SAP PRESS e-books

Print or e-book, Kindle or iPad, workplace or airplane: Choose where and how to read your SAP PRESS books! You can now get all our titles as e-books, too:

- By download and online access
- For all popular devices
- And, of course, DRM-free

Convinced? Then go to www.sap-press.com and get your e-book today.

Integrating Materials Management in SAP S/4HANA®

PRESS

SAP PRESS is a joint initiative of SAP and Rheinwerk Publishing. The know-how offered by SAP specialists combined with the expertise of Rheinwerk Publishing offers the reader expert books in the field. SAP PRESS features first-hand information and expert advice, and provides useful skills for professional decision-making.

SAP PRESS offers a variety of books on technical and business-related topics for the SAP user. For further information, please visit our website: *www.sap-press.com*.

Jawad Akhtar, Martin Murray
Materials Management with SAP S/4HANA:
Business Processes and Configuration (3rd Edition)
2024, 1018 pages, hardcover and e-book
www.sap-press.com/5835

Ghodmare, Bobade, Dua
Integrating Sales and Distribution in SAP S/4HANA
2023, 580 pages, hardcover and e-book
www.sap-press.com/5592

Shailesh Patil, Sudhakar Bandaru
Integrating EWM in SAP S/4HANA
2024, 542 pages, hardcover and e-book
www.sap-press.com/5780

Gautam Bhattacharya, Mehfuze Ali Molla
Integrating Third-Party Logistics with SAP S/4HANA
2024, 455 pages, hardcover and e-book
www.sap-press.com/5717

Stoil Jotev
Configuring SAP S/4HANA Finance (3rd Edition)
2025, 744 pages, hardcover and e-book
www.sap-press.com/5920

Murthy KP

Integrating Materials Management in SAP S/4HANA®

Rheinwerk
Publishing

Editor Megan Fuerst
Acquisitions Editor Emily Nicholls
Copyeditor Melinda Rankin
Cover Design Graham Geary
Photo Credit Shutterstock: 13102567/© Andrea Danti
Layout Design Vera Brauner
Production Kelly O'Callaghan
Typesetting III-satz, Germany
Printed and bound in the United States of America, on paper from sustainable sources

ISBN 978-1-4932-2614-6
1st edition 2025

© 2025 by:
Rheinwerk Publishing, Inc.
2 Heritage Drive, Suite 305
Quincy, MA 02171
USA
info@rheinwerk-publishing.com

Represented in the E.U. by:
Rheinwerk Verlag GmbH
Rheinwerkallee 4
53227 Bonn
Germany
service@rheinwerk-verlag.de

Library of Congress Cataloging-in-Publication Control Number: 2024058112

Contents at a Glance

Contents

3   Master Data

4 Sales and Distribution 189

11   Project System

12   Integration with Other External Systems

Preface

As SAP S/4HANA continues to redefine the landscape of enterprise resource planning (ERP), integrating materials management within this innovative system has become a top priority for businesses striving for operational excellence. This book is the first of its kind on the market to comprehensively address the integration of materials management in SAP S/4HANA, offering readers an in-depth guide tailored to SAP customers, partners, and consultants for mastering this crucial component of their ERP transformation journey.

Integration is a challenging subject, especially within complex ERP environments. In this book, you'll find valuable insights into the intricacies of integration, addressing the critical touchpoints between materials management and other business processes such as production planning, quality management, plant maintenance, finance, Project System, sales, and logistics. Beyond technical configuration, this book provides comprehensive information on the best practices, standard integrated processes, cross-functional organizational structures, and master data. It can serve as your first source of information on the integrated materials management functionalities in SAP S/4HANA.

How This Book Is Organized

This book is organized to provide both foundational and advanced insights into integrating materials management in SAP S/4HANA with a focus on supporting customer-facing business functions and their alignment with materials management functions. Starting with a glance at materials management functions in Chapter 1, one chapter is dedicated to each business function: sales and distribution, quality management, production planning, plant maintenance, extended warehouse management (EWM), transportation management (TM), finance, and Project System. Additional chapters are dedicated to explaining the cross-functional organization structure, core master data and purchasing master data, and integration of materials management with external systems.

It's best to read this book sequentially from Chapter 1 onward; however, if you prefer, you can directly go to any chapter and start reading about that topic. For instance, if you're interested in learning about sales and distribution and materials management integration, you can start reading Chapter 4 without reading previous chapters. Let's review what is covered in each chapter of this book:

- **Chapter 1: Introduction to Integrated Materials Management**
  This chapter introduces the core components of materials management and the key innovations SAP S/4HANA brings to purchasing, inventory management, and logistics invoice verification (LIV) processes. This introductory chapter provides a foundational understanding of materials management in SAP S/4HANA and its critical

role of integration with other business functions. As you move through the following chapters, the groundwork will help you navigate the more detailed and integrated aspects of materials management.

- **Chapter 2: Organizational Structure**
This chapter focuses on the organizational elements of materials management, sales and distribution, financial accounting and controlling, logistics general, logistics execution (including EWM and TM), and plant maintenance in SAP S/4HANA. This chapter sets the stage for understanding how these organizational elements come together in master data management and cross-functional integration.

- **Chapter 3: Master Data**
In this chapter, we explain the significance of core master data such as the material master and business partners, as well as master data specific to materials management in SAP S/4HANA. Along with explaining the step-by-step procedure for setting up the master data, we focus on the configuration setup involved in maintaining the master data. As we move through the following chapters, we'll explain the cross-functional master data details.

- **Chapter 4: Sales and Distribution**
This chapter on the integration of sales and distribution and materials management focuses on the seamless flow of information between these two functions, ensuring that materials, inventory, and procurement data are effectively linked to sales and distribution-related processes. This chapter provides foundational and advanced understanding of best practices, standard integrated processes between the two functionalities, configuration settings to map these integrated processes in SAP S/4HANA, and master data in sales and distribution. In this chapter, we explain the most advanced and essential processes in detail, such as sales and delivery processing, the stock transfer process, inventory management for sales and distribution, third-party processing, and advanced returns management (ARM), focusing all integration points between these two functions.

- **Chapter 5: Quality Management**
This chapter provides a deep understanding of integration between quality management and materials management with a focus on how quality inspection, control, and assurance processes are integrated with procurement, inventory management, and LIV functions. This chapter provides both foundational and advanced understanding of best practices, standard integrated processes between the two functionalities, configuration settings to map these integrated processes in SAP S/4HANA, and master data in quality management. In this chapter, we explain the core quality management functions such as quality inspection, quality certificates, and quality notifications with all integration points, ensuring that materials meet quality standards.

- **Chapter 6: Production Planning**

  This chapter explores how the procurement and inventory management functions of materials management work seamlessly with production planning to ensure the right quantities of materials are available for production at the right time. This chapter provides both foundational and advanced understanding of best practices, standard integrated processes between the two functionalities, configuration settings to map these integrated processes in SAP S/4HANA, and master data in production planning. In this chapter, we explain in detail the material requirements planning (MRP), master production scheduling (MPS), and operational subcontracting processes, as well as their integration with the procurement and inventory management functions of materials management.

- **Chapter 7: Plant Maintenance**

  This chapter explains how plant maintenance activities are seamlessly connected with procurement and inventory management to streamline maintenance processes and efficiently manage spare parts and other components. This chapter provides both foundational and advanced understanding of best practices, standard integrated processes between the two functionalities, configuration settings to map these integrated processes in SAP S/4HANA, and master data in plant maintenance. In this chapter, we explain in detail the preventive maintenance process, corrective maintenance process, refurbishment process, and material planning for maintenance activities, as well as their integration with the procurement and inventory management functions of materials management.

- **Chapter 8: Extended Warehouse Management**

  This chapter focuses on the integration of embedded EWM and materials management, enabling efficient warehouse operations, inventory tracking, and optimized logistics handling for inbound, outbound, and internal warehouse operations. This chapter provides both foundational and advanced understanding of best practices, standard integrated processes between the two functionalities, configuration settings to map these integrated processes in SAP S/4HANA, and master data in embedded EWM. In this chapter, we explain in detail the basic settings in embedded EWM, inbound delivery processing (including putaway strategies), outbound delivery processing (including picking strategies), and internal warehouse movements, as well as their integration with the procurement and inventory management functions of materials management.

- **Chapter 9: Transportation Management**

  This chapter covers the transportation planning, execution, and settlement processes of TM and their integration with the procurement, inventory management, and LIV functions of materials management, ensuring efficient handling of inbound, outbound, and internal logistics. This chapter provides both foundational and advanced understanding of best practices, standard integrated processes between the two functionalities, configuration settings to map these processes in SAP S/4HANA, and master data in TM.

- **Chapter 10: Finance**
  This chapter explains the seamless integration between financial accounting and materials management, ensuring accurate financial accounting, efficient procurement, and robust financial reporting. This chapter provides both foundational and advanced understanding of best practices, standard integrated processes between the two functionalities, configuration settings to map these integrated processes in SAP S/4HANA, and master data in finance. In this chapter, we explain the procurement, inventory management, and LIV process integration with finance.

- **Chapter 11: Project System**
  This chapter we explain how project management activities are tightly integrated with procurement and inventory management, ensuring seamless handling of materials required for projects. This chapter provides both foundational and advanced understanding of best practices, standard integrated processes between the two functionalities, configuration settings to map these integrated processes in SAP S/4HANA, and master data in Project System.

- **Chapter 12: Integration with Other External Systems**
  In this final chapter, we explain electronic data interchange (EDI) for collaborating with suppliers for smoother supply chain transactions. Next, we explain the external tax engine integration with SAP S/4HANA to calculate indirect taxes, supporting purchasing and LIV functions. This chapter provides both foundational and advanced understanding of best practices and standard configuration settings in SAP S/4HANA for integrating external systems.

Acknowledgments

As mentioned previously, a book such as this one is truly the result of significant effort, dedication, and collaboration, and it would not have been possible without the unwavering support and guidance of many remarkable individuals. First, the book would not exist if my lovely wife, Sangamitre KR, and wonderful son, Yashmit Murthy, did not believe in the project and provide their enthusiastic support. To my beloved Mom and Dad, thank you for instilling the values of perseverance and curiosity, always encouraging me to pursue my dreams—this publication is dedicated to your memory, and I hope it brings you pride wherever you are. I extend my sincere gratitude to Emily Nicholls, Megan Fuerst, and the entire Rheinwerk Publishing team for their trust and commitment for bringing this book to life. Your professionalism, guidance, and unwavering support throughout the publishing process have been invaluable.

Next, I am profoundly grateful to the leaders and mentors who have played a significant role in shaping my journey. Their insights and leadership have had a meaningful impact on my professional growth and perspective. In recognition of this, I am truly honored to offer my sincere appreciation and thanks to these leaders whose invaluable guidance, support, and vision have made this work possible: Deb Bhattacharjeee, Frederic Girardeau-Montaut, Jagjeet Singh, Anand Jha, Jerry Hoberman, Anurag Kumar, Christopher Smith, Jonathon Magick, Unmesh Wankhede, Subit J. Mathew, Donna Gray, Maunil Mehta, Merry Kweiter, Biswadeep Sarkar, Niroop Radhakrishnan, Thomas Breuer, Richard A. Swanson, Gautam Gauba, Anurag Jain, Deblina Sharma, Brent Griffith, Zameer Sayed, Harpreet Singh, Sanjib Mukherjee, Bruce Mcquillen, Harish Gopal Kumbhare, Ajay Kulkarni, Jigar Desai, Darshan Shah, Rajesh Tol, Basawaraj Patil, Sanjeev Jain, Balaji Kannapan Ramachandran, and Sunil Kunale.

I would like to extend my gratitude to the clients I have had the privilege to work with over the years. Your trust, collaboration, and challenging real-world scenarios have been instrumental in shaping my understanding of SAP and its practical applications. Finally, I am equally grateful to my friends, colleagues, and peers in the SAP community. Your shared wisdom, real-world experiences, and collaborative spirit have truly inspired and enriched my knowledge.

Conclusion

Integrating materials management in SAP S/4HANA represents a transformative step toward a more streamlined, intelligent, and agile supply chain. Reading this book will provide you with a comprehensive overview of integrated materials management functions. This book explains how to set up cross-functional configuration, organizational structures, and master data to ensure seamless integration across procurement, inventory management, LIV, production planning, quality management, plant maintenance, finance, Project System, sales, and logistics. By mastering these foundational elements, you will be equipped to engage confidently with cross-functional teams during both evaluation and implementation conversations. It will serve as you as both foundational and advanced knowledge source, helping you understand core concepts while also offering deeper insights for optimizing your SAP S/4HANA implementations.

Let's get on board now and proceed to Chapter 1 with a first glance at SAP S/4HANA materials management.

Chapter 1

Introduction to Integrated Materials Management

The primary function of materials management is to manage materials as they flow through the supply chain. Procurement, inventory management, valuation and account determination, and logistics invoice verification are the main functions of materials management. It ensures the right quantities of right materials are available at the right time for the supply chain to function efficiently and cost effectively.

Materials management is a highly integrated function of SAP S/4HANA. Integration with the other functions are built-in capabilities of the standard system. This chapter introduces the integration of materials management with the following functions:

- Production planning
- Sales and distribution
- Quality management
- Plant maintenance
- Extended warehouse management
- Transportation management
- Project System
- Finance

In this introductory chapter, we'll cover core processes within these functions that fall into three buckets: purchasing, inventory management, and logistics invoice verification (LIV).

Materials management manages the end-to-end *purchasing* cycle, which includes purchase requisition creation/generation, purchase order creation, goods receipt processing, and incoming invoice verification. It manages the direct procurement process, indirect procurement process, special procurement processes, service procurement process, and contract and scheduling agreement processes. It supports internal procurement processes that include the intracompany stock transfer process and intercompany stock transfer process and the external procurement processes. The main objective of the procurement function is to ensure that the right quantities of raw materials and other components are available for other supply chain functions at the right time.

Inventory management is another vital function of materials management. It manages the stock of materials in terms of quantity and value. Various goods (material) movements are managed in inventory management, including goods receipts for purchase orders and stock transfer orders, goods issues for customer sales orders and stock transfer orders, and internal movements such as transfer postings, goods issues for reservations, production, plant maintenance, network activities, and so on. It provides different stock types and special stock types to manage inventory. Physical inventory is another important process of inventory management, and the materials stored in the storage locations are counted periodically and compared with the system stock record to ensure inventory accuracy.

The *valuation and account assignment* function of materials management is closely integrated with financial accounting. *Valuation* refers to the determination of inventory value and other material-related costs. This function provides capabilities to determine the value of inventory during material movements in inventory management. Material movements occurring in the inventory management function are accurately valued and posted to the right general ledger accounts in financial accounting. This ensures the automatic general ledger account determination and the posting of the inventory value to the general ledger account in financial accounting. We'll discuss this in detail in Chapter 10, Section 10.3.

The *logistics invoice verification (LIV)* function supports the processing of incoming invoices from suppliers. Incoming invoices are reconciled by performing a three-way match, where invoice line items are matched with corresponding purchase order line items and goods receipt items to determine the quantity, amount, tax, and any discrepancies before payment processing. Evaluated receipt settlement, consignment settlement, and invoicing plan settlement are other features available to support logistics invoice verification functions by automatically generating and posting supplier invoices based on certain conditions, eliminating the need to receive invoices from the suppliers.

1.1   Purchasing

Purchasing is an important function of materials management. It manages the procurement of goods and services for various functions of the organization, including production, plant maintenance, project systems, sales and distribution, and so on, by creating and submitting purchase orders and/or outline agreements for or to suppliers. The purchasing process can be initiated from various other functions in SAP S/4HANA automatically by generating purchase requisitions. A *purchase requisition* is an initial document that can be created manually or automatically to initiate the purchasing process. Purchase requisitions are converted to purchase orders automatically or manually after the purchase requisitions are approved or released. You can have a release strategy or an approval workflow for both purchase requisitions and purchase orders. A best practice recommendation is to have an approval workflow or release strategy for the purchase order. Once released, purchase orders are transmitted to suppliers via email as attachments or electronically via electronic data interchange (EDI). SAP S/4HANA provides application programming interfaces (APIs) to transmit purchase orders.

The purchasing function integrates with production planning, quality management, plant maintenance, Project System, extended warehouse management (EWM), transportation management (TM), and financial accounting. Future chapters cover these integrations in detail.

SAP S/4HANA provides capabilities to manage different purchasing processes using document types, item category references, and account category references:

1. **Standard purchasing process**
   The standard purchasing process is used to procure standard items, and it includes both direct procurement and indirect procurement. A standard item category is used to identify standard items in a purchase order.

2. **Subcontracting process**
   Subcontracting is a special procurement process. It is a special type of service, and the supplier uses the components provided by the buying organization to produce/ assemble semifinished products or to perform certain value-added services, such as packaging, rework, and the like. A subcontracting item category is used to identify subcontracting items in an order.

3. **Consignment process**
   Consignment is a special procurement process. Based on a consignment order, a supplier loans materials to the purchasing company. The purchasing company is not liable to pay to the supplier until it withdraws or directly consumes the consignment stock that belongs to the supplier. A consignment item category is used to identify consignment items in an order.

4. **Stock transfer order process**
   A stock transfer order is a type of purchase order used to procure goods internally from another plant that belongs to the same company code or from another plant that belongs to a different company code in SAP S/4HANA. Based on the stock transfer order, the supplying plant ships the requested materials to the receiving plant to fulfill the stock transfer order. A stock transfer item category is used to procure materials from another plant that belongs to the same company code. A standard item category is used to procure materials from another plant belonging to a different company code.

5. **Third-party process**
   Third-party processes, also called trading processes, are widely used by organizations that sell trade goods. In simple terms, the customer orders will be fulfilled by a third-party supplier instead of the seller who received the customer order. Here the supplier of the trade goods is referred to as a *third party*. The third-party process is initiated from the sales order, and a purchase requisition will be generated for procuring

trading goods with item category third party and with reference to the sales order. The purchase requisition will be converted to a third-party purchase order with a third-party item category to place an order for trading goods to be shipped directly to the customer.

6. **Returns purchase order process**
A returns purchase order is a type of purchase order used to return unwanted, extra, or faulty materials to the supplier. After submitting the returns purchase order to the supplier, a goods issue will be processed to return the goods to the supplier. The supplier will send a credit memo for the returned goods. A returns indicator will be set in the purchase order items that indicate the materials to be returned.

7. **Outline agreements process**
SAP S/4HANA provides outline agreements to formally create contract agreements and scheduling agreements for agreed-upon terms and conditions with suppliers of goods and services. The following are the two types of outline agreement:
   - *Contract agreements* are created with negotiated prices, Incoterms, payment terms, and so on for a set validity period. After the validity period ends, the contract expires. A valid contract can be used to create source lists (see Chapter 3, Section 3.3.1). You can reference a contract agreement to create purchase requisitions and purchase orders. All the pricing details, payment terms, Incoterms, suppliers, materials, and so on are copied over to purchase requisitions and purchase orders from the contract.
   - *Scheduling agreements* are different from contract agreements in terms of functionality, but they are outline agreements created with negotiated prices, Incoterms, payment terms, and so on for a set validity period. Unlike contract agreements, schedule lines can be created with reference to scheduling agreements for the required quantity and expected delivery dates. Schedule lines are transmitted to suppliers, and shipments from suppliers are received against the scheduling agreement.

We'll explain these processes in more detail in the following sections. But first, we'll introduce the key attributes of the purchasing process in SAP S/4HANA.

1.1.1   Purchasing Attributes

Let's discuss how item categories, account assignment categories, and document types control the various purchasing processes.

Item Categories

Item categories are used to distinguish purchasing document items and control the purchasing follow-on processes. The item category is used at the item level in purchase requisitions, purchase orders, and outline agreements. Table 1.1 shows the item categories for purchasing defined in the standard system with their features.

Item Category	Description	Features
(Blank)	Standard	Categorizes standard line items in the purchasing documents (purchase requisitions, purchase orders, and outline agreements). The purchase order is transmitted to the supplier, the supplier ships the goods, and the goods receipt is posted to inventory. This is followed by supplier invoice reconciliation and making a payment to the supplier.
K	Consignment	Categorizes consignment line items in the purchasing documents. The purchase order is transmitted to the supplier. The supplier ships the goods, and the goods receipts will be posted into vendor consignment stock. Upon consumption or withdrawal from consignment stock, consignment settlement and payments to suppliers are processed.
L	Subcontracting	Categorizes subcontracting line items in the purchasing documents. The purchase order is transmitted to the supplier. Components are provided to the suppliers by the purchaser for the subcontracting process. The supplier ships the semifinished or finished product, and the goods receipt is posted to inventory. This is followed by supplier invoice reconciliation and payments to the supplier.
S	Third party	Categorizes third-party line items in the purchasing documents. The purchase order is transmitted to the supplier. The supplier ships the products to the customer directly and sends delivery confirmation. Goods receipts are posted with reference to the sales order account assignment (optional). This is followed by supplier invoice reconciliation and payments to suppliers.
U	Stock transfer	Categorizes stock transfer line items of the stock transfer order. The goods issue from the supplying plant and goods receipt into the receiving plant are posted. The system can be configured for automatic settlement between the supplying company and receiving company for an intercompany stock transfer process. For an intracompany stock transfer process, settlement is not required.
D	Service	Categorizes service line items in the purchasing documents. The purchase order will be transmitted to the supplier. The supplier performs the service, and a service entry sheet will be posted in the system. This is followed by supplier invoice reconciliation and payments to suppliers.

Table 1.1 Item Categories for Purchasing

Item Category	Description	Features
E	Enhanced limits	Categorizes limit items in the purchasing documents. The purchase order will be transmitted to the supplier for periodic payments (e.g., rent). Goods receipt is not performed for these line items, but invoice reconciliation and payment processing are.

Table 1.1 Item Categories for Purchasing (Cont.)

Account Assignment Categories

Account assignment categories associate purchasing documents such as purchase requisitions, purchase orders, outline agreements, and so on with cost objects or accounts in financial accounting. By assigning an account assignment category to purchasing documents at the item level, the received goods and services are directly consumed by the cost center (department), internal order, maintenance order, or network order in the respective business processes. Depending on the account assignment category assigned to the purchasing document item, an **Account Assignment** tab will be added in the purchase order item details. In the **Account Assignment** tab, cost object and general ledger account details are added depending on the account assignment category. During goods receipt and invoice postings with reference to account assigned purchase orders, the costs related to procurement are posted to the respective financial accounts.

Account assignment to purchasing documents is done manually during the creation of purchasing documents. This can happen automatically when the purchase requisitions are generated from production orders, sales orders, maintenance orders, and network activities from Project System. Figure 1.1 shows the account assignment categories defined in the standard system for purchasing.

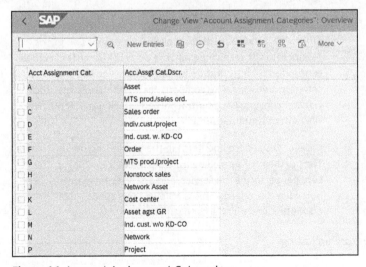

Figure 1.1 Account Assignment Categories

Document Types

A document type is mandatory to create a purchasing document in SAP S/4HANA. Purchasing documents are distinguished based on document types. Every purchasing document is distinguished by a document category, and each document category has multiple document types defined in SAP S/4HANA. Table 1.2 shows the standard document types defined in the system. The system allows you to create a custom document type under every document category.

Document Category	Document Type	Description
A	AN	Request for quotation
B	FO	Framework requisition
B	NB	Purchase requisition
B	RV	Outline agreement requisition
F	FO	Framework order
F	NB	Standard purchase order
F	NBIC	Advanced intercompany purchase order
F	UB	Stock transport order
K	MK	Quantity contract
K	WK	Value contract
L	LP	Scheduling agreement
L	LPA	Scheduling agreement with release document

Table 1.2 Purchasing Document Types

Document category A is defined for request for quotation, category B is defined for purchase requisitions, category K is defined for contract agreements, category F is defined for purchase orders, and category L is defined for scheduling agreements.

As mentioned earlier, the purchase requisition is the initial document for all purchasing processes. Purchase requisition document types are linked to other purchasing document types such as purchase orders, contract agreements, scheduling agreements, and so on. This linkage is required to create the other purchasing documents with reference to purchase requisitions. Allowed item categories can be defined for the document types, and in this way, different purchasing document types can be configured for different purchasing processes.

Let's define the document types and explore how they're used to control different purchasing processes. To access the configuration to define purchase requisition document types, execute Transaction SPRO and navigate to **Materials Management • Purchasing •**

Purchase Requisition • Define Document Types for Purchase Requisitions. Figure 1.2 shows the first screen of the configuration.

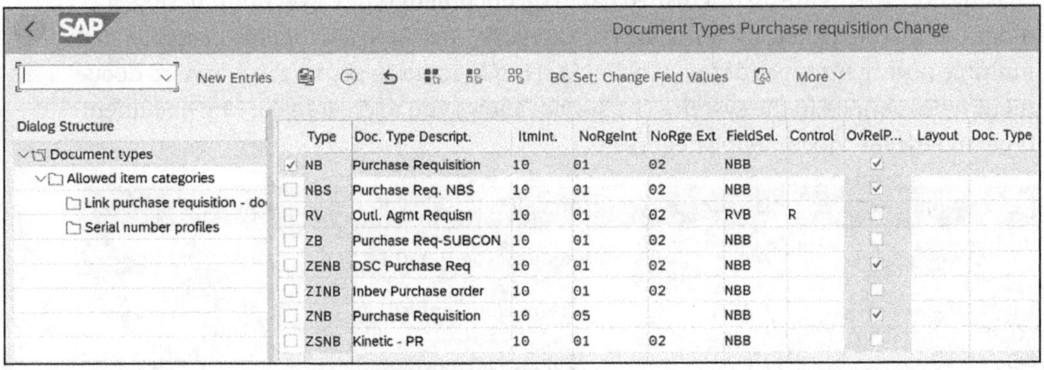

Figure 1.2 Define Purchase Requisition Document Types

The following are the key attributes of purchase requisition document types:

- **Type**
  The purchase requisition document type.

- **ItmInt. (item interval)**
  Controls the purchase requisition line-item number interval. For example, if this value is set to 10, then line-item numbers will be created in the purchase requisition at intervals of 10 (10, 20, 30, etc.).

- **NoRgeInt (number range internal)**
  Controls an internal document number of the purchase requisition.

  To configure the number range, execute Transaction SPRO and navigate to **Materials Management • Purchasing • Purchase Requisition • Define Number Ranges for Purchase Requisitions**. You can reach this configuration directly by executing Transaction SNRO. In this configuration, both internal and external number ranges can be defined against the **BANF** number range object. Figure 1.3 shows the configuration of internal and external number ranges.

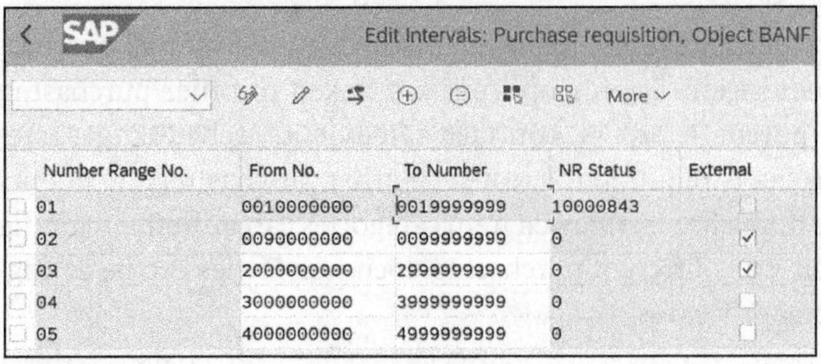

Figure 1.3 Define Purchase Requisition Number Range

- **NoRge Ext (number range external)**
  Controls an external document number of the purchase requisition. An **External** indicator will be set for external number ranges, as shown in Figure 1.3.

- **FieldSel. (field selection key)**
  The field selection key represents the layout of the purchase requisition, including mandatory, optional, and display fields. **NBB** is the standard screen layout for purchase requisition documents.

- **Control**
  This field categorizes different types of purchase requisitions such as standard, stock transfer, subcontracting, and so on.

- **OvRelP... (overall release of purchase requisitions)**
  This indicator controls the release of purchase requisition. If this indicator is set, all line items of the purchase requisition will be released at once.

The next step in the purchase requisition document type configuration is to assign the allowed item categories to the document type. To arrive at this step, select a document type from the document type list and double-click **Allowed item categories** in the **Dialog Structure**. Figure 1.4 shows the allowed item categories for purchase requisition document type **NB**. Item categories in purchasing distinguish among different types of items in purchasing documents.

> **Note**
>
> Best practice is to assign all standard item categories for purchasing to standard document type **NB**. This adds flexibility to the purchasing process, and buyers will have access to this document type. When you create specific document types for every purchasing process, access to a specific document type can be given to a group or individual. For example, planners will have access to create stock transfer orders.

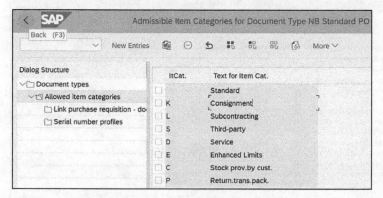

Figure 1.4 Define Purchase Requisition Document Type: Allowed Item Categories for Document Type

The next step in the purchase requisition document type configuration is to link the combination of allowed item categories in the purchase requisition document type to the combination of other purchasing document types and allowed item categories. This configuration controls the creation of a purchase order, requests for quotation (RFQ), and outline agreements with reference to a purchase requisition.

The following are the different purchasing documents that can be created with reference to a purchase requisition:

- RFQ
- Standard purchase order
- Framework order
- Stock transfer order
- Service purchase order
- Subcontract order
- Consignment purchase order
- Third-party purchase order
- Outline agreements: contract agreement
- Outline agreements: scheduling agreement

To configure creating document types for purchase orders, RFQs, and outline agreements, execute Transaction SPRO and navigate to **Materials Management • Purchasing • Purchase Order • Define Document Types for Purchase Orders**. Figure 1.5 shows the first screen of the configuration:

- **Type**
  The document type is the purchase order, RFQ, or outline agreement document type.

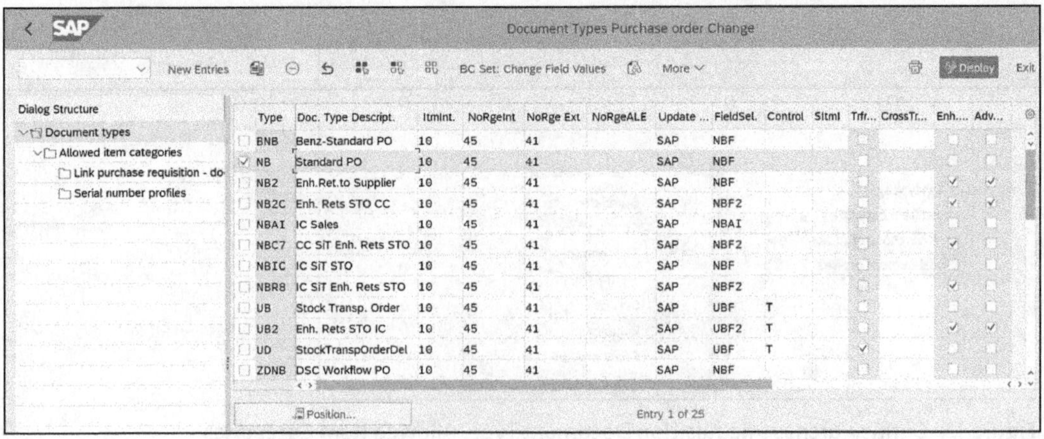

Figure 1.5 Define Purchase Order Document Type

- **ItmInt. (item interval)**
  Controls the purchase document line-item number interval. For example, if this value is set to 10, then line-item numbers will be created in the purchase order in intervals of 10 (10, 20, 30, etc.).

- **NoRgeInt (number range internal)**
  Controls the internal document number of the purchasing document (RFQ, purchase order, or outline agreement).

  To arrive at the number range configuration step, execute Transaction SPRO and navigate to **Materials Management • Purchasing • Purchase Order • Define Number Ranges for Purchasing Documents**. You can directly arrive at this configuration step by executing Transaction SNRO. In this configuration, both internal and external number ranges can be defined against number range object **EINKBELEG**. Figure 1.6 shows the configuration of internal and external number ranges.

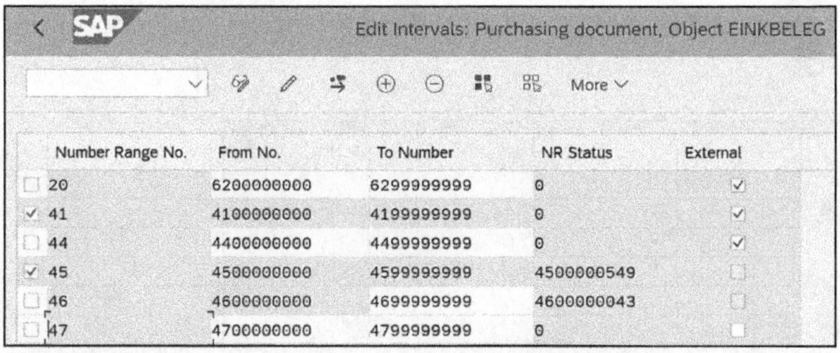

Number Range No.	From No.	To Number	NR Status	External
20	6200000000	6299999999	0	✓
✓ 41	4100000000	4199999999	0	✓
44	4400000000	4499999999	0	✓
✓ 45	4500000000	4599999999	4500000549	
46	4600000000	4699999999	4600000043	
47	4700000000	4799999999	0	

Figure 1.6 Define Purchasing Documents Number Range

- **NoRge Ext (number range external)**
  This controls external document numbers of purchasing documents. The **External** indicator will be set for an external number range, as shown in Figure 1.6.

- **Update ... (update group)**
  This is a logical element that controls how the purchasing documents information is updated in the logistics information system. The logistics information system collects and stores the data from sales, purchasing, plant maintenance, and so on. SAP stores the information in the purchasing area of the logistics information system.

- **FieldSel. (field selection key)**
  The field selection key represents a layout of the purchasing documents, including mandatory, optional, and display fields. **NBF** is the standard screen layout for a purchase order document.

- **Trfr... (stock transfer: take supplier data into account)**
  This indicator controls if a stock transfer order can be created with or without considering the supplier data. If this indicator is set, a vendor as a supplying plant must

be created, and the system reads the supplier information from the vendor master record.

- **Enh... (enhanced store returns)**
  This indicator controls if only outbound deliveries can be created for store returns (stock transfer order) or if both inbound and outbound deliveries are required during the store returns process. If this indicator is set, both inbound and outbound deliveries must be created during the store returns process.

- **Adv... (advanced returns management)**
  This indicator controls if advanced returns management (ARM) is active for the purchasing documents.

The next step in the purchasing document type configuration is to assign the allowed item categories to the document type. To arrive at this step, select a document type from the document type list and double-click **Allowed item categories** from the **Dialog Structure**. Figure 1.7 shows the allowed item categories for purchase order document type **NB (Standard PO)**.

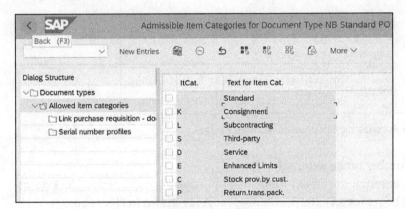

Figure 1.7 Define Purchase Order Document Type: Allowed Item Categories

The third step is to link the purchase requisition document type configuration and purchase order document type such that a purchase order can be created with reference to a purchase requisition. To arrive at this step, select an item category from the list of item categories and double-click **Link purchase requisition—document type** from the **Dialog Structure**. Figure 1.8 shows the linking of purchase requisition document type **NB** to purchase order document type **NB** along with the respective item categories.

To close the loop on purchase requisition document type configuration, the third step is to link the purchase order document type **NB** to the purchase requisition document type **NB** along with the respective item categories, as shown in Figure 1.9.

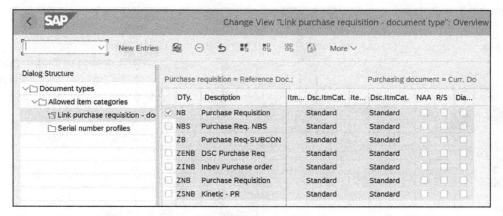

Figure 1.8 Define Purchase Order Document Type: Link Purchase Requisition Document Types

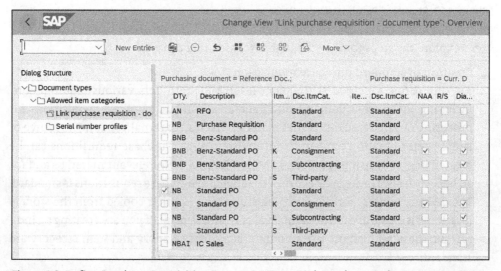

Figure 1.9 Define Purchase Requisition Document Type: Link Purchase Order Document Types

1.1.2 Standard Purchasing Process

Standard purchase orders are used to procure materials (e.g., raw materials) into stock. The process starts with materials planning, and if the material is relevant for material requirements planning (MRP), then the MRP run creates procurement proposals and generates purchase requisitions to initiate the purchasing process. You can create a purchase requisition manually using Transaction ME51N in SAP S/4HANA.

Ad hoc requests and materials that are not relevant for MRP are created manually. Figure 1.10 shows the standard purchase order process.

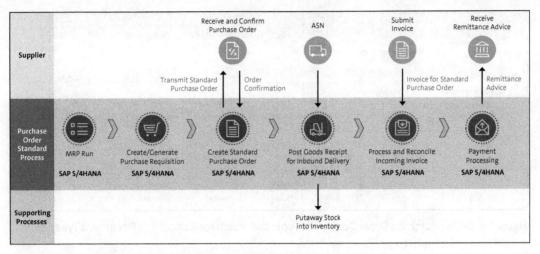

Figure 1.10 Standard Purchase Order Process

Let's explore the steps for the standard purchase order process:

1. **Creation of standard purchase requisition**

   An MRP run is executed at the plant level. MRP considers various requirements, including the current demand, and generates standard purchase requisitions for procuring raw materials and other materials from an external vendor. The source of supply determination happens from the source list. Purchase requisitions can be created manually using Transaction ME51N for non-MRP-relevant materials and for ad hoc requests. The item category of the standard purchase requisitions is standard. The delivery address for the purchase requisition is either copied from the storage location if it exists at the line-item level or from the plant. If you are creating an indirect purchase requisition, the account assignment category and item category are standard.

   Figure 1.11 shows the standard purchase requisition with the standard (blank) item category for **100** EA quantity of material **MS-RAW MATERIAL1** in plant **MJ00**. Source of supply **200019** was determined automatically from the source list. Purchasing info record (PIR) **5700000063** was determined automatically as it was created for material **MS-RAW MATERIAL1** and supplier **200019** in purchasing organization **MJ00**.

2. **Creation of standard purchase order**

   The standard purchase requisition is converted into a standard purchase order manually by the buyer using Transaction ME21N or automatically in SAP S/4HANA using a background job. The line-item details are defaulted into the purchase order, including the source of supply, unit cost, and delivery address. The unit price for the material is automatically copied from the PIR. The purchase order will be transmitted to the supplier based on the output method: email, EDI, or electronic transmission to the supplier on SAP Business Network. The output determination configuration in

SAP S/4HANA drives the purchase order output mechanism. Depending on the confirmation control key, order confirmation from the supplier may be required, and the confirmation control key is defaulted from the PIR. Figure 1.12 shows the standard purchase order created with reference to the purchase requisition with item category standard.

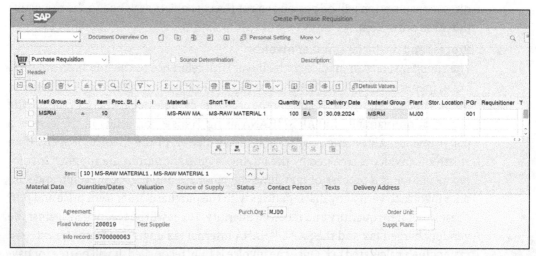

Figure 1.11 Standard Purchase Requisition

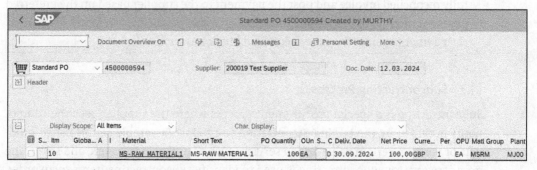

Figure 1.12 Standard Purchase Order

3. **Process goods receipt**

The supplier ships the goods to the buying organization based on the ship-to address provided in the purchase order. It's best practice for the supplier to send an advanced shipping notification (ASN) to the buyer to notify them of the materials, delivery date, serial number, batch, and shipment tracking information. Upon the receipt of the ASN, an inbound delivery is created in SAP S/4HANA automatically, or you can create an inbound delivery manually using Transaction VL31N. Once the shipment physically arrives at the receiving location, a goods receipt is posted for the inbound delivery using Transaction VL32N.

If an ASN is not expected and an inbound delivery creation is not always required, goods receipt can be posted directly with reference to the purchase order using Transaction MIGO. Stock is received into inventory upon goods receipt.

Inbound deliveries are not created for indirect purchase orders with account assignment. For such purchase orders, when they're received using Transaction MIGO, the direct consumption will be posted against the cost object assigned to the respective purchase order item.

4. **Process and reconcile supplier invoice**
   The supplier sends an invoice to the buyer with reference to the purchase order via email (paper invoice), EDI, or electronically via SAP Business Network. The supplier invoice document will be posted by the accounts payable department of the buying organization manually in SAP S/4HANA using Transaction MIRO in the case of a paper invoice. Automatic posting and reconciliation of the invoice happens if an electronic invoice is received by the buying organization from the supplier. Invoice reconciliation is a process of matching the purchase order, goods receipt, and supplier invoice to verify any discrepancies with the purchase order item price and purchase order item quantity and received quantity. Taxes will be reconciled against the vendor charged tax and the SAP S/4HANA internal tax engine or from the external tax engine's calculated tax. Once the invoice is fully reconciled, it will be free for payment processing in the system. The next scheduled payment run will pick up this fully reconciled invoice and post the payment to the supplier based on the preferred payment method of the supplier. Accounts payable personnel can manually process the payment against the fully reconciled invoice as well.

1.1.3   Subcontracting Process

Subcontracting is a special procurement process where the supplier uses the components provided by the buying organization to produce/assemble semifinished products or perform certain value-added services such as packaging, rework, and so on. The item category for subcontracting is used to identify subcontracting items in an order. The process of subcontracting can be triggered from MRP, from a maintenance operation within a maintenance process involving a maintenance order, from a network activity within Project System involving a network activity, or from an external operation within the production process involving processing of the process/production order.

The subcontracting process can be initiated manually by creating a purchase requisition or directly by creating a subcontracting order. However, the production and maintenance processes use the subcontracting process extensively, particularly in manufacturing industries. Figure 1.13 shows all these pieces of the subcontracting process.

Note

Materials management provides the flexibility to procure required components directly from another supplier have the supplier ship the components to the subcontractor directly. Create a standard purchase order and include the subcontractor's business partner number with the supplier role in the **Delivery Address** tab so that the external supplier can ship the components to the subcontractor directly.

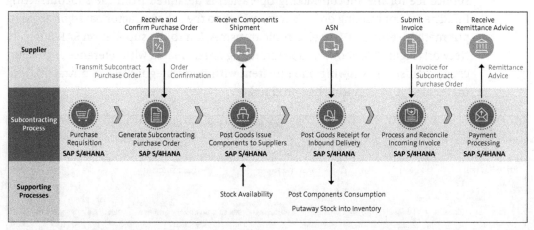

Figure 1.13 Subcontracting Order Process

A bill of materials (BOM) must be created as a prerequisite with the parent material to be produced/assembled/reworked by the subcontractor and with the components details to be provided to the external subcontractor. After this is done, the process steps are as follows:

1. **Create or generate purchase requisition**

   To initiate the subcontracting process, the material planner will run the MRP for planned independent requirements using Transaction MD01N. The system generates subcontracting purchase requisitions for the final product (BOM header) or a subassembly (semifinished product). The system assigns a subcontractor (supplier) as the source of supply from the purchasing master data (source list and subcontracting info record). The item category of the requisition line item will be **L** (subcontracting). The components will be added to the **Material Data** tab of the subcontracting purchase requisition automatically from the BOM.

   Production processes, maintenance processes, Project System, and the like can initiate subcontracting by generating a subcontracting requisition from their respective processes. Later chapters in this book will explain the process steps in detail by providing an integrated view.

   The subcontracting process can also be initiated manually by creating a subcontracting requisition using Transaction ME51N. Select item category L (subcontracting) and then enter the final product to be received from the subcontractor, the quantity, the delivery address (plant address), and so on. The components to be issued to the subcontractor can be manually entered or can be taken from the BOM.

2. **Convert subcontracting purchase requisitions into subcontracting orders**
Subcontracting purchase requisitions will be converted into subcontracting orders either manually by the purchaser via Transaction ME21N or automatically. The line-item details, including the components, are defaulted into the purchase order. The service fee for the subcontracting operation is defaulted from the subcontracting PIR. The subcontracting order is transmitted to the subcontractor based on the output method: email, EDI, or electronic transmission to the supplier on SAP Business Network. Figure 1.14 shows a subcontracting order created with reference to the purchase requisition, showing one line item with item category **L** (subcontracting) for material **SF_001** (semifinished product).

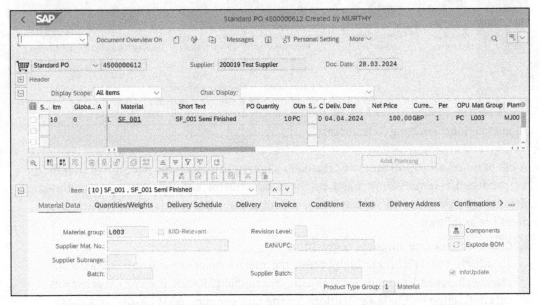

Figure 1.14 Subcontracting Order

The subcontractor will produce **10** PC of the semifinished product **SF_001**. To view the components to be provided to the subcontractor, click the **Components** icon in the **Material Data** tab of the item details.

Figure 1.15 shows the components to be provided to the subcontractor. You can add/remove components manually directly in the purchase order via Transaction ME22N (Purchase Order Change).

3. **Create outbound delivery with reference to subcontract order**
The next step in the process is to create an outbound delivery to issue the components to the subcontractor. Outbound deliveries will be created using Transaction ME2O with reference to the subcontract order.

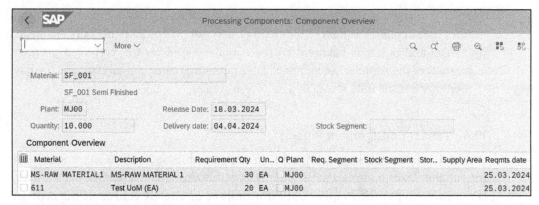

Figure 1.15 Components to Be Provided to Subcontractor

4. **Post goods issue for outbound delivery**

The goods issue will be posted with reference to the outbound delivery after the inventory management functions such as picking and packing of materials are completed. Movement types play an important role in determining the type of inventory movement and posting of the goods issue. Movement type 541 is used for a goods issue for an outbound delivery to ship components to a subcontractor. Movement type 541 ensures the components provided to the subcontracting vendor are tracked and that ownership remains with the company requesting the external operation.

5. **Post goods receipt purchase order**

The subcontractor receives the components and performs the requested operation using the components issued by the buying organization. The subcontractor then ships the produced/assembled final product. Once the shipment from the subcontractor arrives, the goods receipt will be posted manually in SAP S/4HANA with reference to the subcontract order using Transaction MIGO and movement type 101. The stock of the final product received from the subcontractor will be increased in the inventory.

If an ASN is received or the requirement is to create an inbound delivery for processing the goods receipt, then a goods receipt will be posted using Transaction VL32N with reference to the inbound delivery.

Upon posting the goods receipt, consumption of the components provided to the subcontractor will be posted using movement type 543 automatically. This goods movement will reduce the quantities of the components provided to the subcontractor from the inventory.

6. **Post LIV**

The subcontractor sends an invoice for the external operation via email (paper invoice), EDI, or electronically via the SAP Business Network. The supplier invoice document will be posted by the accounts payable personnel manually in the SAP S/4HANA system using Transaction MIRO in the case of a paper invoice. Automatic posting and reconciliation of the invoice happens if an electronic invoice is received from the supplier. The next scheduled payment run will pick up this fully reconciled invoice and post the payment to the supplier based on the preferred payment method of the supplier. Accounts payable personnel can manually process the payment against the fully reconciled invoice as well.

1.1.4 Vendor Consignment Process

Vendor consignment is a special procurement process. Based on a consignment order, the supplier loans materials to the purchasing company, and the purchasing company is not liable to pay the supplier until it withdraws or directly consumes the consignment stock that belongs to the supplier. The consignment item category is used to identify consignment items in an order. Like all special procurement processes (e.g., subcontracting), the consignment process can be triggered from MRP by generating a consignment purchase requisition. Consignment items in the purchase requisition and purchase order are identified by item category K (consignment).

> **Note**
>
> The special procurement key maintained in the **MRP 2** view of the material master plays a vital role in the material requirements planning process. A special procurement type such as consignment, subcontracting, stock transfer, or so on is assigned in the **Special Procurement** field in the **MRP 2** view. Purchase requisitions for stock transfer, subcontracting, and consignment are generated from the MRP run based on this key.

Figure 1.16 shows the vendor consignment process. Let's explore the vendor consignment process steps:

1. **Create or generate purchase requisition**

   To initiate the consignment process, the material planner will run the MRP for planned independent requirements using Transaction MD01N. The system generates consignment purchase requisitions for the materials with the special procurement key maintained as consignment in the **MRP 2** view of the material master. The system determines the supplier from the purchasing master data (source list and subcontracting info record) and assigns it to the consignment purchase requisition as the source of supply. The item category of the requisition line item will be **K** (consignment), and the item category controls the follow-on process steps.

   A consignment info record for the material and supplier combination at the purchasing organization and plant level is required for the consignment process.

2. **Convert consignment purchase requisitions into consignment order**

   A consignment purchase requisition will be converted into a consignment order automatically in SAP S/4HANA via a background job or manually by the buyer using Transaction ME21N. The line-item details including the components are defaulted into the purchase order. The unit price for the material is defaulted from the consignment PIR. The consignment order will be transmitted to the supplier based on the output method: email, EDI, or electronic transmission to the supplier on SAP Business Network. Figure 1.17 shows that purchase order **4500000725** was created for a consignment order of **1.000 EA** of material **1173 (Raw Material 3)** in plant **MJ00**. Item category **K** represents the consignment item.

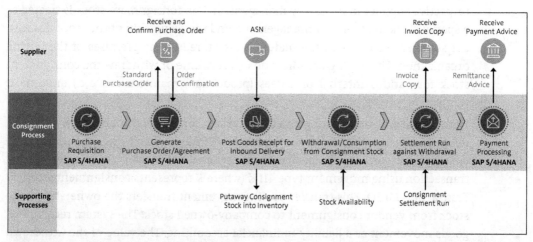

Figure 1.16 Vendor Consignment Process

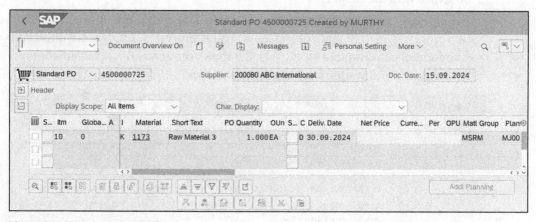

Figure 1.17 Consignment Order

3. **Process goods receipt**

   The supplier ships the consignment goods to the buying organization based on the ship-to address provided in the purchase order. It's best practice for the supplier to send an ASN to the buyer, notifying them of the materials, delivery date, serial number, batch, and shipment tracking information. Upon the receipt of the ASN, an inbound delivery is created in SAP S/4HANA automatically or you can create an inbound delivery manually using Transaction VL31N. Once the shipment physically arrives at the receiving location, the goods receipt is posted for the inbound delivery using Transaction VL32N.

If an ASN is not expected and an inbound delivery creation is not always required, goods receipt can be posted directly with reference to the consignment order using Transaction MIGO. Stock is received into the inventory upon goods receipt.

The received consignment materials are posted to the consignment stock of the respective supplier in inventory management. The consignment stock is treated as a special stock in inventory management, and the ownership of the stock lies with the supplier even though the materials are stored in the premises of the buying organization. The buying organization can consume or withdraw the consignment stock for various internal processes (production, maintenance, etc.) anytime it wants.

4. **Withdrawal and/or consumption of materials from consignment stock**

   Received consignment stock is available for withdrawal and/or consumption. The consignment stock can be posted to unrestricted use stock via a transfer posting transaction using movement type 411 K (where *K* represents consignment stock) in Transaction MIGO. This transfer posting movement transfers the ownership of the stock from vendor consignment to company-owned stock. The system records this goods movement as a liability in financial accounting. The value of the withdrawn stock from consignment to unrestricted use (company-owned) stock is required to be settled with the supplier.

   SAP S/4HANA provides the capability to directly issue consignment stock to production processes, maintenance processes, and other processes. This is called direct consumption of consumption stock. Table 1.3 shows the goods issue movement types for material consumption from consignment stock.

Movement Type	Detailed Description
201 K	Goods issue for a cost center from consignment stock. This movement type is used to issue goods from the consignment stock in the inventory to a cost center for consumption.
221 K	Goods issue for a project from consignment stock. This movement type is used to issue goods from the consignment stock in the inventory to a project for consumption.
261 K	Goods issue for a production order from consignment stock. This movement type is used to issue goods from the consignment stock in the inventory to a production order for consumption.
281 K	Goods issue for a network from consignment stock. This movement type is used to issue goods from the consignment stock in the inventory to a network for consumption.

Table 1.3 Goods Issue Movement Types for Consumption of Consignment Stock

5. **Consignment settlement**
   Withdrawal of materials and consumption/goods issue of materials from consignment stock transfers the ownership of the stock from the supplier to the company. SAP S/4HANA records these stock movements as liabilities in financial accounting. The value of the withdrawn stock from consignment to unrestricted use (company-owned) stock is required to be settled with the supplier. Suppliers will not submit any invoices for the withdrawals and consumption; instead, the company posts an invoice on behalf of the supplier to settle the amount with them. SAP S/4HANA provides Transaction MRRL for consignment settlement. This transaction can be scheduled to run in the background on a regular basis (daily, weekly, etc.). Using this transaction, an invoice can be posted automatically on behalf of the supplier. Best practice is to send a copy of the invoice to the supplier so that they can record it in their books.

The next scheduled payment run will pick up this fully reconciled invoice and post the payment to the supplier based on the preferred payment method of the supplier. Accounts payable personnel can manually process the payment against the fully reconciled invoice as well.

1.1.5 Stock Transfer Process

The stock transfer process is the most important supply chain process. It involves moving the inventory from one storage location to another storage location within the same plant, within the same company code, or across company codes. A stock transfer between storage locations within the same plant is also called a transfer posting. A stock transfer between two plants within the same company code and a stock transfer between two plants across different company codes involve both the sales side and the purchasing side of transactions. The stock transfer process is commonly used to transfer raw materials, semifinished goods, finished goods, and packaging materials between plants/storage locations. It helps to optimize inventory levels across plants, which leads to cost reductions. The stock transfer process mainly integrates materials management, sales and distribution, and finance functions of the organization, though planning, warehouse management, quality management, transportation management, and other functions can also be integrated with this process.

There are two types of stock transfer process in SAP S/4HANA, involving movement of stock/inventory between two plants with materials management and sales and distribution integration:

- **Intracompany stock transfer process**
  Movement of inventory between two plants that belong to the same company code
- **Intercompany stock transfer process**
  Movement of inventory between two plants that belong to different company codes

The stock transfer process is deeply integrated with sales and distribution. Refer to Chapter 4, Section 4.2 for a detailed explanation of the stock transfer process.

1.1.6   Third-Party Process

The third-party process, also called the trading process, is widely used by organizations that sell trade goods. In simple terms, customer sales orders will be fulfilled by a third-party supplier instead of the seller who received the customer order. Here the supplier of the trading goods is referred to as the *third party*. The seller who received the customer order in turn places a purchase order to a third-party supplier to fulfill the customer order. The third-party supplier will ship the goods directly to the customer and bill/invoice the entity that submitted the purchase order. Once the goods are delivered to the customer, the seller bills the customer. Some organizations also sell finished goods to their customers via the third-party process.

The third-party process is initiated from the customer sales order. A third-party purchase requisition is generated automatically from the sales order to procure trade goods. Figure 1.18 shows the third-party purchase requisition created with item category **S** (third-party) and account assignment category **C** (sales order).

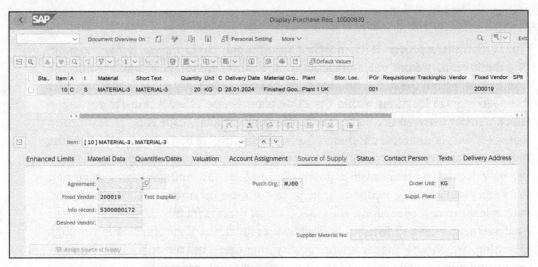

Figure 1.18 Third-Party Purchase Requisition

The third-party process is deeply integrated with sales and distribution. Refer to Chapter 4, Section 4.4 for a detailed explanation of the third-party process.

1.1.7   Returns Purchase Order Process

The returns purchase order process is an important process in the supply chain. It manages supplier returns that occur for various reasons; some of the most common reasons are that the received materials failed to meet quality standards during a quality inspection check, excessive inventory was received, damaged boxes were received, and more. SAP S/4HANA provides the ARM functionality to effectively manage supplier returns. All the documents and process steps are managed automatically and efficiently with full visibility into and control of the end-to-end returns process.

In ARM, the return merchandise authorization (RMA) number plays a vital role. It is a unique identifier for the returns transaction, and it serves as an authorization for returns by the seller or manufacturer. The RMA number for the customer will be provided by the seller to be included in the shipping label generated for returns. Upon the receipt of returned products, the seller uses the RMA number for identifying the goods returned and for the follow-on processes. In this way, the RMA number provides better visibility and tracking of returned products.

Figure 1.19 shows the returns purchase order process. Let's explore the key process steps:

1. **Creation of a returns purchase order**
   Typically, the buyer creates a returns purchase order for the materials to be returned to the supplier using a specific document type that is activated for advanced returns management. Like the stock transfer order, shipping data will be automatically populated at the item level to facilitate the creation of a returns delivery (outbound). At the header level, a **Returns** tab will appear automatically, and the supplier RMA code can be maintained if it's mandatory based on the settings in the business partner. Figure 1.20 shows the creation of a returns purchase order using Transaction ME21N. The **Returns Item** indicator is set for the return purchase order item automatically upon selecting the returns document type.

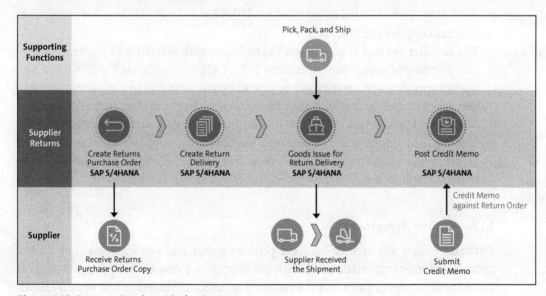

Figure 1.19 Returns Purchase Order Process

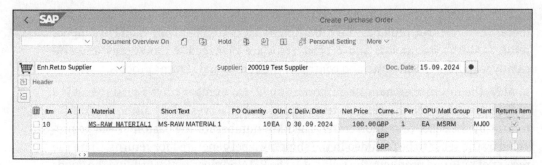

Figure 1.20 Returns Purchase Order

2. **Creation of return delivery with reference to returns purchase order**
Shipping data from the line-item level of the returns purchase order is used to create a return delivery using Transaction VL10B. Best practice is to automate the creation of the returns delivery (outbound) using a batch job (background processing). The return delivery facilitates the movement of materials from the issuing plant mentioned in the returns purchase order item. Picking and packing of materials will be performed within inventory management with reference to the outbound delivery.

3. **Goods issue for return delivery**
After the picking and packing of returned materials, the goods issue will be posted from the issuing plant. Material documents and accounting documents will be updated for the goods movement while updating the quantity and value of the issued stock in the inventory management. Movement types play an important role in determining the type of inventory movement and posting of goods issue. Movement type 161 is used for goods issue for vendor returns delivery.

4. **Process supplier credit memo**
The supplier sends a credit memo to the buyer with reference to the returns purchase order via email (paper invoice), EDI, or electronically via SAP Business Network. Supplier credit memos will be posted by the accounts payable department of the buying organization manually in the SAP S/4HANA system using Transaction MIRO if received via email, mail, or fax. Automatic posting and reconciliation of the credit memo happens if it is received electronically from the supplier. Once the credit memo is fully reconciled and posted in the system, the next scheduled payment run will pick up this fully reconciled credit memo and process it.

1.1.8   Outline Agreements

Purchase orders are used to formally procure goods and services, and they do not promise a long-term relationship with the suppliers. Every time a purchase order is submitted, it doesn't guarantee the required quantity of materials and delivery dates. Prices of materials change over time and require regular updates to purchase prices in info records. Payment terms, Incoterms, and other terms and conditions must be agreed upon with suppliers, and purchase orders do not always guarantee that.

Outline agreements, on the other hand, are long-term agreements between a buying organization and a supplier. Outline agreements promise long-term relationships with suppliers for procuring specific products and services. These agreements are executed and created based on terms agreed upon between the buying organization and the supplier to provide specific products and services at a discounted price over a long period of time. They are beneficial for both businesses (buyer and supplier), and they ensure that buying organizations procure goods in a timely manner. There are two types of outline agreements, which we'll explore in the following sections.

Scheduling Agreements

Scheduling agreements are deeply integrated with MRP. Once executed and created in the system, you can make a scheduling agreement relevant to MRP in the source list. Schedule lines, also called delivery schedules, are created with reference to scheduling agreement. Scheduling agreements are different from contract agreements in terms of functionality, but both types of agreements have the same structure in SAP S/4HANA. Different document types are used to create contract agreements and scheduling agreements.

Scheduling agreements are outline agreements created with negotiated prices, Incoterms, payment terms, and so on, plus a validity period. Schedule lines can be created with reference to scheduling agreements with the required quantity and delivery dates. Schedule lines are transmitted to suppliers, and shipments from suppliers are received against the scheduling agreement. Figure 1.21 shows the scheduling agreement process in SAP S/4HANA.

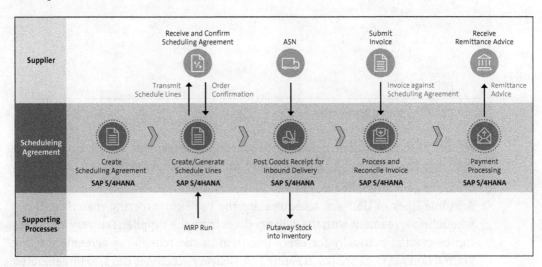

Figure 1.21 Scheduling Agreement Process

Let's explore the key process steps:

1. **Execute and create a scheduling agreement**

   A scheduling agreement can be created with reference to a purchase requisition. Prior to the creation of a scheduling agreement, buyers or commodity managers negotiate terms and conditions with the supplier. A scheduling agreement is created in SAP S/4HANA using Transaction ME31L. Figure 1.22 shows the scheduling agreement header. The scheduling agreement is executed by a purchasing organization with the supplier. You can see both the supplier and the purchasing organization details in the header. The validity period of the agreement, negotiated payment terms, and Incoterms are part of the header details.

   SAP S/4HANA assigns an internal number to the scheduling agreement from the number range assigned to the scheduling agreement document type.

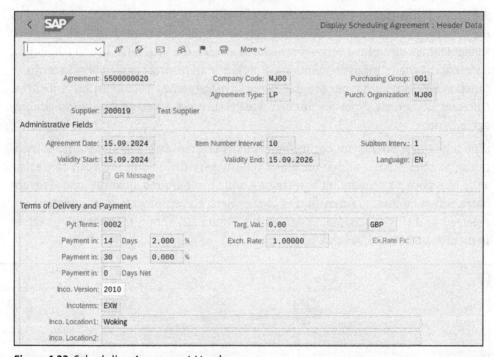

Figure 1.22 Scheduling Agreement Header

   Figure 1.23 shows the item details of the scheduling agreement. It contains the material, negotiated unit price, target quantity, plant, and other details.

2. **Create or generate schedule lines**

   Schedule lines or delivery schedules are the basis for ordering materials in the scheduling agreement with the delivery dates from the supplier. Delivery schedules can be created manually for every line item of the scheduling agreement using Transaction ME38, as shown in Figure 1.24. Delivery schedules can also be generated from the MRP run automatically. A source list must be maintained with the source of supply set as a scheduling agreement, and the MRP relevance must be marked.

Like purchase order transmission, delivery schedules are transmitted to the supplier. The supplier will send confirmation for the delivery schedule and the ASN, depending on the confirmation control key maintained in the scheduling agreement item.

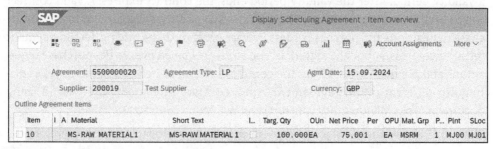

Figure 1.23 Scheduling Agreement Item

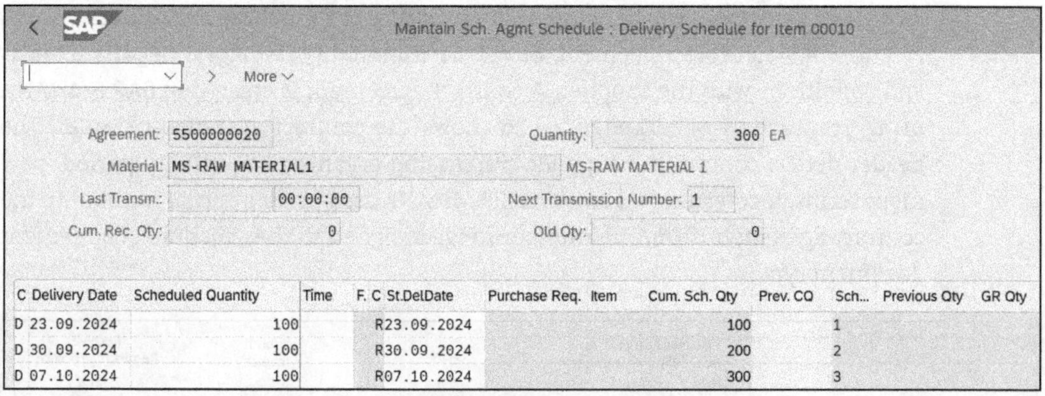

Figure 1.24 Maintain Delivery Schedule for Scheduling Agreement Item

3. **Goods receipt and invoice receipt postings**
   The goods receipt and invoice processing for the scheduling agreement are like the purchase order goods receipt and invoice posting for the purchase order. The goods receipt can be processed using an inbound delivery, or directly using Transaction MIGO if inbound delivery creation is not required based on the confirmation control key.

The incoming invoice from the supplier is posted with reference to the scheduling agreement.

Contract Agreements

Contract agreements are like scheduling agreements in terms of their structure, but they differ in functionality. Delivery dates or delivery schedules can't be created directly for the contract agreement. Instead, release orders (purchase orders with reference to a contract agreement) are created whenever materials are to be procured with reference to the contract. Delivery dates and required quantities are added to the release orders. Contract agreements are executed and negotiated by the purchasing department along with the respective commodity owners. Contract agreements are then created in SAP S/4HANA with a negotiated price, Incoterms, payment terms, and a validity period. After the validity period ends, the contract expires. A valid contract can be used to create source lists. You can reference a contract agreement to create purchase requisitions and purchase orders (release orders). All the pricing details, payment terms, Incoterms, suppliers, materials, and so on are copied over to the purchase requisitions and purchase orders from the contract agreement items. Document types distinguish different outline agreement types. Contract agreements are created using document types **MK** (quantity contract) and **WK** (value contract).

Figure 1.25 shows the contract agreement process in SAP S/4HANA. Let's explore the key process steps:

1. **Execute and create a contract agreement**

   A contract agreement can be created with reference to a purchase requisition. Prior to creation of a contract agreement, buyers or commodity managers negotiate terms and conditions with the supplier. A contract agreement is created in SAP S/4HANA using Transaction ME31K. Figure 1.26 shows the contract agreement header. The header details contain the supplier, purchasing organization, validity period, payment terms, Incoterms, and so on. SAP S/4HANA assigns an internal number to the contract agreement from the number range assigned to the scheduling agreement document type.

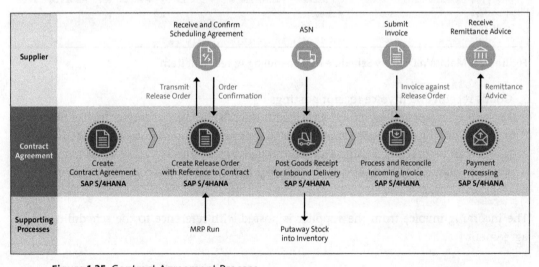

Figure 1.25 Contract Agreement Process

Figure 1.27 shows the item details of the contract agreement. These details include the material, negotiated unit price, target quantity, plant, and more.

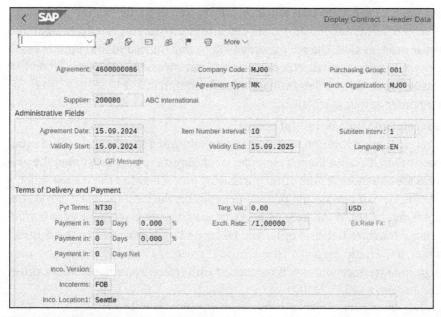

Figure 1.26 Contract Agreement Header

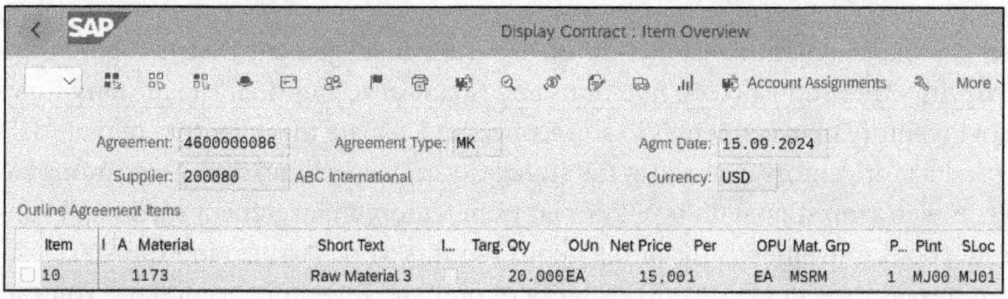

Figure 1.27 Contract Agreement Item

2. **Create release orders**

   Release orders are created with reference to contract agreement items using Transaction ME21N. Release orders are required to formally order materials from a supplier with reference to the contract agreement. Figure 1.28 shows the release order is being created with reference to contract agreement **4600000086** and contract item **10**.

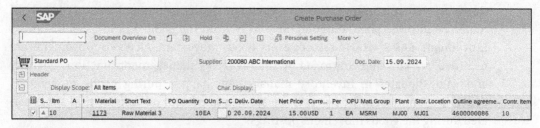

Figure 1.28 Release Order with Reference to Contract Agreement

The unit price for the material is automatically copied from the contract agreement item. Payment terms and incoterms are defaulted from the contract agreement into the release order as well. The release order will be transmitted to the supplier based on the output method: email, EDI, or electronic transmission to the supplier on SAP Business Network. Output determination configuration in SAP S/4HANA drives the purchase order output mechanism.

3. **Goods receipt and invoice receipt postings**

   Goods receipt and invoice processing happens with reference to the release order (purchase order). Because the release order is nothing but a purchase order, the process steps are like those for the other purchasing processes after the release order is created. The type of purchasing process depends on the item category of the contract agreement item. If the item category of the contract agreement item is *standard*, then a standard (release) order will be created with reference to the contract agreement; if the item category of the contract agreement item is *consignment*, then a consignment (release) order will be created with reference to the contract agreement; and so on.

1.2   Inventory Management

Inventory management is a function of materials management that is highly integrated with other supply chain functions and financial accounting. The primary function of inventory management is to store, track, and valuate the inventory of materials and products. Inventory is stored at the storage location level, and storage locations are lower-level organizational units below plants. Inventory management tracks and controls all goods movements, including goods receipts, goods issues, and transfer postings. It always provides real-time visibility of on-hand inventory accurately. You can query the on-hand inventory of a material at a plant level using Transaction MMBE.

Later chapters in this book will provide an integrated view of the inventory management function in other areas in SAP S/4HANA. But here, we'll briefly cover the different features and capabilities of inventory management in materials management.

1.2.1   Goods Movements

Goods movements in inventory management signify incoming inventory (stock) of materials into the plant, outgoing inventory from the plant, and internal inventory movements. SAP S/4HANA records every goods movement. All goods movements are recorded, tracked, and controlled by the inventory management function. Quantity and value updates in inventory management happen during goods movements depending on movement types. There are two important documents that are generated during the goods movement postings:

1. **Material document**

   SAP S/4HANA records every material movement in inventory management with a document called the material document. It records detailed information about the material movement transaction, such as the posting date of the transaction, material number, quantity of the material, movement type, batch information, plant, storage location, use, and who posted the goods movement. Quantity updates in inventory management happen via a material document.

2. **Accounting document**

   SAP S/4HANA records every material movement in inventory management involving financial transactions with an accounting document. It records the detailed information about the financial transaction associated with the goods movement and it is linked to the respective material document. Automatic account determination configuration plays a vital role in the accounting document upon the posting of goods issues, goods receipts, and certain transfer postings. Storage location to storage location goods movement within the same plant doesn't create an accounting document. Value updates in financial accounting happen with the accounting document.

Movement types play an important role in goods movement postings that create material documents and accounting documents. Table 1.4 displays the key movement types used in inventory management for goods receipt, goods issue, physical inventory, and internal inventory movements (transfer postings).

Movement Type	Detailed Description
101	Goods receipt. This movement type is used for standard goods receipts from external vendors and subcontractors as well as goods receipts against production/process orders to receive in-house-produced products into inventory. It increases the stock quantity and value in the receiving storage location. Stock can be received into unrestricted use stock or quality inspection stock.
102	Goods receipt reversal. This movement type is used to cancel the goods receipt posted using movement type 101. Stock will be removed from unrestricted use stock or quality inspection stock upon posting this goods receipt reversal transaction.
103	Goods receipt for purchase order. This movement type is specifically used for goods receipts related to purchase orders into goods receipt blocked stock.
105	Release from blocked stock. This movement type is specifically used for releasing from blocked stock.
501	Goods receipt without purchase order. This movement type is used for goods receipts without a corresponding purchase order. It's typically used for miscellaneous or unplanned receipts.

Table 1.4 Key Movement Types in Inventory Management

Movement Type	Detailed Description
122	Return to vendor. This movement type is used to return materials to the vendor that can't be used in production due to poor quality. Stock will be removed from inventory to return the goods to vendor.
161	Returns purchase order to vendor. This movement type is used to return materials to the vendor with reference to a returns purchase order. Stock will be removed from the inventory to return the materials to the vendor.
301	Transfer posting plant to plant. This movement type is used to transfer materials from one plant to another within the same company code.
309	Transfer posting material to material. This movement type is used to transfer stock of one material to another material within the same plant.
311	Transfer posting from one storage location to another storage location within the same plant. This movement type is used to transfer materials from one storage location to another within the same plant.
321	Transfer posting from quality inspection to unrestricted. This movement type is used to transfer materials which are accepted in the quality inspection into unrestricted use stock.
201	Goods issue for a cost center. This movement type is used to issue goods from inventory to a cost center for consumption.
261	Goods issue for a production order. This movement type is used to issue spare parts and other components from inventory to a production order for consumption.
221	Goods issue for a project. This movement type is used to issue goods from inventory to a project for consumption.
281	Goods issue for a network. This movement type is used to issue goods from inventory to a network for consumption.
531	Goods receipt by-product from production. This movement type is used to post goods receipts for by-products from production against production/process orders to receive them into inventory. It increases the stock quantity and value in the receiving storage location.
541	Goods issue for subcontract supplier. This movement type is used to issue equipment and other components to the subcontractor (service provider) to repair/maintain on behalf of the company that issued the subcontract order. Stock will be transferred from unrestricted use stock to subcontracting stock within inventory management.
543	Consumption from subcontracting. This movement type is used to post the consumption of components provided to the subcontractor automatically upon posting the receipt of the final product produced/assembled by the subcontractor. This goods movement is automatically posted during goods receipt with reference to the subcontract order.

Table 1.4 Key Movement Types in Inventory Management (Cont.)

Movement Type	Detailed Description
545	Goods receipt by-product from subcontracting. This movement type is used to post goods receipts for by-products from subcontracting against a subcontract order.
601	Goods issue for delivery. This movement type is used to issue goods from inventory for an outbound delivery to fulfill a customer order.
641	Goods issue for a stock transport order. This movement type is used to issue goods from inventory for an outbound delivery to fulfill an intracompany stock transfer order.
643	Goods issue for a stock transport order. This movement type is used to issue goods from inventory for an outbound delivery to fulfill an intercompany stock transfer order.
701	Inventory differences in unrestricted use stock. This movement type is used to post differences from unrestricted use stock that arise from the physical inventory process.
703	Inventory differences in quality inspection stock. This movement type is used to post differences from quality inspection stock that arise from the physical inventory process.
707	Inventory differences in blocked stock. This movement type is used to post differences from blocked stock that arise from the physical inventory process.

Table 1.4 Key Movement Types in Inventory Management (Cont.)

1.2.2   Physical Inventory

Physical inventory is an important internal process within inventory management. Businesses largely benefit from physical inventory where it improves the overall efficiency of inventory operations while reducing costs. This is an internal process in which products in the storage locations in inventory management are counted periodically and compared with the system stock records for the same products. Physical inventory can be performed on all stock types (unrestricted use stock, quality inspection stock, and blocked stock). Any difference found between the physical stock and the system stock record will be posted to adjust the stock levels in the system. This difference can be positive or negative, and the difference posting during physical inventory automatically updates the material quantities in inventory management with reference to the material document and the value updates in financial accounting with reference to the accounting document.

Table 1.4 shows the standard movement types for physical inventory processes in inventory management in SAP S/4HANA; these movement types are used to post the differences found during physical inventory based on the stock type.

The frequency (weekly, monthly, quarterly, annually, etc.) of physical inventory depends on multiple criteria: the type of material, business reasons, purposes, stock level monitoring, regulatory reasons, and more. It is recommended to perform physical inventory on all less critical products and materials annually. Figure 1.29 shows the physical inventory process in SAP S/4HANA.

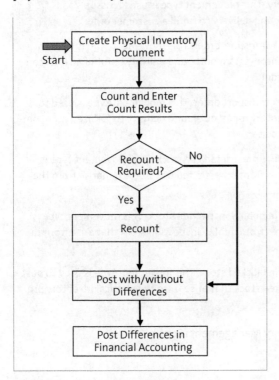

Figure 1.29 Physical Inventory Process

Let's explore the process steps:

1. **Create physical inventory document**
   The process starts with the creation of a physical inventory document in SAP S/4HANA. This document serves as a basis for physical inventory in inventory management and contains the material, plant, storage location, stock type, batch, and other details. Use Transaction MI04 to create the physical inventory document. You can change the physical inventory document to add/remove materials to be counted in physical inventory using Transaction MI05. Figure 1.30 shows the physical inventory document created in SAP S/4HANA.

2. **Counting**
   Counting of the physical stock in inventory management starts after the physical inventory document is created. Inventory management personnel perform this activity based on the physical inventory document details. After the counting is performed at the designated storage location for the materials listed in the physical

inventory document, count results such as actual quantity, counting date and time, and so on are entered manually using Transaction MIO5.

3. **Post physical inventory differences**
   After the actual quantity from the counting results is documented, if there are any differences between the actual quantity and the system record quantity, the differences are posted with reference to the physical inventory document using Transaction MIO7.

4. **Recount and post differences in inventory management**
   Before posting the differences, the system allows you to recount and post the differences. You can use Transaction MIO8 to recount and post the difference in the same transaction.

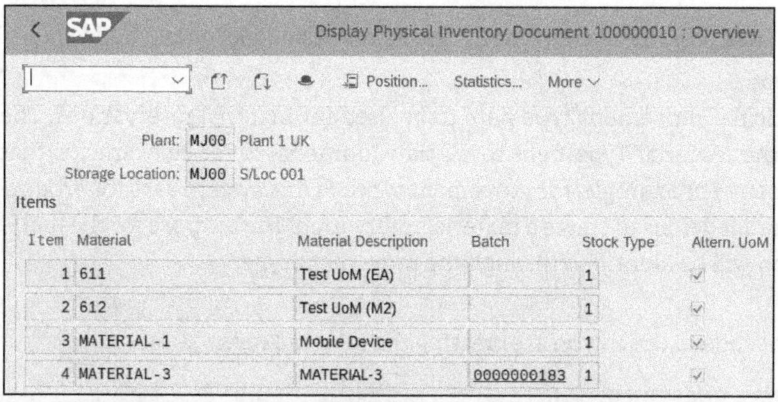

Figure 1.30 Physical Inventory Document

1.2.3   Reservations

Material reservations are requests made to inventory management to reserve required materials for a specific process, typically for a cost center for consumption, a production process, plant maintenance, or network activities. The process for which the reservations are created is defined by the specific movement types used to create reservations in SAP S/4HANA; for example, a reservation with movement type 201 is created for goods issue/consumption for a cost center, movement type 261 is for goods issue/consumption for a production order, and so on. Material reservations ensure the materials are available for the desired purpose and prevent them from being used for another purpose. Material reservations can be created manually using Transaction MB21 and automatically from production planning, maintenance processes, network activity, and so on. Available-to-promise (ATP) checks and MRP runs recognize material reservations, and it's possible to configure both to consider reservations.

Material reservations offer several advantages in production planning, inventory management, and purchasing. Material reservations ensure smooth production operations and facilitate accurate production planning by providing visibility into available materials for future use. The production lead times will be reduced with material reservations and ultimately bring customer satisfaction. On the other hand, material reservations help in effective inventory management and control. They also help prioritize the purchasing process to procure additional stock of reserved materials.

Let's explore the creation of a material reservation in SAP S/4HANA using Transaction MB21. Figure 1.31 shows the initial screen for the creation of a material reservation, where you fill out the following fields:

- **Base Date**
  This is the base date for the reservation. The system uses this date as a requirement date for materials from the inventory.

- **Check Date**
  If you set this indicator, the system checks the factory calendar for any holidays.

- **Movement Type**
  Specify a particular movement type here to be used for material reservations. The system uses the **Material Type** field to display additional fields relevant for that goods movement. For example, for movement type 261, the system asks for production/process order details to create a material reservation. The purpose for the material reservation will be identified through the movement type.

- **Plant**
  Enter the plant code in which you are creating the material reservation.

Figure 1.31 Create Reservations: Initial Screen

Figure 1.32 shows the items screen of the material reservation. For movement type **201** (goods issue for cost center), **Cost Center** and **G/L Account** must be specified to create the reservation.

You can enter multiple line items in one reservation with different materials, quantities, batches, and so on.

Set the **Movement Allowed** indicator to allow goods movement for the reservation. If you deselect this indicator, the system prevents goods movements for this reservation.

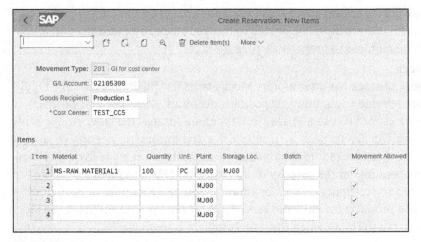

Figure 1.32 Create Reservations: New Items

1.2.4   Stock Types

In SAP S/4HANA, stock types are various categories of stock in inventory management and warehouse management. The stock type controls goods movement, consumption, availability checks, disposition of defective stock, goods issues for customer orders, and goods issues for stock transfer orders. Stock posting between different stock types is managed in inventory management whenever the status of the materials changes. There are three standard stock types defined for business transactions:

1. **Unrestricted use stock**

   This stock category for a material is available for consumption without any restrictions. The stock can be used in sales, production, or direct consumption. You can reserve some of the critical materials stock from unrestricted use stock. This indicates that the stock has passed quality inspection and is certified for use. During goods receipt, you can post the stock directly into unrestricted use stock if quality inspection requirements do not exist for the materials.

2. **Quality inspection stock**

   The status of this material stock category is awaiting quality inspection; a usage decision has not yet been made in quality management. The usage/consumption of this type of stock is restricted. However, some critical materials will be available for the stock availability check depending on the business requirements; necessary settings are required for the availability check in SAP S/4HANA. In certain business scenarios, stock transfers from one plant to another are done from quality inspection stock. During the goods receipt posting with reference to purchase order, the stock will be posted to quality inspection stock automatically if the required quality inspection settings exist and the stock can be posted manually into quality inspection during the goods receipt. Based on the inspection results, the usage decision will be made for the inspection stock. If the stock passes the quality inspection, the stock will be posted to unrestricted use stock automatically. Similarly, if the quality inspection fails, the stock will be posted to blocked stock automatically and follow-up actions are required to properly dispose the stock.

3. **Blocked stock**

This category of stock has a temporary block status for quality reasons, for regulatory reasons, or due to awaiting disposition decisions. Materials and products in blocked stock cannot be used. These stock types are for rejected stock from quality management. Quality inspection results drive the automatic posting of quality inspection stock to blocked stock based on the usage decision made in quality management. Depending on the severity of the problem/defect, blocked stocks are further disposed to scrapping or vendor returns. You can also post certain stock of materials and products into blocked stock temporarily from unrestricted use stock and later remove the block when the issue/problem is resolved.

In addition to the three standard stock types, there are special stock types in SAP S/4HANA that represent a specific status and category. Table 1.5 shows the special stock types defined in SAP S/4HANA.

Special Stock Indicator	Special Stock Text
B	Customer stock
C	Subcontracting customer stock
E	Orders on hand
F	Subcontracting customer order stock
I	Subcontracting returnable transport packaging
J	Subcontracting vendor consignment
K	Supplier consignment
M	Returnable transport packaging vendor
O	Subcontracting stock
P	Pipeline material
Q	Project stock
R	Subcontracting (SC) project stock
T	Stock in transit
V	Returnable packaging with customer
W	Customer consignment
Y	Shipping unit (warehouse)

Table 1.5 Special Stock Types

The usage of stock types and special stock types will be explained in later chapters.

1.3   Logistics Invoice Verification

LIV is a vital function of materials management in SAP S/4HANA. The incoming invoice submitted by a supplier is verified by the accounts payable department personnel for purchase order/scheduling agreement reference, unit price, total amount, tax details, bill-to information, supplier details, and other details of the invoice. The process starts with procurement, where the buying organization issues a purchase order or a scheduling agreement with a delivery schedule to the supplier to procure goods and services. After the fulfillment of purchase order/delivery schedule, the supplier submits an invoice to the buying organization. That triggers the LIV process, mostly handled by the accounts payable department of the buying organization.

Incoming invoices are submitted via different methods:

- **Paper invoice**
  Suppliers submit paper invoices via email in the form of PDF copy or TIF image, by fax, and by mail in the form of a physical copy. The accounts payables team processes paper invoices by manually entering them into SAP S/4HANA using Transaction MIRO to create an LIV document.

- **Electronic invoice**
  Suppliers submit electronic invoices via EDI or from a business-to-business (B2B) platform such as SAP Business Network. Electronic invoices are posted directly in SAP S/4HANA via IDoc processing or via an API, and a LIV document is automatically created. This submission method ensures touchless processing of incoming invoices, eliminating manual errors. Leveraging B2B platforms provides added advantages for both suppliers and buyers with automated and standardized processes. It reduces the invoice-to-pay cycle time with a minimum amount of manual intervention.

The following sections introduce the invoice reconciliation process, which is a vital LIV function.

1.3.1   Invoice Reconciliation

Invoice reconciliation is a process of matching supplier-submitted invoice details with the purchase order details and the goods receipt details to ensure that all transaction details are accurate and complete before posting the invoice for payment. This process, executed by the accounts payable department, helps in reducing discrepancies and ensuring vendors are paid the right amount at the right time for the procured goods and services. Paying the vendors at the right time is crucial for the organizations: it ensures the amount is not overdue and helps prevent penalties, while increasing the cost savings associated with early payment discounts based on agreed-upon payment terms. Thus, reconciliation of incoming invoices and addressing matching exceptions is a vital step of the accounts payable (invoice-to-pay) process. There are two types of matching for purchase order-based invoices, which you can control with standard settings in SAP S/4HANA based on business requirements:

- **Three-way match**
  With three-way matching of an incoming invoice, a purchase order, a receipt, and an invoice are all required to process the invoice. Goods receipt or confirmation is also required from the receiver to process the invoice.

  In general, the invoice is processed for payment without any further approval if the three-way match passes if the invoice purchase order line-item unit price matches the unit price of the respective invoice line item and the overall cumulative invoice total amount is below the purchase order total amount.

  In situations where the goods receipt is missing or not of a sufficient quantity, then the incoming invoice is posted but is blocked for payment processing. To release the payment block, sufficient goods receipt quantities must be posted with reference to the purchase order. The blocked incoming invoice is released for payment once the goods receipt is posted to match the invoice quantity.

- **Two-way match**
  Two-way (purchase order and invoice only) matches are mainly used for low-value, low-risk purchases/categories such as utility charges and the like. A goods receipt with reference to a purchase order is not required to reconcile the incoming invoice. The incoming invoice is directly matched against the purchase order details. If the invoice line item unit price and quantity do not match the respective purchase order item quantity and unit price, the incoming invoice is posted but is blocked for payment processing. To release the payment block, the purchase order item details must be updated to match the invoice, or the invoice must be disputed.

1.3.2 Evaluated Receipt Settlement

In the standard invoice-to-pay process, the supplier submits the invoice to the buying organization, and the accounts payable department receives and processes the incoming invoice. Purchase order, goods receipt, and invoice items are matched to find any discrepancies in price, quantity, tax, and other details. Duplicate invoice checks will also be performed on the received invoice.

Evaluated receipt settlement (ERS) is an automatic invoicing process where the buying organization generates the invoice on behalf of the supplier with reference to purchase order/scheduling agreement and goods receipt. The manual process of reconciling, resolving invoice exceptions, and approving overhead costs is eliminated with evaluated receipt settlement process. The ERS report (Transaction MRRL) is scheduled in the background to run periodically (daily, weekly, biweekly, etc.). This report checks if the supplier is enabled for ERS and if the purchase order line item has the **ERS** flag set. The report extracts the goods receipt items to be invoiced (only) for the purchase order line items that are subjected to ERS and automatically generates a invoice. A copy of the generated invoice will be shared with the respective suppliers for their records. The following are the key features of the ERS process:

- An invoice is generated and reconciled automatically.
- Matching exceptions such as a goods receipt quantity less than the invoice quantity never occur.
- Tax-related invoice exceptions never occur as the tax code is defaulted from the PIR automatically.
- Unit price and quantity discrepancies never occur as the invoice price is defaulted from the purchase order/scheduling agreement, and quantity is defaulted from the goods receipt line item quantity.
- A duplicate check of the invoice is not required.

There are certain prerequisites required to process ERS:

1. The supplier and buying organization must be aligned to enable the ERS process.
2. The goods receipt-based invoice verification indicator (**GR-Bsd IV**) must be set for the supplier (business partner) at the purchasing organization level.
3. The PIR must be created at the purchasing organization and plant level.
4. An ERS flag for the combination of the material and supplier must not be set in the PIR.
5. A tax code must be maintained in the info record as well as the purchase order at the line item level.

Figure 1.33 shows the purchase order line item details. The **Invoice** tab of the purchase order line item shows that the **ERS** flag is enabled, the **GR-Bsd IV** flag is enabled, and an input **Tax Code** is maintained.

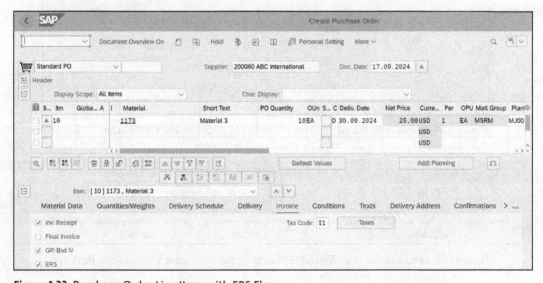

Figure 1.33 Purchase Order Line Item with ERS Flag

This shows that purchase order **4500000726** is subject to ERS. Let's simulate the ERS manually.

Figure 1.34 shows the ERS report that you can execute to perform logistics invoice verification and generate invoice receipts automatically. You can schedule this report to run in the background, which is the best practice. To run ERS manually, execute Transaction MRRL in SAP S/4HANA. **Plant**, **Company Code**, **Supplier** (business partner) number, and **Document Year of Goods Receipt** are the important selection criteria to run the report.

This report can be run in test (simulation) mode manually by selecting the **Test Run** checkbox; doing so will not generate the invoice receipt, but it will display error messages if there are any.

The report fetches the goods receipt items posted with reference to purchase order **4500000726** and automatically posts the invoice against it. Figure 1.35 shows invoice receipt **5105600233** generated automatically for purchase order **4500000726** and with reference to goods receipt material document **5000000572**.

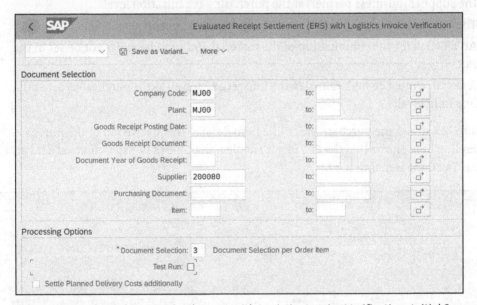

Figure 1.34 Evaluated Receipt Settlement with Logistics Invoice Verification: Initial Screen

Figure 1.35 Evaluated Receipt Settlement with Logistics Invoice Verification: Postings

1.3.3 Consignment Settlement

Consignment settlement is another special LIV process for settling consignment purchase orders based on consumption and withdrawal of consignment stock that belongs to the supplier. With reference to a consignment order, the supplier loans materials to the purchasing company, and the purchasing company is not liable to pay to the supplier until it withdraws or directly consumes the consignment stock that belongs to the supplier.

The consignment stock, once received from the supplier with reference to a consignment order, can be withdrawn/posted to unrestricted use stock via a transfer posting transaction using movement type 411 K (where K represents consignment stock) and using Transaction MIGO. This transfer posting movement transfers the ownership of the stock from vendor consignment to company-owned stock. The system records this goods movement as a liability in financial accounting. The value of the withdrawn stock from consignment to unrestricted use (company-owned) stock is required to be settled with the supplier.

Consignment stock can be directly issued/consumed in production processes, maintenance processes, and other processes using movement types 201 K, 261 K, 281 K, and so on. This is known as *direct consumption* of consumption stock. The system records these goods issue movements as liabilities in financial accounting. The value of the consumed materials from consignment stock is required to be settled with the supplier.

SAP S/4HANA provides a report that is similar to ERS, called the consignment settlement report, which generates an invoice on behalf of the supplier to settle the withdrawn and consumed stock from consignment stock.

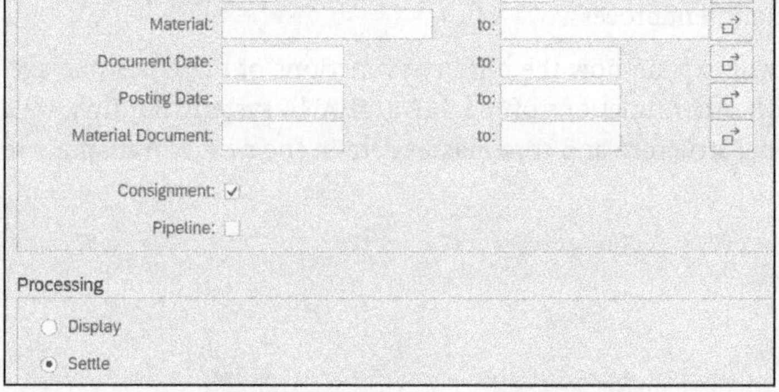

Figure 1.36 Consignment Settlement: Initial Screen

Figure 1.36 shows the consignment settlement report that you can execute to generate invoice receipts automatically. You can schedule this report to run in the background, which is the best practice. To run consignment settlement manually, execute Transaction MRKO in SAP S/4HANA. **Plant**, **Company Code**, **Supplier** (business partner) number, and **Posting Date** of the goods issue (consumption) and stock withdrawal material documents are the important selection criteria to run the report.

1.4   Summary

In this chapter, we introduced the vital functions of integrated materials management, including key master data elements. The functional components explained in this chapter are integrated with other functionalities of an SAP S/4HANA enterprise resource planning (ERP) system. The subsequent chapters of this book provide detailed information about the materials management integration with other functions of the SAP S/4HANA system.

In Section 1.1, we explained the core purchasing process, including standard and special procurement processes such as subcontracting, vendor consignment, stock transfer, and the third-party process. We also explained the returns purchase order process and introduced advanced returns management in SAP S/4HANA. The purchasing process is incomplete without the outline agreements due to their significance for contract agreements and scheduling agreements.

In Section 1.2, we explained inventory management, another vital function of materials management. We explained the various goods movements (material movements) in inventory management with movement types and examples. We also explained physical inventory processes, material reservations, and different stock types in inventory management.

Finally, in Section 1.3, we explained the LIV process in materials management. We explained how to process incoming invoices (paper invoices and electronic invoices) from suppliers with invoice and tax reconciliation. We also explained the ERS concept for generating invoices on behalf of suppliers, and consignment settlement, which is part of the vendor consignment process.

In future chapters, we will explain how the different functions of materials management are integrated with other functions of the SAP S/4HANA system. But first, we'll discuss the organizational structure and core master data in the next two chapters to set a foundation.

Chapter 2
Organizational Structure

The organizational structure, or enterprise structure, is the framework for how an organization's entities, business units, and departments are structured in SAP S/4HANA from the business and legal perspectives. This structure helps manage business processes, transactions, and data within the system. It also ensures the transactional data and master data are accurately captured for the business and that regulatory processes are followed for tracking and reporting.

The organizational structure is the most essential hierarchical structure of business units such as entities, departments, and employee groups managing specific business processes. It also provides a reporting structure within a business area (finance, purchasing, sales, manufacturing, plant maintenance, inventory management, warehouse management, etc.) as well as the entire organization. Implementation of SAP S/4HANA starts with defining the enterprise structure during the blueprinting phase and creating the enterprise structure in the system during the build phase. The hierarchical structure of the organization and its business processes will drive the definition of the organizational structure. An organizational structure is divided into the following areas in SAP S/4HANA:

- Financial accounting and controlling
- Materials management
- Sales and distribution
- Logistics general
- Logistics execution
- Plant maintenance
- Human resources (HR) management

Before exploring organizational elements from all the main functions in SAP S/4HANA, let's discuss the highest-level organizational unit: the client.

2.1 Client

The client is the highest-level organizational unit, and it represents an independent organization in SAP S/4HANA. Each client has its own organizational structure representing a hierarchical structure of business units and entities. Each client has one or more company codes, plants, business units, departments, and so on. The client in SAP S/4HANA is also used as a technical unit to secure master data and transactional data corresponding to a specific organization within the SAP system. Users can log into the SAP S/4HANA system to manage, monitor, and report on business transactions using the client code defined for the organization. The master data and transactional data of the organization or client is protected from other organizations or other clients in SAP S/4HANA.

The client is also used to define configuration settings to control various business processes. The configuration data tables are maintained at the client level as they are applicable to the entire organization. Some of the master data such as basic data views of the material master and classification data are also defined at the client level. The data stored at the client level is accessible by all company codes.

The system landscape of SAP S/4HANA will have a development instance, quality assurance (QA) instance, and a production instance. This system landscape supports the implementation of SAP S/4HANA and management of changes. Each instance of SAP S/4HANA must have at least one client. Technically, business users log into a client to access and transact within the SAP system. The client is created during the setup of the SAP S/4HANA instance (development, quality, or production) using Transaction SCC4. Figure 2.1 shows client **100** defined in SAP S/4HANA.

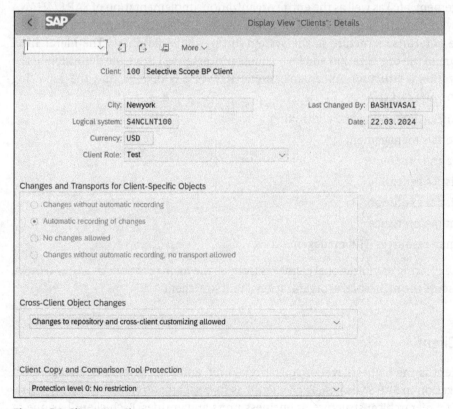

Figure 2.1 Client Details

Now, let's dive deep into the organizational structure for all the functions that integrate with materials management, especially for the flow of materials through the supply chain.

2.2 Financial Accounting and Controlling

The financial accounting and controlling organizational structure is the next level under the client in the organizational hierarchy. It provides the framework for organizing and managing financial processes and financial accounting. This helps to improve and manage an organization's financial data. Let's explore the financial accounting and controlling organizational elements in SAP S/4HANA.

Figure 2.2 shows a typical financial organizational structure. Let's explore the organizational units in financial accounting and controlling in detail.

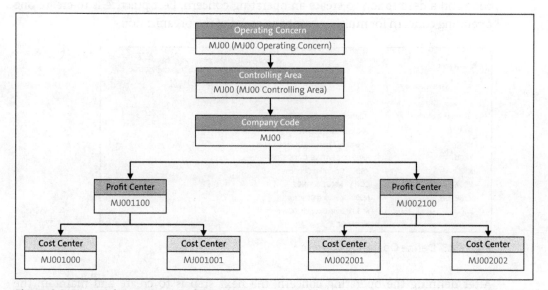

Figure 2.2 Financial Organizational Structure

2.2.1 Operating Concern

The operating concern is the highest-level organizational unit and a key component of profitability analysis. It defines the structure for analyzing profitability data within an organization. The purpose of the operating concern is to perform a detailed analysis of revenues, costs, and margins based on multiple dimensions such as materials, customers, sales area, regions, and so on. The operating concern is structured based on characteristics and value fields. Characteristics define the attributes, based on which the profitability analysis is performed. Some of the characteristic attributes are materials, customers, markets, legal entities, and so on. The value field in the operating concern represents the key performance indicators to be used in profitability analysis. The attributes and value fields of the operating concern are dependent on the business requirements. The operating concern is tightly integrated with finance, materials management, and sales and distribution functions to ensure that profitability analysis data is consistent and accurate across the organization.

We'll explain how to create and maintain key settings for an operating concern in the following sections.

Create Operating Concerns

To define the operating concern, first execute Transaction SPRO and navigate to **Enterprise Structure • Definition • Controlling • Create Operating Concern**.

Figure 2.3 shows the creation of operating concern **MJ00**. Click **New Entries** to create a new operating concern from scratch or copy an existing one to adapt. Enter a four-digit code and a description to create an operating concern. Best practice is to create one operating concern for multiple company codes of the organization.

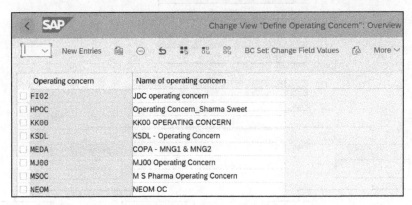

Figure 2.3 Define Operating Concern

After defining the operating concern, the next step is to create and maintain the required characteristics and value fields.

Maintain Characteristics

To maintain characteristics, first execute Transaction SPRO and navigate to **Controlling • Profitability Analysis • Structures • Define Operating Concerns • Maintain Characteristics**.

Figure 2.4 shows the characteristics defined in the standard system. If you require a new characteristic for profitability analysis, you can create one in edit mode by clicking the **Create** icon or pressing F7. Specify the origin table and the original field name from the table for the new field value. A new field value must start with the letters WW, and the maximum length of the field is five characters.

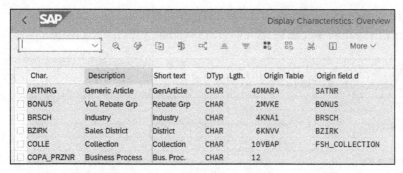

Figure 2.4 Define Operating Concern: Characteristics: Part 1

Figure 2.5 shows the **BZIRK (Sales District)** characteristic as an example. You can start creating the characteristic by defining the characteristic's value and description. Let's explore the field-level settings:

- **Status**
  In this field, you can define the status of the characteristic. You can assign the following values in this field. To use the characteristic in the profitability analysis, the status must be **active**.
  - **A: active**
  - **M: saved**
  - **N: new**
  - **P: partially active**

- **Data element**
  A data element is an ABAP reference object that defines the technical attribute and data type of an object, such as a data table, data field, or structure. Assign the relevant data element that can be used as a reference for this characteristic.

- **Domain**
  The domain is assigned to data elements in SAP. A domain defines the data type and the length of the field.

- **Origin table**
  Enter the table name from the SAP S/4HANA database from which the respective characteristic was chosen.

- **Origin field**
  Enter the field name from the origin table from which the respective characteristic was chosen.

- **Validation**
  You can validate the characteristic values against predefined fixed values with the **Fixed values** radio button, or enter values from a database table. Table T171 stores the

sales districts defined in the system, as shown in Figure 2.6. If you set a **Check table** for validation, only the values stored in table T171 are valid for the characteristic.

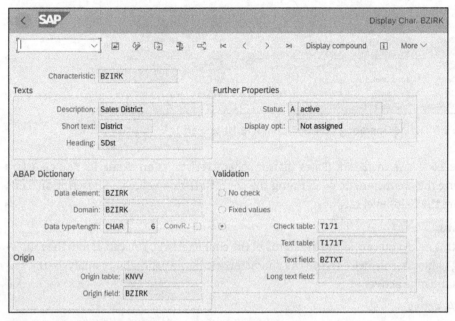

Figure 2.5 Define Operating Concern: Characteristics: Part 2

Table: T171
Displayed Fields: 3 of 3 Fixed Columns:

MANDT	BZIRK	BZTXT
100	US0001	Pacific Northwest
100	US0002	Southwest
100	US0003	Southeast
100	US0004	Northeast
100	US0005	Rockies
100	US0006	Midwest

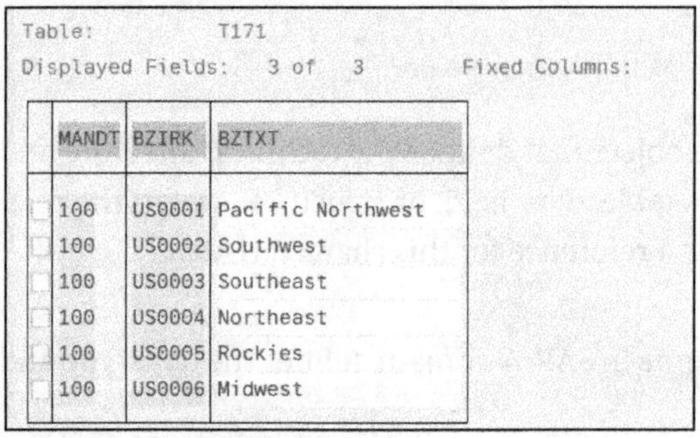

Figure 2.6 Table T171

Maintain Value Fields

To maintain value fields, execute Transaction SPRO and navigate to **Controlling • Profitability Analysis • Structures • Define Operating Concerns • Maintain Value Fields.**

Figure 2.7 shows the value fields defined in the standard system. If you require a new value field for profitability analysis, you can create one in edit mode by clicking the **Create** icon or pressing [F7]. From the popup screen, select an amount field or quantity field from the dropdown list. To select an amount field, set the **Amount** indicator in the popup screen; to select a quantity field, set the **Quantity** indicator in the popup screen.

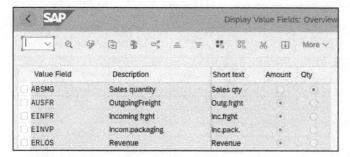

Figure 2.7 Define Operating Concern: Field Values: Part 1

Figure 2.8 shows the **ABSMG** (sales quantity) value field as an example. Let's explore the field-level settings:

- **Status**
  In this field, you can define the status of the value field. You can assign the same values as discussed for characteristics in the previous section. To use the value field in the profitability analysis, the status must be **active**.

- **Value field type: Amount**
  If this indicator is set, it indicates that the respective value field contains amounts.

- **Value field type: Quantity**
  If this indicator is set, it indicates that the respective value field contains quantities.

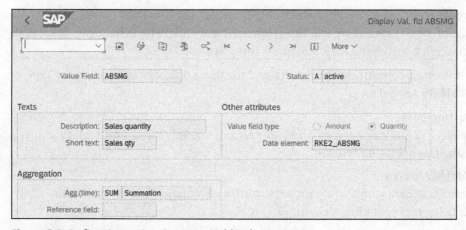

Figure 2.8 Define Operating Concern: Field Values: Part 2

Maintain Operating Concerns

To maintain the operating concern, execute Transaction SPRO and navigate to **Controlling • Profitability Analysis • Structures • Define Operating Concerns • Maintain Operating Concern**.

Figure 2.9 shows the operating concern maintenance screen. Enter "MJOO" for **Operating Concern** to use the operating concern you created earlier and press ⌈Enter⌋.

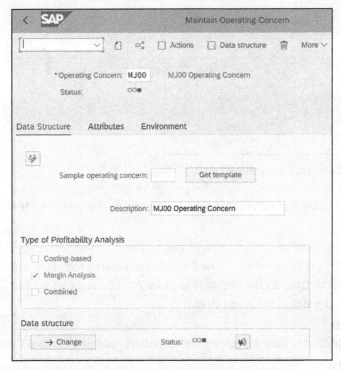

Figure 2.9 Maintain Operating Concern: Data Structure

There are three main tabs for maintaining the operating concern. Figure 2.9 shows the details of the **Data Structure** tab. The green indicator under **Status** indicates the operating concern is active and maintenance is complete.

The operating concern can be maintained for three different options under **Type of Profitability Analysis**:

- **Costing-based**
  Select this indicator if you want to perform costing-based profitability analysis using this operating concern.

- **Margin Analysis**
  Select this indicator if you want to perform a new margin-based profitability analysis using this operating concern.

- **Combined**
  Select this indicator if you want to perform combined costing-based and account-based profitability analysis using this operating concern.

Figure 2.10 shows the details of the **Attributes** tab. The following are the key fields to be maintained in this tab:

- **Operating concern currency**
  The operating concern currency is used in costing-based profitability analysis.

Assigning a currency code to this field indicates the data transferred to profitability analysis is converted and stored in this currency.

- **Fiscal year variant**
  The fiscal year variant in SAP S/4HANA is used to determine the posting period defined in a fiscal year. Assign the relevant fiscal year variant for profitability analysis.

- **Company Code Currency**
  In addition to the operating concern currency, if you want to store the costing-based profitability analysis in the company code currency, set this indicator.

- **OpConcern crcy, PrCtr valuation**
  Set this indicator if you want the profitability analysis to be stored in the operating concern currency with profit center valuation.

- **Comp.Code crcy, PrCtr valuation**
  Set this indicator if you want the profitability analysis to be stored in the company code currency with profit center valuation.

- **Act. 2nd per. type**
  Set this indicator if you want the actual profitability analysis data to be stored by weeks as the second period type.

- **Plan. 2nd per. type**
  Set this indicator if you want the planned profitability analysis data to be stored by weeks as the second period type.

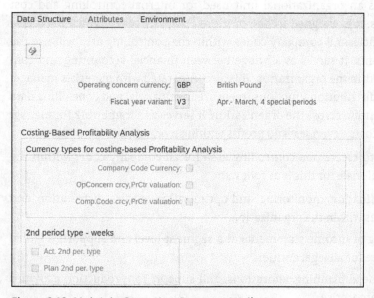

Figure 2.10 Maintain Operating Concern: Attributes

Figure 2.11 shows the details of **Environment** tab. There are two statuses stored in this tab:

- **Cross-client part**

  A green light in this field shows that the cross-client part of the operating concern is active.

- **Client-specific part**

  A green light in this field shows that the client-specific part of the operating concern is active.

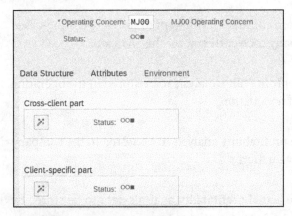

Figure 2.11 Maintain Operating Concern: Environment

2.2.2   Controlling Area

A controlling area is an organizational unit used for internal controlling and cost accounting purposes. It is assigned to one or more company codes, which may have different base currencies. All company codes within the controlling area will use the same chart of accounts. It serves as a bridge between financial accounting and controlling functions within the organization. It is also integrated with materials management and sales and distribution functions to ensure the financial and controlling data is consistent and accurate across the organization. It serves as a framework for managing and controlling costs, revenues, and profits within an organization.

It is recommended to create one controlling area for all company codes within the organization. The rationale for this is as follows:

- It facilitates coordination, monitoring, and optimization of sales, distribution, and purchase processes within the organization.

- It allows reporting of income statements at a segment level and supports a global reporting structure for all legal entities.

- It facilitates overhead planning, allocations, and support for production activities across the entire enterprise.

We'll explain how to define controlling areas and assign them to an operating concern in the following sections.

Definition

To maintain a controlling area, execute Transaction SPRO and navigate to **Enterprise Structure • Definition • Controlling • Maintain Controlling Area**.

Figure 2.12 shows the creation of controlling area **MJ00**. To maintain a new controlling area, click **Maintain Controlling Area Activity** from the **Select Activity** screen. In the next screen, either click **New Entries** or copy an existing controlling area to create a new one. Enter a four-digit code and a description to create a controlling area. Then, fill out the following fields:

- **Person Responsible**
  In this field, you can assign the user responsible for maintaining the controlling area.

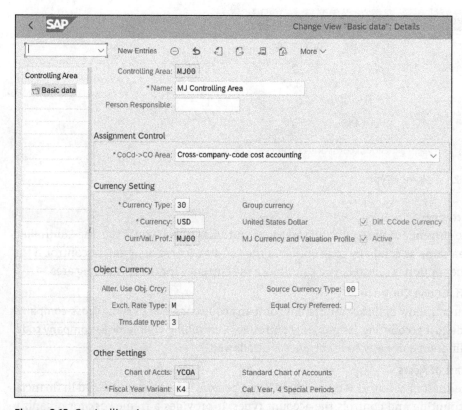

Figure 2.12 Controlling Area

- **Assignment Control: CoCd->CO Area**
  This field controls the assignment of a company code to a controlling area. There are two options:
  - **Controlling area same as company code**
    You can select this option if you want the company code to be the same as the

controlling area. The fiscal year variant, currency code, and chart of accounts can be different for different company codes in this setup.

- **Cross-company-code cost accounting**
  You can select this option if you want to assign multiple company codes to the controlling area. This way, you can have the same fiscal year variant, same currency code, and same chart of accounts for all company codes.

- **Currency Type**
  This field controls the controlling area currency. Table 2.1 shows the list of currency types you can choose from. If you assign the **Controlling area same as company code** option in the **CoCd->CO Area** assignment control, currency type **10** is used by the system by default.

Currency Type	Description
10	Company code currency
30	Group currency
40	Hard currency
50	Index-based currency
60	Global company currency

Table 2.1 Currency Type

- **Currency**
  The currency of the controlling area is set automatically if you assign the **Controlling area same as company code** option in the **CoCd->CO Area** assignment control. If the other option is selected, you can choose the currency for the controlling area.

- **Diff. CCode Currency**
  You can allow or disallow different company code currencies via the **Cross-company-code cost accounting** assignment control. For **Controlling area same as company code**, a different company code currency is disallowed.

- **Chart of Accts**
  The chart of accounts is a structured list of general ledger accounts used in financial accounting and controls the account types. It provides a framework for organizations to report various financial transactions accurately. Depending on the assignment control, you can choose a chart of accounts for controlling area which is applicable to all company codes if assignment control for the controlling area is set to **Cross-company-code cost accounting**.

- **Fiscal Year Variant**
  The fiscal year variant in SAP S/4HANA is used to determine the posting period defined in a fiscal year. Assign the relevant fiscal year variant for profitability analysis.

Assignment

Assignment of an operating concern to a controlling area is required to perform profitability analysis of the controlling area. To arrive at this configuration step to assign an operating concern to a controlling area, execute Transaction SPRO and navigate to **Enterprise Structure • Assignment • Controlling • Assign Controlling Area to Operating Concern**.

Figure 2.13 shows that operating concern **MJ00** is assigned to controlling area **MJ00**. Directly specify the operating concern for the controlling area in the **OpCo** (operating concern) field.

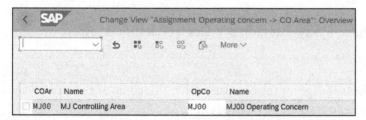

Figure 2.13 Assign Controlling Area to Operating Concern

2.2.3   Company Code

A company code represents an independent legal entity within an organizational structure and is a vital organizational unit for financial accounting. An organization can have one or more company codes. At least one company code is required as financial transactions such as accounts payable, account receivable, asset accounting, and so on are recorded at the company code level. Company codes are also used in controlling activities such as cost center accounting, internal orders, and profitability analysis.

A company code can have several plants, purchasing organizations, and sales organizations. General ledgers are defined at the company code level. Purchasing documents, sales documents, and relevant financial documents are created at the company code level.

Each company code has its own currency code, and the financial transactions are recorded in the company code currency. Financial statements and other legal financial documents are released in the company code currency that includes local tax-related transactions.

Let's explore the company code definition and assignment in SAP S/4HANA.

Definition

To define a company code, execute Transaction SPRO and navigate to **Enterprise Structure • Definition • Financial Accounting • Edit, Copy, Delete, Check Company Code**.

Figure 2.14 displays company code **MJ00**. Click **New Entries** to define new company codes or copy an existing company code to create a new one. Define a four-digit unique alphanumeric or numeric key for the company code and enter a legal name.

A company code has the following additional data:

- **City**
  In this field, enter the name of the city where this company code is registered legally.
- **Ctry/Reg.**
  In this field, enter the ISO country key of the country in which this company code is registered legally.
- **Currency**
  In this field, enter the company code currency. This is typically the respective country's currency.
- **Language**
  In this field, enter the language for the company code. This could be the official language of the respective country.

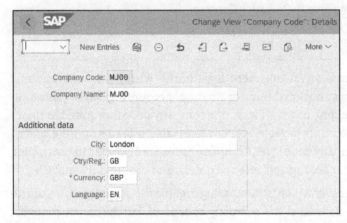

Figure 2.14 Company Code Details

Assignment

Assignment of company codes to a controlling area is required for cost controlling of company codes. To assign company codes to a controlling area, execute Transaction SPRO and navigate to **Enterprise Structure • Assignment • Controlling • Assign Company Code to Controlling Area**.

Click **New Entries** to assign company codes to a controlling area. Figure 2.15 shows that multiple company codes are assigned to controlling area **MJ00** based on the assignment control settings in the controlling area definition.

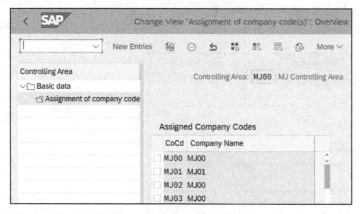

Figure 2.15 Assign Company Codes to Controlling Area

2.2.4 Business Area

A business area represents a specific area of responsibility within the organizational structure. It's an optional organizational unit used for internal reporting and analysis, such as for balance sheets and profit and loss reports for a company code.

To define a business area, execute Transaction SPRO and navigate to **Enterprise Structure • Definition • Financial Accounting • Define Business Area**.

Figure 2.16 shows the business areas. Click **New Entries** to define a new business area for financial accounting.

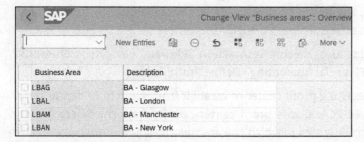

Figure 2.16 Business Areas

Define a four-digit unique key to define a business area and enter a description.

2.2.5 Functional Area

A functional area represents a business unit responsible for a specific function or process within the organizational structure. In financial accounting, functional areas are used in expense reporting.

To define a functional area, execute Transaction SPRO and navigate to **Enterprise Structure • Definition • Financial Accounting • Define Functional Area**.

Figure 2.17 shows the functional areas. Click **New Entries** to define a new functional area for financial accounting.

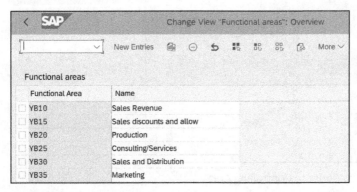

Figure 2.17 Functional Areas

Define a four-digit unique key to define a business area and enter a description.

2.2.6   Profit Center

A profit center is an organizational unit in financial accounting responsible for managing revenues, costs, and profits for a specific product, geographical area (region), or functional area within the organization. The financial performance of a company is generally analyzed at the company code level. Profit centers in SAP S/4HANA provide the ability to analyze financial profits at a more granular level within a company code. Sales, expenses, profits, and return on investment of a product, region, function of an organization, and so on can be analyzed using profit centers.

To maintain a controlling area, execute Transaction SPRO and navigate to **Enterprise Structure • Definition • Financial Accounting • Define Profit Center**.

You can copy from an existing profit center or create it from scratch. To define a new profit center, click the **EC-PCA: Create profit center** activity from the **Select Activity** screen. Specify the controlling area in which you are creating a new profit center. Enter the profit center code and controlling area to create a profit center, then press ⌐Enter⌐. Figure 2.18 shows the general (header) data and basic data of the profit center.

The **General Data** area of the profit center has the following fields:

- **Controlling Area**
  The controlling area is the highest organizational unit in controlling and is used for internal controlling and cost accounting purposes. A profit center is defined within the controlling area.
- **Validity Period**
  The validity period of the profit center is indicated by the from date and to date here for the profit center.

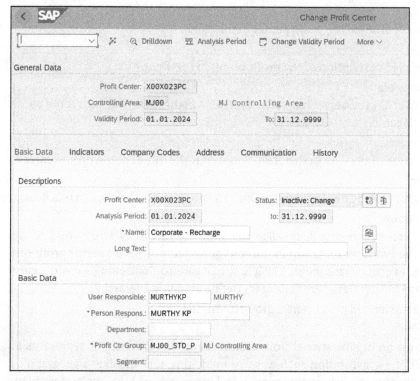

Figure 2.18 Profit Center: Basic Data

The **Basic Data** tab of the profit center has the following fields:

- **Status**
  This field indicates the status of the profit center. To use the profit center in financial accounting, the status must be set to **Active**.
  - **A: Active**
  - **I: Inactive: Create**
  - **U: Inactive: Change**
  - **D: Inactive: Delete**

  Inactive: Create is the status of the profit center during its creation. **Inactive: Change** is the status during a profit center change, but the system also retains the active version of the profit center during the changes. Upon activating the change version of the profit center, the active version is overwritten.

  Inactive: Delete is the status that applies when you set the deletion flag. Upon activating the profit center that is set for deletion, the profit center will be deleted.

- **Analysis Period**
  When the profit center is first created, the analysis period is the same as the validity period. Depending on the business requirements, you can define multiple analysis periods for the profit center that will fall within the validity period.

- **Name**

  Enter a short name for the profit center in this field.

- **Long Text**

  Enter a detailed description of the profit center in this field.

- **User Responsible**

  Enter the user ID of a user stored in SAP S/4HANA in this field. This user is the owner of the profit center.

- **Person Respons.**

  Enter the name of the user in this field. This user is the owner of the profit center.

- **Department**

  Enter the name of the department to which the profit center belongs in this field.

- **Profit Ctr Group**

  The profit center group is also called the standard hierarchy structure. You can maintain the profit center group in the configuration settings. To define profit center groups, execute Transaction SPRO and navigate to **Controlling • Profit Center Accounting • Master Data • Profit Center • Define Standard Hierarchy**.

  Assign the predefined profit center group in this field.

- **Segment**

  A segment is an organizational unit within financial accounting. It represents an area within the organization such as sales, marketing, and the like; a geographical area; or a product that generates revenue and incurs expense. Financial statements can be created against segments.

  Assign the segment to which this profit center belongs to in this field.

Figure 2.19 shows the company codes assigned to the profit center. During the creation of the profit center, the system automatically assigns all company codes that are assigned to the controlling area. But you can control the assignment of profit center to a company code by selecting or deselecting the **Assigned** indicator. If this indicator is set, the profit center is assigned to the respective company code.

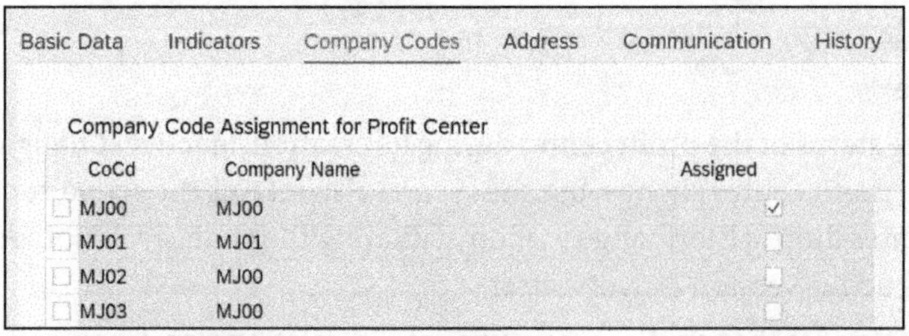

Basic Data	Indicators	Company Codes	Address	Communication	History

Company Code Assignment for Profit Center

CoCd	Company Name	Assigned
MJ00	MJ00	✓
MJ01	MJ01	
MJ02	MJ00	
MJ03	MJ00	

Figure 2.19 Profit Center: Company Codes

2.2.7 Cost Center

A cost center is an organizational unit within a controlling area that represents a defined location where costs are incurred. The cost center could belong to a department, a function, or any other business unit within the controlling organization that incurs costs. Cost centers help to allocate and track costs to different departments and functions of the organization. They are used to track and manage expenses related to indirect procurement activities within an organization.

To create cost centers, navigate to SAP Easy Access menu path **Accounting • Controlling • Cost Center Accounting • Master Data • Cost Center • Individual Processing • KS01—Create**. You can arrive at the same result directly by executing Transaction KS01.

You can copy from an existing cost center or create it from scratch. Enter the cost center code (alphanumeric key), controlling area, and validity start date on the initial screen to create a cost center and press Enter. Figure 2.20 shows the general (header) data and basic data of the cost center.

The general data of the cost center has the following fields:

- **Controlling Area**
  The controlling area is the highest organizational unit in controlling and it is used for internal controlling and cost accounting purposes. A cost center is defined within the controlling area.

- **Valid From**
  This indicates the validity period of the cost center. Enter the valid from and to dates for the cost center.

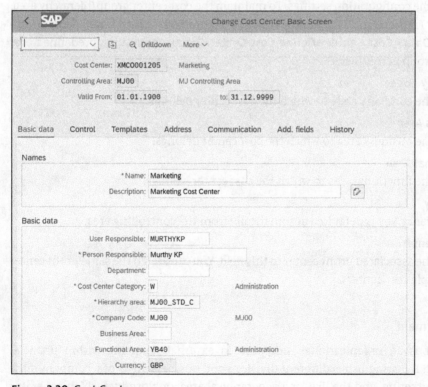

Figure 2.20 Cost Center

The **Basic data** tab of the cost center has the following fields:

- **Name**
  Enter a short name for the cost center in this field.

- **Description**
  Enter a detailed description of the cost center in this field.

- **User Responsible**
  Enter the user ID of a user stored in SAP S/4HANA in this field. This user is the owner of the cost center.

- **Person Responsible**
  Enter the name of the user in this field. This user is the owner of the cost center.

- **Department**
  Enter the name of the department to which the cost center belongs in this field.

- **Cost Center Category**
  You can define cost center categories in the configuration setup. To do so, execute Transaction SPRO and navigate to **Controlling • Cost Center Accounting • Master Data • Cost Center • Define Cost Center Categories**. Assign the predefined cost center category in this field.

- **Hierarchy area**
  The hierarchy area is also called the cost center group. You can define a hierarchy level in the configuration settings. To maintain the cost center group/hierarchy area, execute Transaction SPRO and navigate to **Controlling • Cost Center Accounting • Master Data • Cost Center • Define Cost Center Groups**. Assign the predefined cost center group in this field.

- **Company Code**
  Assign the company code to which the cost center belongs.

- **Business Area**
  Assign the business area to which the cost center belongs.

- **Functional Area**
  Assign the functional area to which the cost center belongs.

- **Currency**
  The currency key is defaulted automatically from the controlling area.

- **Profit Center**
  Assign the associated profit center to this field. You can assign the same profit center to multiple cost centers.

2.2.8   Segment

A segment is an organizational unit used in external reporting within financial accounting. It distinguishes different divisions and business areas of a company such as sales, marketing, and the like, or a geographical area, or a product which generates revenue and incurs expense. Segments allow you to break down financial statements by different divisions of the business and ensure compliance with financial reporting standards.

To define a segment, execute Transaction SPRO and navigate to **Enterprise Structure • Definition • Financial Accounting • Segment**.

Figure 2.21 displays segments. Click **New Entries** to define a new segment for financial accounting.

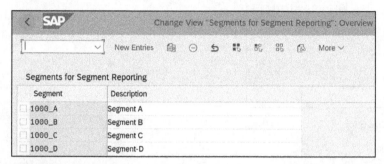

Figure 2.21 Segment

Define an 11-digit unique key to for the segment and enter a description.

2.3   Logistics General

Organizational units defined under *logistics general* are applicable to all modules/functions that deal with logistics. Let's explore the organizational units in logistics general in detail.

2.3.1   Valuation Level

The valuation level plays a vital role in defining at what level a system valuates materials in inventory management. To set the valuation level, execute Transaction SPRO and navigate to **Enterprise Structure • Definition • Logistics General • Define Valuation Level**.

Figure 2.22 shows that the valuation area is a plant, which means the valuation of materials in inventory management will be performed at the plant level. There are two options to choose from:

1. **Valuation area is a plant**
   If this radio button is set, materials are valued at the plant level.

2. **Valuation area is a company code**
   If this radio button is set, materials are evaluated at the company code level.

Best practice is to set the valuation level at the plant level as inventory is managed at the plant level.

Note

Once you set the valuation level either at the plant level or the company code level, it's not recommended to switch the valuation level. If you try to switch the valuation level, it will cause inconsistencies.

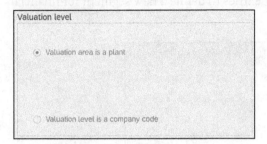

Figure 2.22 Valuation Level

2.3.2   Plant

A plant represents a physical location where manufacturing activities take place or inventory is stored and managed. It is one of the vital organizational units in SAP S/4HANA and serves as a central organizational unit for multiple functions of the organization, such as materials management, production planning, plant maintenance, quality inspection, sales and distribution, warehouse management, and so on. It controls the end-to-end flow of materials within the organization. Entering a plant is mandatory in most of the documents/transactions created in the system: sales orders, purchase orders, material documents for goods receipt and goods issue, maintenance orders, production orders, and so on.

Let's walk through the definition and assignment of plants.

Definition

To define a plant, execute Transaction SPRO and navigate to **Enterprise Structure • Definition • Logistics General • Define Plant**.

Click **New Entries** to define a new plant or copy an existing plant to modify. Define a four-digit unique alphanumeric or numeric key for the plant. Figure 2.23 shows our example plant, **MJ00**.

In the **Name 1** and **Name 2** fields, you can enter a name and an alternate name, respectively, for the plant. Then, complete the following fields in the **Detailed information** area for the plant:

- **Language Key**
  Enter the language of the plant. Texts are entered, displayed, and printed in this language.

- **Street and House No.**
  Enter the street name and house number for where the plant is located.

- **PO Box**
  Enter the PO box (if any) of the plant.

- **Postal Code**
  Enter the postal code of the plant.

- **City**
  Enter the city where the plant is located.

- **Country/Region Key**
  Enter the country where the plant is located.

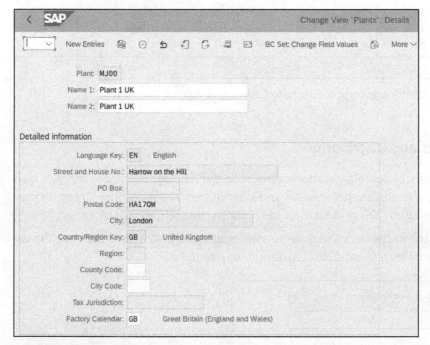

Figure 2.23 Plant

- **Region**
  Enter the region of the country where the plant is located.

- **County Code**
  Enter the county code (if any) for the region where the plant is located.

- **Tax Jurisdiction**
  Enter the tax jurisdiction code of the plant for tax determination within the United States. This is only required for the United States.

- **Factory Calendar**
  A factory calendar defines the working days and nonworking days (holidays) for a country/region, plant, or location (site). Assign a specific factory calendar for the plant.

Assignment

Assignment of plants to the company code is required. You can assign multiple plants to one company code, but the same plant can't be assigned to multiple company codes. To assign plants to a company code, execute Transaction SPRO and navigate to **Enterprise Structure • Assignment • Logistics General • Assign Plant to Company Code**.

Figure 2.24 shows that example plants **MJ00** and **MJ0A** are assigned to company code **MJ00**. Click **New Entries** to assign new plants to the company code.

CoCd	Plnt	Name of Plant	Company Name	Status
MJ00	MJ00	Plant 1 UK	MJ00	
MJ01	MJ01	Plant 2 UK	MJ01	
MJ00	MJ0A	Plant A UK - MJ00	MJ00	

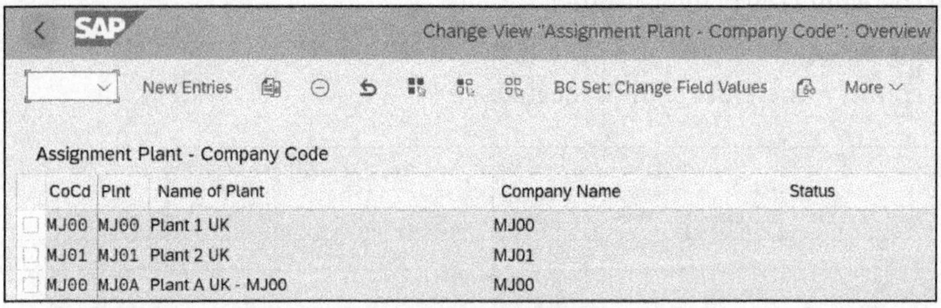

Figure 2.24 Plant Assignment to Company Code

2.4   Sales and Distribution

Sales and distribution organizational structural elements are defined to manage sales operations, customer relationships, and distribution networks. The organizational structure includes sales organization, distribution channels, and divisions.

Figure 2.25 shows the sales and distribution organizational structure. Let's explore the organizational units in sales and distribution in detail.

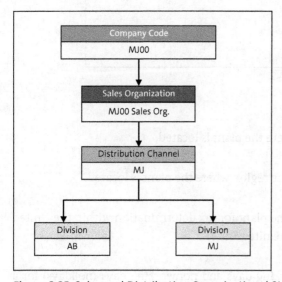

Figure 2.25 Sales and Distribution Organizational Structure

2.4.1   Sales Organization

The sales organization is the highest-level organizational unit in sales and distribution. It is responsible for managing sales activities for specific products or geographical areas. Multiple sales organizations are defined in the system based on the sales strategy and business requirements. Each sales organization is responsible for managing a specific set of customers, products, and sales targets.

Let's walk through the definition and assignment of sales organizations next.

Definition

To define a sales organization, execute Transaction SPRO and navigate to **Enterprise Structure • Definition • Sales and Distribution • Define, Copy, Delete, Check Sales Organization**.

Figure 2.26 displays our example sales organization, **MJ00**. Click **New Entries** to define a sales organization or to copy an existing sales organization to adapt. Define a four-character unique alphanumeric or numeric key to define a sales organization and enter a description.

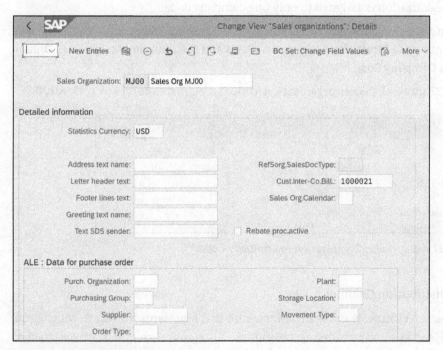

Figure 2.26 Sales Organization

The **Detailed information** area for the sales organization includes the following:

- **Statistics Currency**
  Enter the currency for the sales organization for statistical purposes. The system

defaults this currency when you generate sales statistics for the sales organization. This currency key can be different from the company code currency.

- **RefSorg.SalesDocType (reference sales organization for sales document type)**
  You can assign a reference sales organization for sales document types in this field. Leave this field blank if you want to allow all sales document types for the sales organization.

The **ALE : Data for purchase order** section is used to distribute the sales document data to another logical system to create a purchase order. Application Link Enabling (ALE) technology is used to distribute the data via IDocs from one (SAP S/4HANA) system to another (SAP S/4HANA) system within the same system landscape. This setup is not required if your business requirements do not include distributing the sales document to another system to create a purchase order.

Assignment

The sales organization is assigned to a company code to establish the integration between the sales and distribution and financial accounting and controlling functions. A sales organization is assigned to only one company code.

To assign a sales organization to a company code, execute Transaction SPRO and navigate to **Enterprise Structure • Assignment • Sales and Distribution • Assign Sales Organization to Company Code**.

Figure 2.27 shows that sales organization **MJ00** is assigned to company code **MJ00**.

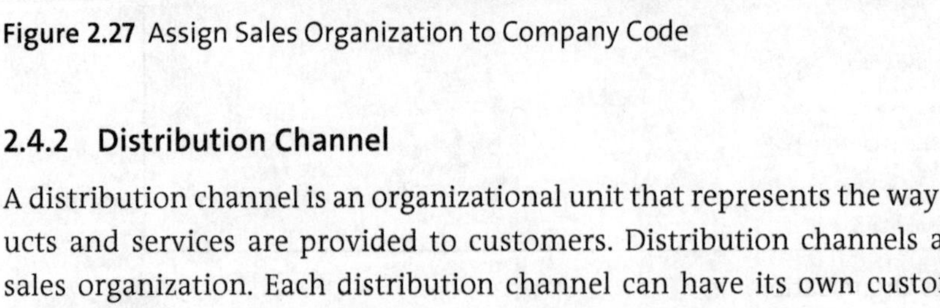

SOrg.	Name	CoCd	Company Name	Status
MJ00	Sales Org MJ00	MJ00	MJ00	
MJ01	Sales Org MJ01	MJ01	MJ01	

Figure 2.27 Assign Sales Organization to Company Code

2.4.2 Distribution Channel

A distribution channel is an organizational unit that represents the way in which products and services are provided to customers. Distribution channels are assigned to sales organization. Each distribution channel can have its own customer segments, products, and pricing strategies. This helps effectively manage and track sales operations.

Let's walk through the definition and assignment of distribution channels.

Definition

To define distribution channels for a sales organization, execute Transaction SPRO and navigate to **Enterprise Structure • Definition • Sales and Distribution • Define, Copy, Delete, Check Distribution Channel**.

Figure 2.28 shows distribution channels **M1**, **M2**, **M3**, and **MJ**. Click **New Entries** to define a new distribution channel or copy an existing distribution channel to edit. Define a two-character unique alphanumeric or numeric key to define a distribution channel and enter a name in the **Name** field.

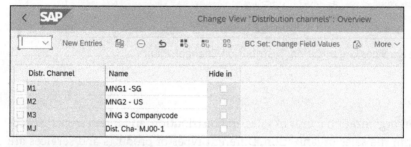

Figure 2.28 Distribution Channel

Assignment

Distribution channels are assigned to sales organizations. Multiple distribution channels can be assigned to one sales organization.

To assign distribution channels to sales organizations, execute Transaction SPRO and navigate to **Enterprise Structure • Assignment • Sales and Distribution • Assign Distribution Channel to Sales Organization**.

Figure 2.29 shows that distribution channel **MJ** is assigned to sales organization **MJ00**.

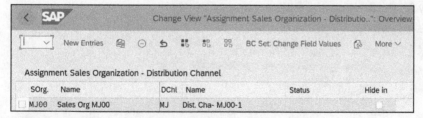

Figure 2.29 Assign Distribution Channel to Sales Organization

Plants are assigned to a combination (or multiple combinations) of the sales organization and distribution channel. To assign a plant to a distribution channel and sales organization combination, execute Transaction SPRO and navigate to **Enterprise Structure • Assignment • Sales and Distribution • Assign Sales Organization – Distribution Channel – Plant**.

Figure 2.30 shows that plants **MJ00**, **MJ01**, and **MJ0A** are assigned to the combination of distribution channel **MJ** and sales organization **MJ00**. Click **New Entries** to assign a distribution channel to a sales organization.

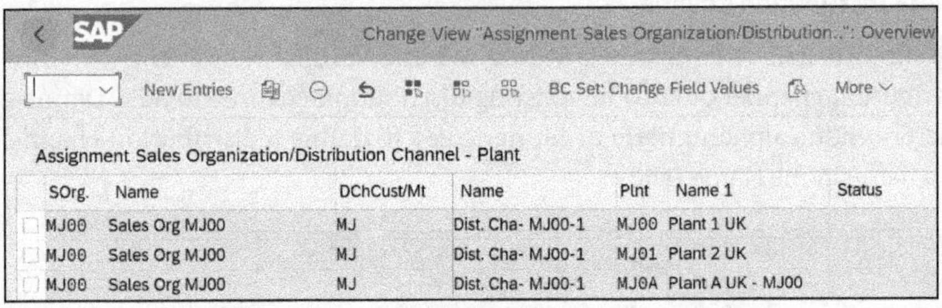

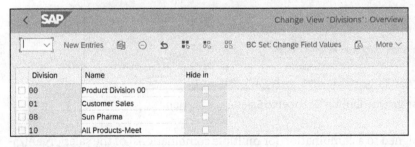

Figure 2.30 Assign Sales Organization: Distribution Channel: Plant

2.4.3   Division

A division is an organizational unit in sales and distribution, and it represents a product range within the sales organization. Different types of products and services are categorized using divisions to manage sales, pricing, and distribution processes. This helps to establish and execute the organization's sales strategies.

We'll explain the definition and assignment of divisions in the following sections.

Definition

To define a division, execute Transaction SPRO and navigate to **Enterprise Structure • Definition • Logistics General • Define, Copy, Delete, Check Division**.

Figure 2.31 shows divisions **00**, **01**, **08**, and **10**. Click **New Entries** to define a new division or copy an existing division to edit. Define a two-character unique alphanumeric or numeric key for the division and enter a name.

Figure 2.31 Division

Assignment

A division is assigned to a sales organization. Multiple divisions can be assigned to a sales organization.

To assign a division to a sales organization, execute Transaction SPRO and navigate to **Enterprise Structure • Assignment • Sales and Distribution • Assign Division to Sales Organization**.

Figure 2.32 shows that divisions **AB** and **MJ** are assigned to sales organization **MJ00**. Click **New Entries** to assign division to a sales organization.

SOrg.	Name	Dv	Name	Status	Hide in
MJ00	Sales Org MJ00	AB	AB Division		
MJ00	Sales Org MJ00	MJ	MJ Division		

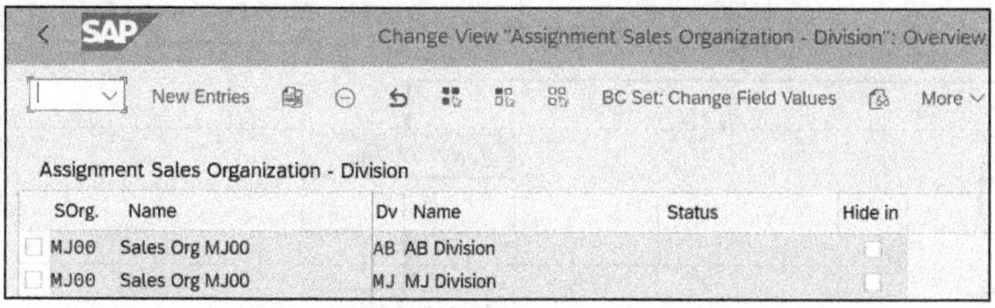

Figure 2.32 Assign Division to Sales Organization

2.4.4   Sales Area Determination

In sales and distribution, a sales area is a combination of a sales organization, distribution channel, and division. A sales area is a unique sales channel to distribute products and services to customers. It helps to define the scope of sales activities, pricing, shipping, and billing processes.

To set up a sales area, execute Transaction SPRO and navigate to **Enterprise Structure • Assignment • Sales and Distribution • Set Up Sales Area**.

Assign the combination of a sales organization, distribution channel, and division in this step. Figure 2.33 displays two sales areas set up for sales organization **MJ00**. Click **New Entries** to set up a new sales area or copy an existing sales area to edit.

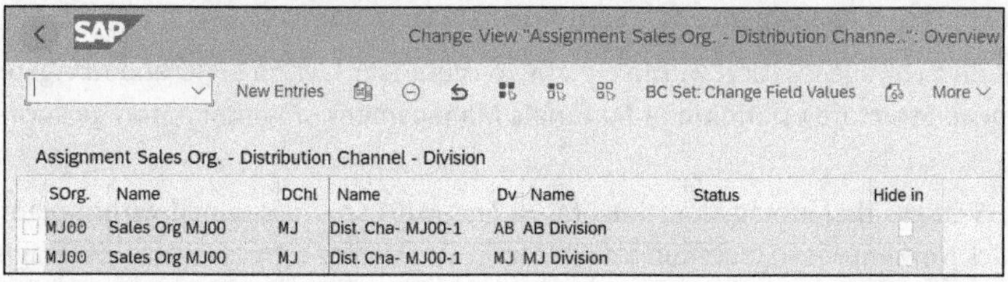

SOrg.	Name	DChl	Name	Dv	Name	Status	Hide in
MJ00	Sales Org MJ00	MJ	Dist. Cha- MJ00-1	AB	AB Division		
MJ00	Sales Org MJ00	MJ	Dist. Cha- MJ00-1	MJ	MJ Division		

Figure 2.33 Set Up Sales Area

2.5   Materials Management

Materials management helps to control and execute procurement activities and inventory management. The organizational structure has four organizational units: plant, storage location, purchasing organization, and purchasing group. Figure 2.34 shows the materials management organizational structure.

Let's explore the organizational units in materials management in detail. (Refer to Section 2.3.2 for the plant details as the plant is a centralized organizational unit that controls all the logistics-related functions.)

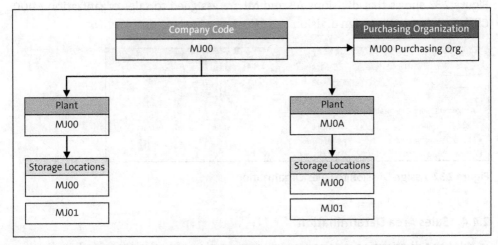

Figure 2.34 Materials Management Organizational Structure

2.5.1   Storage Location

A storage location is a specific physical location within a plant where inventory is stored and managed. It serves as a subdivision of the plant and plays a vital role in inventory management and warehouse management. A storage location is used to track stock levels of materials and to track material movements within a plant and across plants. By using the storage location in SAP, materials can be easily traced within a plant. Entering a storage location is mandatory for any material movement transaction in inventory management and warehouse management.

To maintain a storage location within a plant, execute Transaction SPRO and navigate to **Enterprise Structure • Definition • Materials Management • Maintain Storage Location.**

Figure 2.35 shows that storage locations **MJ00** and **MJ01** are maintained within plant **MJ00.** Click **New Entries** to maintain a new storage location or copy an existing storage location to edit it. Define a four-character unique alphanumeric or numeric key for the storage location.

Fill in the following fields:

- **Location**
  This field holds the four-character storage location ID. You can maintain multiple storage locations within a plant based on your business requirements.

- **Descr. of Storage Loc. (description of storage location)**
  Enter the name or description of the storage location.

- **Validity**

  In this field, you set the status of the storage location to one of the following:

  - **Valid**

    If you set this status, the storage location can be used to post stock movements.

  - **No Longer Valid**

    If you set this status, you cannot use the storage location to post stock movements, and this location is not visible to the user.

  - **Deprecated**

    If you set this status, the storage location cannot be used in the future. If you select a storage location with this status to post the stock movement, the system displays a warning message.

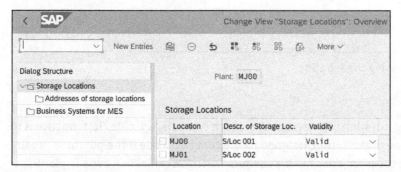

Figure 2.35 Storage Location

Note

If you maintain the address of the storage location, it takes precedence over the plant address in the purchase order, and the storage location address will become the delivery address for the purchase order item.

2.5.2 Purchasing Organization

A purchasing organization is a crucial part of the materials management organizational structure, responsible for procuring goods and services for the organization. The purchasing organization maintains relationships with suppliers; collaborates with suppliers; maintains purchasing master data including pricing conditions, Incoterms, and payment terms; and ensures that the right materials are available at the right time for various functions of the organization such as production, maintenance, sales, and so on.

In the following sections, we'll walk through the definition and assignment of purchasing organizations.

Definition

To maintain a purchasing organization, execute Transaction SPRO and navigate to **Enterprise Structure • Definition • Materials Management • Maintain Purchasing Organization.**

Figure 2.36 shows purchasing organization **MJ00**. Click **New Entries** to maintain a purchasing organization or copy an existing purchasing organization to edit. Define a four-character unique alphanumeric or numeric key for the purchasing organization and enter a description.

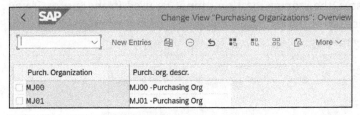

Figure 2.36 Purchasing Organization

Assignment

A purchasing organization can be plant-specific or company-specific. Best practice is to have a purchasing organization assigned to the company code if the purchase organization type is cross-plant-specific. If your purchasing organization type is plant-specific, assign the purchasing organization to the plant. If the purchasing organization is cross-company-code-specific, assign the purchasing organization to the plant and do not assign it to company codes. If a purchase organization is assigned to one company code, it is called a *company-specific purchasing organization*. You can assign one purchasing organization to one or more company codes.

To assign a purchasing organization to company code(s), execute Transaction SPRO and navigate to **Enterprise Structure • Assignment • Materials Management • Assign Purchasing Organization to Company Code.**

Figure 2.37 shows that multiple company codes are assigned to purchasing organization **MJ00**. Directly specify the company code to assign it to the purchasing organization in the **CoCd** (company code) field.

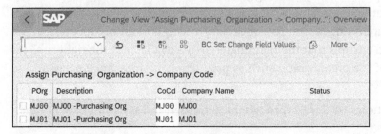

Figure 2.37 Assign Purchasing Organization to Company Code

If you don't choose to assign a purchasing organization to the company code, you can assign a purchasing organization to one or more plants. If a purchase organization is assigned to one plant, it is called a *plant-specific purchasing organization*.

To assign a purchasing organization to one or more plants, execute Transaction SPRO and navigate to **Enterprise Structure • Assignment • Materials Management • Assign Purchasing Organization to Plant**.

Figure 2.38 shows that multiple plants are assigned to purchasing organization **MJ00**. To assign a plant to the purchasing organization, click **New Entries** and specify the plant and purchasing organization.

POrg	Description	Plnt	Name 1	Status
MJ00	MJ00 -Purchasing Org	AKP3	AK Manufacturing Plant3 DU	Company code of plant/purchasing organization are not identical
MJ00	MJ00 -Purchasing Org	MJ00	Plant 1 UK	
MJ00	MJ00 -Purchasing Org	MJ0A	Plant A UK - MJ00	

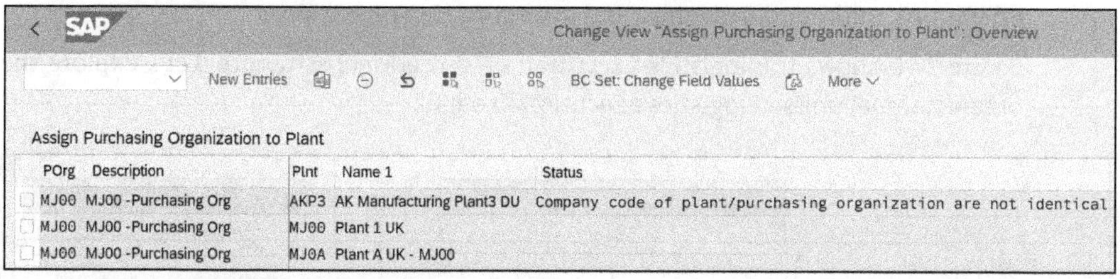

Figure 2.38 Assign Purchasing Organization to Plant

2.5.3 Purchasing Groups

A purchasing group is a subdivision of a purchasing organization, responsible for specific purchasing activities or managing purchasing activities for a specific or multiple product categories. Purchasing groups can have the responsibility of managing procurement activities for a specific region or multiple regions. Purchasing groups are typically assigned to an individual or a group of buyers. Purchasing group information such as contact details are provided to vendors via purchase orders or outline agreements.

To create purchasing groups, execute Transaction SPRO and navigate to **Materials Management • Purchasing • Create Purchasing Groups**.

Figure 2.39 shows purchasing groups **001**, **002**, and **003**. Click **New Entries** to create a new purchasing group or copy an existing purchasing group to edit. Define a three-character unique alphanumeric or numeric key for the purchasing group and enter a description.

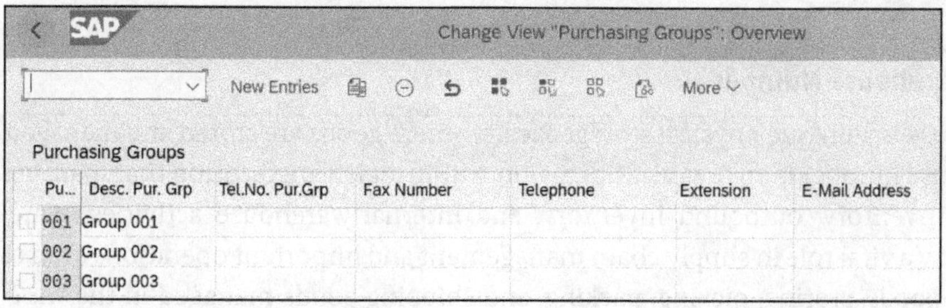

Purchasing Groups

Pu...	Desc. Pur. Grp	Tel.No. Pur.Grp	Fax Number	Telephone	Extension	E-Mail Address
001	Group 001					
002	Group 002					
003	Group 003					

Figure 2.39 Purchasing Groups

You can maintain the telephone numbers, fax numbers, and email addresses of all purchasing groups. Best practice is to share contact details with suppliers by displaying the details of the purchasing group in the purchase order.

2.6   Logistics Execution

The logistics execution function mainly manages warehouse activities, shipping (inbound and outbound), and transportation logistics. It connects procurement and distribution functions of an organization by managing inbound deliveries, warehouse operations, outbound deliveries, and transportation operations.

Figure 2.40 shows the logistics execution organizational structure. Let's explore the organizational units in logistics execution in detail.

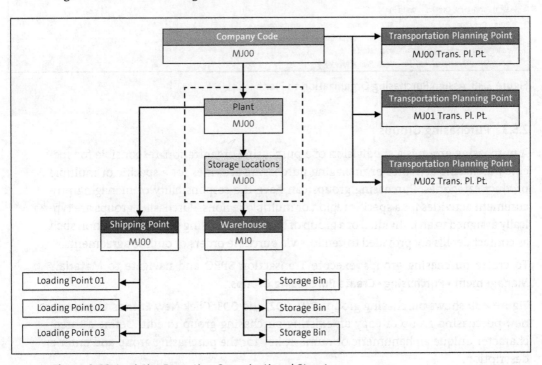

Figure 2.40 Logistics Execution Organizational Structure

2.6.1   Warehouse Number

A warehouse is a unique physical storage facility where goods are stored and managed. A warehouse number is an organizational unit within logistics execution that manages inbound inventory, outbound inventory, and internal warehouse activities. Warehouses play a vital role in supply chain management and important operations such as receiving goods, storing, picking, packing, and shipping goods managed in the warehouse. Each warehouse is identified by a unique warehouse number in SAP S/4HANA.

Let's walk through the definition and assignment for warehouse numbers.

Definition

To define a warehouse number, execute Transaction SPRO and navigate to **Enterprise Structure • Definition • Logistics Execution • Define Warehouse Number**.

Figure 2.41 shows warehouse number **MJ0**. Click **New Entries** to maintain a new warehouse number or copy an existing warehouse number to edit. Define a three-character unique alphanumeric or numeric key for the warehouse number.

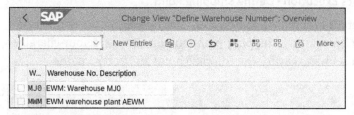

Figure 2.41 Warehouse Number

Assignment

Warehouses are the physical locations within a plant. A plant can have multiple warehouses, and a warehouse number is assigned to the combination of a plant and storage location in SAP S/4HANA.

To assign a warehouse number to the combination of a plant and storage location, execute Transaction SPRO and navigate to **Enterprise Structure • Assignment • Logistics Execution • Assign Warehouse Number to Plant and Storage Location**.

Figure 2.42 shows that warehouse number **MJ0** is assigned to the combination of plant **MJ00** and storage location **MJ00**.

Figure 2.42 Assign Warehouse Number to Plant and Storage Location

Click **New Entries** to assign a warehouse number to the combination of plant and storage location or copy an existing assignment to edit.

2.6.2   Shipping Point

A shipping point is a location within the plant where goods are loaded into transportation vehicles to be shipped to internal and external customers. It represents a final stage in outbound logistics from where goods issue movements are posted in SAP S/4HANA after the picking and packing operations. It is the highest-level organizational unit in shipping.

Let's examine the definition and assignment of shipping points in the following sections.

Definition

To define a shipping point, execute Transaction SPRO and navigate to **Enterprise Structure • Definition • Logistics Execution • Define Shipping Point**.

Figure 2.43 shows the details of shipping point **MJ00**. Click **New Entries** to define a new shipping point or copy an existing shipping point to edit. Define a four-character unique alphanumeric or numeric key for the shipping point and enter a name.

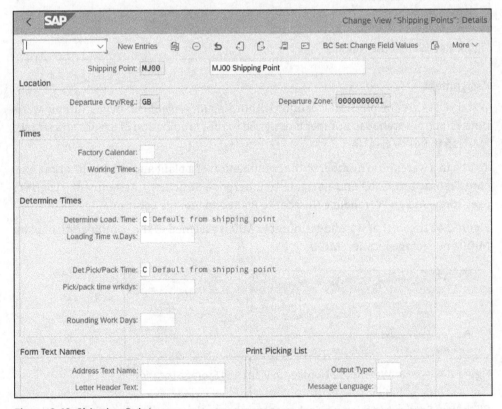

Figure 2.43 Shipping Point

The **Location** area of the shipping point includes the following fields:

- **Departure Ctry/Reg.**
  This field represents the country or region from which goods are shipped to customers. Enter the country key that shows where the shipping point is located.

- **Departure Zone**
  Assign the departure zone, also called the transportation zone, that matches where the shipping point is located. This helps determine the route for transportation. You can define departure zones in the configuration settings. To arrive at the configuration step to define departure zones by country, execute Transaction SPRO and navigate to **Logistics Execution • Shipping • Basic Shipping Functions • Routes • Define Transportation Zones**.

The **Times** area for the shipping point has the following fields:

- **Factory Calendar**
  Assign a specific factory calendar for the shipping point.

- **Working Times**
  Assign a working time for the shipping point. The working time is defined in the factory calendar for working days.

The **Determine Times** area of the shipping point has the following key fields:

- **Determine Load. Time**
  This field controls how the system determines loading time for outbound deliveries. You can set one of the following values in this field:

  - **No loading time determination**
    If there is no value is set, the loading time will not be determined for outbound deliveries for this shipping point.

  - **A: Route dependent**
    If you set this value, the loading time determination is dependent on the route.

  - **B: Route independent**
    If you set this value, loading time determination is independent of the route.

  - **C: Default from shipping point**
    If you set this value, the loading time is determined from the **Loading Time w.Days** field, where you can set the loading time for the shipping point.

- **Loading Time w.Days**
  You can set the loading time here for the shipping point. This field will be available for entry only if value **C** is set in the **Determine Load. Time** field.

- **Det.Pick/Pack Time**
  This field controls how the system determines the picking and packing time for outbound deliveries. You can set one of the following values in this field:

  - **No loading time determination**
    If there is no value is set, the pick/pack time will not be determined for outbound deliveries for this shipping point.

  - **A: Route dependent**
    If you set this value, the pick/pack time determination is dependent on the route.

- **B: Route independent**
  If you set this value, the pick/pack time determination is independent of the route.
- **C: Default from shipping point**
  If you set this value, the pick/pack time is determined from the **Pick/pack time wrkdys** field, where you can set the pick/pack time for the shipping point.
- **Pick/pack time wrkdys**
  You can set the picking and packing time for the shipping point here. This field will be available for entry only if value **C** is set in the **Det.Pick/Pack Time** field.

Under **Print Picking List**, you can set the default values such as output type, output medium, message language, and so on for the shipping point to print the picking list.

Assignment

Shipping points are assigned to the plant. A plant can have multiple shipping points, and a shipping point can be assigned to multiple plants as well.

To assign a shipping point to a plant, execute Transaction SPRO and navigate to **Enterprise Structure • Assignment • Logistics Execution • Assign Shipping Point to Plant**.

Figure 2.44 shows that shipping point **MJ00** is assigned to plants **MJ00** and **MJ01**. It also shows that there are two shipping points assigned to plant **MJ01**.

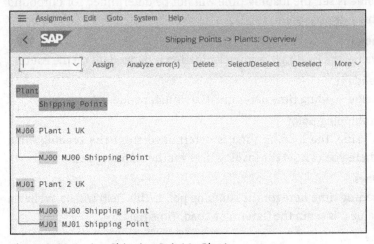

Figure 2.44 Assign Shipping Point to Plant

Click the **Assign** button to assign shipping points to plants.

2.6.3   Loading Point

Loading points are locations within a plant or a warehouse from which goods are loaded onto vehicles for transportation. Loading points are a subdivision of shipping points and are defined with reference to shipping points in SAP S/4HANA.

To maintain loading points for a shipping point, execute Transaction SPRO and navigate to **Enterprise Structure • Definition • Logistics Execution • Maintain Loading Point**.

Enter the shipping point under which the loading points are to be maintained in the initial screen.

Figure 2.45 shows that loading points **01**, **02**, and **03** are maintained for shipping point **MJ00**. Click **New Entries** to define a new loading point or copy an existing loading point to edit. Define a two-character unique alphanumeric or numeric key for the loading point and enter a description.

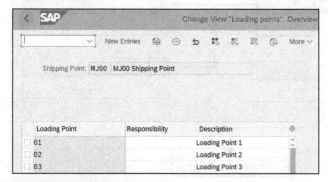

Figure 2.45 Loading Point

2.6.4   Transportation Planning Point

A transportation planning point is a core location within the organization for planning transportation activities. Shipments are scheduled, carriers are selected, and routes are optimized at the transportation planning points. It is the highest-level organizational unit in transportation management. These are defined at the company code level and assigned to a company code.

To define a shipping point, execute Transaction SPRO and navigate to **Enterprise Structure • Definition • Logistics Execution • Maintain Transportation Planning Point**.

Figure 2.46 shows that transportation planning points **MJ00**, **MJ01**, and **MJ02** are maintained and assigned to company code **MJ00**. Click **New Entries** to define a new transportation planning point or copy an existing transportation planning point to create a new one. Define a four-character unique alphanumeric or numeric key to define a transportation planning point and enter a description.

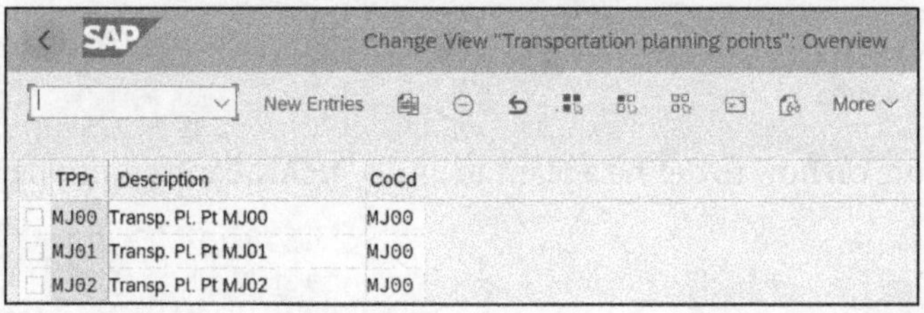

Figure 2.46 Transportation Planning Point

2.7   Plant Maintenance

The plant maintenance organizational structure includes maintenance planning plants, functional locations, and planner groups to effectively manage maintenance operations, improve equipment reliability, and reduce equipment downtime.

Figure 2.47 shows the plant maintenance organizational structure. Let's explore the organizational units in plant maintenance in detail.

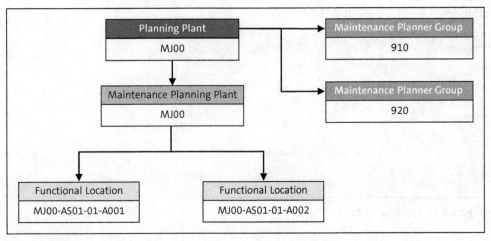

Figure 2.47 Plant Maintenance Organizational Structure

2.7.1   Planning Plant

The planning plant is an organizational unit that provides the planning of resources and materials for a maintenance plant. Planning plants are the logistics plants defined in logistics general (refer to Section 2.3).

Key features of planning plants are as follows:

- A minimum of one planning plant is required for planning purposes.
- Centralized maintenance planning can be performed using one planning plant.
- A maintenance plant is assigned to a planning plant. There can be a one-to-one relationship between a planning plant and a maintenance plant (one planning plant to one maintenance plant) or a one-to-many relationship (one planning plant to multiple maintenance plants).
- A planning plant does not carry any financial impact. Financial impacts only happen at the maintenance plant level.

Refer to Section 2.3.2 for more on how to define a plant in SAP S/4HANA and additional details.

2.7.2   Maintenance Planning Plant

A maintenance planning plant is the highest-level organizational unit in plant mainte-nance. It represents a physical location where maintenance operations are performed. A company's assets such as equipment and the like are defined in the maintenance planning plant.

Key features of a maintenance planning plant are as follows:

- A maintenance planning plant is an extension of a logistics plant.
- A maintenance planning plant is assigned to a planning plant.
- Costs are collected at the maintenance plant level. Financial impacts only happen at the maintenance plant level.

We'll explain the definition and assignment of a maintenance planning plant next.

Definition

To define a maintenance planning plant, execute Transaction SPRO and navigate to **Enterprise Structure • Definition • Plant Maintenance • Maintain Maintenance Planning Plant.**

Figure 2.48 shows maintenance planning plant **MJ00**. Click **New Entries** to maintain a new maintenance planning plant or copy an existing maintenance planning plant to edit.

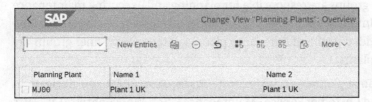

Figure 2.48 Maintenance Planning Plant

Assignment

Maintenance planning plants are assigned to the logistics plant or planning plant. A planning plant can have multiple maintenance planning plants assigned.

To assign maintenance planning plants to a planning plant or maintenance plant or logistics plant, execute Transaction SPRO and navigate to **Enterprise Structure • Assign-ment • Plant Maintenance • Assign Maintenance Planning Plant to Maintenance Plant.**

Figure 2.49 shows that maintenance planning plant **MJ00** is assigned to planning plant **MJ00**. Directly specify the planning plant in the **PlPl** (planning plant) field for every plant relevant for plant maintenance.

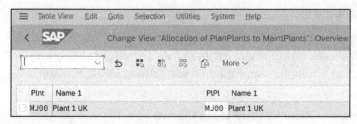

Figure 2.49 Assign Maintenance Planning Plant to Maintenance Plant

2.7.3   Functional Location

A functional location is a specific physical location in the plant where maintenance activities are performed. In functional locations, the company's assets such as equipment and tools are installed and maintained. In SAP S/4HANA, functional locations can be defined in a structural hierarchical format. A functional location can be an assembly line, production line, or a processing location (e.g., paint booth). The functional location helps track and report on maintenance tasks and associated costs.

Functional location details are explained in detail in the plant maintenance master data. Refer to Chapter 7, Section 7.2 for details.

2.7.4   Maintenance Planner Group

Maintenance planner groups are responsible for planning and executing maintenance activities within a maintenance planning plant. Planner groups responsible for maintenance planning are added to the maintenance plan. Maintenance planner groups are defined for planning plants.

To define a maintenance planner group, execute Transaction SPRO and navigate to **Plant Maintenance and Customer Service • Maintenance Plans, Work Centers, Task Lists and PRTs • Basic Settings • Define Maintenance Planner Groups**.

Figure 2.50 shows maintenance planner groups **910** and **920** defined for plant/planning plant **MJ00**.

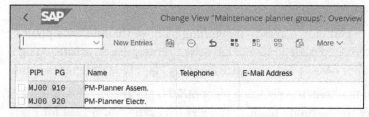

Figure 2.50 Maintenance Planning Group

Click **New Entries** to define a new maintenance planner group or copy an existing maintenance planner group to edit. Define a three-character unique alphanumeric or numeric key to define a maintenance planner group and enter a name. You can also add the telephone number and email ID of the planner group if the maintenance planner is a separate department. In larger organizations, maintenance planner groups are part of a separate department.

2.8 Summary

In this chapter, we explained the significance of the organizational structure and offered best practice examples. We explained the definition of organizational structure elements arranged by functions such as financial accounting and controlling, logistics general, sales and distribution, materials management, logistics execution, and plant maintenance. We also explained the assignment of various organizational units to establish relationships and an enterprise structure.

In future chapters, we will explain how the organizational units are used in maintaining master data and creating transactional data to support various functions of SAP S/4HANA. In the next chapter, we'll discuss the core master data in SAP S/4HANA and the specific master data in materials management.

Chapter 3
Master Data

Master data is a vital component of any ERP system. Master data drives business processes, integrates different functions, and provides the necessary attributes to manage transactions in SAP S/4HANA. It contains internal business-related data such as that for purchasing, production, sales and distribution, finance and controlling, and more, as well as external data such as that for suppliers and customers and regulatory and compliance data.

Master data is the most essential constituent to execute business transactions in SAP S/4HANA. Creation and maintenance of master data is vital for organizations to run their business processes smoothly. Successful SAP S/4HANA implementation projects give the utmost importance to master data, and successful organizations running on SAP S/4HANA have streamlined master data management processes.

There are three types of data in SAP S/4HANA:

- **Master data**
  Master data provides essential data to execute business transactions for a specific set of organizational units. Master data is maintained at different organizational levels based on business context. SAP S/4HANA provides capabilities to define master data and extend to multiple organizational units. For example, material data is defined at the plant level, sales organization level, warehouse level, and so on, and the same material master can be extended to multiple plants, sales organizations, and/or warehouses. Master data consists of multiple data elements, and SAP S/4HANA stores master data using an internal or external number range. The system uses the additional data maintained in the master data record by default when you use master data in a business transaction.

  Master data can be foundational or business function specific. For example, material master data and business partners are foundational, and they are needed to execute transactions of multiple business functions (purchasing, sales, production, maintenance, finance, etc.). Meanwhile, business function-specific master data consists of data attributes to execute a specific business function; for example, purchasing info records (PIRs), source lists, and quota arrangements are purchasing-specific master data.

- **Transactional data**
  Transactional data represents a business transaction of a specific organizational unit. Transactional data is created with reference to one or more master data elements.

A transaction is time-sensitive information often created for internal business transactions such as maintenance, production, inventory management, and so on, and external business transactions such as purchase orders, incoming invoice postings, billing documents to customers, shipments to customers, and the like. Transactional data represent specific business process steps in end-to-end business processes such as procure-to-pay, order-to-cash, forecast-to-stock, and more.

- **Configuration data**
  Configuration data in SAP S/4HANA is used to define business rules to execute business transactions. Configuration data often requires just a one-time setup unless the business rules change. Almost all organizational units use the configuration data to set up master data and execute business transactions. Configuration data controls the screen layout, approval rules to authorize business transactions, automatic financial accounting postings, valuation of stock in inventory management, processing rules for customer orders, and so on.

This chapter focuses on foundational and cross-functional master data that is applicable to most business functions such as material master data, business partner data (vendor data, customer data), and so on. Business function-specific master data is covered in the chapters for each business function.

3.1   Material Master

The *material master* is the foundational data for all logistics processes, and it is centrally maintained to ensure the accuracy of the material data. SAP S/4HANA provides capabilities to maintain material master data for different business functions in terms of separate views, which we will explore in this section. The material master is a material profile that is maintained centrally with different views required for different business functions, such as sales, purchasing, material requirements planning (MRP), inventory management, warehouse management, financial accounting and controlling, and quality management. Material master data is maintained for organizational units depending on the data view. In SAP S/4HANA, the material master can be maintained for multiple sales organizations, plants, storage locations, warehouse numbers, and so on, depending on business requirements. The standard system provides capabilities to extend the material created for a set of organizational units to another set of organizational units. In SAP S/4HANA, different sets of material data attributes can be maintained for different organizational units. For example, purchasing data attributes can be maintained differently for different plants based on business needs.

Here's a summary of material master views:

- **Basic data views**
  Basic data views of material master data are maintained at the client level, which is applicable across all company codes, plants, storage locations, valuation areas,

warehouse numbers, sales organizations, and so on. A basic data view contains general data for a material. Basic data views must be maintained for all types of materials.

- **Sales data views**

  Sales data views of material master data are maintained at the sales organization and distribution channel levels. They contain sales and distribution-related data for the material. Sales data views are mandatory to execute sales transactions (sales order, stock transfer order, etc.) in SAP S/4HANA.

- **Purchasing data view**

  Purchasing views of material master data are maintained at the plant level. They contain the purchasing-related data for the material. A purchasing view is mandatory to execute purchasing transactions (purchase requisition, purchase order, stock transfer order, outline agreements, etc.) in SAP S/4HANA.

- **MRP data views**

  MRP views of material master data are maintained at the plant level. They contain the MRP data for the material. MRP views are mandatory to execute MRP for the respective materials in SAP S/4HANA.

- **Accounting data views**

  Accounting data views of material master data are maintained at the valuation area level. They contain accounting and valuation data for the material. Accounting views are mandatory to execute material valuation in SAP S/4HANA.

- **Storage data views**

  Storage data views of material master data are maintained at the plant and storage location levels. They contain storage-related data for the material for inventory management. Storage data views are mandatory to execute inventory management processes in SAP S/4HANA.

- **Warehouse data views**

  Warehouse data views of material master data are maintained at the warehouse number level. They contain warehouse management-related data for the material. Warehouse data views are mandatory to execute warehouse management processes for the respective materials in SAP S/4HANA.

- **Quality management data view**

  Quality management data views of material master data are maintained at the plant level. They contain quality management-related data for the material required for quality inspection. A quality management data view is mandatory to execute quality management processes for the respective materials in SAP S/4HANA.

In the following sections, we'll explore the material master in more detail, including material groups, material types, and data views. We'll also briefly explain the necessary configuration. But first, let's explore the significance of material master data in SAP S/4HANA.

3.1.1   Importance of Material Master Data

The importance of master data in general is to support business transactions so that the business processes run smoothly. Material master data is a central source for all logistics processes avoiding data redundancy, and it contains the internal product information of the organization. The importance and benefits of material master data are as follows:

- Material master data is a vital component of SAP S/4HANA. It's a central repository of product information to streamline logistics processes.

- It stores the material data in terms of different views to support production planning, purchasing, sales, inventory management, warehouse management, quality management, plant maintenance, and financial accounting processes. SAP S/4HANA uses the subset of relevant data required for a specific business transaction. For example, for purchase order creation, SAP S/4HANA uses the purchasing data of the material master.

- Maintaining material master data helps support the integrated processes that this book is focused on. Tracking a material through the supply chain is possible using material master data.

- Material master data is used to track inventory in inventory management and warehouse management as well as to record various material movements.

- Valuated price information for the material master data is used to evaluate inventory in the inventory management so that actual value of the stock on hand can be determined.

- Material master data in SAP S/4HANA provides flexibility to define additional material characteristics/attributes based on business requirements. A material class can be assigned to the material to maintain additional characteristics.

- Bills of materials (BOMs) for production, maintenance, assembly, and the like are created using material master data. BOMs are crucial master data in production planning and plant maintenance that upcoming chapters will highlight.

- The business process-specific master data such as purchasing master data, quality management master data, sales master data, production planning master data, plant maintenance master data, and the like is created using material master data.

3.1.2   Material Groups

Material groups are used to group materials with similar characteristics/attributes. These materials from the same material group may belong to multiple material types such as raw materials, finished goods, semifinished goods, and so on. A material group is mandatory to create material master data in SAP S/4HANA. Material groups can control security in SAP S/4HANA by defining and assigning authorization objects to them and using these to control access to the material groups. Defining authorizations at the material group level provides flexibility in handling security and controls.

The following configurations explain the definition of material groups and other key configurations using material groups.

Define Material Groups

To define material groups, execute Transaction SPRO and navigate to **Logistics—General • Material Master • Settings for Key Fields • Define Material Groups**.

Figure 3.1 shows the material groups defined in the system. Click **New Entries** to define a new material group. Enter a nine-digit alphanumeric key to define a new material group and a description for the same. The following are the key fields in the material group definition:

- **AGrp**

  In this field, define the authorization group and a four-digit alphanumeric key for the material group. You can define different authorization groups for different material groups, or you can define one authorization group for multiple material groups, based on your business requirements. Authorization groups defined in this configuration are used to control the creation of materials, to change materials, and to display materials by material groups.

- **Description 2 for the Material Group**

  In this field, you can define an additional description for the material group if required.

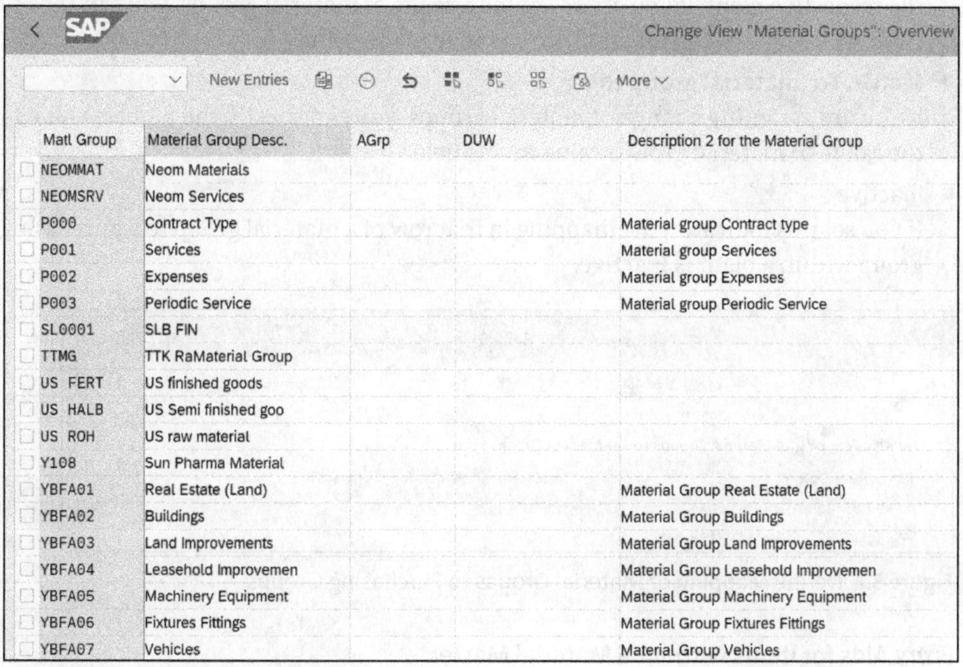

Matl Group	Material Group Desc.	AGrp	DUW	Description 2 for the Material Group
NEOMMAT	Neom Materials			
NEOMSRV	Neom Services			
P000	Contract Type			Material group Contract type
P001	Services			Material group Services
P002	Expenses			Material group Expenses
P003	Periodic Service			Material group Periodic Service
SL0001	SLB FIN			
TTMG	TTK RaMaterial Group			
US FERT	US finished goods			
US HALB	US Semi finished goo			
US ROH	US raw material			
Y108	Sun Pharma Material			
YBFA01	Real Estate (Land)			Material Group Real Estate (Land)
YBFA02	Buildings			Material Group Buildings
YBFA03	Land Improvements			Material Group Land Improvements
YBFA04	Leasehold Improvemen			Material Group Leasehold Improvemen
YBFA05	Machinery Equipment			Material Group Machinery Equipment
YBFA06	Fixtures Fittings			Material Group Fixtures Fittings
YBFA07	Vehicles			Material Group Vehicles

Figure 3.1 Define Material Groups

Define Mapping of Material Groups to Purchasing Group

Purchasing groups can manage purchasing of specific material groups. In this configuration step, you can map material groups to purchasing groups. A purchasing group is assigned to a buyer or a group of buyers who create and manage purchase orders and outline agreements. To assign material groups to purchasing groups, execute Transaction SPRO and navigate to **Materials Management • Purchasing • Purchase Requisition • Define Mapping of Material Groups to Purchasing Group**.

Figure 3.2 shows the mapping of material groups to purchasing groups. Based on the mappings maintained, purchasing group **001** is responsible for purchasing for material group **YBFA02** in plant **MJ00**, whereas purchasing group **MJ1** is responsible for purchasing for all material groups except **YBFA02**. Click **New Entries** to define a new mapping. Specify the following fields to define a new mapping:

- **Plnt (plant)**
  In this field, specify the plant in which you are defining the material groups to purchasing group mapping.

- **PGr (purchasing group)**
  In this field, specify the purchasing group to be mapped to a material group or multiple material groups. A purchasing group is an organizational unit within the purchasing organization.

- **MatGr. Fr. (material group from)**
  In this field, specify the material group to be mapped to the purchasing group within the respective plant. If you are assigning a range of material groups, you can specify the lowest value of the material group range in this field.

- **MatGr. To (material group to)**
  If you are assigning a range of material groups, you can specify the highest value in the range of material groups value in this field.

- **Inactive**
  If you set this indicator, the mapping in that row of a material group to a purchasing group within a plant is inactive.

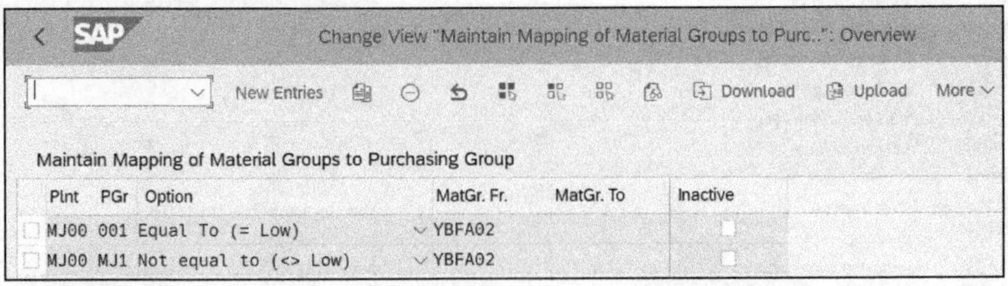

Figure 3.2 Define Mapping of Material Groups to Purchasing Groups

Entry Aids for Items without a Material Master

Indirect purchasing is mostly performed using free text items with material groups and account assignments. An account assignment category is assigned to such purchase order items so that during goods receipt, consumption is directly posted to the respective account assignment category (cost center, internal order, work breakdown structure [WBS], etc.). The general ledger account is also required as an account assignment for such purchase order items for financial account postings during goods receipt and invoice receipt postings. In this configuration, you can assign a valuation class to material groups to determine general ledger accounts.

To assign material groups to purchasing groups, execute Transaction SPRO and navigate to **Materials Management • Purchasing • Material Master • Entry Aids for Items without a Material Master**.

Figure 3.3 shows that valuation classes **1199**, **1231**, and **1276** have been assigned to material groups **148000000**, **153000000**, and **168000000**, respectively. Directly assign the relevant valuation class to the material group in the **ValCl** field. Refer to Chapter 10, Section 10.3 to learn about the valuation class and how it controls the determination of general ledger accounts.

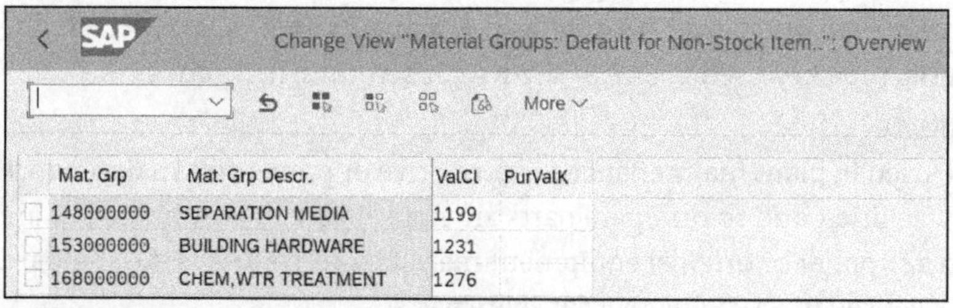

Figure 3.3 Entry Aids for Items without a Material Master

3.1.3 Material Types

While material groups are used to group materials with similar attributes/characteristics, material types are used to decide if a material can be procured, traded, produced in house, used in packaging, used as a spare part, and so on. Material groups can be linked to an external material group to uniquely identify the group of materials. The United Nations Standard Products and Services Code (UNSPSC) is a commonly used coding system for goods and services, and material groups can be mapped to unique UNSPSC codes for identification. Material groups are not created with reference to internal or supplier catalog codes, while material types are used to control internal processes. Material types in SAP S/4HANA control valuation, material master views, and the like, and they are mandatory to create material master data. Material types are for internal purposes; for instance, a buying organization might buy a product that belongs to the raw materials category (material type), but for the supplier's organization, the same product could belong to the finished goods category (material type). SAP S/4HANA provides multiple material types, with the following the most used across industries:

- **Finished goods (FERT)**
  Finished goods are the sellable products to be sold to customers. The finished product

is the final product that a company produces. These products are produced using raw materials and other materials that belong to other material types. These products could have multiple components or different types of materials assembled to produce the finished products. Finished products are the parent materials of BOMs and can't be used as components in a BOM. FERT is the standard material type key used in SAP S/4HANA for finished goods.

- **Raw materials (ROH)**
  Raw materials are used for in-house manufacturing processes to produce semifinished products or finished products. Raw materials are procured from external suppliers or from other plants (stock transfer process). Raw materials can be used as components in a BOM. ROH is the standard material type key used in SAP S/4HANA for raw materials.

- **Semifinished goods (HALB)**
  Semifinished goods are not fully finished products and hence not ready to be sold to customers. Finished products can be produced using one or more semifinished products. Semifinished goods can be used as components in a BOM. HALB is the standard material type key used in SAP S/4HANA for semifinished goods.

- **Spare parts (ERSA)**
  Spare parts are used in plant maintenance for corrective or preventive maintenance of equipment, facilities, and so on. Spare parts are not sellable to customers but are procured from a supplier or original equipment manufacturer. ERSA is the standard material type key used in SAP S/4HANA for spare parts.

- **Trading goods (HAWA)**
  Trading goods are sellable products that are not manufactured in house but procured from an external supplier. Trading goods are mostly used in a direct drop-ship process, where a third-party purchase order is created with reference to the customer sales order and the supplier directly ships the products to the customer. The supplier invoice and customer billing happen separately. Trading goods can be procured, stored in the inventory, and sold to the customer as well. HAWA is the standard material type key used in SAP S/4HANA for trading goods.

- **Nonstock materials (NLAG)**
  Nonstock materials (small products such as nails) are used in maintenance and production processes. These materials are procured from an external supplier, but the inventory is not quantified. These materials can't be reserved for manufacturing or maintenance as the stock is not maintained in SAP S/4HANA, even though it's physically stored at the plant. NLAG is the standard material type key used in SAP S/4HANA for nonstock materials.

- **Configurable materials (KMAT)**
  Configurable materials are complex materials with multiple combinations of parts that go into a product. These are the multiple configurations of a product, such as different paints, chemical compositions, and so on. Instead of creating many material

master data records for every permutation and combination, one configurable product will be created using the KMAT material type. These materials generally have one
super BOM per KMAT material to allow for various configurations. KMAT is the standard material type key used in SAP S/4HANA for configurable materials.

- **Services (SERV)**
  Services are the intangible materials that can't be stocked or transported, such as
  services from an external supplier, and generally require confirmation of services
  from the requester to process an invoice for payment. SERV is the standard material
  type key used in SAP S/4HANA for services.

- **Returnable packaging (LEIH)**
  Returnable packaging materials are a category of packaging materials. When customer shipments are packed using returnable packaging material, the customer is
  legally required to return the packaging material to the seller. These materials can be
  produced in house or procured externally. LEIH is the standard material type key
  used in SAP S/4HANA for returnable packaging materials.

Let's explore the key configuration settings for material types. In this configuration
step, you'll define material types and their attributes. You can select the required material master views for every material type, set the price control, assign a material class to
maintain additional attributes/characteristics of the material, determine field reference
settings, set the material status at the material type level, and so on. To define attributes
of material types, execute Transaction SPRO and navigate to **Logistics—General • Material Master • Basic Settings • Material Types • Define Attributes of Material Types**.

Figure 3.4 shows an overview of material types defined in the system. Click **New Entries**
to define a new material type. Enter a four-digit alphanumeric key to define a new
material type and a description for the same. Double-click the material type key or
select a material type and click the **Details** icon (or press Ctrl + Shift + F2) to display the details screen, where you can maintain the material type attributes.

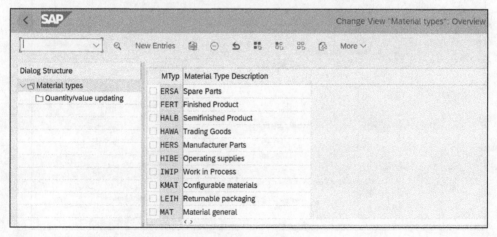

Figure 3.4 Define Attributes of Material Type: Overview

Figure 3.5 shows the first part of the details screen for material type **FERT** (**Finished Product**) defined in the system. Let's explore the key field settings for the material type:

- **Field reference**
  With the field reference key, you can define the mandatory, optional, and display fields of the material master required for the respective material type. Field reference keys for material types can be configured in the system. To define a field reference, execute Transaction SPRO and navigate to **Logistics—General • Material Master • Field Selection • Maintain Field Selection for Data Screens**. Once you define the field reference for the material type, assign the field reference key in the field of the same name.

- **SRef: material type**
  The screen reference for material types groups multiple material types to determine the material master views that can be maintained and the sequence of material master views. During the creation of material master data, the material master data views are available for selection, and you can choose the views you want to maintain for the material.

- **Authorization group**
  In this field, you define the authorization group, a four-digit alphanumeric code for the material type. You can define different authorization groups for different material types, or you can define one authorization group for multiple material types based on business requirements. Authorization groups are used to control access to the materials created for the respective material types.

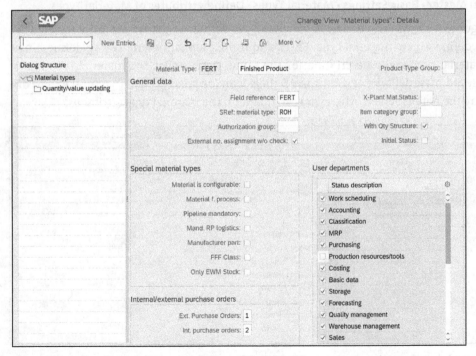

Figure 3.5 Define Material Type Attributes: Details: 1 of 2

> **Note**
>
> SAP S/4HANA provides security controls at the material group and material type levels. You can control the authorization by defining authorization groups for material types and/or material groups based on business requirements.

- **X-Plant Mat.Status**

  In this field, assign a material status that is applicable to all plants. In the supply chain, the material status plays a vital role. It defines whether a material is free to use, in development, or obsolete. Table 3.1 shows the material statuses available in the standard system. Assigning a material status at the material type level defaults that status when you create a material using the corresponding material type.

Material Status	Description
1	Design
2	Design and plan
3	Use
4	Phase-out
5	Obsolete

Table 3.1 Material Statuses

- **Item category group**

  You can assign a default item category group in this field for the material type. The item category group plays a vital role in the determination of item categories for the delivery type. An item category group is assigned in the **Sales: sales org 2** view of the material master. For example, **NORM** is the item category group for a standard item.

- **External no. assignment w/o check**

  If you set this indicator, you can create a material master with an external number arrangement with at least one character without any further checks. If you leave the **Material Number** field blank, the system will assign an internal number range, which is defined for the material type.

- **User departments**

  Under **User departments**, you can select or deselect material master views based on your business requirements. During the creation of the material master, you can maintain these views with reference to the organizational units required to maintain the view.

- **Special material types**

  Under **Special material types**, maintain the special attributes for materials at the material type level. You can set the following indicators based on business requirements:

- **Material is configurable**: If you set this indicator, the material created using the respective material type will be configurable.
- **Material f. process**: If you set this indicator, the material created using the respective material type will be used in a process, and by-products are expected to be produced.
- **Pipeline mandatory**: If you set this indicator, the material created using the respective material type will be handled in a pipeline. Set this indicator for gases, liquids, and so on. You can't create a purchase order to procure this material, and such materials are nonvaluated and not quantifiable.
- **Mand. RP logistics**: If you set this indicator, the material created using the respective material type will be treated as a returnable packaging material and the relevant goods movements will be set in inventory management (e.g., goods issue for outbound delivery with reference to sales order, movement type 601). The system automatically posts the cost to the returnable packaging material account based on valuation and account determination configuration.
- **Manufacturer part**: If you set this indicator, the material created using the respective material type will be treated as a manufacturer part.
- **FFF Class**: If you set the form-fit-function (FFF) class indicator is used for interchangeable parts, and the FFF class groups the interchangeable parts together by material type.
- **Only EWM Stock**: If you set this indicator, the stock of the material created using the respective material type will be maintained in embedded extended warehouse management (EWM) only, not in inventory management. For more details on embedded EWM, see Chapter 8.

■ **Ext. Purchase Orders**
In this field, maintain a key to allow or disallow external purchase orders for the materials created using this material type. Table 3.2 shows the keys that you can set for external purchase orders.

External Purchase Order Key	Description
0	No external purchase orders allowed
1	External purchase orders allowed, but warning issued
2	External purchase orders allowed

Table 3.2 External Purchase Order Allowed

■ **Int. purchase orders**
In this field, maintain a key to allow or disallow internal purchase orders (intercompany stock transfer orders) for the materials created using this material type. Table 3.3 shows the keys that you can set for internal purchase orders.

Internal Purchase Order Key	Description
0	No internal purchase orders allowed
1	Internal purchase orders allowed, but warning issued
2	Internal purchase orders allowed

Table 3.3 Internal Purchase Order Allowed

Figure 3.6 shows the second part of the details screen for material type **FERT** (**Finished Product**) as defined in the system. Let's explore the key field settings for the material type:

- **Classification**
  Under **Classification**, you can assign a **Class type** and the corresponding **Class** to the material type. Using a classification, you can define additional characteristics/attributes for the material based on the class type. The following class types are applicable to the material master, and assigning a class type and class will default the corresponding details into the **Classification** view of the material master when created:
  - **001**: Material class
  - **023**: Batch class
  - **200**: Configurable material class
  - **300**: Variant material class
- **Valuation**
  Under **Valuation**, you can assign a price control and account category reference to the material type:
  - **Price control**
    There are two price control options in SAP S/4HANA—standard price or moving average price—to be assigned to a material type so that the system valuates materials in the inventory management accordingly. The standard price is a fixed price that doesn't vary depending on the stock and value moving in and out, while the moving average price does vary depending on the stock and value moving in and out. The moving average price is mostly assigned to raw materials, and standard price is assigned to finished and semifinished products.
  - **Acct cat. reference**
    The account category reference acts as a bridge between material types and valuation classes. A valuation class is defaulted into the material master data in SAP through the account category reference. The valuation class plays a vital role in automatic account determination during goods movements in inventory management. For more on valuation classes, see Chapter 10, Section 10.3.

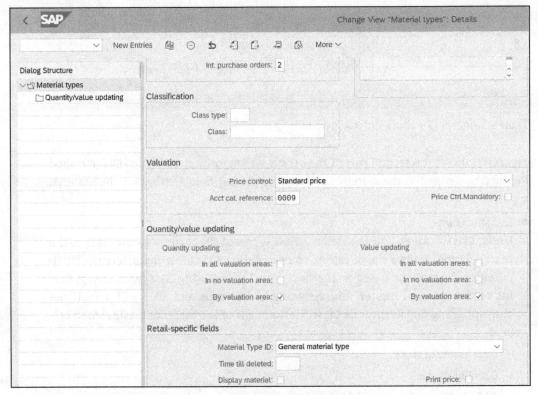

Figure 3.6 Define Material Type Attributes: Details: 2 of 2

- **Quantity/value updating**
  To navigate to the quantity/value updating screen, double-click **Quantity/value updating** in the **Dialog Structure**. Here you can define if the quantity and value are updated in all valuation areas, selected valuation areas, or not for the material type. You can set one of the following checkboxes for **Quantity updating** and **Value updating** separately:

  - **In all valuation areas**: If you set this indicator, the quantity/value update for the materials created using the associated material type happens in all valuation areas (a valuation area can be a plant).

  - **In no valuation area**: If you set this indicator, the quantity/value update for the materials created using the associated material type doesn't happen in any of the valuation areas.

  - **By valuation area**: If you set this indicator, you can select the valuation areas in which the quantity/value update for the materials created using the associated material type will happen.

  Figure 3.7 shows the quantity updating and value updating set for valuation areas **MJ00**, **MJ01**, and **MJ0A** for material type **FERT**. This setting is required if you set the value/quantity updating to happen per valuation area. You can set the indicators for valuation areas based on business requirements.

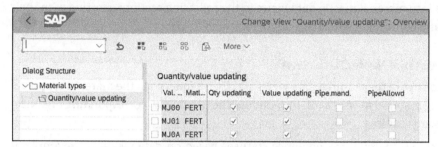

Figure 3.7 Define Material Types: Quantity and Value Update

3.1.4   Material Numbers

In this configuration step, you define the length of material numbers, up to 40 characters. The best practice recommendation is to define a length of 18 characters for the material. To define attributes of material types, execute Transaction SPRO and navigate to **Logistics—General • Material Master • Basic Settings • Define Output Format of Material Numbers**.

Figure 3.8 shows that the length for material numbers was defined as 18 characters.

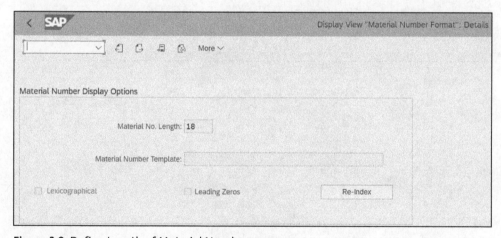

Figure 3.8 Define Length of Material Numbers

3.1.5   Material Master Data Views

Material master data views in SAP S/4HANA separate screen layouts designed based on the business process for ease of maintenance—for example, basic data views, classification views, purchasing views, MRP views, sales views, accounting views, and so on, as we discussed at the beginning of this section. Some process areas like purchasing have only one data view to be maintained, and other process areas like sales, accounting, and MRP have multiple data views to be maintained to create material master data.

We'll dive into material master data views in the following sections, starting with their creation and continuing with an in-depth look at each view.

Creating Material Master Data Views

You can configure the required data views and their sequence at the material type level for creating material master data. To maintain the data views, corresponding organization units are required.

Table 3.4 shows material maintenance statuses based on data views. For example, if you maintain accounting views of a material, the system sets the maintenance status of that material as B. The system keeps track of this for every material master data record to determine which data views are already maintained and which ones are yet to be maintained. The system stores the maintenance status of each material master record in database table MARA.

Maintenance Status	Description
A	Work scheduling
B	Accounting
C	Classification
D	MRP
E	Purchasing
F	Production resources/tools
G	Costing
K	Basic data
L	Storage
P	Forecasting
Q	Quality management
S	Warehouse management
V	Sales
X	Plant stocks
Z	Storage location stocks

Table 3.4 Material Master Maintenance Status

Table 3.5 shows the recommended material master data views for key material types. Maintenance of the material master views is mandatory for executing business transactions.

Material Types	Description	Basic Data: K	Classification: C	Sales: V	Purchasing: E	MRP: D	Forecasting: P	Work Scheduling: A	Warehouse Management: S	Accounting: B	Plant Stocks: X	Storage Location Stocks: Z	Storage: L	Costing: G	Quality Management: Q
PIPE	Pipeline materials	✓	✓		✓					✓	✓	✓			
ROH	Raw materials	✓	✓	✓	✓	✓	✓		✓	✓	✓	✓	✓	✓	✓
SERV	Service materials	✓			✓	✓				✓					
UNBW	Nonvaluated materials	✓	✓		✓	✓	✓		✓		✓	✓	✓		✓
VERP	Packaging	✓		✓	✓	✓	✓		✓	✓	✓	✓	✓		✓
DIEN	Service	✓		✓	✓	✓				✓					
ERSA	Spare parts	✓	✓		✓	✓	✓		✓	✓	✓	✓	✓		✓
FERT	Finished product	✓	✓	✓		✓	✓	✓	✓	✓	✓	✓	✓	✓	✓
HAWA	Trading goods	✓	✓	✓	✓	✓	✓		✓	✓	✓	✓	✓		✓
HERB	Interchangeable part	✓	✓	✓	✓	✓	✓		✓	✓	✓	✓	✓	✓	✓
HERS	Manufacturer part		✓		✓										
LEIH	Returnable packaging	✓		✓	✓	✓			✓	✓	✓	✓	✓		
FRIP	Perishables	✓	✓	✓	✓	✓	✓		✓	✓	✓	✓	✓		✓
HALB	Semifinished product	✓	✓	✓	✓	✓	✓	✓	✓	✓	✓	✓	✓	✓	✓

Table 3.5 Recommended Material Views by Material Type

For instance, a purchasing view is mandatory to create purchasing documents such as purchase requisitions, purchase orders, outline agreements, and so on. So for raw materials that are procured externally, a purchasing view must be maintained. If raw materials need to be transferred across plants and company codes, a sales view must be maintained so that an intercompany stock transfer order or intra company stock transfer order can be created using the material. Accounting views are mandatory if the material is valuated, and goods movement postings can create an accounting entry in the accounting document to record the value. MRP views are mandatory if the material planning is executed using a feature such as MRP, reorder point planning, or the like.

Let's dive deep into the creation and maintenance of material master data views. To create material master data in SAP S/4HANA, execute Transaction MM01. To change material master data, use Transaction MM02. And to display material master data, use Transaction MM03. Figure 3.9 shows the initial screen for creating material master data.

Let's explore the fields in the initial screen:

- **Material**
  Every material master data record will be created with a unique number in SAP S/4HANA. You can enter an external number in the **Material** field based on the external number range assigned to the material type. If you leave this field blank, the system assigns the next available number in the internal number range assigned to the material type.

- **Industry Sector**
  Here, specify the industry sector in which the material is being created. The industry sector controls the screen layout, and industry-specific screens are displayed for the creation of the material master. Mechanical engineering, chemical industry, retail, and services are just some of the different industry sectors.

- **Material Type**
  Specify the material type of the material here. Refer to Section 3.1.3 for more details on material types.

- **Change Number**
  Specify a change number if the material is going through a revision. This is an optional field.

- **Copy from**
  Specify a reference material number here to copy the details of the material data from it. Creation of a material will be easier if you reference another material master data record.

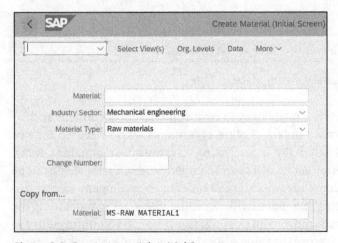

Figure 3.9 Create Material: Initial Screen

The next step in the material master data creation process is to select the views to be created. To select the data views, click the **Select View(s)** button at the top of the screen. Figure 3.10 shows the views selected to create a new raw material.

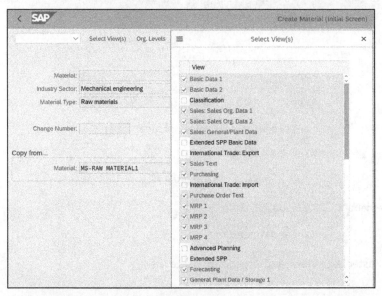

Figure 3.10 Create Material: Select Views

The next step in the material master data creation process is to specify the organizational units in which the material master data is to be created. To do so, click the **Org. Levels** button. Figure 3.11 shows that organizational units such as the plant, storage location, sales organization, division, and warehouse number have been specified to create a new raw material.

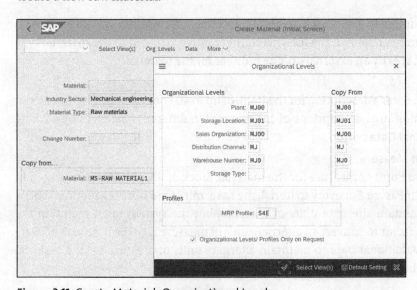

Figure 3.11 Create Material: Organizational Levels

Table 3.6 shows important material data views and the organizational units in which the respective data views are created.

Material Master Data View	Organization Level
Basic data	Client
Purchasing	Plant
MRP	Plant
Costing	Plant
Accounting	Plant (if a plant is the valuation area)
Quality management	Plant
Sales	Sales organization, distribution channel, and plant
Warehouse management	Warehouse number
Storage	Plant and storage location

Table 3.6 Material Master Data Views and Organizational Levels

Basic Data Views

The **Basic data 1** view is displayed after you press `Enter` on the organizational levels screen. From this view, you can navigate to other data views by clicking their tabs at the top. There are two views (screens) for maintaining basic data views of material. Figure 3.12 shows the **Basic data 1** view of the material master. The following are the key fields in this view:

- **Material**
  This value uniquely identifies a material master data record. The system assigns an internal number based on the internal number range assigned to the material type if you don't specify an external unique number for the material.

- **Descr.**
  In this field, enter a description for the item, of up to 40 characters in length. You can maintain additional descriptions of the material in different languages by clicking the **Additional Data** button.

- **Base Unit of Measure**
  Specify a base unit of measure for the material. Stock of the material is managed in this unit of measure. For every material, the base unit of measure is a required entry. You can maintain alternate units of measure, but the system must maintain the conversion factor to convert an alternate unit of measure into the base unit of measure. Click **Additional Data** to maintain alternate units of measure for the material and to set their conversion factors.

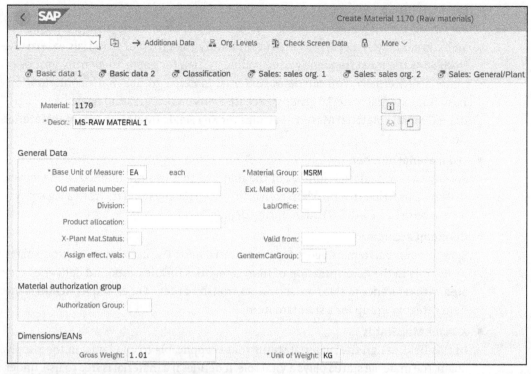

Figure 3.12 Material Master: Basic Data 1 View

- **Division**
  A division is a way to group materials to determine the sales area and business area. Specify a division for the material here. Divisions are defined in the configuration. To define a division, execute Transaction SPRO and navigate to **Logistics—General • Material Master • Settings for Key Fields • Define Divisions**. Figure 3.13 shows the divisions defined in the system.

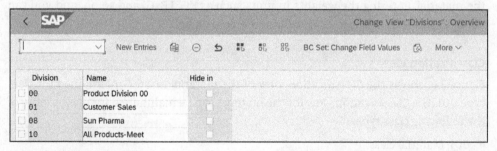

Figure 3.13 Define Divisions

- **Material Group**
  Material groups are used to group materials with similar characteristics/attributes. Specify the material group for the material here.

- **Ext. Matl Group**
  Specify an external material group for the material here. Materials can be linked to an external material group to uniquely identify the group of materials externally. UNSPSC is the most frequently used coding system for goods to identify groups of materials externally. You can define external material groups in the configuration. To define external material groups, execute Transaction SPRO and navigate to **Logistics—General • Material Master • Settings for Key Fields • Maintain External Material Groups**.

- **Old material number**
  Specify the old material number of the material that was maintained in the legacy system. This field is useful when converting material master data from a legacy system to be used in SAP S/4HANA during deployment.

- **GenItemCatGroup**
  Specify a general item category for the material here. The item category group plays a vital role in the determination of item categories for the outbound deliveries created with reference to sales order or stock transfer orders. For example, NORM is the item category group for a standard item.

- **X-Plant Mat.Status**
  In this field, assign the material status that is applicable to all plants. In the supply chain, the material status plays a vital role. It defines if a material is free to use, under development, or obsolete. Enter the **Valid from** date if the material is active.

- **Authorization Group**
  In this field, assign the authorization group, a four-digit alphanumeric code for the material. You can define different authorization groups for material groups and material types. The system will copy the authorization group from the respective material group or material type if it's maintained.

In the **Basic data 1** view, you can maintain the **Gross Weight** and **Net Weight** fields for the material, entering the weights in units such as KG, LB, G, and so on, and you can enter a volume in an appropriate unit in the **Volume** field.

Classification View

Figure 3.14 shows the **Classification** view of the material master, maintained for class type **001**. The **Classification** view for the material can be maintained for one or more of the following class types:

- **001**: Material class
- **023**: Batch class
- **200**: Configurable material class
- **300**: Variant material class

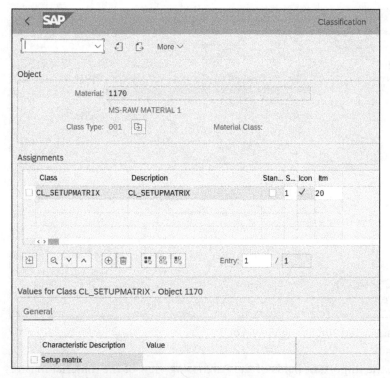

Figure 3.14 Material Master: Classification View

There's only one key field in the **Classification** view: the **Class** field, in which you assign a class to the material. You can assign a material class or batch class in this field. If you want to maintain the material class, assign the class defined for materials with class type **001**. If you are maintaining the classification view for batch class, assign the class defined for class type **023**. A class can be created in SAP S/4HANA in Transaction CLO2 using a specific class type, and you can define the characteristics/attributes to maintain the material master data as well.

The system automatically defaults the characteristics defined in the class assigned to the material. You can maintain the values for the characteristics as needed for your business requirements.

Purchasing View

The **Purchasing** view of the material is maintained at the plant level. You can define the **Purchasing** view of the material for one or more plants. The **Purchasing** view of the material is mandatory to create purchasing documents such as purchase orders or stock transfer orders for the material for a given plant.

Figure 3.15 shows the first part of the **Purchasing** view. **Base Unit of Measure** and **Material Group** are copied from the **Basic data 1** view of the material master automatically.

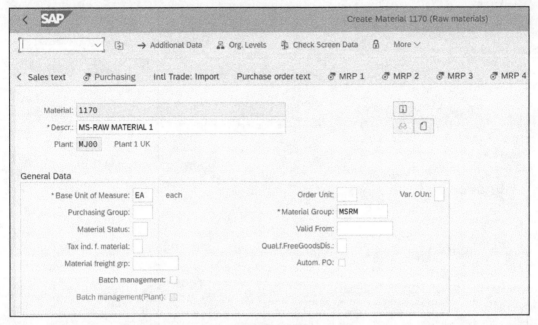

Figure 3.15 Material Master: Purchasing View: 1 of 2

The following are the important fields of the **Purchasing** view of the material master data:

- **Purchasing Group**
  Assign a purchasing group responsible for purchasing this material. This is an optional entry, but if you want a specific buyer/purchasing group to manage purchasing activities of this material, specify a purchasing group. A purchasing group is a subdivision of a purchasing organization, responsible for specific purchasing activities or managing purchasing activities for one specific or multiple product categories. Purchasing groups are typically assigned to an individual or a group of buyers. Purchasing group information such as contact details will be provided via the purchase orders or outline agreements to the vendors.

- **Order Unit**
  Specify a unit of measure in which the material is ordered here. This unit of measure is used in purchasing documents (purchase orders, outline agreements, etc.) to order the material from suppliers. If the order unit is same as the base unit of measure, do not specify the order unit in this field.

- **Material Status**
  In this field, assign the material status that is applicable to the plant in which the purchasing view is being created. In the supply chain, the material status plays a vital role: it defines if a material is free to use, under development, or obsolete. Enter the **Valid From** date for the material status if the respective plant is active. The difference between the X-plant material status from the **Basic data 1** view of material master

data and the status in this field is that the X-plant material status is global, while this field is plant specific.

- **Tax ind. f. material**

  If you need to determine the tax at the time of a purchase order, specify a tax indicator for the material to support the determination of the input tax code during the creation of a purchase order automatically. The following are the most used tax indicators for the material:

  - **0**: Tax exempt
  - **1**: Half tax (50%)
  - **2**: Full tax (100%)

- **Material freight grp.**

  Specify a material freight group for the material here. This is used to determine freight classification and freight code. This information leads to the determination of transportation requirements for the materials and to calculate freight cost.

- **Autom. PO**

  Set this indicator if you want to convert purchase requisitions created for this material into purchase orders automatically using Transaction ME59 or ME59N. You can schedule these two transactions to run in the background.

Figure 3.16 shows the second part of the **Purchasing** view, which includes the following fields:

- **Purchasing value key**

  This key stores the shipping instructions, underdelivery and overdelivery tolerances, and deadline monitoring for purchasing documents. Deadline monitoring includes reminders at regular intervals that can be sent to suppliers for requests for quotation, reminders to ship ordered goods, and so on. Up to three reminders can be set in purchasing value keys.

  Purchasing value keys are defined in the configuration settings. To define purchasing value keys, execute Transaction SPRO and navigate to **Materials Management • Purchasing • Material Master • Define Purchasing Value Keys**.

- **GR processing time**

  Specify the goods receipt processing time in days. This defines the number of days required to process the goods receipt of this material into stock in inventory management.

- **Post to Insp. Stock**

  If you set this indicator, the stock of this material during goods receipt will be posted to quality inspection stock.

- **Source list**

  If you set this indicator, the source list (purchasing master data) is mandatory for external purchasing activities using this material.

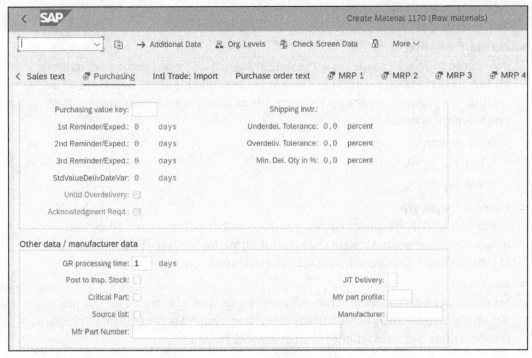

Figure 3.16 Material Master: Purchasing View: 2 of 2

Sales Views

Sales views of a material are maintained at the sales organization, distribution channel, and plant levels. You can define the sales views of the material for one or more sales organization and distribution channel combinations and plants. Sales views of the material are mandatory to create sales documents and stock transfer orders, so it is vital to create sales views for trading goods, certain raw materials, finished products, semifinished products, and so on. There are three sales views for material master data:

- **Sales: sales org. 1**
  Maintained at the sales organization and distribution channel level

- **Sales: sales org. 2**
  Maintained at the sales organization and distribution channel level

- **Sales: General/Plant**
  Maintained at the plant level

Figure 3.17 shows the **Sales: sales org. 1** view of the material master. **Base Unit of Measure**, **Material Group**, and **Division** are copied from the **Basic data 1** view of the material master automatically. However, you can change the division in this view.

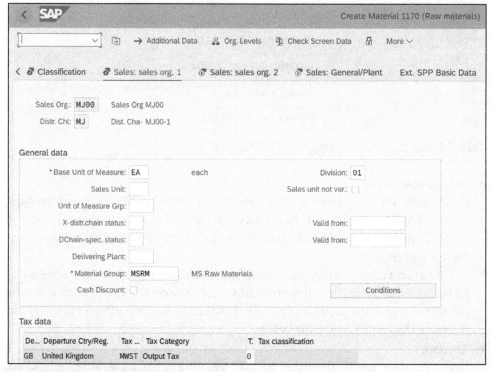

Figure 3.17 Create Material: Sales Org. 1 View

The following are the important fields of the **Sales: sales org. 1** view of the material master data:

- **Sales Unit**
  Specify a unit of measure in which the material is sold. This unit of measure is used in sales documents (e.g., sales orders) to sell the material to customers. If the sales unit is same as the base unit of measure, do not specify the sales unit in this field.

- **Sales unit not var.**
  If you set this indicator and specify a sales unit for the material, the sales order must be created in the sales unit only.

- **Unit of Measure Grp.**
  If you have created a unit of measure group with multiple alternate units of measure to sell the material, specify a unit of measure group applicable for the material to create sales documents (sales order).

- **X-distr.chain status**
  A distribution chain in SAP S/4HANA is a combination of a sales organization and a distribution channel. In this field, specify a material status applicable to all distribution chains, which is a global setting. Enter the **Valid from** date for the X-distribution chain status if the material is active. Table 3.7 shows the distribution chain statuses that can be set for the material.

Distribution Channel Status	Description
01	Discontinued without replacement
02	Discontinued with replacement
03	Technical defect
04	Test part

Table 3.7 Distribution Chain Statuses of the Material

- **DChain-spec. status**
  In this field, specify a material status applicable to the specific distribution chain. Enter the **Valid from** date for the distribution chain status if the material is active. Table 3.7 shows the distribution chain statuses that can be set for the material.

- **Delivering Plant**
  Specify a plant that issues the stock of the material to the customer with reference to an outbound delivery. This allows the sales order to be created in one plant while the plant from which the goods are delivered to the customer is a different plant within the same legal entity (company code).

- **Tax data**
  Under tax data, specify departure country (ship-from), tax condition type, and tax classification (tax indicator) to determine the output tax code in the sales documents.

Figure 3.18 shows the first part of the **Sales: sales org. 2** view. The **Gen. item cat. grp** (general item category group) value is copied from the **Basic data 1** view of the material master automatically. The following are the important fields of the **Sales: sales org. 2** view of the material master data:

- **Matl statistics grp**
  Specify a material statistics group here. The system uses this data to update the sales document information to generate statistics.

- **Material Price Grp**
  Specify a material price group here. The system uses this data to apply the same pricing conditions for price determination in sales documents. You can create the same pricing condition records using the material price group for a group of materials.

- **Acct Assmt Grp Mat.**
  Specify an account assignment group for the material in this field to determine the revenue account in the accounting document during the sales billing document creation.

- **Item Category Group**
  Specify an item category for the material. The item category group plays a vital role in the determination of item categories for the outbound deliveries created with reference to sales orders or stock transfer orders. For example, **NORM** is the item category group for standard items.

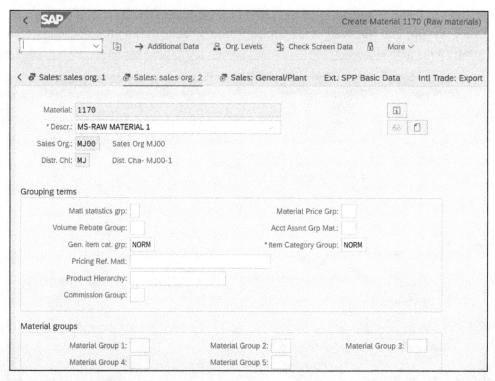

Figure 3.18 Create Material: Sales Org. 2 View

- **Product Hierarchy**

  Different features/characteristics of the materials are grouped together in terms of hierarchy levels. You can group the features in different ways such as by electrical property, mechanical property, color, model, make, and so on. You can define a product hierarchy code of up to 18 digits. For example, an electric water pump can be described by product hierarchy as 000010000200000015, where the first five characters (00001) show that the product belongs to the electrical goods category, characters 6 to 10 (00002) show that the product belongs to the mechanical goods category, and characters 11 to 18 (00000015) show that the product belongs to the water pumps category.

  Specify the product hierarchy of the material as defined in the configuration settings. To define product hierarchies, execute Transaction SPRO and navigate to **Logistics—General • Material Master • Settings for Key Fields • Data Relevant to Sales and Distribution • Define Product Hierarchies**.

- **Material groups**

  There are five material groups field in the **Sales: sales org. 2** view. These are used in sales and distribution only to determine pricing, and in reporting. Assign material groups in these fields based on business requirements.

Figure 3.19 shows the first part of the **Sales: General/Plant** view. **Base Unit of Measure**, **Gross Weight**, and **Net Weight** are copied from the **Basic data 1** view of the material master automatically.

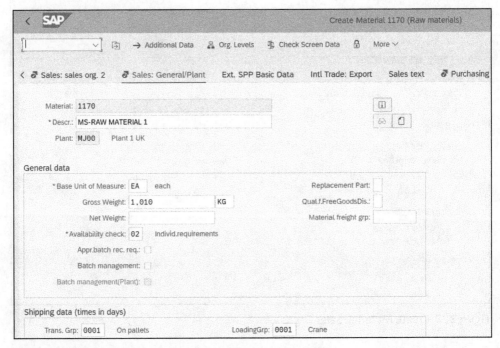

Figure 3.19 Create Material: Sales: General/Plant View

The following are the important fields of the **Sales: General/Plant** view of the material master data:

- **Replacement Part**
  Specify if the material can be a replacement part or not. This indicator for the material is used in sales billing documents.

- **Availability check**
  Availability check groups play a vital role in providing accurate information about stock availability to fulfill customer orders. They are used to configure availability checks in SAP S/4HANA. Specify the availability check group for the material here. See Chapter 4 for more details.

- **Batch management**
  If you activate batch management, the material will be managed and issued in batches in inventory management for outbound deliveries.

- **Trans. Grp**
  Specify a transportation group for the material here for route scheduling for transportation purposes during sales order and delivery processing.

- **LoadingGrp**
  Specify a loading group for the material to automatically determine the shipping point during sales order and delivery processing.

MRP Views

MRP views of a material are maintained at the plant and storage location levels. You can define the MRP views of the material for one or more plants. MRP views of the material are mandatory to perform material planning in SAP S/4HANA. There are four MRP views of material master data:

- **MRP 1**
  Maintained at the plant level
- **MRP 2**
  Maintained at the plant level
- **MRP 3**
  Maintained at the plant level
- **MRP 4**
  Maintained at the plant and storage location levels

Figure 3.20 shows the **MRP 1** view of the material master. **Base Unit of Measure** is copied from the **Basic data 1** view and **Purchasing Group** is copied from the **Purchasing** view of the material master. **Material Status**, if already set in any of the plant-specific views (**Purchasing** view, **Sales: General/Plant** view, etc.), is copied over to the **MRP 1** view.

The following are the important fields of the **MRP 1** view of the material master data:

- **ABC Indicator**
  An ABC indicator classifies materials based on consumption and their significance. This is used in inventory-sampling procedures. Specify an ABC indicator for the material here. Table 3.8 shows the ABC indicators defined in the standard system.

ABC Indicator	Description
A	Significant material and high consumption
B	Material—medium significance and medium consumption
C	Material—low significance and low consumption

Table 3.8 ABC Indicators

- **MRP Group**
  An MRP group is a logical grouping of MRP parameters to control materials planning. The MRP group is assigned in the material master at the plant level. Specify an MRP group here for the material.

- **MRP Type**

  The MRP type determines whether materials are planned in SAP S/4HANA and, if so, how they are planned. Specify an MRP type for the material here. Refer to Chapter 6, Section 6.3.1, for more details about MRP types.

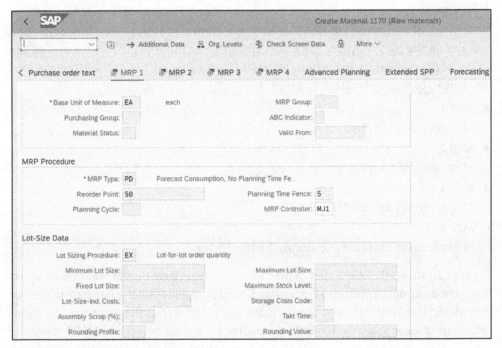

Figure 3.20 Create Material: MRP 1 View

- **Reorder Point**

  A reorder point is required if the material is planned using a reorder point planning procedure, especially for manual reorder point planning. The system automatically determines a reorder point with automatic reorder point planning. If the stock falls below this point, the system automatically creates a necessity for material requirements.

- **Planning Time Fence**

  A planning time fence is a period (in days) in which there will be no automatic changes allowed to the master plan. For example, if the master plan creates procurement proposals within the planning time fence, then no automatic changes to the procurement proposals are allowed. Specify a planning time fence for the material here.

- **Planning Cycle**

  Planning cycles are defined in configuration settings for MRP. Specify a planning cycle here for the material for which the material planning run is performed.

- **MRP Controller**

  MRP controllers are responsible for executing material requirements planning at the plant level. Specify an MRP controller for the material for a specific plant here.

- **Lot-Size Data**
  Specify a lot sizing procedure for the material here. During MRP, the system calcu-lates the lot quantity for procurement or production activities. Specify a **Minimum Lot Size**, **Maximum Lot Size**, or **Fixed Lot Size** for the system to determine the lot size.

Figure 3.21 shows the **MRP 2** view of the material master. It contains procurement-, scheduling-, and net requirements calculation-related attributes of the material.

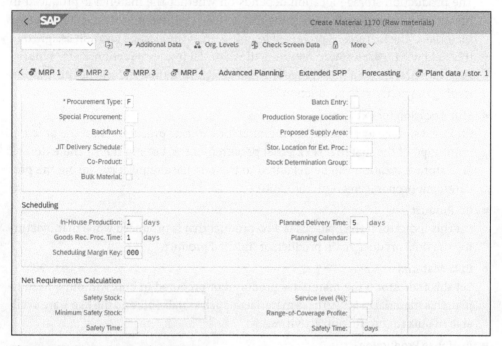

Figure 3.21 Create Material: MRP 2 View

The following are the important fields of the **MRP 2** view of the material master data:

- **Procurement Type**
  Procurement type indicates if the material is produced in house, externally pro-cured, or both. Choose a procurement type for the material from the dropdown menu. Table 3.9 shows the procurement types available in the standard system.

Procurement Type	Description
E	In-house production
F	External procurement
Blank	No procurement
X	Both procurement types

Table 3.9 Procurement Type

- **Special Procurement**
  Special procurement types such as consignment, subcontracting, stock transfer, and so on are defined in the configuration settings at the plant level. Choose a special procurement if any for the material if the procurement type of the material is external procurement or both external and in house.

- **Production Storage Location**
  The production storage location depends on whether the material is produced in house or is a component to be provided to the production process, such as a raw material. If the material is produced in house, then the storage location assigned in this field will be the location receiving the material from production. If the material is a raw material, then the storage location assigned in this field will be the location issuing the material to production.

- **Stor. Location for Ext. Proc.**
  Choose a storage location for the material for external procurement if the procurement type of the material is external procurement or both in house and external. This storage location will be defaulted to the purchase requisition during the procurement proposal creation from MRP.

- **Co-Product**
  Set this indicator if the material is a co-product that is produced while manufacturing the final product, main product, or finished product.

- **Bulk Material**
  Set this indicator if the material is produced or procured in bulk without a need to plan this material in MRP. These materials—such as nails, oil, and the like—are available in bulk for production activities.

- **In-House Production**
  In this field, specify the number of days required to produce the material in house.

- **Planned Delivery Time**
  In this field, specify the number of days required to procure the material externally.

- **Safety Stock**
  Manufacturing facilities keep a certain stock of a material for safety to meet unexpected demand. Specify the safety stock quantity here in the base unit of measure. The **Minimum Safety Stock** field is also used for the same purpose, but for the lowest level of safety stock required to meet unexpected demand.

Figure 3.22 shows the **MRP 3** view of the material master. It contains forecasting-, planning-, and availability check-related attributes of the material.

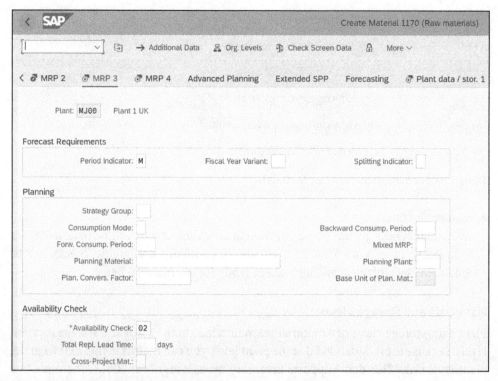

Figure 3.22 Create Material: MRP 3 View

The following are the important fields of the **MRP 3** view of the material master data:

- **Period Indicator**
  Specify a period indicator such as **M** (monthly), **W** (weekly), or **T** (daily) for material forecasting or material planning.

- **Strategy Group**
  A strategy group is used to group multiple planning strategies that can be used for the material: make-to-order, make-to-stock, and so on. Choose a strategy group for the material here.

- **Consumption Mode**
  The consumption mode of the material indicates how the system consumes requirements. Table 3.10 shows the values to choose from. In backward consumption, sales orders, reservations, and the like consume planned independent requirements that fall before the requirement date. In forward consumption, sales orders, reservations, and the like consume planned independent requirement that fall after the requirement date. Specify the forward consumption period or backward consumption period in days (up to 999 days) here, based on the consumption mode set for the material.

Consumption Mode	Description
1	Backward consumption only
2	Backward/forward consumption
3	Forward consumption only
4	Forward/backward consumption
5	Period-specific consumption

Table 3.10 Consumption Mode for Material

- **Availability Check**
  Availability check groups play a vital role in providing accurate information of stock availability to execute MRP. This is used to configure availability checks in SAP S/4HANA. Specify the availability check group for the material here.

Plant Data and Storage Views

Plant data/storage views of a material are maintained at the plant and storage location levels. Because most data is valid at the plant level, you can maintain these views at the plant level only. This data supports inventory management functions. There are two plant data/storage views for material master data:

- **Plant data/stor. 1**
  Maintained at the plant or plant and storage location levels
- **Plant data/stor. 2**
  Maintained at the plant or plant and storage location levels

Figure 3.23 shows the **Plant data/stor. 1** view of the material master. **Base Unit of Measure** is copied from the **Basic data 1** view of the material master. The following are the important fields of the **Plant data/stor. 1** view:

- **Unit of issue**
  If you want to use a goods issue unit of measure for the material that's different from the base unit of measure, specify a unit of issue in this field. The system uses this unit of issue for goods issues, transfer postings, and material reservations.
- **Temp. conditions**
  You can define temperature conditions in the configuration settings for storage purposes. Specify a temperature condition as defined in the configuration to store the material in inventory management in that temperature condition.
- **Storage conditions**
  You can define storage conditions in configuration settings. Specify a storage condition as defined in the configuration to store the material in inventory management

using that storage condition—for example, refrigeration, bulk storage, hot conditions, and so on.

- **Haz. material number**
This number that identifies a material as a hazardous good. Specify a hazardous material number if the material belongs to the hazardous goods category.

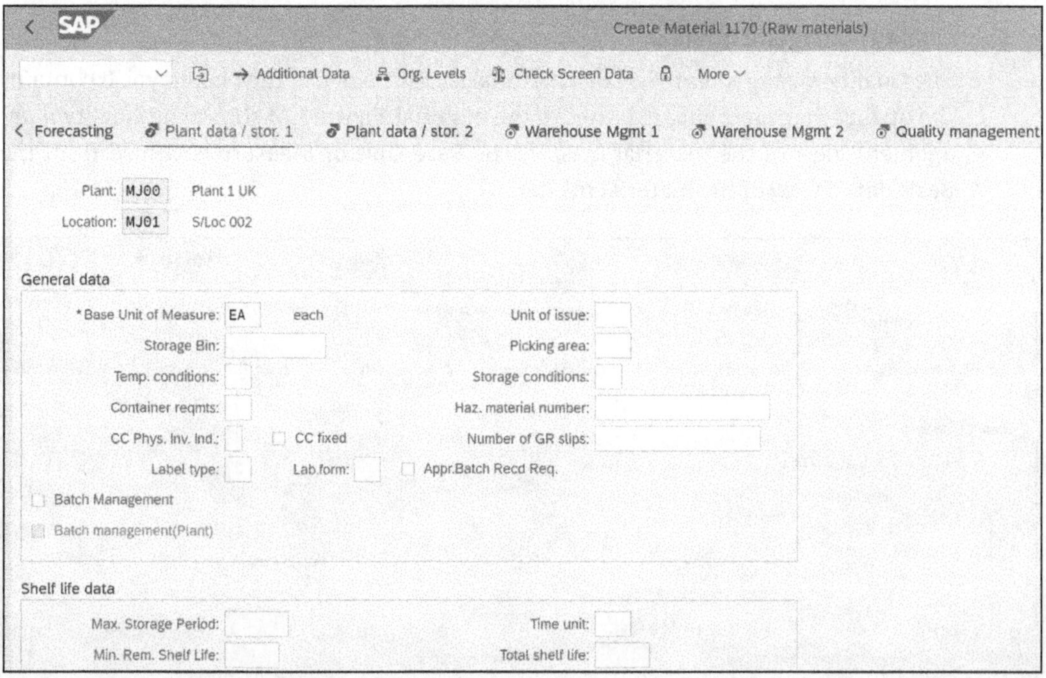

Figure 3.23 Create Material: Plant Data/Stor. 1 View

- **Container reqmts**
Container requirements are defined in configuration settings. If the material needs to be shipped in specific containers, specify the requirements here. This is a regulatory requirement for certain goods.

- **CC Phys. Inv. Ind.**
In this field, specify a cycle-counting indicator, a single-digit alphanumeric or numeric character for cycle counting of products for physical inventory.

- **Batch Management**
If you activate batch management, the material will be managed and issued in batches in inventory management.

- **Max. Storage Period**
Specify a maximum storage period (in days, months, years, etc.) for the material to be stored in inventory management.

- **Min. Rem. Shelf Life**
Specify a minimum remaining shelf life (in days, months, years, etc.) for the material.

This is the minimum remaining time for the material to be allowed in inventory management.

- **Total shelf life**
Specify the total shelf life of the material. This is used to calculate the expiration date of the material from the production date.

Quality Management View

The **Quality management** view of a material is maintained at the plant level. It contains the quality management attributes of the material. Figure 3.24 shows the **Quality management** view of the material master. The **Base Unit of Measure** is copied from the **Basic data 1** view of the material master.

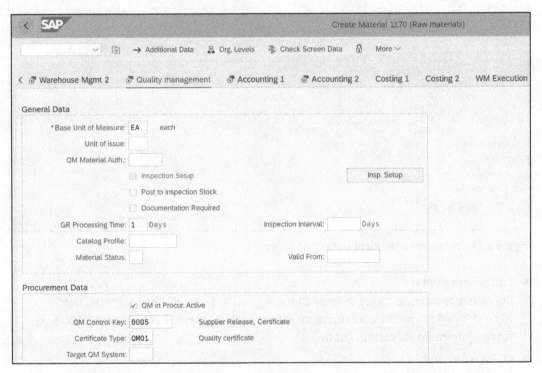

Figure 3.24 Create Material: Quality Management View

The following are the important fields of the **Quality management** view:

- **Inspection Setup**
Assign inspection types and activate them for the material. At least one inspection type must be set up for the material. Inspection types are configured in quality management (see Chapter 5, Section 5.2).

- **Post to Inspection Stock**
If you set this indicator, the stock of this material during goods receipt will be posted to quality inspection stock.

- **QM Material Auth.**
  Assign an authorization group for the material if you want to provide access only to specific users to perform a quality inspection on the material.
- **QM in Procur. Active**
  If you set this indicator, quality management will be active in procurement.
- **QM Control Key**
  This is a unique four-digit numeric value that controls the quality management process in procurement. Assign a four-digit code here to create your own control key based on business requirements.

Accounting Views

Accounting views of a material are maintained at the valuation area level. If the valuation control is set at the plant level, then accounting views are created at the plant level. Accounting views contain the valuation details such as price control, material unit price, attributes for automatic account determination, split accounting details if the material is split valuated, and so on. There are two accounting views for material master data.

Figure 3.25 shows the **Accounting 1** view of the material master. **Base Unit of Measure** is copied from the **Basic data 1** view of the material master. The following are the important fields in this view:

- **Total Stock**
  This is a display field. The system displays the total current stock of the material in the base unit of measure in the plant in which the **Accounting 1** view is created.
- **Valuation Cat.**
  The valuation category is used to determine valuation types for split valuation of a material. Split valuation allows you to configure different valuation categories and valuation types within a valuation area (plant). Assign a valuation category to the material here if you want split valuation.
- **Valuation Class**
  The valuation class is defaulted from the material type. The valuation class controls the general ledger account determination for materials during a goods movement transaction for material valuation.
- **Price Control**
  The price control (not shown; you can access this field by scrolling down) defines how materials are valuated. SAP S/4HANA provides two methods for valuing inventory—standard price and moving average price:
  - **Standard Price (S)**: The standard price is a static price set for a material, and it does not change with procurement activity. The standard price of the material is fixed until you change it manually in the material master.

- **Moving Average Price (V)**: The moving average price is the dynamic price that changes with procurement activities such as goods receipt and invoice receipt. The system calculates the moving average price of material based on the purchase price during goods receipt and invoice receipt postings.

Figure 3.25 shows that price control **S (Standard Price)** was assigned to the material and that the **Standard Price** of the material is maintained in the **Accounting 1** view.

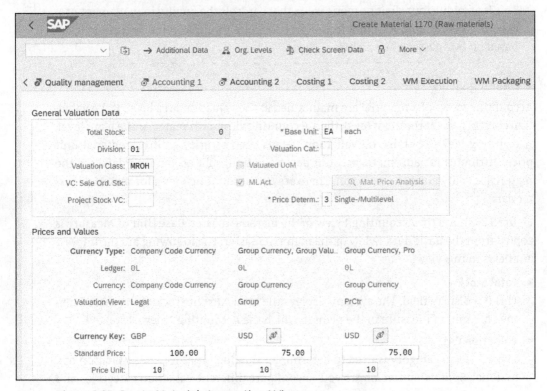

Figure 3.25 Create Material: Accounting 1 View

3.2   Business Partner

Business partners are a new concept in SAP S/4HANA, representing the highest-level data element. The business partner is the most used master data element for almost all functions in SAP S/4HANA. It represents a person, an organization, or a group that has some business interest. A business partner can be a vendor and a customer at the same time. With this new concept, an organization can buy goods and services from and sell their products to the same business partner. Business partners with different roles such as vendors, customers, and so on can be created using one transaction in SAP S/4HANA, which reduces master data duplicity.

We'll dive into business partner master data in the following sections, starting with an introduction to the concept and continuing with business partner roles, customer/vendor integration (CVI), and general configuration.

3.2.1   Business Partner Concept and Advantages

Figure 3.26 illustrates the business partner concept. At the top of the design is the business partner in SAP S/4HANA; next comes the business partner category, which is used in a specific business process. General data, suppliers, and customers are created as business partner roles. Multiple business partner roles are grouped into one business partner group. The system displays the relevant screen based on the business partner grouping at the time of business partner creation. Business partner groupings also help differentiate between internal business partners (e.g., employees) and external business partners (e.g., suppliers and customers). An external number range or an internal number range for the business partner is determined based on the business partner grouping, while the customer and supplier number range are determined from the account group. Account groups are used to classify vendors and customers into different business partner functions.

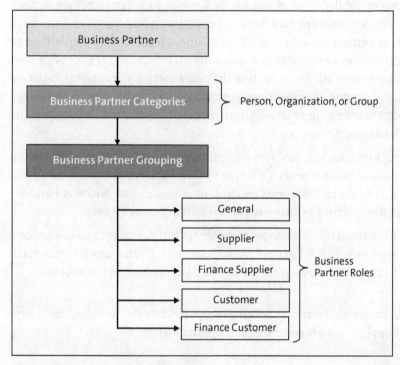

Figure 3.26 Business Partner Concept

The business partner is a simplified concept that makes data management simple. Both vendor and customer roles for an external stakeholder entity and its subsidiaries can be maintained under one business partner record. The business partner concept offers the following advantages:

- Simplified and centralized data management can be achieved for internal and external business partners/stakeholders.

- One business partner can be created with multiple partner roles such as customer, vendor, and so on. Within the customer and vendor roles, multiple vendor and customer partner functions (account groups) can be created.

- It allows for reduced data redundancy with the business partner concept and consolidated data approach.

- It results in improved external stakeholder relationships. The business partner stores parent and subsidiary companies in a hierarchical manner, which makes it easy to manage complex relationships.

3.2.2   Business Partner Roles

Business partner roles are used to define different business partner functions and classify them according to the business partner characteristics and functions. For example, the supplier, finance supplier, general vendor data, goods supplier, customer, finance customer, and so on are different functions of vendors and customers, and the business partner role is defined for every function. Business partner roles are defined to establish a specific business relationship. For example, a purchasing vendor is defined to store purchasing-related attributes such as the order currency, Incoterms, ordering address, and so on to support the creation of purchasing documents, whereas a finance vendor is defined to store payment terms, payment methods, withholding tax, and the like to support the incoming supplier invoice postings.

SAP S/4HANA provides standard business partner roles, and you can add new custom roles based on business requirements. To define business partner roles, execute Transaction SPRO and navigate to **Cross-Application Components • SAP Business Partner • Business Partner • Basic Settings • Business Partner Roles • Define BP Roles**.

Figure 3.27 shows the standard business partner role **FLVN01 (Supplier)**. You can define a new custom business partner role by copying a standard role or can create it from scratch by clicking **New Entries**. The following are the key fields of the business partner role:

- **Title**
  Enter the title for the business partner role in this field. Enter "Supplier" for a supplier role, "FI Supplier" for a finance supplier role, and so on.

- **Description**
  Enter a detailed description for the business partner role.

- **Hide**
  Set this indicator if you don't want to display the business partner role in the selection. You can hide unwanted business partner roles by setting this indicator.

- **BP Role Cat.**
  The business partner role category represents a person, an organization, or a business group. By assigning a business role category to the business partner role, you can create a business partner using the business partner role as a person, a group, or an organization. You can assign multiple business partner roles to a business role category. The system displays all business partner roles assigned to the business partner role category under **Additional BP Roles for BP Role Category**.

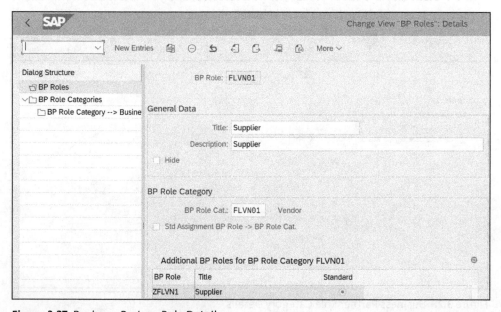

Figure 3.27 Business Partner Role Details

Figure 3.27 also shows the standard business partner category **FLVN01** (**Vendor**). Click **BP Role Categories** in the **Dialog Structure** to display an overview of all business partner categories, as shown in Figure 3.28. You can define a new custom business role category by copying a standard business role category or create one from scratch by clicking **New Entries**. Enter a **Title** and **Description** for the business role category. The following are the key fields of the business role category:

- **Diff.Type**
  The differentiation type is used as logical data to control the screen layout and screens to be displayed during the creation of the business partner. For example, company code data is used to maintain the financial data of the business partner (vendor and customer). The following are some of the differentiation types defined in the standard system:
  - **0: General Data**
  - **1: Company Code Data**
  - **11: Dunning Area Data**

- – 2: Purchasing Organization
- – 3: Sales Area Data
- – 4: Billing Group Data
- ■ **Possible Business Partner Categories**
  Select the possible business partner categories (person, organization, and/or business group) for the business partner role. During the creation of a business partner, the system displays those categories you selected, and you can create the business partner as one of those if you have selected all business partner categories.

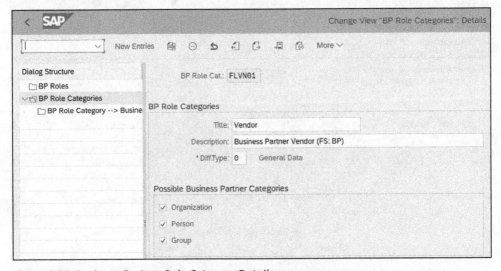

Figure 3.28 Business Partner Role Category Details

Business partner groupings are defined to group relevant business partner roles together. The system displays the business partner views according to the business partner groupings during the creation of the business partner. To define business partner role groupings, execute Transaction SPRO and navigate to **Cross-Application Components • SAP Business Partner • Business Partner • Basic Settings • Business Partner Roles • Define BP Role Groupings**.

Figure 3.29 shows the details of business partner role grouping **ZVEND**, defined for vendor role groupings. You can define a new business role grouping by copying an existing one or create one from scratch by clicking **New Entries**. Enter a **Title** and **Description** for the business role grouping.

There's only one key field for the business role grouping: **Role Group.Cat.**. Define a role group category under **BP Role Grouping Categories** in the **Dialog Structure**, then assign it in this field. This is used for grouping multiple business partner role groupings logically.

Figure 3.30 shows the business partner roles assigned to business partner role grouping **ZVEND**. Click **New Entries** to assign a new business partner role grouping.

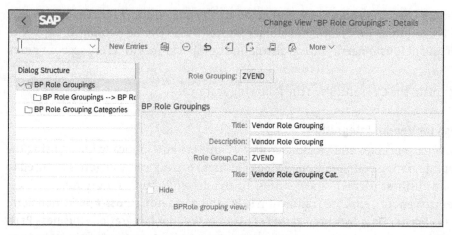

Figure 3.29 Business Partner Role Grouping Details

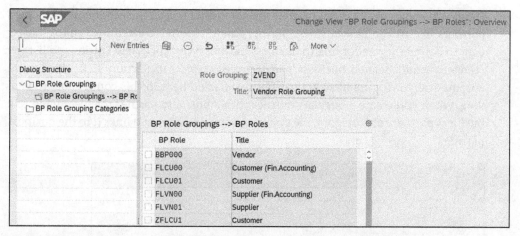

Figure 3.30 Business Partner Role Assignments to Business Partner Role Groupings

3.2.3   Customer/Vendor Integration

Vendor account groups play a vital role in defining the integration between a vendor and business partner and a customer and business partner. Vendor account groups and customer account groups are defined per partner function: vendor, invoicing party, ordering address, ship-to party, sold-to party, and so on.

You can define vendor account groups in the configuration settings. To define vendor account groups, execute Transaction SPRO and navigate to **Financial Accounting • Accounts Receivable and Accounts Payable • Supplier Accounts • Master Data • Preparations for Creating Supplier Master Data • Define Account Groups with Screen Layout (Vendors)**. You can also define number ranges for every account group here.

Similarly, you can define customer account groups in the configuration settings. To define customer account groups, execute Transaction SPRO and navigate to **Financial**

Accounting • Accounts Receivable and Accounts Payable • Customer Accounts • Master Data • Preparations for Creating Customer Master Data • Define Account Groups with Screen Layout (Customers). You can also define number ranges for every account group here.

We'll explain the CVI settings in the following sections.

Settings for Vendor Integration

Business partner roles and vendor account groups are automatically linked using the vendor integration configuration settings. Business partner role categories are used to integrate business partners and vendors. To enable vendor integration from business partners to vendors, execute Transaction SPRO and navigate to **Cross-Application Components • Master Data Synchronization • Customer/Vendor Integration • Business Partner Settings • Settings for Vendor Integration • Set BP Role Category for Direction BP to Vendor.**

Figure 3.31 shows the business partner role categories to link the vendors and business partners. To list the new business partner role categories created for vendor master data to define the vendor link to business partner roles, click **New Entries** and list the relevant vendor-related business partner roles created in Section 3.2.2. The system automatically links the business partner roles listed here in this config table to vendors. When you create a business partner with a business partner role belonging to a business partner role category, the system creates a vendor and links it to the business partner.

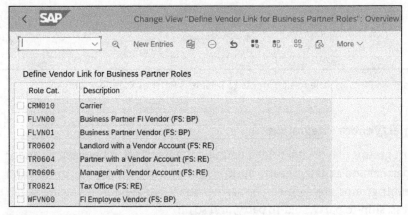

Figure 3.31 Settings for Vendor Integration

Settings for Customer Integration

Business partner role categories are used to integrate business partners and customers. To enable vendor integration from business partner to vendors, execute Transaction SPRO and navigate to **Cross-Application Components • Master Data Synchronization • Customer/Vendor Integration • Business Partner Settings • Settings for Customer Integration • Set BP Role Category for Direction BP to Customer.**

Figure 3.32 shows the business partner role categories to link the customers and business partners. To list the new business partner role categories created for customer master data to define the vendor link to business partner roles, click **New Entries** and list the relevant customer-related business partner roles created in Section 3.2.2. The system automatically links the business partner roles listed here in this config table to customers. When you create a business partner with a business partner role belonging to a business partner role category, the system creates a customer and links it to the business partner.

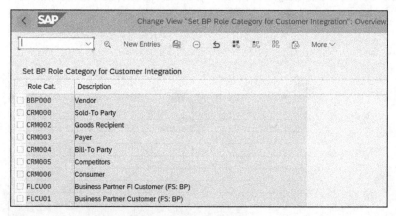

Figure 3.32 Settings for Customer Integration

3.2.4   Business Partner

To understand the configuration settings and the business partner concept more, let's explore the business partner master data in SAP S/4HANA. A business partner is created as a person, a group, or an organization using Transaction BP.

Figure 3.33 shows the initial screen for the creation of business partner. You can create a business partner as a person, group, or organization by clicking the corresponding buttons at the top of the screen. You must click one of those buttons (business partner categories) to create the business partner. In our example, we'll create a business partner as an organization by clicking the **Organization** button.

Figure 3.33 Business Partner: Initial Screen

Next, we will explore the creation of general data for the business partner, which is mandatory for all business partners.

Business Partner: General Role

Figure 3.34 shows that the business partner general role (**Business Partner (Gen.)**) was defaulted automatically after clicking **Organization** on the initial screen.

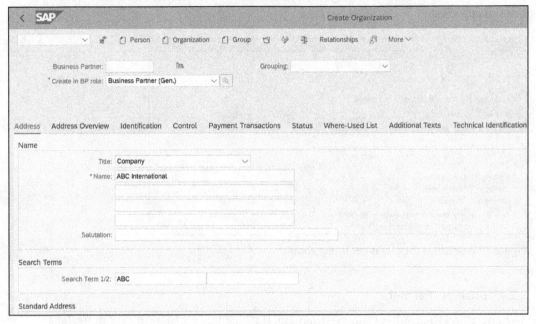

Figure 3.34 Business Partner General Role: Address Data

The general role of the business partner data has nine different views to maintain the general data of the business partner; we'll walk through the main ones in this section.

The **Address** view is shown in Figure 3.34 and includes the following key fields:

- **Title**
  Select a title for the business partner from the dropdown menu. Titles are configurable based on the business partner category. For example, for the person business partner category, you can configure titles such as Mr., Ms., and the like; for an organization, you can configure titles such as company, and so on.

- **Name**
  SAP S/4HANA provides four fields to maintain business partner name(s) flexibly. You can enter long names using multiple fields or you can maintain alternate names of the business partner.

- **Search Term**
  Search terms are useful for searching business partners in the foreground and in the

background. It's always recommended to enter a search term—typically, the first name of the business partner.

- **Standard Address**

  Here you can maintain the full address of the business partner. This address is used in communication with the business partner and in tax reconciliation and compliance reporting.

  You can maintain communication details in the **Address** tab that include the street/house number, postal code, city, country, and region.

Figure 3.35 shows the **Address Overview** view (tab) for the general data of the business partner. The system displays the main address of the business partner maintained in the **Address** view here. In this view, you also can maintain additional addresses for the business partner.

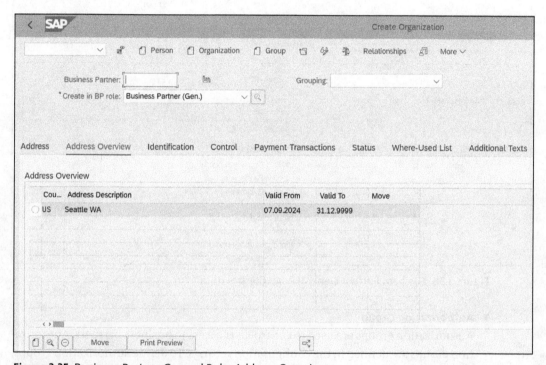

Figure 3.35 Business Partner General Role: Address Overview

To maintain additional business partner addresses, click the **Create** icon at the bottom of the screen. You can enter the complete address here, including the street/house number, postal code, city, country, and region. For every additional address you maintain, you can add communication details such as preferred language, telephone number, mobile phone, fax, and email address.

Figure 3.36 shows the **Control** view of the general role of the business partner. The control parameters for the business partner are maintained in this view. The following are the key fields of this view:

- **BP Type**
  Business partner types are used to control the screen layout (fields to be displayed) during the creation of a business partner. Assign a business partner type based on the business partner record you are creating. Business partner types are defined in the configuration settings. To define business partner types, execute Transaction SPRO and navigate to **Cross-Application Components • SAP Business Partner • Business Partner • Basic Settings • Business Partner Types • Define Business Partner Types**.

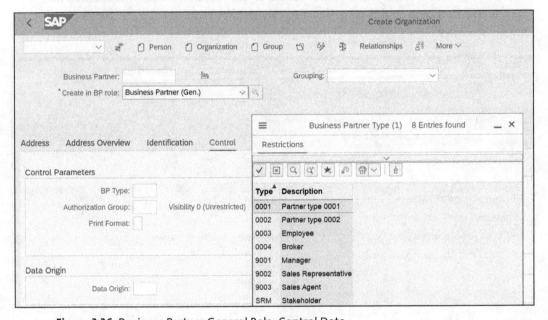

Figure 3.36 Business Partner General Role: Control Data

- **Authorization Group**
  Authorization groups are used to control user access to business partners. Authorization groups are defined in the configuration settings. To define authorization groups, execute Transaction SPRO and navigate to **Cross-Application Components • SAP Business Partner • Business Partner • Basic Settings • Authorization Management • Maintain Authorization Groups for Business Partner**.

- **Data Origin**
  The data origin is the source of the business partner data. You can define entries for the data origin in configuration settings. This shows if the business partner data is converted from a legacy system, derived from a supplier onboarding process, and so on. It is useful for reporting purposes.

Figure 3.37 shows the **Payment Transactions** view of the general role of the business partner. You can maintain a list of bank accounts for a business partner and procurement card details if you are using the payment card functionality. The following are the key fields of bank details:

- **Bank Key**
  A bank key is used to store the bank details such as legal address of the bank, country of origin, and other control data. Specify a bank key for every bank account you maintain.

- **Bank acct**
  Specify the bank account number for the business partner in this field.

- **Control Key**
  Specify if the bank account type is a savings account, checking account, and so on in this field.

- **IBAN**
  An international bank account number (IBAN) is used for both domestic and international payment transactions. Specify an IBAN in this field.

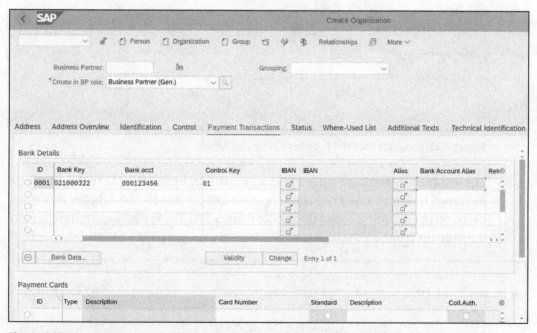

Figure 3.37 Business Partner General Role: Payment Transactions

After maintaining the required attributes for the general role of the business partner, click the **Save** button. The system assigns an internal number automatically and saves the business partner general role details.

Business Partner: Supplier Role

Once the business partner with a general role is created, you can maintain additional business partner roles for the same business partner. To maintain additional business partner roles, execute Transaction BP and open the previously created business partner with the general role in change/edit mode. Figure 3.38 shows that the **Supplier (New)** business partner role was selected manually from the business partner general role view. The system displays the views corresponding to the supplier role. As you can see, the **Purchasing** button is available on the screen to maintain purchasing data for the supplier.

Purchasing data views contain the vital attributes of the supplier for the creation of purchasing documents such as purchase requisitions, purchase orders, outline agreements, requests for quotation, and so on. These attributes are used to create purchasing master data (source list, PIR, etc.) as well.

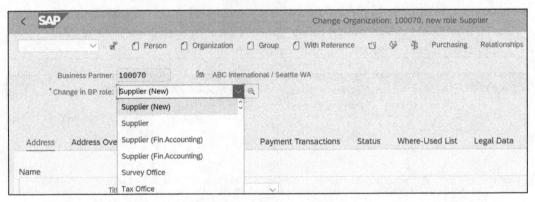

Figure 3.38 Business Partner: Supplier Role Initial Screen

The **Vendor: General Data** tab is added automatically upon selecting the supplier role, as shown in Figure 3.39. An internal number assignment for the supplier was added based on the account group of the vendor (here, **SUPL**). The system determined the vendor account group automatically.

Figure 3.39 Business Partner: Supplier Role: Vendor: General Data

From the general data view, click **Purchasing** to maintain purchasing data for the supplier role. Figure 3.40 shows the **Purchasing Data** view of the supplier role for the business partner. Purchasing data is created at the purchasing organization level.

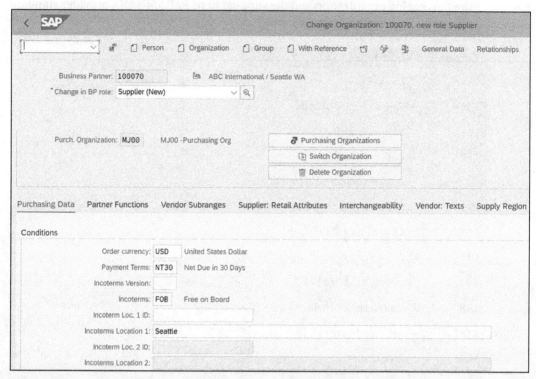

Figure 3.40 Business Partner Supplier Role: Purchasing Data: 1 of 3

As shown in Figure 3.40, the **Purchasing Data** view was maintained for purchasing organization **MJ00**. Typically, purchasing data for suppliers is collected via a supplier onboarding process. Only approved suppliers are entered into the system as business partners.

Let's explore the key fields of the **Purchasing Data** view of the business partner supplier role. The **Conditions** section has the following key fields:

- **Order currency**
  The order currency is the supplier's preferred currency for purchase orders and outline agreements. Maintain the order currency preferred by the supplier and approved during supplier onboarding.

- **Payment Terms**
  Payment terms contain details about discount calculation and due date calculation for incoming (supplier) invoices. Payment terms are agreed-upon terms between the purchasing organization and the supplier. Those from the supplier role are copied to the purchasing documents and then to the incoming invoices when they're posted in the system. Maintain payment terms in this field.

- **Incoterms**
  Incoterms (international commerce terms) are trading terms/rules for suppliers and buyers for the sale/purchase of goods. Maintain the Incoterms agreed on between the purchasing organization and the supplier here. SAP S/4HANA provides standard Incoterms, as listed in Table 3.11.

Incoterms	Description
EXW	Ex works
FAS	Free alongside ship
FOB	Free on board
CFR	Cost and freight
CIF	Cost, insurance, and freight
CPT	Carriage paid to
DES	Delivered ex ship
DEQ	Delivered ex quay
DAF	Delivered at frontier
DDP	Delivered duty paid
DDU	Delivered duty unpaid
FCA	Free carrier
CIP	Carriage and insurance paid to

Table 3.11 Incoterms

- **Incoterm Location**
  Incoterm locations are used in transportation planning. These locations indicate where the goods are picked up from for transportation.

Figure 3.41 shows the **Control Data** section of the purchasing data. The following are the key fields:

- **ABC indicator**
  ABC indicators classify suppliers based on their significance to the purchasing organization. A has the highest significance, B is of medium significance, and C is of low significance. Maintain the ABC indicator here according to the significance of the supplier.
- **Shipping Conditions**
  Shipping conditions play an important role in route determination for transportation. Maintain a shipping condition if you manage transportation of goods from a supplier.

Business Partner: 100070
ABC International / Seattle WA
*Change in BP role: Supplier (New)
Purchasing Data
Partner Functions
Vendor Subranges
Supplier: Retail Attributes
Interchangeability
Vendor: Texts
Supply Region
Control Data
ABC indicator:
Shipping Conditions: 01
GR-Based Invoice Verif.:
Grant Free Goods:

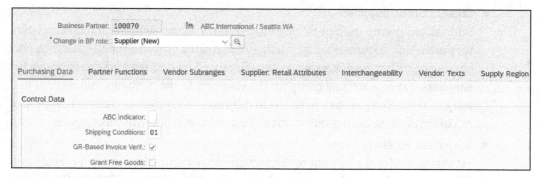

Figure 3.41 Business Partner Supplier Role: Purchasing Data: 2 of 3

- **GR-Based Invoice Verif.**
  Set this indicator to enable goods receipt-based invoice verification. During the incoming invoice posting, the system displays a separate item for every goods receipt item posted with reference to the purchase order or scheduling agreement. The invoice items are first matched against goods receipt items. For evaluated receipt settlements (auto invoice posting functionality), setting this indicator is mandatory.

Figure 3.42 shows the **Additional Purchasing Data** section of the purchasing data record. The following are the key fields:

- **Delete flag for Purch Org.**
  Set this indicator to flag the supplier for deletion for the purchasing organization. If you have this supplier created for multiple purchasing organizations, SAP S/4HANA provides the flexibility to set this flag only for a specific purchasing organization.

Business Partner: 100070
ABC International / Seattle WA
*Change in BP role: Supplier (New)
Purchasing Data
Partner Functions
Vendor Subranges
Supplier: Retail Attributes
Interchangeability
Vendor: Texts
Supply Region
Additional Purchasing Data
Settlement Management:
Delete flag for Purch Org:
Schema Group Supplier:
Automatic Purchase Order:
Pricing Date Control:
No Control
Sort criterion materials:
By VSR sequence number
Purchasing block:
Purchase block POrg level:
Confirmation Control:
Returns supplier:

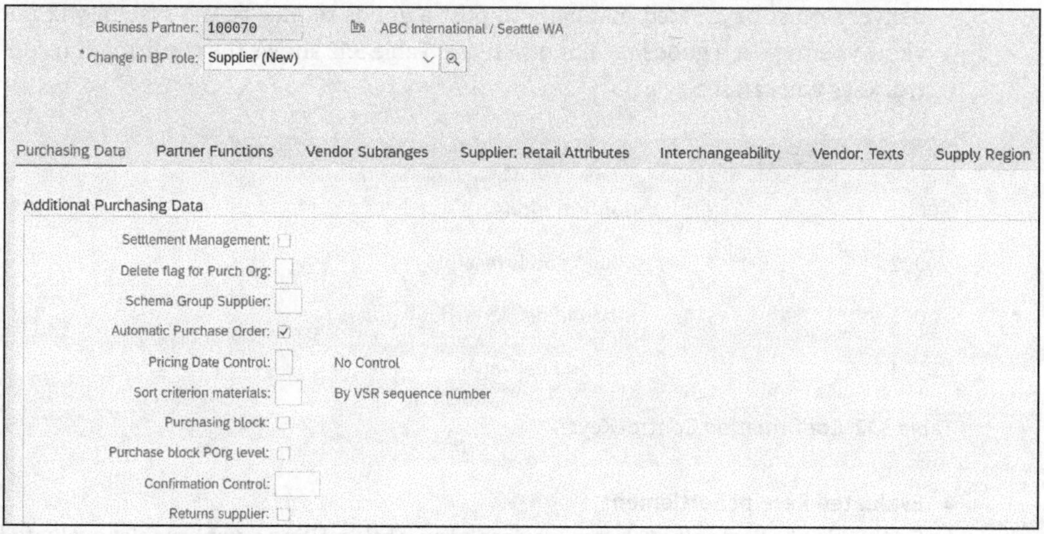

Figure 3.42 Business Partner Supplier Role: Purchasing Data: 3 of 3

- **Schema Group Supplier**

  Schema groups are used in pricing procedures. You can define a pricing procedure for purchasing documents (e.g., purchase order) for a group of suppliers or for a purchasing organization. If you have a pricing procedure defined for a group of specific suppliers, define a schema group for the suppliers in the configuration settings and assign the schema group of suppliers in this field. The system defaults the respective pricing procedure during the creation of a purchase order with this supplier.

- **Automatic Purchase Order**

  Set this indicator if you want to convert purchase requisitions created for this supplier into purchase orders automatically using Transactions ME59 or ME59N. You can schedule these two transactions to run in the background.

- **Purchasing block**

  Set this indicator if you want to block the suppliers for all purchasing organizations. The purchasing block prevents you from creating purchase orders for and submitting them to this supplier.

- **Purchasing block POrg level**

  Set this indicator if you want to block the suppliers for a specific purchasing organization. The purchasing block prevents you from creating a purchase order for and submitting it to this supplier in the respective purchasing organization.

- **Confirmation Control**

  Confirmation control is a vital component of purchasing data. It implies that confirmation and/or advanced shipping notifications (ASNs) are expected from the supplier. Suppliers send an order confirmation to confirm the delivery quantity and delivery date for purchase order items. If an inbound delivery is expected, an ASN must be received electronically or manually (via email); in manual cases, an inbound delivery must be created manually to post a goods receipt as the system will not allow you to post a goods receipt otherwise. Table 3.12 shows the confirmation control keys you can set.

Confirmation Control Key	Description
0001	Confirmations
0002	Rough goods receipt
0003	Inbound delivery/rough goods receipt
0004	Inbound delivery

Table 3.12 Confirmation Control Keys

- **Evaluated Receipt Settlement**

  Setting this indicator (not shown; you can view it by scrolling down) enables evaluated receipt settlement (ERS) for the supplier in the respective purchasing organization.

ERS enables the buying organization to post an invoice on behalf of the vendor with reference to the goods receipt items in agreed-upon terms with the supplier. The supplier doesn't submit invoices for purchases of goods and services; instead, the purchasing organization automatically posts the invoice receipt based on the goods receipt quantity with reference to purchase order. The tax code for the invoice receipt will be defaulted from the PIR. SAP S/4HANA provides a report (Transaction MRRL) that can be scheduled to run in the background to post accounts payable invoices automatically. You must set the **GR-Bsd IV** indicator as well along with this indicator to enable ERS (see Chapter 1, Section 1.3.2).

Figure 3.43 shows the **Partner Functions** view of the supplier role of the business partner. This view is also maintained at the purchasing organization level. Partner functions are the different roles and responsibilities of business partners. Table 3.13 shows the various partner functions of suppliers.

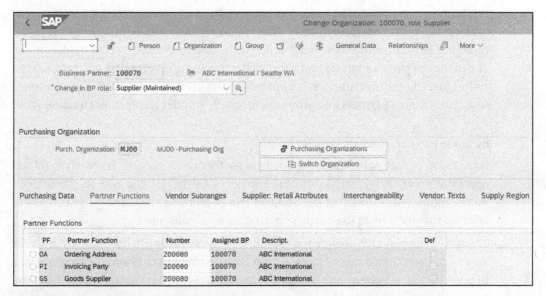

Figure 3.43 Business Partner: Supplier Role: Partner Functions

Partner Functions	Description
CP	Contact person
AZ	Alternative payment recipient
OA	Ordering address
FS	Freight supplier
VN	Supplier

Table 3.13 Partner Functions

Partner Functions	Description
PI	Invoicing party
CR	Forwarding agent
CA	Contract address
GS	Goods supplier

Table 3.13 Partner Functions (Cont.)

The supplier (VN) is the main vendor, and this partner function is used in the creation of purchase orders and outline agreements. The ordering address function receives purchase orders and prioritizes deliveries, and this partner function is used in the purchasing function. The goods supplier picks, packs, and ships the requested goods, and this partner function is used in the intrastate and import process. The forwarding agent is responsible for transportation of shipped goods from the issuing location to the receiving location, and this partner function is used in transportation management. The invoicing party generates and submits the incoming invoice to the buying organization, and this partner function is used in logistics invoice verification. You can create separate business partners and group them together under the **Partner Function** view.

Business Partner: Finance Supplier Role

Once more, to maintain additional business partner roles, execute Transaction BP and open the previously created business partner with the general role in change/edit mode. Figure 3.44 shows that the **Supplier (Fin.Accounting)** business partner role was selected manually from the business partner general role view.

The system displays the views corresponding to the supplier (finance accounting) role. As you can see, the **Company Code** button is available here to maintain the supplier's finance data. The system also adds the **Vendor: General Data** and **Vendor: Tax Data** tabs, as shown in Figure 3.44.

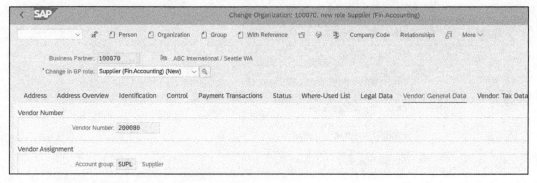

Figure 3.44 Business Partner: Finance Supplier Role: Initial Screen

From the **Vendor: General Data** view, click **Company Code** at the top of the screen to maintain financial accounting data for the finance supplier role. The supplier's finance data is created at the company code level. Figure 3.45 shows the **Vendor: Account Management** view of the finance supplier role of the business partner.

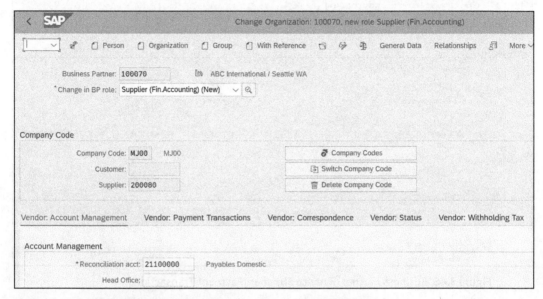

Figure 3.45 Business Partner: Finance Supplier: Account Management

There's one key field in the **Vendor: Account Management** view: **Reconciliation acct.** Reconciliation accounts are special types of general ledger accounts that integrate subledgers such as accounts payable and accounts receivable into the general ledger accounts. They ensure that all financial transactions recorded in the subledgers are reflected automatically in the general ledger accounts. These accounts are configured in SAP S/4HANA to maintain the integrity of financial data.

It's mandatory to assign a vendor reconciliation account in this field. All financial transactions related to suppliers are recorded in the vendor reconciliation accounts. During the supplier invoice posting, SAP S/4HANA updates the accounts payable subledger with the outstanding amount owed to the supplier and automatically posts the entry to the vendor reconciliation account in the general ledger.

Figure 3.46 shows the **Vendor: Payment Transactions** view of the finance supplier role of the business partner.

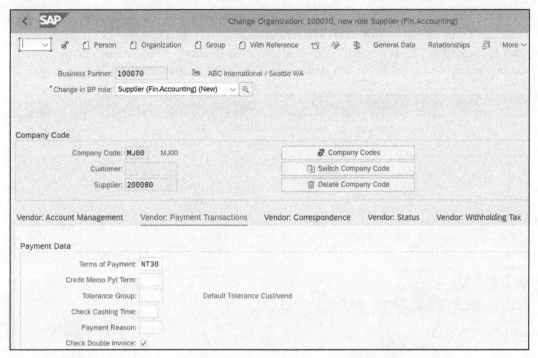

Figure 3.46 Business Partner: Finance Supplier: Payment Transactions

The following are the key fields of this view:

- **Terms of Payment**
  Payment terms contain details about discount calculation and due date calculation for incoming (supplier) invoices. Payment terms are agreed-upon terms between the buying organization and the supplier. Payment terms from the finance supplier role are copied over to the supplier invoice documents. Maintain payment terms in this field.

- **Check Double Invoice**
  It is a best practice to check this indicator, ensuring that you perform duplicate checks during incoming invoice postings (logistics invoice verification). A duplicate invoice check is performed during logistics invoice verification based on the following to find any potential duplicates:
  - Company code
  - Supplier
  - Currency
  - Gross invoice amount
  - Supplier's invoice reference number
  - Invoice date

> **Note**
>
> A supplier role maintained at the purchasing organization level is mandatory to create purchasing documents (purchase orders, outline agreements, etc.), whereas the finance supplier role maintained at the company code level is mandatory to post incoming invoices with a supplier.

3.3   Master Data in Materials Management

This section covers the purchasing master data. Purchasing master data is maintained using material master data and business partners in SAP S/4HANA with purchasing-specific attributes to support procurement. Let's dive deep into purchasing master data.

3.3.1   Source List

The source list is vital for determining the source of supply for procurement and during MRP. A source list is an essential procurement master data element for a valid source of supply to be determined for purchase requisitions and for automatic source determination for purchase orders. A source list is also called an approved supplier list. You can maintain all sources of supply such as business partners (suppliers) and outline agreements for a material in a specific plant using the source list. Preferred suppliers for a material can also be maintained using a source list.

The following are the prerequisites for maintaining the source list:

1. Material master data must be created in the same plant in which the source list is being maintained, and a purchasing view must exist.
2. A business partner with a supplier role must be created in the same purchasing organization to which the plant belongs.
3. If a source of supply is one of the plants (supplying plant) for the material, then material master data must be extended to that plant.
4. To maintain MRP relevance for a source of supply, the material master data must be maintained with MRP views.
5. If you want to list an outline agreement (supplier contract agreement or scheduling agreement) as a source of supply, the outline agreement item must be created for the same material and plant combination.

Source lists are maintained in SAP S/4HANA using Transaction ME01; use Transaction ME03 to display a source list. Figure 3.47 shows the source list maintained for the material **MS-RAW MATERIAL1** and plant **MJ00**. Press [Enter] to display the source list maintenance screen.

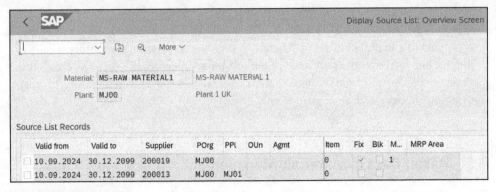

Figure 3.47 Source List

A source list is maintained for a combination of a material and a plant. The following are the source list key fields:

- **Valid from**
  Every source of supply that you list in the source list must have a valid from date. Enter the current date as the valid from date.

- **Valid to**
  Every source of supply that you list in the source list must have a valid to date. Enter a valid to date in this field. Typically, a valid to date is maintained as 12.31.9999 or 12.31.2099.

- **Supplier**
  Enter a valid supplier created as a business partner with the supplier role in this field. The supplier ID is copied over to the purchasing documents from the source list.

- **POrg (purchasing organization)**
  A purchasing organization is an organizational unit responsible for procurement. It maintains relationships with suppliers. Assign the purchasing organization that is linked to the plant in which the source list is being maintained here.

- **PPl (procurement plant)**
  A procurement plant is an internal source of supply, also called a supplying plant, that can belong to the same company code of the plant in which the source list is being created or belong to a different company code. Using a procurement plant, a stock transfer requisition and a stock transfer order can be created so that the material can be procured internally.

- **OUn (order unit)**
  Specify a unit of measure in which the material is ordered. This unit of measure is copied to the purchase requisition and purchase order.

- **Agmt (agreement)**
  Specify an outline agreement number (contract agreement or a scheduling agreement) created for the material and plant. The system creates schedule lines during

the MRP run if a scheduling agreement is specified in this field as the source of supply. If a contract agreement is specified in this field, it gets copied to the purchase requisition item.

- **Item (agreement item)**
  Specify the corresponding agreement line item number in this field.

- **Fix (fixed source of supply)**
  Set this indicator if the source of supply is preferred among the options. The system suggests the fixed source as the priority for purchase requisitions and purchase orders.

- **Blk (blocked)**
  Set this indicator to block a source of supply in the source list. The system doesn't suggest this source of supply for purchase requisitions and purchase orders.

- **M... (MRP)**
  Choose the MRP usage key in this field. It controls the usage of a source list in the MRP run. Table 3.14 shows the MRP usage keys.

MRP Usage	Description
Blank	Source list record not relevant to MRP
1	Record relevant to MRP
2	Record relevant to MRP; schedule lines generated automatically

Table 3.14 MRP Usage

Note

MRP usage key **2** (**Record relevant to MRP. Sched. lines generated automatically**) can be set only if the material is relevant for MRP in the plant and a scheduling agreement is assigned as a valid source of supply in the source list.

Let's simulate how the system assigns a source of supply during the purchase requisition creation. Figure 3.48 shows a purchase requisition being created for material **MS-RAW MATERIAL1** in plant **MJ00**.

After clicking the **Assign Source of Supply** button, fixed vendor **200019** was assigned to the purchase requisition automatically as this vendor is listed as a preferred (fixed) source of supply in the source list for material **MS-RAW MATERIAL1** in plant **MJ00**.

The purchase requisition header data contains a **Source Determination** indicator, as shown in Figure 3.49. If you set this indicator and there is only one source of supply listed for the material and plant combination in the source list, then that source of supply is assigned to the purchase requisition automatically.

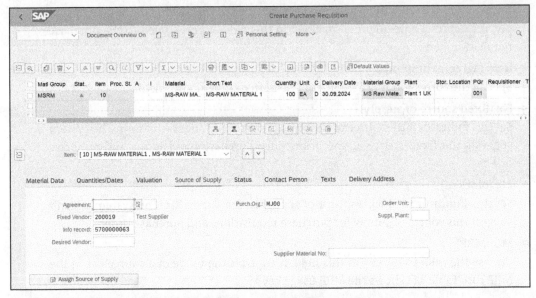

Figure 3.48 Create Purchase Requisition: Assign Source of Supply

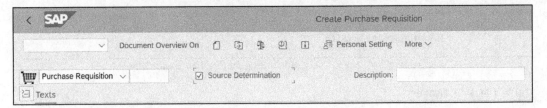

Figure 3.49 Create Purchase Requisition: Source Determination

3.3.2   Purchasing Info Record

The purchasing info record (PIR) is another important purchasing master data element used in different types of purchase orders. It contains commercial information for procuring a material from a supplier (business partner) that includes the unit price, planned delivery time, standard order quantity, minimum order quantity, underdelivery and overdelivery tolerances, Incoterms, supplier part number, confirmation control, and so on. The **Purchasing** view of the material master data, maintained at the plant level, contains procurement-related attributes of the material, and the business partner (supplier role) contains the supplier-specific attributes for procurement, maintained at the purchasing organization level. PIRs, created with a combination of a material and a business partner (supplier) at the purchasing organization and plant levels, contain refined purchasing data specific to the material and supplier combination. If a supplier supplies multiple materials for a purchasing organization, then the commercial attributes could be different for different combinations of material and supplier. There are four types of PIRs:

1. **Standard**

   Standard PIRs are used in standard purchase orders. You can create standard PIRs with a combination of a material group and a supplier for indirect procurement. This type of record is used in standard purchase order line items with the standard item category.

2. **Subcontracting**

   Subcontracting PIRs are used in subcontracting orders. The subcontracting process is a special procurement process in SAP S/4HANA where certain operations (assembly, packaging, manufacturing ops, etc.) are outsourced by the purchasing organization to external suppliers and the required components provided to the external suppliers to perform the operation. This type of record is used in subcontracting order line items with the subcontracting item category. Creation of a subcontracting PIR is mandatory to create a subcontracting order.

3. **Consignment**

   Consignment PIRs are used in supplier consignment orders. Supplier consignment is a type of external procurement where the supplier loans materials with reference to the consignment purchase order to the purchasing company. The purchasing company is not liable to pay to the supplier until it withdraws or directly consumes the consignment stock that belongs to the supplier. This type of record is used in supplier consignment order line items with the consignment item category. Creation of a consignment PIR is mandatory to create a supplier consignment order.

4. **Pipeline**

   Pipeline PIRs contain information about materials that are supplied by suppliers via a pipeline directly to the sites of buying organizations.

Let's explore a standard PIR in SAP S/4HANA. Execute Transaction ME11 to create a PIR, Transaction ME12 to change it, and ME13 to display it. Figure 3.50 shows the creation of a PIR for supplier **200019** and material **MS-RAW MATERIAL1** in purchasing organization **MJ00** and plant **MJ00**. Press [Enter] on the initial screen to display the general data view of the PIR.

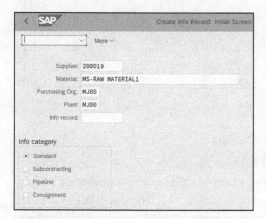

Figure 3.50 Create Purchasing Info Record: Initial Screen

Unlike source lists, either PIRs are assigned an internal number range, or you can specify an external number in the **Info record** field.

Select an **Info category** (info record type) option for the PIR. **Standard** was selected in the example to create a standard PIR.

> **Note**
>
> PIRs can be created at the purchasing organization level alone. Creation of a PIR at the plant level is required for ERS and consignment settlement processes.

You can navigate between several views at the top of the screen. Figure 3.51 shows the general data of the PIR.

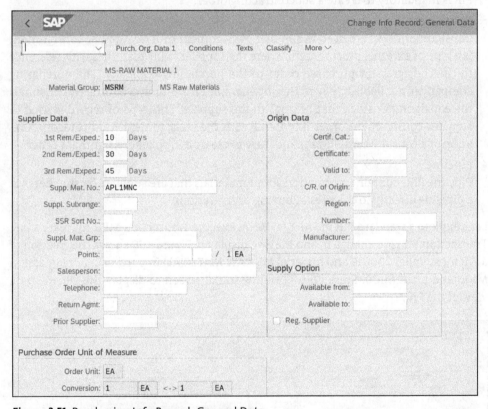

Figure 3.51 Purchasing Info Record: General Data

The following are the key fields of the general data:

- **1st Rem./Exped., 2nd Rem./Exped., and 3rd Rem./Exped.**
  These three fields are copied from the purchasing value key maintained in the purchasing view of the material master data (see Section 3.1.5). These reminders are used for deadline monitoring of purchasing documents. Deadline monitoring includes

reminders at regular intervals that can be sent to suppliers for requests for quotation, reminders to ship ordered goods, and so on. Up to three reminders can be set in purchasing value keys. You can change and update the default values of these three fields directly in the PIR.

- **Supp. Mat. No.**
  This is an important field of the general data of the PIR. It stores the corresponding material number stored in the supplier's ERP system. If you are required to send the purchase order to the supplier via EDI or to an B2B application such as SAP Business Network, sending the supplier part number for every purchase order line item is preferable to exchange documents electronically with the supplier.

- **Order Unit**
  Specify a unit of measure in which the material is ordered. This unit of measure is copied to the purchase order.

- **Conversion**
  If you specify an **Order Unit** different from the base unit of measure of the material, specify a conversion factor here to convert the order unit in terms of the base unit of measure.

Figure 3.52 shows the **Purch. Organization Data 1** view of the PIR. The following are the key fields of this view:

- **Pl. Deliv. Time (planned delivery time)**
  Specify the number of days it takes to receive the materials from the supplier from the purchase order date.

- **Tol. Underdl. (underdelivery tolerance)**
  Specify the underdelivery tolerance as a percentage. The system accepts the delivery and sets the delivery completion indicator if the received quantities from the supplier are less than the ordered quantity and within the underdelivery tolerance limit.

- **Tol. Overdl. (overdelivery tolerance)**
  Specify the overdelivery tolerance as a percentage. The system accepts the delivery and sets the delivery completion indicator if the received quantities from the supplier are greater than the ordered quantity and within the overdelivery tolerance limit.

- **Standard Qty**
  The standard order quantity of a material from a supplier should be greater than the minimum order quantity. The standard quantity is used as a basis for scale prices. Specify a standard order quantity in this field.

- **Minimum Qty**
  This is the minimum order quantity of a material from a supplier. The system will not allow you to create the purchase order if the purchase order quantity of the material is less than the minimum order quantity maintained in the info record. Specify a minimum order quantity in this field.

- **Ackn. Rqd (acknowledgement required)**
  If you set this indicator, it will be copied over to the **Confirmation** tab of the purchase order item. Using this indicator, you can communicate to the supplier that they must acknowledge the receipt of the purchase order using a desired communication method (email, letter, etc.).

- **Conf. Ctrl (confirmation control)**
  Confirmation control is a vital component of purchasing data. It implies that an order confirmation and/or ASN is expected from the supplier. Table 3.15 shows the confirmation control keys defined in the standard system; choose one of them based on business requirements.

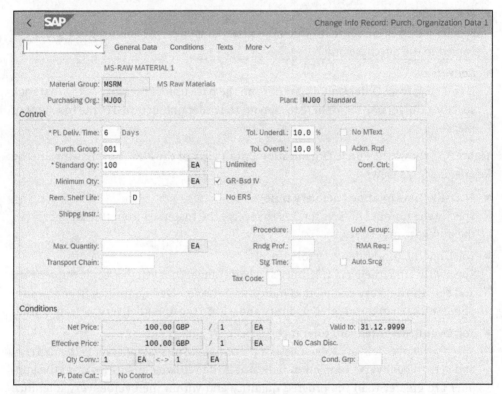

Figure 3.52 Purchasing Info Record: Purchasing Org. Data

Confirmation Control Key	Description
0001	Confirmations
0002	Rough goods receipt
0003	Inbound delivery/rough goods receipt
0004	Inbound delivery

Table 3.15 Confirmation Control Keys

- **No ERS**

  ERS is a process that enables a buying organization to post an invoice on behalf of a vendor with reference to the goods receipt items. If an ERS indicator is set in the supplier role of the business partner in the purchasing view, then the supplier doesn't submit invoices for purchases of goods and services. Instead, the purchasing organization automatically posts the invoice receipt based on the goods receipt quantity with reference to purchase order.

  If the supplier is set for ERS, but you do not want to enable this automated process for this specific material, then set the **No ERS** indicator in the info record.

- **Auto.Srcg. (autosourcing)**

  If you set this indicator, the system considers this info record during the MRP run for the material in the same plant in which the info record exists. Purchase requisitions are created with reference to the info record during the MRP run.

- **Conditions**

  In this section, you can directly maintain the **Net Price** (price after discounts and surcharges) for the first time at the time of creation of the info record. To revise or update the price from the next time on, use the **Conditions** view and maintain a revised pricing condition with validity dates. The system automatically derives a value for **Effective Price** from the pricing conditions.

Figure 3.53 shows the **Conditions** view of the info record. In our example, a condition price of 100.00 GBP per 1 EA was maintained for condition type **PB00** in purchasing organization **MJ00**.

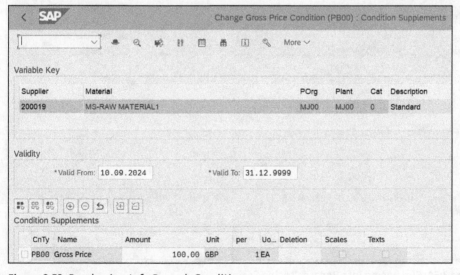

Figure 3.53 Purchasing Info Record: Conditions

To revise or edit the condition price, execute Transaction ME12 to display the PIR in edit mode. Enter the material, supplier, purchasing organization, and plant in the initial screen. Press ⏎Enter on the initial screen to display the general data of the PIR. From there, navigate to the **Conditions** view of the PIR. In our example, we'll reduce the **Valid To** date to the current date minus one day, and add a new condition record with the **Valid To** date as the current date and the **Valid From** date as 12.31.9999.

Pricing scales can be maintained in the PIR using condition type **PB00**. You can define a pricing scale based on order quantity. The scale basis can be configured in the condition type configuration. To navigate to maintain scale prices for the condition record, click the **Scales** icon or press ⎗F2 on your keyboard from the **Conditions** view of the PIR, Figure 3.54 shows the different scale prices maintained for the material based on quantity. The following examples describe how pricing scales work when you create a purchase order:

- If you create a purchase order with quantity 90, the system automatically derives the unit price as 100.00 GBP.
- If you create a purchase order with quantity 150, the system automatically derives the unit price as 95.00 GBP.
- If you create a purchase order with quantity 299, the system automatically derives the unit price as 90.00 GBP.
- If you create a purchase order with quantity 500, the system automatically derives the unit price as 80.00 GBP.

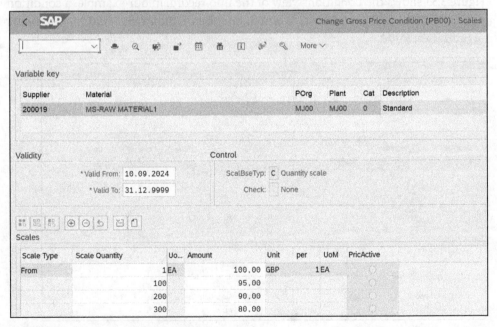

Figure 3.54 Purchasing Info Record: Pricing Scales

Let's consider how a PIR supports the creation of purchasing documents by simulating a purchase order. Figure 3.55 shows a purchase order being created for material **MS-**

RAW MATERIAL1 in plant **MJ00**, purchasing organization **MJ00**, and supplier **200019**. A net price of 90.00 GBP was derived automatically from the scale pricing maintained in the condition record in PIR for the quantity of 250 EA. As shown in Figure 3.55, a scale price of 90.00 was maintained for a quantity of 200 EA to 299 EA.

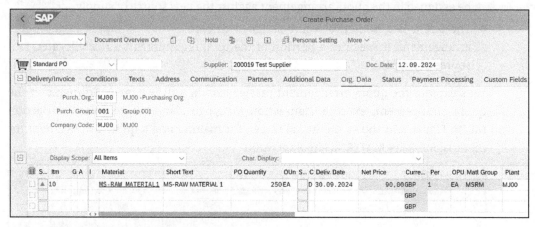

Figure 3.55 Create Purchase Order: Scale Price

3.3.3   Quota Arrangement

A quota arrangement is a powerful feature in materials management. It divides the total requirements for procuring materials among multiple sources (suppliers) over a period based on the quota assigned to each source of supply. A quota arrangement is master data in materials management and is created for a material in the purchasing plant with multiple sources of supply (suppliers). It has a validity period that you can set based on business requirements. If a quota arrangement is maintained for the material in the respective plant, the system automatically considers the quota arrangement in source determination for purchasing. During the source determination for a purchasing requirement (e.g., purchase requisition) for the material, the system determines the next source of supply from which the material requirement can be purchased from among the sources of supply maintained in the quota arrangement. For purchase requisitions that are created manually, you can press the **Assign Source of Supply** button in the **Source of Supply** tab of the purchase requisition line item details to determine the source of supply using a quota arrangement. The system calculates the quota rating for all sources of supply maintained in the quota arrangement using the following formula, and the source of supply that has the lowest quota rating is suggested as the source of supply for the material requirement:

Quota rating = (Quota allocated quantity + Quota base quantity) ÷ Quota

Let's look at the elements of this formula:

- The quota allocated quantity is the total requirements (quantity) amount already allocated to the source of supply from the quota arrangement. The quota allocated quantity includes the purchase requisitions, purchase orders, and delivery sched-ules for scheduling agreements, allocated to the respective source of supply.

- The quota base quantity is used only when a new source of supply is included in the quota arrangement with no quantity allocation so far. The quota base quantity can be assigned in the quota arrangement against the new source of supply.

- The quota is a numerical value assigned to each source of supply in the quota arrangement. It specifies a portion of the total requirements to be allocated to a source of supply.

Let's explore the quota arrangement master data in SAP S/4HANA. To maintain the quota arrangement, execute Transaction MEQ1; to display it, execute Transaction MEQ3. Figure 3.56 shows the initial screen for maintaining a quota arrangement for material **MS-RAW MATERIAL1** in plant **MJ00**.

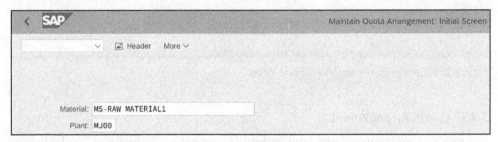

Figure 3.56 Maintain Quota Arrangement: Initial Screen

The quota arrangement is maintained for a material and plant combination. Enter the material for which you want to maintain the quota arrangement and the plant in which you want to maintain it. Press ⟨Enter⟩ to display the maintenance view of the quota arrangement period overview.

Figure 3.57 shows the overview of the quota arrangement periods for material **MS-RAW MATERIAL1** and plant **MJ00**.

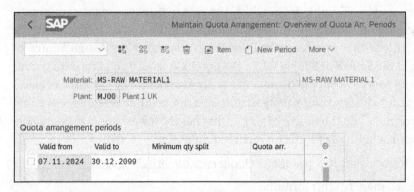

Figure 3.57 Maintain Quota Arrangement: Overview of Quota Arrangement Periods

The following are the key fields to be maintained for quota arrangement periods:

- **Valid from**
  Specify a valid from date in this field. The quota arrangement becomes valid from this date.

- **Valid to**
  Specify a valid to date in this field. The quota arrangement will be valid until this date.

- **Minimum qty split**
  Specify a minimum quantity here for the quota arrangement to split the material requirements among the sources of supply.

- **Quota arr.**
  This field is automatically populated with the number of the quota arrangement once the quota arrangement master data is saved.

Next, click the **Item** icon at the top or press F7 on your keyboard to display the maintenance view of the quota arrangement item overview in order to maintain the sources of supply for the respective validity period. Figure 3.58 shows the overview of the quota arrangement items for material **MS-RAW MATERIAL1** and plant **MJ00**. Sources of supply are listed as items in the quota arrangement. The header section of the view also shows the validity period. You can directly list the sources of supply and maintain the quota arrangement-relevant data by clicking **New Entries** to list the sources of supply/suppliers. The **QAI** (quota arrangement item) field is populated automatically.

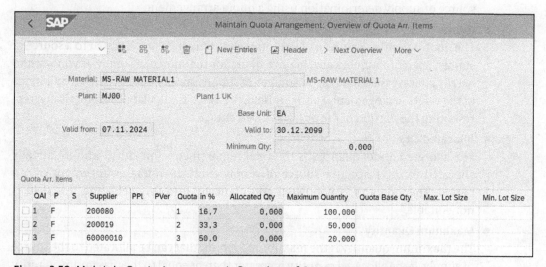

Figure 3.58 Maintain Quota Arrangement: Overview of Quota Arrangement Items

The following are the key fields to be maintained for quota arrangement items:

- **P (procurement type)**
  The procurement type indicates if the material is procured internally or externally.

Choose a procurement type for the material from the dropdown menu. Table 3.16 shows the procurement types available for quota arrangements.

Procurement Type	Description
E	Internal procurement
F	External procurement

Table 3.16 Procurement Type

- **S (special procurement)**
  Special procurement types such as consignment, subcontracting, stock transfer, and the like are available for selection from the dropdown menu if you have chosen external procurement for the procurement type. Choose a relevant special procurement type applicable for the source of supply.

- **Supplier**
  Specify a valid supplier created as a business partner with supplier role in this field for external procurement. The supplier added in this field is available for source of supply determination using quota arrangement.

- **PPl (procurement plant)**
  Specify a valid supply/procurement plant created as a plant (organizational unit) in this field for internal procurement if the procurement type for the quota arrangement item is internal procurement. The plant added in this field is available for source of supply determination using a quota arrangement.

- **Quota**
  The quota specifies a portion of the total requirements to be allocated to a source of supply. Specify a numerical value as a quota for the source of supply. If you want to evenly allocate the quota for all sources of supply, specify quota value 1 for all items of the quota arrangement. For the **%** field, the value is calculated automatically, representing the portion of the quota as a percentage.

- **Allocated Qty**
  The allocated quota quantity is the total requirements (quantity) amount already allocated to the respective source of supply from the quota arrangement over a period. This quantity is populated automatically into this field, and this field is noneditable.

- **Maximum Quantity**
  The maximum quantity is the maximum amount that can be allocated to the source of supply. If the allocated quantity is greater than or equal to the maximum quantity maintained in this field, the respective source of supply is not determined.

- **Quota Base Qty**
  The quota base quantity is one of the attributes for the quota rating, and you can update it to calculate the quota rating without changing the quota for the respective source of supply.

3.4 Summary

In this chapter, we introduced the core master data in SAP S/4HANA, which is widely used by all supply chain functions, financial accounting and controlling, and extended warehouse management (EWM) and transportation management (TM). This chapter explained all the key attributes and material master and business partner data, its significance, and configuration settings with best practice examples. In this chapter, we also explained the master data in materials management to support purchasing and its integration with other functions.

In subsequent chapters, we will explain how material master data, business partners, and master data in materials management support different functions of SAP S/4HANA, EWM, and TM. In the next chapter, we'll discuss sales and distribution and its integration with materials management.

Chapter 4
Sales and Distribution

Sales and distribution and materials management are core areas of SAP S/4HANA and the most important components for any organization to operate. Sales and distribution handles processing customer orders and delivery. Its seamless integration with materials management ensures the smooth flow of a material through the supply chain to the customer.

Sales order processing, shipping, outbound delivery processing, and customer billing are the key processes within sales and distribution. The materials management functions integrate with all the main processes within sales and distribution, ensuring the fulfillment of customer orders.

This chapter covers all the key integration points with sales and distribution, including the functionality and configuration in SAP S/4HANA. Starting with sales order creation for a finished or trading good, this chapter covers how this integration helps with the availability check for goods within inventory management that drives the determination of delivery date for the order. It also covers the stock transfer process for transferring materials from one plant to another within the same company code and across company codes. We then discuss the inventory management functionality within materials management that ensures the processing of sales orders and deliveries, updating inventory, determining and updating general ledger accounts, and processing of customer returns. Next, we cover the third-party purchasing processes within materials management that support the fulfillment process by procuring trading/finished goods from a third party, with materials directly fulfilled from a third-party vendor for the customer against the customer order. Finally, we'll cover advanced returns management (ARM), a new and powerful feature for both the customer returns and supplier returns processes.

4.1   Sales Order and Delivery Processing

Sales orders capture actual customer demand in terms of specific order quantities, delivery dates, and other details. Demand and forecast planning enable organizations to predict the future demand, optimize inventory management, and run effective production planning. SAP S/4HANA offers various tools and functionality to generate demand forecasts. In SAP S/4HANA, material requirements planning (MRP) plays a vital role in managing the inventory, production planning, and procurement activities. Demand forecasts are one input in the MRP process. MRP also considers the inventory levels, actual sales order demand, stock transfers, production schedules, and more to determine the material requirements to fulfill customer demand.

SAP S/4HANA also performs an availability check of materials during sales order processing to ensure that requested quantities of materials are available in inventory. After the availability check, the system confirms the order if sufficient stock is available in inventory and automatically allocates the requested quantities to the sales order. If enough material is not available in the inventory to fulfill sales order, the system automatically creates a backorder to track the outstanding demand. When the enough stock becomes available, a backorder is confirmed and processed for delivery.

Let's dive deep into sales orders and delivery processing to get more insights into the integration between sales and distribution and materials management. We'll then walk through how to set up key integration points for MRP, available-to-promise (ATP), and automatic account determination during delivery processing.

4.1.1   Integrated Sales Order and Delivery Process

Sales order and delivery processing functionalities of sales and distribution have direct integration with materials management—specifically, MRP, inventory management, and purchasing. Table 4.1 shows all the key processes impacted.

Area	Processes Impacted	Description
Sales and distribution	Sales order processing	This is one of the core processes in sales and distribution that focuses on creating and managing customer orders.
	Availability check	This function helps ensure enough material is available to fulfill customer demand/sales order.
	Create and manage customer pricing conditions	This process supports sales order processing by creating and maintaining pricing, discount, freight, and tax conditions as well as scale pricing.
	Simulate and calculate order pricing	This process supports the simulation and calculation of sales order pricing.
	Create and manage customer invoice	This process deals with creating and managing customer invoice/billing as well as transmitting the billing documents to the customer.

Table 4.1 Process Impact: Sales Order and Delivery Process

Area	Processes Impacted	Description
Materials management	MRP run	This process supports the materials requirement planning run for make to order strategy.
	Post goods issue for outbound delivery	This process supports the processing of the goods issue for the outbound delivery.

Table 4.1 Process Impact: Sales Order and Delivery Process (Cont.)

Figure 4.1 shows the process flow of the end-to-end sales order and delivery process. Let's take a closer look at the process steps:

1. **Creation of sales order**
   Customer orders requesting goods or services will be converted into a sales order in SAP S/4HANA using Transaction VA01. Creation of sales orders can be fully automated via electronic data interchange (EDI). (See Chapter 12, Section 12.1 for details about EDI.) Sales orders can also be created with reference to the following documents:
   – An *inquiry* helps to keep track of customer interactions.
   – A *quotation* helps ensure consistency with pricing and other terms.
   – Creating a *sales order* with reference to another sales order helps to copy the details over and maintain consistency with pricing and other details.
   – A common practice is to have a customer *contract* created for a long term. Creating a sales order with reference to a contract helps ensure consistency with pricing and other terms as well as contractual agreements.
   – *Scheduling agreements* ensure recurring deliveries to customers over a set period. Creating a sales order with reference to a scheduling agreement helps to keep compliance with delivery schedules and improve overall efficiency.
   – Sales orders with reference to a *billing document* help in scenarios where customers have special billing agreements and payment terms.

   The system allows you to enter the business partner ID of the customer created in the SAP S/4HANA system, materials, requested quantity, requested delivery date, and pricing details during the creation of a sales order.

2. **Availability check**
   An availability check is the most important functionality of the sales order process to ensure that a sufficient quantity of the materials requested by the customer is available in the stock to fulfill the sales order for the requested delivery date. An availability check is triggered automatically during the creation of a sales order using a checking rule. A checking rule is a predefined set of rules to perform an availability check. Performing availability checks ensures timely delivery of products to the customers and optimizes inventory management while improving customer satisfaction. After a successful availability check, the system confirms the sales order and automatically allocates the requested quantities to the sales order to deliver the requested quantities of the material on the requested date.

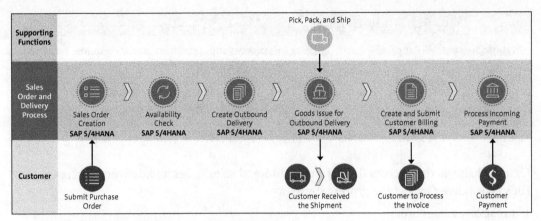

Figure 4.1 Sales Order and Delivery Processing

3. **Create outbound delivery**

   The next step in the sales order processing, after the requested quantities are available to fulfill the customer order, is to create an outbound delivery. Outbound deliveries are created using Transaction VL01N with reference to the sales order. Outbound deliveries help organizations to prepare for the shipments of goods to the customer and contain specific quantities of the materials to be shipped and delivery dates for scheduling shipments. Inventory and warehouse management activities such as picking of materials and packing activities are performed with reference to outbound deliveries.

   In addition, outbound deliveries will have a document flow to trace and track the subsequent documents such as a goods issue document, shipment document, and billing document. A shipment document can be created after the outbound deliveries are created, and the outbound deliveries for one or more customers can be assigned to the shipment document.

4. **Goods issue for outbound delivery**

   A goods issue will be posted with reference to the outbound delivery after the inventory management functions such as picking and packing of materials are completed. This is an important functionality of inventory management, which removes the materials from inventory and issues them for outbound delivery to ship them to the customer. Movement types play an important role in determining the type of inventory movement and posting of goods issue. Movement type 601 is used for a goods issue for an outbound delivery to ship the goods to the customer. A goods issue for outbound delivery can be posted using Transaction VL02N.

5. **Create and submit customer billing document**

   Customer billing/invoice documents can be created with reference to the sales order or outbound delivery document. The best practice is to create the billing document after the shipment (goods issue for outbound delivery) is completed. Customer billing documents can be created using Transaction VF01. The system calculates the output tax and determines an output tax code based on the country of origin, shipping address, and products/services delivered. The output tax will be added to the billing document and collected from the customer. The seller will pay the taxes on behalf of the customer to the local government authority of the shipping county and state. Billing documents are transmitted to the customer via email or EDI.

6. **Process incoming payment**
   Incoming payments or payments received from customers against the outstanding billing documents will be processed in SAP S/4HANA as part of the accounts receivable function. Payment receipts are entered into the system using Transaction F-28. A reference document (check, eCheck, etc.), payment amount, payment date, and payment method will be recorded in the payment receipt document. This process can be automated as well.

Sales and delivery processing impacts user roles from both materials management and sales and distribution functions within the organization.

The following are the key user roles impacted within sales and distribution:

- **Sales representatives**
  Responsible for creation of sales orders and interacting with customers to address their queries

- **Sales manager**
  Responsible for managing the sales team and monitoring the sales order process

- **Accounts receivable clerk**
  Responsible for creating and maintaining customer billing document while ensuring accuracy and compliance with company and regulatory policies

- **Master data management team**
  Responsible for maintaining the pricing condition records for the transfer price

The following are the key user roles impacted within materials management:

- **Logistics/shipping coordinator**
  Responsible for coordinating the shipments to customers based on outbound deliveries

- **Inventory manager**
  Responsible for picking and packing of materials from inventory based on availability for the goods issue for outbound deliveries

4.1.2   Master Data Requirements

Sales order and delivery processing require the following master data to be maintained in the SAP S/4HANA system:

- **Material master data**
  Maintaining certain views of material master data are mandatory for the sales order and delivery process. The other material master views are required based on business requirements. For example, if the same material must be procured internally

using the stock transfer process, then the purchasing view must be maintained as well. Table 4.2 shows the material master views required for the sales order and delivery process.

Maintenance Status	Material Master Views
K	Basic data
D	MRP
P	Forecasting
L	Storage
Q	Quality management
V	Sales
B	Accounting

Table 4.2 Material Master Maintenance Status

- **Business partner (customer)**
  A business partner with a customer role is required for customers for the sales order and delivery processing. Customer-specific settings are maintained in the business partner.

For more information on the material master and business partner configuration, see Chapter 3.

4.1.3   Available-to-Promise Check

ATP checks are a critical functionality that allow the business to check availability of materials to fulfill customer orders. Availability checks happen at the item level of the sales order. The system checks the availability of the requested quantity on the requested date for every material in the sales order. After the availability check, the system confirms the sales order if sufficient stock is available in the inventory and automatically allocates the requested quantities to the order. If enough is not available in inventory to fulfill the sales order, the system automatically creates a backorder to track the outstanding demand. When enough stock becomes available, a backorder will be confirmed and processed for delivery.

ATP checks can be configured based on your business requirements in SAP S/4HANA to consider current stock levels, reservations, open receipts, production orders, and purchase orders. The system calculates the available stock of the material using the following formula:

Available stock at the plant = Stock on hand + Supply – Demand

Let's break down the parts of this formula:

- **Stock on hand**
  Available stock of material in inventory management. This is typically unrestricted-use stock. However, the system can be configured to consider safety stock, quality inspection stock, blocked stock, and more.

- **Supply**
  The system considers open receipts, open production orders, stock transfer orders, and open purchase orders to calculate the total supply of material.

- **Demand**
  The system considers sales orders and reservations for the material to calculate the total demand.

ATP checks in SAP S/4HANA offer several benefits to organizations:

- **Customer satisfaction**
  With the real-time visibility of the available stock, ATP checks provide accurate delivery dates for customer orders for the requested quantity of materials and products. Businesses can provide accurate delivery commitments to the customers.

- **Optimized inventory management**
  ATP checks will optimize the inventory levels of not only the finished products but also the components by accurately calculating and confirming sales orders. This will reduce surplus inventory of materials.

- **Reduced backorders**
  The checks help to identify stock shortages, and businesses can proactively act to reduce backorders.

- **Improved production planning**
  ATP checks integrate directly with production planning processes. They help to generate planned orders that will be converted to purchase requisitions and/or production orders depending on the business requirements and settings in SAP S/4HANA.

- **Improved supply chain efficiency and accuracy**
  By accurately predicting the availability of materials and products, ATP checks help businesses to make the right decisions to optimize inventory levels, optimize production capacity, and procure products internally or externally.

4.1.4 Delivery Scheduling

On-time delivery of customer orders is the number one priority for organizations to enhance customer satisfaction. Delivery scheduling is the most important functionality and part of the sales order processing. It determines the appropriate delivery date for the order considering various factors, including material availability, picking and packing lead times, goods issue lead time, and transit lead time, while prioritizing customer requirements. Delivery scheduling begins as soon as the customer order is entered into the system using Transaction VA01 (Create Sales Order) or a customer order is received via EDI.

Before scheduling the delivery, the system performs an availability check considering all the control parameters that are configured in the system based on business requirements. Once the requested product quantities are confirmed by the availability check and promised to the sales order, the system calculates the picking and packing lead time, loading lead time, goods issue lead time, and transportation lead time to calculate the accurate delivery date for the confirmed items of the sales order. The system also considers the shipping point, delivery priority, and transportation planning to calculate the delivery date.

There are two scenarios to explore for delivery scheduling:

- **Forward scheduling**
  The availability of material for fulfilling the sales order is the key data point for the system for delivery scheduling. If the material availability date is in the past—that is, before the requested delivery date of the sales order item—then the system automatically confirms the earliest possible delivery date for the sales order item. Table 4.3 illustrates a forward scheduling example.

Description	Value
Ordered date	2/24/24
Current date	2/24/24
Ordered quantity	100 EA
Requested delivery date of the order	3/15/24
Available quantity in inventory	500 EA
Transportation planning time	5 Days
Picking and packing time	3 Days
Loading time	1 Day
Transit time	3 Days

Table 4.3 Forward Scheduling Example

In this example, a sufficient quantity of the requested product stock is available in the inventory. The system calculates the delivery date as follows:

Delivery date for the order line item = Material available date (i.e., current date) + Transportation planning time, including picking and packing time + Loading time + Transit time

The ordered date will be considered the baseline date for forward scheduling. The system calculates the goods issue date as 3/1/24 and the delivery date as 3/4/24, which is earlier than the customer-requested delivery date of 3/15/24. Picking and packing will be done in parallel to transportation planning.

Figure 4.2 shows how forward scheduling works in SAP S/4HANA.

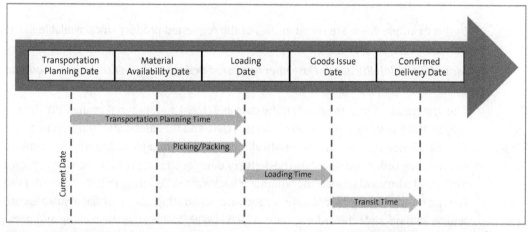

Figure 4.2 Forward Scheduling

- **Backward scheduling**

  The customer-requested delivery date will be considered the baseline to calculate the material availability for backward scheduling. The system calculates the material availability date backward so that the requested quantities of the product must be available on a particular date to meet the on-time delivery for the customer. Table 4.4 illustrates a backward scheduling example.

Description	Value
Ordered date	2/24/24
Current date	2/24/24
Ordered quantity	100 EA
Requested delivery date of the order	3/15/24
Available quantity in inventory	0 EA
Replenishment lead time	24 Days
Transportation planning time	5 Days
Picking and packing time	3 Days

Table 4.4 Backward Scheduling Example

Description	Value
Loading time	1 Day
Transit time	3 Days

Table 4.4 Backward Scheduling Example (Cont.)

In this example, there are no quantities of the requested product stock available in the inventory. The system calculates the material availability date in this scenario as follows:

Material availability date = Customer requested delivery date – Transit time – Loading time – Picking and packing time

The system calculates the goods issue date as 3/12/24 and material availability date as 3/8/24. But the stock is not available on this date and the sales order item can't be confirmed. Hence the system automatically considers the replenishment lead time to confirm the order and calculate the delivery date for the sales order item. The system switches to forward scheduling whenever backward scheduling fails. Considering the baseline date as the ordered date for forward scheduling, the material available date will be 24 days after the ordered date, which is 3/19/24. Because the picking and packing lead time is three days, the system calculates the loading date as 3/22/24. And as the loading lead time is one day, the system calculates the goods issue date as 3/23/24. Because the transit lead time is three days, the system calculates the delivery date as 3/26/24. Figure 4.3 shows how backward scheduling works in SAP S/4HANA.

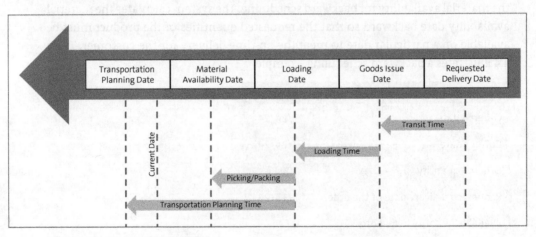

Figure 4.3 Backward Scheduling

4.1.5 Configuration of ATP Checks for Sales Orders and Deliveries

Let's explore how we can meet business requirements for ATP checks for sales orders and deliveries through configuration in the SAP S/4HANA system. We'll provide the step-by-step instructions for the ATP check setup in the following sections.

Define Availability Checking Group

Availability checking groups play a vital role in providing accurate information about stock availability to fulfill customer orders. Let's discuss the availability checking group and how to configure it.

To arrive at this configuration step, execute Transaction SPRO and navigate to **Sales and Distribution • Basic Functions • Availability Check and Transfer of Requirements • Availability Check • Availability Check with ATP Logic or Against Planning • Define Availability Checking Group**. Click **New Entries** from the initial screen to define a new availability checking group. Figure 4.4 shows the availability checking group configuration.

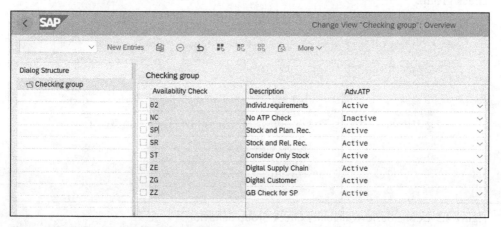

Figure 4.4 Define Availability Checking Group

Let's consider the fields to be configured:

- **Availability Check**
  This is a two-digit alphanumeric value. Along with a checking rule, this group is used to determine the availability checking procedure in the SAP S/4HANA system. Availability checks are used to determine whether sufficient stock of material is available in the inventory to fulfill a customer sales order or a stock transfer order. The availability checking rule is assigned to the material master record in the **Sales: General/ Plant** data view.

- **Adv.ATP**
  Advanced ATP is a powerful feature in SAP S/4HANA in sales and distribution that provides enhanced functionality for checking the stock availability in inventory management to fulfill customer orders and stock transfer orders. This functionality integrates with various other supply chain functions to dynamically determine the accurate stock availability of materials efficiently. Select **Active** or **Inactive** from the dropdown list to either activate or deactivate advanced ATP for the checking group for the availability check. By activating it for the checking group, the system considers inventory levels, planned orders, production orders, stock in transit, planned receipts, and the entire supply chain network to determine the stock availability.

Define Basic Availability Check Group Behavior

The reason to have multiple checking groups is to define how the availability checks are performed differently for different types of materials and products. In this configuration step, we define the behavior of the availability check group.

To arrive at this configuration step, execute Transaction SPRO and navigate to **Sales and Distribution • Basic Functions • Availability Check and Transfer of Requirements • Availability Check • Availability Check with ATP Logic or Against Planning • Define Basic Availability Check Group Behavior**. Click **New Entries** from the initial screen to define a new availability checking group. Figure 4.5 shows the settings for each availability check group.

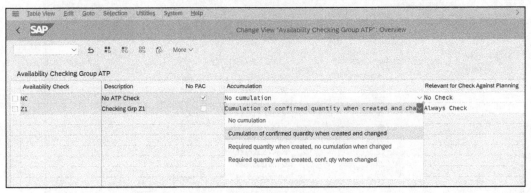

Figure 4.5 Define Basic Availability Check Group Behavior

Let's consider the fields to be configured:

- **No PAC**
  Set this indicator if you do not want to apply the ATP check logic to the checking group. For materials and products that are assigned with the availability checking group but for which no product availability check (PAC) indicator is set, the system will not perform an ATP check. The other settings of this configuration node will become irrelevant when this indicator is turned on.

- **Accumulation**
  This is an important basic setting for the ATP check. Availability checks are performed by the system during the creation of a sales order and during change order processing. The system performs the ATP check and promises the available quantities to the sales order so that the sales order can be confirmed. This will reduce the ATP quantities for the new sales orders created after the ATP quantities are assigned to the previous order. This setting will control the system behavior to commit the ATP quantities to the sales orders. The following values can be assigned to this field:
  - **No cumulation:** If this value is assigned to this field, the system will not consider the cumulated ATP quantities. This may cause inconsistencies as the available quantities are automatically confirmed to the new orders and the system does not consider the previously committed quantities to the sales orders created in the past.

- **Cumulation of confirmed quantity when created and changed**: If this value is assigned to this field, the system cumulates the confirmed quantities when the sales orders are created and changed. The system will consider the previously confirmed quantities when calculating the ATP quantities during a new sales order creation and during change order processing. This way, the system avoids inconsistencies, and the ATP quantities are confirmed for a new sales order if a sufficient quantity is available considering the previously confirmed quantities.

- **Required quantity when created, no cumulation changed**: If this value is assigned to this field, the system cumulates the required quantities from all open sales orders and doesn't cumulate the quantities when sales orders are changed. The required quantities are the requested quantities in a sales order.

- **Required quantity when created, confirmed quantity when changed**: If this value is assigned to this field, the system cumulates the required quantities from all open sales orders and cumulates only the confirmed quantities when sales orders are changed. The required quantities are the requested quantities in a sales order. Confirmed quantities are the ATP quantities for the sales order item.

- **Relevant for Check Against Planning**
  Checking for the product availability against planning refers to validating against planned production, procurement, and stock transfer replenishments. The following values can be assigned to this field:
  - **No Check**: If this value is assigned to this field, the system will not check the product availability against planning.
  - **Always Check**: If this value is assigned to this field, the system will always check the product availability against planning.
  - **Only Check If Dummy**: If this value is assigned to this field, the system will check the product availability against planning only if it's a dummy product. Dummy products are often used in testing and quality assurance.

Configure Default Values for Availability Checking Group

In this configuration step, we can assign a default availability check group to a combination of plant and material type. This setting will ensure that the default availability check group is defaulted when a material master is created with this combination.

Execute Transaction SPRO and navigate to **Sales and Distribution • Basic Functions • Availability Check and Transfer of Requirements • Availability Check • Availability Check with ATP Logic or Against Planning • Configure Default Values for Availability Checking Group**.

Click **New Entries** from the initial screen to add a new entry for a combination of plant and material type. Figure 4.6 shows the assignment of the availability check group **SR** to the combination of material type **FERT** (**Finished Product**) and plant **MJ00**. Click **New Entries** to assign an availability check group for the combination of material type and plant. This setting helps to maintain availability check criteria by plant and material type.

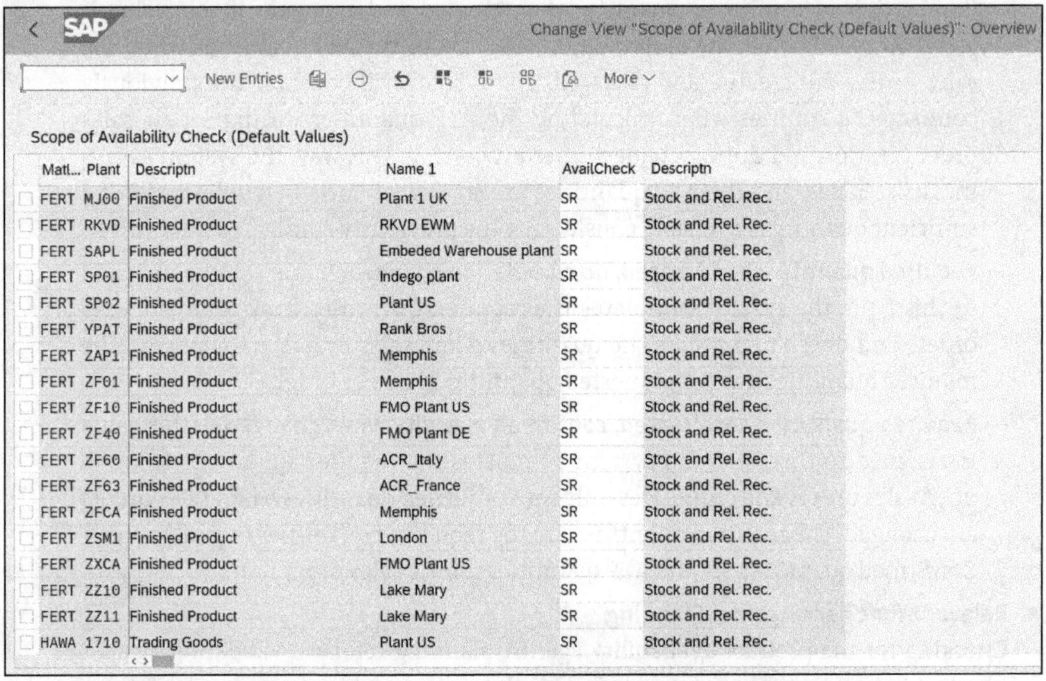

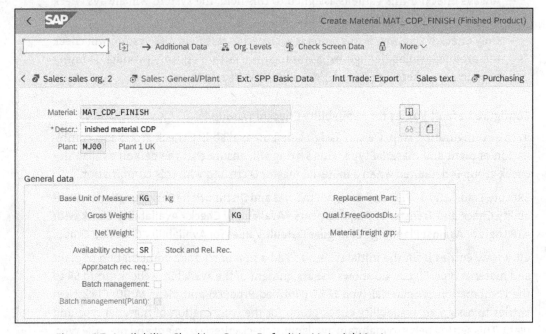

Figure 4.6 Configure Default Values for Availability Checking Group

Figure 4.7 shows that availability checking group **SR** was defaulted while creating a material with type **FERT** in plant **MJ00**.

Figure 4.7 Availability Checking Group Default in Material Master

Configure Scope of Availability Check

In this configuration, we can define the scope of the availability check. This configuration is set with reference to a checking rule (checking rules are configured in inventory management and purchasing; see Section 4.2.3 for details) and an availability checking group. The scope involves the type of stock of the product to be checked for availability (e.g., quality inspection stock), the future supply scope to be considered or not (e.g., including confirmed stock transfer orders), and so on. You can also define the requirements to be considered for the availability check such as the requirements arising from sales orders, reservations, deliveries, and so on.

To arrive at this configuration step, execute Transaction SPRO and navigate to **Sales and Distribution • Basic Functions • Availability Check and Transfer of Requirements • Availability Check • Availability Check with ATP Logic or Against Planning • Configure Scope of Availability Check**. Click **New Entries** from the initial screen to add a new scope of availability check for a combination of checking rule and availability checking group, and press Enter. Figure 4.8 shows the settings for a scope of availability check for checking rule **A** and availability checking group **SR**.

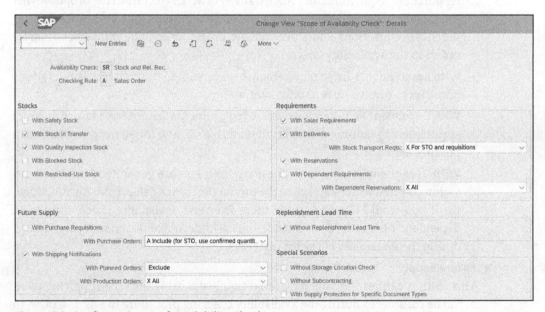

Figure 4.8 Configure Scope of Availability Check

Let's explore the configuration settings:

- **Stocks**
  This section defines whether the following stock types will be considered to perform an availability check in SAP S/4HANA to fulfill the stock transfer order:
  - **With Safety Stock**
  - **With Stock in Transfer**

- With Quality Inspection Stock
- With Blocked Stock
- With Restricted-Use Stock

Set the indicator against each stock type if, based on business requirements, you want to consider these stock types during the availability checking procedure.

- **Requirements**
  This section defines if certain requirements are to be considered during the availability checking procedure in SAP S/4HANA to fulfill the stock transfer order. Set the indicator or make a selection for the following requirements to consider them during the availability check:

  - **With Sales Requirements**: These requirements are generated from open (unfulfilled) customer sales orders.

  - **With Deliveries**: These requirements are generated from open outbound deliveries.

  - **With Stock Transport Reqts**: These requirements are generated from stock transfer purchase requisitions and stock transfer orders. Select from the dropdown list if you want to exclude them from availability checks, include both stock transfer purchase requisitions and stock transfer orders, or include only stock transfer orders in the availability check.

  - **With Reservations**: These requirements are generated from material reservations (requests to reserve materials for later use).

  - **With Dependent Requirements**: These requirements are generated from material requirements planning from bills of materials (BOMs) for all the components to produce a final product.

  - **With Dependent Reservations**: These requirements are generated from the reservations created with reference to the production order. Select from the dropdown list if you would like to exclude them from the availability check, include all dependent reservations, or include only withdrawable dependent reservations during the availability check for materials.

- **Future Supply**
  This section defines if planned material supplies coming into the supplying plant are to be considered during the availability checking procedure in SAP S/4HANA to fulfill the stock transfer order. Set the indicator or make a selection for the following supplies to consider them during the availability check:

  - **With Purchase Requisitions**: This field defines if existing open purchase requisitions are to be considered during the availability check for materials.

  - **With Purchase Orders**: This field defines if existing open stock transfer orders are to be considered during the availability check for materials. Select from the dropdown list if you would like to exclude them from the availability check, include order quantities from stock transfer orders, or include only confirmed quantities from stock transfer orders during the availability check to fulfill stock transfer orders.

– **With Shipping Notifications**: This field defines if existing open shipping notifications/inbound deliveries are to be considered during the availability check for materials.

– **With Planned Orders**: This field defines if planned orders are to be considered during the availability check for materials. Select from the dropdown list if you would like to exclude them from the availability check, include all planned orders, include only the firmed planned orders, or include only fully confirmed planned orders during the availability check to fulfill stock transfer orders.

Note

Firmed planned orders are those which will not be changed during the MRP run. *Confirmed planned orders* are those for which all components are available for production of the finished goods.

– **With Production Orders**: This field defines if production orders are to be considered during the availability check for materials. Select from the dropdown list if you would like to exclude them from the availability check, include all production orders, or include only the released production orders during the availability check.

- **Replenishment Lead Time**
  This indicator controls if replenishment lead times must be considered during the availability check for materials to fulfill stock transfer orders. Do not set this indicator if you want the system to consider replenishment lead times during the availability check. If the delivery date for the stock transfer order is within the replenishment lead time, the system confirms the delivery after the replenishment lead time. If the delivery date for the stock transfer order falls after the replenishment lead time, the system automatically confirms the delivery with the same delivery date.

Note

The replenishment lead time in SAP S/4HANA is calculated based on the planned delivery time for the material and goods receipt processing of the material master record. The following data must exist in the material master for the system to accurately check replenishment lead times and confirm deliveries. If this data doesn't exist in the material master, the system automatically confirms the deliveries without checking the replenishment lead times:

- Goods receipt processing time (**MRP 2** view of the material master)
- Goods receipt processing time (**Purchasing** view of the material master)
- Planned delivery time (**MRP 1** view of the material master)

- **Special Scenarios**
  The following are the special scenarios to be considered during the availability check for materials to fulfill stock transfer orders:
  - **Without Storage Location Check**: If this indicator is set, the system carries out the availability check at the plant level and not the storage location level.
  - **Without Subcontracting**: If this indicator is set, the system carries out the availability check without considering subcontracting stock.
  - **With Supply Protection for Specific Document Types**: Specific document types such as sales orders, stock transfer orders, and so on will get supply protection when this indicator is set. If these documents are already confirmed and quantities are already assigned, the follow-on documents of these documents will also be protected from the availability check.

Define an Availability Check Procedure for Each Schedule Line Category

Activating availability for schedule line categories enables MRP and availability checks for every schedule line in the sales documents. This setup is applicable only for sales documents. In this configuration, we define for every schedule line category whether the availability check will be carried out and whether the transfer of requirements to planning will be carried out during the creation of sales documents.

A schedule line category is a prerequisite for this configuration. You can find instructions for making the following configuration settings for the sales document creation setup in SAP S/4HANA in Section 4.4.3:

1. Define a sales order type
2. Define an item category
3. Define the item category determination
4. Define a schedule line category
5. Define the schedule line category determination

To arrive at this configuration step, execute Transaction SPRO and navigate to **Sales and Distribution • Basic Functions • Availability Check and Transfer of Requirements • Availability Check • Availability Check with ATP Logic or Against Planning • Define Availability Check Procedure for Each Schedule Line Category**.

Figure 4.9 shows the configuration settings for this node. If the availability check must be carried out and the requirements are not required to be transferred to material planning, set the availability check (**AvC**) and product allocation (**All.**) indicators.

Let's explore the settings of this configuration:

- **AvC (availability check for sales)**
  If this indicator is set, the availability check will be carried out for the sales order item with the respective schedule line category.

- **Rq. (transfer of requirements)**

  If this indicator is set, the sales demand will be transferred to MRP. Transfer of requirements to MRP can trigger creation of planned orders or purchase requisitions depending on the setup. Planned orders will be transferred to a manufacturing plant to produce additional inventory. If the distribution center has sufficient inventory, stock transfer purchase requisitions will be created.

- **All. (product allocation)**

  If this indicator is set, confirmed quantities from the availability check will be allocated to the sales order items. Product allocation ensures the stock is distributed efficiently to fulfill customer orders.

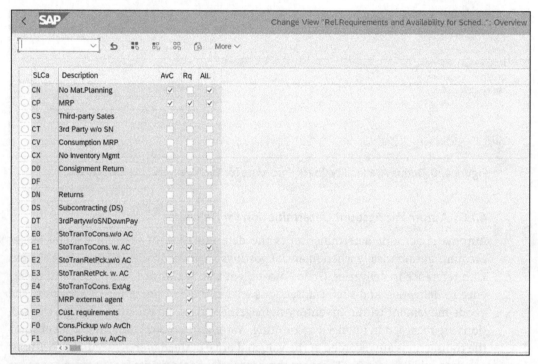

Figure 4.9 Define Availability Check Procedure for Each Schedule Line Category

Define an Availability Check Procedure for Each Delivery Line Category

In this configuration, we turn off the availability check for certain delivery line categories—for example, returns and nonstock items.

To arrive at this configuration step, execute Transaction SPRO and navigate to **Sales and Distribution • Basic Functions • Availability Check and Transfer of Requirements • Availability Check • Availability Check with ATP Logic or Against Planning • Define Availability Check Procedure for each Delivery Line Category**. Figure 4.10 shows the availability check was turned off for delivery line category **CB10 (Returnables Item)**.

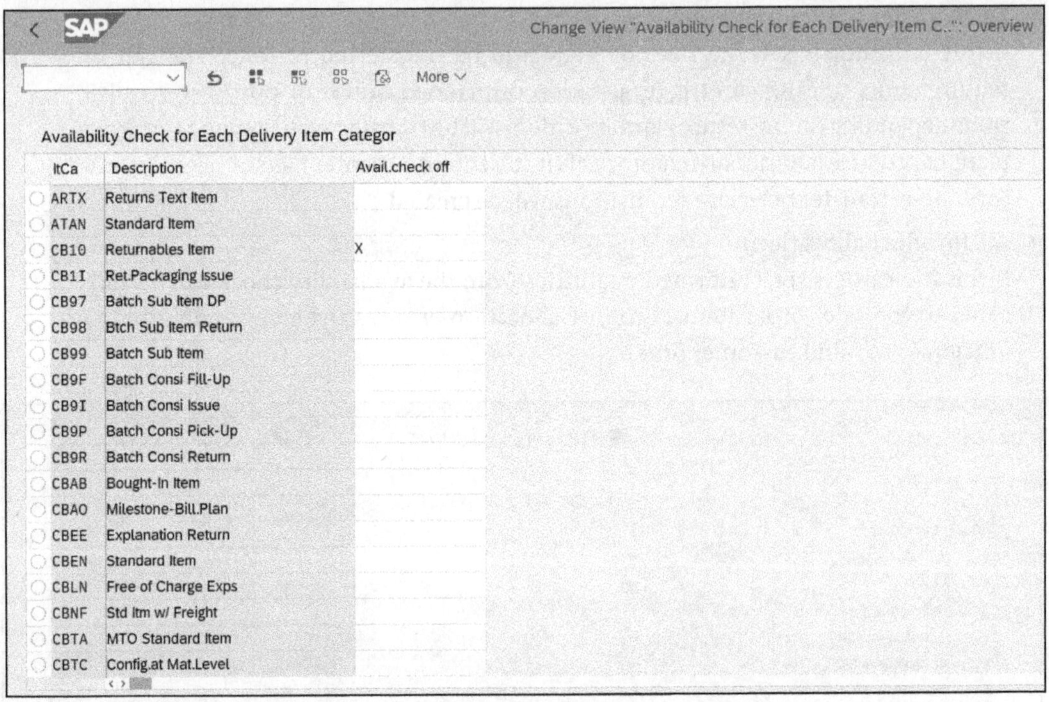

Figure 4.10 Define Availability Check Procedure for Each Delivery Line Category

4.1.6   Automatic Account Determination for Deliveries

Automatic account determination is the determination of relevant general ledger accounts automatically when financial postings are made in the SAP S/4HANA system with reference to deliveries. Goods movement transactions will be posted with reference to deliveries, and such transactions trigger the determination of accounts. The goods movement within inventory management must be valuated, and the transactions are recorded in financial accounting. Automatic account determination ensures the transactions are recorded at the valuated price of the materials accurately without any manual intervention. There are three important data points for automatic account determination:

- **Material**
  The material master record is the main driver for the automatic account determination. The valuation class and price control indicator assigned to the material master record at the plant level are used to determine the general ledger account during goods movement transactions such as goods issues and goods receipts. A valuation class acts as a link between inventory management and finance.

- **Company code and plant**
  A valuation area plays a vital role in the automatic determination of accounts. A valuation area is typically a plant, but a company code can also be defined as a valuation area if all plants belong to one company code.

- **Goods movement**
  The goods movement transaction also controls the account determination. The movement type of the goods movement transaction plays a crucial role in the determination of accounts automatically.

The system creates a material document and accounting document to record the goods movement transaction in SAP S/4HANA. The material document records the type of goods movement details such as posting date, document reference, material, issued quantity/received quantity, movement type, customer, plant, storage location, and so on. For goods issue for an outbound delivery, the reduction of the material stock from the inventory management will be recorded in the material document. The system creates a material document for all goods movements in inventory management.

An accounting document will be created from all goods receipt and goods issue transactions, but for certain transfer posting goods movements—such as storage location to storage location transfer within the same plant, posting stock from quality inspection stock to unrestricted-use stock, and so on—no accounting document will be created in the system. An accounting document records the financial accounting data for goods movement transactions, such as stock value that is coming into the inventory or that is reduced from the inventory in the case of goods issue, corresponding general ledger accounts that are determined automatically, date posted, and so on.

To determine the account for goods movements with reference to outbound delivery to the customer or return delivery from the customer, materials management provides valuation and account assignment settings that can be configured based on business requirements. We'll discuss valuation and account determination in Chapter 10, Section 10.3.

4.2 Stock Transfer Process

Stock transfer is a widely used and important supply chain process. It involves moving inventory from one storage location to another storage location within the same plant or same company code, or across company codes. The stock transfer between storage locations within the same plant is also called a *transfer posting*. Stock transfer between two plants within the same company code and stock transfer between the two plants across different company codes involve both the sales side and purchasing side of transactions. The stock transfer process is commonly used to transfer raw materials, semifinished goods, finished goods, and packaging materials between plants/storage locations. It helps to optimize inventory levels across plants, which leads to cost reduction. The stock transfer process mainly integrates materials management, sales and distribution, and finance functions of the organization, while planning, warehouse management, quality management, transportation management, and other functions can also be integrated with this process.

Let's dive deep into the stock transfer process in SAP S/4HANA and understand the process steps involved in the movement of inventory between plants within the same company code and across company codes. We'll then cover relevant master data requirements before walking through the configuration instructions.

4.2.1   Integrated Stock Transfer Process

A seamless integration of the stock transfer process between sales and distribution and materials management ensures visibility and control over the end-to-end stock transfer process. Table 4.5 shows all the key processes impacted.

Area	Processes Impacted	Description
Sales and distribution	Create and maintain stock transfer order	This process supports the creation and maintenance of stock transfer order.
	Create and process replenishment delivery	This process supports the creation of replenishment delivery/outbound delivery against the stock transfer order.
	Create intercompany billing document (sales side)	This process supports the creation of intercompany billing document on the sales side (supplying plant).
Materials management	MRP run	This process supports the MRP run to generate stock transfer purchase requisitions.
	Create and maintain stock transfer requisition	This process supports the creation and maintenance of stock transfer purchase requisitions.
	Create and maintain stock transfer order	This process supports the creation and maintenance of stock transfer order.
	Post goods issue for outbound delivery	This process supports the processing of goods issue for the outbound delivery.
	Create and process inbound delivery	This process supports the creation of inbound delivery against the stock transfer order.
	Post goods receipt for inbound delivery	This process supports the processing of goods receipt for the inbound delivery.
	Receive and process intercompany invoice (purchasing side)	This process supports the creation and reconciliation of intercompany invoices on the purchasing side (receiving plant).

Table 4.5 Process Impact: Stock Transfer Process

There are two types of stock transfer process in SAP S/4HANA involving movement of stock/inventory between two plants, with materials management and sales and distribution integrations:

- **Intracompany stock transfer process**

  This is for movement of inventory between two plants belonging to the same company code. Both the receiving (purchasing) plant and the supplying plant belong to the same legal entity (i.e., same company code), so there will be no settlement that happens for internal selling and purchasing unlike in an intercompany stock transfer process. But the intracompany stock transfer process in SAP S/4HANA supports the shipping process using outbound deliveries; the receiving process using inbound deliveries; picking, packing, and goods issue via embedded extended warehouse management (EWM); putaway of received goods via inbound deliveries into the warehouse; and transportation management (TM). Intracompany stock transfer thus is a widely used feature in SAP S/4HANA. Figure 4.11 shows all the process steps.

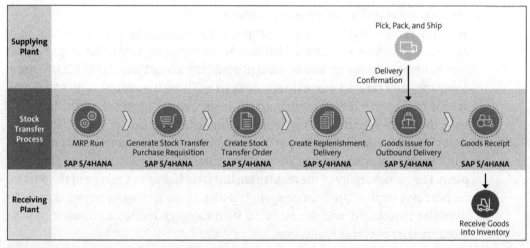

Figure 4.11 Intracompany Stock Transfer Process

- **Intercompany stock transfer process**

  This is movement of inventory between two plants belonging to different company codes. Figure 4.12 shows all the process steps. One notable difference between intercompany and intracompany stock transfer process is that intercompany billing and settlement is required for the intercompany stock transfer process.

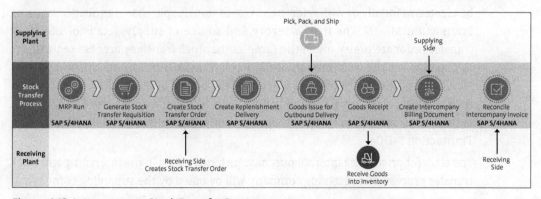

Figure 4.12 Intercompany Stock Transfer Process

Let's dive deep into every subprocess involved in the stock transfer process:

1. **MRP run**

   MRP is one of the key functionalities within materials management. This process ensures the right quantities of materials are available at the right time for the production process to meet the sales demand. The MRP run in SAP S/4HANA calculates the material requirements based on sales demand, planned receipts, and inventory levels. Purchase requisitions and planned orders will be the output of MRP run. Planned orders will be converted to production orders, while purchase requisitions are converted to purchase orders. The MRP run creates stock transfer purchase requisitions as well based on the inventory availability at the supplying plant. An MRP run can be executed manually using Transaction MD06.

2. **Create stock transfer purchase requisitions**

   The MRP run generates stock transfer purchase requisitions in the receiving plant automatically based on stock availability at the supplying plant. The stock transfer purchase requisition can also be created manually using Transaction ME51N. Intracompany stock transfer purchase requisitions will have the supplying plant ID as the source of supply, whereas intercompany stock transfer purchase requisitions will have the business partner of the supplying plant as the source of supply. The valuation price of the stock transfer purchase requisition in most cases will be the standard price or moving average price of the material maintained in the supplying plant. The item category of the requisition line item indicates the type of stock transfer purchase requisition; item category U indicates the intracompany stock transfer purchase requisition, and the standard item category indicates an intercompany stock transfer purchase requisition.

3. **Create stock transfer order**

   The stock transfer order is the main driver of the stock transfer process, and this document is the combination of a sales order and a purchase order in the same document. With reference to the stock transfer order, an outbound delivery and goods issue and intercompany billing document will be created on the sales and distribution side. On the purchasing side, an inbound delivery, goods receipt, and intercompany invoice verification and settlement will be processed. Stock transfer orders can be created manually or with reference to stock transfer purchase requisition(s) using Transaction ME21N. The item category and source of supply details of the stock transfer order are pretty much the same as the stock transfer purchase requisition, as explained in the walkthrough of the stock transfer purchase requisition process step. Unlike the other purchase order types, the output of the stock transfer order to the supplying plant is not required as it is an internal document; the SAP S/4HANA system automatically creates a stock requirement list that can be displayed using Transaction MD04.

   The transfer price is the internal purchase unit price used in the intercompany stock transfer process. The receiving company will be billed by the supplying company at this unit price. An intercompany stock transfer order will be created by the receiving plant using the transfer price. The transfer price for the material can be maintained in the purchasing info records (PIRs).

4. **Create replenishment delivery (outbound delivery)**

   Shipping data from the line item level of the stock transfer order is used to create a replenishment delivery using Transaction VL10B. SAP S/4HANA performs an availability check before the outbound delivery can be created. The outbound delivery facilitates the movement of materials from the supplying plant to the receiving plant. Picking and packing of materials will be performed within the inventory management against the outbound delivery. This helps to manage inventory movements efficiently.

5. **Goods issue for outbound delivery**

   Goods issue is an important functionality of inventory management that removes materials from inventory for transfer in a stock transfer process. Posting of a goods issue in SAP S/4HANA on the supplying side updates the material document and accounting documents while reducing the quantities in inventory. Movement types play an important role in determining the inventory movement type and the goods issue posting. Movement type 641 is used for the intracompany stock transfer process and 643 for the intercompany stock transfer process. Goods issue for replenishment delivery moves the stock from the supplying plant into transit.

6. **Goods receipt**

   Goods receipt against a stock transfer order facilitates the movement of materials from stock in transit to the inventory of the receiving plant. Goods receipt posting in SAP S/4HANA increases the stock level in inventory management. Movement type 101 is used for goods receipt postings for both intracompany and intercompany stock transfer processes.

   An inbound delivery, also called an advanced shipping notification (ASN), can be created prior to goods receipt posting. Creation of an inbound delivery is controlled by the confirmation control key maintained at the item level of the stock transfer order. An inbound delivery is mandatory when handling units are involved, EWM integration is required, and the shipments are required to be tracked in advance. The putaway process will be performed with reference to the inbound delivery followed by the goods receipt posting.

7. **Create intercompany billing document (sales side)**

   An intercompany billing document is required only for an intercompany stock transfer process involving two different legal entities. After the goods issue is posted from the supplying plant, SAP S/4HANA allows the creation of the intercompany billing document with reference to the outbound delivery using Transaction VF01 on the sales (supplying plant) side. This document will reference both the outbound delivery and the stock transfer order. The transfer price from the stock transfer order will be copied over to the billing document. Best practice is to send the billing document to the business partner of the receiving plant. An EDI message can be triggered as an output so that a logistics invoice verification (LIV) document can be posted automatically on the receiving (purchasing) side.

8. **Post and reconcile intercompany invoice (purchasing side)**
   This step is required only for an intercompany stock transfer process involving two different legal entities. The intercompany invoice is also called the LIV document, created with reference to the intercompany stock transfer order using Transaction MIRO in SAP S/4HANA. The system matches the invoice quantity and price with the received quantities against the transfer price and quantities of the stock transfer order. If there are any discrepancies during the invoice reconciliation (three-way match), the system blocks the invoice from payment processing. Those discrepancies will be handled by the exception handlers (accounts payable and buyer/planner). A fully reconciled invoice with no exceptions will be cleared for payment (intercompany settlements).

The stock transfer process impacts user roles from both materials management and sales and distribution within an organization.

The following are the key user roles impacted within sales and distribution:

- **Sales representatives**
  Responsible for initiating a stock transfer process based on customer demand, sales forecasts, and inventory requirements
- **Accounts receivable clerk**
  Responsible for creating and maintaining intercompany billing documents while ensuring accuracy and compliance with company and regulatory policies
- **Master data management team**
  Responsible for maintaining the pricing condition records for the transfer price

The following are the key user roles impacted within materials management:

- **Buyers/planners**
  Responsible for initiating stock transfer requests based on demand, production requirements, and inventory levels
- **Inventory manager**
  Responsible for picking materials from inventory based on availability for the stock transfer process
- **Receiving clerk**
  Responsible for verifying material quantities and the condition of the received goods and for receiving incoming stock at the receiving location into inventory
- **Accounts payable personnel**
  Responsible for receiving and processing of intercompany invoices while ensuring accuracy and compliance with company and regulatory policies

- **Master data management team**
  Responsible for maintaining the purchasing master data for the stock transfer process

4.2.2 Master Data Requirements

The stock transfer process requires the following master data to be maintained in the SAP S/4HANA system:

- **Material master data**
  Maintaining the material master data views listed in Table 4.6 is mandatory for the stock transfer process. The other views of the material master are required based on other business processes in which the same material is being used.

Maintenance Status	Material Master Views
K	Basic data
D	MRP
P	Forecasting
E	Purchasing
L	Storage
Q	Quality management
V	Sales
B	Accounting

Table 4.6 Material Master Maintenance Status

- **Business partner (customer)**
  A business partner with a customer role is required to determine shipping data, and this business partner will be assigned to the receiving plant in the stock transfer order configuration. Our discussion in Section 4.2.3 about defining shipping data for plants explains the importance of a business partner (customer).

- **Business partner (supplier)**
  A business partner with a vendor role is required for an intercompany stock transfer process. The supplying plant will be assigned to the business partner to determine the supplying plant for the process. Figure 4.13 shows the business partner (supplier) marked as **Plant relevant** and that the supplying plant was assigned to it.

  A business partner (supplier) is required for the intercompany stock transfer process and not required for the intracompany stock transfer process. This is because the intercompany stock transfer process involves the creation of intercompany billing on the sales side and invoice verification on the purchasing side, followed by an intercompany settlement process. This must be created in the receiving company code and the purchasing organization of the receiving plant.

The **Plant relevant** indicator must be set and the supplying plant must be assigned in the **Vendor: General Data** tab of the business partner. This is how the system determines the supplying plant for the intercompany stock transfer process so that goods issues for the replenishment delivery can be processed.

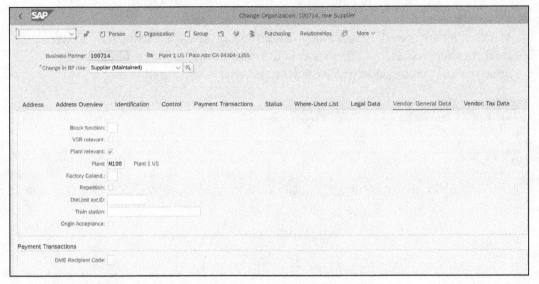

Figure 4.13 Assign Supplying Plant to Business Partner (Supplier)

- **Source list**
  A source list needs to be created in the receiving plant with the supplying plant information to procure materials internally using the stock transfer process.

- **PIR**
  A PIR needs to be created for the combination of a material and the supplier record of the supplying plant with the transfer price and other details.

For more information on master data configuration for the material master, business partners, source lists, and PIRs, see Chapter 3.

4.2.3   Configuration of Stock Transfer Orders

This section explains the key configuration settings for executing the stock transfer process in SAP S/4HANA and the key integration points between materials management and sales and distribution. These settings drive the end-to-end process while considering all the business requirements. Let's dive deep into the configuration settings and their importance.

Define Shipping Data for Plants

The definition of shipping data of plants ensures that the key components of shipping data are available for the creation of a stock transfer order, for replenishment delivery, for goods issue processing, and for creation of an intercompany billing document. The receiving plant in the stock transfer process will be technically treated as a customer to execute the sales and distribution side of the process. The following shipping data is required for those subprocesses:

- **Shipping point**
  A shipping point is a physical location within the plant from which the goods are issued for outbound deliveries and are shipped to the receiving plant/customers. Shipping points are assigned to plants in configuration.

- **Loading point**
  A loading point is a physical location within the plant from which the goods are loaded into a vehicle for transportation. Loading points are assigned to shipping points.

- **Shipping conditions**
  A shipping condition represents the condition in which the goods are shipped to the receiving plant/customer: standard shipping, express shipping, returns, and so on. Shipping conditions play a vital role in the determination of loading lead time, priority, and freight calculation.

- **Incoterms**
  Incoterms define the terms of a sale and the responsibilities of both parties—the seller and the buyer—in the sense of who owns the transportation costs, customs clearance, insurance, and other risks associated with the delivery of goods.

- **Transportation-specific data**
  Route determination, shipping point determination, and freight forwarders (forwarding agents) are also equally important. These elements are discussed in Chapter 9.

As a prerequisite, we need a business partner (customer) to be created in the sales area belonging to the receiving plant. SAP S/4HANA determines the shipping data of the supplier from the business partner (customer). Figure 4.14 displays the business partner (customer) record of the receiving plant.

To arrive at this configuration step, execute Transaction SPRO and navigate to **Materials Management • Purchasing • Purchase Order • Set Up Stock Transport Order • Define Shipping Data for Plants**.

Figure 4.15 shows the initial screen for maintaining shipping data for plants. Select the receiving plant for which the shipping data is to be maintained, and click the **Details** button at the top or press $\boxed{\texttt{Ctrl}}$+$\boxed{\texttt{Shift}}$+$\boxed{\texttt{F2}}$ to display the details of the settings, as shown in Figure 4.16.

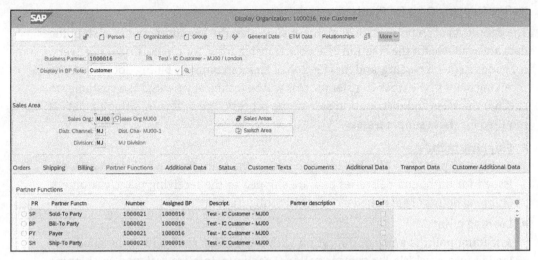

Figure 4.14 Business Partner (Customer) Record of Receiving Plant

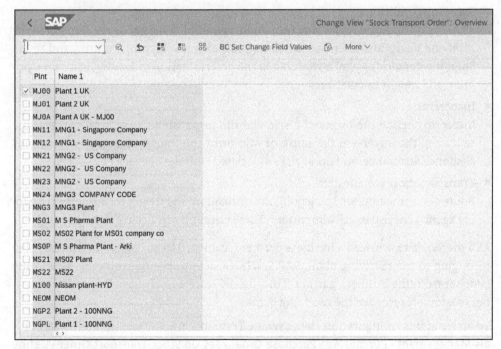

Figure 4.15 Maintain Shipping Data for Plants: Initial Screen

Let's dive deep into some of the key configuration settings for maintaining the shipping data for plants. Figure 4.16 shows the fields to be configured:

- **Customer No. Plant (customer number of plant)**
  A business partner (customer record) created for the plant for the purpose of a stock transfer process needs to be assigned to this field in the configuration. This record

will serve as the main driver for determining the shipping data during the creation of a stock transfer order and the creation of an outbound delivery.

- **SlsOrg.Int.B (sales organization for intercompany billing)**
  Assign a sales organization that belongs to the same sales area in which the business partner (customer record) for the plant was created in this field. SAP S/4HANA uses this sales organization for the creation of intercompany billing documents.

- **DistChannelB (distribution channel for intercompany billing)**
  Assign a distribution channel that belongs to the same sales area in which the business partner (customer record) for the plant was created in this field. SAP S/4HANA uses this distribution channel for the creation of intercompany billing documents.

- **Div.Int.Billing (division for intercompany billing)**
  Assign a division that belongs to the same sales area in which the business partner (customer record) for the plant was created in this field. SAP S/4HANA uses this division for the creation of intercompany billing documents.

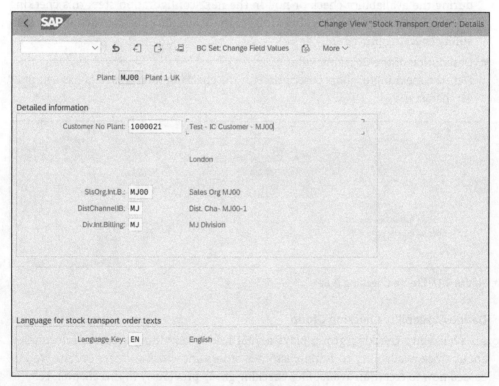

Figure 4.16 Maintain Shipping Data for Plants: Details Screen

Create Checking Rule

Availability checks, which we discussed at length in Section 4.1, are also important during the stock transfer process. SAP S/4HANA checks the availability of the materials in the supplying plant at the time of replenishment delivery or outbound delivery creation during the stock transfer process. A checking rule is a two-digit alphanumeric value, logically used to configure the availability check. It defines the sequence and type of check to be performed for the stock availability.

To arrive at this configuration step, execute Transaction SPRO and navigate to **Materials Management • Purchasing • Purchase Order • Set up Stock Transport Order • Create Checking Rule**. Click **New Entries** from the initial screen to add the new checking rules for the stock transfer order process. Figure 4.17 shows the fields to be configured to add new checking rules for the stock transfer process.

Let's examine the fields to be configured:

- **CRl (checking rule for the availability check)**
  This is a two-digit alphanumeric value used to determine the availability checking procedure in the SAP S/4HANA system. For example, using this checking rule, the system can be configured to check if the quality inspection stock is to be considered during the availability check or not. In the next configuration step, this checking rule will be assigned to a checking group of the availability check to define the availability checking procedure.

- **Description of the Checking Rule**
  Define a meaningful short description for the checking rule. This field has a limit of 40 characters.

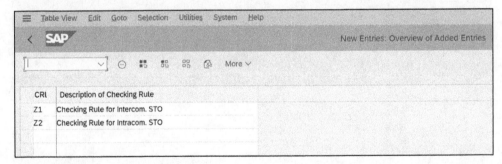

Figure 4.17 Create Checking Rule

Define Availability Checking Group

An availability checking group plays a vital role in providing accurate information about stock availability to fulfill customer orders and stock transfer orders. We discussed how to create an availability checking group previously in this chapter; refer to Section 4.1.5 for instructions.

Define Checking Rule

In this configuration step, we define the availability checking procedure based on the availability checking group and the checking rule defined in the previous step. For the complete instructions, see our discussion of defining the ATP check scope in Section 4.1.5.

Define Delivery Type for Stock Transfer Process

A delivery type plays a vital role in the creation of a replenishment delivery or outbound delivery. In general, in SAP S/4HANA, every document has its own document type and number ranges, reference documents and so on. The delivery type controls the number range, picking location determination, reference document to be used to create the delivery document, output determination, route determination for transportation management, and so on. SAP S/4HANA provides two standard delivery types for the stock transfer process:

- **NL**
  Delivery type for the intracompany stock transfer process

- **NLCC**
  Delivery type for the intercompany stock transfer process

Standard delivery types can be used to create outbound delivery documents with reference to a stock transfer order. Based on business requirements, if the standard delivery type settings need to be changed, best practice is to copy the standard delivery types, **NL** and **NLCC**, to create new delivery types. Custom delivery types must start with the letter Z.

To arrive at this configuration step, execute Transaction SPRO and navigate to **Logistics Execution • Shipping • Deliveries • Define Delivery Types**. From the initial screen, select delivery types **NL** and **NLCC** and click the **Copy As** icon to configure custom delivery types **ZNL** and **ZNLC**, respectively. Figure 4.18 and Figure 4.19 show the delivery type settings for the intracompany and intercompany stock transfer processes.

Let's dive deep into some of the important settings for delivery types:

- **Document Cat.**
  This setting defines the document category of the delivery type. It defines and controls the functionality of the document such as inquiry, scheduling agreement, invoice, returns, and so on. For the intracompany and intercompany stock transfer delivery types, assign **J** (**Delivery**) as the document category.

- **Number Systems**
  This section of the delivery type configuration defines the internal and external number ranges for the delivery type. The system automatically chooses the next available number in the internal number range while creating the delivery document in SAP S/4HANA. External numbers can be manually entered and must be within the defined external number range.

  To arrive at the number range configuration step, execute Transaction SPRO and navigate to **Sales and Distribution • Sales • Sales Documents • Sales Document Header • Define Number Ranges for Sales Documents**. You can arrive at this configuration step directly by executing Transaction VN01. In this configuration step, both internal and external number ranges can be defined against number range object RV_BELEG.

- **Order Required**

  In this field, define the order basis for the creation of an outbound delivery. In a stock transfer process, the stock transfer order will be the basis for creating a replenishment delivery or outbound delivery. Hence, assign the value **Purchase order required** for both intracompany stock transfer delivery type **ZNL** and intercompany stock transfer delivery type **ZNLC**.

- **Storage Location Rule**

  In this field, assign a picking location/storage location determination rule. Based on this rule, the system determines a storage location or picking location for the outbound delivery within the inventory management. There are three standard rules available in SAP S/4HANA to determine the picking location:

  - **MALA**: Shipping point + plant + shipping condition
  - **RETA**: Plant + shipping condition + situation
  - **MARE**: First, the system checks for rule MALA; if not possible to determine the picking location, it then checks rule RETA

  Because the business partner (customer record) of the receiving plant contains the shipping condition, and storage location-specific shipping points can be defined, assign **MALA** for both intracompany stock transfer delivery type **ZNL** and intercompany stock transfer delivery type **ZNLC**. This is also the standard SAP setting for default delivery types **NL** and **NLCC** for intracompany and intercompany stock transfer processes.

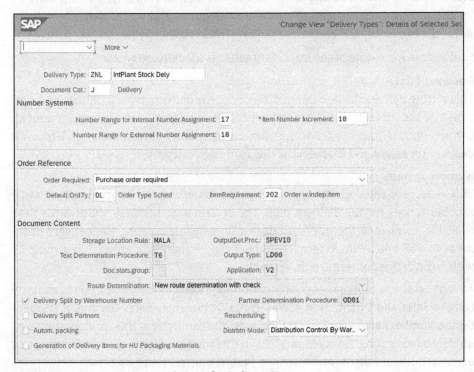

Figure 4.18 Intracompany Stock Transfer Delivery Type

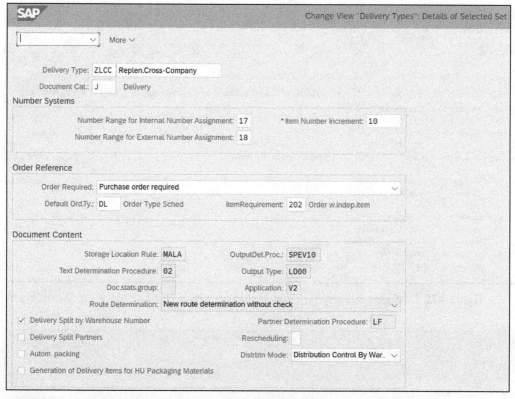

Figure 4.19 Intercompany Stock Transfer Delivery Type (Replenishment Delivery)

Configure Delivery Type and Availability Check Procedure by Plant

In this configuration, a delivery type, a stock transfer order type, and a checking rule (for material availability) are assigned to the supplying plant and stock transfer order document type combination. Refer to our discussion of the schedule line category in Section 4.4.3 for the configuration of stock transfer order document types. Typically, document type **UB** is used for an intracompany stock transfer order and document type **NB** is used for an intercompany stock transfer order. Through this configuration setting, the system determines the delivery type when a delivery is created with reference to the stock transfer order.

To arrive at this configuration step, execute Transaction SPRO and navigate to **Materials Management • Purchasing • Purchase Order • Set up Stock Transport Order • Configure Delivery Type & Availability Check Procedure by Plant**. Figure 4.20 shows the initial screen for configuring the delivery type and availability check procedure per supplying plant. Click **New Entries** to add new entries for this configuration.

Figure 4.21 shows the two supplying plants added to this configuration. Supplying plant **M100** is configured for the intercompany stock transfer process, and supplying plant **MJ01** is configured for the intracompany stock transfer process.

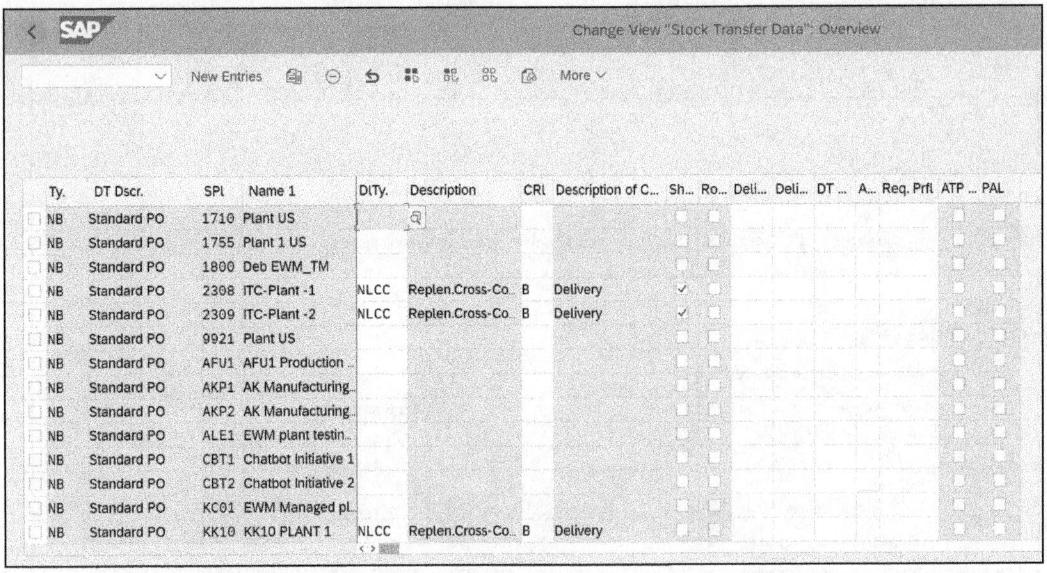

Figure 4.20 Configure Delivery Type and Availability Check Procedure by Plant: Initial Screen

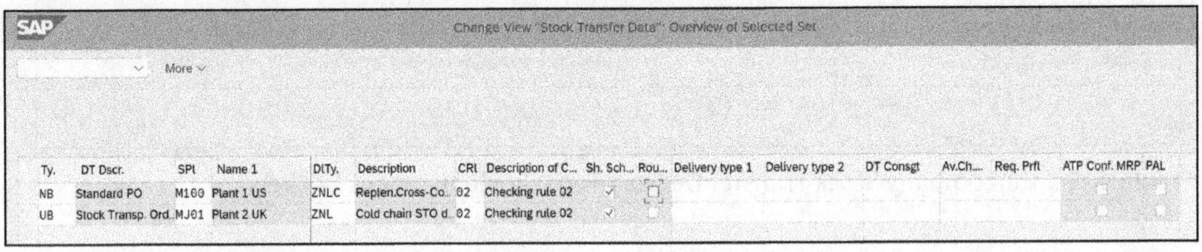

Ty.	DT Dscr.	SPl	Name 1	DlTy.	Description	CRl	Description of C...	Sh...	Ro...	Deli...	Deli...	DT ...	A...	Req. Prfl	ATP ...	PAL
NB	Standard PO	1710	Plant US													
NB	Standard PO	1755	Plant 1 US													
NB	Standard PO	1800	Deb EWM_TM													
NB	Standard PO	2308	ITC-Plant -1	NLCC	Replen.Cross-Co..	B	Delivery	✓								
NB	Standard PO	2309	ITC-Plant -2	NLCC	Replen.Cross-Co..	B	Delivery	✓								
NB	Standard PO	9921	Plant US													
NB	Standard PO	AFU1	AFU1 Production...													
NB	Standard PO	AKP1	AK Manufacturing...													
NB	Standard PO	AKP2	AK Manufacturing...													
NB	Standard PO	ALE1	EWM plant testin...													
NB	Standard PO	CBT1	Chatbot Initiative 1													
NB	Standard PO	CBT2	Chatbot Initiative 2													
NB	Standard PO	KC01	EWM Managed pl...													
NB	Standard PO	KK10	KK10 PLANT 1	NLCC	Replen.Cross-Co..	B	Delivery									

Figure 4.21 Configure Delivery Type and Availability Check Procedure by Plant: New Entries

Ty.	DT Dscr.	SPl	Name 1	DlTy.	Description	CRl	Description of C...	Sh. Sch...	Rou...	Delivery type 1	Delivery type 2	DT Consgt	Av.Ch...	Req. Prfl	ATP Conf. MRP PAL
NB	Standard PO	M100	Plant 1 US	ZNLC	Replen.Cross-Co..	02	Checking rule 02	✓							
UB	Stock Transp. Ord..	MJ01	Plant 2 UK	ZNL	Cold chain STO d..	02	Checking rule 02	✓							

Let's dive deep into some of the important settings of this configuration:

- **Ty. (purchasing document type)**
  The purchasing document type controls the type of the stock transfer process: intra-company or intercompany. It also plays an important role in controlling the screen layout, pricing procedures, allowed item categories, types of purchase requisition document reference, and so on. Every purchasing document controls different purchasing processes and rules. Assign document type **UB** for the intracompany stock transfer process and document type **NB** for the intercompany stock transfer process in this field.

- **SPl (supplying plant)**
  Assign the supplying plant of the stock transfer order in this field. A supplying plant that belongs to a different company must be assigned for purchasing document type **NB**, and the supplying plant that belongs to the same company code must be assigned for the purchasing document type **UB**. We can have a plant producing materials or procuring them from a vendor and supplying all the other plants within

the organization. If inventory optimization is the sole requirement, all plants within the organization can supply others as well as receive materials from one another. This configuration controls both the scenarios.

- **DlTy. (delivery type)**
  The delivery type plays a vital role in the creation of a replenishment delivery or outbound delivery against the stock transfer order. Assign the delivery types configured for the intracompany stock transfer process to the combination of intracompany stock transfer document type **UB** and the supplying plant that belongs to the same company code. Assign the delivery types configured for the intercompany stock transfer process to the combination of intracompany stock transfer document type **NB** and the supplying plant that belongs to a different company code. Delivery types **ZNL** and **ZNLC** were configured in the previous section for the intracompany stock transfer process and intercompany stock transfer process respectively.

- **CRl (checking rule)**
  Maintain the checking rule configured earlier in this field for the system to perform an availability check for materials for the stock transfer order in the supplying plant.

Assign Document Type, One-Step Procedure, Underdelivery Tolerance

In this configuration, assign a purchasing document type for the stock transfer order for every combination of supplying plant and receiving plant for the intracompany stock transfer and intercompany stock transfer processes. Assign document type **UB** for the combination for the intracompany stock transfer process and document type **NB** for the combination for the intercompany stock transfer process. This configuration also controls the underdelivery tolerance and the one-step stock transfer procedure.

To arrive at this configuration step, execute Transaction SPRO and navigate to **Materials Management • Purchasing • Purchase Order • Set Up Stock Transport Order • Assign Document Type, One-Step Procedure, Underdelivery Tolerance**. Figure 4.22 shows the initial screen of this configuration step. Click **New Entries** to add purchasing document types for a new combination of supplying plant and receiving plant, as shown in Figure 4.23.

Figure 4.23 shows that two combinations of supplying plant and receiving plant have been added to the respective purchasing document type based on the intracompany and intercompany stock transfer process.

To assign the purchasing document type for a new combination of a supplying plant and receiving plant, click **New Entries** and specify a supplying plant, receiving plant, and relevant purchasing document type. If the supplying plant and receiving plant combination belong to the same legal entity (i.e., company code), assign purchasing document type **UB**; if the supplying plant and receiving plant combination belong to the different legal entities (i.e., company codes), assign purchasing document type **NB**.

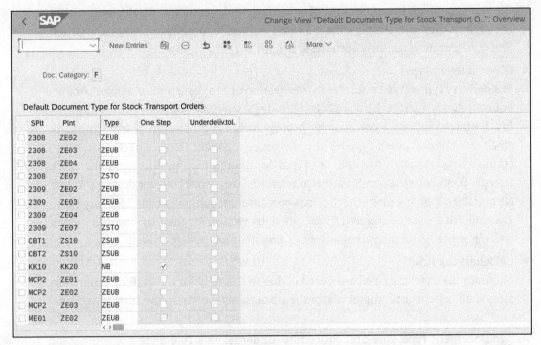

Figure 4.22 Assign Document Type, One-Step Procedure, Underdelivery Tolerance: Initial Screen

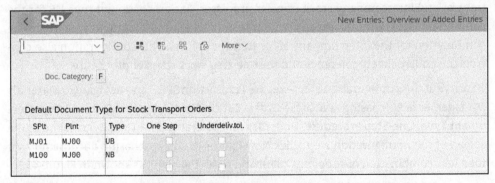

Figure 4.23 Assign Document Type, One-Step Procedure, Underdelivery Tolerance: New Entries

This configuration controls the following features:

- **One Step (one-step procedure)**

  If this indicator is set, the goods issue from the supplying plant and the goods receipt into the receiving plant happens at the same time using one goods movement. Best practice is to use a two-step procedure for the stock transfer process, where the goods issue from the supplying plant moves the stock into transit for the receiving plant and the receiving plant receives the goods into inventory using movement type 101.

- **Underdeliv.tol. (underdelivery tolerance)**
  If this indicator is set, the goods issue from the supplying plant will consider an underdelivery tolerance set for the item in the stock transfer order. If the total quantities issued are within the underdelivery tolerance set, then the system automatically sets the **Delivery Complete** indicator at the item level of the stock transfer order.

Goods Issue Movement Type Determination for Outbound Delivery for Stock Transfer Process

Determination of the correct movement type is vital for any goods movement. Each movement type represents a specific business process and a transaction. There are specific movement types in SAP S/4HANA for every stock transfer process. For example, the movement type for goods issues for intracompany stock transfer orders is different from the movement type for goods issues for intercompany stock transfer orders. Similarly, the movement type for goods issues for intracompany stock transfer orders using a one-step procedure is different from the movement type for goods issues for intracompany stock transfer orders using a two-step procedure.

Let's discuss the logic behind the determination of movement types:

- **Item Category Determination for Deliveries**
  In this configuration step, an item category will be assigned to the delivery type. To arrive at this configuration step, execute Transaction SPRO and navigate to **Logistics Execution • Shipping • Deliveries • Define Item Category Determination in Deliveries**. Figure 4.24 shows that the delivery types created for the stock transfer process have been assigned to a standard item category:

  - **NLN** is the standard item category (**ItmC**) defined in the SAP S/4HANA system for the intracompany stock transfer process.

  - **NLC** is the standard item category (**ItmC**) defined in the SAP S/4HANA system for the intercompany stock transfer process.

  - The item category group (**ItCG**) plays a vital role in the determination of item categories for the delivery type. An item category group is assigned in the **Sales: sales org 2** view of the material master. For example, **NORM** is the item category group for standard items.

  - The item usage (**Usge**) indicator further simplifies the item category determination for the delivery types. For example, configuring a specific usage indicator such as **V** (purchase order) ensures that the item category determination is made with reference to a stock transfer order.

- **Schedule Line Category Determination for Item Categories**
  In this configuration step, a schedule line category will be assigned to the item category. To arrive at this configuration step, execute Transaction SPRO and navigate to **Sales and Distribution • Sales • Sales Documents • Sales Document Item • Schedule Lines • Assign Schedule Line Categories**.

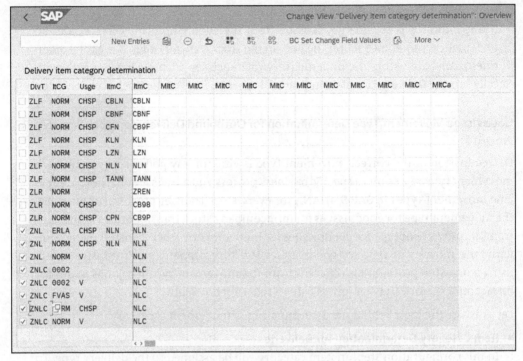

Figure 4.24 Define Item Category Determination in Deliveries

Figure 4.25 shows that schedule line category **NN** is assigned to the standard item category **NLN** for the intracompany stock transfer process.

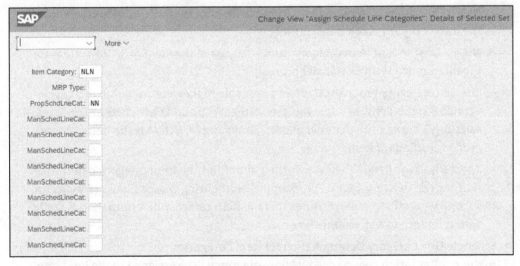

Figure 4.25 Assign Schedule Line Category for Item Category NLN

Figure 4.26 shows that schedule line category **NC** is assigned to the standard item category **NLC** for the intercompany stock transfer process.

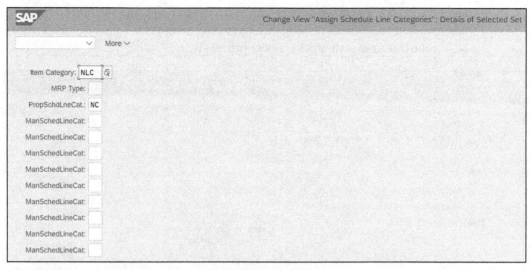

Figure 4.26 Assign Schedule Line Category for Item Category NLC

- **Assign Movement Types to the Schedule Line Category**

    In this configuration step, movement types for goods issues for outbound delivery in the stock transfer process will be assigned to the schedule line categories. To arrive at this configuration step, execute Transaction SPRO and navigate to **Sales and Distribution • Sales • Sales Documents • Sales Document Item • Schedule Lines • Define Schedule Line Categories**.

    Figure 4.27 shows that goods issue movement types **641** and **647** for the two-step and one-step procedures, respectively, have been assigned to schedule line category **NN** for the intracompany stock transfer process.

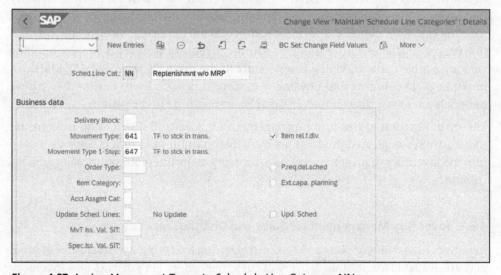

Figure 4.27 Assign Movement Types to Schedule Line Category NN

Figure 4.28 shows that goods issue movement types **643** and **645** for the two-step and one-step procedures, respectively, have been assigned to schedule line category **NC** for the intercompany stock transfer process.

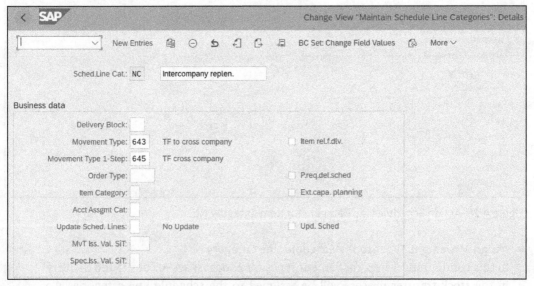

Figure 4.28 Assign Movement Types to Schedule Line Category NC

4.3   Inventory Management

Inventory management is a critical and key function of the supply chain. It stores, tracks, and controls various goods movements, including inbound inventory and outbound inventory, as well as internal storage movements. It provides real-time visibility into inventory levels at the plant and the storage location. Optimizing inventory levels, meeting customer demand, and minimizing costs are the main objectives of organizations in enterprise resource planning. Ensuring inventory accuracy is vital for fulfilling customer orders and providing components to the production orders. SAP S/4HANA provides cycle counting and physical inventory functionality to ensure that critical materials are physically counted, and results are captured in the system.

The main function of inventory management for sales and distribution is to ensure that customer-requested products are available to fulfill sales orders. Let's dive deep into the inventory management process, focusing on its integration with sales and distribution.

4.3.1   Inventory Management for Sales and Distribution

Inventory management plays a vital role in ensuring that requested quantities of products are available for on-time delivery of customer orders. Let's discuss how inventory management integrates with sales and distribution processes:

1. **Stock availability to fulfill customer orders**
   The system performs an availability check during the creation of a sales order in SAP S/4HANA. An availability check happens at the item level of the sales order, meaning that the system checks the availability of the requested quantity on the requested date for every material in the sales order. After the availability check, the system confirms the sales order if sufficient stock of materials is available in inventory and automatically allocates the requested quantities to the order.

2. **Stock allocation and reservation**
   Requested quantities of the products are allocated to the sales order after the stock availability check ensures that enough stock is available in inventory. Allocating the available stock to the sales order ensures that the allocated quantities of the products are reserved for the pertinent sales order and not available for another sales order.

3. **Backorder processing**
   During the stock availability check, if enough material is not available in inventory to fulfill the sales order, then the system automatically creates a backorder to track the outstanding demand. When enough stock becomes available, the backorder will be confirmed and processed for delivery.

4. **Goods issue for outbound delivery**
   Goods issue will be performed with reference to the outbound delivery after the inventory management functions such as picking and packing of materials are completed. This is an important functionality in inventory management that removes the materials from inventory and issues an outbound delivery to ship them to the customer. Goods issue posting will also automatically trigger account determination and posting to update the material ledger for the goods movement in order to accurately reflect the reduction of stock levels.

5. **Inventory replenishment**
   Based on actual customer demand and forecast demand, MRP generates proposals for internal procurement or external procurement or production. The stock transfer process is used in internal procurement to replenish the stock to fulfill the customer demand. The system triggers proposals for in-house production to replenish the inventory and, in certain cases, the materials will be procured from another third-party vendor to fulfill customer demand.

6. **Returns processing**
   Return order processing involves processing customer returns and integrating them into inventory. SAP S/4HANA allows you to process returns orders and returns deliveries. The returned products will be received into inventory with reference to the returns delivery. Effective management of receiving returns into inventory optimizes inventory levels and improves overall operational efficiency while maintaining customer satisfaction.

There are two important documents in SAP S/4HANA to record detailed information about the goods movements within inventory management:

- **Material document**

  SAP S/4HANA records every material movement in inventory management with the material document. It records detailed information about the material movement transaction, such as the posting date, material number, quantity of the material, movement type, batch information, plant and storage location, and user who posted the goods movement.

  Figure 4.29 shows the material document created in the system after posting a goods issue for an outbound delivery. The following are some important details of the material document:

  - Material documents will be assigned with a unique number based on the number range defined in the system for material documents.

  - They record the posting date and reference document information (purchase order, outbound delivery, etc.).

  - The movement type plays an important role in determining the type of material movement, such as goods receipt, goods issue, or transfer posting.

  - They record information about the material movement, including the issuing location and the customer details.

  - They record batch details and serial number information if the material is managed in batches and a serial number profile is assigned to the material.

  - They record the issuing location (plant and storage location) and the receiving location (plant and storage location if the goods movement happens within inventory management; i.e., a storage location to storage location transfer).

  - Material documents are linked with accounting documents if a specific goods movement involves value updates in financial accounting.

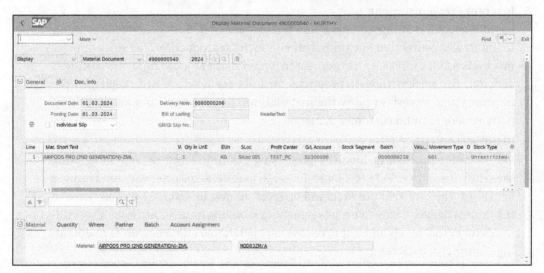

Figure 4.29 Material Document Created for Goods Issue for Outbound Delivery

- **Accounting document**

  SAP S/4HANA records every material movement in inventory management involving financial transactions in the accounting document. It records detailed information about the financial transaction associated with the goods movement and is linked to the respective material document. Automatic account determination configuration plays a vital role in the accounting document posting upon the posting of a goods issue, goods receipt, or certain transfer postings. Storage location to storage location goods movement within the same plant doesn't create an accounting document.

  Figure 4.30 shows the accounting document created in the system automatically after posting a goods issue for an outbound delivery. Refer to Chapter 10, Section 10.3 to learn how the system determines the accounts automatically.

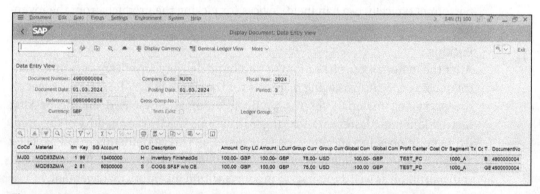

Figure 4.30 Accounting Document Created for Goods Issue for Outbound Delivery

4.3.2   Goods Issue for Outbound Deliveries

Goods issue is the process of removing material stock from inventory to fulfill sales orders, stock transfer orders, consumption for production, and internal consumption. In sales and distribution, goods issue for outbound deliveries is a common process that involves issuing of materials and products with reference to an outbound delivery to fulfill either a customer order or a stock transfer order. The goods issue posting creates both a material document and an accounting document in the system to record the material movement and the valuation of the stock in financial accounting, respectively.

The goods issue for outbound delivery involves the following steps (refer to Section 4.1.1 and Section 4.2.1 for the detailed end-to-end processes):

1. **Creation of outbound delivery**

   The process starts with the creation of an outbound delivery with reference to a sales order or stock transfer order. This document serves as the basis for picking and packing of goods from the inventory. It contains the different materials or products to be issued to the customer or another plant/warehouse, the quantity to be issued, shipping details, and delivery dates for every item. Outbound deliveries can be manually created or automatically created using background jobs in SAP S/4HANA. The creation of outbound deliveries considers the availability of stock in inventory management.

2. **Picking**

   After the outbound delivery is created, the inventory/warehouse personnel will be notified based on the priority, delivery dates, and picking and packing lead times. The required materials or products will be physically picked by the warehouse personnel considering various factors such as batch numbers, shelf-life expiration date, production date, and so on. SAP S/4HANA automatically suggests the materials to be picked and the exact location of the materials in inventory management. Inventory personnel typically use handheld devices for picking the products from the inventory.

3. **Packing**

   After the materials are picked for the outbound delivery, inventory personnel pack them using specific packaging materials for the shipments. SAP S/4HANA suggests the packaging materials for the products automatically based on the packing instructions and master data. Packing instructions specify the quantity of the product to be shipped in a specific packaging material. This combination in SAP is called a *handling unit*. Handling unit management is an important aspect of shipping and transportation as well as managing products within inventory management.

4. **Post goods issue**

   After the picking and packing of materials/products from inventory management, the next step is to issue the goods with reference to the outbound deliveries for shipping and transportation. Posting the goods issue can be done manually or automatically upon the completion of packing of the goods to be shipped in SAP S/4HANA. The goods issue reduces the stock of material from inventory management and updates the accounts in financial accounting with the valuation of the stock issued for the outbound delivery. The system creates a material document and accounting documents to record the quantity and value updates upon the posting of goods issue.

5. **Shipping and transportation**

   Shipping and transportation activities begin after the goods issue posting is completed. The system creates a shipment document if the transportation management is active in the SAP S/4HANA system. Shipments can be tracked until the products are delivered to the customer or receiving plant in case of stock transfer orders.

The goods issue is an important step to not only record the goods movement transaction but also process the customer billing or intercompany billing documents. Billing documents will typically be created after the goods issue posting in the system. The goods issue for outbound deliveries helps in reporting and tracking the process as well.

4.3.3   Goods Receipt for Returns Deliveries

Goods receipt for returns deliveries involves the processing of customer returns and stock transfer returns. Goods receipt into inventory management will be performed with reference to the return deliveries. This process ensures the returned materials are accurately updated in inventory management and are free from any damages. A goods receipt posting with reference to a return delivery creates both a material document and an accounting document in the system to record the material movement and the valuation of the stock in financial accounting, respectively. The goods receipt for return delivery involves the following steps:

1. **Creation of returns delivery**
   The process starts with the creation of a returns delivery with reference to a return order or stock transfer order for returns. The return delivery document contains the materials/products, reason for return, quantities to be returned, and return dates. Return deliveries can be manually created or created automatically using background jobs in SAP S/4HANA.

2. **Return authorization**
   To check that return deliveries are complaint with the company policy and regulatory policies, return deliveries are typically authorized by the selling organizations. This process will be supported by an approval workflow in SAP S/4HANA. An approval workflow can be configured in the system based on business requirements considering materials/products, values of goods, regulatory policies, company policies, and so on.

3. **Physically receiving the returned goods**
   Once the returned materials/products are physically arrived at the receiving location of the plant, the receiving personnel checks the quantities received and the product details for accuracy with reference to a slip that comes along with the shipment before receiving them into inventory management. The condition of the goods will be checked for any damages.

4. **Post goods receipt**
   Once the returned materials/products are verified, the goods receipt will be posted manually in SAP S/4HANA. The posting of a goods receipt increases the stock of materials in inventory management and updates the accounts in financial accounting with the valuation of stock received with reference to returns delivery. The system creates a material document and accounting documents to record the quantity and value updates upon the posting of a goods receipt. Typically, stock of returned materials will be posted into quality inspection stock in inventory management.

5. **Quality inspection**
   In certain cases, the stock returned into inventory management undergoes a quality inspection before the returned products can be issued to another outbound delivery. Quality management is a function within the supply chain that helps with quality inspection and control of the products and materials (we'll discuss this in detail

in Chapter 5). Quality inspection of the returned products will be based on the quality procedures, and the inspection results are recorded in the SAP S/4HANA system. Based on the inspection results, one of the following actions will be taken:

- Post the stock from quality inspection stock to unrestricted use stock. Once this step is done, materials/products become available for another customer order.
- Decide to send the materials/products for repairs to another facility or a third-party vendor.
- Scrap returned goods if they are damaged and unusable.

■ Upon posting a goods receipt into inventory with reference to a returns delivery, a customer credit memo or intercompany credit memo will be processed in the system. The system records the transactions for reporting and tracking purposes.

4.3.4   Standard Inventory Movement Types

A movement type in SAP S/4HANA is a three-digit code that represents a specific material movement in inventory management. Inventory management has the following processes:

1. **Goods receipt**
   This increases the stock of the material involved in this process in inventory management.

2. **Goods issue**
   This reduces the stock of the material involved in this process in inventory management.

3. **Transfer posting**
   This process neither increases the stock nor decreases the stock of the material involved in this process in inventory management. However, it changes the characteristics of the stock such as the stock type, special stock type, batch number, plant, or storage location.

4. **Physical inventory**
   This process either increases or decreases the stock of the material involved in this process in inventory management depending on the gain or loss from the physical inventory.

These inventory management processes are associated with a specific set of movement types defined in the standard system. A movement type controls one of these processes. Movement types in SAP S/4HANA are predefined and available to use by standard. Movement types can be customized based on business requirements. Movement types play a key role in the account determination for the material movements. Chapter 10, Section 10.3 explained how the movement types control the automatic account determination.

Let's explore some of the important movement types defined in the standard system. There are many other movement types, and you can display all movement types in tables T156 and T156T using Transactions SE16 or SE16N.

Table 4.7 shows the standard movement types for inventory.

Process	Movement Type	Detailed Description
Goods receipt	101	Goods receipt. This movement type is used for standard goods receipts from external vendors. It increases the stock quantity and value in the receiving storage location. Stock can be received into unrestricted-use stock or quality inspection stock.
	102	Goods receipt reversal. This movement type is used to cancel the goods receipt posted using movement type 101.
	103	Goods receipt for purchase order. This movement type is specifically used for goods receipts related to purchase orders into goods receipt blocked stock.
	105	Release from blocked stock. This movement type is specifically used for releasing from the blocked stock
	501	Goods receipt without purchase order. This movement type is used for goods receipts without a corresponding purchase order. It's typically used for miscellaneous or unplanned receipts.
Transfer posting	301	Transfer posting plant to plant. This movement type is used to transfer materials from one plant to another within the same company code.
	309	Transfer posting material to material. This movement type is used to transfer stock of one material to another material within the same plant.
	311	Transfer posting from one storage location to another storage location within the same plant. This movement type is used to transfer materials from one storage location to another within the same plant.
	321	Transfer posting from quality inspection stock to unrestricted-use stock. This movement type is used to transfer materials from quality inspection stock to unrestricted stock within the same storage location.

Table 4.7 Inventory Movement Types

Process	Movement Type	Detailed Description
Material consumption	201	Goods issue for a cost center. This movement type is used to issue goods from the inventory to a cost center for consumption.
	221	Goods issue for a project. This movement type is used to issue goods from the inventory to a project for consumption.
	261	Goods issue for a production order. This movement type is used to issue goods from the inventory to a production order for consumption.
	281	Goods issue for a network. This movement type is used to issue goods from the inventory to a network for consumption.
Outbound delivery	601	Goods issue for delivery. This movement type is used to issue goods from the inventory for an outbound delivery to fulfill customer order.
	641	Goods issue for a stock transport order. This movement type is used to issue goods from the inventory for an outbound delivery to fulfill an intracompany stock transfer order.
	643	Goods issue for a stock transport order. This movement type is used to issue goods from the inventory for an outbound delivery to fulfill an intercompany stock transfer order.
Physical inventory	701	Inventory differences in unrestricted use stock. This movement type is used to post differences from unrestricted use stock that arising from the physical inventory process.
	703	Inventory differences in quality inspection stock. This movement type is used to post differences from quality inspection stock that arise from the physical inventory process.
	707	Inventory differences in blocked stock. This movement type is used to post differences from blocked stock that arise from the physical inventory process.

Table 4.7 Inventory Movement Types (Cont.)

4.3.5   Configuration of Inventory Movement Types

This section explains the key configuration settings for movement types in SAP S/4HANA. Movement types control the automatic account determination, quantity, and value updates of stock and the handling of special stocks. Let's explore the settings.

In SAP S/4HANA, movement types are defined in the standard system. You can change the settings for the existing movement types or copy and create new movement types. New custom movement types can start with Z. For example, if you need a new custom movement type based on standard movement type 601, copy the standard movement type 601 and create a custom movement type starting with Z.

To arrive at this configuration step, execute Transaction SPRO and navigate to **Materials Management • Inventory Management and Physical Inventory • Movement Types • Copy, Change Movement Types**. In the initial screen for copying and creating a new movement type, select the movement type (**MvT**) checkbox and either click the **Apply** icon or press ⌜Enter⌟. This opens a popup screen in which to enter the movement type you want to change or copy. Do so to create a new custom movement type.

Figure 4.31 shows the overview screen for changing the existing movement type settings or copying the existing standard movement type and creating a custom movement type. Select the standard movement type (**601**) and click the **Copy As** icon to copy and create a new custom movement type. Figure 4.32 shows the movement type details.

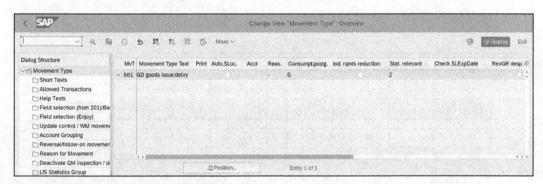

Figure 4.31 Change, Copy Movement Type: Overview

Let's dive deep into the important field-level settings for copying the existing standard movement type and creating a custom movement type:

- **Movement Type**
  A movement type in SAP S/4HANA is a three-digit code that represents a specific material movement in inventory management. Best practice is to assign a code that starts with a Z for a custom movement type. Figure 4.32 shows the new custom movement type **Z71** created by copying standard movement type **601**.

- **Print item**
  This field controls if an item from the material document that corresponds to the respective movement type can be printed. Table 4.8 shows the possible values for this field.

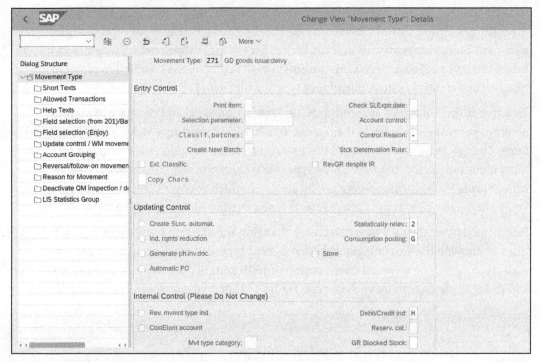

Figure 4.32 Movement Type Details

Print Item Value	Description
Blank	No document printout
1	Material document printout
2	Return delivery
3	Goods receipt/goods issue slip for subcontracting
4	Inventory list without sales price
5	Inventory list with sales price
6	Material document printout for goods receipt/goods issue
7	Transfer posting

Table 4.8 Printing of Material Document Item: Possible Values

- **Create SLoc. automat. (automatic creation of storage location allowed)**
  When this indicator is set in the configuration, the system creates storage location data in the material master record used in the goods movement. This indicator should be set for all putaway movement types such as **101, 301, 561, 501**, and so on so that the system will automatically create storage location data for them based on the storage location the stock is moving into for the very first time.

- **Account control**

  This field controls whether the **G/L Account** field appears in the material document. There are four possible values that can be assigned to this field, as shown in Table 4.9. Maintain a blank value if you want to enter a general ledger account for certain transactions.

Account Control	Description
Blank	Entry in this field is optional.
+	Entry in this field is required.
-	Field is suppressed.
.	Entry in this field is optional.

Table 4.9 General Ledger Account Field Control

- **Control Reason**

  This field controls whether it's mandatory, optional, or not required to enter a reason for the specific goods movement transaction. There are four possible values that can be assigned to this field, as shown in Table 4.10. For returns movements, it's a standard practice to enter a reason for the movement.

Reason Control	Description
Blank	Entry in this field is optional.
+	Entry in this field is required.
-	Field is suppressed.
.	Entry in this field is optional.

Table 4.10 Reason for Movement Control

- **Consumption posting**

  This field controls consumption updates in inventory management for all stock withdrawals or goods issue postings. Table 4.11 shows the possible entries for this field. For example, for goods issue for outbound delivery, assign **G** as the consumption posting indicator as it is a planned withdrawal of stock with reference to an outbound delivery.

Consumption Posting	Description
Blank	No consumption update
G	Planned withdrawal (total consumption)

Table 4.11 Consumption Posting Values

Consumption Posting	Description
R	Planned, if referenced to a reservation; otherwise unplanned
U	Unplanned withdrawal (unplanned consumption)

Table 4.11 Consumption Posting Values (Cont.)

- **Check SLExpir.date (check best-before/production date)**
  This field controls the validation of the shelf-life expiration date and production date for the goods movement transaction. Table 4.12 shows the possible entries for this field. For a goods issue for outbound delivery, the shelf-life expiration date must not be checked; in that case, maintain a blank value for the movement type.

Consumption Posting	Description
Blank	No check
1	Enter and check
2	Enter only
3	No check at goods issue

Table 4.12 Check Best-Before Date/Production Date: Possible Values

- **RevGR despite IR (reversal of goods receipt allowed despite invoice)**
  This corresponds to the procurement scenario in which purchasing of the materials is done with reference to a purchase order. This indicator controls if a reversal of the goods receipt can be posted without reversing an invoice reconciliation document (created using Transaction MIRO). This is not recommended as it creates an imbalance between the goods receipt and invoice receipt.

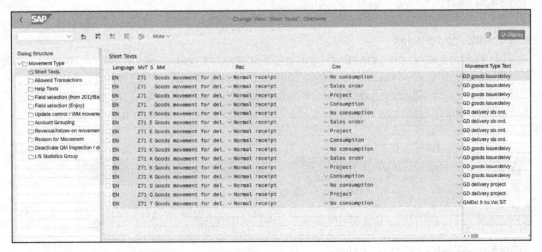

Figure 4.33 Maintain Movement Type Description.

Navigating through the **Dialog Structure,** the next step in the configuration of a new movement type is to define the description in different languages. Typically, descriptions are maintained in English and German. Figure 4.33 shows the descriptions maintained for movement type **Z71**.

The next step in the configuration of a new movement type is to assign all allowed transaction codes to the movement type. Figure 4.34 shows the allowed transaction codes for custom movement type **Z71**. In this step, assign all relevant transaction codes for create, change, and display transactions to the movement types. If you have a custom transaction code to create and process an outbound delivery, assign those transactions in this step.

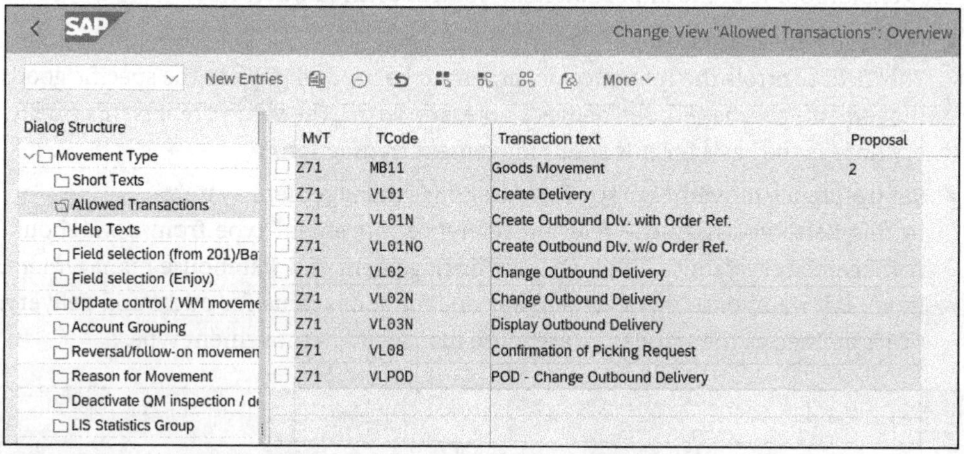

Figure 4.34 Maintain Allowed Transactions for Movement Types

The next important step in the configuration of a new movement type is to maintain field selections for the goods movement transaction corresponding to the movement type. Figure 4.35 shows the material document field names assigned to movement type **Z71**.

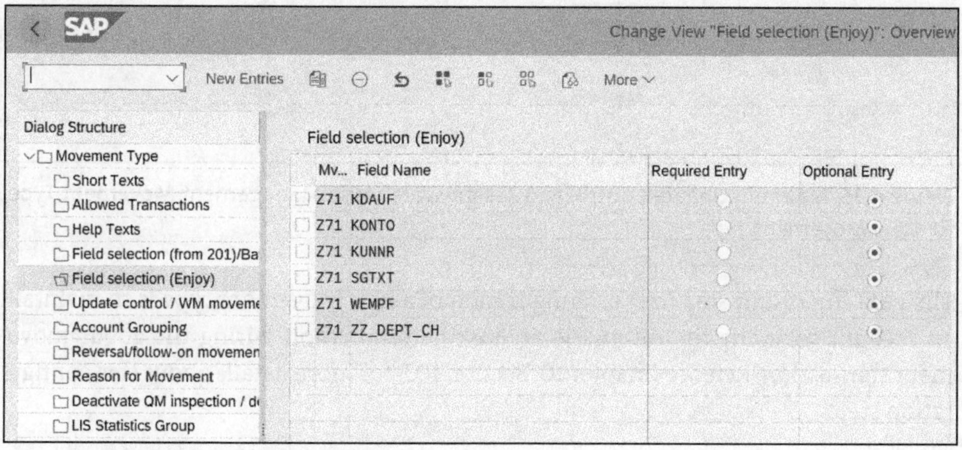

Figure 4.35 Maintain Field Selection for Movement Type

If you create a new custom field at the material document level and it is relevant for the goods movement transaction corresponding to the movement type, maintain it in this configuration step. If the field selection is a required entry during the specific goods movement, set the **Required Entry** radio button; otherwise, set the **Optional Entry** radio button.

The next important step in the configuration of a new movement type is to maintain quantity update and value update controls for the goods movement transaction, including the transaction involving special tasks. Figure 4.36 shows the update control (**Val** [value control] and **Qua** [quantity control]), movement indicator (**Mvt**), and warehouse management movement type reference (**Ref**) assigned to movement type **Z71**. Let's explore some of the important fields for this configuration step:

- **Mvt (movement indicator)**
  This field controls the reference document to be used to process the specific goods movement: purchase order, delivery note, and so on. The value **L** represents a delivery note as the basis for this goods movement transaction.

- **Ref (reference movement type for warehouse management system)**
  In this field, we assign the relevant reference movement type from a warehouse management system to link inventory management and warehouse management. In an SAP warehouse management system, for goods issue for outbound delivery, **601** is defined as the standard warehouse management movement type.

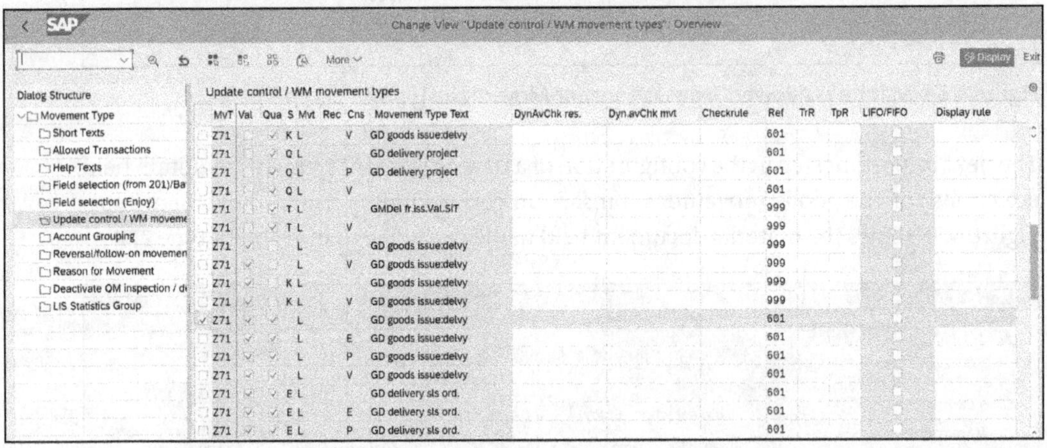

Figure 4.36 Maintain Update Control and Assign Warehouse Management Movement Type for New Movement Type

The next important step in the configuration of a new movement type is to maintain an account grouping for automatic account determination during the goods movement transaction. Refer to Chapter 10, Section 10.3 for more details about this configuration step.

The next important step in the configuration of a new movement type is to assign a reversal or follow-on movement type. Figure 4.37 shows reversal movement type **602** assigned to the new custom movement type **Z71**.

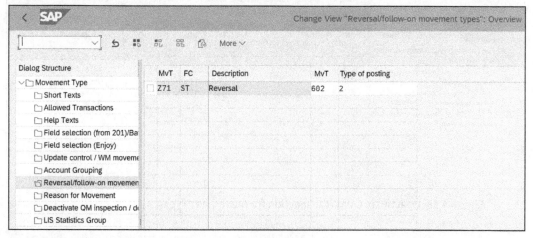

Figure 4.37 Maintain Reversal or Follow-On Movement Types

The **Type of posting** field controls if the quantity for the reversal or follow on movement can be entered manually or not. The possible values are shown in Table 4.13. If you assign a posting value of **2**, then during the reversal posting, the quantity cannot be entered manually.

Type of Posting	Description
Blank	Neither return delivery nor reversal
1	Return delivery (quantity can be entered; no generation of match code)
2	Reversal (quantity cannot be entered; generation of match code)
3	Transfer posting (quantity can be entered; no generation of match code)

Table 4.13 Type of Posting

The next important step in the configuration of a new movement type is to activate or deactivate the quality inspection for the goods movement. Figure 4.38 shows the configuration for activating or deactivating the quality inspection for the movement type. If the **QM not active** indicator is set, quality inspection will be deactivated, and the system will not create any inspection lots.

Because this custom movement type is for goods issue for outbound delivery to fulfill a customer order, assigning it to the relevant schedule line category in the configuration will ensure the determination of the movement type automatically during the goods issue for outbound delivery. Refer to Section 4.2.3 for schedule line category configuration.

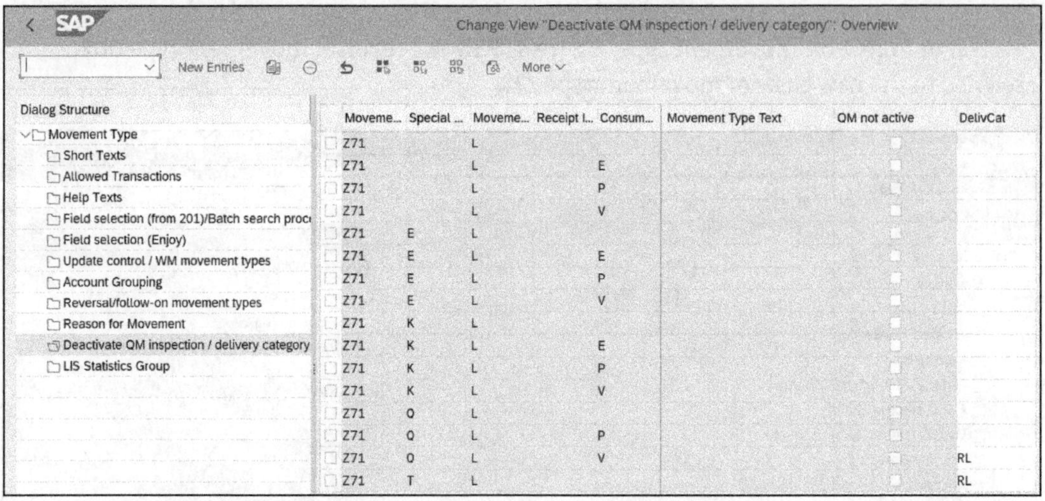

Moveme...	Special ...	Moveme...	Receipt I...	Consum...	Movement Type Text	QM not active	DelivCat
Z71		L					
Z71		L		E			
Z71		L		P			
Z71		L		V			
Z71	E	L					
Z71	E	L		E			
Z71	E	L		P			
Z71	E	L		V			
Z71	K	L					
Z71	K	L		E			
Z71	K	L		P			
Z71	K	L		V			
Z71	Q	L					
Z71	Q	L		P			
Z71	O	L		V			RL
Z71	T	L					RL

Dialog Structure: Movement Type — Short Texts — Allowed Transactions — Help Texts — Field selection (from 201)/Batch search proc — Field selection (Enjoy) — Update control / WM movement types — Account Grouping — Reversal/follow-on movement types — Reason for Movement — Deactivate QM inspection / delivery category — LIS Statistics Group

Figure 4.38 Deactivate Quality Inspection for Movement Types

4.4   Third-Party Order Processing

The third-party process, also called the trading process, is widely used by organizations that sell trading goods. In simple terms, the customer orders will be fulfilled by a third party instead of the seller who received the customer order. Here, the supplier of the trading goods is referred to as a third party. The seller who received the customer order in turn places a purchase order to a third-party supplier to fulfill the customer order. The third-party supplier will ship the goods directly to the customer and bill/invoice the entity that submitted the purchase order. Once the goods are delivered to the customer, the seller bills the customer. Some organizations also sell finished goods to their customers via a third-party process.

Let's dive deep into the third-party process and understand the integration points between materials management and sales and distribution.

4.4.1   Integrated Third-Party Process

This process is a testimony to the tight integration between sales and distribution and materials management. Table 4.14 shows the key subprocesses impacted.

Area	Processes Impacted	Description
Sales and distribution	Sales order processing	This is one of the core processes in sales and distribution that focuses on creating and managing customer orders.

Table 4.14 Process Impact

Area	Processes Impacted	Description
Sales and distribution (Cont.)	Create and manage customer pricing conditions	This process supports sales order processing by creating and maintaining pricing, discounts, freight, and tax conditions as well as scale pricing.
	Simulate and calculate order pricing	This process supports the simulation and calculation of sales order pricing.
	Create and manage customer invoice	This process deals with creating and managing customer invoices/billing as well as transmitting billing documents to customers.
Materials management	Sourcing materials and services	This process supports the sourcing of a best source of supply to procure trading goods.
	Create and maintain purchasing master data	This process supports the creation and maintenance of purchasing master data such as source lists and PIRs to support the purchasing of trading goods from a third-party supplier.
	Create and maintain purchasing documents	This process focuses on creation of purchasing documents such as purchase requisitions and purchase orders to procure trading goods from a third-party vendor.
	Create and manage outline agreements	This process supports the creation of outline agreements such as contracts with a third-party supplier to procure trading goods for the third-party process.
	Supplier invoice management	This process supports the processing of incoming invoices from the third-party supplier.

Table 4.14 Process Impact (Cont.)

Figure 4.39 shows all the steps of third-party processing. One of the notable differences between the regular sales order processing and the third-party drop-ship process is that the seller doesn't have to maintain the inventory of the goods. The fulfillment of the customer order is driven by materials management functions such as purchasing and logistics invoice verification. This also helps the seller to identify best and multiple sources of supply for the materials.

Let's walk through these process steps:

1. **Processing of sales order for trading goods**
   Customer orders containing trading goods or other third-party material will be converted into a sales order in SAP S/4HANA using Transaction VA01. Creation of a sales order can be fully automated via EDI. (Refer to Chapter 12, Section 12.1 for EDI details.) A sales order can have one or more line items depending on the customer order. The line items of the sales order can be a mix of third-party trading materials and other materials that are directly fulfilled by the seller.

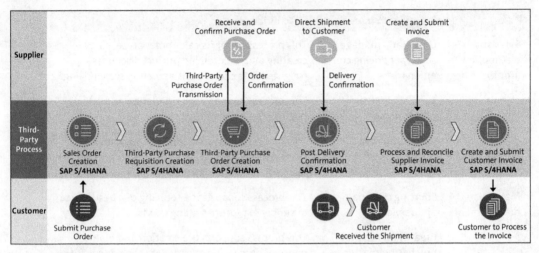

Figure 4.39 Third-Party Process

2. **Creation of third-party purchase requisition**

   SAP S/4HANA triggers the creation of third-party purchase requisitions from the sales order automatically with the item category reference set as *third-party* and the account assignment category set as *sales order*. Section 4.4.3 explains the configuration settings behind the triggering of automatic purchase requisition creation in SAP S/4HANA. The delivery address for the items in the purchase requisition will have the customer address and the customer ID (business partner ID) of the customer stored in SAP S/4HANA, which are defaulted into the **Delivery Address** tab at the item level of the purchase requisition. Source of supply determination happens during the creation of a purchase requisition from a source list (purchasing master data).

3. **Creation of third-party purchase order**

   A third-party purchase requisition will be converted into a third-party purchase order manually by the purchaser using Transaction ME21N or automatically in SAP S/4HANA. The line item details are defaulted into the purchase order, including the source of supply, unit cost, and delivery address. The unit price for the material is defaulted from the PIR (purchasing master data). The purchase order will be transmitted to the third-party supplier based on the output method: email, EDI, or electronic transmission to the supplier on SAP Business Network. Output determination configuration in SAP S/4HANA drives the purchase order output mechanism. Depending on the confirmation control key, order confirmation from the supplier may be required, and the confirmation control key is defaulted from the PIR.

4. **Delivery confirmation and goods receipt**

   The supplier ships the goods directly to the customer and sends a delivery or shipping notification to the seller. It's best practice to send a shipping notification to the seller, with the third-party supplier notifying them of the materials, delivery date, serial number, batch, and shipment tracking information. Upon the receipt of the

ASN, the seller performs goods receipt in the SAP S/4HANA system using Transaction MIGO. Inventory will not be updated upon goods receipt; instead, the stock will be received against the sales order account assignment.

5. **Process and reconcile supplier invoice**

   The supplier sends an invoice to the seller against the fulfilled purchase order via email (paper invoice), via EDI, or electronically via SAP Business Network. The supplier invoice document will be posted by the accounts payable department of the seller manually in the SAP S/4HANA system using Transaction MIRO for a paper invoice. Automatic posting and reconciliation of the invoice happens if an electronic invoice is received by the seller from the supplier. Invoice reconciliation is a process of matching a purchase order, goods receipt, and supplier invoice to verify any discrepancies with a purchase order item price and purchase order item quantity and received quantity. Taxes will be reconciled against the vendor charged tax and the SAP S/4HANA internal tax engine or the external tax engine's calculated tax. Once the invoice is fully reconciled, it will be free for payment processing in the system. The next scheduled payment run will pick up this fully reconciled invoice and post the payment to the supplier based on the preferred payment method of the supplier. Accounts payable personnel can manually process the payment against the fully reconciled invoice as well.

6. **Create and submit customer invoice (customer billing)**

   This is the final step in the third-party process. The seller submits a billing document against the sales order, also called the customer invoice, to the customer. This step can't be completed until after the goods receipt against the third-party purchase order has been completed or the supplier invoice was received and processed. The item category and copy control configuration in sales and distribution will drive the billing relevance depending on whether the billing document can be created after the goods receipt was posted or the invoice receipt was posted against the third-party purchase order as well as whether the item quantities were copied from the goods receipt or from the supplier invoice document. Section 4.4.3 explains the configuration of item categories. Billing documents are transmitted to the customer via email or EDI.

> **Note**
>
> Step 4 of the third-party process is optional. This process can be completed without performing goods receipt against the third-party purchase order as there is no inventory movement within the selling organization. The item category and copy setting configuration drives this step, as explained in step 6.

The third-party drop-ship process impacts user roles from both materials management and sales and distribution within the seller's organization.

The following are the key user roles impacted within sales and distribution:

- **Sales representatives**
  Responsible for creating and maintaining a sales order against the customer order. They also collaborate with buyers to fulfill the sales order in a timely manner.

- **Accounts receivable personnel**
  Responsible for creating and maintaining customer billing document and answering any billing-related queries from the customer while ensuring accuracy and compliance with company and regulatory policies.

- **Master data management team**
  Responsible for maintaining the master data required for the third-party process from the sales side, such as the customer info record and pricing condition records.

The following are the key user roles impacted within materials management:

- **Category managers or sourcing team**
  Responsible for sourcing the best source of supply for the trading goods and other materials that are sold to the customers through the third-party process

- **Buyers/purchasers**
  Responsible for creating and maintaining third-party purchase orders with reference to the third-party purchase requisition

- **Master data management team**
  Responsible for maintaining the purchasing master data required for the third-party process from the purchasing side, such as source list and PIR

- **Accounts payable personnel**
  Responsible for receiving and processing of the supplier invoice while ensuring accuracy and compliance with company and regulatory policies

4.4.2   Master Data Requirements

The third-party process requires the following master data to be maintained in the SAP S/4HANA system:

- **Material master data**
  Maintaining the material master data views listed in Table 4.15 is mandatory for the trading goods to perform the third-party process. The other views of the material master are required based on other business processes in which the same material is being used.

Maintenance Status	Material Master Views
K	Basic data
E	Purchasing
V	Sales
B	Accounting

Table 4.15 Material Master Maintenance Status

In the **Sales: sales data 2** view of the material master, it is required to maintain the item category group as BANS for the third-party process.

- **Business partner (customer)**
  A business partner with a customer role is required for customers to perform the third-party process.

- **Business partner (vendor)**
  A business partner with a vendor role is required for third-party suppliers to perform the third-party process.

- **Source list**
  A source list needs to be created for the trading goods and the third-party supplier so that the source determination happens during the automatic purchase requisition creation.

- **PIR**
  A PIR needs to be created for the combination of trading goods and third-party supplier so that the unit price of the material and other details are defaulted into the purchase order.

4.4.3   Configuration of Third-Party Processing

This section explains the key configuration settings for performing a third-party process from end to end in SAP S/4HANA from both materials management and sales and distribution. These settings ensure smooth execution of the process and represent business requirements. Let's dive deep into the various settings and understand how the third-party process works in SAP S/4HANA.

Sales Order Type

Sales order types in SAP S/4HANA are used to distinguish between various sales transactions. Each sales order type represents a specific business transaction and requirement. There are standard sales order types available in the system and those types are further configured based on business requirements. The sales order type is part of the sales order document header and controls various aspect of sales order processing such as the transaction flow, pricing procedure determination, availability check, delivery date, and so on. Let's dive deep into some of the key settings within the sales order type configuration.

To create a new sales order document type, execute Transaction SPRO and navigate to **Sales and Distribution • Sales • Sales Documents • Sales Document Header • Define Sales Document Types**. Figure 4.40 shows the initial screen for maintaining sales order types.

To create a new sales order document type, click **New Entries** and specify a four-digit alphanumeric key and a description. You can also copy an existing sales order type and create a new one (which is recommended). To display an existing sales order type's details, select a sales order type (e.g., **OR**) and click the **Details** icon at the top or press [Ctrl]+[Shift]+[F2]. Figure 4.41 and Figure 4.42 show the details of sales order type **OR**.

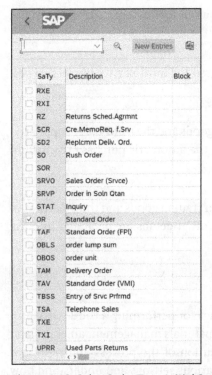

Figure 4.40 Sales Order Type Initial Screen

Let's dive deep into some of the key configuration settings for the sales order types:

- **SD Document Category**
  This setting defines the category of the sales order type. It defines and controls the functionality of the sales document such as the sales order, scheduling agreement, inquiry, returns, and so on. For the third-party order process, set the **SD Document Category** to **C**—that is, sales order.

- **Number systems**
  This section of the sales order type configuration defines the internal and external number ranges for the sales order document. The system automatically chooses the next available number in the internal number range while creating the sales order document in SAP S/4HANA. External numbers can be manually entered and must be within the defined external number range.

  To arrive at the number range configuration step, execute Transaction SPRO and navigate to **Sales and Distribution • Sales • Sales Documents • Sales Document Header • Define Number Ranges for Sales Documents**. You can arrive at this configuration step directly by executing Transaction VNO1. In this configuration step, both internal and external number ranges can be defined against number range object RV_BELEG.

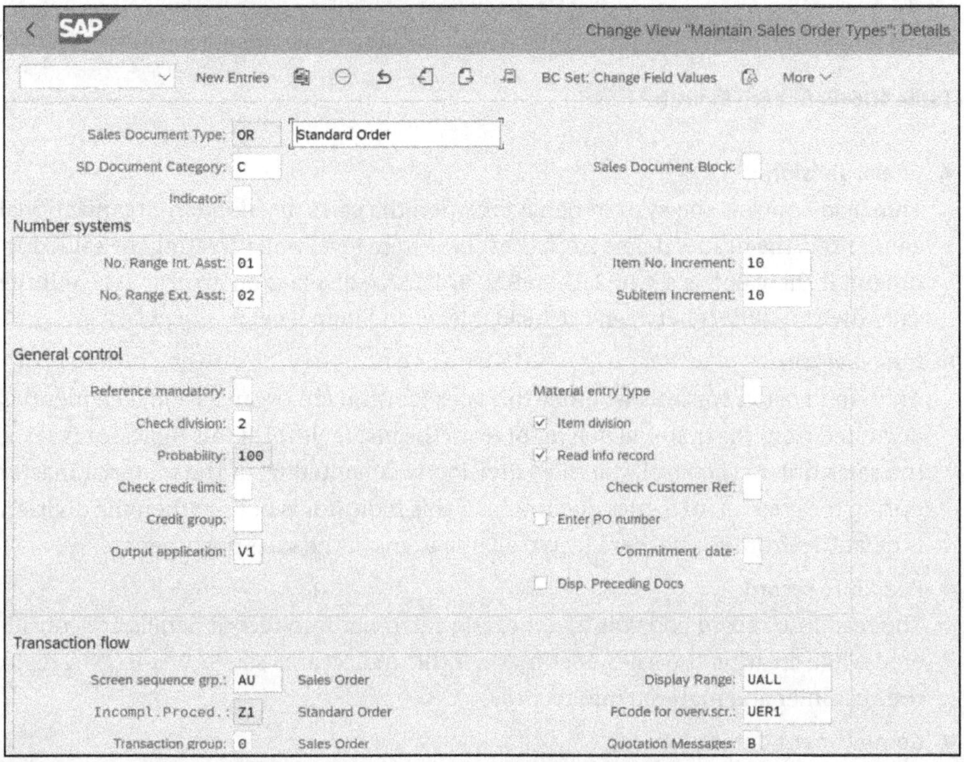

Figure 4.41 Sales Order Type: 1 of 2

- **Reference mandatory**
  This field controls whether another reference document is required to create the sales document in SAP S/4HANA. Table 4.16 shows the different values that can be set in this field according to the business requirements. For example, if the value in this field is set as **A**, then the sales order document can be created with reference to an inquiry only.

Reference Mandatory Indicator	Description
Blank	No reference required
A	With reference to an inquiry
B	With reference to a quotation
C	With reference to a sales order
E	Scheduling agreement reference

Table 4.16 Reference Indicator

Reference Mandatory Indicator	Description
G	With reference to a quantity contract
M	With reference to a billing document

Table 4.16 Reference Indicator (Cont.)

- **Check division**
  This field controls the system behavior when the sales division (an organizational element) in the header defers from that at the item level while creating the sales document. If the value is set to **2**, then SAP S/4HANA displays an error message if the sales division differs between the header level and item level.

- **Item division**
  An item division controls whether the sales division (an organizational element) is defaulted from the material master or from the header level. If this indicator is set at the sales order type level, the sales division is defaulted from the material master during the creation of a sales document. If this indicator is not set, then the division is defaulted from the header level to all line items of the sales document.

- **Read info record**
  The read info record controls whether the customer info record details are considered while creating the sales document. If this indicator is set, SAP S/4HANA reads the customer and material info records.

- **Commitment date**
  Commitment dates for the schedule lines are calculated based on the value set in this field in the sales order type configuration in the respective sales document. For a standard order type, used in the third-party process, commitment dates typically are not maintained. Table 4.17 shows the different values that can be maintained in this field.

Commitment Date	Description
Blank	Do not maintain commitment date
A	Consider agreed delivery time only
B	First confirmation date
C	Best confirmation date, saved up until now

Table 4.17 Commitment Date Indicators for Schedule Lines

- **Screen sequence grp**
  The screen sequence group controls the screen sequence based on the sales order type selected during the creation of sales document. The system displays different screens

for different types of sales documents based on this setting in the respective sales order type. For example: If the screen sequence group value is set as **AU**, SAP S/4HANA displays the screens relevant for sales order creation.

- **Incompl. Proced.**
  This field typically controls the mandatory information being entered or not during the sales document creation. It creates necessary logs if the sales document is incomplete. For example, if the incomplete procedure value is set to **11**, SAP S/4HANA checks for the incompleteness of the sales order document.

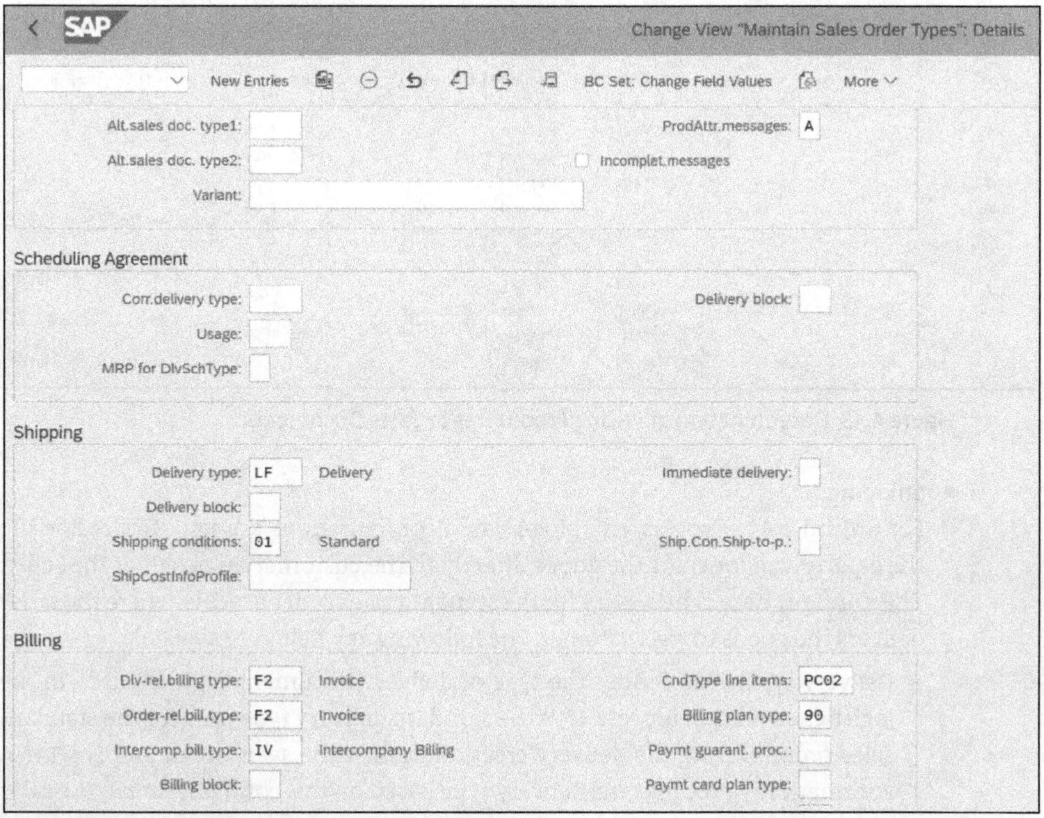

Figure 4.42 Sales Order Type: 2 of 2

- **Doc. Pricing Proc. (document pricing procedure)**
  This field (not shown; you can scroll down to view this field) defines the classification of the sales document for the determination of the pricing procedure. Figure 4.43 shows how the value maintained in this field is used in the determination of the pricing procedure for the respective sales document.

To arrive at the pricing procedure determination for sales documents configuration step, execute Transaction SPRO and navigate to **Sales and Distribution • Basic Functions • Pricing • Pricing Control • Define and Assign Pricing Procedures**.

Change View "Det. of Pricing Procedures in Sales Docs.": Overview

New Entries BC Set: Change Field Values More

Sales Organiza...	Distribut...	Division	Doc....	Cust...	Pricing Proced...	Pricing Procedure	Con...	Condition Type for Fast Entry
MJ00	MJ	MJ			RVAA01	ndard	PR00	Price
MJ00	MJ	MJ	A		RVAA01	Standard	PR00	Price
MJ00	MJ	MJ	A	01	RVAA01	Standard	PR00	Price
MJ00	MJ	MJ	A	02	RVAA01	Standard	PR00	Price
MJ00	MJ	MJ	A	I	ZCAA01	Intercompany Billing		
MJ00	MJ	MJ	O1		RVAA01	Standard	PR00	Price
MJ00	MJ	MJ	S1		RVAA01	Standard	PR00	Price
MJ00	MJ	MJ	Y1		RVAA01	Standard	PR00	Price
MJ00	MJ	MJ	Y1	01	RVAA01	Standard	PR00	Price
MJ01	MJ	MJ	A		ZVAA01	Standard-MJ00	PR00	Price
MJ01	MJ	MJ	A	I	ZCAA01	Intercompany Billing		
MNG1	M1	M1	A	01	RVAA01	Standard	PR00	Price
MNG1	M1	M1	A	02	RVAA01	Standard	PR00	Price
MNG2	M2	M2	A	01	RVAA01	Standard	NB00	Price
MNG2	M2	M2	A	02	RVAA01	Standard	NB00	Price
MSOS	MS	MS	A	01	RVAA01	Standard	PR00	Price
NL01	10	10	A	01	ZJASM1	Standard	PR00	Price
RAR	RA	R1			RVAA01	Standard		
UKSO	L1	L1	A	01	ZCWSPP	Computer world standar		
US01	10	10	A	01	ZJASM1	Standard	PR00	Price
US01	92	10	A	01	ZJASM1	Standard		

Figure 4.43 Determination of Pricing Procedures in Sales Documents

- **Shipping**
  For a third-party process, shipping-related configurations are not required as the third-party vendor ships the goods directly to the customer on behalf of the seller. This configuration is necessary for the standard sales order process, where the seller delivers the goods to the customer. The following key fields are available:
  - **Delivery type**: This defines the type of delivery document typically used in the logistics execution process. **LF** is the standard delivery type used for the standard sales order. During the delivery creation against the sales order, SAP S/4HANA automatically proposes the delivery type based on this configuration in the sales order type.
  - **Shipping conditions**: This defines the shipping condition applicable to the sales order type. Here, **01** is the standard shipping condition used for the standard sales order. During the creation of a sales document, SAP S/4HANA automatically defaults the shipping condition based on this configuration in the sales order type.
- **Order-rel.bill.type (order-related billing type)**
  This is an important setting for the third-party process. It defines the type of billing document such as an invoice, credit memo, and the like. For the standard sales order type, define the value as **F2** so that the SAP S/4HANA system automatically proposes

this billing type during the creation of a billing document against the sales document.

Item Category

Item categories are an essential part of the sales document and are used to distinguish between different types of sales document items such as standard items, free of charge items, returnable items, packaging items, third-party sales, service items, and so on. They play a crucial role in determining how the specific sales documents are processed in SAP S/4HANA. A sales document can have the same or different item categories. Each item category in the system provides specific control mechanisms and drives the determination of the pricing relevance, schedule line, outbound delivery relevance, and so on.

To arrive at this configuration step, execute Transaction SPRO and navigate to **Sales and Distribution • Sales • Sales Documents • Sales Document Item • Define Item Categories**. Executing this configuration node displays the initial screen of item categories. Select item category **TAS** and click the **Details** button to display the important settings, as shown in Figure 4.44.

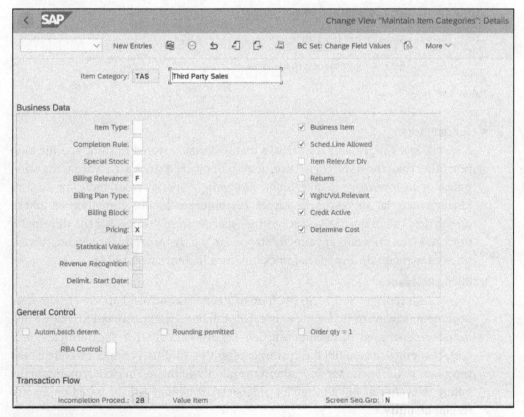

Figure 4.44 Maintain Item Category

TAS is the standard item category for third-party sales. To master this process, a deep understanding of item category **TAS** is essential. Let's dive deep into some of the important item category configurations and find out how this item category controls the third-party process:

- **Item Type**
  The item type configuration setting further distinguishes the type of the item in the sales document. Based on this setting, SAP S/4HANA processes the item in a sales document differently. For example, the delivery might be required for the standard item type rather than for a text item. Similarly, pricing could be different for different item types. Table 4.18 shows different values that can be assigned to the item type field.

Item Type	Description
Blank	Standard item
A	Value item
B	Text item
C	Packing item (will be generated)
D	Material not relevant
E	Packaging item (external)

Table 4.18 Item Types

- **Special Stock**
  A special stock represents a stock of a material that is stored in some specific location other than the unrestricted use, quality, or blocked stock areas within the warehouse or in inventory management. Assigning a special stock indicator such as pipeline material, subcontracting stock, customer stock, consignment stock, and so on restricts the usage of the stock to that specific special stock. For the third-party process, as the materials are directly shipped to the customer by the vendor, there is no need to maintain any special stock indicator in item category **TAS**.

- **Billing Relevance**
  This is an important setting in the item category. It controls if an item in the sales document is relevant for billing or not. This can be further restricted to create a billing document based on an outbound delivery or a sales order. In the third-party process, best practice is to bill the customer after the third-party vendor has delivered the goods to the customer and submit the invoice to the seller. Hence, maintain the billing relevance as **F: Relevant for order-related billing—Status based on invoice receipt quantity**.

- **Billing Plan Type**

  This field controls how SAP S/4HANA determines the billing date and amount to be billed for the goods delivered or service performed by the seller to the customer. The following are the billing plan types:

  - In periodic billing, a customer is billed periodically (weekly, monthly, quarterly etc.) for the entire amount.

  - In milestone billing, a customer is billed based on milestones. The billing date and time vary depending on the milestone. This type of billing is seen in service industries, such as consulting, auditing, and so on.

- **Billing Block**

  A billing block prevents an item in the sales document from creating customer billing. If the billing block is set at the item level, then the respective item in the sales document will be blocked for billing/invoicing.

- **Pricing**

  The pricing relevance indicator controls if an item category is relevant for pricing or not. For third-party order processing, the item category is relevant for standard pricing.

- **Sched.Line Allowed**

  This indicator controls if the item category is relevant for delivery or not. All item categories that are relevant for delivery require schedule lines. For the third-party process, even though the delivery is made by the third party, creation of the schedule line is always required against the item. Through the schedule lines, SAP S/4HANA triggers the creation of a purchase requisition against the third-party supplier.

- **Item Relev.for Dlv**

  This indicator controls if the item in the sales document is to be considered during delivery creation or not. In the third-party process, the seller doesn't create/process a delivery, and the goods will be shipped directly by the vendor to the customer. It is not required to set this indicator for item category **TAS**.

- **General Control**

  For the third-party item category **TAS**, the following indicators are not required to be set:

  - **Autom.batch determ.**: If this indicator is set for the item category, SAP S/4HANA determines the batches automatically based on customer requirements.

  - **Rounding permitted**: If this indicator is set for the item category, SAP S/4HANA rounds off the quantities at the item level of the sales document based on the rounding profile.

  - **Order qty = 1**: If this indicator is set for the item category, SAP S/4HANA restricts the sales order item quantity to one unit.

Item Category Determination

The item category group plays an important role in the automatic determination of item categories based on the material and the sales order type. Before we discuss item category determination during sales order creation, let's consider the item category group.

The item category group is a logical grouping similar to a material solely for the purpose of determination of item categories within the sales document. Figure 4.45 shows that item category group **BANS** was assigned to material **MATERIAL-3** in the **Sales: sales org. 2** view of the material master.

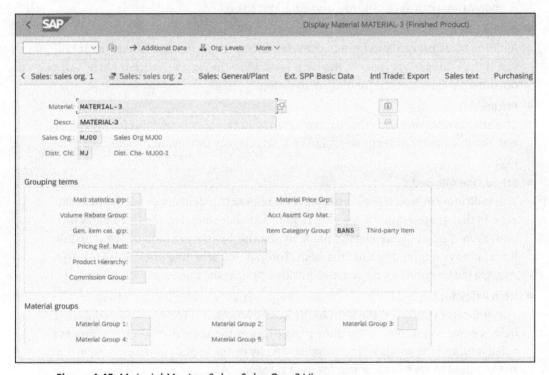

Figure 4.45 Material Master: Sales: Sales Org. 2 View

Item category groups are predefined in the SAP S/4HANA system. Additional ones can be defined in the system based on business requirements. To arrive at this configuration step, execute Transaction SPRO and navigate to **Sales and Distribution • Sales • Sales Documents • Sales Document Item • Define Item Category Groups**.

Note

BANS is the standard item category group defined in the system for the third-party process.

Determination of item categories for the sales order type happens through the configuration. To arrive at this configuration step, execute Transaction SPRO and navigate to

Sales and Distribution • Sales • Sales Documents • Sales Document Item • Assign Item Categories. Figure 4.46 shows that item category **TAS** is assigned to the standard order type **OR** and item category group **BANS**.

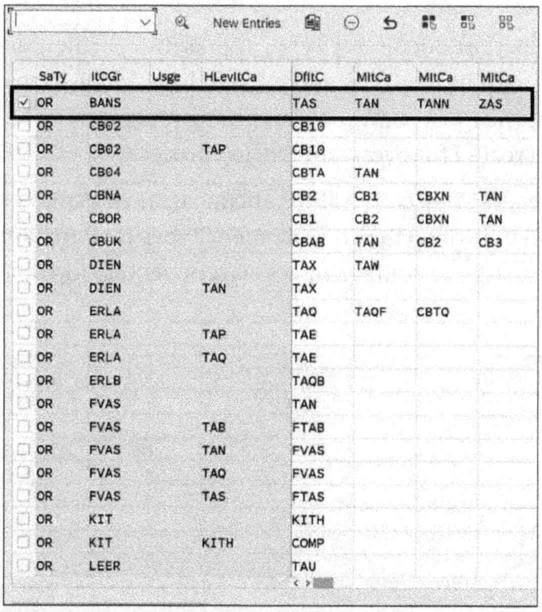

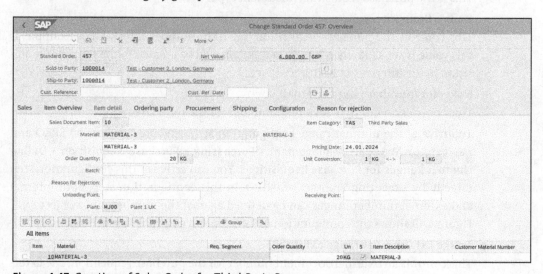

	SaTy	ItCGr	Usge	HLevItCa	DfItC	MltCa	MltCa	MltCa
✓	OR	BANS			TAS	TAN	TANN	ZAS
	OR	CB02			CB10			
	OR	CB02	TAP		CB10			
	OR	CB04			CBTA	TAN		
	OR	CBNA			CB2	CB1	CBXN	TAN
	OR	CBOR			CB1	CB2	CBXN	TAN
	OR	CBUK			CBAB	TAN	CB2	CB3
	OR	DIEN			TAX	TAW		
	OR	DIEN	TAN		TAX			
	OR	ERLA			TAQ	TAQF	CBTQ	
	OR	ERLA	TAP		TAE			
	OR	ERLA	TAQ		TAE			
	OR	ERLB			TAQB			
	OR	FVAS			TAN			
	OR	FVAS	TAB		FTAB			
	OR	FVAS	TAN		FVAS			
	OR	FVAS	TAQ		FVAS			
	OR	FVAS	TAS		FTAS			
	OR	KIT			KITH			
	OR	KIT	KITH		COMP			
	OR	LEER			TAU			

Figure 4.46 Item Category Determination

Figure 4.47 shows how the system determines the item category during the creation of a sales order for material **MATERIAL-3**. The system automatically defaults the item category **TAS** based on the sales order type **OR** and the item category group **BANS**. The system reads the item category group **BANS** from the material master.

Figure 4.47 Creation of Sales Order for Third-Party Process

Purchase Requisition

Before we configure the schedule line category for the third-party process, it is vital to master the purchasing document type settings. Let's dive deep into the purchase requisition configuration in SAP S/4HANA first.

Like sales order types, purchase requisition document types represent specific purchasing processes based on business requirements. Standard purchase requisitions, subcontracting, and stock transfers are the most common purchasing processes. A specific purchase requisition type can be created for every purchasing process.

To arrive at this configuration step, execute Transaction SPRO and navigate to **Materials Management • Purchasing • Purchase Requisition • Define Document Types for Purchase Requisitions**. Figure 4.48 shows the first step of configuring a purchase requisition document type.

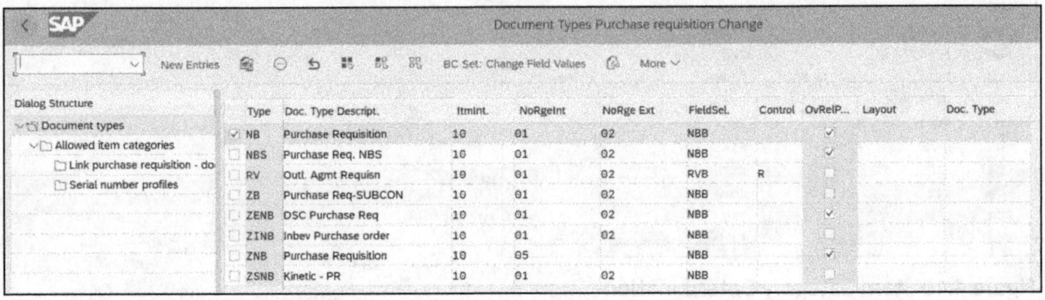

Figure 4.48 Define Purchase Requisition Document Type: 1 of 3

The following are the details of the screen fields seen here:

- **Type (document type)**
  This is the purchase requisition document type.

- **ItmInt. (item interval)**
  This controls the purchase requisition line item number interval. For example, if this value is set as **10**, then line item numbers will be created in the purchase requisition in the intervals of 10 (10, 20, 30, etc.).

- **NoRgeInt (number range internal)**
  This controls an internal document number for the purchase requisition.

  To arrive at the number range configuration step, execute Transaction SPRO and navigate to **Materials Management • Purchasing • Purchase Requisition • Define Number Ranges for Purchase Requisitions**. You can arrive at this configuration step directly by executing Transaction SNRO. In this configuration step, both internal and external number ranges can be defined against the number range object BANF. Figure 4.49 shows the configuration of internal and external number ranges.

- **NoRge Ext (number range external)**
  This controls an external document number for the purchase requisition. The **External** indicator will be set for an external number range.

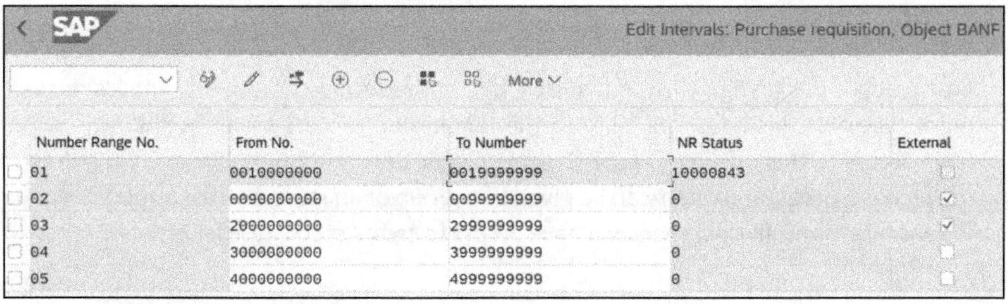

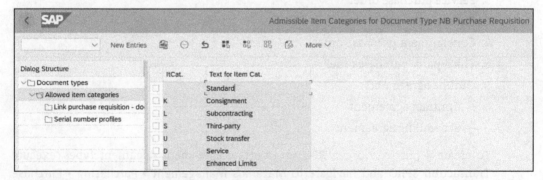

Figure 4.49 Define Purchase Requisition Number Range

- **FieldSel. (field selection key)**
  The field selection key represents a layout of the purchase requisition, including mandatory, optional, and display fields. **NBB** is the standard screen layout for purchase requisition documents.

- **Control**
  This field categorizes different types of purchase requisitions such as standard, stock transfer, subcontracting, and so on.

- **OvRelP (overall release of purchase requisitions)**
  This indicator controls the release of a purchase requisition. If this indicator is set, all line items of the purchase requisition will be released at once.

The next step in the purchase requisition document type configuration is to assign the allowed item categories to the document type. To arrive at this step, select a document type from the document type list and double-click **Allowed item categories** from the **Dialog Structure**. Figure 4.50 shows the allowed item categories for purchase requisition document type **NB**.

Figure 4.50 Define Purchase Requisition Document Type: 2 of 3

Like item categories in sales and distribution, item categories in purchasing distinguish between different types of items in the purchasing documents. To review the item categories for purchasing, look back to Table 1.1 in Chapter 1, Section 1.1.1.

> **Note**
>
> Best practice is to assign all standard item categories for purchasing to standard document type **NB**. This adds flexibility to the purchasing process for buyers. Buyers will have access to this document type. Create specific document types for every purchasing process; necessary access can be given to a group or an individual for a specific document type. Typically, planners will have access to create stock transfer orders.

The next step in the purchase requisition document type configuration is to link the combination of allowed item categories of the purchase requisition document type to the combination of purchase order document type and allowed item category. This configuration controls the creation of purchase orders, requests for quotation, and outline agreements with reference to a purchase requisition.

Purchase Order

Now that we're here, mastering the purchase order document type is necessary to understand the purchasing process better. Let's dive deep into the purchase order document types before completing the **Link purchase requisition—document type** step. Refer to Chapter 1, Section 1.1.2 for an overview of the purchasing process. The following are the different purchasing documents that can be created with reference to a purchase requisition:

1. Request for quotation (RFQ)
2. Standard purchase order
3. Framework order
4. Stock transfer order
5. Service purchase order
6. Subcontract order
7. Consignment purchase order
8. Third-party purchase order
9. Outline agreements:
   - Contract agreement
   - Scheduling agreement

To create a purchase order, RFQ, and outline agreement document types, execute Transaction SPRO and navigate to **Materials Management • Purchasing • Purchase Order • Define Document Types for Purchase Orders**. Figure 4.51 shows the first step for configuration of a purchase requisition document type. The following are the details of the screen fields:

- **Type**
  This is the purchase order, RFQ, or outline agreement document type.

- **ItmInt (item interval)**

  This controls the purchase document line item number interval. For example, if this value is set to **10**, then line item numbers will be created in the purchase order in the intervals of 10 (10, 20, 30, etc.).

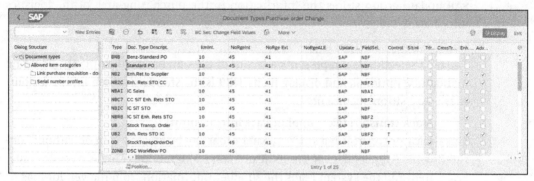

Figure 4.51 Define Purchase Order Document Type: 1 of 3

- **NoRgeInt (number range internal)**

  This controls the internal document number for the purchasing documents (RFQ, purchase order, and outline agreements).

  To arrive at the number range configuration step, execute Transaction SPRO and navigate to **Materials Management • Purchasing • Purchase Order • Define Number Ranges for Purchasing Documents**. You can arrive at this configuration step directly by executing Transaction SNRO. In this configuration step, both internal and external number ranges can be defined against number range object EINKBELEG. Figure 4.52 shows the configuration of internal and external number ranges.

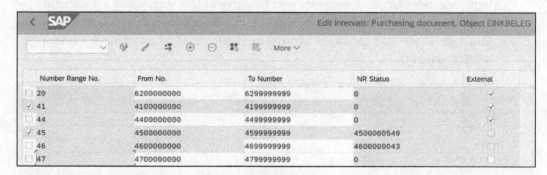

Figure 4.52 Define Purchasing Documents Number Range

- **NoRge Ext (number range external)**

  This controls an external document number for the purchasing documents. The **External** indicator will be set for an external number range, as shown in Figure 4.52.

- **Update... (update group)**
  This is a logical element that controls how the purchasing document information is updated in the logistics information system. The logistics information system collects and stores the data from sales, purchasing, plant maintenance, and so on. The SAP update group stores the information in the purchasing area of the logistics information system.

- **FieldSel. (field selection key)**
  The field selection key represents a layout of the purchasing documents including mandatory, optional, and display fields. **NBB** is the standard screen layout for purchase requisition documents.

- **Trfr... (stock transfer: take supplier data into account)**
  This indicator controls if a stock transfer order can be created with or without considering the supplier data. If this indicator is set, a supplying plant as a vendor must be created, and the system reads the supplier information from the vendor master record.

- **Enh... (enhanced store returns)**
  This indicator controls if only an outbound delivery can be created for store returns (stock transfer order) or if both inbound and outbound deliveries are required during the store returns process. If this indicator is set, both inbound and outbound deliveries must be created during the store returns process.

- **Adv... (advanced returns management)**
  This indicator controls if advanced returns management is active for purchasing documents.

The next step in the purchasing document type configuration is to assign the allowed item categories to the document type. To arrive at this step, select a document type from the document type list and double-click **Allowed item categories** in the **Dialog Structure**. Figure 4.53 shows the allowed item categories for purchase document type **NB (Standard PO)**.

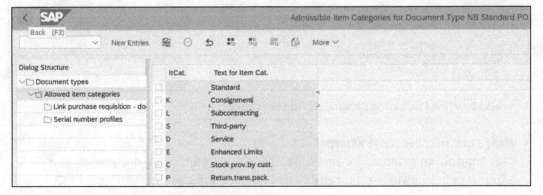

Figure 4.53 Define Purchase Order Document Type: 2 of 3

Document Types

The third step is to link the purchase requisition document type configuration and purchase order document type such that a purchase order can be created with reference to a purchase requisition. To arrive at this step, select an item category from the list of item categories and double-click **Link purchase requisition—document type** from the **Dialog Structure**. Figure 4.54 shows the linking of purchase requisition document type **NB** to purchase order document type **NB** along with the respective item categories.

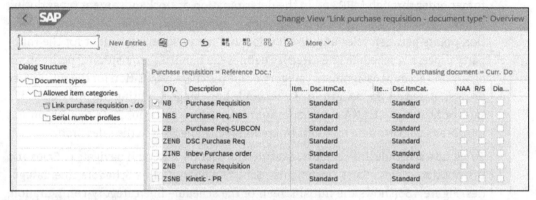

Figure 4.54 Define Purchase Order Document Type: 3 of 3

To close the loop on purchase requisition document type configuration, the third step is to link purchase order document type **NB** to the purchase requisition document type **NB** along with the respective item categories. Figure 4.55 shows the linking of purchase order document type **NB** to the purchase requisition document type **NB** along with the respective item categories.

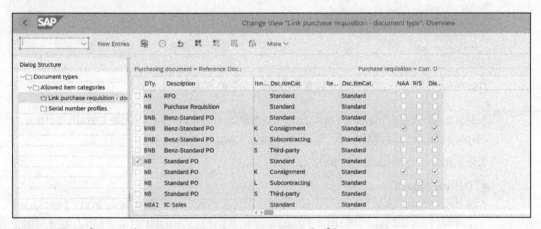

Figure 4.55 Define Purchase Requisition Document Type: 3 of 3

Schedule Line Category

Now you should understand the configuration of the purchase requisition and purchase order document types. Next, let's master the settings for the schedule line category and understand how the third-party process works in SAP S/4HANA and the integration of sales and distribution and materials management.

Schedule line categories control the delivery quantities and delivery dates of sales order items. This plays a vital role in ensuring the availability of materials against a customer order through MRP and in the determination of stock movement types during delivery processing if the item is relevant for delivery. This is one of the main integration points between sales and distribution and materials management for the third-party process. A schedule line category defines the purchase requisition type, item category, and account assignment category of the purchase requisition for the third-party process. Based on the definition of a schedule line category, during the processing of a sales order, SAP S/4HANA automatically creates a third-party purchase requisition and proposes a delivery date for items by creating schedule lines in the sales order.

To set up a schedule line category, execute Transaction SPRO and navigate to **Sales and Distribution • Sales • Sales Documents • Schedule lines • Define Schedule Line Categories**. Figure 4.56 shows the initial screen of the schedule line category configuration. The system displays a predefined, standard list of schedule line categories with alphanumeric, two-character codes.

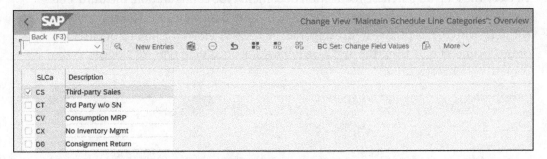

Figure 4.56 Maintain Schedule Line Category: Initial Screen

Select a schedule line category from the initial screen and click the **Details** button to display the details of the configuration, as shown in Figure 4.57.

The important fields are as follows:

- **Delivery Block**
  A delivery block set at the schedule line category level blocks the delivery processing of that specific schedule line. For example, if the system determines multiple schedule line categories in the sales order, delivery processing will be blocked for those schedule lines where the delivery block was set. For the third-party process, if a delivery block is set at the schedule line category level, then the system doesn't trigger the creation of a third-party purchase requisition.

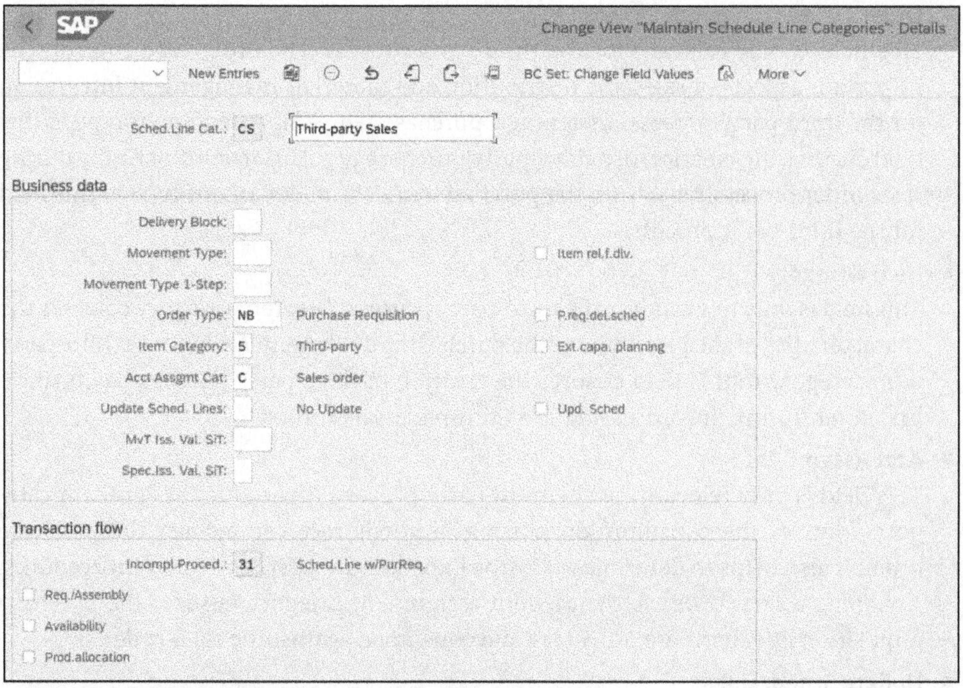

Figure 4.57 Maintain Schedule Line Category

- **Movement Type**

  The movement type in the schedule line category controls the goods movement during delivery processing. Movement type 601 is used for most of schedule line categories that are relevant for customer delivery. However, for the third-party process, direct delivery happens from a third party, so assigning a movement type in the schedule line category is not required.

- **Movement Type 1-Step**

  This controls the one-step procedure of a goods movement involving a supplying location and a receiving location; that is, the goods issue from the supplying location and goods receipt at the receiving location happen at the same time, and SAP S/4HANA creates a single material document for such goods movements involving a movement type with a one-step procedure. These movement types for the one-step procedure are configured in the schedule line categories meant for the stock transfer process:

  - **NL:** This is the standard schedule line category for the intercompany stock transfer process. Movement type 645 is assigned in the **Movement Type 1-Step** field.

  - **NN:** This is the standard schedule line category for the intracompany stock transfer process. Movement type 647 is assigned in the **Movement Type 1-Step** field.

  However, for the third-party process, a direct delivery happens from a third party, so assigning a movement type in this field for the schedule line category is not required.

- **Order Type**

  This field is only relevant for the third-party process. The order type controls the important aspects of sales and distribution and materials management integration for the third-party process. Assigning a purchase requisition document type in this field ensures the creation of a third-party purchase requisition automatically during sales order processing. Assign the purchase requisition document type in this field for the third-party process.

- **Item Category**

  This field is only relevant for the third-party process. The item category controls the characteristics of the line items of the purchasing document. Assigning a third-party item category to this field ensures the creation of third-party purchase requisition line items during the processing of a third-party sales order.

- **Acct Assgmt Cat**

  This field is only relevant for the third-party process. An account assignment category controls the consumption posting of goods received against the purchase order. It also helps to determine how the expenses are posted to different accounts. Assigning a sales order as the account assignment category ensures the purchase requisition line items are processed and consumed against the sales order.

- **Update Sched. Lines**

  This field is also relevant for the third-party process (see the possible values in Table 4.19). The system updates the schedule lines in the sales order upon posting the goods receipts against the third-party requisition if the value **1** was assigned to this field. If you assign the value **2**, the system updates the schedule lines upon receiving the shipping notification as well as when the goods receipt is posted. No updates to schedule lines will be made if there is no value assigned.

Update Schedule Lines	Description
Blank	No update
1	Only for goods receipt
2	For goods receipt and shipping notification

Table 4.19 Update Schedule Line Values

- **Item rel.fr.dlv. (item relevant for delivery)**

  This indicator controls if the schedule line in the sales document is to be considered during delivery creation or not. In the third-party process, the seller doesn't create/process any delivery, and the goods will be shipped directly by the vendor to the customer, so it is not required to set this indicator in the schedule line category.

- **Upd. Sched. (update schedule line from purchase order)**

  This indicator is also relevant for the third-party process. Setting this indicator in the schedule line category configuration enables the schedule line updates in the sales order whenever an element of the corresponding third-party purchase order is changed, especially the delivery date.

- **Ext.capa. planning (external capacity planning)**
  This indicator is also relevant for the third-party process. To accurately determine the fulfillment date of the third-party purchase order by the third-party vendor, you must consider the capacity of the vendor and plan the delivery dates. When this indicator is set, external planning is possible in SAP Supply Chain Management (SAP SCM; a prerequisite for this functionality). If SAP SCM is not part of the landscape, don't set this indicator.

- **P.req.del.sched. (purchase requisition with delivery scheduling)**
  This indicator is also relevant for the third-party process. Best practice in the third-party process is for the supplier to directly ship the goods to the customer. However, based on your business requirements, if you want the goods from the supplier to be shipped to the seller, set this indicator.

Schedule Line Category Determination

The previous section explained the importance of the schedule line category in the third-party process. Automatic determination of a schedule line category is vital in sales order processing.

Determination of a schedule line category happens through configuration. To arrive at this configuration step, execute Transaction SPRO and navigate to **Sales and Distribution • Sales • Sales Documents • Sales Document Item • Schedule lines • Assign Schedule Line Categories**. Figure 4.58 shows that schedule line category **CS** is assigned to standard item category **TAS**.

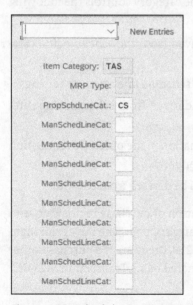

Figure 4.58 Schedule Line Category Determination

The MRP type is another criterion to determine the schedule line category during sales order processing. But a third-party sales order is not relevant for MRP, so you can leave the **MRP Type** field blank for third-party-related schedule line category determination configuration.

Figure 4.59 shows that schedule line category **CS** was determined automatically for item category **TAS**, and schedule lines were created against the line items of the sales order for the third-party process. Creation of schedule lines will also trigger the creation of a purchase requisition.

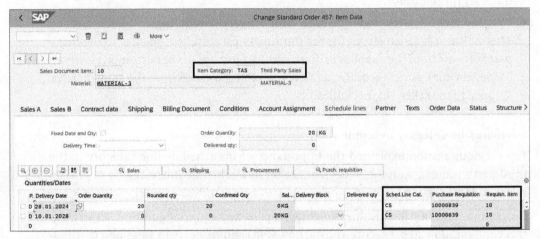

Figure 4.59 Schedule Line Category Determination during Sales Order Processing

Figure 4.60 shows the purchase requisition created from the sales order based on the schedule line category configuration. The schedule line category controls the following fields in the purchase requisition:

- The purchase requisition document type was derived from the schedule line category settings.
- Item category **S** (third-party) was derived from the schedule line category settings.
- Account assignment category **C** (sales order) was derived from schedule line category settings.
- The source of supply (third-party vendor) determination in the purchase requisition happens through the source (list purchasing master data).
- The unit price of the purchase requisition item will be derived from the PIR.

The next step in the third-party process after the creation of purchase requisition is to create a purchase order and transmit it to the third-party supplier. The buyer can manually create the purchase order using Transaction ME21N. Creation of a purchase order with reference to a purchase requisition can be automated using Transaction ME59N.

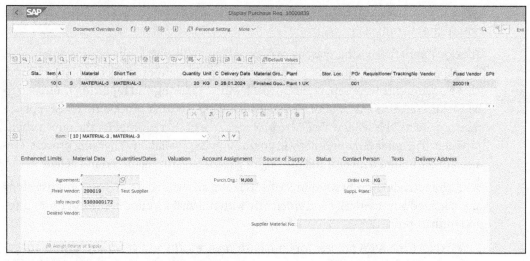

Figure 4.60 Third-Party Purchase Requisition

The supplier will fulfill the order and ship the goods to the customer directly. The seller processes the supplier invoice and creates a customer invoice/billing document. The system will not allow the seller to submit an invoice until the goods are delivered to the customer or the supplier invoice is processed, depending on the settings.

4.5 Advanced Returns Management

Returns management is an important process in the supply chain, which manages the returned products efficiently and effectively while enhancing customer satisfaction. The returns process involves customer returns, supplier returns, and internal stock transfer order returns. Efficiently managing returns from customers is vital in the supply chain, and it involves processing return receipts for the retuned products, quality inspections of returned products, restocking of the products in inventory management, and processing of customer credit memos.

In advanced returns management (ARM), which we introduced in Chapter 1, Section 1.1.7, the return merchandise authorization (RMA) is a unique identifier for the returns transaction, and it serves as an authorization for returns by the seller or manufacturer.

Quality inspection of returned products is an important step in the process, enabling you to make the decision to scrap the products, restock them, or send them for repair. Returned products will be received into quality inspection stock, and after the quality inspection, a usage decision will be made.

Based on the terms agreed upon with the customer, a replacement product will be sent in certain cases free of charge or in others with some nominal charges. If there's no replacement product that can be sent to the customer, a credit memo will be created to refund the customer.

The document flow visibility was a major concern in the past as the returns order, quality inspection for the returned products, and replacement order were not interlinked. Those documents were separate from each other. Tracking and reporting thus used to be major concerns for organizations/sellers. If the returns process involves replacement products, the seller will need to process a free of charge order for the replacement products. The replacement order will go through the availability checking process, creation of outbound delivery, and posting of goods issue for the outbound delivery.

With the introduction of ARM in SAP S/4HANA, all the documents and process steps are managed automatically and efficiently with full visibility and control of the end-to-end returns process.

Let's explore the ARM process for both customer returns and vendor returns management. We'll then walk through the master data, learn about supplier and customer returns, and provide step-by-step instructions for configuration.

4.5.1   Integrated Advanced Returns Management

The ARM process involves subprocess steps from materials management and sales and distribution for both vendor returns and customer returns. Table 4.20 shows the key subprocesses impacted by vendor returns.

Area	Processes Impacted	Description
Sales and distribution	Create and process returns delivery (outbound)	This process supports the creation of a returns delivery/outbound delivery with reference to the returns purchase order.
Materials management	Create and maintain purchasing master data	This process supports the creation and maintenance of purchasing master data required for the vendor returns process.
	Create and maintain returns purchase order	This process supports the creation and maintenance of a returns purchase order.
	Post goods issue for returns delivery	This process supports picking, packing, and goods issue posting with reference to a returns delivery.
	Create and maintain replacement purchase order	This process supports the creation and maintenance of a replacement purchase order for the replacement item, typically a free of charge item.

Table 4.20 Process Impact: Vendor Returns

Area	Processes Impacted	Description
Materials management (Cont.)	Create and process inbound delivery	This process supports the creation of an inbound delivery with reference to the replacement purchase order.
	Post goods receipt for inbound delivery	This process supports posting of a goods receipt and putaway into stock in inventory management.

Table 4.20 Process Impact: Vendor Returns (Cont.)

Table 4.21 shows the key subprocesses impacted by customer returns.

Area	Processes Impacted	Description
Sales and distribution	Create and maintain returns order	This process supports the creation of a customer returns order.
	Create and process returns delivery (inbound)	This process supports the creation of a returns delivery with reference to the returns order.
	Create and maintain replacement order	This process supports the creation of a replacement order to deliver a replacement product to the customer.
	Create and process outbound delivery	This process supports the creation of an outbound delivery with reference to the replacement order.
Materials management	Post goods receipt for returns delivery	This process supports posting of a goods receipt and putaway into stock in inventory management.
	Post goods issue for outbound delivery	This process supports picking, packing, and goods issue posting with reference to an outbound delivery.

Table 4.21 Process Impact: Customer Returns

The ARM process impacts user roles from both materials management and sales and distribution functions.

The following are the key user roles impacted within sales and distribution:

- **Sales representatives**
  Responsible for creating and maintaining returns order and replacement order for customer returns. These representatives coordinate with customers to manage the returns process efficiently while enhancing customer satisfaction.

- **Master data management team**
  Responsible for maintaining the master data required process for customer returns.

The following are the key user roles impacted within materials management:

- **Buyers/purchasers**
  Responsible for creating and maintaining returns purchase orders and replacement purchase orders.
- **Logistics/shipping coordinator**
  Responsible for coordinating shipments to customers based on outbound deliveries for replacement products.
- **Inventory manager**
  Responsible for picking and packing materials from inventory based on availability for the goods issue for outbound deliveries for replacement products. Also responsible for receiving the returned products and sending them for quality inspection.
- **Master data management team**
  Responsible for maintaining the purchasing master data required for vendor returns.

4.5.2   Master Data Requirements

The ARM process requires the following master data to be maintained in the SAP S/4HANA system:

- **Material master data**
  Maintaining the material master data views listed in Table 4.22 is mandatory for the ARM process. The other views of the material master are required based on other business processes in which the same material is being used.

Maintenance Status	Material Master Views
K	Basic data
E	Purchasing
V	Sales
L	Storage
Q	Quality management
B	Accounting

Table 4.22 Material Master Maintenance Status

- **Business partner (customer)**
  A business partner with a customer role is required for customers to perform the ARM process.

- **Business partner (vendor)**

  A business partner with a vendor role is required to perform the ARM process.

  In the purchasing data at the purchasing organization level, set the **Supplier RMA Required** indicator if the supplier requires RMA for processing returns, as shown in Figure 4.61.

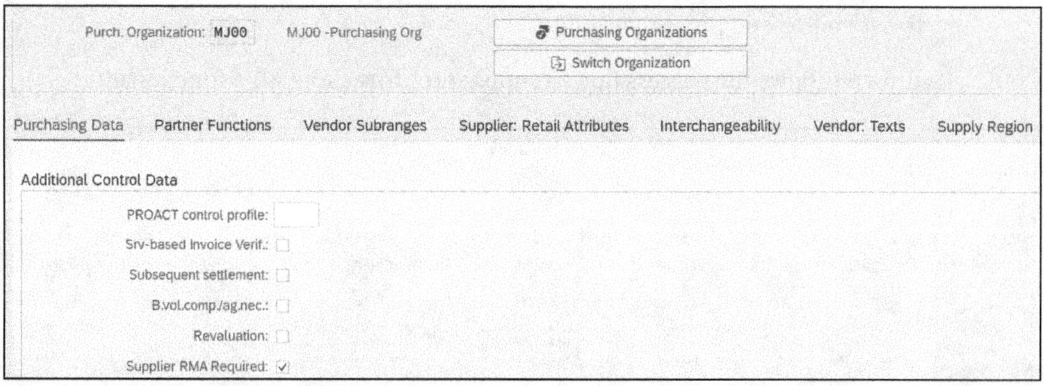

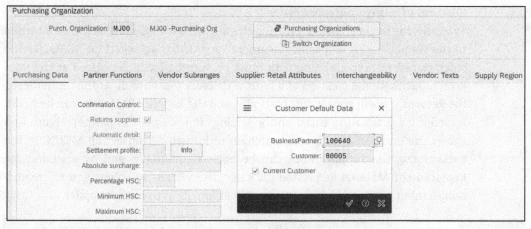

Figure 4.61 Maintain Business Partner (Supplier): 1 of 2

Set the **Returns supplier** indicator within the **Purchasing Data** tab, as shown in Figure 4.62. In the popup window, the corresponding customer ID created for the business partner using the customer business partner role must be assigned to the **Customer** field. Based on this setup, the system controls the returns processing to be carried out via the shipping process with reference to an outbound delivery.

Figure 4.62 Maintain Business Partner (Supplier): 2 of 2

4.5.3 Supplier Returns Management

Processing supplier returns using ARM provides the following business benefits:

- Materials can be returned to the external vendor via the shipping process with an outbound delivery created with reference to the returns purchase order.

- ARM can be used for stock transfer order returns for both intracompany and intercompany stock transfer orders.
- The end-to-end returns process can be tracked and reported using an RMA code.
- ARM supports both credit memo scenarios and obtaining replacement materials.
- Replacement material orders can be tracked along with the returns order processing to get end-to-end process visibility.

Figure 4.63 shows the process flow of supplier returns using ARM functionality.

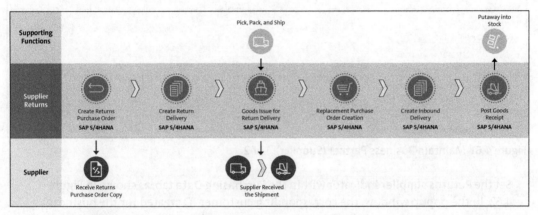

Figure 4.63 Advanced Returns Management: Supplier Returns

Let's walk through the process steps:

1. **Creation of returns purchase order**
   Typically, the buyer creates a returns purchase order for the materials to be returned to the supplier using a specific document type that is activated for ARM. Like the stock transfer order, shipping data must be automatically populated at the item level to facilitate the creation of a returns delivery (outbound). At the header level, the **Returns** tab will appear automatically, and the supplier RMA code can be maintained if it's mandatory based on the settings in the business partner. Figure 4.64 shows the creation of a returns purchase order using Transaction ME21N. In the **Returns** tab, the supplier RMA can be entered manually, and by selecting the **Replacement Material Requested** indicator, the system will allow you to create a replacement purchase order to get a replacement item from the supplier.

Figure 4.64 Creation of Returns Purchase Order

2. **Creation of returns delivery with reference to returns purchase order**
   Shipping data from the line item level of the returns purchase order is used to create a returns delivery using Transaction VL10B. Best practice is to automate the creation of returns delivery (outbound) using a batch job (background processing). A returns delivery facilitates the movement of materials from the issuing plant mentioned in the returns purchase order item. Picking and packing of materials will be performed within inventory management with reference to the outbound delivery.

3. **Goods issue for return delivery**
   After the picking and packing of returned materials, a goods issue will be posted from the issuing plant. Material documents and accounting documents will be updated for the goods movement while updating the quantity and value of the issued stock in inventory management. Movement types play an important role in determining the type of inventory movement and posting of the goods issue. Movement type 161 is used for goods issue for a vendor returns delivery.

4. **Creation of replacement material purchase order**
   A replacement material purchase order can be created using Transaction MSR_VRM_GR with reference to the original returns purchase order created in the first step of the process. As a prerequisite, in the original returns purchase order, the **Replacement Material Requested** indicator must be set, as shown in Figure 4.64. The replacement material and quantity must be mentioned in the replacement material purchase order.

5. **Creation of inbound delivery for replacement material purchase order**
   With reference to the replacement material purchase order, an inbound delivery can be created using Transaction VL31N. The confirmation control key assigned at the item level of the purchase order controls the creation of the inbound delivery. An inbound delivery can be created automatically upon receiving the ASN message via EDI.

6. **Goods receipt for inbound delivery**
   A goods receipt against the inbound delivery will be posted upon receiving the replacement materials from the supplier using Transaction VL32N. Goods receipt posting in SAP S/4HANA increases the stock level in inventory management. Movement type 101 is used for goods receipt postings for both the intracompany and intercompany stock transfer processes.

4.5.4   Customer Returns Management

Processing customer returns using ARM provides the following business benefits:

- ARM streamlines the returns process, making it faster and more efficient. This can lead to customer satisfaction as it makes it convenient and easy for customers to return products to the seller.

- ARM integrates seamlessly with other functions such as inventory management, quality management, and finance.

- An integrated quality inspection process enables easy disposition of the returned products to restock the material, scrap the material, or send it for repairs.
- A replacement material can be sent to the customer with reference to a replacement order, which makes it easy to replace a faulty item with a new one.
- It provides enhanced visibility and control of the returns process, which helps in quick decision-making.

Figure 4.65 shows the process flow of customer returns using ARM functionality.

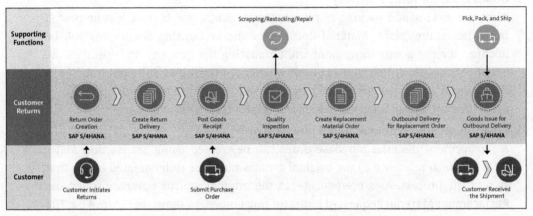

Figure 4.65 Advanced Returns Management: Customer Returns

Let's walk through the process steps:

1. **Creation of returns order**

   A returns order to process customer returns can be created using Transaction VA01 to receive the returned products from the customer. Figure 4.66 shows the returns order created in SAP S/4HANA to process customer returns. **MATERIAL-1** is the material to be returned by the customer, and the order quantity is the returned quantity of the material.

   The **Return Reason, Follow-Up Act.** (after receiving the materials from the customer), and **Next Follow-Up** fields must be all be completed in the return order. For **Refund Type**, you can select either **Credit Memo** or **Replacement Material**. If the latter, set the **Replacement Material from Supplier Requested** indicator.

   A replacement material can be different from the original material provided to the customer based on the variant configuration. The replacement quantity must be the same as the return order quantity. The system automatically generates the RMA number for the return order upon saving the document.

2. **Creation of returns delivery with reference to return order**

   A returns delivery to process the receipt of the returned products will be created as the next step using Transaction VL01N. Best practice is to automate the creation of the returns delivery (inbound) using a batch job (background processing). The returns delivery facilitates the putaway of returned products into the quality inspection facility within inventory management.

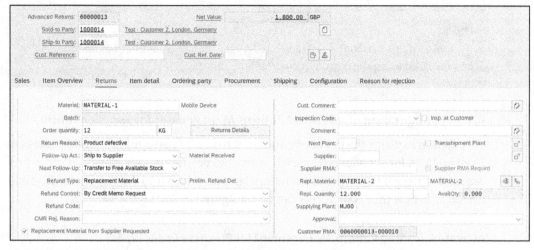

Figure 4.66 Creation of Returns Order for Customer Returns

3. **Goods receipt for return delivery**

    The goods receipt against the return delivery will be posted upon receiving the returned products from the customer using Transaction VLO2N. The returned products will be received into the quality inspection stock to perform a quality check of the returned products.

4. **Quality inspection of the returned products**

    The next step after receiving the returned products into the quality inspection stock is to carry out the quality inspection checks to handle the products accordingly. After the quality inspection, the returned products will be restocked, scrapped, or sent for repairs. The quality inspection results can be added to the returns order using Transaction MSR_INSPWH with reference to the returns delivery.

5. **Creation of replacement material order**

    After entering the quality inspection results using Transaction MSR_INSPWH with reference to the returns delivery, a replacement material order can be created automatically by setting the indicator to release the subsequent delivery free of charge in the same Transaction MSR_INSPWH and by saving the transaction. The system performs an availability check for the replacement material.

6. **Creation of outbound delivery for the replacement material order**

    An outbound delivery will be created as a next step with reference to the replacement material order. You can create an outbound delivery manually using Transaction VLO1N or this can be automated. An outbound delivery facilitates picking and packing of replacement materials from inventory management.

7. **Goods issue for outbound delivery**

    A goods issue will be posted for the outbound delivery to ship the replacement materials to the customer.

> **Note**
>
> If a replacement material is not requested by the customer, a refund of up to 100% can be processed using the customer credit memo process.

4.5.5   Configuration of Advanced Returns Management for Supplier Returns

This section explains the key configuration settings for performing an ARM process in SAP S/4HANA for supplier returns. These settings ensure smooth execution of the process and represent business requirements. Let's dive deep into the various settings and understand how the ARM process can be configured in SAP S/4HANA.

Activate Advanced Returns Management for Purchase Order Types

In this configuration, ARM will be activated for specific document types. It is recommended to create separate document types for returns and to activate ARM. Refer to Section 4.4.3 for more details about the creation of purchasing document types.

To arrive at this configuration step, execute Transaction SPRO and navigate to **Materials Management • Purchasing • Purchase Order • Returns Order • Advanced Returns Management • Activate Advanced Returns Management for Purchase Order Types**. Figure 4.67 shows the configuration settings for activating ARM for specific purchase order document types.

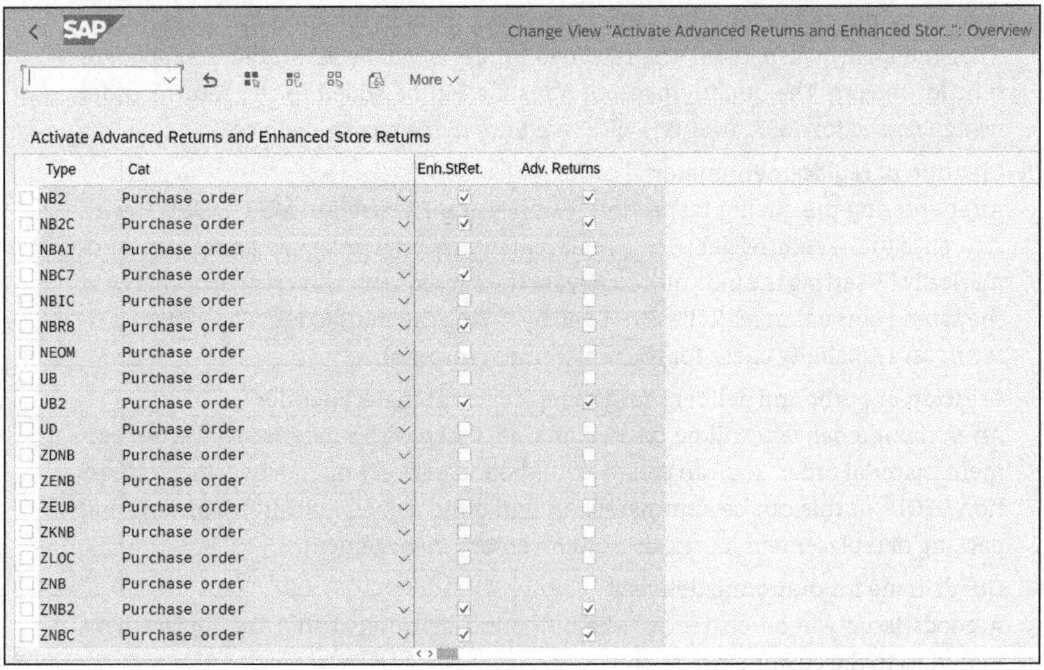

Type	Cat		Enh.StRet.	Adv. Returns
NB2	Purchase order	∨	✓	✓
NB2C	Purchase order	∨	✓	✓
NBAI	Purchase order	∨	☐	☐
NBC7	Purchase order	∨	✓	☐
NBIC	Purchase order	∨	☐	☐
NBR8	Purchase order	∨	✓	☐
NEOM	Purchase order	∨	☐	☐
UB	Purchase order	∨	☐	☐
UB2	Purchase order	∨	✓	✓
UD	Purchase order	∨	☐	☐
ZDNB	Purchase order	∨	☐	☐
ZENB	Purchase order	∨	☐	☐
ZEUB	Purchase order	∨	☐	☐
ZKNB	Purchase order	∨	☐	☐
ZLOC	Purchase order	∨	☐	☐
ZNB	Purchase order	∨	☐	☐
ZNB2	Purchase order	∨	✓	✓
ZNBC	Purchase order	∨	✓	✓

Figure 4.67 Activate Advanced Returns Management for Purchase Order Types

Let's explore the settings:

- **Type (purchasing document type)**

  This controls the various attributes and processing rules for creation of specific types of purchase order. It's used to distinguish among various types of purchase orders, such as the following:

  – Standard purchase order

  – Returns order

  – Stock transfer order

  You can configure new purchase order document types and activate ARM based on business requirements. Document types such as **NB** (for a standard purchase order), **NB2** (for an enhanced returns purchase order), **UB** (for an intracompany stock transfer order), and more exist in the standard system.

- **Cat (purchasing document category)**

  This categorizes different purchasing documents based on the business process and functionality. Each purchasing document category can have multiple document types. The following are the different purchasing document categories in SAP S/4HANA. Select the purchase order as the document category for this configuration:

  – Request for quotation

  – Purchase requisition

  – Purchase order

  – Contract

  – Scheduling agreement

  – Source list

  – PIR

  – Service entry sheet

  – Central contract

  – Central purchase order

  – Central request for quotation

  – Central supplier quotation

- **Enh.St.Ret. (enhanced store returns)**

  This indicator refers to stock transfer order returns. By activating it, the stock transfer returns will be carried out through outbound delivery and inbound delivery processes for the respective stock transfer order for those returns created using the document type with enhanced store returns activated. A goods issue for return delivery (outbound) will be processed, and a goods receipt will be processed with reference to an inbound delivery.

- **Adv. Returns (advanced returns)**

  This indicator, when set, activates the ARM process. When you create a supplier returns order using a document type with advanced returns activated, the system automatically allows you to perform all functions explained in Section 4.5.3.

Define Number Ranges for Advanced Returns Process IDs

In this configuration, the number range for the ARM process will be defined. To arrive at this configuration step, execute Transaction SPRO and navigate to **Materials Management • Purchasing • Purchase Order • Returns Order • Advanced Returns Management • Define Number Ranges for Advanced Returns Process IDs**.

Figure 4.68 shows the number ranges defined for the ARM process. You can arrive at this configuration step directly by executing Transaction SNRO. In this configuration step, both internal and external number ranges can be defined against number range object `MSR_ID`.

Based on this configuration, the system assigns a unique number/ID for every ARM process initiated by the creation of a returns purchase using a specific document type to track the end-to-end process.

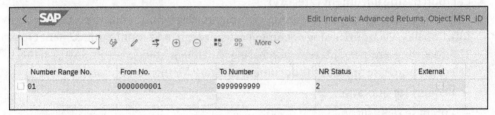

Figure 4.68 Define Number Ranges for Advanced Returns Process IDs

Define Return Reasons for Supplier Returns

In this configuration, you can define reasons for supplier returns based on business requirements. To arrive at this configuration step, execute Transaction SPRO and navigate to **Materials Management • Purchasing • Purchase Order • Returns Order • Advanced Returns Management • Define Return Reasons for Supplier Returns**.

A return reason can be entered in the **Returns** tab in the header of the returns purchase order. This information will be sent to the supplier when a returns purchase order is transmitted. Figure 4.69 shows the definition of return reasons.

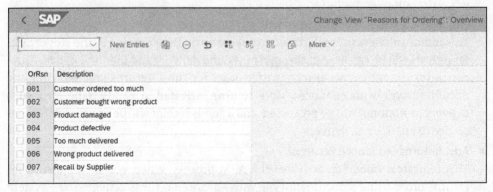

Figure 4.69 Define Return Reasons for Supplier Returns

Click the **New Entries** button to define your return reasons based on business requirements. Define a four-digit return reason code (**OrRsn**) and enter the **Description**.

Define Rejection Reasons for Supplier Returns

In this configuration, you can define a rejection reason for supplier returns based on business requirements. To arrive at this configuration step, execute Transaction SPRO and navigate to **Materials Management • Purchasing • Purchase Order • Returns Order • Advanced Returns Management • Define Rejection Reasons for Supplier Returns.**

Figure 4.70 shows the definition of rejection reasons. Click the **New Entries** button to define your return reasons based on business requirements. Define a two-digit rejection reason (**RejReason**) code and enter a **Description**. A rejection reason can be entered in the **Returns** tab in the header of a returns purchase order. This happens after the returns purchase order is created and a supplier rejects the return because of regulatory, company policy, or compliance issues. The system stops further processing of a return when the rejection reason is updated in the returns purchase order.

SAP
Change View "Rejection Reasons": Overview
New Entries
More
Rejection Reasons
RejReason
Description
01
Not in Warranty - Vendor rejected
02
Not under Quality check - Vendor rejected
03
Company revert
04
Customer induced revert

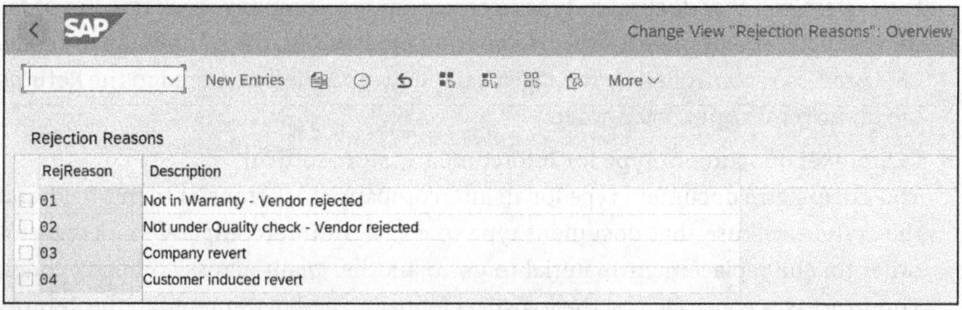

Figure 4.70 Define Rejection Reasons for Supplier Returns

Specify Settings for Replacement Materials from Supplier

In this configuration, you can specify the purchase order document type for the creation of a replacement material purchase order using Transaction MSR_VRM_GR. To arrive at this configuration step, execute Transaction SPRO and navigate to **Materials Management • Purchasing • Purchase Order • Returns Order • Advanced Returns Management • Specify Settings for Replacement Materials from Supplier.** Figure 4.71 shows the configuration settings for specifying a purchase order document type for the creation of a replacement material purchase order against every returns purchase order type for ARM.

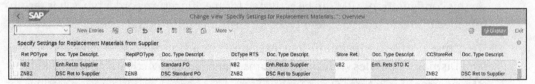

SAP
Change View "Specify Settings for Replacement Materials..": Overview
New Entries
More
Display
Exit
Specify Settings for Replacement Materials from Supplier
Ret POType
Doc. Type Descript.
ReplPOType
Doc. Type Descript.
DcType RTS
Doc. Type Descript.
Store Ret.
Doc. Type Descript.
CCStoreRet
Doc. Type Descript.
NB2
Enh.Ret.to Supplier
NB
Standard PO
NB2
Enh.Ret.to Supplier
UB2
Enh. Rets STO IC
ZNB2
DSC Ret to Supplier
ZENB
DSC Standard PO
ZNB2
DSC Ret to Supplier
ZNB2
DSC Ret to Supplier

Figure 4.71 Specify Settings for Replacement Materials from Supplier

Let's explore the settings:

- **Ret POType (returns purchase order type)**
  This is the returns purchase order document type with ARM activated. This is the key element for this configuration.

- **ReplPOType (replacement purchase order type)**
  You can assign a purchase order type to this field, and the system will use that document type to create a replacement purchase order for the replacement material from the supplier.

- **DcType RTS (purchasing document type for follow-up return to supplier)**
  You can assign a purchase order type to this field, and the system will use that document type to create a returns purchase order for the replacement material to be returned to a supplier. This process is controlled by the logistical follow-up activity entered in the **Returns** tab of the returns purchase order.

- **Store Ret. (document type for intracompany store return)**
  You can assign a document type for an intracompany stock transfer to this field, and the system will that document type to create an intracompany stock transfer order for the replacement material to go to another plant within the same company code. This process is controlled by the logistical follow-up activity entered in the **Returns** tab of the returns purchase order.

- **CCStoreRet. (document type for intercompany store return)**
  You can assign a document type for an intercompany stock transfer to this field, and the system will use that document type to create an intercompany stock transfer order for the replacement material to go to another plant across company codes. This process is controlled by the logistical follow-up activity entered in the **Returns** tab of the returns purchase order.

4.5.6   Configuration of Advanced Returns Management for Customer Returns

This section explains the key configuration settings for performing the ARM process in SAP S/4HANA for customer returns. These settings ensure smooth execution of the process and represent business requirements. Let's dive deep into the various settings and understand how the ARM process can be configured in SAP S/4HANA.

Activate Advanced Returns Management for Returns Order Types

In this configuration, ARM can be activated for specific returns order types. It is recommended to create separate order types for customer returns and to activate ARM.

To arrive at this configuration step, execute Transaction SPRO and navigate to **Sales and Distribution • Sales • Advanced Returns Management • Activate Advanced Returns Management for Returns Order Types**. Figure 4.72 shows the configuration settings for activating ARM for specific returns order types.

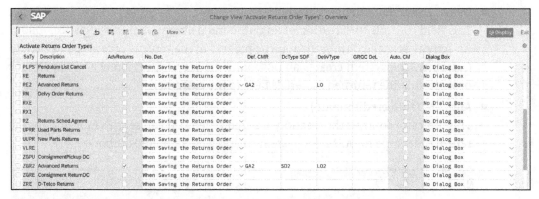

Figure 4.72 Activate Advanced Returns Management for Returns Order Types

Let's explore the settings:

- **SaTy (sales document type)**
  The sales document type controls the various attributes and processing rules for creation of specific types of sales orders or a customer returns order.

- **AdvReturns (advanced returns)**
  This indicator, when set, activates the ARM process for the respective sales document type. When you create a customer returns order using the document type with advanced returns activated, the system automatically allows you to perform all functions explained in Section 4.5.4.

- **No. Det. (determination of returns order number)**
  In this field, you can specify at what step in the creation of a returns order using ARM the system creates a returns order document number. You can select one of the following values:

  - When saving the returns order

  - Immediately when creating the returns order

  - On user request

- **Def. CMR (default credit memo request order type)**
  This field controls the credit memo document creation with reference to the returns order. The document type specified in this field will be defaulted automatically when a credit memo is created with reference to the returns order that was created using the corresponding returns order type.

- **DCType SDF (sales document type for subsequent delivery free of charge)**
  This field controls the creation of subsequent delivery of a free of charge order with reference to the returns order. The order type specified in this field will be defaulted automatically when a subsequent delivery of a free of charge order is created using Transaction MSR_INSPWH with reference to the returns order that was created using the corresponding returns order type.

- **Deliv.Type (default delivery type)**

  This field controls the creation of an outbound delivery when the logistical follow-up activity selected during the returns process in the returns order is **Send Back to Customer**. The system creates an outbound delivery with reference to the returns order to ship the returned materials back to the customer.

- **Auto CM (auto credit memo)**

  This indicator controls the creation of a credit memo. If you set this indicator for the return order type for ARM, the system creates a credit memo automatically.

> **Note**
>
> The system creates a credit memo for zero dollars even if a subsequent delivery free of charge order exists for the returns order when the **Auto CM** indicator is set.

Activate Advanced Returns Management for Credit Memo Request Order Types

In this configuration, ARM can be activated for specific credit memo request order types.

To arrive at this configuration step, execute Transaction SPRO and navigate to **Sales and Distribution • Sales • Advanced Returns Management • Activate Advanced Returns Management for CMR Order Types**. Figure 4.73 shows the configuration settings for activating ARM for specific credit memo request order types.

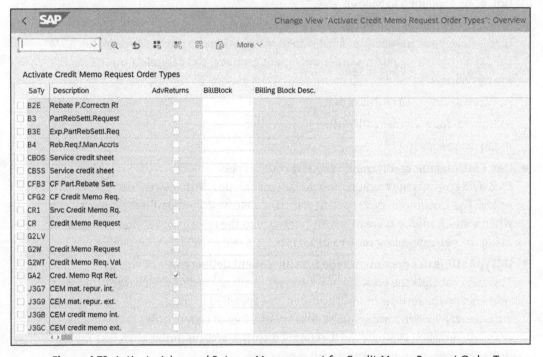

Figure 4.73 Activate Advanced Returns Management for Credit Memo Request Order Types

Let's explore the settings:

- **SaTy (sales document type)**
  This is the sales document type that represents the credit memo request.

- **AdvReturns (advanced returns)**
  This indicator, when set, activates the ARM process for the associated credit memo request order type.

- **BillBlock (billing block)**
  If a billing block is specified, the credit memo will be blocked when it's created. It can be released later.

Activate Advanced Returns Management for Subsequent Delivery Free of Charge Order Types

In this configuration, ARM can be activated for specific subsequent delivery free of charge order types. To arrive at this configuration step, execute Transaction SPRO and navigate to **Sales and Distribution • Sales • Advanced Returns Management • Activate Advanced Returns Management for SDF Order Types**. Figure 4.74 shows the configuration settings for activating ARM for specific subsequent delivery free of charge order types.

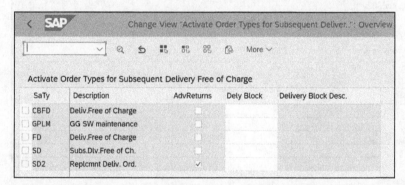

Figure 4.74 Activate Advanced Returns Management for Subsequent Delivery Free of Charge Order Types

Let's explore the settings:

- **SaTy (sales document type)**
  This is the sales document type that represents the subsequent delivery free of charge order.

- **AdvReturns (advanced returns)**
  This indicator, when set, activates the ARM process for the respective subsequent delivery free of charge order type.

- **Dely Block (delivery block)**

  If a delivery block is specified, the subsequent delivery free of charge order will be blocked when it's created. It can be released later.

Activate and Rename Follow-Up Activities

In this configuration, you can activate the logistical follow-up activities you want to use in the ARM process.

To arrive at this configuration step, execute Transaction SPRO and navigate to **Sales and Distribution • Sales • Advanced Returns Management • Activate and Rename Follow-Up Activities**. Figure 4.75 shows the configuration settings for activating and renaming logistical follow-up activities for ARM.

In the standard system, the logistical follow-up activities are predefined. You can activate them for ARM, and the active ones will appear in the follow-up activity and next follow-up activity fields of the returns order.

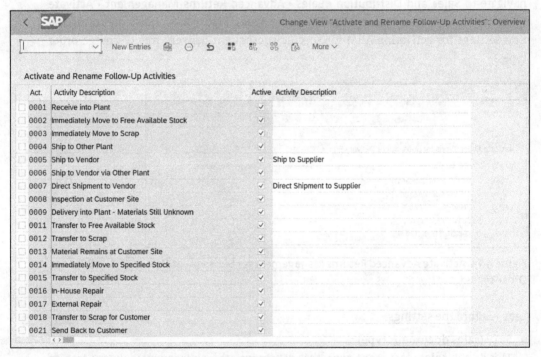

Figure 4.75 Activate and Rename Follow-Up Activities

Activate and Rename Refund Control

In this configuration, you can activate refund controls to use in the ARM process.

To arrive at this configuration step, execute Transaction SPRO and navigate to **Sales and Distribution • Sales • Advanced Returns Management • Activate and Rename**

Refund Control. Figure 4.76 shows the configuration settings for activating and renaming refund control for ARM.

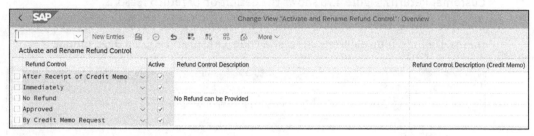

Figure 4.76 Activate and Rename Refund Control

In the standard system, the refund controls are predefined. You can activate them for ARM, and the active ones will appear in the **Refund Control** field of the returns order.

Define Number Ranges for Advanced Returns Process IDs

In this configuration, the number range for the ARM process will be defined.

To arrive at this configuration step, execute Transaction SPRO and navigate to **Sales and Distribution • Sales • Advanced Returns Management • Define Number Ranges for Advanced Returns Process IDs**.

Figure 4.77 shows the number range defined for the ARM process. You can arrive at this configuration step directly by executing Transaction SNRO. In this configuration step, both internal and external number ranges can be defined against number range object MSR_ID.

Figure 4.77 Define Number Ranges for Advanced Returns Process IDs

Based on this configuration, the system assigns a unique number/ID for every ARM process initiated by the creation of a returns purchase order using a specific document type to track the end-to-end process.

Define Return Reasons for Customer Returns

In this configuration, you can define reasons for customer returns based on business requirements.

To arrive at this configuration step, execute Transaction SPRO and navigate to **Sales and Distribution** • **Sales** • **Advanced Returns Management** • **Define Return Reasons for Customer Returns**. Figure 4.78 shows the definition of return reasons.

Click the **New Entries** button to define your return reasons based on business requirements. Define a four-digit return reason code (**OrRsn**) and enter a **Description**.

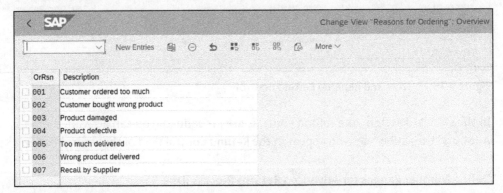

Figure 4.78 Define Return Reasons for Customer Returns

Define Returns Refund Codes

In this configuration, you can define refund codes for customer returns based on business requirements.

To arrive at this configuration step, execute Transaction SPRO and navigate to **Sales and Distribution** • **Sales** • **Advanced Returns Management** • **Define Return Refund Codes**. Figure 4.79 shows the definition of return reasons.

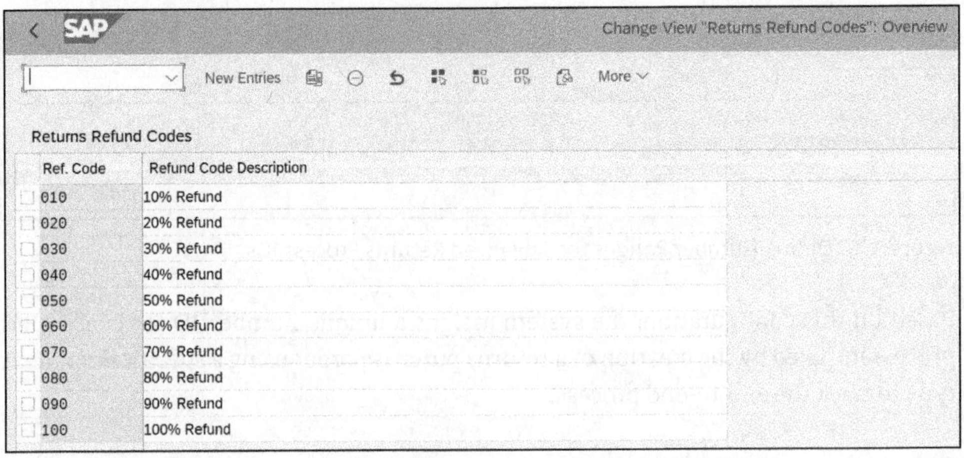

Figure 4.79 Define Returns Refund Codes

Click the **New Entries** button to define your refund codes based on business requirements. Define a three-digit code (**Ref. Code**) and enter a description.

Configure Default Values for Returns Refund Codes

In this configuration, based on the combination of the quality inspection result and a return reason, you can assign a specific refund code based on business requirements.

To arrive at this configuration step, execute Transaction SPRO and navigate to **Sales and Distribution • Sales • Advanced Returns Management • Configure Default Values for Returns Refund Codes**. Figure 4.80 shows the configuration settings for maintaining the default values for return refund codes. Click **New Entries** to maintain the default values for return refund codes. Select an **Inspection Code**, **Return Reason**, and **Refund Code** from the dropdown menus.

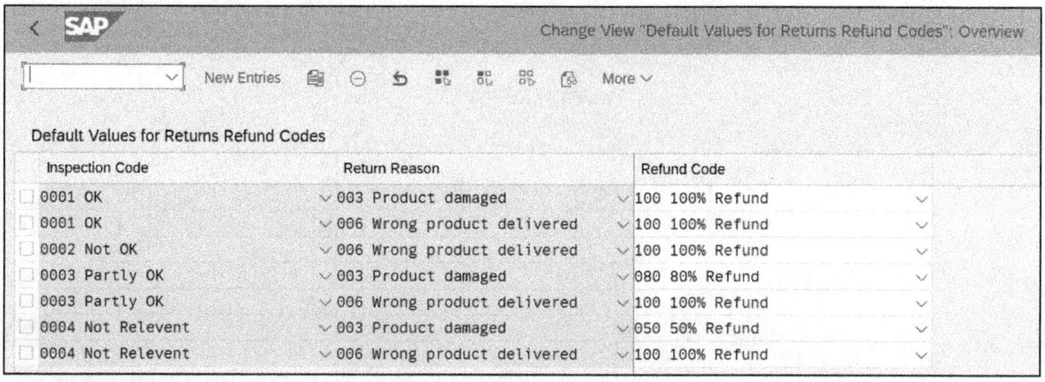
SAP
Change View "Default Values for Returns Refund Codes": Overview
New Entries
More
Default Values for Returns Refund Codes
Inspection Code
Return Reason
Refund Code
0001 OK
003 Product damaged
100 100% Refund
0001 OK
006 Wrong product delivered
100 100% Refund
0002 Not OK
006 Wrong product delivered
100 100% Refund
0003 Partly OK
003 Product damaged
080 80% Refund
0003 Partly OK
006 Wrong product delivered
100 100% Refund
0004 Not Relevent
003 Product damaged
050 50% Refund
0004 Not Relevent
006 Wrong product delivered
100 100% Refund

Figure 4.80 Configure Default Values for Returns Refund Codes

Based on these settings, the system automatically determines the return refund codes upon entering the quality inspection results in Transaction MSR_INSPWH.

4.6 Summary

In this chapter, we introduced the integration of sales and distribution functions with materials management and offered best practice examples and processes. In Section 4.1, we explained the master data, availability checks, the MRP process, and automatic account determination for sales order and delivery processing with configuration settings in SAP S/4HANA. In Section 4.2, we explained the stock transfer process and introduced the concept of creating stock transfer orders that include sales order and purchase order details in one document with master data and configuration settings. In Section 4.3, we explained inventory management for sales and distribution processes, including goods issue for outbound deliveries, goods receipt for customer return deliveries, important standard movement types, and configuration of inventory movement types and other related settings. In Section 4.4, we explained third-party order processing and the materials management integration with master data requirements and configuration settings. Finally, in Section 4.5, we explained the ARM feature of SAP S/4HANA for both customer returns management and supplier returns management with master data requirements and configuration settings.

In the next chapter, we'll discuss the integration between quality management and materials management to enforce quality standards throughout the purchasing and inventory management processes. We'll also explain how the integrated quality management process can lead to improved product quality, compliance, and customer satisfaction.

Chapter 5
Quality Management

Quality management in SAP S/4HANA provides a comprehensive list of tools and capabilities to support quality control and assurance across different business processes. The seamless integration of quality management with the procurement and inventory management functions of materials management enables organizations to maintain high quality standards for incoming materials while ensuring compliance with regulatory requirements.

Materials are procured externally from external suppliers and internally from other plants via the stock transfer process. At the time of goods receipt into inventory, the system can trigger a quality inspection check automatically based on a predefined criterion such as a vendor, material type, or purchase order. Quality management of incoming materials optimizes the efficiency and effectiveness of the procurement process.

This chapter covers quality management processes, integration points between materials management and quality management functionalities, quality management-specific master data, and key configuration settings in SAP S/4HANA. The quality management process in general starts with quality planning to ensure that quality consistently meets all business and compliance requirements. During the quality inspection of incoming materials and products, the quality plan will be executed and incoming materials and products inspected. Quality notifications can be created manually or generated automatically after the quality inspection in case of product defects found during the inspection. Quality certificates can be generated in the quality management functionality to certify that the materials and products meet the quality standards and compliance requirements.

SAP S/4HANA quality management supports issuing quality certificates to customers and receiving quality certificates from suppliers. It provides analytical reports to monitor the quality management processes and quality control. Quality audits are performed internally as well as externally to evaluate the effectiveness of the quality management processes and find any discrepancies with reference to quality standards, compliance requirements, and regulatory requirements. Quality management supports quality audits and helps document the audit findings.

Quality collaboration with suppliers can provide the following features:

- **Quality notifications**

  Buyers and suppliers can collaborate on deviations using quality notifications. These can be issued electronically by the external manufacturer (subcontractor) to the buyer or can be issued by the buyer to the supplier during quality inspection of received materials and products.

- **Quality inspections**

  Suppliers can send quality certificates along with inspection results electronically to the buyers. Suppliers can also send quality certificates to the buyers along with advanced shipping notifications (ASNs).

- **Quality reviews**

  Quality reviews can be initiated by the buyer's quality management users to collaborate with suppliers on process and material change requests, for reviews of product complaints, and in other general use cases.

Let's dive deep into the master data requirements in quality management. We'll then walk through the step-by-step configuration for quality management integration points with materials management—specifically, in the procurement and inventory management processes.

5.1   Master Data in Quality Management

Material master data is the central repository for all material information related to purchasing, sales, materials requirement planning (MRP), accounting, storage, warehouse management, and quality management. Refer to Chapter 3, Section 3.1 for more details. This chapter covers quality management-specific master data.

Master data plays a vital role in executing and controlling the quality management processes. The following sections discuss some of the key master data elements used in SAP S/4HANA quality management.

5.1.1   Inspection Characteristics

Quality inspection characteristics, also called master inspection characteristics (MICs), are the features of a material or product that are testable or measurable during quality inspection. These characteristics include dimensions, weights, chemical properties, and other relevant parameters to perform quality inspection of products or materials. These are broadly categorized into two types:

- **Qualitative characteristics**

  Qualitative characteristics are the attributes of a material or a product that are assessed based on their qualitative properties: color, taste, odor, and the like. Qualitative characteristics are typically assessed using subjective judgements rather than precise measurements. For example, a visual inspection of products for any scratches and other visual defects, a visual inspection of color, sensing the odor of a product, and so on. The inspection results of qualitative characteristics can be recorded descriptively as OK or not OK, pass/fail, and so on.

- **Quantitative characteristics**
  Quantitative characteristics are the attributes of a material or a product that can be measured precisely, and the results can be documented numerically using measuring instruments. For example, the physical dimension of a product, the product weight, the chemical composition of products, and so on. Inspection results for quantitative characteristics are recorded numerically and compared against specified acceptance criteria.

Both qualitative and quantitative characteristics can be managed in inspection plans to effectively conduct quality inspection of products and materials. SAP S/4HANA quality management provides functionalities to inspect and control both types of characteristics.

MICs are created in the system at the plant level using Transaction QS21. Let's explore the various attributes of MICs.

Figure 5.1 shows the initial screen for the creation of MICs. Fill out the following fields:

- **Plant**
  Enter the plant code in this field for the plant in which you want to create MICs.

- **Master Insp. Charac. (master inspection characteristic)**
  In this field, enter a unique name for the MIC you want to create.

- **Valid From**
  Enter the date from which the MIC is valid for the plant here.

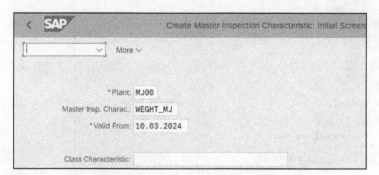

Figure 5.1 Create Master Inspection Characteristics: Initial Screen

Figure 5.2 shows the general data screen for the creation of MICs. Press ⌷Enter⌷ on the initial (previous) screen to display the general data screen. Let's explore some of the key fields:

- **Control Data**
  Specify if the MIC is a quantitative or qualitative characteristic. For example, if you want to create a MIC for measuring the moisture content of a material, which is measured as a percentage, select **Quantitative Charc** as the control data. For a MIC for the color of a material, select **Qualitative Charc** as the control data.

This setting controls the control parameters for the MICs. Control indicators are explained later in this list.

- **Status**

  This field controls the processing status of the MIC. The following are the different statuses for the MICs:

  - **Being Created**: This is the status by default at the time of creation. An MIC with this status can't be used.

  - **Released**: The MIC processing status must be set to **Released** to use it to test the quality of products and materials. You must set this status to use it in your quality inspection plan.

  - **Can No Longer Be Used**: This status can be set when you no longer want to use this MIC.

  - **Deletion Flag**: Set this processing status when you want to archive this MIC during the next archiving transaction.

  - **Archived**: This status will be set after archiving the MIC, and the record is deleted from the system.

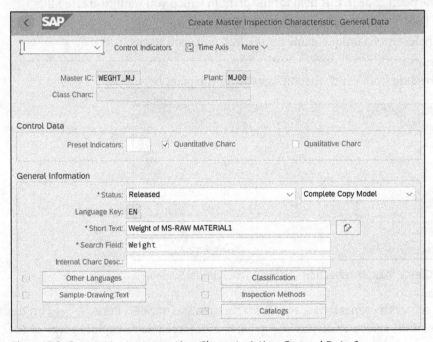

Figure 5.2 Create Master Inspection Characteristics: General Data Screen

- **Complete Copy Model (copy model/reference characteristic)**

  By default, the **Complete Copy Model** value is set in this field. This field controls how the MIC's details are referenced or copied over to the quality inspection plan. There are three standard settings:

- **Incomplete Copy Model**: Select this copy model if the MIC has some missing details and the inspection plan can't be saved until you manually add the missing data.
- **Complete Copy Model**: Select this copy model if the MIC has all required data specific to the master record and all the data from the MIC will be copied over to the inspection plan. You can still update certain data for the MIC in the inspection plan.
- **Reference Characteristic**: Select this value if you do not want users to make any changes to the MIC in the inspection plan. Any changes required for the MIC will be made within the MIC via Transaction QS22.

- **Control Indicators**

  Based on the inspection characteristic type, different control indicators appear when you click the **Control Indicators** button at the top of the screen or press `Shift`+`F1`. Figure 5.3 displays the control indicators for quantitative characteristics. Because quantitative characteristics are measured and the results are documented numerically, you can set a lower specific limit, upper specific limit, and/or a specific target value to be checked:

  - **Lower Specif. Limit**: This is the lowest specification limit for the characteristic. If the test results fall below this value, the testing can be unsuccessful.
  - **Upper Specif. Limit**: This is the highest specification limit for the characteristic. If the test results fall above this value, the testing can be unsuccessful.
  - **Check Target Value**: The system automatically checks the target value if the lower specification limit or upper specification limit is set. The system checks if the target value lies within the upper limit and lower limit tolerance set.

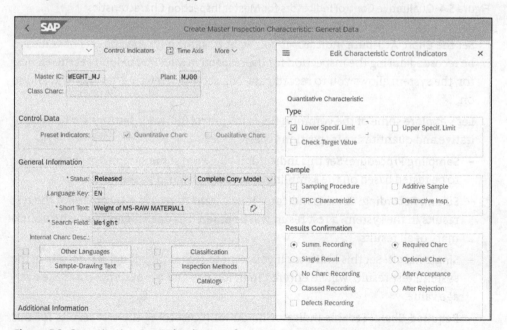

Figure 5.3 Quantitative Control Indicators for Master Inspection Characteristics

Figure 5.4 shows the control indicators for qualitative characteristics. Because qualitative characteristics are visually inspected and cannot be measured and the results are not documented numerically.

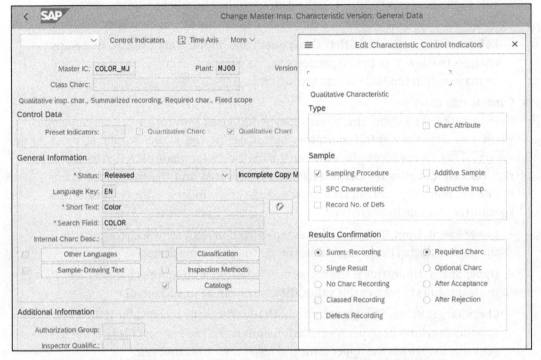

Figure 5.4 Qualitative Control Indicators for Master Inspection Characteristics

Set the **Charc Attribute** indicator if you want to allow only specific sets of values to be recorded during the inspection for the respective MIC. If you don't set this indicator, the system allows you to record pass/fail, accepted/rejected, OK/not OK, and so on.

Let's explore some of the common fields in control indicator settings for both qualitative and quantitative inspection types:

– **Sampling Procedure**: Set this indicator if you want some samples of the entire lot to be tested based on the sampling procedure assigned to the MICs.

– **Summ. Recording**: Set this indicator if you want to record a summary of the test results in the system; a test result will be entered as one single result with a summary of all results.

– **Single Result**: Set this indicator if you want to record individual test results in the system; test results will be entered individually, and the system shows the average value.

– **Required Charc**: Set this indicator if you want recording the test results for this MIC to be mandatory.

- **Optional Charc:** Set this indicator if you want recording the test results for this
  MIC to be optional.

- **Quantitative Data**
  This tab appears only when you select a quantitative characteristic type. Click the
  Quantitative Data tab to specify a tolerance key, lower limit and upper tolerances,
  and a unit of measure (this is the key data for this). The upper limit, lower limit, and
  target value will be mandatory if you set **Lower Specific Limit**, **Upper Specific Limit**,
  and **Target Value** indicators in the **Control Indicators** tab (see Figure 5.5).

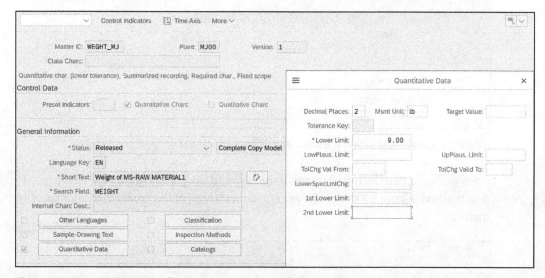

Figure 5.5 Quantitative Data for Master Inspection Characteristics

5.1.2   Inspection Methods

In quality management, inspection methods define how an inspection characteristic
can be tested with a set of work instructions and testing procedures. It specifies the
sequence of steps to be followed during the material inspection process, including how
to prepare for tests, execute tests, and record the test results, including acceptance criteria. The purpose of creating inspection methods is to ensure materials and products
meet required quality standards and regulatory requirements. Inspection methods are
used to inspect incoming inventory, perform a finished product inspection, and so on.

Inspection methods are created in the system at the plant level using Transaction QS31.
Inspection methods are assigned to MICs. You can assign an inspection method to an
MIC using Transaction QS23.

Let's explore the various attributes of an inspection method. Figure 5.6 shows the initial screen for the creation of an inspection method, where you can fill out the following fields:

- **Plant**
  Enter the plant code in this field for the plant in which you want to create inspection method.
- **Insp. Method**
  In this field, enter a unique name for the inspection method you want to create.
- **Valid From**
  Enter the date from which the inspection method is valid for the plant.

Figure 5.6 Create Inspection Method: Initial Screen

Press Enter from the initial screen to navigate to the general data view of the inspection method. Figure 5.7 shows the general data screen for the creation of an inspection method.

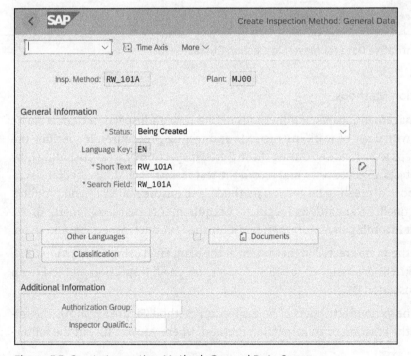

Figure 5.7 Create Inspection Method: General Data Screen

Let's explore some of the key fields:

- **Status**
  This field controls the processing status of the inspection method. The following are the different statuses for the inspection method:

  - **Being Created**: This will be the status by default at when the inspection method is created. An inspection method with this status can't be used.
  - **Released**: The inspection method processing status must be set to **Released** to use it to test the quality of products and materials.
  - **Can No Longer Be Used**: This status can be set when you no longer want to use this inspection method.
  - **Deletion Flag**: Set this processing status when you want to archive this inspection method during the next archiving transaction.
  - **Archived**: This status will be set after archiving the inspection method, and the record will be deleted from the system.

- **Documents**
  By clicking the **Documents** button, you can add external reference documents for work instructions and testing procedures that include the sequence of steps to be followed during the material inspection process, including how to prepare for the tests, execute the tests, and record the test results, including the acceptance criteria. You can add multiple documents to the inspection method. Figure 5.8 shows the documents added to the inspection method.

Figure 5.8 Inspection Method: Documents

- **Authorization Group**
  If you have defined specific authorization groups to control the access for quality management master data, you can assign an authorization group. Only users with access to the assigned authorization group can use the inspection method.

5.1.3   Sampling Procedure

In quality management, a sampling procedure defines the methodology to select samples from the stock of a material being received into inventory or used during the manufacturing process. It determines the inspection lot to perform quality inspection. It plays a vital role in the quality inspection process and determines a lot size to be tested/inspected without having to inspect every single unit.

Sample determination in SAP S/4HANA happens during specific events/transactions such as goods receipt against a purchase order, goods receipt against a production order, and so on.

The attributes of the sampling procedure include the name, description, inspection lot size, acceptance quality level (AQL), sampling plan, sampling method, and inspection level. The sampling procedure is created using Transaction QDV1. Let's explore how to create a sampling procedure and set the size for the inspection lot to be tested/ inspected.

Figure 5.9 shows the initial screen for the creation of a sampling procedure. Enter a unique name in the **Sampling Procedure** field for the sampling procedure you want to create.

Figure 5.9 Create Sampling Procedure: Initial Screen

Press Enter on the initial (previous) screen to navigate to the assignments view of the sampling procedure. Figure 5.10 shows the assignment screen for the creation of a sampling procedure. Let's explore some of the key fields:

- **Sampling Type**

  A sampling type is a predefined methodology for determining the sampling size to perform quality inspection for the entire lot of materials and products to be tested/ inspected. You can select one of the following, based on your requirements:

  - **Fixed sample**: Select this sampling type if the inspection lot size will be fixed irrespective of the quantity of stock subjected to quality inspection.

  - **100% inspection**: Select this sampling type if the inspection lot size will be 100% of the quantity of stock subjected to quality inspection.

  - **Percentage sample**: Select this sampling type if the inspection lot size will be a percentage of the quantity of stock subjected to quality inspection.

  - **Use sampling scheme**: A sampling scheme is a predefined set of rules to determine the sampling size for quality inspection. Select this if you want to assign a predefined set of rules to determine sampling size.

> **Note**
>
> A sampling procedure can be assigned to the material master record in the **Quality management** view at the plant level. See Chapter 3, Section 3.1.5 for more details.

- **Valuation Mode**

  The valuation mode determines how the system evaluates the results of inspections conducted based on a set of rules. It is crucial to select a valuation mode to accept or reject a sample during quality inspection. The following are some of the key valuation modes you can set:

  - **Attributive inspection nonconf. units**: If you set this valuation mode in the sampling procedure, the system compares the number of nonconforming units to the acceptance number in an inspection plan to determine if the inspection characteristic can be accepted or rejected,

  - **Attr. insp. nonconforming units / manual**: If you set this valuation mode in the sampling procedure, the system first compares the number of nonconforming units to the acceptance number in an inspection plan to determine if the inspection characteristic can be accepted or rejected. If this comparison is systematically not possible, it switches to manual valuation mode.

  - **Attributive inspection number of defects**: If you set this valuation mode in the sampling procedure, the system first compares the number of defects to the acceptance number in an inspection plan to determine if the inspection characteristic can be accepted or rejected.

  - **Attr. insp. number of defects / manual**: If you set this valuation mode in the sampling procedure, the system first compares the number of defects to the acceptance number in an inspection plan to determine if the inspection characteristic can be accepted or rejected. If this comparison is systematically not possible, it switches to manual valuation mode.

  - **Variable inspection s-Method (one limit)**: If you set this valuation mode in the sampling procedure, a sampling plan is used for variable inspection where the mean value is checked against the tolerance specifications of either the upper limit or lower limit to determine if the inspection characteristic can be accepted or rejected.

  - **Variable insp. s-Method (two limits)**: If you set this valuation mode in the sampling procedure, a sampling plan is used for variable inspection where the mean value is checked against both the upper limit and lower limit of the tolerance specifications to determine if the inspection characteristic can be accepted or rejected.

- **Manual Valuation:** If you set this valuation mode in the sampling procedure, a valuation to determine if the inspection characteristic can be accepted or rejected will be performed manually.

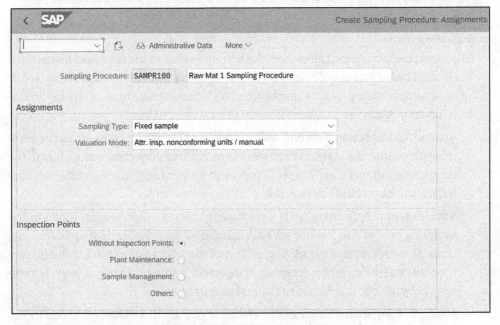

Figure 5.10 Create Sampling Procedure: Assignment Screen

After selecting the sampling type and valuation mode, press ⌈Enter⌉ to display the **Special Conditions** view of the sampling procedure.

Click the **Sample** button from the special conditions view to specify the sample size for the sampling procedure. Figure 5.11 shows the special conditions view and the popup window to enter the sampling size and acceptance criteria:

- **Sample Size**
  You can enter a sample size for the sampling procedure depending on the sampling type selected. If you have selected a fixed sample as the sampling type, you can enter a numerical value, and if you have selected a percentage sample as the sampling type, you can enter a percentage value.

- **Acceptance Number**
  You can specify the highest number of defects or nonconforming units that are allowed during the quality inspection check to accept the inspection characteristic. If the number of defects is more than this number, the inspection characteristic will be rejected.

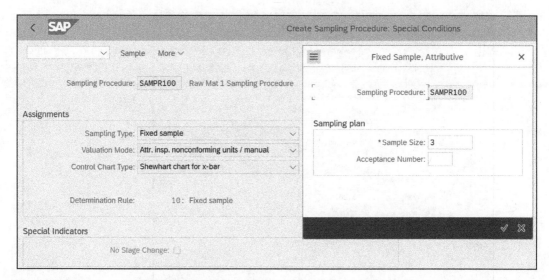

Figure 5.11 Create Sampling Procedure: Sample Size

5.1.4 Inspection Plan

In quality management, the inspection plan defines the sequence of inspection operations, inspection characteristics to be tested/inspected, and the sampling procedure to be followed during the quality inspection of materials and products. For every material to be inspected, it includes what characteristics are tested and how.

The attributes of the inspection plan include a name, description, inspection type, inspection stages, inspection characteristics, sampling procedure, and acceptance criteria. In SAP S/4HANA, the inspection plan is created using Transaction QP01 for a combination of a material and a plant.

Let's explore how to create an inspection plan. Figure 5.12 shows the initial screen for the creation of inspection plan, where you fill out the following fields:

- **Material**
  In this field, enter the material master record to create the inspection plan.

- **Plant**
  Enter the plant code in this field for the plant in which you want to create inspection plan.

- **Key Date**
  Enter date from which the inspection plan is valid for the plant.

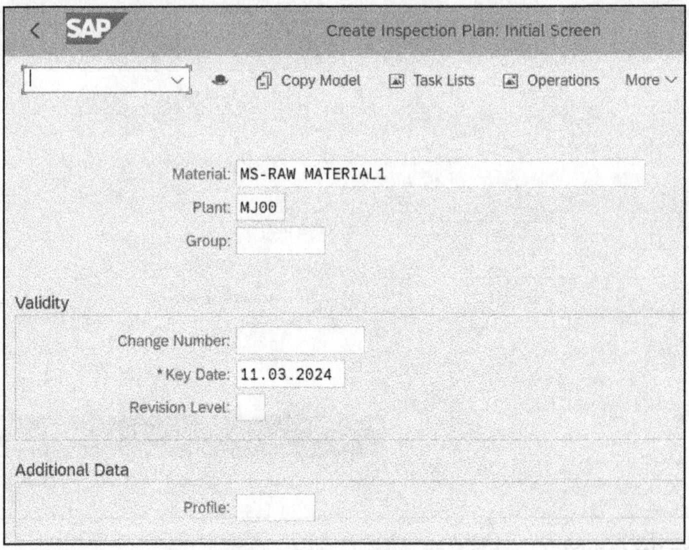

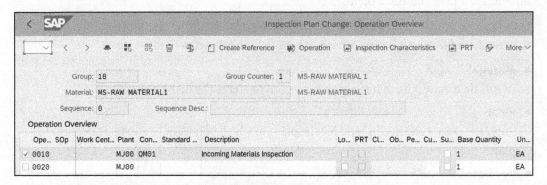

Figure 5.12 Inspection Plan: Initial Screen

Press Enter on the initial (previous) screen to navigate to the operations overview screen of the inspection plan. Figure 5.13 shows the operations overview screen to add operations for the inspection plan. An inspection plan can have one or more operations, and each operation refers to each stage of the inspection plan. You can assign specific inspection characteristics to each operation.

The key field is the **Con...** (control key) column. The control key controls how every operation will be handled in an inspection plan. To use MICs, select **QM01** as the control key. **Base Quantity** and **Un...** (unit of measure) are prepopulated by default, which you can change.

Figure 5.13 Inspection Plan: Operations Overview

To navigate to the characteristics overview screen to add MICs to the operation, select the quality operation from the operations overview screen and click the **Inspection Characteristics** button at the top or press F7. Figure 5.14 shows the characteristic overview screen to assign MICs to an operation within an inspection plan. On this screen, do the following:

- Select and assign MICs defined at the plant level in the **Master Ins...** column.
- Enter a short text for the inspection characteristic you are assigning.
- Select and assign an inspection method for the inspection characteristic in the **Insp...** column.
- Select and assign the sampling procedure for the inspection characteristic in the **Sampling...** column.

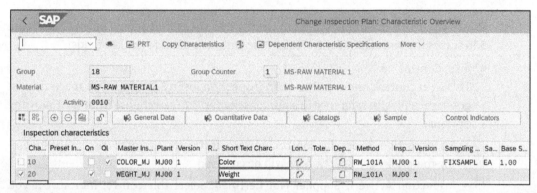

Figure 5.14 Inspection Plan: Characteristic Overview

5.1.5   Quality Info Record

A quality info record (QIR) is an important quality management master data element that contains quality-related information for materials and vendors. The QIR plays a vital role in ensuring that procured materials from vendors meet the required quality standards.

A QIR is created for a combination of a material and a vendor. The attributes of the QIR include the material, vendor, inspection type, frequency, sampling procedure, tolerance limits, inspection method, and so on. In SAP S/4HANA, QIRs are created using Transaction QI01.

Let's explore the creation of QIRs. Figure 5.15 shows the initial screen for the creation of a QIR, where you fill out the following fields:

- **Material**
  Enter the material master record in this field for which to create the QIR.
- **Supplier**
  Enter the supplier/source of supply for the material in this field for which to create the QIR.
- **Plant**
  Enter the plant code in this field for the plant in which you want to create the QIR.

Figure 5.15 Create Quality Info Record: Initial Screen

Press [Enter] on the initial screen to navigate to the details screen of the QIR. Figure 5.16 shows the **Inspection Control** tab of the QIR. Let's explore the key fields:

- **Insp. Control**

  This field controls the creation of an inspection lot in SAP S/4HANA at the time of goods receipt with reference to a purchase order. You can choose from the following options:

  - **No Inspection**: Select this control indicator if you don't want to inspect the incoming material. No inspection lot will be created.

  - **Inspection Active If Supplier Not Certified**: Select this control indicator if you want to inspect the incoming material only if the supplier is not certified. If any overrides are active that control the creation of an inspection lot, an inspection lot will not be generated at the time of goods receipt.

  - **Inspection Active Regardless of Supplier Certificate**: Select this control indicator if you want to inspect the incoming material regardless of the supplier certificate.

Figure 5.16 Create Quality Info Record: Inspection Control

- **Source Inspection**

  A source inspection can be performed upon request at the source with participation from the buyer, the supplier, or a third-party vendor authorized by the buyer. Select an appropriate value for the source inspection of the material at the vendor's site.

Table 5.1 shows the source inspection types defined in the standard system for inspection lot origin **01** (goods receipt). Assign inspection type **01** if you want to inspect the incoming materials with reference to the purchase order at the source.

Inspection Lot Origin	Source Inspection Type	Inspection Type Description
01	01	Goods receipt inspection for purchase order
01	101	Model inspection at goods receipt for purchase order
01	130	Receiving inspection from external processing
01	102	Source inspection

Table 5.1 Source Inspection Types

- **Lead Time (in Days)**
  This field controls the timing of the creation of an inspection lot for source inspection. The system considers the delivery date of the item in the purchase order to calculate the date for the creation of an inspection lot for material inspection before the scheduled delivery date. Assign the lead time in days.

- **Source Insp.—No GR**
  Set this indicator if you don't want to create an inspection lot at the time of goods receipt as the source inspection is active.

There are two additional tabs to fill out:

- **Release**
  Set the released until date for the QIR within the **Release** tab. This date indicates that the material can be purchased from the respective supplier until that date for the plant for which the QIR exists.

- **Quality Agreement**
  You can add a document for the quality agreement with the supplier if there is one in the **Quality Agreement** tab.

5.1.6 Catalogs

In quality management, catalogs are generic master data used across all plants to define qualitative data and to categorize defects. You can assign catalogs to qualitative characteristics to evaluate the characteristics during quality inspection and record results. A catalog will be assigned to an MIC. Codes are used to record test results for the inspection characteristic during quality inspection of materials and products.

In SAP S/4HANA, catalogs are created using Transaction QS41. Let's explore the creation of catalogs. Figure 5.17 shows the initial screen for the creation of catalogs, where you fill out the following fields:

- **Catalog**
  This field defines the category of the catalog. For a quality catalog to be assigned to a MIC, assign value **1** (**Characteristic Attributes**).

- **Code Group**
  A code group is the unique name for a group of codes, such as colors, surfaces, and on.

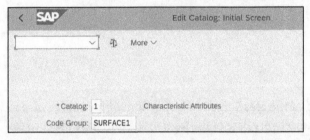

Figure 5.17 Create Catalog: Initial Screen

Press Enter on the initial (previous) screen to navigate to the code group overview screen of the catalog, shown in Figure 5.18. You can define multiple code groups and codes in a catalog. Fill out the following fields:

- **Short Text**
  Enter a short text to define the code group.

- **Status of Code Group**
  The status of the code group must be set to **Released** to use the catalog in quality inspection.

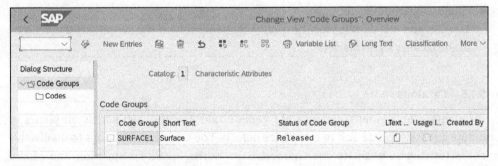

Figure 5.18 Create Catalog: Code Group

Navigate to **Codes** in the **Dialog Structure** area to arrive at the screen shown in Figure 5.19, which shows the codes assigned to the code group. You can add new entries to define codes.

A code is a unique value that defines one of the attributes of the characteristic, a defect, or a cause. For example, to inspect the condition of a metal surface, you can define codes such as rough, smooth, scratches, dents, and so on.

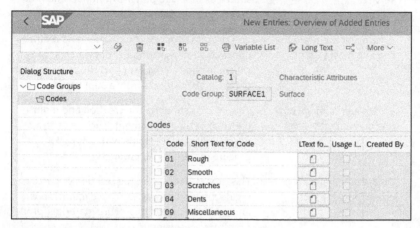

Figure 5.19 Create Catalog: Codes

5.2   Quality Management in Procurement

Quality management in procurement is crucial in ensuring the procured materials and products meet the required quality standards. Quality management in procurement is a continuous improvement area as it enhances quality performance and the feedback from quality management is used to improve procurement processes. Supplier onboarding or supplier selection involves assessing the supplier's quality standards, certificates, and compliance with regulatory requirements. The buying organization also communicates to the vendors if there are any specific quality requirements for the materials and products to be procured. Supplier collaboration is essential in the quality management of procured materials and products. Collaboration with suppliers ensures best practices are followed and processes are standardized and enhanced.

We'll walk through key quality management processes that integrate with materials management, specifically procurement, in the following sections, including step-by-step instructions for configuration. First, we'll provide a high-level overview of the integration.

5.2.1   Integrated Quality Management Process

In this section, let's explore the integration between quality management and materials management. Table 5.2 shows all the key processes impacted.

Area	Processes Impacted	Description
Materials management	Sourcing materials and services	This process supports the sourcing of a best source of supply to procure materials and products from a supplier.
	Create and maintain purchasing master data	This process supports the creation and maintenance of purchasing master data such as a source list and a PIR to support the purchasing of materials/products from a supplier.
	Create and manage outline agreements	This process supports the creation and maintenance of outline agreements such as contracts or scheduling agreements with a supplier to procure materials and products.
	Create and maintain purchasing documents	This process focuses on the creation of purchasing documents such as purchase requisitions and purchase orders to procure materials/products from a supplier.
	Create and maintain stock transfer orders	This process supports the creation and maintenance of stock transfer orders for both intracompany and intercompany stock transfer processes.
	Post goods receipt	This process supports the posting of procured materials into quality inspection stock in inventory management.
	Automatic stock posting based on usage decision	This process supports the automatic stock posting from quality inspection stock to unrestricted stock, blocked stock, or scrap based on the usage decision.
	Process incoming invoice	This process supports the creation and reconciliation of incoming invoices received from the supplier.
Quality management	Maintain quality master data	This process supports the creation and maintenance of quality management master data such as MICs, inspection methods, sampling procedures, inspection plans, QIRs, and so on to perform quality inspection of incoming materials and products procured from suppliers.
	Perform quality inspection against procurement	This process supports quality inspection of incoming materials/products from a supplier, including inspection lot generation, verifying or recording supplier quality certificates, inspection results recording, and making usage decisions.

Table 5.2 Process Impact

Area	Processes Impacted	Description
	Perform stock transfer quality inspection	This process focuses on creation of purchasing documents such as purchase requisitions and purchase orders to procure trading goods from a third-party vendor.

Table 5.2 Process Impact (Cont.)

Figure 5.20 shows the process flow of quality management process for procured materials and products. Let's walk through the process steps:

1. **Creation of purchase order**
   A purchase order will be created by the purchaser/buyer manually using Transaction ME21N with reference to a purchase requisition or contract agreement. If the item is subjected to quality inspection, the stock type will be defaulted as *quality inspection* and the quality assurance (QA) control key will be defaulted from the material master data. The purchase order will be transmitted to the supplier based on the output method: email, electronic data interchange (EDI), or electronic transmission to the supplier on SAP Business Network. Output determination configuration in SAP S/4HANA drives the purchase order output mechanism. Depending on the confirmation control key, an order confirmation from the supplier may be required, and the confirmation control key is defaulted from the PIR.

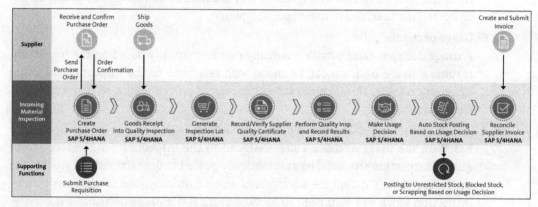

Figure 5.20 Quality Inspection of Procured Materials

2. **Post goods receipt into quality inspection**
   Once the shipment from the supplier arrives at the receiving location within the plant, goods receipt will be posted manually in SAP S/4HANA with reference to the purchase order using Transaction MIGO. If an inbound delivery exists, goods receipt will be posted with reference to the inbound delivery. Based on the quality information from the purchase order line item, the stock will be posted to quality inspection stock.

3. **Generation of inspection lot automatically**

   Based on the settings, if the material is flagged for quality inspection during the goods receipt of incoming materials from the supplier, the system generates the inspection lot automatically. The inspection lot size will be dependent on the sampling procedure. You can navigate to the inspection lot details from the material document (Transaction MIGO). The inspection lot will have the quality inspection-related information such as inspection type, material to be inspected, quantity, and so on.

4. **Print/record/verify quality certificate received from the supplier**

   If the buyer has requested a quality certificate from the supplier, it can be received before the physical receipt of the goods, at the time of goods receipt, or after the goods receipt. To receive the certificate, quality management in procurement in the material master record must be active, a control key with certificate required must be set in the material master, and the certificate type must be specified in the material master. The certificate receipt can be recorded, verified, and printed in the system.

5. **Perform quality inspection and record inspection results**

   Quality inspection of the incoming materials procured from the supplier will be carried out against the sample size identified in the inspection lot. The results will be recorded in the system using Transaction QA32 with reference to the inspection lot generated at the time of goods receipt. Based on the inspection plan, the results recording will be carried out against every qualitative and quantitative inspection characteristic specified in the inspection plan.

6. **Usage decision**

   A usage decision can be made manually or automatically based on the inspection results. A usage decision can be made with reference to the inspection lot using Transaction QA32. This includes accepting, rejecting, or returning received materials from the suppliers.

7. **Automatic stock posting based on usage decision**

   Based on the usage decision codes assigned to the inspection lot, the stock from quality inspection stock will be automatically posted to unrestricted use stock if the usage decision is *accepted—unrestricted stock*. Similarly, the stock from quality inspection stock will be posted to blocked stock if the usage decision is *rejected—blocked stock*.

8. **Post and reconcile supplier invoice**

   A supplier sends an invoice to the buyer with reference to the purchase order via email (paper invoice), via EDI, or electronically via SAP Business Network. A supplier invoice document will be posted by the accounts payable department of the seller manually in SAP S/4HANA using Transaction MIRO in the case of paper invoice. Automatic posting and reconciliation of the invoice happens if an electronic invoice is received by the seller from the supplier. Invoice reconciliation is a process of matching the purchase order, goods receipt, and supplier invoice to verify any discrepancies with purchase order item prices and purchase order item quantities and received quantities. Taxes will be reconciled against the vendor charged tax and the SAP S/4HANA internal tax engine, or from the external tax engine calculated tax. Once the invoice is fully reconciled, it will be free for payment processing in the system. The next scheduled payment run will pick up this fully reconciled invoice and post the payment to the supplier based on the preferred payment method of the supplier. Accounts payable personnel can manually process the payment against the fully reconciled invoice as well.

> **Note**
>
> You can block a vendor because of product quality reasons at both the company code level and the purchasing organization level. You can unblock the blocked vendor later.

5.2.2   Quality Inspection

Quality inspection is the assessment, testing, and inspection of incoming materials procured from suppliers and a company's own products to ensure they meet required quality standards for use in production processes and to sell them to customers. Product characteristics are measured, tested, and inspected using predefined methods and procedures during the quality inspection. With procurement integration, quality inspections are done either at the source (a *source inspection*) or during goods receipt into inventory. Let's explore the various stages of quality inspection of incoming materials procured from suppliers:

1. **Quality inspection planning**

   Inspection planning involves defining the criteria and procedure for quality inspection of materials and products. It defines a step-by-step procedure to execute quality inspection of materials and products during production processes, goods receipt, goods issue, and so on. Quality plans are created for the material at the plant level.

   It defines a set of operations to be performed sequentially during quality inspection. Each operation can have multiple qualitative and quantitative inspection characteristics to be inspected based on specific requirements of the materials and products to be inspected.

   Sampling procedures can be specified in the inspection plan, which determine how samples for quality inspection will be selected from the inspection lot.

   Inspection methods can be specified in the inspection plan, which define the techniques and procedures to be used to perform quality inspection.

   Quality inspection plans can be defined using the following transactions:
   - Transaction QP01: Create Inspection Plan
   - Transaction QP02: Change Inspection Plan
   - Transaction QP03: Display Inspection Plan

Quality inspection plans bring consistency and standardization to the quality inspection process.

2. **Inspection lot creation**
   The next step in the quality inspection process is to create the inspection lot. This represents a batch of materials to be measured, checked, and inspected in quality management. Inspection lots can be created during the goods receipt of the incoming materials procured from the suppliers or during the production order confirmation. The system checks if the material is subjected to quality inspection based on material data and the quality inspection plan.

   During certain events, the system automatically creates inspection lots to perform quality inspection. Such events include goods receipt with reference to a purchase order, production order confirmation, and certain goods movements.

   Inspection lots can be created manually in certain situations, such as when external manufacturers send some random sample to test and confirm. Inspection lots can be created using the following transactions:

   – Transaction QA01: Create Inspection Lot
   – Transaction QA02: Change Inspection Lot
   – Transaction QA03: Display Inspection Lot

   Inspection lots are crucial in the quality management process, and they facilitate the quality inspection process.

3. **Inspection results recording**
   During the quality inspection, results recording is a crucial step to document the inspection results, which leads to usage decision-making. Inspection lots are assessed based on predefined procedures, and the results are recorded for every inspection characteristic defined in the inspection plan. Typically, quantitative characteristic testing results are numerical, and qualitative characteristic test results are entered as text (accepted/rejected, OK/not OK, etc.).

   Results recordings are done manually in most cases. But if you have an interface between SAP S/4HANA and the measuring device, results can be captured automatically.

   Quality inspection results can be recorded manually using the following transactions:

   – Transaction QA32: Change Inspection Lot (Worklist)
   – Transaction QE51N: Results Recording
   – Transaction QE72: For all Inspection Lots

   Accurate results recording is the most essential part of the quality inspection to ensure required quality standards are met and to ensure compliance with regulatory requirements.

4. **Usage decision**

   A usage decision is the final step of the quality inspection process. It involves deciding whether to accept or reject the inspection lot based on the quality inspection results. The system performs an appropriate goods movement posting automatically when the usage decision is made.

   If the usage decision is made to accept the inspection lot, the respective material stock moves from quality inspection to unrestricted use stock. If the usage decision is made to reject the inspection lot, the respective material stock moves from quality inspection to blocked stock. If the usage decision is made to rework, return, or repair the inspection lot, the appropriate follow-up function will be performed.

   A usage decision for the inspection lot can be made manually using the following transactions:

   – Transaction QA32: Change Inspection Lot (Worklist)
   – Transaction QA11: Record
   – Transaction QA13: Display

   A usage decision ensures that only those materials that meet the quality standards can be used, and the rejected materials can be sent for repair, rework, or scrapping.

5.2.3 Quality Certificates

A quality certificate provides assurance of the quality of products and materials, and it certifies that the materials and products meet the required quality standards and regulatory requirements. The quality certificate can be issued by suppliers to their customers after the quality inspection of products and materials they are supplying. A quality certificate is created in SAP S/4HANA after the inspection with reference to a delivery, batch, or inspection lot. There are two types of certificates that are most common:

- **Certificate of assurance (COA)**
  A document that provides detailed information of testing and analysis of products. Test conditions and specifications of the products are included in the COA certificate.
- **Certificate of conformance (COC)**
  A document that declares the product meets certain quality standards or regulations. This certificate doesn't include test results.

In certain industries, a COA certificate is issued, while other industries issue a COC certificate.

We'll walk through the processes for both incoming and outgoing quality certificates in the following sections.

Incoming Quality Certificate

To receive an incoming quality certificate from the supplier, a quality management control key and certificate type must be defined in the system. The required certificate must be set in the quality management control key. You can receive the certificate via EDI directly from the supplier.

The content of the quality certificate is as follows:

1. Material details, including batch numbers, quantity, and so on
2. Testing methods used and inspection sample details
3. Inspection characteristics and the test results
4. Any regulatory compliance details
5. Signature from the approving authority

The quality certificate can also be received from the supplier in a PDF form via email or a paper form along with the shipped materials and products. If you receive a paper copy physically or a PDF copy via email, the receipt of the quality certificate (incoming certificate) can be confirmed by creating the incoming quality certificate using Transaction QC51. You can set the status of the incoming quality certificate as expected or received.

Outgoing Quality Certificate

Outgoing quality certificates are issued to customers to certify that the supplied products and materials meet the required quality standards and compliance requirements. Outgoing quality certificates are created with reference to an outbound delivery at the time of shipment if the customer requires a quality certificate for specific materials and products. Customers may have specific requirements for quality certificates, and the required content must be printed in the output form. Customer-specific requirements are included in the custom form and require custom development in the SAP S/4HANA system. The output determination can be configured in the system to send the quality certificate via EDI directly to the customer, via email as a PDF copy attachment, or printed and sent with the shipment in a paper form. You can create outbound quality certificates using Transactions QC20 (for a delivery note), QC21 (for an inspection lot), and QC22 (for a material batch).

Creation of outgoing quality certificates requires the following settings in the system:

1. **Quality certificate profile**
   You can create and maintain certificate profiles using Transactions QC01 (Create), QC02 (Change), QC03 (Display), and QC06 (Delete). Figure 5.21 shows the certificate profile header data. To navigate to the header data of the quality certificate profile, from the initial screen, enter the name of the certificate profile and certificate type **QM01**, which is the standard quality certificate profile type. Then press ⌈Enter⌋.

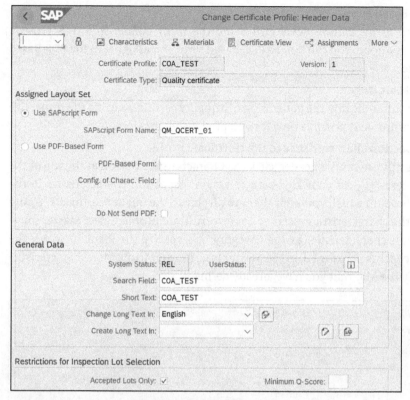

Figure 5.21 Quality Certificate Profile: Header

The following are the vital settings in the certificate profile:

- **Use SAPscript Form**

  An SAP script form is used to generate business documents such as purchase orders, invoices, delivery notes, quality certificates, and so on in a specific format based on business requirements. These forms are called during the output determination to print, send via EDI, or send via email as PDF attachments to business partners such as vendors, customers, and internal partners. Script forms in SAP are created in SAP S/4HANA using Transaction SE71.

  In the standard system, **QM_QCERT_01** exists as a standard form to print quality certificates. You can build a custom form to print quality certificates based on business requirements.

> **Note**
>
> SAPscripts are created using Transaction SE71. SAPscript forms can also be created using Transaction SMARTFORMS. You can create these SAPscript forms using either transaction, but Transaction SMARTFORM offers advanced features such as smart forms that are client-independent, and you can create multiple page formats. It takes comparatively less effort to create SAPscript forms in Transaction SMARTFORMS.

- **Use PDF-Based Form**

  PDF-based forms can be developed using the form builder in SAP S/4HANA, which uses Adobe. You can build a PDF-based form and assign it in the **PDF-Based Form** field.

- **Accepted Lots Only**

  If you set this indicator, the generation of a quality certificate using the respective certificate profile is possible only for accepted inspection lots.

2. **Assign inspection characteristics to the certificate profile**

   From the certificate profile header, click the **Characteristics** button at the top of the screen or press F5 on your keyboard to assign inspection characteristics in the sequential order in which you want them to display in the quality certificate. Figure 5.22 shows the characteristics assigned to the certificate profile. Select **Master** (master inspection characteristic) as the **Category** and specify the **Plant** in which the inspection check is performed. In the **Master Ins...** (master inspection characteristic) field, specify the MIC that you have already defined.

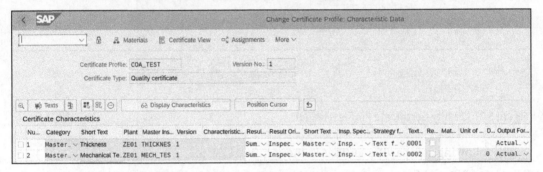

Figure 5.22 Quality Certificate Profile: Characteristics

3. **Assign to material, material group, or customer/material**

   To generate the quality certificate automatically, or manually from a delivery note, you must assign the certificate profile to a material, material group, or customer/material based on the business requirements. For example, if a customer has specific requirements for the quality certificate, use the customer and material option to assign a specific certificate profile. You can perform this activity using Transactions QC15 (Create), QC16 (Change), and QC17 (Display). From the initial screen of Transaction QC15, select one of the following key combinations to assign the quality certificate to the material/customer, material, or material group, respectively, then press Enter :

   - **Material/Customer**

   - **Material**

   - **Material Group**

Figure 5.23 shows the certificate profile assigned to material group. To assign a new material group to the quality certificate profile, directly specify the material group and profile in the **Matl Group** (material group) and **Profile** fields, respectively.

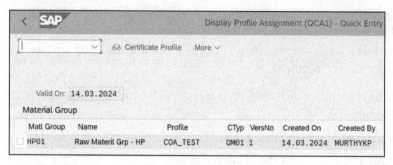

Figure 5.23 Quality Certificate Profile Assignment

4. **Create output condition record for certificate recipient**

Using output condition type LQCB (quality certificate for sold-to party), create and maintain condition record for recipients of a quality certificate using Transactions VV21 (Create), VV22 (Change), and VV23 (Display). An output type must be created using a condition technique as a prerequisite to generate output for the quality certificate.

From the initial screen, after entering the output type, press ⌈Enter⌋ to display the condition record details screen. Figure 5.24 shows the condition record for printing/sending a quality certificate to a customer. Specify the sales organization, customer, and output parameters to trigger the output of the quality certificate for the customer.

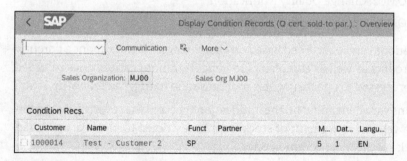

Figure 5.24 Quality Certificate Recipients

5.2.4   Quality Notifications

Quality notifications in SAP S/4HANA are used to capture and analyze defects during quality inspection so that corrective action or preventive action can be taken to improve the quality of the goods. This helps in continuous improvement of the quality of products and materials. The quality notification process is as follows:

1. Describe the problem/defect in the quality notification related to the quality of materials/products.
2. Analyze the problem/defect and find the root cause.
3. Find a solution based on the root cause so that a corrective action or a preventive action can be taken.
4. Implement the tasks for corrective action or preventive action and update the inspection plan/method if required.

Quality notification types distinguish different types of problems/defects based on the origin of the problem:

- Customer complaint about the quality of goods
- Complaint against a vendor about the quality of goods
- Internal problem report

Quality notifications record all the details of the defect/problem, such as the affected material, type of inspection (customer complaint, vendor issues, etc.), inspection characteristics, severity, root cause of the defect, and so on. Depending on the quality notifications, follow-up actions can be taken, such as supplier communications, supplier returns, and so on.

The status of the quality notifications can be updated in the system to track if the defect is investigated, resolved, and so on. Quality notifications can be created/updated in the system using the following transactions, and a quality notification can be created with reference to a purchase order and goods receipt material document:

- Transaction QM01: Create Quality Notification
- Transaction QM02: Change Quality Notification
- Transaction QM03: Display Quality Notification

Quality notifications can be recorded based on customer complaints about a potential defect with a product as well as complaints to vendors. A notification, once recorded, can be further processed to update it with a root cause and to trigger a follow-up action.

Follow-up actions vary depending on the notification type and the severity of the problem. Returns to vendor, movement of stock from unrestricted to blocked, and scrapping of the stock are the most common ones.

5.2.5   Vendor Returns

When quality inspection of incoming materials from the supplier fails, a quality notification will be created, which triggers the supplier returns process.

Figure 5.25 shows the process flow for quality inspection-initiated vendor returns.

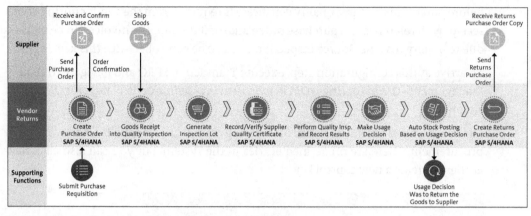

Figure 5.25 Quality Inspection Initiated Vendor Returns

The process steps for the creation of a purchase order and carrying out a quality inspection are largely the same as discussed in Section 5.2.1. The differences are as follows:

- **Automatic stock posting based on usage decision**
  Based on the usage decision codes assigned to the inspection lot, the stock from quality inspection stock will be posted to blocked stock if the usage decision is *rejected—blocked stock*. A quality notification will be created to trigger vendor returns.

- **Creation of returns purchase order**
  Typically, the buyer creates a returns purchase order for the materials to be returned to the supplier using a specific document type that is activated for advanced returns management. Like the stock transfer order, shipping data must be automatically populated at the item level to facilitate the creation of a returns delivery (outbound). At the header level, the **Returns** tab will appear automatically, and the supplier RMA code can be maintained if it's mandatory based on the settings in the business partner. See Chapter 4, Section 4.5 for more details.

5.2.6 Configuration of Quality Management in Procurement

This section explains the key configuration settings for performing quality inspection of procured materials and products in SAP S/4HANA. These settings drive the smooth execution of quality management process in procurement and represent business requirements. Let's dive deep into the various settings and understand how quality management in procurement can be configured in the system.

Define Control Keys for Quality Management in Procurement

In this configuration, control keys for quality management in procurement can be configured. The control key plays a vital role in determining if supplier release is required before purchasing goods, a quality certificate is required from the vendor during goods receipt with reference to a purchase order, and/or if the supplier invoice can be blocked due to quality reasons. Source inspection can also be controlled with this key.

To arrive at this configuration step, execute Transaction SPRO and navigate to **Quality Management • QM in Logistics • QM in Procurement • Define Control Keys for QM in Procurement**.

Figure 5.26 shows the initial screen for defining control keys for quality management in procurement. There are predefined entries in the system, but you can also click **New Entries** to create a new control key.

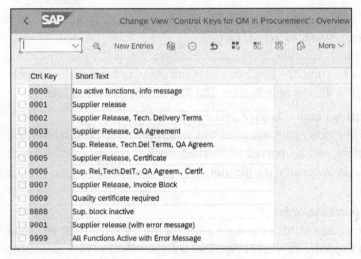

Figure 5.26 Control Keys for Quality Management in Procurement: Initial Screen

Figure 5.27 shows the details screen for the configuration. Let's explore the control indicators:

- **QM Control Key**
  A unique four-digit numeric value that controls the quality management process in procurement. Assign a four-digit code to create your own control key based on business requirements.

- **Document Control**
  This controls if the technical delivery terms and quality assurance documents are required. If you set the **Tech. Delivery Terms** indicator, technical delivery terms must be stored in the **Quality management** view of the material master, and the terms will be defaulted into the purchasing documents and sent to the supplier via purchasing document transmission. If you set the **Q-Agreement Reqd** indicator, the QA agreement must be stored in the purchasing info record (PIR) and sent to the supplier via purchasing document transmission.

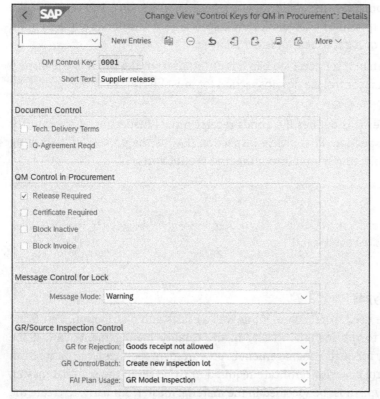

Figure 5.27 Control Keys for Quality Management in Procurement: Details Screen

- **QM Control in Procurement**
  There are four control indicators to control quality management in procurement:
  - **Release Required:** If this indicator is set, the PIRs for the supplier, material, and plant must exist before the purchasing activity.
  - **Certificate Required:** If this indicator is set, a quality certificate must be received from the supplier during the goods receipt. The certificate requirement will be sent to the supplier with the purchasing document output.
  - **Block Inactive:** If this indicator is set, it deactivates the supplier block set in the supplier master data (business partner) and in the PIR, which are set because of quality reasons.
  - **Block Invoice:** If this indicator is set, the invoice document will be blocked for quality inspection when an inspection lot exists, and the inspection is still in progress. The block is applied even if the usage decision was taken to reject the inspection lot. Goods receipt-based invoice verification must be active in the purchase order item as a prerequisite for this block to function.
- **GR/Source Inspection Control**
  There are three control indicators to control goods receipt in quality management:

- **GR for Rejection**: If the inspection lot was rejected at the source, you can control if the goods receipt can be posted, if a new inspection lot creation is required upon goods receipt, or if goods receipt is not allowed.
- **GR Control/Batch**: If the received batch is different from the batch of the inspection lot, you can control if a new inspection lot must be created or not during the goods receipt.
- **FAI Plan Usage**: This defines if a goods receipt model inspection is relevant for first article inspection. First article inspection (FAI) is the process of inspecting materials by selecting samples from the first production run.

Note

A control key for quality management in procurement will be assigned to the material master, typically at the plant level.

Define Certificate Types

Certificate types are used in procurement as well as sales and distribution. Certificate types are assigned to the material master in the **Quality management** view. If the assigned control key for quality management in procurement requires a certificate from the supplier, a control key with certificate required must be set in the material master, and the certificate type must be specified in the material master. Similarly, to create the certificate profile for the outbound certificate, you must specify a certificate type.

In this configuration, you can define the quality certificate types. If the **Certificate Required** indicator is set in the control key for quality management in procurement, you must assign the certificate type in the material master.

To arrive at this configuration step, execute Transaction SPRO and navigate to **Quality Management • QM in Logistics • QM in Procurement • Define Certificate Types**.

Figure 5.28 shows the initial screen for defining certificate types. There are predefined entries in the system. Click **New Entries** to create a new certificate type. Specify a four-digit alphanumeric key and a description to create a new certificate type. You can copy an existing certificate type to create a new one.

Figure 5.29 shows the details screen for the configuration. Let's explore the settings:

- **Certificate for Each PO Item**
  If this indicator is set, a certificate is required per purchase order line item during the certificate goods receipt posting.
- **Certificate for Each GR Item**
  If this indicator is set, a certificate is required per goods receipt line item during the certificate goods receipt posting.

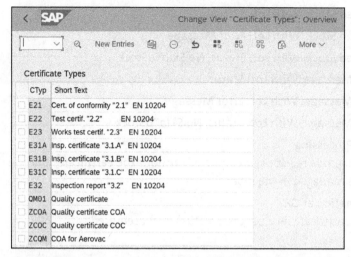

Figure 5.28 Define Certificate Types: Initial Screen

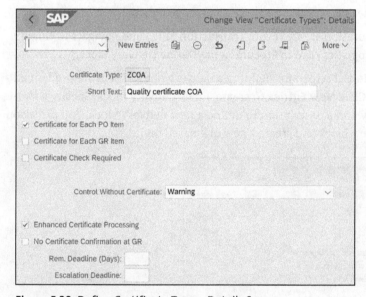

Figure 5.29 Define Certificate Types: Details Screen

- **Certificate Check Required**
  During the goods receipt, the system checks for the certificate. If the certificate is not provided, the system behaves as defined in the **Control Without Certificate** field.

- **Control Without Certificate**
  If the required quality certificate is not provided during the goods receipt, you can set one of the following controls:
  - **Warning**
  - **InformationNo Message**

- – **Without Lot: Blocked Stock, With Lot: Status**
- – **Without Lot: Blocked Stock, With Lot: Error Message**
- – **Without Lot: Blocked Stock, With Lot: Status, No SkipToStock**
- – **Without Lot: Error Message, With Lot: Status**
- – **Without Lot: Error Message, With Lot: Error Message**
- – **Without Lot: Error Message, With Lot: Status, No SkipToStock**

- **Enhanced Certificate Processing**
  If this indicator is set, goods receipt can be posted without the certificate, and the quality certificates are managed in the inbox.

- **No Certificate Confirmation at GR**
  If this indicator is set, certificate receipt confirmation is not required.

Define Delivery Block

In this configuration, you can define a delivery block to be assigned at the supplier master record level or in the QIR.

To arrive at this configuration step, execute Transaction SPRO and navigate to **Quality Management • QM in Logistics • QM in Procurement • Define Delivery Block**.

Figure 5.30 shows the initial screen for defining a delivery block. There are predefined entries in the system. Click **New Entries** to create a new delivery block. Specify a two-digit alphanumeric key and a description to define a new delivery block. You can also copy an existing delivery block and create a new one based on it.

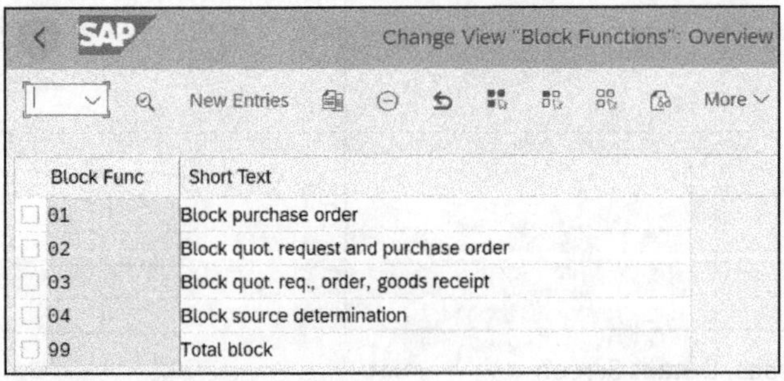

Figure 5.30 Define Delivery Block: Initial Screen

Click the **Details** button at the top or double-click a delivery block to display the details screen. Figure 5.31 shows the details screen for the configuration for editing existing delivery block **01**. The main settings are in the **Control Indicators for Block Function** section.

These indicators control the block functionality of specific documents when a delivery block is set in the corresponding master data (business partner or QIR):

- **Quotation Request**

  If this indicator is set, the creation of a request for quotation is not allowed if the delivery block is set in the corresponding master data (business partner or QIR).

- **Source Determination**

  If this indicator is set, the supplier master record will not be determined during the source determination for purchasing documents.

- **Purchase Order**

  If this indicator is set, the creation of a purchase order is not allowed if the delivery block is set in the corresponding master data (business partner or QIR).

- **Goods Receipt**

  If this indicator is set, the goods receipt for the supplier and material combination is not allowed if the delivery block is set in the QIR.

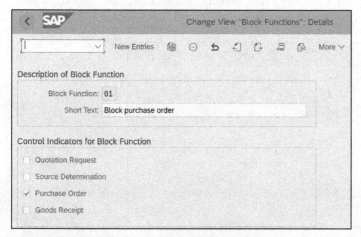

Figure 5.31 Define Delivery Block: Details Screen

Note

If the delivery block is assigned at the business partner (supplier) level, all materials supplied by the supplier will be impacted. If you assign the delivery block at the QIR level, the specific material supplied by the supplier will be impacted.

5.2.7 Configuration of Quality Planning

This section explains the key configuration settings for quality planning in SAP S/4HANA. Inspection planning involves defining the criteria and procedures for quality inspection of materials and products. Let's dive deep into the various settings and understand the configuration settings for quality planning.

Define Defect Classes

In this configuration, you can define defect classes for defect result recording during the quality inspection of materials and products against the inspection lot. A defect class is a two-digit numeric value that is used in result recording to classify the severity of the defect in terms of minor defect, major defect, critical defect, and so on. Each defect class can control the follow-up action after defect result recording such as quality notification updates, triggering a workflow during defect result recording, or no action.

To arrive at this configuration step, execute Transaction SPRO and navigate to **Quality Management • Quality Planning • Basic Data • Catalogs • Define Defect Classes**. Figure 5.32 shows the initial screen for defining defect classes. There are predefined entries in the system. Click **New Entries** to create a new defect class. Specify a two-digit alphanumeric key and a description to define a new defect class. You can also copy an existing defect class and create a new one from it.

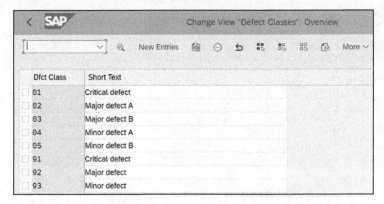

Figure 5.32 Define Defect Classes: Initial Screen

In our example, we'll edit existing defect class **05** to explain the settings. Figure 5.33 shows the details screen for the configuration. Let's explore the settings:

- **Quality Score**
  A quality score is used in vendor evaluation and to determine the quality of the inspection lot. A quality score is high for minor defects and low for major defects. It is defined based on the severity of the defect. Enter the quality score as per the severity of the defect.

- **Defect Class Allowed**
  Set this indicator if you want to record the defect results for this class even though the inspection characteristic attribute is accepted during quality inspection. If you don't set this indicator, you can only record defects to this class if the inspection characteristic attribute is rejected during quality inspection.

- **Next Functions in Results Recording**
  Set one of the following options:
  - **Workflow Link**: When you record a defect for this class for the inspection lot, the system automatically triggers a workflow with event **defectscreated.**
  - **Activation Q-Notif.**: When you record a defect for this class for the inspection lot, the system automatically activates the quality notification.
  - **No next function**: When you record a defect for this class for the inspection lot, the system neither triggers a workflow nor activates a quality notification.

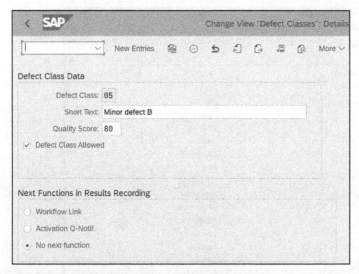

Figure 5.33 Define Defect Classes: Details Screen

Define Catalog Types

In this configuration, you can define catalog types to create catalogs to record inspection results for qualitative characteristics. Catalogs are the generic master data used across all plants to define qualitative data as well as to categorize defects.

To arrive at this configuration step, execute Transaction SPRO and navigate to **Quality Management • Quality Planning • Basic Data • Catalogs • Define Catalog Types**.

Figure 5.34 shows the initial screen for defining catalog types. There are predefined entries in the system. Click **New Entries** to create a new catalog type. Specify a one-digit unique letter or number and a description to define a new catalog type. You can also copy an existing catalog type and create a new one from it.

In our example, we'll edit existing catalog type **1** to explain the settings. Figure 5.35 shows the details screen for the configuration. Let's explore the settings:

- **Selected Sets Allowed**
  Set this indicator if you want to create selected sets from the code list from the catalog.

- **Valuation Required**
  Set this indicator if you want to record the results of the inspection characteristic as accepted or rejected.

- **Deactivate**
  Set this indicator if you do not want to use the catalog codes during results recording.

- **Follow-Up Action**
  Set this indicator if you want to assign a follow-up action to the catalog codes to be triggered after the results recording.

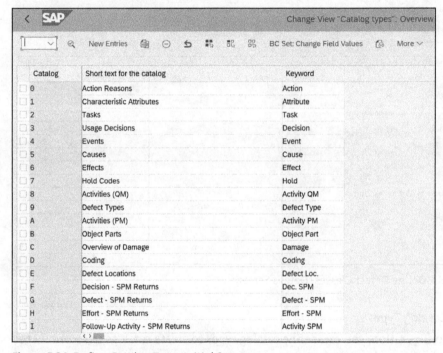

Figure 5.34 Define Catalog Type: Initial Screen

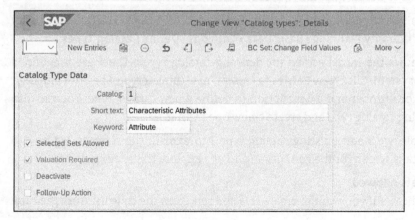

Figure 5.35 Define Catalog Type: Details Screen

Define Control Keys for Inspection Operations

In this configuration, control keys for inspection operations can be configured to manage inspection planning. A control key plays a vital role in determining how quality inspections are conducted, which characteristics are inspected, when inspections are conducted, and so on.

To arrive at this configuration step, execute Transaction SPRO and navigate to **Quality Management • Quality Planning • Inspection Planning • Operation • Define Control Keys for Inspection Operations**.

Figure 5.36 shows the initial screen for defining control keys for inspection operations. There are predefined entries in the system. Click **New Entries** to create a new four-character alphanumeric control key for inspection operations.

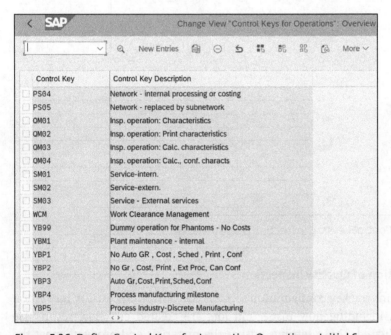

Figure 5.36 Define Control Keys for Inspection Operations: Initial Screen

In our example, we'll edit the existing predefined and standard control key **QM01** to explain the configuration settings. Figure 5.37 shows the details screen for the configuration. Let's explore the settings:

- **Scheduling**
  Set this indicator if you want to schedule inspection operations and suboperations in the inspection plan.

- **Det. Cap. Reqmnts (determine capacity requirements)**
  If you have already set the scheduling indicator, you can set this indicator to determine the capacity requirements of inspection operations and suboperations.

- **Insp. Char. Required (inspection characteristics required)**
  Set this indicator if you want to assign inspection characteristics to the inspection operation in the inspection plan.
- **Automatic GR**
  This setting is not relevant for procurement; it is only applicable to the production planning (process industries) process that involves production orders.
- **Rework**
  Set this indicator if you want to execute rework operations when this control key is assigned to the inspection plan.

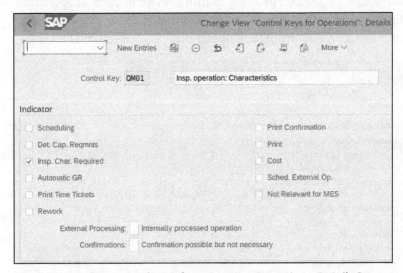

Figure 5.37 Define Control Keys for Inspection Operations: Details Screen

5.2.8   Configuration of Quality Inspection

This section explains the key configuration settings for performing quality inspection in SAP S/4HANA, including how inspection lots are created and how results are recorded and evaluated. These settings ensure the materials and products meet the required quality standards and regulatory requirements. Let's dive deep into the various settings and understand how the quality inspection process can be configured in the system.

Maintain Inspection Types

In this configuration, you can maintain inspection types for the quality inspection of materials and products against the inspection lot.

To arrive at this configuration step, execute Transaction SPRO and navigate to **Quality Management • Quality Inspection • Inspection Lot Creation • Maintain Inspection Type**.

Figure 5.38 shows the initial screen for maintaining inspection types. There are predefined entries in the system. Click **New Entries** to maintain a new inspection type.

Specify a two-character alphanumeric key and a description to define a new inspection type. You can also copy an existing inspection type and create a new one from it.

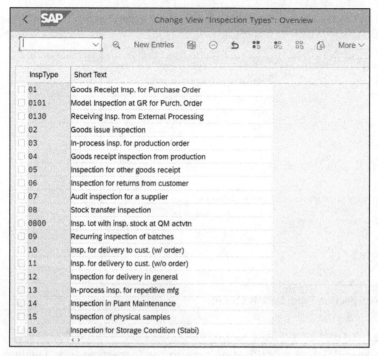

Figure 5.38 Maintain Inspection Type: Initial Screen

In our example, we'll edit existing inspection type **01** to explain the configuration settings. Figure 5.39 shows the details screen for the configuration. Let's explore the important settings for inspection type:

- **Sample Type**
  The sample type determines the origin of the material sample for quality inspection. For goods receipt with reference to a purchase order, the inspection lot will be created automatically, and the system determines the origin of the material sample automatically. You don't need to assign a sample type to this field.

- **UD Selected Set**
  If you have defined a selected set for the usage decision at the plant level, you can assign a usage decision selected set to this field. The system allows you to select one of the usage decision codes from this list when you use this inspection type to create an inspection lot.

- **Recording View**
  This field controls how you can record the results after the inspection. By default, **Single Values and Summarized Results** is set, and the system will allow you to enter single values as inspection results and show summarized results.

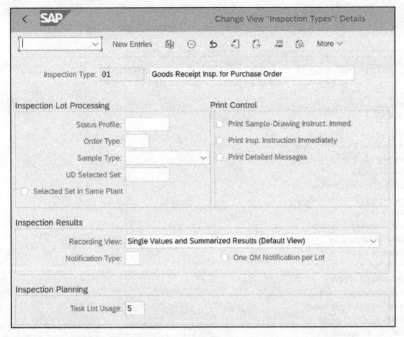

Figure 5.39 Maintain Inspection Type: Details Screen

- **Notification Type**
  You can assign a specific quality notification type to the inspection type. Quality notification types were introduced in Section 5.2.4, and their configuration will be discussed in Section 5.2.9.

- **One QM Notification per Lot**
  Set this indicator if you want to combine all defects and problems during the quality inspection and create one quality notification.

Maintain Inspection Lot Origins and Assign Inspection Types

In this configuration, you can assign specific inspection types to the inspection lot origins. Inspection lot origins are predefined, and you can't define your own origin. Assigning inspection types to inspection lot origins will help in the automatic creation of inspection lots. To arrive at this configuration step, execute Transaction SPRO and navigate to **Quality Management • Quality Inspection • Inspection Lot Creation • Maintain Inspection Lot Origins and Assign Inspection Types**.

Inspection lots can originate from goods receipt against a purchase order, goods receipt against a production order, goods issue against an outbound delivery, and so on. In SAP S/4HANA, these origins are predefined.

Figure 5.40 shows the inspection lot origins that are predefined in the system. Let's explore the key fields:

- **No (number range)**

  Create a number range for the inspection lot creation. To create the number ranges, execute Transaction SPRO and navigate to **Quality Management • Quality Inspection • Inspection Lot Creation • Define Number Ranges for Inspection Lots**. Maintain the number range for object QLOSE.

- **DCr (dynamic modification criteria)**

  To determine the scope of the quality inspection dynamically, dynamic modification criteria are used. These are defined by some of the key fields. For example, for goods receipt inspection with reference to a purchase order, the material, supplier, and manufacturer are used as dynamic modification criteria. This determines the quality level at the time of inspection lot creation.

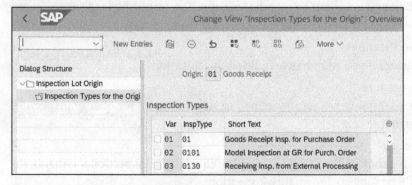

Origin	No	DCr	TLType	Status	Lot Origin Text
01	01	005	Q	4	Goods Receipt
03	03	001		4	Production
04	04	001	Q	4	Goods Receipt from Production
08	08	002	Q	4	Stock Transfer
10	10	002	Q	4	Delivery to Customer with Sales Order
17	17	001	Q	4	Extended Warehouse Inspection
89	89	001	Q	4	Miscellaneous

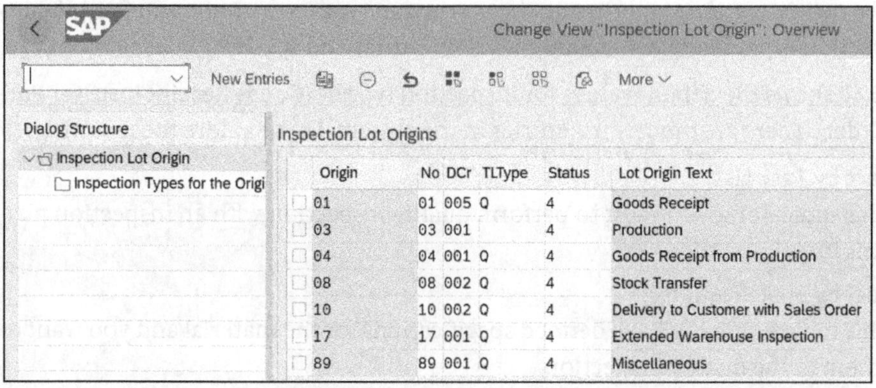

Figure 5.40 Inspection Lot Origins

In the next step, navigate in the **Dialog Structure** area to **Inspection Types for the Origin** to assign the inspection type to inspection lot origin. Figure 5.41 shows multiple inspection types that are assigned to inspection lot origin **01** (**Goods Receipt**). Click **New Entries** to assign the relevant inspection types to the inspection lot origin. The **Var** (variant) field is used to add different variants as inspection types to the inspection type origin in increments of **01**.

Origin: **01** Goods Receipt

Inspection Types

Var	InspType	Short Text	
01	01	Goods Receipt Insp. for Purchase Order	
02	0101	Model Inspection at GR for Purch. Order	
03	0130	Receiving Insp. from External Processing	

Figure 5.41 Assign Inspection Types to Inspection Origin

Define Default Values for Inspection Type

The values from these configuration settings are defaulted at the time of creation of new material master data. Inspection types are assigned to the **Quality management** view of the material master. You can update these default values in the material master later.

To arrive at this configuration step, execute Transaction SPRO and navigate to **Quality Management • Quality Inspection • Inspection Lot Creation • Define Default Values for Inspection Type**.

From the initial screen, **Define Default Values for Inspection Type: Overview**, select an inspection type and click the **Details** button at the top or press $\boxed{\text{Ctrl}}$+$\boxed{\text{Shift}}$+$\boxed{\text{F2}}$ to display the details screen to define default values for the inspection type. If you have defined a new inspection type, to add the inspection type to the **Define Default Values for Inspection Type: Overview** screen, click **New Entries** and specify the inspection type.

Figure 5.42 shows the default values for inspection type 01 (**Goods Receipt Insp. for Purchase Order**). There are predefined entries in the system. Let's explore these settings:

- **Inspect by Task List**
  Set this indicator if you want to perform quality inspection with an inspection plan for this inspection type.

- **Inspect by Specification**
  Set this indicator if you have defined specifications for the material and you want to use them in the quality inspection.

- **Inspect According to Config.**
  Set this indicator if you want to use the quality inspection settings from the purchase order document, production order, or sales order.

- **Assign Specification Automatically**
  Set this indicator if you have already set either the **Inspect by Task List** or **Inspect by Specification** indicator; the system automatically assigns an inspection plan or material specification automatically.

- **Record Characteristic Results**
  Set this indicator if you want to record the inspection results by inspection characteristic.

- **Multiple Specifications**
  Set this indicator if you want to use multiple specifications for quality inspection.

- **Sampling Procedure**
  If you have a sampling procedure defined to select samples for quality inspection, you can assign it in this field.

- **100% Inspection**
  If you set this indicator, the sample size will be equal to the inspection lot; that is, the entire lot will be subjected to quality inspection.

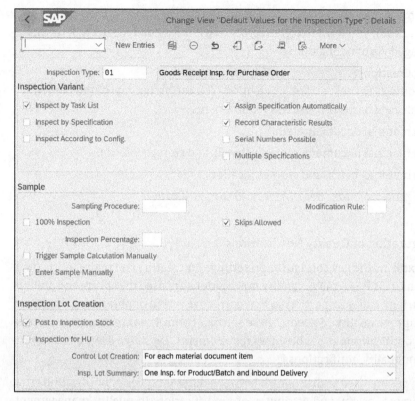

Figure 5.42 Define Default Values for Inspection Type

- **Skips Allowed**
  Set this indicator if you want to skip certain inspection characteristics during quality inspection of lots.

- **Inspection Percentage**
  If you don't have an inspection plan or a material specification for the quality inspection, you can assign a percentage in this field for the inspection sample calculation from the lot.

- **Trigger Sample Calculation Manually**
  If you want to calculate the sample size for quality inspection from the lot manually, set this indicator.

- **Enter Sample Manually**
  Set this indicator if you want to enter the sample size for quality inspection manually.

- **Post to Inspection Stock**
  Set this indicator if you want to automatically post the goods receipt against the purchase order/production order to quality inspection stock in inventory management.

- **Inspection for HU**
  Set this indicator if you want to create inspection lots for handling units, if the material is managed via handling units.

- **Control Lot Creation**
  This field controls the inspection lot creation. You can assign one of the predefined values in this field based on business requirements:
  - For each material document item
  - For each material document, material, batch, and storage location
  - For each material, batch, and storage location
  - For each purchase order item, batch, and storage location

5.2.9   Configuration of Quality Notifications

This section explains the key configuration settings for creating and maintaining quality notifications in SAP S/4HANA. Quality notifications are used to capture and analyze the defects during quality inspection so that corrective action or preventive action can be taken to improve quality of goods. These settings control the creation and processing of quality notifications as per business requirements. Let's dive deep into the various settings for quality notifications.

Notification types are used to distinguish between different types of notifications that define the purpose, origin, and attributes of a notification. In quality management, notifications are used to report, track, and resolve an inspection defect or a problem. There are three important quality notification types to distinguish among different types of problems/defects based on the origin of the problem:

- Customer complaint about the quality of goods
- Complaint against a vendor about the quality of goods
- Internal problem report

To define notification types, execute Transaction SPRO and navigate to **Quality Management • Quality Notifications • Notification Creation • Notification Type • Define Notification Types**.

Figure 5.43 shows the initial screen for defining quality notification types. There are predefined entries in the system. Click **New Entries** to create a new quality notification type. Let's explore the quality notification type settings using notification type **02**, for a complaint against a supplier. Select the notification type and click the **Details** button at the top or press Ctrl + Shift + F2 to display the details screen to define the attributes of the notification type.

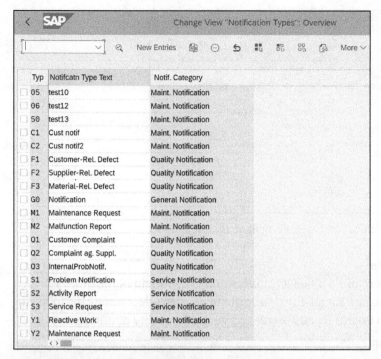

Figure 5.43 Define Notification Types: Initial Screen

Figure 5.44 shows the notification type for a complaint against a supplier. This type is used to record defects during quality inspection of incoming materials and products. Let's explore the key fields:

- **Notif. Category**
  Notification categories are used to distinguish between notifications from different application components like quality, maintenance, services, and so on. Table 5.3 shows the notification categories defined in the standard system. For quality notifications, use notification category **02**.

Category	Notification Category Text
01	Maintenance notification
02	Quality notification
03	Service notification
04	Claim
05	General notification

Table 5.3 Notification Categories

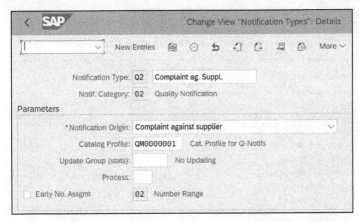

Figure 5.44 Define Notification Types: Compliant against Supplier

- **Notification Origin**
  The notification origin for a quality notification is the origin of the quality inspection defect or problem. For quality notifications, you can select among the following three notification origins if you are creating your own quality notification type:
  - **Customer complaints**
  - **Complaint against supplier**
  - **Internal problem notifications**
- **Catalog Profile**
  The catalog profile contains a list of codes and code groups that can be used to create quality notifications in the system. You must create catalogs for the purpose of creation of quality notifications as a prerequisite before the creation of quality notifications. In the standard system catalog profile, **QM0000001** has been defined for quality notifications.
- **Number Range**
  A number range must be defined and assigned to the notification type. To create the number ranges, execute Transaction SPRO and navigate to **Quality Management • Quality Notifications • Notification Creation • Notification Type • Define Number Ranges for Notification Type**. Maintain the number range for object QMEL_NR.

5.3   Inventory Management

Quality management and inventory management functions are closely integrated within SAP S/4HANA. Effective coordination and management of both functions is essential in the supply chain to optimize inventory levels and maintain quality of materials and products.

Incoming material inspection needs to be performed during the goods receipt from supplier. The system automatically determines the quality inspection requirements and posts the stock to quality inspection automatically. Quality inspection of incoming materials ensures the materials and products meet quality standards that are pre-defined as well as regulatory requirements.

Quality inspection results help to move the inventory within inventory management. The inspected materials that pass the quality inspection will be posted to unrestricted use stock for consumption, and the materials that fail the quality inspection will be placed based on the severity of issues, business requirements, and regulatory require-ments. The failed inspection lots will be moved either to scrap or to blocked stock loca-tions for further action.

Quality management drives continuous improvement of the quality of incoming materials and products. It provides tools and processes to analyze defects during qual-ity inspection so that corrective action or preventive action can be taken to improve quality of goods. You describe a problem or defect in the quality notification related to the quality of materials/products. This helps in the end-to-end process of analyzing a problem or defect, determining its root cause, and finding a solution to the problem.

In this section, let's explore the integration between inventory management (a func-tion within materials management) and quality management. We'll walk through key integration point configuration for stock types, movement types, batch management, usage decisions, and more.

5.3.1   Stock Types

In SAP S/4HANA, stock types are various categories of stock in inventory management and warehouse management. Stock types control various goods movement, consump-tion, availability checks, disposition of defective stock, goods issue for customer orders, and goods issue for stock transfer orders. Stock posting between different stock types is managed in inventory management whenever the status of the materials changes. There are three standard stock types defined for business transactions:

- **Unrestricted use stock**
  This category of stock of a material is available for consumption without any restric-tions. The stock can be used in sales, production, or direct consumption. You can reserve some of the critical materials stock from unrestricted use stock. This indi-cates that the stock has passed quality inspection and is certified for use. During goods receipt, you can post the stock directly into unrestricted use stock if quality inspection requirements do not exist for the materials.

- **Quality inspection stock**
  This category of stock of a material is awaiting quality inspection; a usage decision has not yet been made in quality management. The usage/consumption of this type of stock is restricted. However, some critical materials will be available for stock

availability checks depending on the business requirements; necessary settings are required for availability checks in SAP S/4HANA. In certain business scenarios, stock transfers from one plant to another are done from quality inspection stock. During the goods receipt posting with reference to a purchase order, the stock will be posted to quality inspection stock automatically if the required quality inspection settings exist and the stock can be posted manually into quality inspection during the goods receipt. Based on the inspection results, a usage decision will be made regarding the inspection stock. If the stock passes the quality inspection, the stock will be posted to unrestricted use stock automatically. Similarly, if the quality inspection fails, the stock will be posted to blocked stock automatically, and follow-up actions are then required to properly dispose the stock.

- **Blocked stock**
  This category of stock has a temporary block status for quality reasons, for regulatory reasons, or due to disposition decisions. Materials and products in blocked stock cannot be used. These stock types are the rejected stock from quality management. Quality inspection results drive the automatic posting of quality inspection stock to blocked stock based on usage decisions made in quality management. Depending on the severity of the problem/defect, blocked stocks are further disposed to scrap or vendor returns. You can also post certain stocks of materials and products into blocked stock temporarily from unrestricted use stock and remove the block when the issue or problem is resolved.

In addition to the three main stock types, in SAP S/4HANA there are other stock types that represent a specific status and a category, called *special stock types*. Table 5.4 shows the special stocks defined in the standard SAP S/4HANA system.

Special Stock Indicator	Special Stock Text
B	Customer stock
C	Subcontracting customer stock
E	Orders on hand
F	Subcontracting customer order stock
I	Subcontracting returnable transport packaging
J	Subcontracting vendor consignment
K	Supplier consignment
M	Returnable transport packaging vendor
O	Subcontracting stock

Table 5.4 Special Stock Types

Special Stock Indicator	Special Stock Text
P	Pipeline material
Q	Project stock
R	Subcontracting (SC) project stock
T	Stock in transit
V	Returnable packaging with customer
W	Customer consignment
Y	Shipping unit (warehouse)

Table 5.4 Special Stock Types (Cont.)

Each special stock type serves a special purpose and status and represents a specific business transaction. All stock types, including special stock types, of a material can be displayed at the plant level using Transaction MMBE. Figure 5.45 shows the stock overview of a material in inventory.

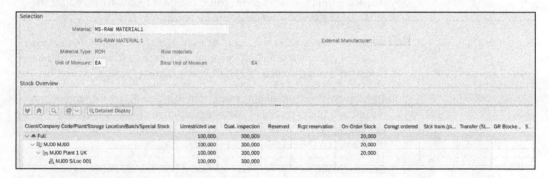

Figure 5.45 Stock Overview

5.3.2   Movement Types for Stock Postings

Movement types are used in quality management for goods movement postings into quality inspection as well as for quality inspection stock movement to other stock types. There are specific movement types for every goods movement transaction into quality inspection and from quality inspection stock. The materials going through the quality inspection process must be taken out of quality inspection stock after the inspection process to a specific stock type depending on the test results. Once the usage decision is made for an inspection lot, the system determines a specific movement type based on the usage decision code and moves the stock into a specific stock type within inventory management.

Table 5.5 shows the movement types used in quality management.

Movement Type	Detailed Description
101	Goods receipt. This movement type is used for standard goods receipts from external vendors. It increases the stock quantity and value in the receiving storage location. Stock can be received into unrestricted use stock or quality inspection stock.
102	Goods receipt reversal. This movement type is used to cancel the goods receipt posted using movement type 101. Stock will be removed from unrestricted use stock or quality inspection stock upon posting this goods receipt reversal transaction.
122	Return to vendor. This movement type is used to return materials to vendor because of poor quality. Stock will be removed from the quality inspection stock to return the goods to the vendor.
321	Transfer posting from quality inspection to unrestricted. This movement type is used to transfer materials which are accepted in quality inspection into unrestricted use stock.
323	Transfer posting from quality inspection to quality inspection stock within plant. This movement type is used to transfer materials which are undergoing quality inspection from one storage location to another storage location within the same plant. This movement type will not change the quality inspection stock type of the materials.
331	Withdrawal of samples from quality inspection stock. This movement type is used to withdraw samples from quality inspection stock for sample consumption. Samples will be taken from inventory for sample inspection.
350	Transfer posting from quality inspection to blocked. This movement type is used to transfer materials that are rejected in quality inspection into blocked stock. The stock posted to blocked stock will be dispositioned further based on the severity of the problem.
553	Goods issue to scrapping. This movement type is used to issue goods for scrapping directly from quality inspection.

Table 5.5 Movement Types Used in Quality Management

5.3.3   Batch Management

Batch management is the management of materials and products in batches through their lifecycle. It helps with quality assurance and tracking of the flow of a material through the supply chain. Materials managed in batches can be easily traced and reported. Batch management provides better visibility and brings efficiency into inventory management processes.

A batch consists of the following attributes in SAP S/4HANA:

- Material or product
- Batch number (alphanumeric)

- Batch status
- Manufacture/production date
- Expiration date

To create, change, and display a batch master record, execute Transactions MSC1N, MSC2N, and MSC3N, respectively. Specify the material, batch number, plant, and storage location, and then press Enter to display the attributes of the batch master record. Figure 5.46 shows the batch master record in SAP S/4HANA. The status of the batch controls the usage of batches: *unrestricted* batches can be used and *restricted* batches can't be used. If the batch of the material is in one of the following stock types, batches are restricted, and it's not possible to change the status of the batch by posting any goods movements manually:

- Quality inspection stock
- Blocked stock
- Stock in transit

Batch management must be activated in the material master to manage the materials in batches. During the goods receipt with reference to a purchase order, the system automatically creates batches and assigns a batch number to each received item.

If a batch is undergoing quality inspection, it is set to **Restricted** status. Once the batch is accepted in quality inspection and an appropriate usage decision is set, the stock is posted to unrestricted use stock in inventory management automatically, and the corresponding batch status will be set to **Unrestricted** automatically.

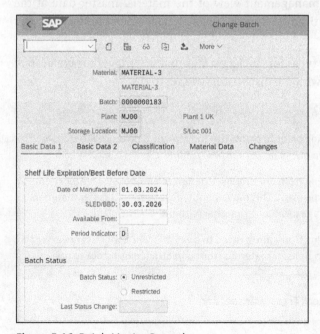

Figure 5.46 Batch Master Record

You can define the inspection lot creation during goods receipt based on the batch of materials and with reference to a purchase order with one of the following criteria:

- For each material document, material, batch, and storage location
- For each material, batch, and storage location
- For each purchase order item, batch, and storage location

This criterion can be set in the inspection type configuration; refer to Section 5.2.8 for the inspection lot creation settings in the configuration. Best practice is to create an inspection lot per batch at the time of goods receipt if the materials are managed in batches.

Batch management is seamlessly integrated with quality management. During the goods receipt inspection, batches undergo quality inspection. Accepted batches in quality management are unrestricted for usage, and rejected batches in quality management will be disposed based on the severity of the problem/defect.

Quality certificates include batch information if the materials are produced and managed in batches. You can create outgoing quality certificates by batch number using Transaction QC22.

See Chapter 6, Section 6.5.5 for more details on batch management functionality.

5.3.4   Stock Transfer Inspection Process

Stock transfer inspection can be activated by assigning inspection type 08 (stock transfer inspection) in the **Quality management** view of the material master data at the plant level. The system automatically creates an inspection lot when the stock transfer is triggered.

Table 5.6 shows the movement types for stock postings that will trigger the creation of an inspection lot for materials with the inspection type set to 08.

Movement Type	Detailed Description
301	Transfer posting plant to plant. This movement type is used to transfer materials from one plant to another within the same company code.
311	Transfer posting from one storage location to another storage location within the same plant. This movement type is used to transfer materials from one storage location to another within the same plant.
322	Transfer posting from unrestricted to quality inspection. This movement type is used to transfer materials from unrestricted use stock to quality inspection.

Table 5.6 Movement Types for Stock Transfer Inspection

Movement Type	Detailed Description
323	Transfer posting from quality inspection to quality inspection stock within plant. This movement type is used to transfer materials which are undergoing quality inspection from one storage location to another storage location within the same plant. This movement type will not change the quality inspection stock type of the materials.
349	Transfer posting from blocked to quality inspection. This movement type is used to transfer materials from blocked stock to quality inspection stock.
309	Transfer posting material to material. This movement type is used to transfer stock of one material to another material within the same plant.

Table 5.6 Movement Types for Stock Transfer Inspection (Cont.)

Figure 5.47 shows the transfer posting for a stock transfer from one storage location to another storage location using movement type 311. As prerequisites, quality management must be active, the indicator to post to inspection stock must be set, and the inspection type 08 must be assigned in the material master at the plant level.

Quality management master data must be created for the material for quality inspection. Refer to Section 5.1 for quality management master data that supports the end-to-end quality inspection process.

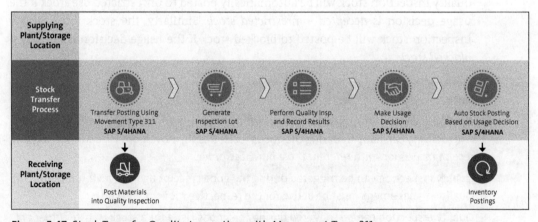

Figure 5.47 Stock Transfer Quality Inspection with Movement Type 311

Let's explore the various stages of stock transfer inspection:

1. **Transfer posting of stock into quality inspection**
   Table 5.6 shows the movement types that are used in stock transfer inspection. These are transfer postings within the plant or across plants. Based on the material settings, whenever a transfer posting is triggered, the system automatically posts the stock into quality inspection.

2. **Automatic generation of inspection lot**

   Based on the settings, if the material is flagged for quality inspection during the transfer postings for stock transfers, the system generates the inspection lot automatically. The inspection lot size will be dependent on the sampling procedure. You can navigate to the inspection lot details from the material document (Transaction MIGO/MB03). The inspection lot will carry quality inspection-related information such as the inspection type, material to be inspected, quantity, and so on.

3. **Perform quality inspection and record inspection results**

   Quality inspection of the incoming materials procured from the supplier will be carried out against the sample size identified in the inspection lot. The results will be recorded in the system using Transaction QA32 with reference to the inspection lot generated at the time of goods receipt. Based on the inspection plan, the results recording will be carried out against every qualitative and quantitative inspection characteristic specified in the inspection plan.

4. **Usage decision**

   A usage decision can be made manually or automatically based on the inspection results. Usage decisions can be made with reference to the inspection lot using Transaction QA32. This includes accepting, rejecting, or returning the inventory of received materials from the suppliers.

5. **Automatic stock posting based on usage decision**

   Based on the usage decision codes assigned to the inspection lot, the stock from quality inspection stock will be automatically posted to unrestricted use stock if the usage decision is *accepted—unrestricted stock*. Similarly, the stock from quality inspection stock will be posted to blocked stock if the usage decision is *rejected—blocked stock*.

> **Note**
>
> Quality inspection for intracompany stock transfer and intercompany stock transfer with delivery can happen during the goods receipt using movement type 101, like goods receipt inspection with reference to a purchase order.
>
> Quality inspection can be triggered during the goods receipt against customer returns. Quality management must be active for the respective movement type.
>
> Quality inspection can be triggered during the goods issue for a stock transfer order or customer order.

5.3.5   Usage Decision

A usage decision is an important step in the quality inspection process. It removes materials from the quality inspection stock. A user decision authorization will be given to quality assurance managers, who thoroughly examine the testing results and make an appropriate usage decision for the inspection lot. Accurate results recording of qualitative and quantitative inspection characteristics is vital in making the right usage decision. The system performs an appropriate goods movement posting automatically when the usage decision is made.

If the usage decision is made to accept the inspection lot, the respective material stock moves from quality inspection to unrestricted use stock. If the usage decision is made to reject the inspection lot, the respective material stock moves from quality inspection to blocked stock. If the usage decision is made to rework, return, or repair the inspection lot, the appropriate follow-up function will be performed.

Usage decisions for the inspection lot can be made manually using the following transactions:

- Transaction QA32: Change Inspection Lot (Worklist)
- Transaction QA11: Record
- Transaction QA13: Display

A usage decision ensures that only those materials that meet the quality standards can be used, and the rejected materials can be sent for repair, rework, or scrapping.

You can make usage decisions in SAP S/4HANA in the following different ways:

- Make a usage decision against a single inspection lot.
- Make a usage decision against multiple inspection lots.
- Make a usage decision automatically when certain conditions are met, such as no defects being found during quality inspection for the lot and none of the inspection characteristics being rejected. A background job must be scheduled to perform an automatic usage decision for the inspection lot.

5.3.6   Configuration of Inventory Management for Quality Management

This section explains the key configuration settings to integrate quality management and inventory management in SAP S/4HANA. For the smooth flow of materials through the quality inspection process, these settings are vital. These settings ensure automation and standardization of quality management processes and that required quality standards for materials and products are met. Let's explore the important settings for inventory management that correspond to quality management.

Inspection for Goods Movements

Movement types control the inspection lot creation for quality inspection. When the goods movement transactions are posted, the system determines the inspection lot origin, and from that, it determines the inspection types. In this configuration, we can activate or deactivate quality management for movement types. If you activate quality management for a movement type, you must maintain an inspection lot origin.

As a prerequisite to this step, inspection types are assigned to inspection lot origins. Refer to Section 5.2.8 for more details about the link between the inspection lot origin and inspection type.

To arrive at this configuration step, execute Transaction SPRO and navigate to **Quality Management • Quality Inspection • Inspection Lot Creation • Inspection for Goods Movements**.

This configuration step has two activities, and you can select the following configuration activities from the **Activities** screen:

- **Movement Type/Update Control/Inspection Lot Origin**
  Double-click the **Movement Type/Update Control/Inspection Lot Origin** activity from the activities screen to assign the inspection lot origin to a movement type. Figure 5.48 shows that inspection lot origin (**Inspect.lot origin**) 01 is assigned to movement type **101**.

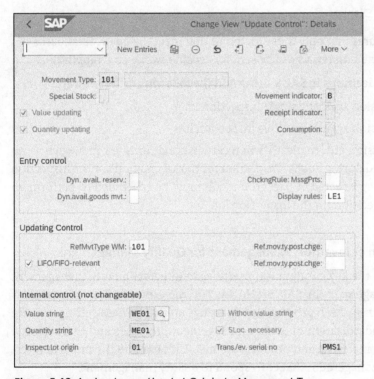

Figure 5.48 Assign Inspection Lot Origin to Movement Type

- **Deactivate Quality Inspection for a Movement Type**
  Double-click the **Deactivate Quality Inspection for a Movement Type** activity from the **Activities** screen to deactivate quality inspection for a movement type. Figure 5.49 shows that the **QM not active** indicator is not set for movement type **101**. If you want to deactivate quality management for any movement type, set the **QM not active** indicator.

Change View "Deactivate Movement Types for QM Inspection..": Overview

New Entries More

Deactivate Movement Types for QM Inspection Processing

MvT	S	Movement ind.	Receipt Ind.	Consumpt.	Movement Type Text	QM not active
101		Goods movement for purchase order	Normal receipt	No consumption	GR goods receipt	
101		Goods movement for purchase order	Normal receipt	Asset	GR goods receipt	
101		Goods movement for purchase order	Normal receipt	Sales order	GR goods receipt	
101		Goods movement for purchase order	Normal receipt	Project	GR goods receipt	
101		Goods movement for purchase order	Normal receipt	Consumption	GR goods receipt	
101		Goods movement for purchase order	Stock transport order	No consumption	GR goods receipt	
101		Goods movement for purchase order	Stock transport order	Asset	GR goods receipt	
101		Goods movement for purchase order	Stock transport order	Consumption	GR goods receipt	
101		Goods movement for production order	Normal receipt	No consumption	GR goods receipt	
101		Goods movement for production order	Normal receipt	Asset	GR goods receipt	
101		Goods movement for production order	Normal receipt	Consumption	GR goods receipt	
101	B	Goods movement for purchase order	Normal receipt	No consumption	GR goods receipt	
101	B	Goods movement for purchase order	Normal receipt	Consumption	GR goods receipt	
101	B	Goods movement for production order	Normal receipt	No consumption	GR goods receipt	
101	E	Goods movement for purchase order	Normal receipt	No consumption	GR goods receipt	
101	E	Goods movement for purchase order	Normal receipt	Sales order	GR goods receipt	
101	E	Goods movement for purchase order	Normal receipt	Project	GR goods receipt	
101	E	Goods movement for purchase order	Normal receipt	Consumption	GR goods receipt	

Figure 5.49 Activate/Deactivate Quality Management for Movement Types

Define Inventory Postings

In quality management, usage decisions are made using Transaction QA11 against the inspection lot based on inspection results. The usage decision codes control various stock postings from quality inspection within inventory management. To arrive at this configuration step, execute Transaction SPRO and navigate to **Quality Management • Quality Inspection • Inspection Lot Completion • Define Inventory Postings.**

Figure 5.50 shows the configuration of stock postings for usage decisions. These are predefined entries in the system, and the SAP recommendation is not to change this configuration.

Change View "QM: Inventory Postings with Usage Decision": Overview

New Entries More

Posting UD	Origin	Quantity Description	Q Stock	UnrestrStk	GR Blocked
TRANSFER01		Stock Transfers (Plant)	323	301	
TRANSFER02		Stock Transfers (StLoc)	323	311	
VMENGE01		To Unrestricted Use	321	311	105
VMENGE02		To Scrap	553	551	
VMENGE03		To Sample Consumption	331	333	333
VMENGE04		To Blocked Stock	350	344	
VMENGE05		To Reserve Samples	350	344	
VMENGE06		New Material	321	309	
VMENGE07		Return Posting	122	122	124

Figure 5.50 Define Inventory Postings

Let's explore the key fields:

- **Posting UD (posting proposal in usage decision)**
  When a usage decision is made against the inspection lot, the posting proposal linked to the usage decision will be triggered automatically and the respective stock posting will be performed in the background. These posting proposals are predefined in the system.

- **Q Stock (quality stock), UnrestrStk (unrestricted stock), and GR Blocked (blocked stock)**
  These are movement types for stock postings from quality inspection. Assign a relevant movement type for the posting proposal so that an automatic goods movement will be posted in the background. Assign or change the movement type carefully so that the relevant stock postings can trigger in the background.

5.4   Summary

In this chapter, we introduced the integration of quality management functions with materials management and offered best practice examples and processes. In Section 5.1, we explained the master data in quality management with procedures to maintain quality management-specific master data in SAP S/4HANA. In Section 5.2, we explained the integrated quality management process for purchasing materials from suppliers, including quality inspection, quality certificate, and quality notification processes. We also explained the various configuration settings to support the quality management of received materials from suppliers. Finally, in Section 5.3, we explained how the inventory management functionality of materials management is integrated with quality management, including batch management and usage decision postings in quality management after the inspection process.

In the next chapter, we'll discuss the integration between production planning and materials management to optimize manufacturing process and inventory levels, with best practice examples, master data requirements, and configuration settings.

Chapter 6
Production Planning

Production planning in SAP S/4HANA provides a comprehensive list of tools and procedures to create, manage, and execute production plans in order to produce finished products and meet customer demand, optimize production processes and resources, and achieve operational excellence. Production planning and materials management integration ensures that required materials and components are made available for production processes on time.

Production planning creates a bill of materials (BOM) with a list of materials, components, and assemblies to produce a finished product. Materials management functions ensure the required materials, components, and assemblies are available to produce the finished product. The quantities of the finished products to be produced are driven by customer demand in terms of sales orders and demand forecasts based on historical data and other criteria. The material requirements planning (MRP) function of production planning calculates the material requirements to produce finished products, while the inventory management function of materials management issues the available materials to the production process, and the procurement function of materials management ensures that unavailable materials are procured to support the production process.

This chapter covers production planning processes, integration points between production planning and materials management functionalities, production planning master data, and the key configuration settings in SAP S/4HANA. First, let's understand the different manufacturing processes across industries.

6.1 Integrated Production Planning Processes

Before we get into the details, we'll explore the different manufacturing processes:

1. **Discrete manufacturing**
   Discrete industries produce unique products from their production lines. The products produced in discrete industries mainly consist of different components that are assembled into a finished product. Examples of such products include cars, mobile devices, laptops, toys, home appliances, television sets, cameras, and so on. Such products can be disassembled, recycled, and reassembled.

In discrete manufacturing, the demand and supply vary with time, and so the frequency of production of finished goods varies with time. Discreate manufacturers produce not only finished products, but also semifinished products and spare parts.

The master data required in SAP S/4HANA for discrete manufacturing is as follows:

- Material master.
- Bill of materials.
- Work center.
- Routing.
- Production version.
- Production orders are the main document for discrete manufacturing, and the required materials to manufacture the products are issued against a production order.

2. **Process manufacturing**

The final products are produced in batches out of a master recipe, a chemical mixture, or a specific composition of ingredients. Manufacturing processes of process industries are highly regulated, and compliance requirements are followed and validated from procuring raw materials to producing finished products and through storage in inventory. The final products undergo thorough quality inspections to check quality standards and compliance with regulatory requirements. Examples of such products include packaged food and beverages, dairy products, fertilizers, pharmaceuticals, and so on.

In process manufacturing, final products can be produced via continuous production or discontinuous production depending on the customer demand and forecasts. Process manufacturers produce mainly finished products only.

The master data required in SAP S/4HANA for process manufacturing is as follows:

- Material master.
- Bill of materials.
- Resource (called a work center in discrete manufacturing).
- Master recipe (called a routing in discrete manufacturing).
- Production version.
- Process orders are the main document for process manufacturing, and the required materials are either continuously supplied or discontinuously supplied.

3. **Repetitive manufacturing**

Repetitive manufacturing industries produce large amounts of the same final products repetitively and continuously. The manufacturing processes in this industry are mostly automated so that a large quantity of the same product can be produced with a shorter lead time. Examples of such products are smartphones, cars, toys, and so on.

The master data required in SAP S/4HANA for repetitive manufacturing is as follows:

– Material master.
– Bill of materials.
– Production line (called a work center in discrete manufacturing).
– Rate routing (called a routing in discrete manufacturing).
– Production version.
– Production orders or process orders are used for manufacturing depending on the final products. The required raw materials must be procured and maintained in inventory for continuous supply to production.

There are two distinct production planning processes in SAP S/4HANA that are common across industries. The first is the *make-to-order* (MTO) process, as shown in Figure 6.1, which is a strategy used in manufacturing industries for production planning and to produce finished products based on a customer order. In the MTO process, the finished products are produced to fulfill a customer order after it is received.

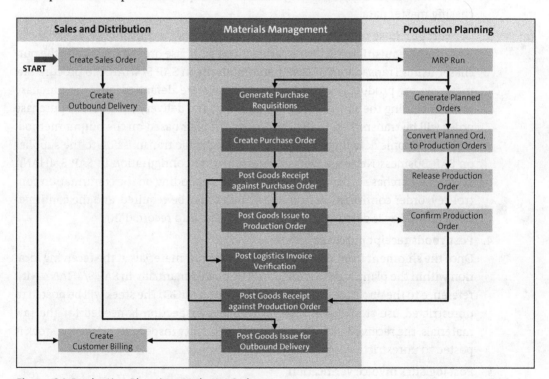

Figure 6.1 Production Planning: Make-to-Order

The planning and production process in SAP S/4HANA is triggered by the creation of a sales order for the products requested by the customer. Because the customer demand varies over time, the manufacturers must consider the average unit cost of manufacturing, procurement lead times for raw materials procurement, and manufacturing lead times to produce finished products while considering capacity planning for resources/work centers. Examples of MTO products include aircrafts, large ships, certain luxury cars, and so on.

Let's walk through these process steps:

1. **Sales order creation**

   The MTO process starts when a customer order is received for products. Customer orders requesting finished products will be converted into sales orders in SAP S/4HANA using Transaction VA01. An availability check will be performed during the sales order creation and an email notification will be triggered for the material planner about any unavailable finished products needed to fulfill the sales order.

2. **MRP run**

   The material planner will run MRP with reference to a sales order using Transaction MD50 for MTO planning. The system generates a planned order for the finished product manufacturing and purchase requisitions for raw materials if not available in the inventory. The system assigns a source of supply for raw materials from purchasing master data.

3. **Convert purchase requisitions into purchase orders**

   A purchase requisition will be converted into a purchase order manually by the purchaser using Transaction ME21N or automatically in SAP S/4HANA to procure raw materials for production. The line item details are defaulted into the purchase order, including the source of supply, unit cost, and delivery address. The purchase order will be transmitted to the third-party supplier based on the output method: email, electronic data interchange (EDI), or electronic transmission to the supplier on SAP Business Network. Output determination configuration in SAP S/4HANA drives the purchase order output mechanism. Depending on the confirmation control key, order confirmation from the supplier may be required, and the confirmation control key is defaulted from the purchasing info record (PIR).

4. **Post goods receipt purchase order**

   Once the shipment from the supplier arrives for raw materials at the receiving location within the plant, a goods receipt will be posted manually in SAP S/4HANA with reference to the purchase order using Transaction MIGO. The stock will be posted to unrestricted use stock upon receipt. If quality inspection is needed for the raw materials, the received materials will undergo quality inspection before the stock is posted to unrestricted use stock.

5. **Post logistics invoice verification**

   A supplier sends an invoice for the purchase order via email (paper invoice), EDI, or electronically via SAP Business Network. The supplier invoice document will be posted by accounts payable personnel manually in SAP S/4HANA using Transaction MIRO in the case of a paper invoice. Automatic posting and reconciliation of the invoice happens if an electronic invoice is received from the supplier. The next scheduled payment run will pick up this fully reconciled invoice and post the payment to the supplier based on the preferred payment method of the supplier. Accounts payable personnel can manually process the payment against the fully reconciled invoice as well.

6. **Convert planned orders into production orders**
Planned orders are the results of the MRP run, and a planned order represents materials that are not available in the inventory to fulfill a customer order. A planned order is converted into a production order to produce finished goods using Transaction CO01. A production order consists of finished products to be produced, the quantity to be produced, and the production dates.

7. **Release production order**
A production order must be released to make it available for the shop floor to start the production activities. This is done using Transaction CO02.

8. **Post goods issue raw materials to production**
The received raw materials and other required materials/components will be issued to production with reference to production order using Transaction MIGO. This uses movement type 261.

9. **Confirm production order**
Production orders will be confirmed by the production workers with quantities produced. This confirmation also processes the goods issue of raw materials that are consumed during production and are not issued manually.

10. **Post goods receipt with reference to production order**
The produced finished products will be posted into inventory by performing goods receipt with reference to production order using Transaction MIGO. The stock will be posted to unrestricted use stock upon posting goods receipt. This uses movement type 101.

11. **Create outbound delivery with reference to sales order**
Once the finished goods are available in stock to fulfill the sales order, an outbound delivery will be created manually using Transaction VL01N. This can be automated as well. Outbound deliveries help organizations to prepare for the shipments of goods to the customer and contain specific quantities of the materials to be shipped and delivery dates for scheduling shipments. Inventory and warehouse management activities such as picking of materials and packing activities are performed with reference to outbound deliveries.

12. **Post goods issue for outbound delivery**
Goods issue will be performed with reference to the outbound delivery after the inventory management functions such as picking and packing of materials are completed. This is an important functionality of inventory management that removes the materials from inventory and issues an outbound delivery to ship them to the customer. Movement types play an important role in determining the type of inventory movement and posting the goods issue. Movement type 601 is

used for goods issue for an outbound delivery to ship goods to a customer. Goods issue for outbound delivery can be posted using Transaction VLO2N.

13. **Create customer billing document**

Goods customer billing/invoice documents can be created with reference to sales orders or outbound delivery documents. Best practice is to create the billing document after the shipment (goods issue for outbound delivery) is completed. The customer billing document can be created using Transaction VFO1. The system calculates output tax and determines an output tax code based on the country of origin, shipping address, and product/services delivered. Output tax will be added to the billing document and collected from the customer. The seller will pay the taxes on behalf of the customer to the local government authority of the shipping county and state. The billing document will be transmitted to the customer via email or EDI.

14. **Process incoming payments**

Incoming payments or payments received from customers against the outstanding billing documents will be processed in SAP S/4HANA as part of the accounts receivable function. Payment receipts are entered into the system using Transaction F-28. A reference document (check, eCheck, etc.), payment amount, payment date, and payment method will be recorded in the payment receipt document.

The second production planning process, the *make-to-stock* (MTS) process, is a strategy used in manufacturing industries to production plan and produce finished products based on forecast. In MTS, the finished products are produced in advance to fulfill a future demand. The planning and production process in SAP S/4HANA is triggered by the forecast. The system provides tools and procedures to generate forecasts based on historical sales demand, current trend, seasonal products, and so on. Customer orders are fulfilled faster in MTS than in MTO. Examples of MTS products include packaged food and beverages, consumer goods, clothes, and so on.

Figure 6.2 shows the process flow for the MTS scenario. Let's walk through the process steps:

1. **Create planned independent requirements**

The MTS process starts with the creation of a forecast demand for the finished products. Planned independent requirements without reference to sales order and with reference to forecasts will be created using Transaction MD61. The planned independent requirements will be the input for MRP.

2. **MRP run**

The material planner will run MRP for planned independent requirements using Transaction MD01N. The system generates a planned order for the finished product manufacturing and purchase requisitions for raw materials if not available in the inventory. The system assigns a source of supply for raw materials from purchasing master data.

3. **Convert purchase requisitions into purchase order**

   The purchase requisition will be converted into a purchase order manually by the purchaser using Transaction ME21N or automatically in SAP S/4HANA to procure raw materials for production. The line item details are defaulted into the purchase order, including the source of supply, unit cost ,and delivery address. The purchase order will be transmitted to the third-party supplier based on the output method: email, EDI, or electronic transmission to the supplier on SAP Business Network.

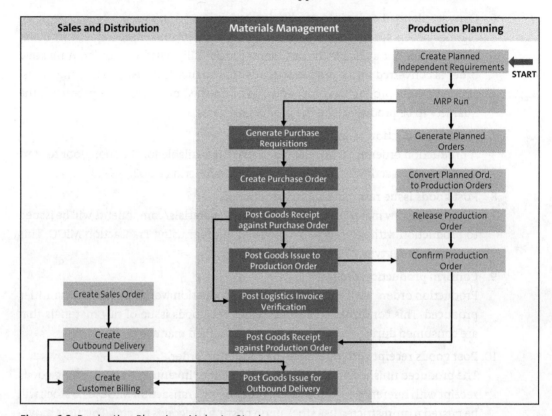

Figure 6.2 Production Planning: Make-to-Stock

4. **Post goods receipt purchase order**

   Once the shipment from the suppliers arrives for raw materials at the receiving location within the plant, goods receipt will be posted manually in SAP S/4HANA with reference to the purchase order using Transaction MIGO. The stock will be posted to unrestricted use stock upon receipt. If quality inspection is required for the raw materials, the received materials will undergo quality inspection before the stock is posted to unrestricted use stock.

5. **Post logistics invoice verification**

   The supplier sends an invoice for the purchase order via email (paper invoice), via EDI, or electronically via SAP Business Network. The supplier invoice document will be posted by the accounts payable personnel manually in SAP S/4HANA using Transaction MIRO in the case of a paper invoice. Automatic posting and reconciliation of the invoice happens if an electronic invoice is received from the supplier. The next scheduled payment run will pick up this fully reconciled invoice and post the payment to the supplier based on the preferred payment method of the supplier. Accounts payable personnel can manually process the payment against the fully reconciled invoice as well.

6. **Convert planned order into production order**
Planned orders are the results of the MRP run, and a planned order represents materials that are not available in the inventory to fulfill a customer order. A planned order is converted into a production order to produce finished goods using Transaction COO1. A production order consists of finished products to be produced, the quantity to be produced, and the production dates.

7. **Release production order**
A production order must be released to make it available for the shop floor to start the production activities. This is done using Transaction COO2.

8. **Post goods issue raw materials to production**
The received raw materials and other required materials/components will be issued to production with reference to a production order using Transaction MIGO. This uses movement type 261.

9. **Confirm production order**
Production orders will be confirmed by the production workers with the quantities produced. This confirmation also processes the goods issue of raw materials that are consumed during production and are not issued manually.

10. **Post goods receipt with reference to production order**
The produced finished products will be posted into inventory by performing goods receipt with reference to a production order using Transaction MIGO. The stock will be posted to unrestricted use stock upon posting the goods receipt. This uses movement type 101.

MTS will increase the inventory of finished goods for future sales orders. Both MTO and MTS have their own pros and cons. Depending on the products, customer demand, and seasonal demand, organizations will choose a strategy to execute.

Table 6.1 shows all the key processes impacted in production planning and materials management.

Area	Processes Impacted	Description
Production planning	Maintain production master data	This process creates and maintains production master data such as BOMs, work centers, routings, and production versions.
	Perform MRP	This process supports MRP to generate planned orders for finished product manufacturing and purchase requisitions for raw materials if not available in inventory.
	Manage production operations	This process supports the monitoring of production activities that are in progress.
	Manage external operations	This process supports the external operations performed by a subcontractor (external vendor) during the production of finished goods.
Materials management	Sourcing materials and services	This process supports the sourcing of a best source of supply to procure raw materials for production.
	Create and maintain purchasing master data	This process creates and maintains purchasing master data such as the source list and PIR to support the purchasing of raw materials from suppliers.
	Create and manage outline agreements	This process supports creation of outline agreements such as contracts with suppliers to procure raw materials for production.
	Create and maintain purchase requisition	This process supports the creation, generation, and maintenance of purchase requisitions that originate from MRP.
	Create and maintain purchasing documents	This process focuses on the creation of purchasing documents to procure raw materials from suppliers.
	Post goods receipt	This process supports the goods receipt of finished products from production into inventory.
	Post goods issue materials to production	This process supports the issuing of raw materials and components to production with reference to a production order.
	Supplier invoice management	This process supports the processing of incoming invoices from suppliers.

Table 6.1 Process Impact: Production Planning

6.2   Master Data in Production Planning

The material master is the central repository for all material information related to purchasing, sales, MRP, work scheduling, accounting, storage, warehouse management, and quality management. Refer to Chapter 3, Section 3.1 for more details. This section covers production planning-specific master data. The work scheduling view of the material master is required for production planning to control scheduling of production/process orders.

Master data plays a vital role in the production planning processes. The following sections will discuss the key master data components used in SAP S/4HANA production planning.

6.2.1   Bill of Materials

A BOM is a list of materials, components, subassemblies, and semifinished products in specific quantities to produce a finished product. The BOM is the most essential part of the production process. During MRP, the BOM is exploded to determine all dependent requirements—that is, the required quantities of materials and components to produce finished products. The BOM helps to optimize the inventory of raw materials required for production in inventory management. This helps with procurement of the right raw materials in the right quantities for production consumption. It also helps in managing production activities effectively.

Let's explore the different types of BOMs in SAP S/4HANA:

- **Production BOM**
  This type of BOM is used in a production order to produce finished products, semifinished products (if the selling organizations sell those), and spare parts. It lists the raw materials, components, and semifinished products, and their specific quantities to produce the final products. The final products in a BOM are listed as parent items, and the raw materials and components are listed as child items.

- **Sales BOM**
  This type of BOM is used in sales and distribution to sell final products to the customers. Typically, a sales BOM consists of a parent item, which is the finished product, and the associated components as child items. An example sales BOM could be for a television set, where the television is listed as the parent item, while the remote control, HDMI cable, power cord, and so on are listed as child items.

- **Assembly BOM**
  This type of BOM is used in sales and distribution to sell final products to the customers. Unlike a sales BOM, an assembly BOM consists of the final finished product as its parent item and the required components assembly as the child items. The components are assembled into a finished product. A classic example is a bicycle,

which is sold in a box with subassembly components and other components. Usually, consumers assemble the child items into a finished product.

- **Template BOM**
This type of BOM is used in sales and distribution as well as in production. Typically, a template BOM contains a finished product as the parent item and a list of optional components as child items. A template BOM provides flexibility in selecting and removing items from the sales order, production order, and so on. An example of a template BOM is for a car that comes with components like leather seats, premium audio, and a sunroof as upgrade options.

In the following sections, we'll explain how to create a BOM and walk through the configuration.

Create Bills of Materials

Let's discuss the creation of BOMs in SAP S/4HANA. BOMs are created in the system using Transaction CS01.

Figure 6.3 shows the initial screen for creating a BOM for a finished product. Let's explore the fields in the initial screen:

- **Material**
In this field, enter the material number of the finished good or the component to be produced. This is the parent item of the BOM.

- **Plant**
Enter the plant code in this field for the plane in which the BOM will be used to produce the finished product or a component.

- **BOM Usage**
In this field, enter a function or a business unit within the organization in which the BOM will be used. The following are the functions within the organization in which a BOM can be used:
  - **1: Production**
  - **3: Universal**
  - **4: Plant Maintenance**
  - **5: Sales and Distribution**
  - **6: Costing**

For production, select BOM usage code **1**.

- **Alternative BOM**
If you create additional BOMs for the same material in the same plant, you can assign an alternative BOM in this field.

- **Valid From**
In this field, enter the date from which the BOM is valid.

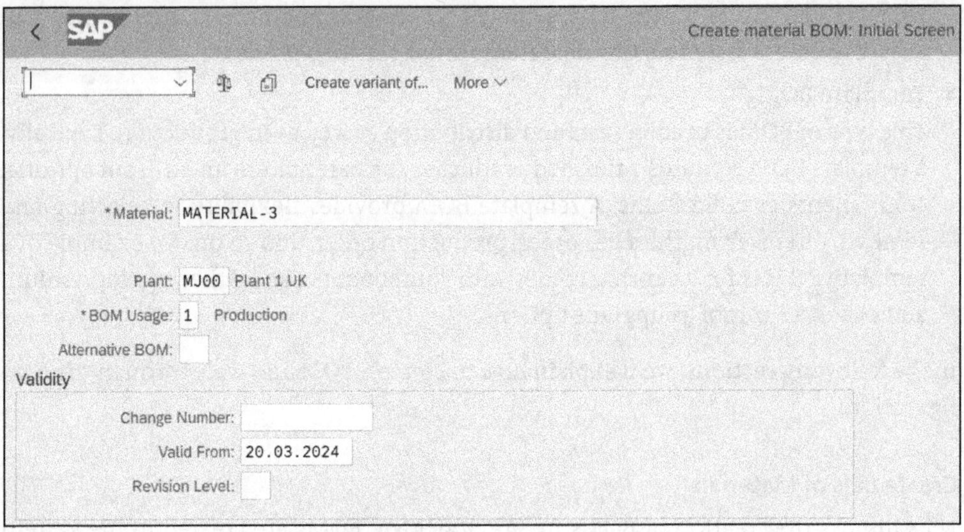

Figure 6.3 Create Material BOM: Initial Screen

Press ⌈Enter⌉ after specifying the plant, materials, BOM usage indicator, and the valid from date to navigate to the general item overview screen from the initial screen. Figure 6.4 shows the **General Item Overview** screen for the BOM. In this screen, you can list all raw materials, components, and subassemblies used in the production of the parent item of the BOM. Let's explore some of the key fields:

- **ICt (item category)**
  The item category is used to distinguish between different categories of BOM items. The system uses the value in this field to control activities during the BOM usage. The following are the different item categories defined for BOM items:
  - Document item (**D**): Use this item category if you want to list the design document or another specification as an item in the BOM for reference.
  - Class item (**K**): If you have a component or a material that is a configurable material, you can enter a material class with one or more characteristics for the BOM item.
  - Stock item (**L**): Use this item category if the material or component is stored in the inventory. This material or component can be issued to production using movement type 261.
  - Nonstock item (**N**): Use this item category to list a nonstock item as a BOM item. A nonstock item can have a material number or just a short text (without a material number).
  - Variable-size item (**R**): Use this item category to list a variably sized material as a BOM item—for example, a metallic sheet.
  - Text item (**T**): Use this item category to enter a long text in between different BOM items.

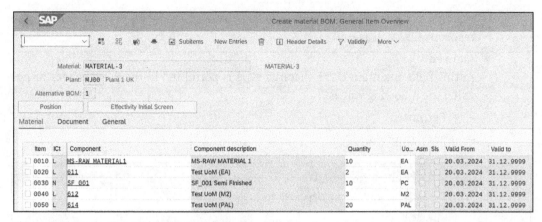

Figure 6.4 Create Material BOM: Item Overview

- **Component**
  In this field, enter a material master ID for relevant item categories such as stock item, variable-size item, and so on. The system automatically determines the unit of measure (**UoM**) for the components with the material ID.

- **Quantity**
  These are the component quantities required to produce a unit of the finished product or the parent item of the BOM. If you are listing a by-product (a material that is created during the production of another material or parent material), then it must be listed in negative quantities as a separate item of the BOM.

- **Asm (assembly)**
  Set this indicator if the component listed in the BOM is an assembly. An assembly item is an inventory item made up of multiple components, but identified as a single item.

Define Material Types Allowed for BOM Header

In this configuration step, you can define allowed material types and disallowed material types for BOM header data. This configuration controls if the material for the parent item of the BOM is allowed or not.

To arrive at this configuration step, execute Transaction SPRO and navigate to **Production • Basic Data • Bill of Materials • General Data • Define Material Types Allowed for BOM Header**. Click **New Entries** from the initial screen to define a new material type to be allowed or disallowed in the BOM header. This configuration setting is added based on the BOM usage indicator.

Figure 6.5 shows the following configuration settings:

- **Mat type headr (material type header)**
  Define the material type you want to allow or disallow in the BOM header (parent

material of the BOM). The value * indicates that all material types are allowed for the respective BOM usage.

- **Allowed**
  In this field, maintain if the material type is permitted in the BOM header or not using the following values:
  - **+**: Permitted
  - **-**: Not permitted

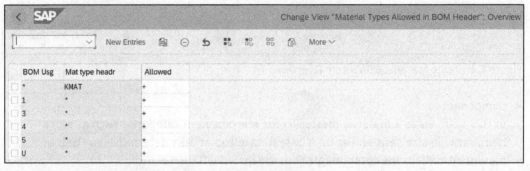

BOM Usg	Mat type headr	Allowed
*	KMAT	+
1	*	+
3	*	+
4	*	+
5	*	+
U	*	+

Figure 6.5 Define Material Types for BOM Header

Define Material Types Allowed for BOM Items

In this configuration step, you can define allowed material types and disallowed material types for BOM items/component data. This configuration controls if the material for the item data of the BOM is allowed or not.

To arrive at this configuration step, execute Transaction SPRO and navigate to **Production • Basic Data • Bill of Materials • Item Data • Define Material Types Allowed for BOM Items**. Click **New Entries** from the initial screen to define a new material type to be allowed or disallowed in BOM items. This configuration setting is added based on the **BOM Usg** (BOM usage) indicator and **Mat Type headr** (material type header).

Figure 6.6 shows the following configuration settings:

- **Mtype BOM item (material type BOM item)**
  Define the material type you want to allow or disallow in the BOM items (components of the BOM). The value * indicates that all material types are allowed for the respective BOM usage.

- **Allowed**
  In this field, maintain if the material type is permitted it the BOM items or not using the following values:
  - **+**: Permitted
  - **-**: Not permitted

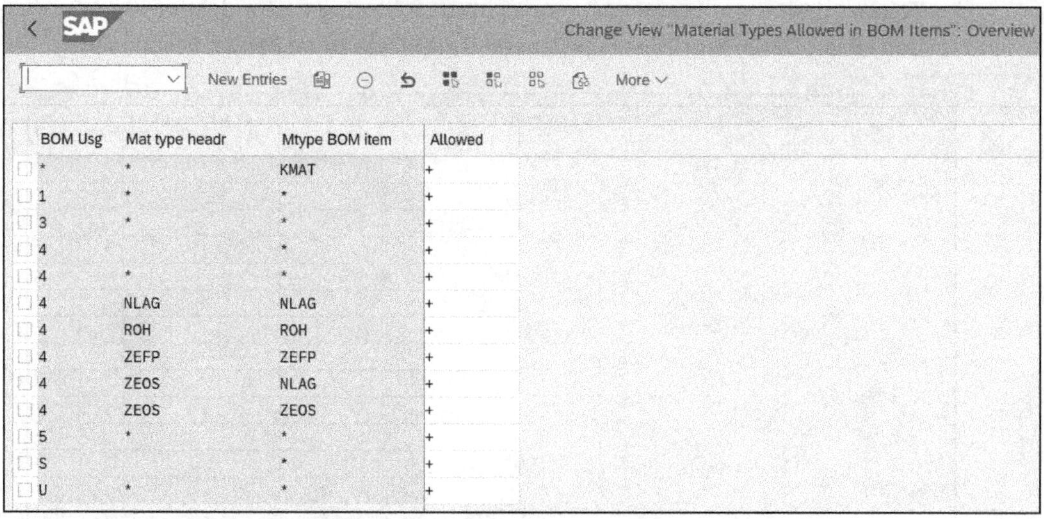

BOM Usg	Mat type headr	Mtype BOM item	Allowed
*	*	KMAT	+
1	*	*	+
3	*	*	+
4		*	+
4	*	*	+
4	NLAG	NLAG	+
4	ROH	ROH	+
4	ZEFP	ZEFP	+
4	ZEOS	NLAG	+
4	ZEOS	ZEOS	+
5	*	*	+
S		*	+
U	*	*	+

Figure 6.6 Define Material Types for BOM Items

6.2.2 Work Center

A work center is the location within a manufacturing plant where the production activities will be performed. In production planning, the capacity of the work centers is defined in terms of the number of units they produce over a period and the number of hours they can operate in a day. The capacity details of work centers help in scheduling operations for production orders.

Let's understand the creation and configuration of work centers in SAP S/4HANA.

Initial View

Work centers are created in the system using Transaction CR01. Figure 6.7 shows the initial screen for creating work centers. Let's explore the fields in the initial screen:

- **Plant**
  Enter the plant in which the work center is located.
- **Work Center**
  Enter the name of the work center, up to eight characters.
- **Work Center Category**
  This is the categorization of the work centers: production work centers, maintenance work centers, and so on. The following are the different work center categories defined for production planning:
  - **0001**: Machine
  - **0003**: Labor
  - **0005**: Plant maintenance
  - **0007**: Production line

You can define work center categories in the following customization step: **Produc-tion • Basic Data • Work Center • General Data • Determine Person Responsible**.

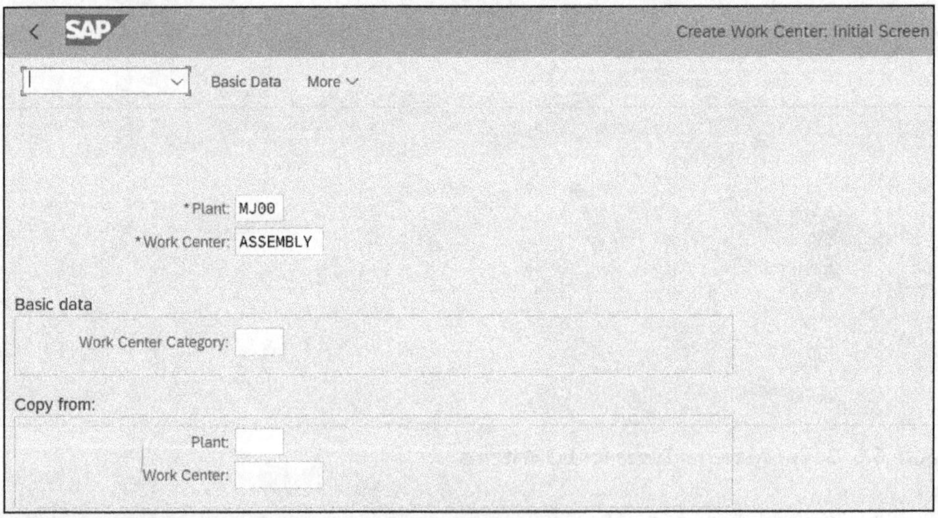

Figure 6.7 Create Work Center: Initial Screen

After filling in these fields, click **Basic Data** at the top to display the **Basic Data** view of the work center.

Basic Data View

Figure 6.8 shows the **Basic Data** view for the creation of a work center. Let's explore the key fields:

- **Person Responsible**
  In this field, assign the person or group responsible for supervising the respective work center. You can define the person or group responsible in the customization step via menu path **Production • Basic Data • Work Center • General Data • Determine Person Responsible**.

- **Supply Area**
  This is an interim storage area for raw materials and components to be supplied to the respective work center for production. Define a production supply area if the materials are issued to production from extended warehouse management (EWM) or the like.

- **Usage**
  Assign a key for task list usage. The relevant keys are as follows:
    - **003**: Only networks
    - **004**: Only maintenance task lists
    - **008**: Master recipe plus process order
    - **009**: All task list types

- **Backflush**
  If you set this indicator, the specific operations that take place in this work center will automatically backflush materials and components assigned to the operation, and the materials withdrawals will be posted automatically.

- **Standard Value Key**
  Standard value keys are defined in the configuration settings for work centers, where you can define up to six default values for work centers, such as standard values for operation execution, maintenance rules that control if a standard value is mandatory or optional, default unit of measure, costing-specific data in work centers, and so on. Choose a predefined standard value key from the dropdown list.

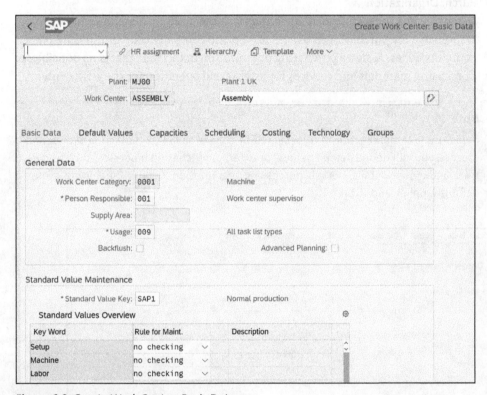

Figure 6.8 Create Work Center: Basic Data

Default Values View

Next, navigate to the **Default Values** view for the creation of a work center, as shown in Figure 6.9. Default values defined here will be copied over to routings when you reference the work center in a routing. Let's explore the key fields:

- **Control Key**
  This is a four-digit alphanumeric key that defines the business application this work center belongs to. By setting the reference indicator (**Ref. Ind.**) next to this field, the

system defaults the control key value into the routing. The following are the important control key values defined in the standard system:

- PP01: Routing/Ref. op. set—internal proc.

- PP02: Routing/Ref. op. set—external proc.

- **Purch. Group**
  A purchasing group is an organizational unit in materials management responsible
  for buying specific products and services within a purchasing organization. Assign a
  purchasing group in this field that is responsible for purchasing materials and services for an external operation for the work center.

- **Purch. Organization**
  A purchasing organization is an organizational unit in materials management
  responsible for purchasing activities for a plant/company code or multiple plants/
  company codes. Assign a purchasing organization in this field that is responsible for
  purchasing materials and services for an external operation for the work center.

Note

If you use external operations represented by control key **PP02** during the production
process, you can define default values for external purchasing processes, such as contract agreement details, default purchasing organization, default purchasing group,
default supplier, and so on.

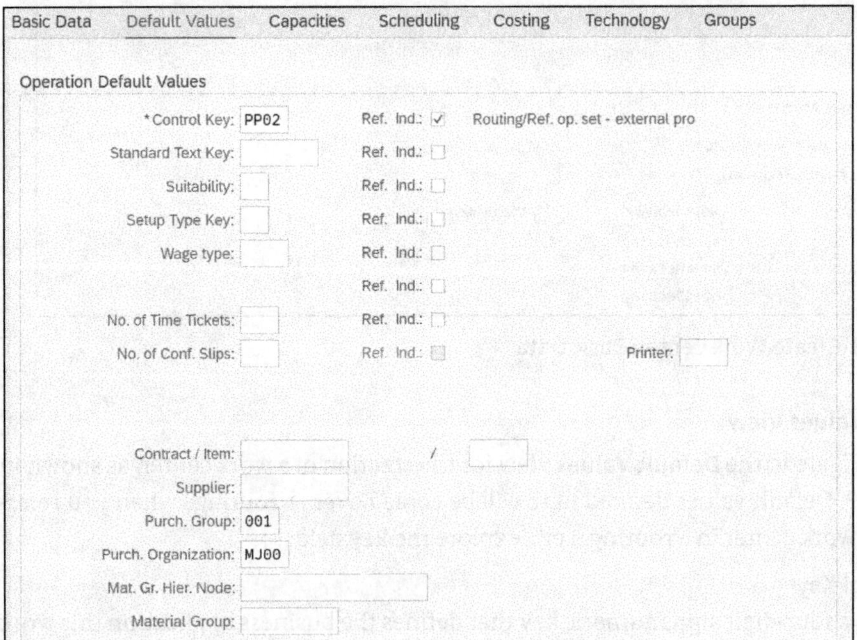

Figure 6.9 Create Work Center: Default Values

Capacities View

Next, navigate to the **Capacities** view for the creation of a work center, as shown in Figure 6.10.

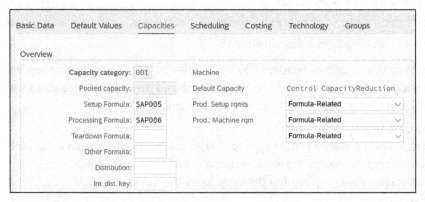

Figure 6.10 Create Work Center: Capacities

This information is crucial for the capacity planning of work centers for production. Let's explore the key fields:

- **Capacity category**

  This field categorizes the capacity. It defines if the capacity category of the work center is a machine or type of labor. The following are the capacity categories defined in the standard system:

  - **001**: Machine
  - **002**: Person
  - **008**: Processing unit
  - **011**: Warehouse
  - **012**: Warehouse plus processing unit

  Assign a value that is relevant to calculate the capacity of the work center.

- **Pooled capacity**

  If you have defined the pooled capacity at the plant level and at the capacity category level, you can assign a pooled capacity for the work center. If your production process runs with several work centers and all work centers share a common capacity, assign a pooled capacity.

- **Formulas**

  In the standard system, there are formulas defined to calculate the setup capacity requirements, processing capacity requirements, and so on. These formulas trigger ABAP code in the background to calculate capacity, internal processing time, processing costs, and so on for the work center tasks. The following formulas are available:

  - The setup formula is a calculation used to determine the duration of the setup operation of a work center.

- The processing formula is a calculation used to determine the duration of a processing operation of a work center.
- The teardown formula is a calculation used to determine the duration of a teardown operation of a work center.

The execution time of a work center is determined as follows:

Setup time + Processing time + Teardown time

Next, navigate to the capacities header view by double-clicking the capacity category from the **Capacities** view or pressing F2 . Figure 6.11 shows the capacities header view for the creation of a work center. Let's explore the key fields:

- **Capacity Responsible**
  If you have defied a capacity planner group in the configuration, you can assign the planner group responsible for the work center. A capacity planner group is the group of planners responsible for planning the capacity of the work center.

- **Factory Calendar**
  Assign the factory calendar in this field. The system uses the factory calendar to determine capacity requirements of the work center.

- **Capacity Base Unit**
  Define a measurement unit for capacity calculations. If you define **MIN**, the system calculates the capacity of work center in minutes, as shown in Figure 6.11.

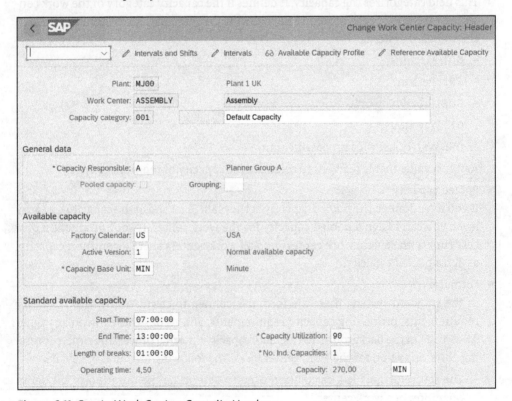

Figure 6.11 Create Work Center: Capacity Header

- **Start Time**
  Enter the shift start time for the work center. This is used in calculating the capacity of the work center.

- **End Time**
  Enter the shift end time for the work center. This is used in calculating the capacity of the work center.

- **Capacity Utilization**
  This is an important setting. If you define the capacity utilization, the system calculates the total shift time as the work center capacity as defined in the **Start Time** and **End Time** fields. For example, if utilization capacity is 100%, the start time is 07:00:00 hours, the end time is 16:00:00 hours, and the shift break is 01:00:00 hour, the total shift time of the work center is eight hours, and the capacity of the work center is 480 minutes (as the capacity base unit is minutes [**MIN**]).

Scheduling View

Let's move on to the **Scheduling** view of the work center, as shown in Figure 6.12. This information is crucial for the capacity planning of work centers for production. Let's explore the key fields:

- **Duration of Setup**
  Assign a formula to determine the setup time of an operation at the work center.

- **Processing Duration**
  Assign a formula to determine the processing time of an operation at the work center.

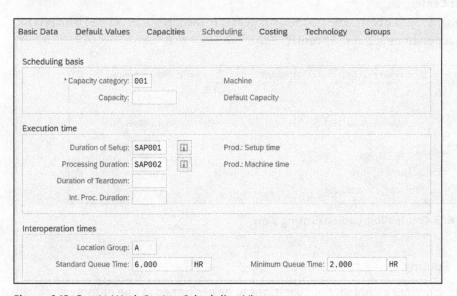

Figure 6.12 Create Work Center: Scheduling View

- **Location Group**
  Assign the physical location of the work center in the plant. You must define this value in the configuration. To arrive at this configuration step to define location groups, execute Transaction SPRO and navigate to **Production • Capacity Requirements Planning • Master Data • Work Center Data • Define Move Time Matrix**.

- **Standard Queue Time**
  Enter a standard time duration for production orders to queue before being processed at the work center.

- **Minimum Queue Time**
  Enter a minimum time duration for production orders to queue before being processed at the work center.

Costing View

Finally, navigate to the **Costing** view of the work center, as shown in Figure 6.13. This information is crucial for calculating costs associated with production operations and for incurring costs. Let's explore the key fields:

- **Validity**
  Enter the **Start date** and **End Date**. This defines the validity period of costing calculations.

- **Controlling Area**
  This is an organizational unit and defined for a company code. Typically, a controlling area has a 1:1 relationship with a company code.

- **Cost Center**
  Assign a specific cost center to the work center to incur costs associated with production operations at the work center.

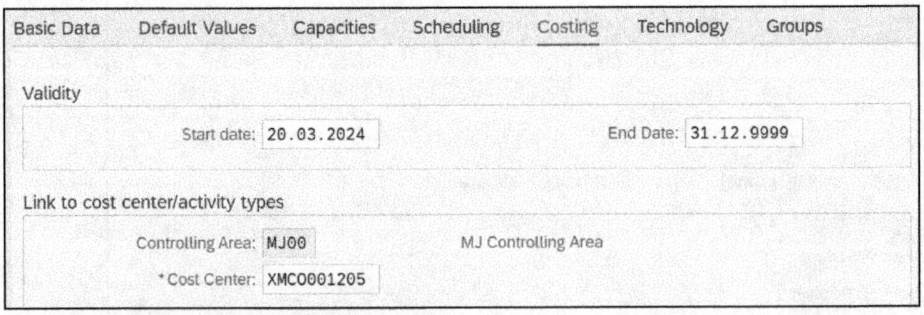

Basic Data	Default Values	Capacities	Scheduling	Costing	Technology	Groups

Validity

Start date: 20.03.2024   End Date: 31.12.9999

Link to cost center/activity types

Controlling Area: MJ00   MJ Controlling Area

*Cost Center: XMCO001205

Figure 6.13 Create Work Center: Costing View

6.2.3 Routing

A routing defines a sequence of operations to be performed at one or more work centers for production of finished products or components. Both external and internal operations are defined in the routing to manufacture the finished product, and the components to manufacture the finished products will be assigned to the respective operation of the routing.

A BOM for the material and work centers in the plant must be created as a prerequisite before a routing is created for the material in the same plant.

Let's discuss the creation of routings in SAP S/4HANA. Routings are created in the system using Transaction CA01.

Figure 6.14 shows the initial screen for creating routing. Let's explore the fields in the initial screen:

- **Material**
  In this field, enter the material number of the finished good or the component for which the routing is to be created. This is the same parent item from the BOM.

- **Plant**
  Enter the plant code in this field for the plant in which the routing will be used to produce the finished product or a component.

- **Key Date**
  Enter the key date for the routing. The routing is valid from this date, and this routing will be used in production orders from this date.

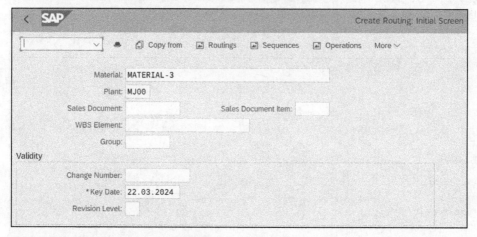

Figure 6.14 Create Routing: Initial Screen

After filling in these fields, click the **Header** icon at the top or press F6 to navigate to the routing header. Figure 6.15 shows the header details of the routing. Let's explore the key fields:

- **Usage**
  In this field, define the function within the organization in which the routing is used. For production planning usage, enter value **1**. The following are the values that can be assigned to this field:

- **1**: Production
- **2**: Engineering/design
- **3**: Universal
- **4**: Plant maintenance
- **5**: Goods receipt
- **51**: Goods receipt model inspection
- **6**: Goods issue
- **9**: Material check

- **Overall Status**

  At the time of creation of the routing, the system assigns value **1** by default. The processing status of the routing must be set to **4** to be used in the production order. The following are the values that can be assigned to this field:

  - **1**: Created
  - **2**: Released for order
  - **3**: Released for costing
  - **4**: Released (general)

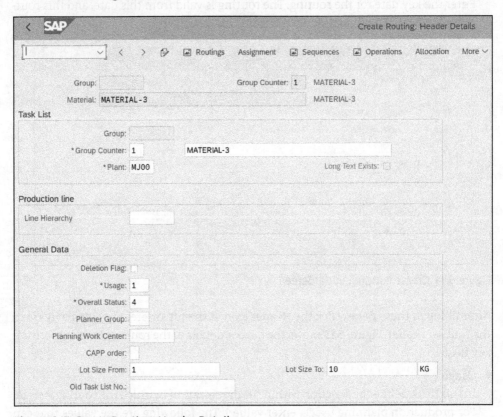

Figure 6.15 Create Routing: Header Details

- **Planner Group**
  If you have defined a person responsible or a group to maintain the routing, assign it here.

- **Planning Work Center**
  If you have a separate planning work center defined, you can assign it here.

- **Lot Size From/To**
  Enter a minimum lot size in the **Lot Size From** and maximum lot size value in the **Lot Size To** fields. This range is applicable to the routing and defines the minimum and maximum lot size it can produce.

Next, click **Operations** at the top of the header details screen to navigate to the operations details. Figure 6.16 shows the operations details of the routing. Let's explore the key fields:

- **Work Cent... (work center)**
  Assign the work center belonging to the plant where the corresponding operation will be performed.

- **Con... (control key)**
  This is a four-digit alphanumeric key that defines the business application this routing belongs to. The following are the important control key values defined in the standard system:

  - **PP01: Routing/Ref. op. set—internal proc.**

  - **PP02: Routing/Ref. op. set—external proc.**

  Assign **PP01** for internal manufacturing operations and **PP02** for external operations.

- **Base Qu... (base quantity)**
  This is the quantity of the components produced during the operations within the defined time in the **Setup** and **Machine** (standard time) fields.

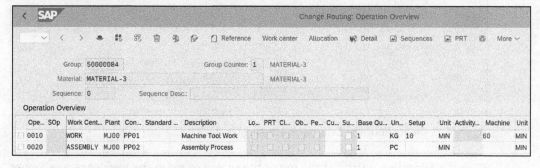

Figure 6.16 Create Routing: Operations

Next, maintain the external processing parameters for an external operation. To navigate to the external processing parameters of an external operation, double-click the operation or press F6. Figure 6.17 shows the external processing parameters assigned to the external operation of the routing. Let's explore the key fields:

- **Subcontracting**
  Set this indicator for external processing to be done by a subcontracting vendor.

- **Purchasing Info Rec.**
  The PIR is an important purchasing master data element that defines the purchasing-related data for a combination of a material and a supplier within a purchasing organization (see Chapter 3, Section 3.3 for more information). If you set the **Subcontracting** indicator, you must create a subcontracting PIR and assign the document number in this field. The system automatically defaults the purchasing group, supplier, planned delivery time, net price, and so on from the PIR.

- **Purchas. Organization**
  The purchasing organization is the organization element that controls purchasing activities of the plant and company code.

- **Cost Element**
  Assign a cost element to incur costs associated with external processing. A cost element defined in the same controlling area of the plant must be assigned here. This cost element is used in financial accounting and reporting.

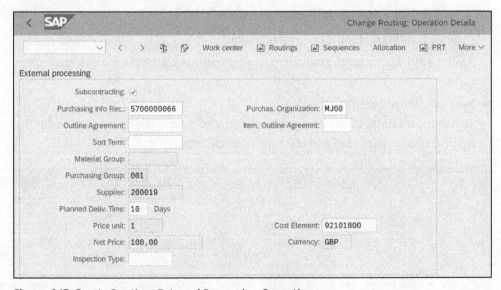

Figure 6.17 Create Routing: External Processing Operation

Click the **Allocation** button at the top of the screen to reach the view shown in Figure 6.18, which shows the components that have been defaulted from the BOM.

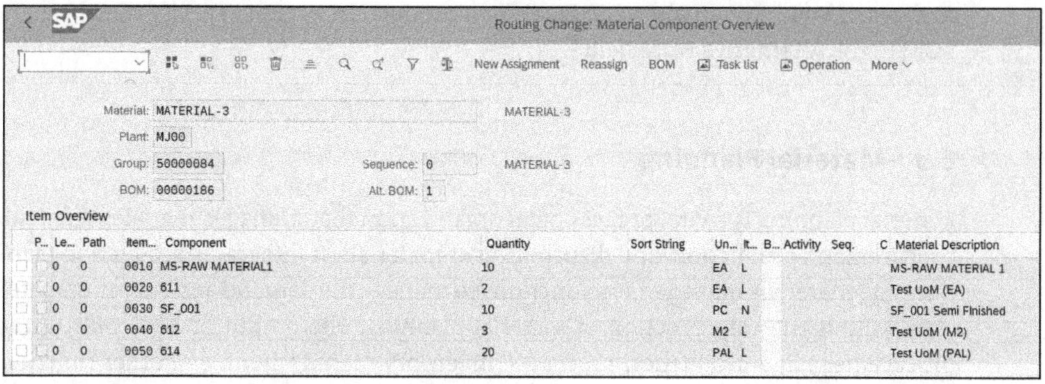

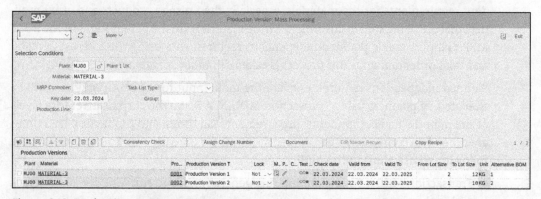

Figure 6.18 shown above the following figure is referenced.

Figure 6.18 Create Routing: Material Component Overview

6.2.4   Production Version

The production version defines different methods of producing a finished product or a semifinished product. A material can have multiple BOMs and routings, and production versions can be created for every BOM and routing combination. A production version is mandatory in SAP S/4HANA.

Let's discuss the creation of a production version in SAP S/4HANA. Production versions are created in the system using Transaction C223. Directly specify the **Material**, **Plant**, and **Key date** on the initial screen, then press ⌈Enter⌋ to display the production version details in the same screen.

SAP S/4HANA provides a simple method to define different ways of producing the same material via production versions. Figure 6.19 shows that **MATERIAL-3** in plant **MJ00** has two production versions as there are two BOMs that exist for the material.

Figure 6.19 Production Version: Mass Processing

Production versions are maintained for the parent item of the BOM. Specify the parent item of the BOM to define the production version and specify the valid-from and valid-to dates. If you have an alternate BOM for a production version, specify the alternate BOM in the **Alternative BOM** field.

6.3   Material Planning

Material planning is a vital process of an organization that manages the sales demand, procurement, production, and distribution of materials. It integrates production planning and materials management functions to manage the demand and supply of materials. MRP is the core function of material planning. The output of MRP will trigger procurement and production processes. Inventory management is another function that is integrated with MRP as it considers available inventory levels of materials as an important input for this process. Let's explore this vital functionality in detail, including the configuration instructions.

6.3.1   Material Requirements Planning

MRP keeps demand and supply in sync for an organization. It converts independent requirements arising from the sales demand and forecast demand into dependent requirements. It ensures the right quantities of the right material are available for the production process at the right time so that organizations always keep optimized inventory levels. MRP controllers run MRP frequently to address ever-changing demand, and during the MRP run the system displays exception messages for the planners to take necessary follow-up actions such as changing or cancelling purchase orders or schedule lines of scheduling agreements, changing delivery dates of purchase order items or schedule lines, and so on.

Figure 6.20 shows the MRP process. The process starts with defining planned independent requirements for the materials, especially the finished products and other products that are being sold to customers. Those materials are the parent items of the BOM. MRP planners create planned independent requirements using the forecast demand and sales order demand. This process is called *demand management* in SAP S/4HANA.

Planned independent requirements are the main input for MRP. An MRP run will be executed by planners using Transaction MD01N. A BOM, the current inventory levels of the material and components, procurement lead times, and production lead times are other data components considered as inputs for the MRP run. You can also use the following transactions to run MRP:

- Transaction MD01: Total planning run, generally used to run MRP for all materials within the planning scope
- Transaction MD02: Single-item and multilevel planning, generally used to run MRP for materials with multilevel BOMs.
- Transaction MD03: Single-item and single-level planning, generally used to run MRP for materials without a BOM

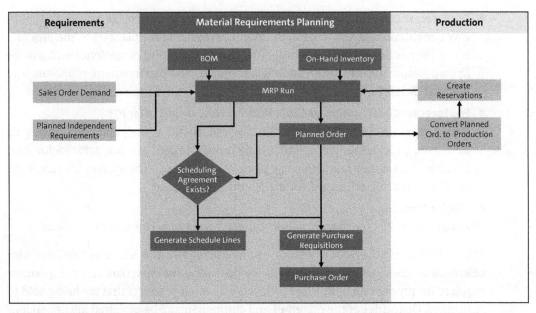

Figure 6.20 Material Requirements Planning Process

MRP types control the MRP parameters and maintain the material master MRP views. These MRP types are assigned to lower-level materials and components of the BOM. The following are the MRP types used for MRP:

- **PD: Forecast Consumption, No Planning Time Fence**
  If you maintain this MRP type in the material master at the plant level, the system considers this material during the MRP run and creates procurement proposals and planned orders for new requirements for shortages.

- **P1: Forecast Consumption, Auto Firming, New Orders after PTF**
  If you maintain this MRP type in the material master at the plant level, the system considers this material during the MRP run and creates procurement proposals and planned orders for new requirements for shortages. The existing procurement proposals and planned orders are firmed automatically if they are within the planned time fence. The new procurement proposals and planned orders created during the planning run will be moved to the end of the planning time fence and will not be firmed.

- **P2: Forecast Consumption, Auto Firming, No New Orders in PTF**
  If you maintain this MRP type in the material master at the plant level, the system considers this material during the MRP run and the existing procurement proposals and planned orders are firmed automatically if they are within the planned time fence. But there are no new procurement proposals or planned orders created during the planning run.

- **P3: Forecast Consumption, Manual Firming, No New Orders in PTF**
  If you maintain this MRP type in the material master at the plant level, the procurement proposals and planned orders within the planning time fence will not be firmed automatically. MRP run will not create new procurement proposals and planned orders for new requirements if they fall within the planning time fence.

- **P4: Forecast Consumption, Manual Firming, New Orders after PTF**
  If you maintain this MRP type in the material master at the plant level, during an MRP run, procurement proposals and planned orders are not created for new requirements if they fall within the planning time fence. The system considers the requirements that fall outside of the planning time fence.

- **ND: No Planning**
  Assign this MRP type to materials if you don't want to perform MRP for them.

MRP runs explode the BOM to generate purchase requisitions and schedule lines with reference to the scheduling agreement for procuring the materials and components required for production of finished products and other products that are being sold to customers. Quantities of the materials and components to be procured are dependent on the available inventory on hand. The delivery dates of the schedule lines and purchase requisitions are dependent on the procurement lead times set in the PIR. Certain organizations have set up their SAP S/4HANA systems to generate planned orders as the outcome of MRP runs. MRP planners can adjust the dependent requirements in the planned order before releasing them to generate schedule lines and purchase requisitions.

The results of an MRP run can be displayed in SAP S/4HANA using Transaction MD04 (Stock/Requirements List). This transaction displays the following statuses of the material:

- MRP parameters such as MRP type, material number, plant, material type, and so on at the header level
- Required quantity and available quantity of the material in the plant for customer orders
- Planned consumption details for production reservations
- Planned goods issues for outbound deliveries
- Planned goods receipts for purchase requisitions, purchase orders, scheduling agreements, planned orders, production orders, and so on

The planned order will be converted to a production order. Once done, material and quantity adjustments cannot be made in the production order. The required raw materials can be reserved for production for future withdrawal/consumption. The procured raw materials will be received into inventory from internal/external suppliers and later issued to production with reference to a production order.

Another crucial result of the MRP run is the exception messages that the system triggers if there's a change in demand. These messages enable planners to adjust procurement proposals such as updating purchase requisitions to adjust delivery dates and quantities.

MRP creates a lot size for production and defines the quantities of the finished products and other components to be produced in the production process. In processing industries, lot sizing is also called *batching*. Lot sizing helps optimize production and inventory of raw materials and components while reducing the production costs and improving overall efficiency.

> **Note**
>
> The planning time fence is the period (future time) withing materials planning in which the master plan is protected and the changes to certain planning parameters are restricted. This will stabilize the production schedules. The planning time fence is defined in the material master's **MRP 1** view. When you run subsequent MRP runs within the planning time fence, the existing plan will not be affected.

6.3.2   Master Production Scheduling

MRP runs daily to determine the raw materials and components required for production and creates dependent requirements by exploding the BOM. The customer demand and forecasts for certain finished products vary over time. While MRP creates procurement proposals and inventory requests for raw materials, subassemblies, and other components for production, master production scheduling (MPS) is aimed at concentrating on direct demand from customers for critical products that bring profits.

Figure 6.21 shows the MPS process, which is one of the inputs for MRP. MPS, another kind of MRP, is typically executed less frequently than MRP and plans for direct independent requirements that come from customer demand and forecast demand. Customer demand and forecast demand for finished products and other sellable goods flow directly into MPS. While MRP plans for raw materials and components of the BOM, MPS plans for the parent item (finished product) of the BOM. In MPS, extra care will be given to those critical finished products that bring profits to the organization. Execution of MSP for those critical products will be done prior to the MRP run so that attention will be given to those critical finished goods during manufacturing.

An MPS run can be executed in SAP S/4HANA using the following transactions:

- Transaction MD40: Master production scheduling online
- Transaction MD41: Master production scheduling for single item and multilevel planning

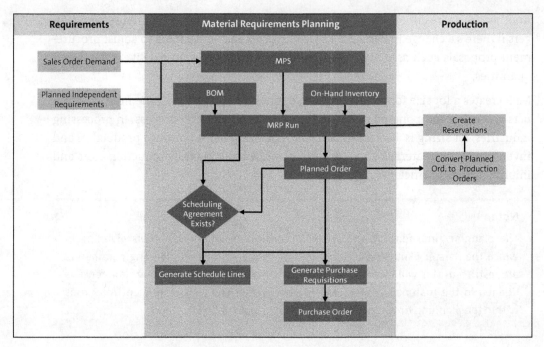

Figure 6.21 Master Production Scheduling Process

MRP types control the MRP parameters and maintain the material master MRP views. These MRP types are assigned to the critical finished goods, and those materials are the parent items of the BOMs. The following are the MRP types used for MPS:

- **M0: MPS, FCST Consumption, No Firming**
  If you maintain this MRP type in the material master at the plant level, the system considers this material during the MPS run and creates procurement proposals and planned orders based on requirements. The procurement proposals and planned orders will not be firmed automatically.

- **M1: MPS, FCST Consumption, Auto Firming, New Orders after PTF**
  If you maintain this MRP type in the material master at the plant level, the system considers this material during the MPS run and creates procurement proposals and planned orders for new requirements for shortages. The existing procurement proposals and planned orders are firmed automatically if they are within the planned time fence. The new procurement proposals and planned orders created during the planning run will be moved to the end of planning time fence and will not be firmed.

- **M2: MPS, FCST Consumption, Auto Firming, No New Orders in PTF**
  If you maintain this MRP type in the material master at the plant level, the system considers this material during the MPS run, and the existing procurement proposals and planned orders are firmed automatically if they are within the planned time

fence. But there are no new procurement proposals and planned orders created during the planning run.

- **M3: MPS, FCST Consumption, Manual Firming, New Orders after PTF**
  If you maintain this MRP type in the material master at the plant level, the procurement proposals and planned orders within the planning time fence will not be firmed automatically. The MPS run will not create new procurement proposals and planned orders for new requirements if they fall within the planning time fence.

- **M4: MPS, FCST Consumption, Manual Firming, No New Orders in PTF**
  If you maintain this MRP type in the material master at the plant level, during the MPS run, procurement proposals and planned orders are not created for new requirements if they fall within the planning time fence. The system considers the requirements that fall outside of the planning time fence.

> **Note**
>
> MRP types **M0**, **M1**, **M2**, **M3**, and **M4** are used in MPS and are assigned to the critical finished materials that are planned at the individual material level (parent item of the BOM).
>
> MRP types **PD**, **P1**, **P2**, **P3**, and **P4** are used in MRP and are assigned to the lower-level materials and components of the BOM as well as noncritical finished products.

Once the independent requirements of critical finished materials are planned in MPS, the MRP run will be executed more frequently to plan for dependent requirements of the BOM for finished materials and other products. The MRP run explodes the BOM to generate purchase requisitions and schedule lines with reference to the scheduling agreement for procuring the materials and components required for production of finished products and other products that are being sold to the customers. Quantities of the material and components to be procured are dependent on the available inventory on hand. The delivery dates of the schedule lines and purchase requisitions are dependent on the procurement lead times set in the PIR. Certain organizations have set up their SAP S/4HANA systems to generate planned orders as the outcome of MRP runs. MRP planners can adjust the dependent requirements in the planned order before releasing them to generate schedule lines and purchase requisitions.

As with MRP, the results of an MPS run can be displayed in SAP S/4HANA using Transaction MD04 (Stock/Requirements List).

The planned order will be converted to a production order. Once done, the material and quantity adjustments cannot be made in the production order. The required raw materials can be reserved for production for future withdrawal/consumption. The procured raw materials will be received into inventory from internal/external suppliers and later issued to production with reference to a production order.

6.3.3   Example Material Requirements Planning Run

Let's simulate MRP and MSP scenarios to understand how the system supports MRP.

As a prerequisite, maintain planning master data as defined in Section 6.2.

Figure 6.22 shows the **MRP 1** view of the material master for **MATERIAL-3**, which is a finished product and the parent item of the BOM. **MRP Type M1** is maintained, and this indicates that MPS is to be executed. The **Planned Time Fence** field is maintained as **5** days.

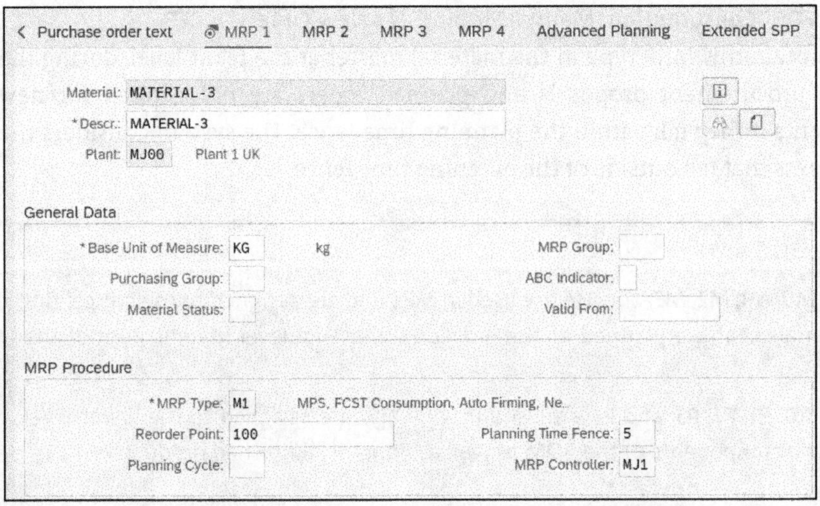

Figure 6.22 MRP Data for Finished Product: Parent Item of BOM

Figure 6.23 shows the **MRP 1** view of the material master for **MS-RAW MATERIAL1**, which is a raw material and the component item of the BOM. **MRP Type P1** is maintained, and this indicates that MRP is to be executed. The **Planned Time Fence** field is maintained as **5** days.

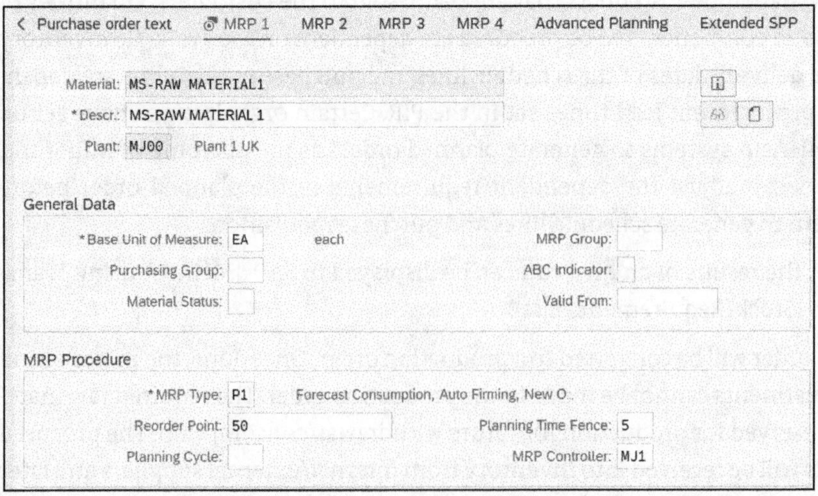

Figure 6.23 MRP Data for Raw Material: Component Item of BOM

Figure 6.24 shows the sales order created against the customer order for 200 KG of finished product **MATERIAL-3**, with a requested delivery date of March 25, 2024.

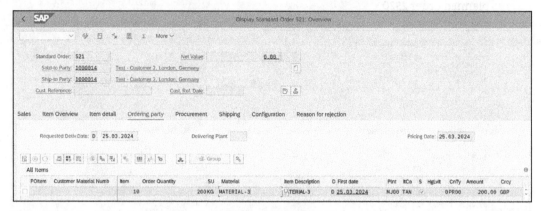

Figure 6.24 Customer Demand for Finished Product MATERIAL-3

Execute Transaction MD40 to run MPS for **MATERIAL-3**, which is a finished product and the parent item of the BOM.

Figure 6.25 shows the stock/requirements list (Transaction MD04) for **MATERIAL-3**. Customer sales order **521** was listed, and planned order **2520** was created.

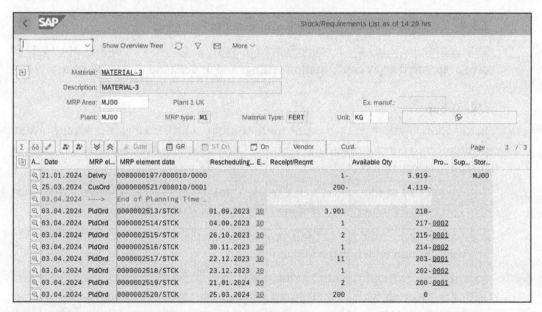

Figure 6.25 Stock Requirements List for MATERIAL-3

Next, execute Transaction MD40 to run MRP for **MS RAW MATERIAL1**, which is a raw material and the component of the BOM.

Figure 6.26 shows the stock/requirements list (Transaction MD04) for **MS RAW MATE-RIAL1**. Dependent requirements for the component were created with reference to planned order **2520**.

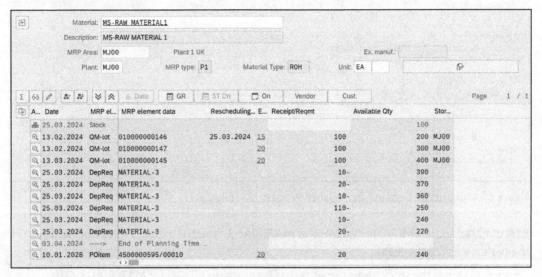

Figure 6.26 Stock/Requirements List for MS RAW MATERIAL1

6.3.4   Configuration of Material Requirements Planning

This section explains the key configuration settings for MRP in SAP S/4HANA. Let's dive deep into the key configuration settings and understand those that control MRP.

MRP Group

In this configuration, you can define the MRP group, which is a logical grouping of MRP parameters to control materials planning. The MRP group is assigned in the material master at the plant level.

To arrive at this configuration step, execute Transaction SPRO and navigate to **Production • Material Requirements Planning • MRP Groups • Define MRP Controllers**.

Figure 6.27 shows MRP group **0001** defined for the plant **MJ00**. Enter a four-digit key and a description to define the MRP group.

For materials without the MRP group, the system uses plant parameters to perform MRP. For materials with an MRP group assigned, the system uses MRP group parameters to perform MRP.

Material Requirements Planning (MRP) Group
Overview
More
* Plant: MJ00
Create
Maintain
Delete
Copy
Create MRP Group
Plant: MJ00 Plant 1 UK
* MRP Group: 0001 MRP Group for MJ00
Create
Cancel

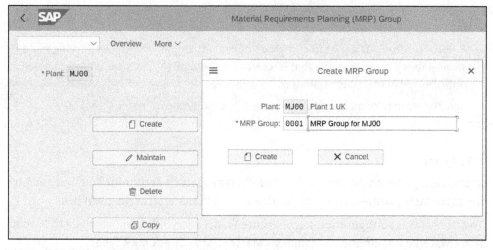

Figure 6.27 Define MRP Group at Plant Level

MRP Controllers

In this configuration, you can maintain MRP controllers who are responsible for executing MRP within the organization. MRP controllers are defined at the plant level.

To arrive at this configuration step, execute Transaction SPRO and navigate to **Production • Material Requirements Planning • Master Data • Define MRP Controllers.**

Figure 6.28 shows the overview of MRP controllers defined at the plant level. You can define multiple MRP controllers based on business requirements. Click **New Entries** to define a new MRP controller for the plant. Specify a plant in the **Plnt** field and a unique three-digit alphanumeric key in the **MRP Cont.** (MRP controller) field to define a new MRP controller at the plant level.

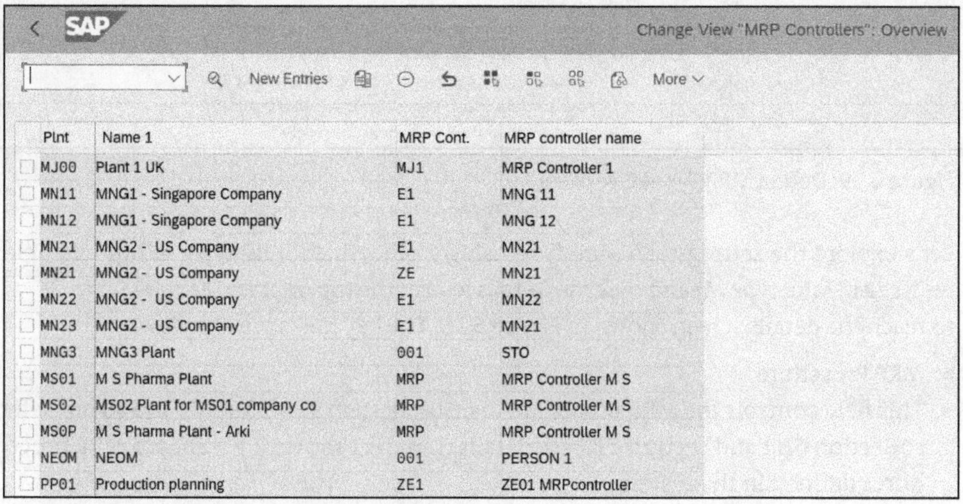

Change View "MRP Controllers": Overview
New Entries
More
Plnt | Name 1 | MRP Cont. | MRP controller name
MJ00 | Plant 1 UK | MJ1 | MRP Controller 1
MN11 | MNG1 - Singapore Company | E1 | MNG 11
MN12 | MNG1 - Singapore Company | E1 | MNG 12
MN21 | MNG2 - US Company | E1 | MN21
MN21 | MNG2 - US Company | ZE | MN21
MN22 | MNG2 - US Company | E1 | MN22
MN23 | MNG2 - US Company | E1 | MN21
MNG3 | MNG3 Plant | 001 | STO
MS01 | M S Pharma Plant | MRP | MRP Controller M S
MS02 | MS02 Plant for MS01 company co | MRP | MRP Controller M S
MS0P | M S Pharma Plant - Arki | MRP | MRP Controller M S
NEOM | NEOM | 001 | PERSON 1
PP01 | Production planning | ZE1 | ZE01 MRPcontroller

Figure 6.28 Define MRP Controllers at Plant Level

MRP Types

In this configuration, you can maintain MRP types, which determine whether and how the material is planned. This is maintained for each material at the plant level.

To arrive at this configuration step, execute Transaction SPRO and navigate to **Produc-tion • Material Requirements Planning • Master Data • Check MRP Types.**

Figure 6.29 shows the overview of MRP types defined in the standard system. Click **New Entries** to maintain a new MRP type.

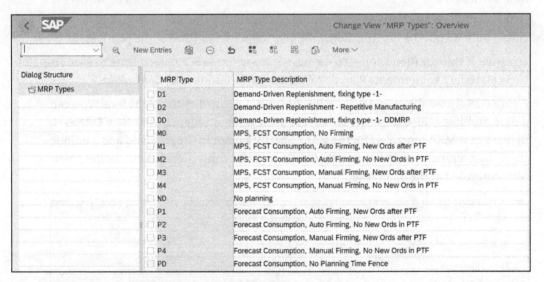

Figure 6.29 Define MRP Types Overview

Let's explore the settings of a MRP type using a predefined MRP type. In our example, we'll select MRP type **M1** and click the **Details** icon at the top or press Ctrl + Shift + F2 to reach the details screen shown in Figure 6.30. The key fields are as follows:

- **MRP Procedure**
  This field controls the MRP procedure to be used, such as MPS, MRP, and so on. Refer to Section 6.3.1 and Section 6.3.2 for details. Table 6.2 shows the standard MRP proce-dures defined in the system.

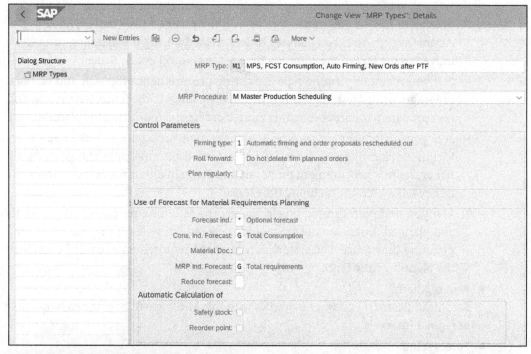

Figure 6.30 Define MRP Types Details

MRP Procedure	Description
D	MRP
M	MPS
B	Reorder point planning (ROP)
C	Demand-driven replenishment (DD)
S	Forecast-based planning
N	No MRP

Table 6.2 MRP Procedures

- **Firming type**

  This field controls how procurement proposals and planned orders are firmed. Firmed requirements in MRP are the planned orders and procurement proposals that cannot be changed by the MRP run. The following firming types are defined in the standard system:

  - **0**: If you maintain firming type **0**, the procurement proposals and planned orders will not be firmed automatically.

- **1**: If you maintain firming type **1**, existing procurement proposals and planned orders are firmed automatically if within the planned time fence. The new procurement proposals and planned orders created during the planning run will be moved to the end of the planning time fence and will not be firmed.

- **2**: If you maintain firming type **2**, existing procurement proposals and planned orders are firmed automatically if within the planned time fence. But there are no new procurement proposals and planned orders created during the planning run.

- **3**: If you maintain firming type **3**, procurement proposals and planned orders within the planning time fence will not be firmed automatically. The MRP run will not create new procurement proposals and planned orders for new requirements if they fall within the planning time fence.

- **4**: If you maintain firming type **4**, during the MRP run, procurement proposals and planned orders are not created for new requirements if they fall within the planning time fence. The system considers the requirements that fall outside of the planning time fence.

- **Plan regularly**
  Set this indicator if you want materials with this MRP type to be planned frequently at regular intervals.

- **Forecast ind.**
  Set the relevant value if you want to include the forecasts of materials with this MRP type in the MRP run. The following values are allowed:
  - Blank: No forecast
  - **+**: Obligatory forecast
  - *****: Optional forecast

- **Cons. Ind. Forecast**
  Set the relevant consumption value to be considered for forecasts. The following values are allowed:
  - **U**: Unplanned consumption: Unplanned consumption is the quantities of materials taken from inventory without a reservation.
  - **G**: Total consumption: Total consumption is the sum of planned (quantities of materials taken from inventory with a reservation) and unplanned consumption.

- **MRP Ind. Forecast**
  Set the relevant value for the system to consider the forecast values during MRP. The following values are allowed:
  - Blank: Not to be included in planning
  - **U**: Unplanned requirements from unforeseen demand
  - **G**: Total requirements, the sum of planned requirements (planned reservations, planned receipts, planned consumption, etc.) and unplanned requirements

- **Safety stock**
  Set this indicator if you want to calculate the safety stock automatically during MRP.

- **Reorder point**

  Set this indicator if you want to calculate the reorder point automatically during MRP.

Check MRP Lot-Sizing Procedure

In this configuration, you can maintain lot-sizing procedures for MRP. The system uses lot-sizing procedures within materials planning to calculate the quantity to be procured or produced.

To arrive at this configuration step, execute Transaction SPRO and navigate to **Production • Material Requirements Planning • Planning • Lot-Size Calculation • Check Lot-Sizing Procedures**.

Figure 6.31 shows the overview of MRP lot-sizing procedures defined in the standard system. Click **New Entries** to maintain a new MRP lot-sizing procedure.

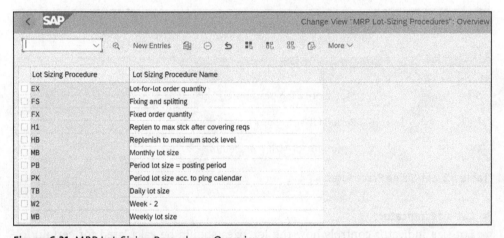

Figure 6.31 MRP Lot-Sizing Procedures Overview

Let's explore the settings of a MRP lot-sizing procedure using one of the predefined options. For our example, select **EX** and click the **Details** icon at the top or press Ctrl + Shift + F2 to navigate to the details of the MRP lot-sizing procedure. Figure 6.32 shows the details of MRP lot-sizing procedure **EX**.

MRP lot-sizing procedures are assigned in the material master at the plant level. The key fields are as follows:

- **Lot-sizing procedure**

  Three lot-sizing methods are defined in the standard system for MRP. The system calculates the lot size for procurement proposals and planned orders based on the lot-sizing procedure; that is, the procurement quantity and planned order quantity of the materials are calculated based on the lot-sizing procedure. Table 6.3 lists the lot-sizing procedures.

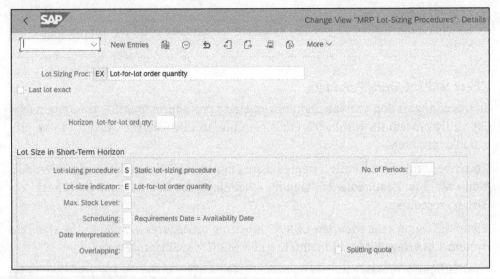

Figure 6.32 MRP Lot-Sizing Procedures Details

Lot-Sizing Procedure	Description
S	Static lot-sizing procedure
P	Period lot-sizing procedure
O	Optimum lot-sizing procedure

Table 6.3 Lot-Sizing Procedures

- **Lot-size indicator**

  Lot-size indicator controls how the lot sizes should be calculated for certain lot-sizing procedures during the MRP run. Table 6.4 shows the lot-size indicators defined in the standard system.

Lot-Size Indicator	Description
W	Period lot size: weekly lot/dynamic lot sizing: least unit cost
F	Fixed lot size
E	Lot-for-lot order quantity
H	Replenish to maximum stock level
T	Daily lot size
M	Monthly lot size

Table 6.4 Lot-Size Indicators

Lot-Size Indicator	Description
P	Period lot sizes according to posting periods
D	Dynamic lot size creation
G	Groff reorder procedure
S	Static lot size: fixed w. splitting/dynamic lot size: part pd bal.
K	Period lot size as in PPC planning calendar

Table 6.4 Lot-Size Indicators (Cont.)

6.4   Operational Subcontracting

Subcontracting is a special procurement process in SAP S/4HANA where certain operations are outsourced by organizations to external suppliers and the required components provided to the external suppliers for the operation. The external supplier, the *subcontractor*, will perform the operation on behalf of the organization. The subcontractor will ship the semifinished product back to the organization and charge a service fee. This process involves tight integration between production planning and materials management processes. There are generally two types of subcontracting:

- **Operational subcontracting**
  In this scenario, during the in-house production of finished products and other products, specific manufacturing operations will be performed by the subcontractor (external supplier) and the required components will be provided to the subcontractor. The subcontractor will perform the requested production operation, ship the semifinished products to the organization, and charge a fee for the operation.

- **Assembly subcontracting**
  In this scenario, all the required components will be provided by the organization to the subcontractor to assemble a finished product or other sellable product on behalf of the organization. The assembled product will be shipped to the organization. The subcontractor will charge a service fee for the assembly operation.

Let's dive deep into the subcontracting process and understand the integration points between production planning and materials management, followed by step-by-step configuration instructions.

6.4.1   Subcontracting Manufacturing

Table 6.5 shows all the key subprocesses impacted in production planning and materials management for the external subcontracting process.

Area	Processes Impacted	Description
Production planning	Maintain production master data	This process creates and maintains production master data such as the BOM, work center, routing, and production version.
	Perform MRP	This process supports MRP to generate planned orders for the finished product manufacturing and purchase requisitions for raw materials if not available in the inventory.
	Manage production operations	This process supports the monitoring of production activities that are in progress.
	Manage external operations	This process supports the external operations performed by a subcontractor (external vendor) during the production of finished goods.
Materials management	Sourcing materials and services	This process supports the sourcing for a subcontracting vendor for operational subcontracting.
	Create and maintain purchasing master data	This process creates and maintains purchasing master data such as subcontracting PIRs to support the subcontracting process.
	Create and maintain subcontracting purchase requisition	This process supports the creation, generation, and maintenance of subcontracting purchase requisitions originating from MRP and/or production process.
	Create and maintain subcontract order	This process supports the creation and maintenance of a subcontracting purchase order.
	Provide components to subcontractor	This process supports the issuing of components to a subcontractor for external operations.
	Post goods receipt	This process supports the goods receipt of semifinished products manufactured/assembled by the subcontractor into inventory.
	Post component consumption	This process supports the automatic consumption of components provided to the subcontracting vendor for external operations.
	Supplier invoice management	This process supports the processing of incoming invoices from the supplier for the operational subcontracting charges.

Table 6.5 Process Impact: Subcontracting Manufacturing

The subcontracting process can be triggered from MRP or from an external operation within the production process that involves processing of the process/production order.

Let's break down the prerequisites for the MRP-driven subcontracting process:

1. Maintain procurement type **F** (external procurement) and special procurement type **30** (subcontracting) in the **MRP 2** view of the material master for the final product to be produced/assembled via the external operation from a subcontractor. Figure 6.33 shows the final material setup for subcontracting process. For more details, see Chapter 3, Section 3.1.5.

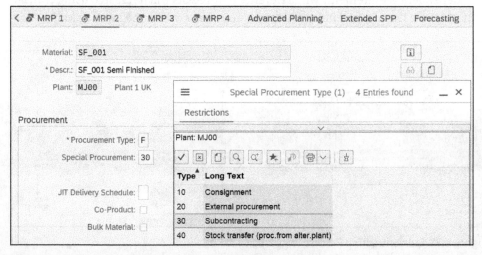

Figure 6.33 Material Master Setup for Subcontracting Process from MRP

2. Maintain the BOM for the final product to be produced/assembled via the external operation from a subcontractor. Assign the components required to produce/assemble the final product from the external operation for BOM items. Figure 6.34 shows the BOM setup for the subcontracting process (see Section 6.2.1 for instructions on creating the BOM).

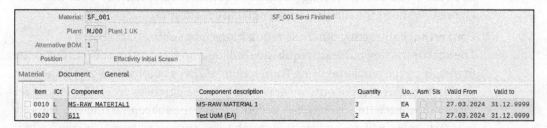

Figure 6.34 BOM Setup for Subcontracting Process

3. Finally, set up the PIR for subcontracting for the combination of the final product to be produced/assembled and the external subcontractor (supplier). Maintain the service fee for the subcontracting process in the PIR. For more information, see Chapter 3, Section 3.3.2.

Figure 6.35 shows the subcontracting process triggered from the MRP process. The process steps for the subcontracting process triggered by MRP are as follows:

1. **Create planned independent requirements**
   Once the necessary master data is maintained, the subcontracting process starts with the creation of a forecast demand for the final product (BOM header). Planned independent requirements with reference to forecasts will be created using Transaction MD61. The planned independent requirements will be the input to MRP.

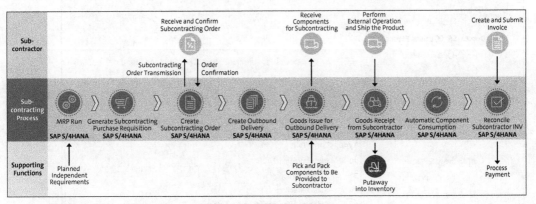

Figure 6.35 Subcontracting Process Triggered by MRP

2. **MRP run**
   The material planner will run MRP for planned independent requirements using Transaction MD01N. The system generates subcontracting purchase requisitions for the final product (BOM header). The system assigns a subcontractor (supplier) as the source of supply from purchasing master data. The item category of the requisition line item will be **L** (subcontracting). The components will be added to the **Material Data** tab of the subcontracting purchase requisition automatically from the BOM. Figure 6.36 shows the subcontracting purchase requisition.

3. **Convert subcontracting purchase requisitions into subcontract orders**
   The subcontracting purchase requisition will be converted into a subcontract order manually by the purchaser using Transaction ME21N or automatically. The line item details including components are defaulted into the purchase order. The service fee for the subcontracting operation is defaulted from the subcontracting PIR. The subcontract order will be transmitted to the subcontractor based on the output method: email, EDI, or electronic transmission to the supplier on SAP Business Network.

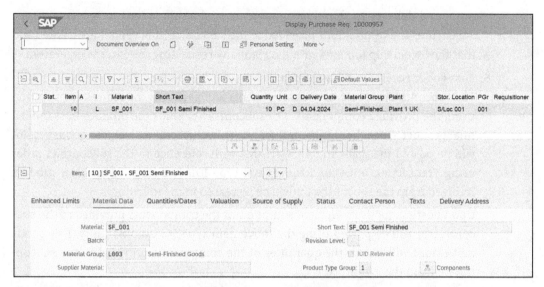

Figure 6.36 Subcontracting Purchase Requisition Generated from MRP

4. **Create outbound delivery with reference to subcontract order**

   The next step in the process is to create an outbound delivery to issue the components to the subcontractor. Outbound deliveries will be created using Transaction ME2O with reference to the subcontract order. Inventory and warehouse management activities such as picking of materials and packing activities are performed with reference to outbound deliveries. Figure 6.37 shows the subcontracting monitor used to create an outbound delivery to issue components to the subcontractor.

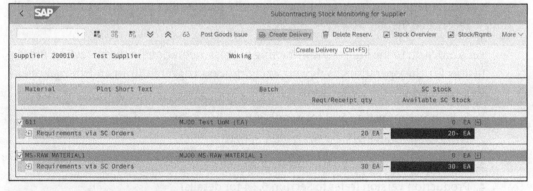

Figure 6.37 Subcontracting Stock Monitor (Transaction ME2O)

5. **Post goods issue for outbound delivery**

   Goods issue will be posted with reference to the outbound delivery after the inventory management functions such as picking and packing of materials are completed. Movement types play an important role in determining the type of inventory movement and posting of goods issue. Movement type 541 is used for goods issue for outbound delivery to ship the components to the subcontractor. Movement type 541 ensures the components provided to the subcontracting vendor are tracked and that the ownership remains with the company requesting the external operation.

6. **Post goods receipt purchase order**

The subcontractor receives the components and performs the requested operation using the provided components. The subcontractor ships the produced/assembled final product. Once the shipment from the subcontractor arrives, the goods receipt will be posted manually in SAP S/4HANA with reference to the subcontract order using Transaction MIGO and movement type 101. The stock of the final product received from the subcontractor will be increased in the inventory.

Upon posting goods receipt, consumption of the components provided to the subcontractor will be posted using the movement type 543 automatically. This goods movement will reduce the quantities of the components provided to the subcontractor from the inventory.

7. **Post logistics invoice verification**

The subcontractor sends an invoice for the external operation via email (paper invoice), via EDI, or electronically via SAP Business Network. The supplier invoice document will be posted by the accounts payable personnel manually in the SAP S/4HANA system using Transaction MIRO in the case of a paper invoice. Automatic posting and reconciliation of the invoice happens if an electronic invoice is received from the supplier.

The next scheduled payment run will pick up this fully reconciled invoice and post the payment to the supplier based on the preferred payment method of the supplier. Accounts payable personnel can manually process the payment against the fully reconciled invoice as well.

Let's move on to the other subcontracting process: operational subcontracting. Prerequisites for the operational subcontracting process triggered from an external operation from production order are as follows:

1. BOM (refer to Section 6.2.1 for details)

2. Work center (refer to Section 6.2.2 for details)

3. Routing (refer to Section 6.2.3 for details)

4. Production version (refer to Section 6.2.4 for details)

Figure 6.38 shows the operational subcontracting process where the external operation is triggered from the processing of a production order in discrete manufacturing.

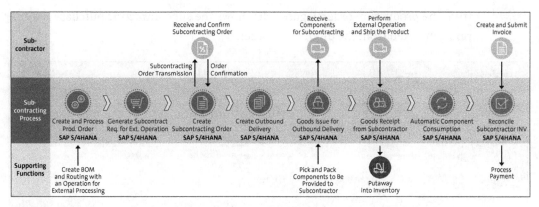

Figure 6.38 Operational Subcontracting Process Triggered by Production Operation

Process steps for the operational subcontracting process triggered from an external operation from a production order are as follows:

1. **Create and process production order**

   The process starts with the creation of a production order for the header material of the BOM. All the master data will be copied over to the order based on material and plant details. BOMs will be copied to the **Components** tab and the production operations will be copied from the routing.

   Figure 6.39 shows a production order with the first operation to be performed in house and the second operation, with control key **PP02**, to be performed externally by the subcontractor.

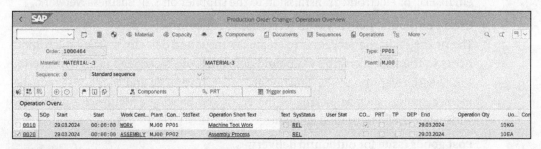

Figure 6.39 Production Order Created with External Operation

2. **Generation of subcontracting purchase requisition for external operation**

   The system automatically generates subcontracting purchase requisitions for the second operation of a production order based on external processing parameters set in the routing. The system assigns a subcontractor (supplier) as the source of supply from external processing parameters. The item category of the requisition line item will be **L** (subcontracting), and the account assignment category will be **F** (order). The components will be added to the **Material Data** tab of the subcontracting purchase requisition automatically from the components assigned to the external operation from the production order. Figure 6.40 shows the subcontracting purchase requisition generated from the production order.

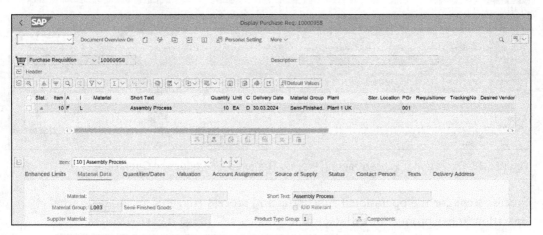

Figure 6.40 Subcontracting Purchase Requisition Generated from Production Order

3. **Convert subcontracting purchase requisitions into subcontract orders**
   The subcontracting purchase requisition will be converted into a subcontract order manually by the purchaser using Transaction ME21N or automatically. The line item details including components are defaulted into the purchase order. The service fee for the subcontracting operation is defaulted from the subcontracting PIR. Subcontract orders will be transmitted to the subcontractor based on the output method: email, EDI, or electronic transmission to the supplier on SAP Business Network.

4. **Create outbound delivery with reference to subcontract order**
   The next step in the process is to create an outbound delivery to issue the components to the subcontractor. Outbound deliveries will be created using Transaction ME2O with reference to the subcontract order. Inventory and warehouse management activities such as picking of materials and packing activities are performed with reference to outbound deliveries.

5. **Post goods issue for outbound delivery**
   Goods issue will be posted with reference to the outbound delivery after the inventory management functions such as picking and packing of materials are completed. Movement types play an important role in determining the type of the inventory movement and posting of goods issue. Movement type 541 is used for goods issue for outbound delivery to ship the components to the subcontractor. Movement type 541 ensures that the components provided to the subcontracting vendor are tracked and that ownership remains with the company requesting the external operation.

6. **Post goods receipt purchase order**

   The subcontractor receives the components and performs the requested operation using the provided components. The subcontractor ships the assembled final product. Once the shipment from the subcontractor arrives, the goods receipt will be posted manually in SAP S/4HANA with reference to the subcontract order using Transaction MIGO and movement type 101. The stock of the final product received from the subcontractor will be directly consumed by the production order as the account assignment category of the subcontract order is F (order).

   Upon posting goods receipt, consumption of the components provided to the subcontractor will be posted using movement type 543 automatically. This goods movement will reduce the quantities of the components provided to the subcontractor from the inventory.

7. **Post logistics invoice verification**

   The subcontractor sends an invoice for the external operation via email (paper invoice), via EDI, or electronically via the SAP Business Network. The supplier invoice document will be posted by the accounts payable personnel manually in SAP S/4HANA using Transaction MIRO in the case of a paper invoice. Automatic posting and reconciliation of the invoice happens if an electronic invoice is received from the supplier. The next scheduled payment run will pick up this fully reconciled invoice and post the payment to the supplier based on the preferred payment method of the supplier. Accounts payable personnel can manually process the payment against the fully reconciled invoice as well.

6.4.2   Configuration of Subcontracting Process

In this section, let's explore the configuration settings for the subcontracting process from production planning and materials management in SAP S/4HANA. Let's dive deep into the key configuration settings and understand the configuration settings that control the subcontracting process.

Define External Procurement

In this configuration, you can define default values for external procurement for materials relevant to MRP. Default values are maintained at the plant level and the MRP group level.

To arrive at this configuration step, execute Transaction SPRO and navigate to **Production • Material Requirements Planning • Planning • Procurement Proposals • Define External Procurement**.

When the procurement proposals are created during the execution of MRP, the system gets these default values to create purchase requisitions and purchase orders. Figure 6.41 shows the initial screen for maintaining default values for external procurement.

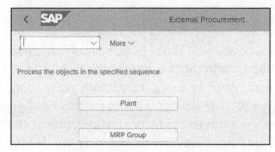

Figure 6.41 Define External Procurement: Initial Screen

Figure 6.42 shows the default values for external procurement triggered by MRP at the plant level:

- **Purch. Process. Time (purchase processing time)**
  Maintain the purchase processing time that the company's purchasing department typically takes to convert the purchase requisition into a purchase order. During the MRP run, the system considers the purchase processing time to determine the delivery date for the requisition item.
- **Sub. Purch. Group (substitute purchasing group)**
  Maintain a default purchasing group in this field that is responsible for converting purchase requisitions originating from MRP into purchase orders. The system uses this default value only if the purchasing group can't be determined from the purchasing master data or from the material master.
- **Unknown Acct Assignment**
  Maintain a default account assignment category for the purchase requisition line items that contain nonvaluated materials.

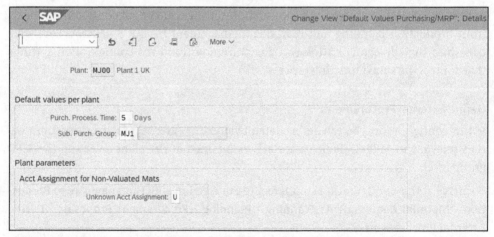

Figure 6.42 Default Values for External Procurement for MRP at Plant Level

Figure 6.43 shows the default values for external procurement triggered by MRP at the MRP group level.

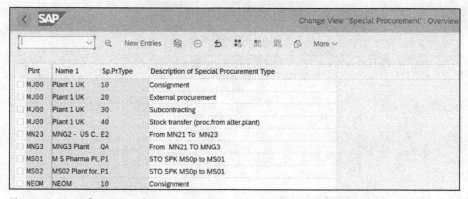

Figure 6.43 Default Values for External Procurement for MRP at MRP Group Level

In this configuration, relevant purchase requisition document types can be directly assigned to the combination of MRP group and plant. Maintain **DocType SPO** (default purchase requisition document type) for a standard purchase requisition, **DocType SC** (subcontracting purchase requisition document type), and **DocType ST** (stock transfer purchase requisition document type) for the combination of MRP group and plant.

> **Note**
>
> If you don't maintain a default purchase requisition document type in this configuration step, the system sets document type NB by default.

Define Special Procurement Type

In this configuration, you can define special procurement types at the plant level. A special procurement type is assigned in the **MRP 2** view of the material master values for external procurement for materials relevant for MRP.

To arrive at this configuration step, execute Transaction SPRO and navigate to **Materials Management • Consumption-Based Planning • Master Data • Define Special Procurement Types.**

Figure 6.44 shows the special procurement types defined for plant **MJ00**. Special procurement type **30** is defined for subcontracting.

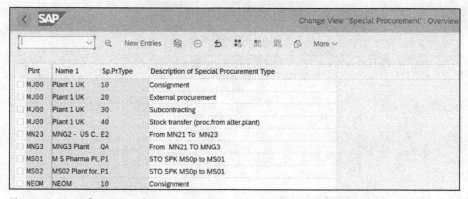

Figure 6.44 Define Special Procurement Type: Initial Screen

Next, maintain the attributes of special procurement type 30. Select the line for plant **MJ00** and special procurement type **30**, and then click the **Details** icon or press Ctrl + Shift + F2 to navigate to the details screen. Figure 6.45 shows the special procurement type details for type **30** (subcontracting):

- **Procurement Type**
  Every material relevant for MRP should be assigned a procurement type. Assign **F** (**External Procurement**) for special procurement type **30** (subcontracting). In the standard system, there are two procurement types:
  - **F: External Procurement**
  - **X: In-House Production**

  For subcontracting process with an external supplier, select procurement type **F**.

- **Special procurement**
  In this field, set the special procurement value relevant for the special procurement type. Table 6.6 shows the possible options. For subcontracting, assign the value **L**.

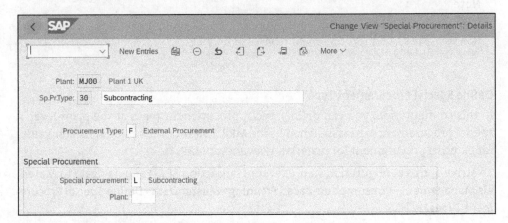

Figure 6.45 Define Special Procurement Type: Details

Procurement Type	Special Procurement	Description
E	E	In-house production
E	P	Prod. other plant
F		Initial value: external
F	K	Consignment
F	L	Subcontracting
F	U	Stock transfer

Table 6.6 Special Procurement Indicators

Define Purchase Requisition Document Type for Subcontracting

In this configuration, you can define a purchase requisition document type that represents a specific purchasing process based on business requirements. Standard purchase requisition, subcontracting, and stock transfer are the most common purchasing processes. A specific purchase requisition type can be created for every purchasing process for better control and visibility.

To arrive at this configuration step, execute Transaction SPRO and navigate to **Materials Management • Purchasing • Purchase Requisition • Define Document Types for Purchase Requisitions.**

Refer to Chapter 4, Section 4.4.3 for the detailed steps for this configuration.

The document type you create must allow item category **L** (**Subcontracting**) to create the subcontract purchase requisition, as shown in Figure 6.46.

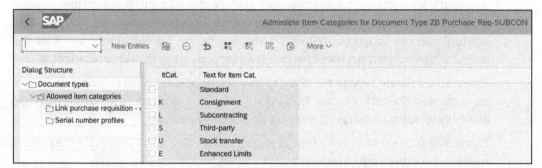

Figure 6.46 Purchase Requisition Document Type Settings for Subcontract Purchase Requisition

Define Document Type for Subcontracting

In this configuration, you can define a purchase order document type that represents a specific purchasing process based on business requirements. Standard purchase order, subcontract order, and stock transfer order are the most common purchasing processes.

To arrive at this configuration step to define a subcontract order document type, execute Transaction SPRO and navigate to **Materials Management • Purchasing • Purchase Order • Define Document Types for Purchase Orders.**

Refer to Chapter 4, Section 4.4.3 for the detailed steps for this configuration.

The document type you create must allow item category **L** (**Subcontracting**) to create the subcontract order, as shown in Figure 6.47.

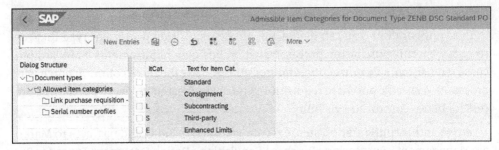

Figure 6.47 Purchase Order Document Type Settings for Subcontract Order

Set Up Subcontract Order

In the subcontracting process, components are provided by the company that placed the subcontract order. Components stock can be provided to the vendor in two ways, one with a sales and distribution delivery and another without. Standard best practice is to ship the components via the sales and distribution delivery process. In this configuration, you can assign the relevant delivery type for the creation of an outbound delivery for a subcontract order to issue the components to the subcontractor.

To assign a delivery type to the plant to issue components to the subcontract vendor via outbound delivery, execute Transaction SPRO and navigate to **Materials Management • Purchasing • Purchase Order • Set Up Subcontract Order**.

Figure 6.48 shows that delivery type **LB** is assigned to plant **MJ00** to provide the components to the subcontract vendor.

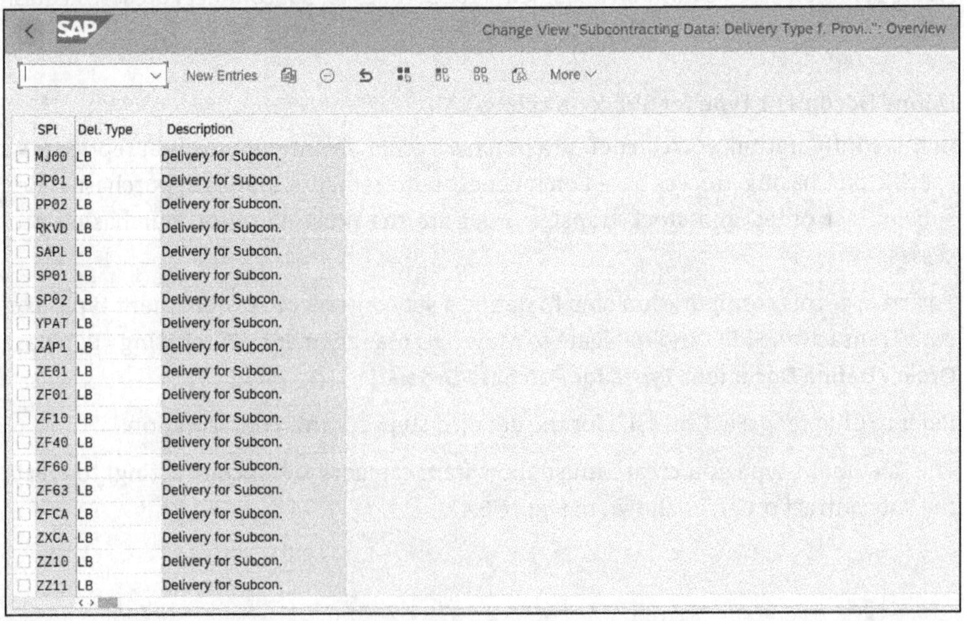

Figure 6.48 Assign Delivery Type for Subcontract Order at Plant Level

The system determines item category LBN for delivery type LB. Schedule line category LB is determined for item category LBN. The system determines movement type 541 for the goods issue stock to the subcontractor during the creation of an outbound delivery with reference to the subcontract order.

Figure 6.49 shows movement type **541** assigned to schedule line category **LB**.

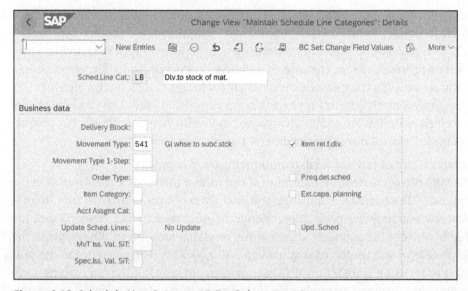

Figure 6.49 Schedule Line Category LB For Subcontract Process

6.5   Inventory Management

Production planning and inventory management are tightly integrated functions in SAP S/4HANA. Inventory management ensures the required materials and components required to manufacture the company's finished products are adequately supplied to production operations. The finished products manufactured from production planning are received into specific locations within inventory management, from which finished products are issued to customers with reference to sales orders.

Inventory management provides visibility into the inventory on hand and location information for where the materials are stored in inventory. During MRP, the system generates procurement proposals for any shortage of inventory to produce finished products based on the demand forecast and customer demand. The system performs an availability check for materials and components based on the BOM to produce finished products.

The inventory management function tracks all material movements within the production process that include goods issue of raw materials for internal operations and external operations and goods receipt of finished products from production.

This helps optimize the production process, reducing costs and meeting customer demand in a timely fashion. Let's explore the integration between inventory management and production planning functions.

6.5.1   Goods Receipt Postings

Goods receipt for production planning involves receiving manufactured products from the in-house production function and from external manufacturers (subcontractors). Goods receipt into inventory management is performed with reference to a production or process order in the case of in-house production and with reference to a subcontract order in the case of external manufacturing. Goods receipt also involves receiving raw materials from suppliers that are planned in MRP. This process ensures the received materials are accurately updated in inventory management. For production planning, the following are the different goods receipt processes:

1. **Goods receipt of raw materials/components from suppliers**
   The MRP run generates purchase requisitions for raw materials if not available in the inventory. The purchase requisition will be converted into a purchase order to procure raw materials for production operations from an external vendor. Once the shipment from the suppliers arrives at the receiving location within the plant, the goods receipt will be posted manually in SAP S/4HANA with reference to the purchase order using Transaction MIGO and movement type 101. The raw materials will be directly consumed by production if the production order is assigned to the purchase order line item.

2. **Goods receipt of finished products from inhouse production**
   The produced finished products will be posted into inventory by performing goods receipt with reference to the production order/process order using Transaction MIGO. The stock will be posted to unrestricted use stock upon posting the goods receipt. This uses movement type 101.

3. **Goods receipt of semifinished products from external manufacturer**
   The subcontracting process is a special procurement process in SAP S/4HANA where certain operations will be outsourced by an organization to external suppliers and the required components will be provided to the external supplier for the operation. The external supplier, the subcontractor, will perform the operation on behalf of the organization. The subcontractor will ship the semifinished product back to the organization and charge a service fee. Once the shipment from the subcontractor arrives at the receiving location of the plant, a goods receipt will be posted manually with reference to the subcontract order using Transaction MIGO and movement type 101.

6.5.2   Goods Issue Postings

Goods issue for production planning is a process of removing raw materials and components stock from inventory for consumption for production. The goods issue posting creates both a material document and an accounting document in the system (depending on the movement types used) to record the material movement and the valuation of the stock in financial accounting. For production planning, the following are the different goods issue processes:

1. **Goods issue of raw materials/components to in-house manufacturing**
   Raw materials and other required materials/components will be issued to production with reference to a production order or process order using Transaction MIGO and movement type 261. MRP ensures that the required materials and components to produce the finished products are available at the right time. Inventory management provides a material reservation feature to reserve materials and components for production in advance. The goods issue ensures that the required materials are supplied to production for consumption during manufacturing of the finished products.

2. **Goods issue of raw materials/components to external manufacturing**
   Raw materials and other required materials/components are issued to an external manufacturer (subcontractor) with reference to a subcontract order using Transaction ME2O and movement type 541. Goods issue can be posted with reference to a (sales and distribution) outbound delivery. Movement types play an important role in determining the type of the inventory movement and posting of goods issue. Movement type 541 is used for a goods issue for an outbound delivery to ship the components to the subcontractor. Movement type 541 ensures the components provided to the subcontracting vendor are tracked and that ownership remains with the company requesting the external operation.

6.5.3   Movement Types and Stock Postings

Production planning in SAP S/4HANA governs the MRP, in-house manufacturing, and external manufacturing processes. Managing material movements to support these processes is crucial for the organization. The availability of raw materials and other components required for the production process is essential to effectively managing the production process. There are specific movement types defined in the standard system for material movements to support production planning processes.

Table 6.7 shows the movement types used in production planning processes.

Movement Type	Detailed Description
101	Goods receipt. This movement type is used for standard goods receipts from external vendors and subcontractors as well as goods receipt against production/process orders to receive in-house produced products into inventory. It increases the stock quantity and value in the receiving storage location. Stock can be received into unrestricted use stock or quality inspection stock.

Table 6.7 Movement Types Used in Production Planning-Related Processes

Movement Type	Detailed Description
102	Goods receipt reversal. This movement type is used to cancel the goods receipt posted using movement type 101. Stock will be removed from unrestricted use stock or quality inspection stock upon posting this goods receipt reversal.
122	Return to vendor. This movement type is used to return materials to the vendor that can't be used in production due to poor quality. Stock will be removed from inventory to return the goods to vendor.
161	Returns purchase order to vendor. This movement type is used to return materials to the vendor with reference to a returns purchase order. Stock will be removed from the inventory to return the materials to vendor.
321	Transfer posting from quality inspection to unrestricted. This movement type is used to transfer materials that are accepted in quality inspection into unrestricted use stock.
201	Goods issue for a cost center. This movement type is used to issue goods from the inventory to a cost center for consumption.
261	Goods issue for a production order. This movement type is used to issue spare parts and other components from the inventory to a production order for consumption.
531	Goods receipt by-product from production. This movement type is used to post goods receipt for by-products from production against production/process orders to receive them into inventory. It increases the stock quantity and value in the receiving storage location.
541	Goods issue for subcontract supplier. This movement type is used to issue equipment and other components to the subcontractor (service provider) to repair/maintain on behalf of the company that issued the subcontract order. Stock will be transferred from unrestricted use stock to subcontracting stock within the inventory management.
543	Consumption from subcontracting. This movement type is used to post the consumption of components provided to the subcontractor automatically upon posting the receipt of the final product produced/assembled by the subcontractor. This goods movement is automatically posted during the goods receipt with reference to the subcontract order.
545	Goods receipt by-product from subcontracting. This movement type is used to post goods receipt for by-products from subcontracting against the subcontract order.

Table 6.7 Movement Types Used in Production Planning-Related Processes (Cont.)

6.5.4 Reservations

Material reservations are requests made to inventory management to set aside required quantities of specific materials for a particular purpose, typically for production or maintenance. The purpose is defined by specific movement types used to create reservations in SAP S/4HANA. Material reservations ensure that materials are available for a particular purpose and prevent them from being used for another purpose. Material reservations can be created manually using Transaction MB21 and automatically from production planning during the creation of the production order. The available-to-promise (ATP) check and MRP run recognize material reservations, and it is possible to configure both the availability check and MRP run to consider reservations.

Material reservations offer several advantages in production planning, inventory management, and purchasing. Material reservations ensure smooth production operations and facilitate accurate production planning by providing visibility into available materials for future use. The production lead times will be reduced with material reservations and ultimately bring customer satisfaction. On the other hand, material reservations help in effective inventory management and control. They also help in prioritizing the purchasing process to procure additional stock of reserved materials.

In production planning, the following movement types are most used in the creation of material reservations:

- **201**
  Used to reserve materials for a cost center to issue materials to a cost center at a future date. The production department cost center will be used to create reservations.

- **261**
  Used to reserve materials for a production/process order to issue materials for production consumption at a future date.

We'll walk through the steps to create and configure reservations in the following sections.

Create Material Reservations

Let's explore the creation of a material reservation in SAP S/4HANA. You can create a material reservation using Transaction MB21. Figure 6.50 shows the initial screen for creation of a material reservation, where you fill out the following fields:

- **Base Date**
  This is the base date for the reservation. The system uses this date as a requirement date for materials from the inventory.

- **Check Date**
  If you set this indicator, the system checks the factory calendar for any holidays.

- **Movement Type**
Specify a particular movement type here to be used for material reservations. The system uses the **Material Type** field to display additional fields relevant for that goods movement. For example, for movement type 261, the system asks for production/process order details to create a material reservation. The purpose for the material reservation will be identified through movement types.

- **Plant**
Specify the plant in which you are creating the material reservation.

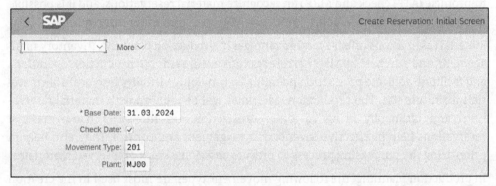

Figure 6.50 Create Reservations: Initial Screen

After specifying those fields, press ⌜Enter⌝ to navigate to the **New Items** screen. Figure 6.51 shows the items screen of material reservation. For movement type **201** (goods issue for cost center), the **Cost Center** and **G/L Account** fields must be filled to create the reservation.

You can enter multiple line items in one reservation with different materials, quantities, batches, and so on. The **Movement Allowed** indicator allows goods movement for the reservation. If you deselect this indicator, the system prevents goods movement for this reservation.

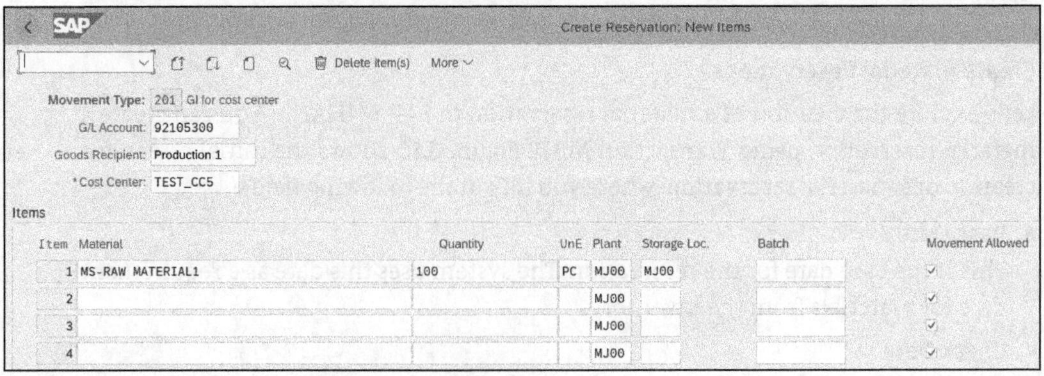

Figure 6.51 Create Reservations: New Items

Define Number Assignment for Reservations

In this configuration, you can define the number range for reservations. The system uses this number range to create reservations and assign internal numbers to reservations automatically.

To arrive at the number range configuration step, execute Transaction SPRO and navigate to **Materials Management • Inventory Management and Physical Inventory • Number Assignment • Define Number Assignment for Reservations**. You can arrive at this configuration step directly by executing Transaction SNRO. In this configuration step, both internal and external number ranges can be defined against number range object RESB.

Figure 6.52 shows the configuration of internal and external number ranges for reservations.

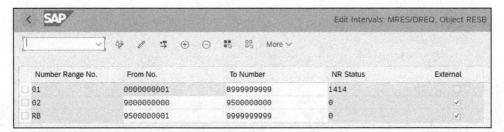

Number Range No.	From No.	To Number	NR Status	External
01	0000000001	8999999999	1414	☐
02	9000000000	9500000000	0	✓
RB	9500000001	9999999999	0	✓

Figure 6.52 Define Number Range for Reservations

Define Default Values

In this configuration, you can define default values for reservations at the plant level.

To arrive at the configuration step, execute Transaction SPRO and navigate to **Materials Management • Inventory Management and Physical Inventory • Reservation • Define Default Values**. The key fields are as follows (see Figure 6.53):

- **Mvt (movement allowed)**
  This indicator can be set at the plant level so that during the material reservation creation, this indicator is defaulted at the line item level of the reservation. However, you can select/deselect this indicator during the creation of a reservation using Transaction MB21. If this indicator is set in the reservation item, the system allows the material movement for the reservation.

- **Days m (number of days movement allowed indicator is set automatically)**
  Specify the number of days from the required date in the reservation from which the system allows the goods movement for the reservation. For example, if the current date is March 30, 2024, and the number of days for which movement is allowed is set as 10 days, then the system sets this indicator automatically for all reservations created with requirement dates less than or equal to April 9, 2024.

- **Rete (retention period, in days, for reservation items)**
  Specify the number of days for the system to retain reservations. Reservation items will be deleted automatically after a specified number of days of retention. For example, if the current date is March 30, 2024, the requirement date is March 30, 2024, and the retention period is set as 30 days, then the system deletes the unused reservation immediately after 30 days from the requirement date.

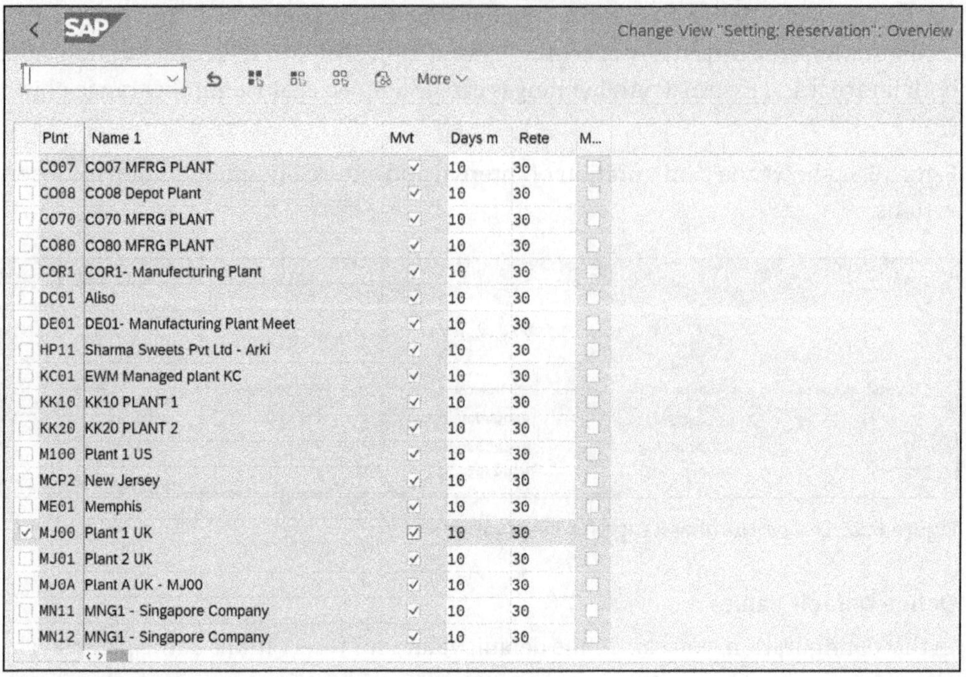

Plnt	Name 1	Mvt	Days m	Rete	M...
CO07	CO07 MFRG PLANT	✓	10	30	☐
CO08	CO08 Depot Plant	✓	10	30	☐
CO70	CO70 MFRG PLANT	✓	10	30	☐
CO80	CO80 MFRG PLANT	✓	10	30	☐
COR1	COR1- Manufacturing Plant	✓	10	30	☐
DC01	Aliso	✓	10	30	☐
DE01	DE01- Manufacturing Plant Meet	✓	10	30	☐
HP11	Sharma Sweets Pvt Ltd - Arki	✓	10	30	☐
KC01	EWM Managed plant KC	✓	10	30	☐
KK10	KK10 PLANT 1	✓	10	30	☐
KK20	KK20 PLANT 2	✓	10	30	☐
M100	Plant 1 US	✓	10	30	☐
MCP2	New Jersey	✓	10	30	☐
ME01	Memphis	✓	10	30	☐
MJ00	Plant 1 UK	☑	10	30	☐
MJ01	Plant 2 UK	✓	10	30	☐
MJ0A	Plant A UK - MJ00	✓	10	30	☐
MN11	MNG1 - Singapore Company	✓	10	30	☐
MN12	MNG1 - Singapore Company	✓	10	30	☐

Figure 6.53 Define Default Values for Reservations

Configure Scope of Availability Check

In this configuration, we can define the scope of the availability (ATP) check. This configuration will be set with reference to a checking rule and availability checking group. The scope involves the type of stock of the product to be checked for availability (e.g., quality inspection stock), the future supply scope to be considered or not (e.g., include confirmed stock transfer orders), and so on. You can also define the requirements to be considered for the availability check such as the requirements arising from sales orders, reservations, deliveries, and so on.

To arrive at the configuration step, execute Transaction SPRO and navigate to **Materials Management • Inventory Management and Physical Inventory • Reservation • Configure Scope of Availability Check**.

In this configuration step, you can define a checking rule, define a scope of check, and assign a checking rule to specific transaction codes. Click **Define Checking Rule** from the initial screen to navigate to the screen to define the scope of availability check.

Figure 6.54 shows the scope of availability check. Set the **With Reservations** indicator for the system to consider reservations during the availability check.

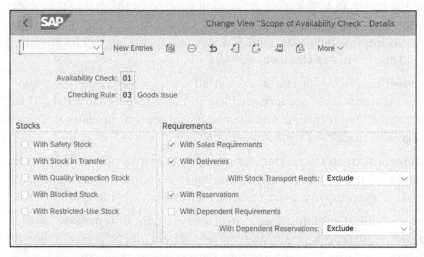

Figure 6.54 Scope of Availability Check for Reservations

The next step in this configuration is to assign the availability checking rule to the transaction codes for reservation creation and reservation change. Click **Transaction code** from the initial screen to navigate to the screen to assign the availability checking rule to the transaction codes for reservation creation and reservation change. Figure 6.55 shows the configuration settings for assigning an availability checking rule to the **Create Reservation** and **Change Reservation** transaction codes. The system performs an availability check for the items in the reservation during the execution of these transactions.

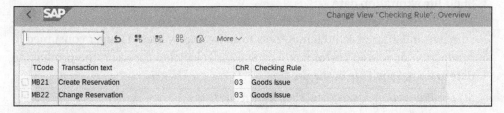

Figure 6.55 Assign Checking Rule to Transaction Codes for Reservation Creation and Change

6.5.5 Batch Management

In certain industries, especially in process manufacturing industries, materials are produced in batches. Batch management plays a vital role in production planning and is an important part of production, and it provides traceability, quality control, and regulatory compliance of batches of materials produced in house and via external manufacturing. Batch information is captured during the goods receipt transaction for tracking and tracing purposes in inventory management. Specific batch information such as batch number, material, quantity, production date, expiration date, and other attributes will be captured during the goods receipt. This helps in goods issue transactions to determine the materials with earlier production dates to be shipped to customers rather than newly produced batches. In regulated industries such as pharma, batch management ensures that regulatory compliance requirements are met by providing traceability and documentation for batches.

Batch management is used in life sciences and healthcare, food and beverage production, chemical industries, milk product production, and aerospace industries. These industries are also highly regulated, and the batch management provides necessary visibility and control to monitor regulatory compliance.

Batch management also helps in making sure unexpired raw materials are issued from inventory management for in-house manufacturing and external manufacturing. The expired batches in inventory management will be disposed of accordingly.

In batch management, the following date attributes are critical to manage batches. They provide the necessary traceability and control different functions within the supply chain. These dates are updated the first time the batches are received:

1. **Production date**
   The production date is a critical attribute of the batch of a material. It indicates the date on which the materials of the batch are produced in house or externally. The production date is captured at the time of goods receipt, and it is recorded in the batch master record. The production date helps in inventory management for issuing materials with an earlier production date, called the first in, first out (FIFO) strategy. The production date is vital to calculate the shelf-life expiration date (SLED) and the expiration date of the batch.

2. **Shelf-life expiration date**
   The SLED of a batch is the date until which the materials of the batch are usable/consumable and after which the materials from inventory are no longer consumed in production or sold to customers. The SLED is calculated based on the production date and total shelf life of the product:

   SLED = Production date + Total shelf life (in days)

   The SLED is continuously monitored in inventory management to know the remaining shelf life of the materials and batches. Warehouse and inventory management picking strategies are established based on the SLED to issue materials to customers based on business requirements. The expired materials will be disposed of from inventory.

Let's explore the key configuration settings for batch management.

Specify Batch Level

In SAP S/4HANA, batches can be maintained at the client level, plant level, and material level. Best practice is to create and maintain batches at the material level. During the goods receipt of the materials for the first time into inventory, a batch number will be created with production date and expiration date attributes.

In this configuration, you can maintain the batch level. To arrive at this configuration step, execute Transaction SPRO and navigate to **Logistics—General** • **Batch Management** • **Specify Batch Level and Activate Status Management**.

Figure 6.56 shows the batch level is defined at the material level by selecting the **Batch unique at material level** option.

Figure 6.56 Define Batch Level

Activate Internal Batch Number Assignment

In this configuration, you can activate the internal number range for batches.

To arrive at this configuration step, execute Transaction SPRO and navigate to **Logistics—General** • **Batch Management** • **Batch Number** • **Activate Internal Batch Number Assignment**.

As shown in Figure 6.57, select **Active** to activate the internal batch number assignment for batch master creation.

Figure 6.57 Activate Internal Batch Number Assignment

Maintain Internal Batch Number Assignment Range

In this configuration, you can maintain an internal number range for internal batch number assignment.

To arrive at the number range configuration step, execute Transaction SPRO and navigate to **Logistics—General** • **Batch Management** • **Batch Number** • **Maintain Internal**

Batch Number Assignment Range. You can arrive at this configuration step directly by executing Transaction SNRO.

In this configuration step, both internal and external number ranges can be defined against number range object BATCH_CLT. Figure 6.58 shows the configuration of internal number ranges.

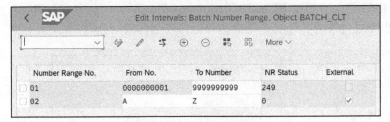

Figure 6.58 Maintain Internal Batch Number Assignment Range

Define Batch Creation for Production Order/Process Order

Batches can be created automatically at the time of creation of a production order or process order or during the release of the production order or process order.

This configuration setting is performed in the production scheduling profile. To arrive at this configuration step, execute Transaction SPRO and navigate to **Production Planning for Process Industries • Process Order • Master Data • Define Production Scheduling Profile for Process Manufacturing**.

Figure 6.59 shows that the production scheduling profile is defined for the plant. To define a new production scheduling profile for the plant, click **New Entries** and specify the plant and a six-digit alphanumeric key in **PS Prof.** (production scheduling profile).

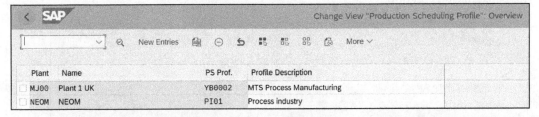

Figure 6.59 Batch Management Settings for Production Order or Process Order: 1 of 2

To maintain the batch management settings for the combination of a plant and production scheduling profile, select the combination from the overview screen and click the **Details** icon at the top or press Ctrl + Shift + F2 . Figure 6.60 shows the batch management settings for the production scheduling profile. Let's explore the key configuration fields:

- **Auto Batch Creation**
  In this field, you can define if the automatic batch creation should happen at the time the production order or process order is created, at the time of the release of

the production order or process order, or not at all during the creation or release of the production/process order. The following values are possible:

– No automatic batch creation in production order or process order

– Automatic batch creation at time of production order or process order creation

– Automatic batch creation at time of production order or process order release

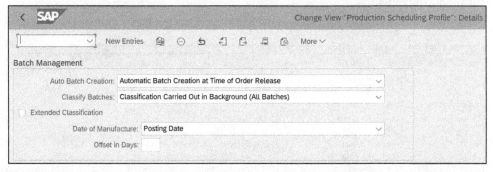

Figure 6.60 Batch Management Settings for Production Order or Process Order: 2 of 2

- **Classify Batches**
  Classification of batches is an important activity to define additional attributes based on characteristics such as qualitative data (color, odor, etc.) and quantitative data (weight, viscosity, etc.). Later in this section, we'll discuss how to define characteristics and classes for batches of material. The classification of batches will be carried out at the time of creation and change of the batch master record.

  In this field, you can define if the classification of batches should happen in the background, foreground, or not at all. The classification of batches will be carried out in the background or foreground based on this setting. The following values are possible:

  – No branching to batch classification

  – Classification carried out in background for all batches

  – Classification only in foreground for mandatory characteristics for all batches

  – Classification always invoked in foreground for all batches

  – Classification always invoked in foreground for new batches only

- **Date of Manufacture**
  In this field, you can define if the production date for the batch master record should be derived from the posting date, confirmation date, or system date. The following values are possible:

  – No default

  – Posting date

  – Confirmation date (start of execution)

  – Confirmation date (end of execution)

  – System date

Define Batch Creation for Goods Movements

In this configuration, you can define how the new batch records are created for every movement type.

To arrive at the configuration step, execute Transaction SPRO and navigate to **Logistics—General • Batch Management • Creation of New Batches • Define Batch Creation for Goods Movements**.

Figure 6.61 shows the definition of a new batch creation for every movement type.

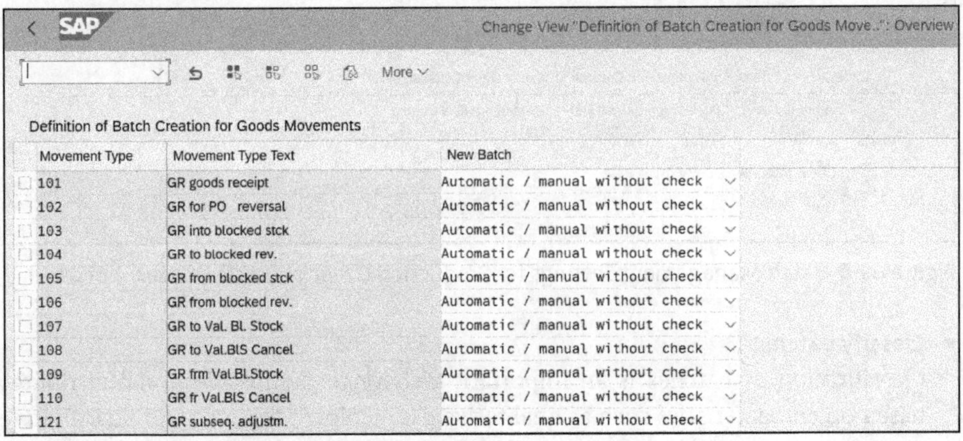

Movement Type	Movement Type Text	New Batch
101	GR goods receipt	Automatic / manual without check
102	GR for PO reversal	Automatic / manual without check
103	GR into blocked stck	Automatic / manual without check
104	GR to blocked rev.	Automatic / manual without check
105	GR from blocked stck	Automatic / manual without check
106	GR from blocked rev.	Automatic / manual without check
107	GR to Val. Bl. Stock	Automatic / manual without check
108	GR to Val.BlS Cancel	Automatic / manual without check
109	GR frm Val.BLStock	Automatic / manual without check
110	GR fr Val.BlS Cancel	Automatic / manual without check
121	GR subseq. adjustm.	Automatic / manual without check

Figure 6.61 Definition of Batch Creation for Goods Movements

The following values are possible to set in the **New Batch** field based on business requirements:

- Automatic / manual without check
- Automatic / manual and check against external number range
- Automatic / manual and check in user exit
- Automatic / no manual creation
- Manual without check
- Manual and check against external number range
- Manual and check in user exit
- No creation

Define Batch Class and Characteristics

SAP S/4HANA provides batch classifications to define additional attributes for batches of materials using characteristics. You can define multiple characteristics and multiple classes for material batches. One or more characteristics can be assigned to a class, and the class is assigned to the material in the classification view. Batch classification brings flexibility to batch management, and organizations can define batch classes and characteristics. Using the batch classification feature, you can define different attributes for different batches of materials. Batches can be searched and reported using batch classification. Batch classification can also help with defining regulatory compliance-related attributes for reporting and tracking.

Let's explore the definition of batch classification for materials:

1. **Create and maintain characteristics for batch classification**

   Batch characteristics can be defined using Transaction CT04. Figure 6.62 shows batch characteristic **Z_COLOR** defined for batch classification. Let's explore the key field settings:

   – **Status:** In this field, you can define the processing status of batch characteristic. At the time of creation of the characteristic, **In Preparation** will be assigned as the status. To use this characteristic in the batch classification, the status must be set to **Released.**

   – **Data Type:** In this field, you can define the standard data format to define the attribute values for this characteristic in the batch classification. One of the following values can be set:

     • **Character Format**
     • **Currency Format**
     • **Date Format**
     • **Numeric Format**
     • **Time Format**

   – **Value Assignment:** You can define how you should be defining an attribute value for the characteristic in the batch classification. You can set this to **Single Value** or **Multiple Values** for the characteristic.

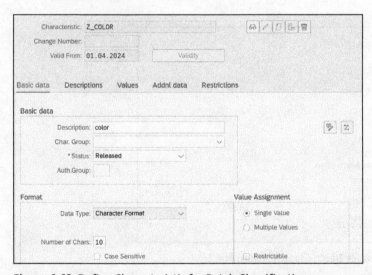

Figure 6.62 Define Characteristic for Batch Classification

   – **Values:** In this tab, you can define the values for the characteristic. If you define values for the characteristic in this setting, you can select from the list of values during the batch creation for batch classification.

– **Restrictions**
In this tab, you can define if this characteristic can be used in different class types. The following are a few class types defined in the standard system:

- **023**: Batch
- **001**: Material Class
- **002**: Equipment Class
- **005**: Inspection Characteristics

2. **Create and maintain a class for batch classification**
A batch class can be defined using Transaction CL02. Figure 6.63 shows batch class **Z_BATCH_SEL** defined for batch classification. Let's explore the key field settings:

– **Class type**
In this field, specify a specific class type for the class. For batch classification, assign **023** as the class type.

– **Status**
In this field, you can define the processing status of a batch characteristic. At the time of creation of the characteristic, **In Preparation** will be assigned as the status. To use this characteristic in the batch classification, the status must be set to **Released**.

– **Valid From/Valid to**
Define a **Valid From** date and **Valid to** date for the batch class. To assign this class to materials, the status of the class must be released and should be within the validity range.

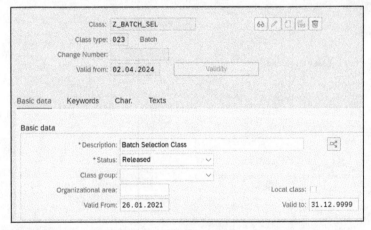

Figure 6.63 Define Class for Batch Classification: 1 of 2

Next, assign the predefined batch characteristics to the class. Click the **Char.** tab to navigate to the batch characteristics, as shown in Figure 6.64. Assign the batch characteristics defined for batch classification in this tab by directly specifying the batch characteristics in a sequence.

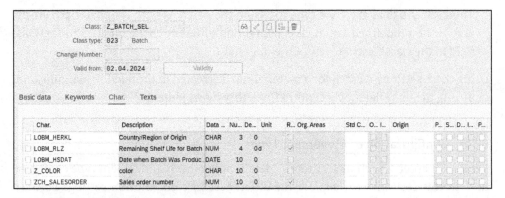

Figure 6.64 Define Class for Batch Classification: 2 of 2

3. **Assign batch class to material master data**

   In this setting, assign the defined class in the **Classification** view of the material master for batch classification, as shown in Figure 6.65. The characteristics defined in the class will be defaulted automatically.

   To assign the batch class to the material, execute Transaction MM02 and specify the material to which the batch class is to be assigned. Then in the **Select Views** dialog box, select the **Classification** view and press [Enter].

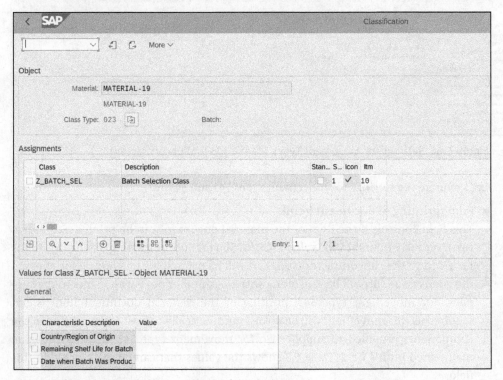

Figure 6.65 Assign Batch Class to Material Master Data

Then select **023** as the class type to assign the batch class to the material. Directly assign the predefined batch class containing different characteristics to the **Class** field in the classification view of the material.

Using these settings, a batch for a material can be created in production planning and/or during the material movements in inventory management.

6.5.6   Configuration of Movement Types

Refer to Chapter 4, Section 4.3.5 for the details of inventory movement type configuration. In this section, we'll explore the differences in configuration settings between movement types 541 (goods issue components for subcontract supplier) and 543 (consumption posting of components from subcontracting). These two movement types support the subcontracting process. Movement type 541 supports the goods issue of components to the subcontractor to produce/assemble the final product. But the ownership of the components still stays with the company, unlike other goods issue postings. The consumption of these components happens at the time of goods receipt of the final product from the subcontractor. Figure 6.66 shows the differences in configuration settings between movement types **541** and **543**.

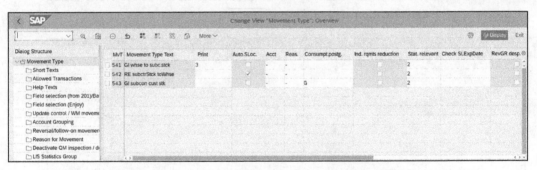

Figure 6.66 Settings for Movement Types 541 and 543 for Subcontracting

Let's explore the settings for movement types **541** and **543**:

- **Print (printing of document item)**
  This field controls if the item of the material document can be printed. For movement type **541**, maintain the value **3** (**GR/GI Slip for Subcontracting**) so that the items provided to the subcontractor (supplier) will be printed in the goods issue slip and the goods issue slip will be sent along with the shipment of components to the supplier. For movement type **543**, printing of the material document item is not required as this movement will be posted automatically to post consumption of the components provided to supplier. So for movement type **543**, there is no value maintained in this field. Table 6.8 shows the values that can be maintained in this field.

Printing of Doc. Item Control	Description
Blank	No document printout
1	Material document printout
2	Return delivery
3	Goods receipt/goods issue slip for subcontracting
4	Inventory list without sales price
5	Inventory list with sales price
6	Material document printout for goods receipt/goods issue
7	Transfer posting

Table 6.8 Printing of Document Item for Material Movements

- **Consumpt.postg. (consumption posting)**
  This field controls the consumption posting for specific movement types. During the consumption posting, goods issue of the materials will be posted and the materials will be removed from the inventory. For movement type **541**, there is no value maintained in this field as the materials provided to the subcontractor will be physically shipped, but the component stock ownership still lies with the company that requested the subcontracting operation. So in SAP S/4HANA, the components stock will not be removed from the inventory during this material movement; instead the materials will be transferred to subcontracting stock for tracking purposes. For movement type **543**, value **G** (**Planned withdrawal (total consumption)**) is maintained in this field as the system posts the consumption of the components stock automatically during the goods receipt from the subcontractor. Table 6.9 shows the values that can be maintained in this field.

Consumption Posting	Description
Blank	No consumption update
G	Planned withdrawal (total consumption)
R	Planned, if reference to reservation, otherwise unplanned
U	Unplanned withdrawal (unplanned consumption)

Table 6.9 Consumption Posting Indicators for Material Movements

6.6   Summary

In this chapter, we introduced the integration of production planning functions with materials management and offers best practice examples and processes. In Section 6.1, we explained the integrated production planning processes. In Section 6.2, we explained the master data in production planning with step-by-step procedures to maintain production planning-specific master in SAP S/4HANA. In Section 6.3, we explained the integrated MRP and MPS processes for materials planning for production processes, including key configuration settings. In Section 6.4, we explained the operational subcontracting process. Finally, in Section 6.5, we explained how the inventory management functionality of materials management integrates with production planning, including goods receipts, goods issues, reservations, and batch management.

In the next chapter, we'll discuss the integration between plant maintenance and materials management to streamline the maintenance processes and ensure efficient management of spare parts with best practice examples, master data requirements, and configuration settings.

Chapter 7
Plant Maintenance

Plant maintenance in SAP S/4HANA provides a comprehensive list of tools and procedures for facility management, equipment maintenance, and more. It optimizes maintenance processes and reduces downtime of assets such as equipment, machinery, and facilities while improving their overall performance. The integration between plant maintenance and materials management ensures the on-time supply of materials required to perform maintenance activities.

The main objective of the plant maintenance function is to extend the life of a company's assets, prevent failures, and make sure that equipment and machinery function efficiently. Maintenance planning is a crucial activity within a plant (manufacturing plant, distribution plant, storage plant, etc.) and is performed by maintenance planners. SAP S/4HANA provides tools and procedures to plan maintenance tasks, manage equipment master data, and manage maintenance orders. Maintenance planning involves creation and scheduling of maintenance tasks, including inspection tasks. Maintenance order management involves managing maintenance tasks, material requirements, and resource requirements to execute maintenance tasks. Equipment management involves managing equipment master data, technical specifications, and other documentation while retaining maintenance history.

Notification management is a standard feature available in plant maintenance that allows timely reporting of equipment failures and other maintenance incidents that can help prevent equipment downtimes. Maintenance notifications can trigger the creation of maintenance orders.

Maintenance planning within plant maintenance triggers the material requirements needed to execute maintenance tasks. Critical maintenance, repair, and operations (MRO) items are planned in materials management using consumption-based planning, particularly reorder point planning. Material requirements from plant maintenance and reorder point planning generate purchase requisitions to procure required materials. Purchase requisitions are converted to purchase orders and sent to external vendors. The received materials will be stored in inventory and issued to maintenance tasks when required using movement type 261. Materials management functions ensure the required materials are available on time for maintenance activities. Coordination between plant maintenance and materials management is vital for organizations to prevent downtimes and achieve overall operational efficiencies.

This chapter covers the plant maintenance processes, integration points between plant maintenance and materials management, plant maintenance master data, and key configuration settings in SAP S/4HANA. First, we'll provide an overview of the integrated plant maintenance processes and how they work cross-functionally with materials management.

7.1   Integrated Plant Maintenance Processes

There are two maintenance strategies applied based on the industry best practices and the business requirements: preventive maintenance and corrective maintenance. Let's start with an overview of preventive maintenance.

Preventive maintenance involves regular scheduled maintenance of equipment, machinery, and facilities. It involves regular inspection of assets and performing necessary maintenance activities regularly such as lubricating, changing filters, and comparing the inspection results against predefined lists of value ranges. Preventive maintenance prevents downtime caused by sudden failure of equipment. Preventive maintenance tends to incur regular maintenance costs while it increases the lifespan of a company's assets.

Figure 7.1 shows the preventive maintenance process flow, illustrating the integration between plant maintenance and materials management.

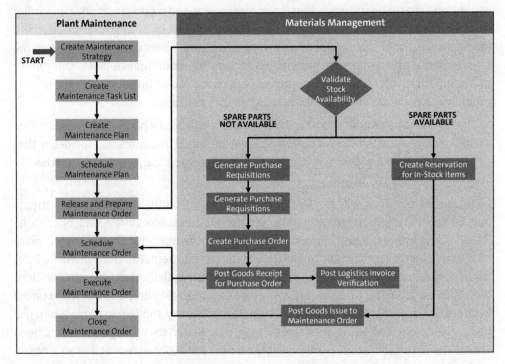

Figure 7.1 Integrated Maintenance Planning Process: Preventive Maintenance

The main objective of preventive maintenance (also called *proactive maintenance*) is to avoid equipment breakdown. Maintenance plans are used in preventive maintenance that contain the maintenance tasks, the scope and time of execution of maintenance tasks, equipment and functional location details, frequency of maintenance task execution, and so on. The following are the different types of preventive maintenance:

- **Time-based preventive maintenance**
  Maintenance tasks are scheduled at regular intervals—for example, performing maintenance activities for a critical asset every two months. The system automatically generates maintenance plans every two months in this case.

- **Performance-based preventive maintenance**
  Maintenance tasks are scheduled based on performance values of technical objects, equipment, or functional locations—for example, performing maintenance activities after running a production unit for 1,000 hours. In this case, the system automatically generates maintenance tasks and uses real-time data to schedule maintenance tasks.

- **Condition-based preventive maintenance**
  Maintenance tasks are scheduled based on measuring points. For example, imagine that a production unit normally runs at a temperature of 150 degrees Fahrenheit and currently is running at over 200 degrees Fahrenheit. In this case, the system automatically generates maintenance tasks and uses real-time data to schedule maintenance tasks.

Let's explore the process steps from plant maintenance and materials management involved in preventive maintenance:

1. **Create maintenance strategy**
   A maintenance strategy is vital in case of preventive maintenance, and it defines the rules for the sequence and cycles/frequencies of planned maintenance activities based on time, performance, or condition. Maintenance strategies, once defined, can be used in maintenance plans and task lists. You can create maintenance strategies using Transaction IP11.

2. **Create maintenance task list**
   A maintenance task list defines the sequence of tasks to be performed during preventive maintenance. You can define operations in a sequence for equipment while defining the task list. You can create a maintenance task list using Transaction IA01.

3. **Create maintenance plan**
   The next step in the process is to create a maintenance plan for a technical object using a task list. The maintenance plan defines the date and scope of the maintenance tasks to be performed on a piece of equipment or a functional location. You can create maintenance plans using Transaction IP41.

4. **Schedule maintenance plan**
   Scheduling of maintenance plans is typically carried out automatically using a

scheduled background job. This activity triggers the creation of maintenance order to perform maintenance activities defined in the maintenance plan. You can schedule maintenance plans manually using Transaction IP10. Scheduling the maintenance plans generates the maintenance order.

5. **Release and preparation for maintenance order**
Once the maintenance order is generated automatically, it will be released, and the status will be set to *in preparation* to start preparing to perform maintenance activities. The maintenance planners will validate the resource availability, materials (spare parts) availability, and services to be performed.

6. **Create material reservations if spare parts are available in inventory**
During the availability check for spare parts, if the spare parts are available in the inventory, material reservations will be created for spare parts for goods issue for a maintenance order on a specific date and for a specific quantity. Material reservations are requests made to inventory management to keep aside required quantities of specific materials for a particular purpose, typically for production or maintenance. Material reservations for plant maintenance will be created using movement type 261 for issue of spare parts against a maintenance order.

7. **Generate purchase requisitions if spare parts are not available in inventory**
During the availability check for spare parts, if the spare parts are not available in the inventory, the system generates purchase requisitions automatically from a maintenance order to procure spare parts from a third-party vendor. The system assigns a source of supply (a third-party vendor) for spare parts from purchasing master data on the materials management side.

8. **Convert purchase requisitions into purchase order**
The purchase requisition will then be converted into a purchase order manually by the purchaser using Transaction ME21N or automatically in SAP S/4HANA to procure spare parts for maintenance activities. The line item details are defaulted into the purchase order, including the source of supply, unit cost, and delivery address. The purchase order will be transmitted to the third-party supplier based on the output method: email, electronic data interchange (EDI), or electronic transmission to the supplier on SAP Business Network. Output determination configuration in SAP S/4HANA drives the purchase order output mechanism. Depending on the confirmation control key, an order confirmation from the supplier may be required, and the confirmation control key is defaulted from the purchasing info record (PIR).

9. **Post goods receipt for purchase order**
Once the shipment from the suppliers arrives for spare parts at the receiving location within the plant, goods receipt will be posted manually in SAP S/4HANA with reference to the purchase order using Transaction MIGO. The stock will be issued directly to the maintenance order as the account assignment of the purchase order has the reference of the maintenance order.

The maintenance order will be scheduled to start the maintenance activities after all the required resources are available to carry out maintenance tasks.

10. **Post logistics invoice verification**

    The supplier sends an invoice for the purchase order via email (paper invoice), via EDI, or electronically via SAP Business Network. The supplier invoice document will be posted by the accounts payable personnel manually in the SAP S/4HANA system using Transaction MIRO in the case of a paper invoice. Automatic posting and reconciliation of the invoice happens if an electronic invoice is received from the supplier. The next scheduled payment run will pick up this fully reconciled invoice and post the payment to the supplier based on the preferred payment method of the supplier. Accounts payable personnel can manually process the payment against the fully reconciled invoice as well.

11. **Post goods issue spare parts to maintenance order**

    For all the in-stock materials (spare parts), goods issue will be posted with reference to a reservation and/or directly without reference to a material reservation maintenance order. Goods issue will be posted manually using Transaction MB1A and movement type 261.

    The maintenance order will be scheduled to start the maintenance activities after all the required resources are available to carry out maintenance tasks.

12. **Execute maintenance order**

    In this step, the actual execution of maintenance operations will be carried out. Maintenance and repair tasks will be performed as per the maintenance plan. Technicians will perform preliminary operations and main operations and record the results.

13. **Close maintenance order**

    The results from the execution step are reviewed by the supervisors, who confirm the completion of maintenance tasks. The status of the order will be set to *technically complete*.

Table 7.1 shows the key processes impacted in plant maintenance and materials management to support corrective maintenance of equipment.

Area	Processes Impacted	Description
Plant maintenance	Maintain plant maintenance master data	This process creates and maintains plant maintenance master data such as functional location, functional location BOM, equipment, equipment BOM, work center, and so on.
	Create and manage maintenance strategy	This process supports the creation and maintenance of maintenance strategies that contain maintenance cycles for preventive maintenance.
	Create maintenance plan	This process supports the creation of detailed maintenance tasks list to carry out preventive maintenance.

Table 7.1 Process Impact: Preventive Maintenance

Area	Processes Impacted	Description
Plant maintenance (Cont.)	Schedule maintenance plan	This process supports scheduling of maintenance plans to start maintenance and deadline monitoring.
	Maintenance order preparation	This process supports the release of a maintenance order, validates resources and spare parts availability, and aids procurement activities.
	Maintenance order scheduling	This process supports the scheduling of maintenance orders and assigning of resources to maintenance operations.
	Maintenance order execution	This process supports the execution of maintenance operations and recording of maintenance time and readings.
	Maintenance order completion	This process supports the technical closure of a maintenance order.
Materials management	Manage material reservations	This process supports the creation of material reservations for maintenance orders.
	Sourcing materials and services	This process supports the sourcing of a best source of supply to procure spare parts for maintenance.
	Create and maintain purchasing master data	This process creates and maintains purchasing master data such as source lists and PIRs to support the purchasing of spare parts from suppliers.
	Create and manage outline agreements	This process supports the creation of outline agreements such as contracts with suppliers to procure spare parts for production.
	Create and maintain purchase requisition	This process supports the creation, generation, and maintenance of purchase requisitions originating from plant maintenance orders.
	Create and maintain purchasing documents	This process focuses on creation of purchasing documents to procure spare parts from suppliers.
	Post goods receipt	This process supports the goods receipt of finished products from production into inventory.
	Post goods issue materials to production	This process supports the issuing of spare parts and components to maintenance operations with reference to a maintenance order.
	Supplier invoice management	This process supports the processing of incoming invoices from suppliers.

Table 7.1 Process Impact: Preventive Maintenance (Cont.)

Let's move on to the second plant maintenance strategy, *corrective maintenance*. This is also known as *reactive maintenance* and involves taking corrective actions such as repairs when equipment is broken, resulting in equipment failure. This strategy is applied to equipment that does not incur high costs of maintenance when it fails or is completely broken down. Some of noncritical assets fall into this category. Corrective maintenance tends to incur a lower maintenance cost, but it may cause unplanned downtime and could result in high one-time costs for repairs.

Figure 7.2 shows the corrective maintenance process flow, illustrating the integration between plant maintenance and materials management.

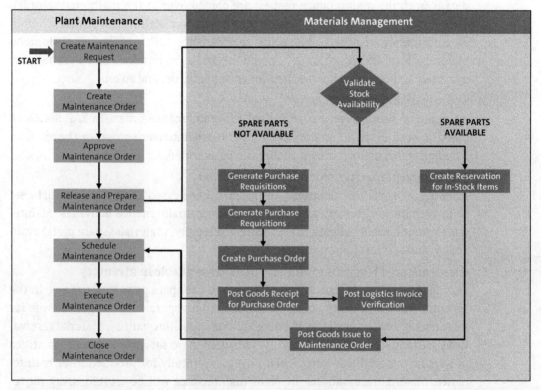

Figure 7.2 Integrated Maintenance Planning Process: Corrective Maintenance

During the regular maintenance of a company's assets, such as equipment, a routine checkup will be performed by a technician. If the technician finds a faulty spare part, failure, or malfunction of equipment, corrective maintenance is required. Let's explore the process steps:

1. **Create maintenance request**
   A maintenance request is a plant maintenance notification formally submitted by the technicians upon spotting faulty or defective equipment for corrective maintenance. It contains a detailed description of the problem, details of the technical object such as functional location and equipment on which the maintenance tasks

are to be performed, and all the relevant information to perform maintenance activities. Maintenance requests are created in SAP S/4HANA using Transaction IW26.

Maintenance requests are screened, updated, and accepted by the supervisor before a maintenance order can be created.

2. **Create maintenance order**
   Maintenance planners create maintenance orders either directly without reference to a plant maintenance notification or with reference to a plant maintenance notification such as a maintenance request. Maintenance orders with reference to maintenance requests can be created in SAP S/4HANA using Transaction IW34. The details from the maintenance request are copied over to the maintenance order, and maintenance planners can assign additional tasks. Maintenance orders contain all information required to execute maintenance tasks on faulty equipment, including start and finish dates, operations and suboperations to be carried out, materials and components required for each operation, and so on.

3. **Approve maintenance order**
   An approval process can be configured for maintenance orders in SAP S/4HANA based on business requirements. Only approved maintenance orders can be released, based on which maintenance activities can be executed.

4. **Release and preparation for maintenance order**
   After the approval, the maintenance order will be released and the status will be set to *in preparation* to start preparing to perform maintenance activities. Maintenance planners will validate the resource availability, materials (spare parts) availability, and services to be performed.

5. **Create material reservations if spare parts are available in inventory**
   During the availability check for spare parts, if the spare parts are available in the inventory, material reservations will be created for spare parts for goods issue for maintenance orders on a specific date and for a specific quantity. Material reservations are requests made to inventory management to set aside required quantities of specific materials for a particular purpose, typically for production or maintenance. Material reservations for plant maintenance will be created using movement type 261 for issue of spare parts against maintenance order.

6. **Generate purchase requisitions if spare parts are not available in inventory**
   During the availability check for spare parts, if the spare parts are not available in the inventory, the system generates purchase requisitions automatically from a maintenance order to procure spare parts from a third-party vendor. The system assigns a source of supply (a third-party vendor) for spare parts from the purchasing master data on the materials management side.

7. **Convert purchase requisitions into purchase order**
   A purchase requisition will then be converted into a purchase order manually by the purchaser using Transaction ME21N or automatically in SAP S/4HANA to procure spare parts for maintenance activities. The line item details are defaulted into

the purchase order that includes a source of supply, unit cost, and delivery address. The purchase order will be transmitted to the third-party supplier based on the output method: email, EDI, or electronic transmission to the supplier on SAP Business Network. Output determination configuration in SAP S/4HANA drives the purchase order output mechanism. Depending on the confirmation control key, order confirmation from the supplier may be required and the confirmation control key is defaulted from the PIR.

8. **Post goods receipt for purchase order**

   Once the shipment from the suppliers arrives for spare parts at the receiving location within the plant, goods receipt will be posted manually in SAP S/4HANA with reference to the purchase order using Transaction MIGO. The stock will be issued directly to the maintenance order as the account assignment of the purchase order has the reference of the maintenance order.

   The maintenance order will be scheduled to start the maintenance activities after all the required resources are available to carry out maintenance tasks.

9. **Post logistics invoice verification**

   The supplier sends an invoice for the purchase order via email (paper invoice), via EDI, or electronically via SAP Business Network. The supplier invoice document will be posted by the accounts payable personnel manually in SAP S/4HANA using Transaction MIRO in the case of a paper invoice. Automatic posting and reconciliation of the invoice happens if an electronic invoice is received from the supplier. The next scheduled payment run will pick up this fully reconciled invoice and post the payment to the supplier based on the preferred payment method of the supplier. Accounts payable personnel can manually process the payment against the fully reconciled invoice as well.

10. **Post goods issue spare parts to maintenance order**

    For all the in-stock materials (spare parts), goods issue will be posted with reference to a reservation and/or directly without reference to a material reservation maintenance order. Goods issue will be posted manually using Transaction MB1A and movement type 261.

    The maintenance order will be scheduled to start the maintenance activities after all the required resources are available to carry out maintenance tasks.

11. **Execute maintenance order**

    In this step, the actual execution of maintenance operations will be carried out. Maintenance and repair tasks will be performed as per the maintenance plan. Technicians will perform preliminary operations and main operations and record the results.

12. **Close maintenance order**

    The results from the execution step are reviewed by the supervisors, who confirm the completion of maintenance tasks. The status of the order will be set to *technically complete*.

Table 7.2 shows the key processes impacted in plant maintenance and materials management to support corrective maintenance of equipment.

Area	Processes Impacted	Description
Plant maintenance	Maintain plant maintenance master data	This process supports creation and maintenance of plant maintenance master data such as functional location, functional location BOM, equipment, equipment BOM, work center, and so on.
	Maintenance request initiation	This process supports the creation, submission, and prioritization of maintenance request with failure modes, detection methods, and effects.
	Maintenance request screening	This process supports the approval of maintenance requests.
	Maintenance order planning	This process supports the creation and planning of maintenance orders.
	Maintenance order preparation	This process supports the release of a maintenance order, validation of resources and spare parts availability, and procurement activities.
	Maintenance order scheduling	This process supports the scheduling of maintenance orders and assigning of resources to maintenance operations.
	Maintenance order execution	This process supports the execution of maintenance operations and recording of maintenance time and readings.
	Maintenance order completion	This process supports the technical closure of maintenance order.
Materials management	Manage material reservations	This process supports the creation of material reservations for the maintenance orders.
	Sourcing materials and services	This process supports the sourcing of a best source of supply to procure spare parts for maintenance.
	Create and maintain purchasing master data	This process creation and maintenance of purchasing master data such as source lists and PIRs to support the purchasing of spare parts from suppliers.
	Create and manage outline agreements	This process supports creation of outline agreements such as contracts with suppliers to procure spare parts for production.
	Create and maintain purchase requisition	This process supports the creation, generation, and maintenance of purchase requisitions originating from plant maintenance orders.

Table 7.2 Process Impact: Corrective Maintenance

Area	Processes Impacted	Description
Materials management (Cont.)	Create and maintain purchasing documents	This process focuses on the creation of purchasing documents to procure spare parts from suppliers.
	Post goods receipt	This process supports the goods receipt of finished products from production into inventory.
	Post goods issue materials to production	This process supports the issuing of spare parts and components to maintenance operations with reference to a maintenance order.
	Supplier invoice management	This process supports the processing of an incoming invoice from the supplier.

Table 7.2 Process Impact: Corrective Maintenance (Cont.)

7.2   Master Data in Plant Maintenance

There are no specific plant maintenance views in material master data. However, the other material master views must be maintained based on the business requirements to support the plant maintenance process. For example, the **Purchasing** view must be maintained for procuring materials required for plant maintenance tasks. Refer to Chapter 3, Section 3.1 for more details on how material master data drives the plant maintenance process.

This section covers plant maintenance-specific master data. The following are the different master data elements in plant maintenance:

- **Work center**
  A work center is the location within a manufacturing plant where the production activities are performed. In plant maintenance, a work center is used as a reference in maintenance orders so that the work centers can be maintained. Refer to Chapter 6, Section 6.2.2 for more details on work centers.

- **Functional location**
  A functional location is a physical location within a plant where the maintenance activities are performed. This is a plant maintenance-specific master data, and it provides a structural way to identify a company's assets to perform maintenance tasks.

- **Equipment**
  Equipment in plant maintenance includes physical assets such as machinery, vehicles, and tools that require maintenance within the plant. Each piece of equipment is a master data element in plant maintenance, containing all the information about it from a maintenance standpoint.

- **Equipment BOM**
  An equipment bill of materials in plant maintenance is a list of spare parts and components to assemble a piece of equipment. It helps in planning maintenance activities and procuring required materials for maintenance activities.

Let's dive deep into the master data in plant maintenance.

7.2.1   Functional Location

In functional locations, the company's assets such as equipment and tools are installed and maintained. In SAP S/4HANA, functional locations can be defined in a structural hierarchical format. A functional location can be an assembly line, production line, or a processing location. Functional locations help in tracking and reporting of maintenance tasks and associated cost.

Figure 7.3 shows the functional locations in a plant, created with four levels of hierarchy structure to easily track and trace a specific location. The first level represents a plant such as a manufacturing plant, distribution plant, and so on. The second level represents a department within the plant such as assembly, operations, inspection, and so on. The third level represents a subdepartment such as production line 1, assembly line 1, painting, and so on. And the fourth level represents an operating area where the equipment is installed and maintained.

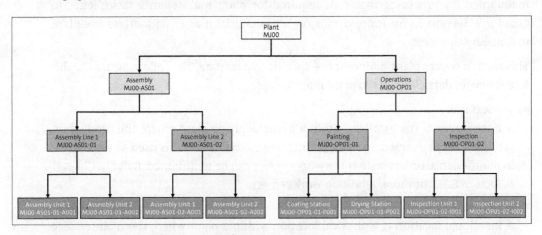

Figure 7.3 Functional Location in Plant Maintenance

The levels defined in the functional location are controlled by a structural indicator defined in the configuration settings, which we'll discuss next. We'll then explore how to create a functional location in SAP S/4HANA and the important attributes of a functional location. The functional location can be created individually or in bulk using the hierarchical structure.

Structure Indicator

This setting determines the generic structure of the functional locations in a plant using the **Edit mask** option. A structural indicator is used as a reference to create the functional locations.

To arrive at this configuration step, execute Transaction SPRO and navigate to **Plant Maintenance and Customer Service • Master Data in Plant Maintenance and Customer Service • Technical Objects • Functional Locations • Create Structure Indicator for Reference Locations/Functional Locations**.

The functional location structural indicator can be defined up to 10+ levels. Typically, this structure is defined up to five levels and with a minimum of two levels. From the initial screen of this configuration node, which displays an overview of all existing structure indicators, click **New Entries**, specify a five-digit alphanumeric key in the **StrInd** field, and specify the description of structure indicator in the **StructIndText** field to define a new structure indicator.

Figure 7.4 shows the functional location structural indicator defined with four hierarchy levels. Let's explore the fields of the structural indicator:

- **Edit mask**
  This field controls the generic structure of functional location. You can define the length and allowed characters of each level of the functional location. In the standard system, the following edit masks are used to control the allowed characters of functional locations:
  - **X**: Allows alphanumeric characters; that is, both letters and numbers are allowed.
  - **N**: Only numbers are allowed.
  - **A**: Only letters are allowed.
  - **S**: Special characters, letters, and numbers are allowed.

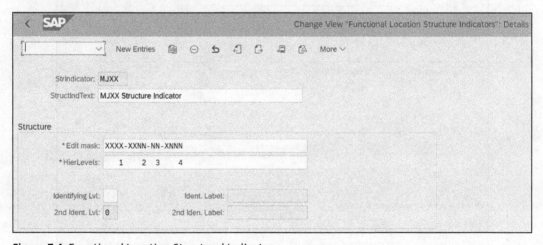

Figure 7.4 Functional Location Structural Indicator

- **HierLevels**

  In this field, define the hierarchy levels for functional locations. Levels are represented by numbers from 0 to 9. If you want to define hierarchy levels higher than 10, define the 10th level with 0 and start the 11th level with 1, 12th level with 2, and so on. The hierarchy level is dependent on the edit mask and defines the hierarchy level at the end of each edit mask block, as shown in Figure 7.4.

Initial View

To create a functional location, execute Transaction IL01. You'll arrive at the initial screen, as shown in Figure 7.5. Enter the structural indicator **MJXX** defined in the configuration for the functional location structure, and the system will automatically default the edit mask and hierarchy levels defined for the structural indicator.

Let's explore the fields in the initial screen:

- **Functional Loc.**

  In this field, define the functional location per the structural indicator definition. Enter only allowed characters and the allowed length of characters for each hierarchy level. The system will not allow you to create the functional location if the allowed characters and the defined length of the hierarchy level vary.

- **FunctLocCat**

  The functional location category is used to distinguish among different functional locations. Table 7.3 shows some of the functional location categories defined in the standard system.

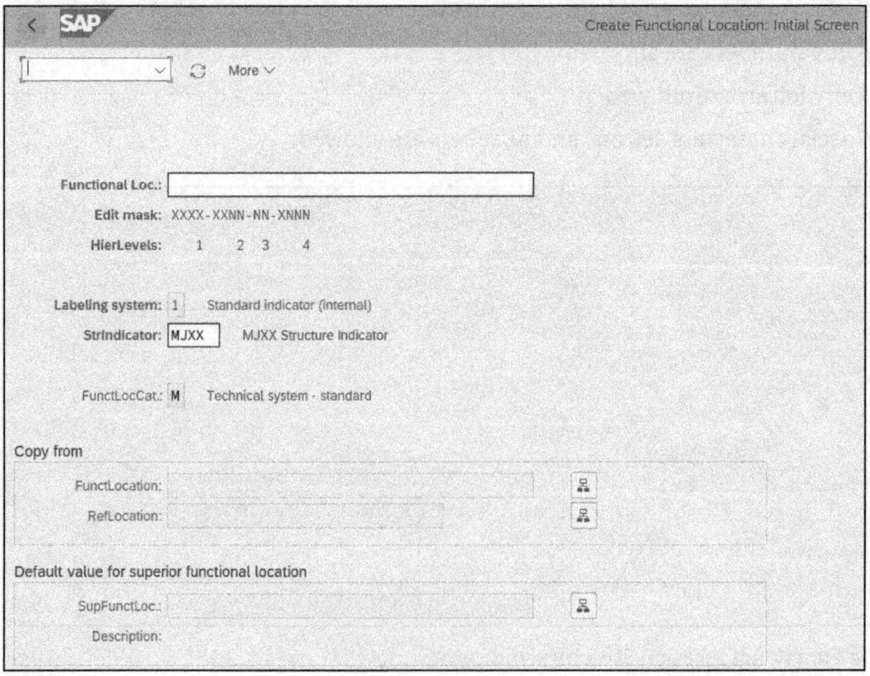

Figure 7.5 Create Functional Location: Initial Screen

Functional Location Category	Description
L	Linear functional location
M	Technical system—standard
P	Plant and machinery
S	Customer location

Table 7.3 Functional Location Categories

- **Copy from**

  Functional locations can be created with reference to another functional location or by using a reference location. Best practice is to create a reference location using Transaction IL11 if you want to create multiple functional locations with the same structure.

- **SupFunctLoc**

  A superior functional location is the top hierarchy structural level of the functional location. You can default the values from the superior functional location into the functional location. For example, MJ00-AS01 is the superior functional location for functional location MJ00-AS01-01-A001.

General Data Tab

Press [Enter] to go to the **General** view, as shown in Figure 7.6. Let's explore the fields:

- **Class**

  In this field, you can assign a predefined class for the functional location. You can define a class for the functional location using Transaction CL02 and class type 003. By defining a class, you can define additional attributes for the functional location.

- **Object Type**

  An object type can be defined to group similar equipment and technical objects together to categorize the functional locations. In this field, you can assign a predefined technical object.

  To define technical object types, execute Transaction SPRO and navigate to **Plant Maintenance and Customer Service • Master Data in Plant Maintenance and Customer Service • Technical Objects • General Data • Define Types of Technical Objects**.

- **AuthorizGroup**

  If you have defined specific authorization groups to control the access for a functional location in plant maintenance, you can assign the authorization group in this field. Only users with access to the assigned authorization group can use the functional location to perform maintenance activities.

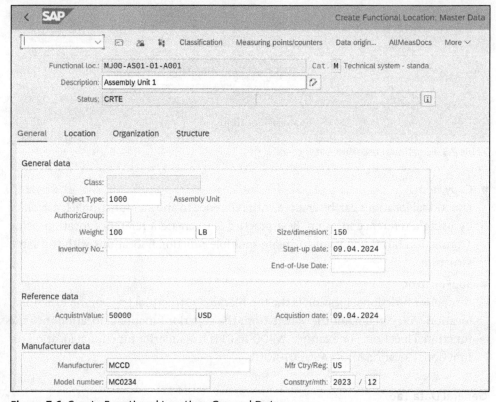

Figure 7.6 Create Functional Location: General Data

- **Inventory No.**

  An inventory number is defined in asset accounting to identify and monitor fixed assets. If you have created a specific asset in asset accounting for the piece of equipment to be maintained at the functional location, assign that number in this field. An inventory number, along with the main asset number, uniquely identifies a company's asset.

- **Start-up date**

  This defines the date on which the operation of a piece of equipment started at this functional location.

- **End-of-Use Date**

  This defines the planned date on which the operation of a piece of equipment will be ended at this functional location.

- **Reference data**

  You can define the acquisition value of the asset and the acquisition date of the asset for reference.

- **Manufacturer data**

  You can define manufacturer data such as the manufacturer, country/region of manufacture, model number, manufacturer part number, and manufacturer serial number for the equipment for reference.

Location Tab

Now navigate to the **Location** tab, as shown in Figure 7.7. Let's explore the fields:

- **MaintPlant**
  A maintenance plant is the physical plant in which the maintenance activities on equipment, technical objects, and machinery are carried out. Assign the relevant plant defined in the organization structure in this field.

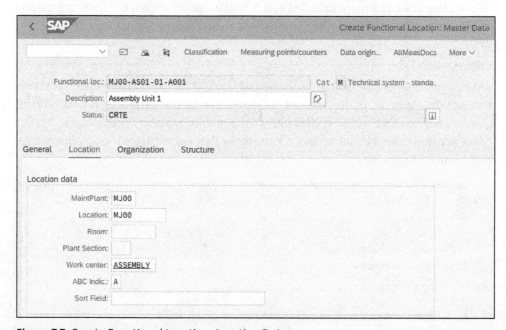

Figure 7.7 Create Functional Location: Location Data

- **Location**
  Location data helps to identify where the functional location is located within a physical plant. A location is an organizational unit and defined in the enterprise structure configuration. Assign the relevant location defined in this field.

- **Plant Section**
  The plant section further divides the maintenance plant into multiple sections to trace equipment easily for performing maintenance tasks. You can define the plant section in configuration for the maintenance plant. To arrive at this configuration step, execute Transaction SPRO and navigate to **Plant Maintenance and Customer Service • Master Data in Plant Maintenance and Customer Service • Technical Objects • General Data • Define Plant Sections**.

- **Work center**
  A work center is the location within a manufacturing plant where the production activities are performed. In plant maintenance, a work center is used as a reference in maintenance orders so that the work centers can be maintained. Assign the relevant

work center defined in the production planning master data in this field (see Chapter 6, Section 6.2.2) to carry out maintenance tasks.

- **ABC Indic.**
  ABC analysis is mostly used in inventory management and in resource allocation based on the criticality of items. A items are the most valuable, B items are fairly valuable, and C items are the least valuable. Assigning the value **A** in this field indicates that the equipment installed at this functional location is the most critical and that important resources must be allocated to perform maintenance activities.

Organization Tab

Proceed to the **Organization** tab, as shown in Figure 7.8. You can assign assets, cost centers, or WBS elements under **Account assignment** in this tab. The account assignment objects assigned here will be copied over to the maintenance order automatically. The costs incurred from the maintenance activities will be settled using the account assignments entered here.

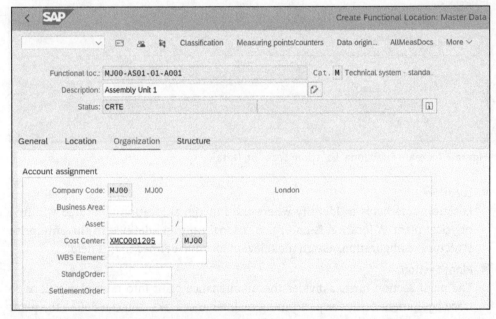

Figure 7.8 Create Functional Location: Organization Data

Note

Reference functional locations can be very useful if you create functional locations in bulk. Reference functional locations can be created using Transaction IL11, which do not represent actual functional locations in a plant but contain data that can be copied over to the functional locations.

Structure Tab

Navigate to the **Structure** tab to display the equipment assigned to the functional location. In the next section, we'll explain how to assign a functional location to equipment.

7.2.2   Equipment

In SAP plant maintenance, equipment is a physical asset of an organization. A piece of equipment represents an individual asset such as machinery, a tool, or a technical object and is installed at a functional location. The equipment master data in SAP S/4HANA contains the equipment type, functional location, detailed specifications, and other attributes of a piece of equipment.

An equipment master record can be part of maintenance plans for preventive maintenance tasks. Maintenance activities are performed, tracked, and reported for every piece of equipment in SAP S/4HANA.

In the following sections, we'll explain how to create equipment categories, followed by equipment master data creation and configuration.

Creating Equipment Categories

An equipment category is an important attribute to create an equipment master record, and it distinguishes physical assets of a plant based on their usage. Based on business requirements, you can define equipment categories in the configuration.

To define equipment categories, execute Transaction SPRO and navigate to **Plant Maintenance and Customer Service • Master Data in Plant Maintenance and Customer Service • Technical Objects • Equipment • Equipment Categories • Maintain Equipment Category**.

Figure 7.9 shows the equipment categories defined in the configuration. The equipment category (column **C**) is a unique one-character code. To define a new equipment category, click **New Entries** and specify a one-character alphanumeric key in the **C** column and its description in the **Equipment CatDesc.** column.

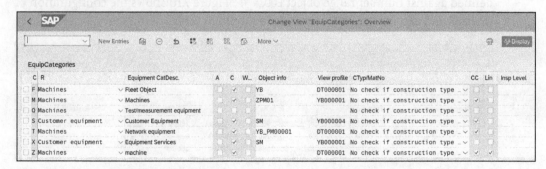

Figure 7.9 Define Equipment Categories

The key fields are as follows:

- **R (equipment reference category)**

  You can assign equipment reference categories to one or more equipment categories in the **R** (equipment reference category) column. The following equipment reference categories are available in the standard system:

  - **Machines**
  - **Production resources/tools**
  - **Customer equipment**

  If you want to change the equipment category for an equipment master, the system will only allow you to change it to one of the equipment categories created with the same equipment reference category. For example, if the equipment master was created with **Machines** as the equipment reference category, then you can only change it to one of the equipment categories created with **Machines** as the equipment reference category.

- **C (change documents)**

  If you set this indicator, change documents will be updated for changes to the equipment that uses this category.

- **W... (workflow event)**

  If you set this indicator, one of the following workflow events will be triggered during the maintenance of the equipment master data:

  - `PieceOfEquipment.Created`
  - `PieceOfEquipment.Changed`
  - `PieceOfEquipment.LocationChanged`

You can define number ranges for equipment categories. To arrive at the number range configuration step, execute Transaction SPRO and navigate to **Plant Maintenance and Customer Service • Master Data in Plant Maintenance and Customer Service • Technical Objects • Equipment • Equipment Categories • Define Number Ranges for Equipment Categories**. You can arrive at this configuration step directly by executing Transaction SNRO. In this configuration step, both internal and external number ranges can be defined against number range object `EQUIP_NR`. Figure 7.10 shows the configuration of internal and external number ranges for equipment categories.

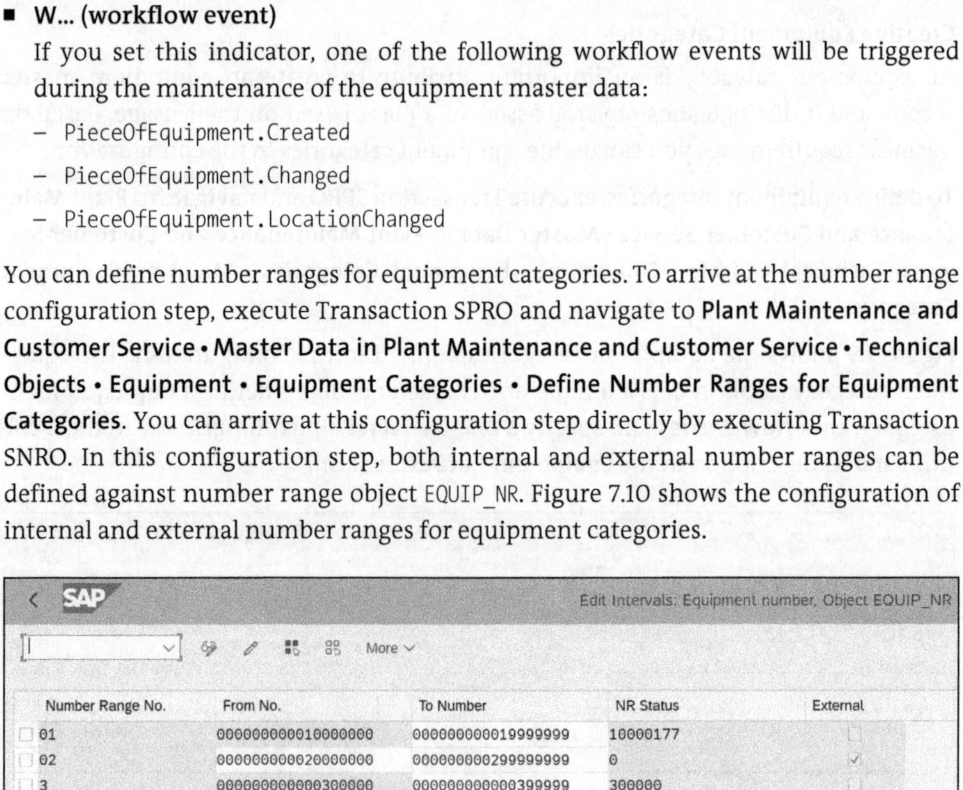

Number Range No.	From No.	To Number	NR Status	External
01	000000000010000000	000000000019999999	10000177	☐
02	000000000020000000	000000000299999999	0	☑
3	000000000000300000	000000000000399999	300000	☐

Figure 7.10 Define Number Ranges for Equipment Categories

Creating Equipment

Let's explore how to create an equipment master in SAP S/4HANA and the important attributes of a piece of equipment. Use Transaction IE01 to create the equipment master.

Figure 7.11 shows the initial screen of equipment master data creation, where you fill out the following fields:

- **Equipment**
  If you have defined an internal number range for the respective equipment category used in the creation of equipment master, leave this field blank. If you have defined an external number range, assign a number to this field as per the external number range definition. The equipment master will be stored in the database by this number.

- **Valid On**
  Enter the valid from date for the equipment master.

- **Equipment category**
  You must enter an equipment category defined in the configuration to create an equipment master. Enter the relevant equipment category here.

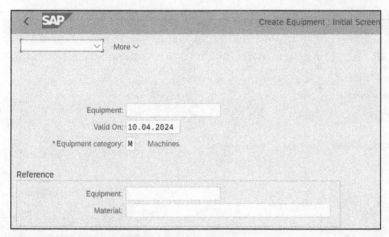

Figure 7.11 Create Equipment: Initial Screen

You can create an equipment master with reference to another piece of equipment or a material under the **Reference** section.

General Data Tab

Press [Enter] to go to the **General** view. Figure 7.12 shows the **General data** area of the equipment master data, which includes the following fields (similar to what we described in Section 7.2.1 for the functional location):

- **Class**

  In this field, you can assign a predefined class for the equipment. You can define a class for the functional location using Transaction CLO2 and class type OO3. By defining a class, you can define additional attributes for the equipment.

- **Object Type**

  An object type can be defined to group similar equipment and technical objects together to categorize the equipment. In this field, you can assign a predefined technical object. You can define an object type in configuration.

- **AuthorizGroup**

  If you have defined specific authorization groups to control access to the equipment in plant maintenance, you can assign the authorization group in this field. Only users with access to the assigned authorization group can use the equipment to perform maintenance activities.

- **Inventory No.**

  An inventory number is defined in asset accounting to identify and monitor fixed assets. If you have created a specific asset in asset accounting for a piece of equipment, assign that number in this field. The inventory number, along with the main asset number, uniquely identifies a piece of equipment.

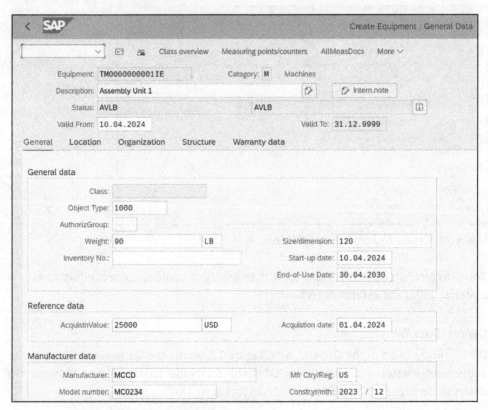

Figure 7.12 Create Equipment: General Data

- **Start-up date**
  This defines the date on which the operation of the equipment started.
- **End-of-Use Date**
  This defines a date on which the operation of the equipment is planned to end.
- **Reference data**
  You can define the acquisition value and date of the equipment here.
- **Manufacturer**
  Specify the manufacturer's name or supplier number (business partner number in the system) of the equipment in this field.
- **Model number**
  Specify the manufacturer's model number for the equipment in this field.
- **Mfr Ctry/Reg (manufacturer country/region)**
  Specify the manufacturer's country or region in this field.
- **Constr.yr/mth**
  Specify the year and month of the construction of the equipment by the manufacturer.

Location Tab

Let's move on to the **Location** tab of the equipment master, as shown in Figure 7.13, which includes the following fields:

- **MaintPlant**
  The maintenance plant is the physical plant in which the maintenance activities for the equipment are carried out. Assign the relevant plant defined in the organization structure in this field.

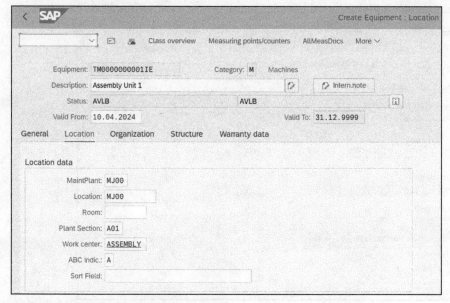

Figure 7.13 Create Equipment: Location Data

- **Location**

  Location data helps to identify where the equipment is located within a physical plant. The location is an organizational unit and is defined in the enterprise structure configuration. Assign the relevant location in this field.

- **Plant Section**

  A plant section further divides the maintenance plant into multiple sections to trace equipment easily for performing maintenance tasks. You can define the plant section in configuration for the maintenance plant.

- **Work center**

  A work center is the location within a manufacturing plant where the production activities are performed. In plant maintenance, the work center is used as a reference in maintenance orders so that the work centers can be maintained. Assign the relevant work center defined in the production planning master data in this field to carry out maintenance tasks. See Chapter 6, Section 6.2.2 for more details.

- **ABC Indic.**

  ABC analysis is mostly used in inventory management and in resource allocation based on the criticality of items. Assigning the value **A** in this field indicates the equipment is the most critical and that important resources must be allocated to perform maintenance activities.

Organization Tab

Next, navigate to the **Organization** tab, as shown in Figure 7.14. Similar to the functional location described in Section 7.2.1, you can assign an asset, cost center, or WBS element under **Account assignment** in this tab. The account assignment objects assigned here will be copied over to the maintenance order automatically. The costs incurred from the maintenance activities will be settled using the account assignments entered here.

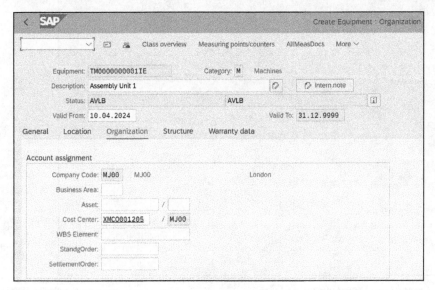

Figure 7.14 Create Equipment: Organization Data

Structure Tab

Figure 7.15 shows the **Structure** tab of the equipment master. Assign the functional location where the respective piece of equipment is installed in the **Functional loc.** field by pressing the **Change InstLoc** icon on the right-hand side. Upon assigning the functional location in this field, the status of equipment will be changed to **INST** (installed).

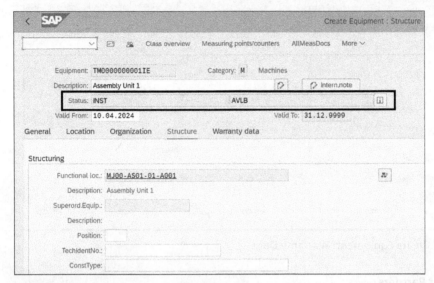

Figure 7.15 Create Equipment: Structure Data

Warranty Data Tab

Finally, navigate to the **Warranty data** tab, as shown in Figure 7.16. A warranty is an assurance given by the manufacturer that the equipment will function in the expected way for a certain period. If the equipment fails during the warranty period, the manufacturer will replace it or repair it free of charge. Depending on the terms agreed upon between the seller and the purchaser at the time of acquiring the equipment, some warranties cover only repair costs, not the shipping and handling costs.

Let's explore the fields in the **Warranty data** tab:

- **Warranty Start**
  This is the date from which the warranty period starts for the equipment.

- **Warranty end**
  This is the end date for the warranty period for the equipment. The repair or replacement of the equipment will be accepted by the manufacturer if the equipment fails between the warranty start date and warranty end date.

- **InheritWarranty**
  Set this indicator if a superior equipment exists for the piece of equipment and the warranty details of the superior equipment should be copied over to the equipment master.

- **Pass on warrnty**

  Set this indicator if the equipment master being created is the superior equipment and subequipment exists for the equipment being created. The warranty details of the current equipment being created will be passed on to the subequipment.

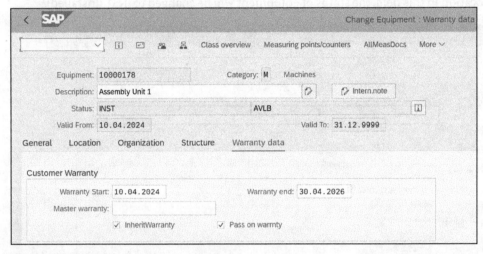

Figure 7.16 Create Equipment: Warranty Data

Equipment Partners

Next, maintain the equipment partners who are responsible for equipment maintenance. Click the **Partners** icon at the top or press ⌈Ctrl⌉+⌈F2⌉ to navigate to the **Partners** screen. Figure 7.17 shows the partners for the equipment master. Partners are the people, departments, or organizations involved in business transactions, such as vendors, customers, sold-to parties, ship-to parties, planners, buyers, and so on. Equipment partners specifically are the people, departments, and/or organizations associated with the equipment, such as persons responsible for the equipment, departments responsible for the equipment, and/or manufacturers/vendors of the equipment.

The following are the key fields of this view:

- **Funct (partner function)**

  Partners are defined by partner function. Select a partner function defined for plant maintenance in the standard system. The following are the partner functions that you can select from the dropdown list:

  - **Department resp.**
  - **Supplier**
  - **Person respons.**

- **Partner**

  Select a partner based on the partner function in this field. The department responsible is defined as an organizational unit within plant maintenance, the person responsible is defined in plant maintenance, and suppliers are defined as business partners.

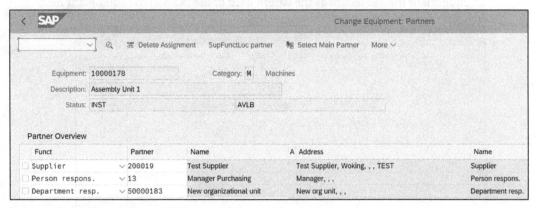

Figure 7.17 Create Equipment: Partners

7.2.3   Equipment Bill of Materials

An equipment BOM in plant maintenance is a list of spare parts and components to assemble and maintain a piece of equipment. The header or parent item of the BOM is the equipment, and spare parts and other components are part of the BOM items. BOM items contain detailed information about spare parts and components such as the part number, description, quantity, unit of measure, and so on. Equipment BOMs are integrated with materials management functions, allowing seamless integration with procurement and inventory management. There are three types of BOM used in plant maintenance:

- **Equipment BOM**
  Equipment BOMs used in plant maintenance are lists of spare parts and components to assemble and maintain a piece of equipment. Equipment BOMs are created using Transaction IB01.

- **Functional location BOM**
  Functional location BOMs used in plant maintenance are lists of components to maintain a functional location. Functional location BOMs are created using Transaction IB11.

- **Assembly BOM**
  Assembly BOMs consist of a final finished product as the parent item and the required components assembly as child items. The components are assembled into a finished product. Assembly BOMs are created using Transaction CS01.

As a prerequisite for creating an equipment BOM, spare parts must be created in the same plant in which the equipment is located using the material type defined for spare parts in Transaction MM01.

Figure 7.18 shows the initial screen for creating an equipment BOM. The fields are nearly the same as provided for materials in Chapter 6, Section 6.2.1, except that you can fill in an equipment master record as the parent item of the BOM in the **Equipment** field.

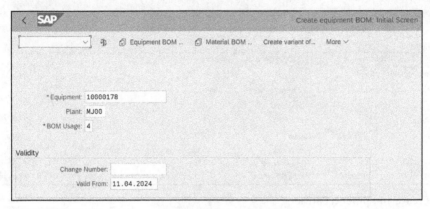

Figure 7.18 Create Equipment BOM: Initial Screen

After specifying the equipment master, plant, and BOM usage, press ⎡Enter⎤ to navigate to the **General Item Overview** screen. Figure 7.19 shows the general item overview screen for the equipment BOM. In this screen, you can list all spare parts and components used to maintain the equipment (parent item of the BOM). The key fields are the same as described in Chapter 6, Section 6.2.1.

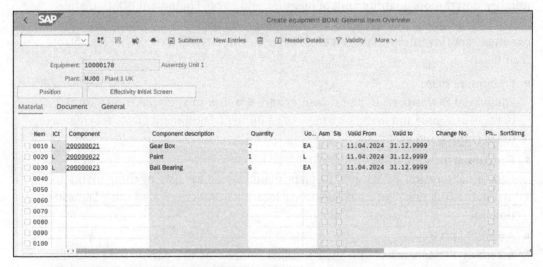

Figure 7.19 Create Equipment BOM: Item Overview

> **Note**
>
> You can maintain a functional location BOM like an equipment BOM. The parent item of the BOM will be the functional location, and you can list all spare parts and components used to maintain the functional location in the functional location BOM.

7.3   Maintenance Planning

In preventive maintenance, a *maintenance plan* is a crucial step of the maintenance process. It is used to manage routine maintenance activities on a company's critical assets such as equipment and tools. Maintenance plans define maintenance tasks, schedules, and resources to maintain technical objects (equipment and tools) and thus optimize the overall performance. In SAP S/4HANA, a maintenance plan is created against a specific maintenance strategy based on business and compliance requirements. You can predefine task lists for maintenance plans. Maintenance plans can reference a task list that contains each task in a sequence, and each task in a task list contains material and resource requirements to carry out maintenance of equipment. Maintenance plans, once created, can be scheduled to automatically generate maintenance orders. Maintenance activities are carried out with reference to maintenance orders.

Maintenance strategy and equipment maintenance task lists are the prerequisites for maintenance plan and maintenance order processing, especially in preventive maintenance. After walking through these prerequisites, we'll dive deep into maintenance order processing and how the integration between plant maintenance and materials management functions cooperates to keep the equipment and other assets operating for a long time with maximum efficiency.

7.3.1   Maintenance Strategies

Maintenance planning begins with the creation of a relevant maintenance strategy for the equipment and or functional location maintenance. It includes scheduling information that can be used in maintenance plans. You can define a maintenance strategy using Transaction IP11.

Figure 7.20 shows the maintenance strategy initial screen, which also shows the details. You can create a new strategy by clicking the **New Entries** button or by copying an existing strategy.

Let's explore the details of the maintenance strategy:

- **Scheduling indicator**
  The scheduling indicator controls the type of maintenance cycle or scheduling of maintenance plans. There are two types of scheduling indicators that can be set in the maintenance strategy.

  In the *time-based* maintenance plan, the next planned date for the maintenance activity will be scheduled based on time. Within the time-based plan, you can set the scheduling indicator based on calendar days, the key date, or the factory calendar. Maintain the scheduling indicator **Time** for calendar days, **Time—key date** for scheduling the next maintenance cycle on a specific day of every month, or **Time—factory caldr** for scheduling the next maintenance cycle based on the factory calendar of the maintenance plant. For example:

- **Scheduling indicator** set to **Time**: If you define the monthly maintenance cycles in the maintenance strategy and assume the current date as April 14, 2024, the next maintenance plan dates are calculated as follows:
    - First maintenance plan date or next maintenance cycle date = 4/14/24 + 30 = 5/13/24
    - Second maintenance plan date or next maintenance cycle date = 5/13/24 + 30 = 6/11/24
- **Scheduling indicator** set to **Time—key date**: If you define the monthly maintenance cycles in the maintenance strategy and assume the current date as April 14, 2024, the next maintenance plan dates are calculated as the 14th day of every month:
    - First maintenance plan date or next maintenance cycle date = 5/14/24
    - Second maintenance plan date or next maintenance cycle date = 6/14/24
- **Scheduling indicator** set to **Time—factory caldr**: If you define the monthly maintenance cycles in the maintenance strategy and assume the current date as April 14, 2024, the next maintenance plan dates are calculated after 30 working days defined in the factory calendar:
    - First maintenance plan date or next maintenance cycle date = 5/14/24
    - Second maintenance plan date or next maintenance cycle date = 6/14/24
- **Scheduling indicator** set to **Activity**: If you want to define performance-based scheduling, select **Activity** as the scheduling indicator. Unlike time-based scheduling, *performance-based* scheduling on maintenance plans depends entirely on the defined performance indicators such as hours of operation, temperature, and so on. The system determines the next planned date based on current readings, averages, and the like.

- **Strategy unit**
  A strategy unit is the base unit for the calculation of next planned date for the maintenance cycle. It is recommended to use **DAY** as the strategy unit even if you are planning to schedule maintenance cycle every year.

- **Call horizon**
  The call horizon is a percentage value that determines the timing of the creation of a maintenance order automatically for the maintenance plan. The value maintained in this field will be used by the system to calculate the date for the creation of the maintenance order.

  Let's consider an example. If the first maintenance plan date or next maintenance cycle date is 5/14/24, the call horizon is 10%, and the duration of the cycle is 30 days, then the date for the creation of maintenance order is calculated as $30 \div 100 \times 10$ (= 3 days) + 5/14/24 = 5/16/24. The maintenance order will be created automatically for the first cycle of maintenance on 5/16/24.

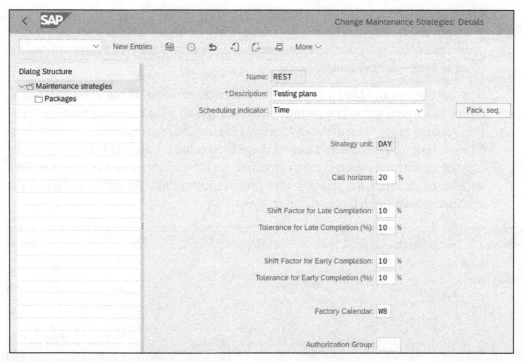

Figure 7.20 Maintenance Strategy: Details

- **Shift Factor for Late Completion/Tolerance for Late Completion**
  A shift factor for late completion is a percentage value and is applicable for the calculation of the next planned maintenance cycle date. If the completion of the current maintenance cycle is delayed and the variance is outside **Tolerance for Late Completion**, the system calculates the next planned maintenance cycle date based on the **Shift Factor for Late Completion** percentage value. The next planned maintenance cycle date is shifted forward (late) by N number of days. For example, if the shift factor for late completion is 10% and the duration of the cycle is 30 days, the next planned maintenance cycle date is shifted by 30 ÷ 100 × 10 = 3 days.

- **Shift Factor for Early Completion/Tolerance for Early Completion**
  Shift factor for early completion is a percentage value and it is applicable for the calculation of the next planned maintenance cycle date. If the completion of the current maintenance cycle is earlier than the planned date and the variance is outside of **Tolerance for Early Completion**, the system calculates the next planned maintenance cycle date based on the **Shift Factor for Early Completion** percentage value. The next planned maintenance cycle date is shifted backward by N number of days. For example, if the shift factor for early completion is 10% and the duration of the cycle is 30 days, the next planned maintenance cycle date is shifted backward (early) by 30 ÷ 100 × 10 = 3 days.

- **Factory Calendar**
  Assign the factory calendar in this field. The system uses the factory calendar to determine the next planned maintenance cycle date if the scheduling indicator is set as **Time—factory caldr.**

Let's navigate in the **Dialog Structure** to **Packages** to arrive at the screen shown in Figure 7.21, which shows the maintenance packages of a maintenance strategy. Packages are used to schedule maintenance tasks in a sequence with a cycle length (in days) for each task. You can create a new package by clicking the **New Entries** button or by copying an existing package. In our example, we have maintained a testing plan for a time-based maintenance strategy with four testing cycles with different **Cycl.length** (cycle length) and with **Day** as the **Unit.**

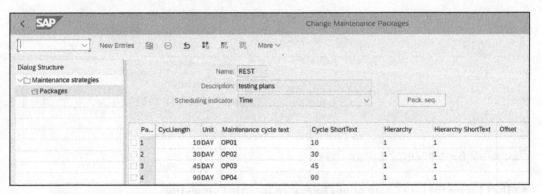

Figure 7.21 Maintenance Strategy: Maintenance Packages

Hierarchy defines the testing plan's place in the hierarchy; for example, if all maintenance cycles are due at the same time, maintain the same hierarchy for all testing cycles. If you require a maintenance cycle to replace the other when both are due at the same time, assign a higher value for that cycle in the **Hierarchy** field.

You can define multiple packages for preventive maintenance tasks with different cycle lengths. Let's explore the details of packages:

- **Pa... (package number)**
  The package number defines the consecutive number of maintenance tasks.

- **Cycl.length (cycle length)**
  The cycle length defines the interval of the maintenance cycle. For example, in a time-based maintenance plan, cycle length 10 defines that the respective maintenance task shall be performed every 10 days.

- **Hierarchy**
  The hierarchy defines the priority of the maintenance task. If several tasks are due at the same time, you can define the priority of each task so the system prioritizes the initiation of the work order for the task having highest priority.

The package sequence will be calculated automatically by the system upon saving the maintenance strategy. This defines the sequence of maintenance tasks by considering the cycle length. Click the **Pack. seq.** button to reach the screen shown in Figure 7.22, where the maintenance package sequence appears for the defined maintenance strategy. For example, task #1 or the first package has a cycle length of 10 days in a time-based plan, so the maintenance task will be performed every 10 days.

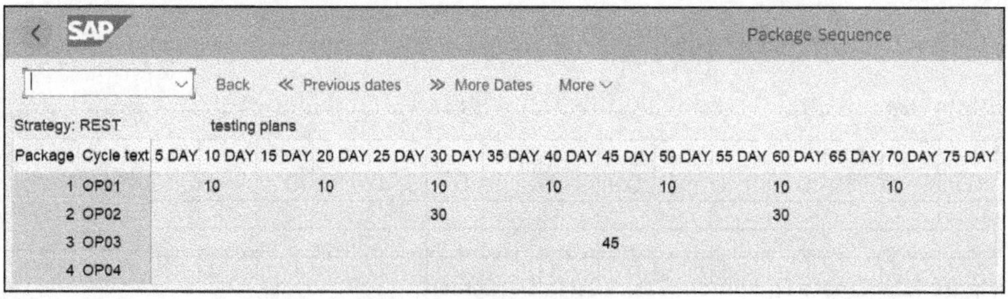

Package	Cycle text	5 DAY	10 DAY	15 DAY	20 DAY	25 DAY	30 DAY	35 DAY	40 DAY	45 DAY	50 DAY	55 DAY	60 DAY	65 DAY	70 DAY	75 DAY
1	OP01		10		10		10		10		10		10		10	
2	OP02						30						30			
3	OP03									45						
4	OP04															

Strategy: REST   testing plans

Figure 7.22 Maintenance Strategy: Package Sequence

7.3.2   Maintenance Task List

A maintenance task list defines the sequence of maintenance tasks to be performed repeatedly to maintain a piece of equipment or a functional location. It defines the plant maintenance operations in a sequence as tasks. All the maintenance tasks to be performed to maintain a company's assets have been grouped in a task list. There are three types of task lists:

- **Equipment task list**
  This is the set of sequential maintenance tasks to maintain a piece of equipment.

- **Functional location task list**
  This is the set of sequential maintenance tasks to maintain a functional location.

- **General maintenance task list**
  This is used for general maintenance tasks in a plant.

Let's understand the task list by creating an equipment task list. You can create an equipment task list using Transaction IA01. Figure 7.23 shows the initial screen for creating an equipment task list, where you fill out the following fields:

- **Equipment**
  Enter the equipment master created in the planning plant in this field to create an equipment task list (refer to Section 7.2.2).

- **Profile**
  You can define default settings for internal maintenance and external maintenance in the profile in the configuration. These values will be defaulted into the maintenance routings. Assign the predefined profile in this field.

- **Key Date**

  Enter the key date for an equipment task list. The equipment task list is valid from this date, and the task list can be used in a maintenance plan from this date.

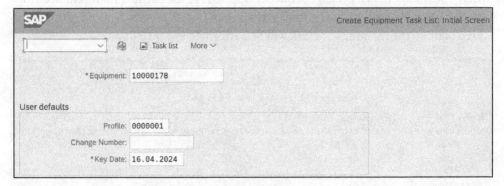

Figure 7.23 Create Equipment Task List: Initial Screen

After specifying the equipment, profile, and key date, press ⌨Enter⌨ to navigate to the header details of the equipment task list, as shown in Figure 7.24. Let's explore the key fields:

- **Planning Plant**

  Maintenance activities will be carried out in the planning plant. Assign the planning plant where the equipment maintenance tasks are executed.

- **Work Center**

  The work center is the location within a manufacturing plant where the production activities are performed. In an equipment task list, the work center is the location where the equipment is installed and the maintenance tasks are to be executed. Assign the relevant work center.

- **Usage**

  In this field, enter a function or a business unit within the organization in which the equipment task list will be used. The following are the functions within the organization in which an equipment task list can be used:

  - **1: Production**
  - **3: Universal**
  - **4: Plant Maintenance**
  - **5: Sales and Distribution**
  - **6: Costing**

  For equipment maintenance, select usage code **4**.

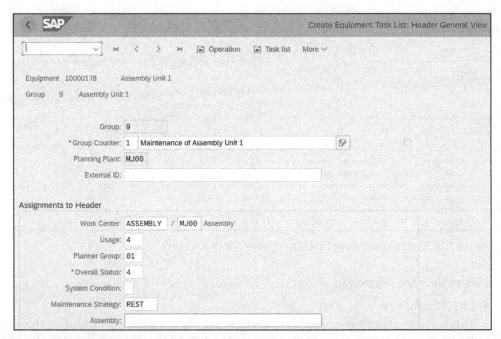

Figure 7.24 Create Equipment Task List: Header Details

- **Planner Group**

  A planner group is responsible for maintaining the equipment task list. You can define a planner group for the plant in configuration. Assign the predefined planner group responsible for maintaining the equipment task list.

- **Overall Status**

  At the time of creation of the equipment task list, the system assigns value **1** by default. The processing status of the equipment task list must be set to **4** to be used in the maintenance plan:

  - **1**: Created
  - **2**: Released for order
  - **3**: Released for costing
  - **4**: Released (general)

- **Maintenance Strategy**

  A maintenance strategy contains the scheduling information for the maintenance plan. Assign the predefined maintenance strategy for the equipment task list.

The next step in the creation of an equipment task list is to define the operations or activities for the maintenance of equipment. Click the **Operation** button at the top of the screen to arrive at the screen in Figure 7.25, which shows the operations overview of the equipment task list. Here, the activity number (**Act.**) is displayed in increments of 10 by default. The work center and plant values are copied from the header details.

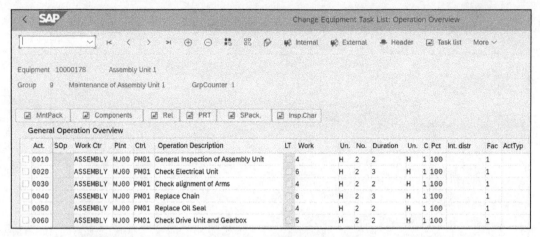

Figure 7.25 Create Equipment Task List: Operations Overview

Let's explore the key fields:

- **Ctrl (control key)**
  This is a four-digit alphanumeric key that defines the business application this routing or the maintenance activity belongs to. The following are the important control key values defined in the standard system:

  - **PM01** (plant maintenance—internal): Assign **PM01** to carry out the specific maintenance activity internally.

  - **PM02** (plant maintenance—external): Assign **PM02** to carry out the specific maintenance activity or operation externally; the respective operation will be performed by an external vendor. Using the external parameters such as order quantity, net price, cost center, material group, and source of supply, the system automatically generates a purchase requisition to procure spare parts from the external vendor.

  - **PM03** (plant maintenance—external [services]): Assign **PM03** if an external vendor should be performing the respective maintenance service. This will be followed by a service entry sheet, invoice verification, and payment to an external party. External processing parameters for external maintenance service need to be entered like for **PM02** operation. But in the case of **PM03**, service procurement-related details need to be entered.

Note

For all types of operations (internal or external), you can procure required spare parts or components for the maintenance operation from an external vendor. The procured spare parts and components will then be issued for maintenance with reference to the maintenance order. Order quantity and net price for the spare parts are required to create purchase order. Enter the order quantity and net price for the operation.

- **Work**
  In this field, enter the amount of work involved in hours (standard) to perform the activity.

- **No.**
  Enter the number of employees/resources required to perform the respective maintenance activity.

- **Duration**
  In this field, enter the duration in hours to perform the maintenance activity. The duration can be less than the work if multiple employees perform the task.

- **Pct (percentage of work)**
  In this field, enter the percentage of work involved in carrying out this maintenance activity. The system uses this field to calculate the capacity, operating time of the work center, and so on. By default, the system uses 100% if no percentage value is specified.

The next step is to assign packages to the operations. Packages will be copied from a maintenance strategy. To navigate to the packages overview screen, click the **MntPack** button at the top of the **Operations Overview** screen. Figure 7.26 shows the maintenance packages overview for maintenance tasks.

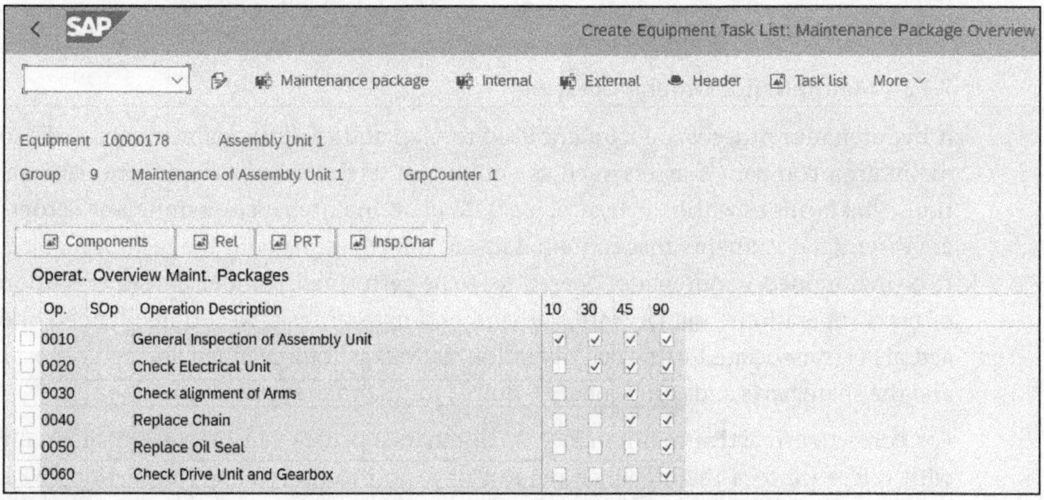

Figure 7.26 Create Equipment Task List: Maintenance Package Overview

You can define the interval for maintenance operation based on the maintenance strategy. Intervals **10, 30, 45,** and **90** are copied from the maintenance strategy. Set the indicator under these intervals per operation if you want to perform that maintenance operation during those intervals.

The next step is to assign components/spare parts to the maintenance operations. This indicates the spare part requirements to perform the maintenance operation. To assign the components to the maintenance operation, go back to the **Operations Overview** screen and then select the operation for which you want to maintain the components. Click the **Components** button at the top to reach the screen shown in Figure 7.27.

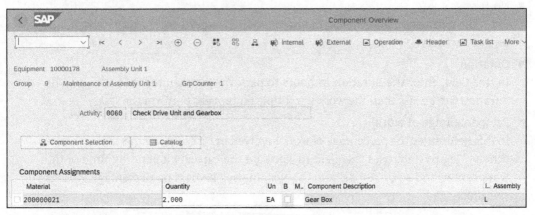

Figure 7.27 Create Equipment Task List: Assign Components to Operations

Assign the spare parts (note that spare parts are created using material type ERSA) as components to the maintenance operations that require spare parts to complete the task.

7.3.3   Maintenance Order Processing

A maintenance order is a document used to plan and execute maintenance tasks to maintain a company's assets such as equipment, technical objects, functional locations, and facilities within a physical plant. In plant maintenance, maintenance orders are essential documents that contain detailed information about the technical object to be maintained, maintenance operations to be performed, planned date of execution of tasks (operations), actual start date and end date of task execution, planned and actual costs associated with each operation, as well as total planned and actual costs, and the spare parts and components required for maintenance activities.

For corrective/reactive maintenance, maintenance orders can be created manually with reference to a maintenance request or a notification using Transaction IW34. Maintenance orders can be created directly without reference to a maintenance request or maintenance order manually using Transaction IW31.

For preventive/proactive maintenance, maintenance orders are generated automatically based on the maintenance plan. There are three types of maintenance plans in plant maintenance:

- **Single-cycle plan**
  In a single-cycle plan, maintenance activities are planned and performed on a technical object at regular intervals. For example, in a time-based maintenance plan,

routine inspections are performed every month on a piece of equipment to ensure the equipment runs efficiently for a longer period, preventing sudden breakdown. Whereas in a performance-based maintenance plan, general maintenance activities are performed after every 100 hours of operation of a piece of equipment. In a single-cycle plan, either a time-based maintenance plan or performance-based maintenance plan can be used for preventive maintenance. Single-cycle plans can be created using Transaction IP41.

- **Multiple-counter plan**
  A multiple-counter plan provides a comprehensive approach to maintain critical tools and equipment. It uses both time-based and performance-based maintenance plans to schedule and execute maintenance activities. Multiple-counter plans will have maintenance cycles with multiple dimensions. For example, an assembly unit in the production line can be maintained every three months or 1,000 hours of operation. Multiple-counter plans can be created using Transaction IP43.

- **Strategy plan**
  A strategy plan involves a maintenance strategy and packages to schedule and execute maintenance activities. In a strategy plan, multiple operations are performed based on a package defined in the maintenance strategy, as explained earlier. For example, in a strategy plan, a regular inspection is performed on a piece of equipment every 15 days, filters and gaskets are changed every three months, a gear box is checked every six months, and so on. Strategy plans can be created using Transaction IP42.

To understand the integration between plant maintenance and materials management, let's consider and create a strategy plan to maintain a piece of equipment. Figure 7.28 shows the initial screen for the creation of a maintenance strategy plan. A maintenance strategy is mandatory to create a maintenance strategy plan (Section 7.3.1). Let's explore the fields in the initial screen:

- **Maintenance plan**
  If you have defined an internal number range assignment for a maintenance plan, leave this field blank.

  To arrive at the configuration step to define number ranges for maintenance plans, execute Transaction SPRO and navigate menu path **Plant Maintenance and Customer Service • Maintenance Plans, Work Centers, Task Lists and PRTs • Maintenance Plans • Define Number Ranges for Maintenance Plans**. You can directly arrive at this configuration step by executing Transaction SNRO. In this configuration step, both internal and external number ranges can be defined against number range object MPLA_NR.

- **Maint. plan cat.**
  The maintenance plan category controls the document to be generated when the maintenance call is due. You can generate the following documents:

- – Maintenance order
- – Maintenance notification
- – Inspection lots
- – Service order

Select maintenance order (**PM Order**) to generate a maintenance order automatically when the maintenance call is due.

- ■ **Strategy**
  Assign the predefined maintenance strategy in this field.

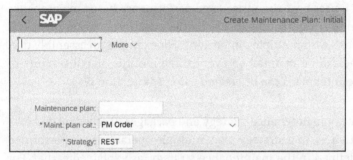

Figure 7.28 Create Maintenance Strategy Plan: Initial Screen

Press ⌈Enter⌋ from the initial screen to enter the maintenance plan details. Figure 7.29 shows the **Item** tab of the maintenance plan. Let's explore the key fields:

- ■ **Reference object**
  In this section, assign the technical object on which the maintenance activities are required to be carried out. You can assign equipment, functional location, and assembly as reference objects.

- ■ **Planning Data**
  Enter the following planning data:

  - – **Planning Plant**: This is the plant where the maintenance activities are planned and executed. Enter the planning plant where the functional location and equipment are located.

  - – **Order Type**: This is the categorization of the different types of orders. Enter the relevant order type. You can enter standard order type **YA02** for proactive maintenance. The system uses this order type to generate a maintenance order.

  - – **Main WorkCtr**: Assign the work center where the production planning activities are carried out and the location where the equipment is installed.

- ■ **Task List**
  You can assign the equipment task list or functional location task list (Section 7.3.2). The operations, packages, components, and external processing parameters will be copied over to the maintenance order from the task list.

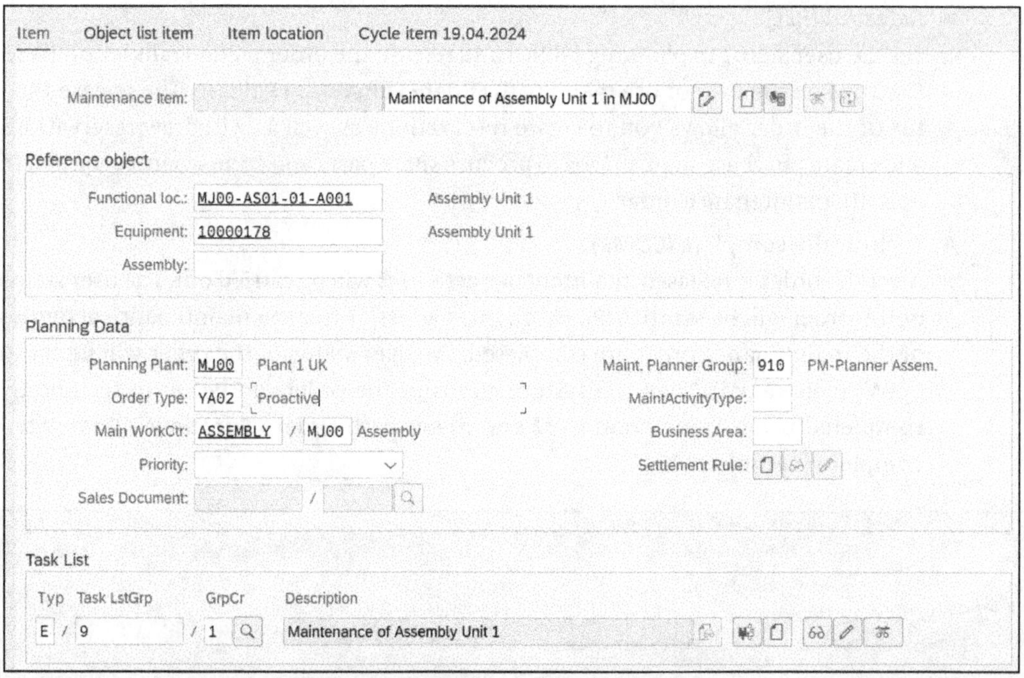

Figure 7.29 Create Maintenance Strategy Plan: Item Overview

Upon saving the maintenance plan, the system generates two documents and assigns a number to them. The first one is the maintenance plan (header) and the second is the maintenance item. You can assign one or more items to the maintenance plan and the system generates the maintenance item number for each item.

After creating the maintenance plan, you can schedule the maintenance operation using Transaction IP10. In the initial screen of the schedule maintenance plan, specify the maintenance plan created. This activity will automatically generate the maintenance order, as shown in Figure 7.30. Item details such as the functional location and equipment have been copied over to the maintenance order from the maintenance plan.

The planned start date, finish date, and due date for the maintenance activities are displayed in the maintenance order header along with the planner group responsible for the maintenance activities.

Status management is used for the maintenance order to recognize the status of the order. The following are the key statuses of maintenance order:

- **Created (CRTD)**
  When the maintenance order is created, the system status of the order is set to **CRTD**, and the user status of the order is set to **INTL** (initial). Reservations can be created for the maintenance order, but not withdrawn if the status of the order is **CRTD**.

- **Released (REL)**

  Set the user status to planning (**PLNG**) and release the order using Transaction IW32. Use the **Release** icon or use menu path **Order • Functions • Release**. The released status of the order allows you to create reservations as well as withdraw reservations and create purchase requisitions to procure spare parts and components required to execute maintenance order.

- **Technically complete (COMP)**

  After the order is released, maintenance activities will be carried out. The user status of the order will be set to **WIPR** (work in progress). After the maintenance activities of the maintenance order are completed, the user status of the order will be set to **COMP** (completed). Then the system status of the order can be set to technically completed using Transaction IW32 and menu path **Order • Functions • Complete • Complete (Technically)**.

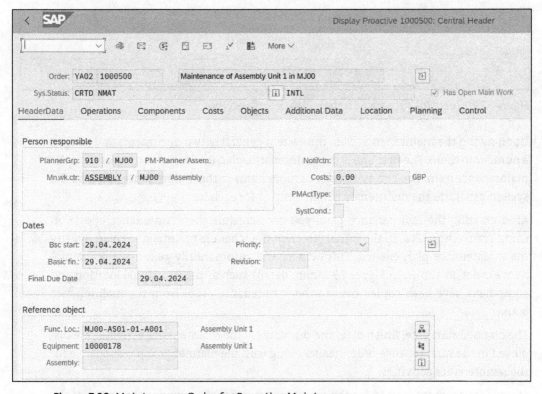

Figure 7.30 Maintenance Order for Proactive Maintenance

Figure 7.31 shows the completed activity/operation of the maintenance order. **PM01** control key represents internal maintenance activity performed by the planning group responsible for the execution of the maintenance tasks. Status **COMP** indicates the maintenance activities are completed. System status **CLSD** indicates the order is closed.

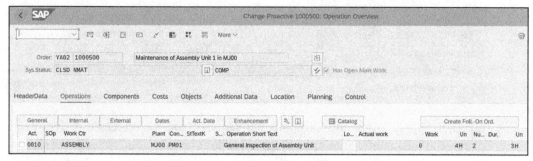

Figure 7.31 Maintenance Order: Operation Overview

7.3.4   Material Planning

Material planning for plant maintenance ensures the necessary materials, spare parts, and components are available to perform maintenance activities. Plant maintenance integrates with MRP and purchasing to automatically create procurement proposals to procure materials for maintenance activities in time. Inventory management helps to track and manage spare parts and components required for maintenance activities. BOMs play a vital role in planning of materials for plant maintenance. The equipment BOM, assembly BOM, and functional location BOM provide the list of components/spare parts required for maintenance operations. The materials required in plant maintenance are the components of the BOM. You can also assign components directly to the operations of a maintenance order.

Purchasing and inventory management functions are directly integrated with material planning for plant maintenance. The following item categories of the assembly, equipment, and functional location BOM are relevant for purchasing:

- **Stock item (L)**
  Use this item category if the material or component is stored in the inventory. This material or component can be issued to production using movement type 261.

- **Non-stock item (N)**
  Use this item category to list a nonstock item as a BOM component. A nonstock item can have a material number or just a short text (without material number).

- **Variable-size item (R)**
  Use this item category to list a variable-sized material as a BOM component. An example variable-sized material is a metallic sheet.

- **Text item (T)**
  Use this item category to enter a long text in between different BOM components.

Note

Stock items and variable-size items required for maintenance activities are managed in inventory. Material reservations will be created automatically for the components, and

---

materials can be withdrawn with reference to reservations from inventory for maintenance activities, whereas nonstock items that are required for maintenance activities are procured from suppliers. Procurement proposals/purchase requisitions are generated to procure nonstock items from a maintenance order based on external processing parameters.

Materials (spare parts and components) required for maintenance activities are planned in the maintenance order at the operations level by directly assigning components required for the respective maintenance activity/operation. Required materials/components are copied from equipment tasks list to the maintenance order automatically. However, you can assign components directly to the maintenance order in both corrective maintenance and preventive maintenance.

Let's discuss the material planning process for maintenance orders with a corrective maintenance process. Refer to Section 7.3.3 for the details of preventive maintenance and material planning for the same.

Maintenance Request

The corrective maintenance process starts with the creation of a maintenance request, or maintenance notification. You can create a maintenance request by using Transaction IW21 and setting the notification type as **M1** or by using Transaction IW26 directly.

In our example, we'll select **M1** and press ⌈Enter⌋ to arrive at the notification details, as shown in Figure 7.32. In the maintenance request, you can assign the equipment, functional location, and assembly details to perform maintenance.

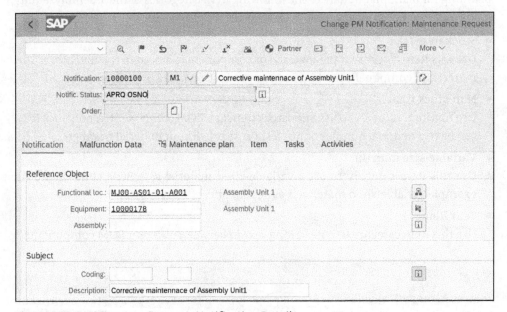

Figure 7.32 Maintenance Request: Notification Details

You can define malfunction data in the **Malfunction Data** tab, such as malfunction start date and time of the technical object. If it's a complete breakdown, you can indicate that in that tab.

You can assign a maintenance task list under the **Maintenance plan** tab (see Section 7.3.2 to create the task list). By assigning the task list, the operations, components details will be copied over to the maintenance order when created using this maintenance request.

A maintenance request must be approved to carry out the corrective maintenance of the technical object by creating a maintenance order.

Maintenance Order

Maintenance orders with reference to a maintenance request can be created using Transaction IW34. Figure 7.33 shows the initial screen for the creation of a maintenance order. Fill out the following fields:

- **Order Type**
  The order type distinguishes different types of maintenance orders. Assign the appropriate order type for the creation of the maintenance order. **YA01** is an order type defined for corrective/reactive maintenance. The following are the standard maintenance order types:
  - **PM01**: Maintenance order
  - **PM03**: Maintenance order/notification
  - **PM04**: Refurbishment order
  - **PM05**: Calibration order
  - **PM06**: Capital investment order
  - **SM01**: Service order
  - **SM02**: Service order (with revenues)
  - **SM03**: Repair service

- **Notification**
  Enter the maintenance request created for the corrective maintenance in this field. All the details from the maintenance request will be copied over to the maintenance order.

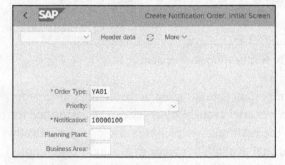

Figure 7.33 Create Maintenance Order for Notification

After specifying the order type and notification, press [Enter] to navigate to the **HeaderData** tab of the maintenance order. From there, click the **Operations** tab to arrive at the **Operations Overview** screen. Figure 7.34 shows the maintenance order **Operations** overview, showing the planned operations for the maintenance activities based on the maintenance request.

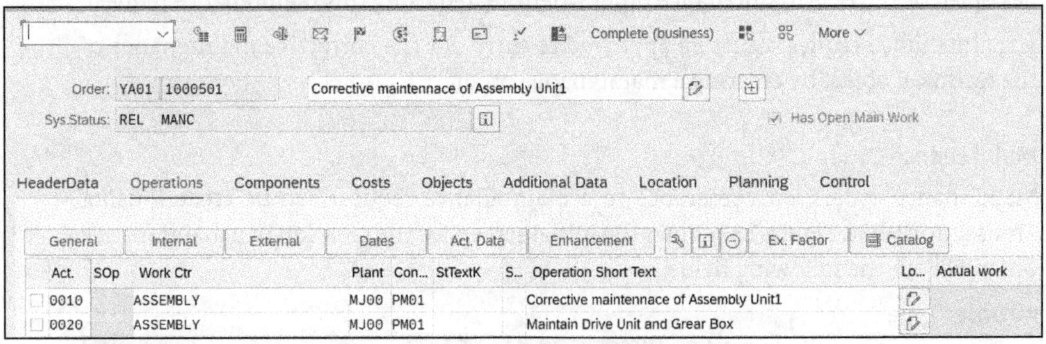

Figure 7.34 Create Maintenance Order for Notification: Operations

Next, click the **Components** tab to view the components planned for the maintenance activities, as shown in Figure 7.35. A component can be a stock item, nonstock item, or text item. You can assign all components required to maintain the technical object. Assign the required quantity and the unit of measure for every component. You can list the components required for the maintenance operation by specifying the materials one by one in the **Components** column. Specify the **Reqmt Qty** (requirement quantity) and **UM** (unit of measure) for every component listed in the **Components** tab.

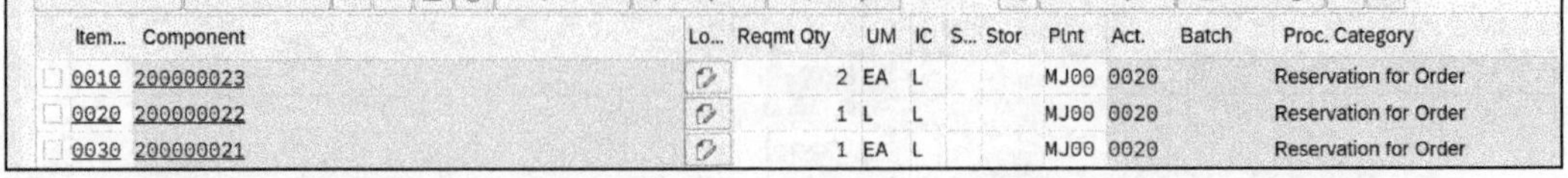

Figure 7.35 Create Maintenance Order for Notification: Components

Once the required spare parts and components are assigned to the maintenance order to execute maintenance operations, a material availability check can be performed for the maintenance order manually by the maintenance planners and/or automatically by the system. You can perform material availability check manually if the maintenance order is not yet released. Click the **Material Availability, Overall** icon at the top or press $\boxed{\text{Ctrl}}$+$\boxed{\text{F9}}$ to perform the availability check manually. At the time of maintenance order release (via Transaction IW32), the system performs the material availability check automatically.

Figure 7.36 shows the material reservations created automatically by the system after the maintenance order was released. Materials can be withdrawn from the inventory with reference to the material reservation to carry out the maintenance operations.

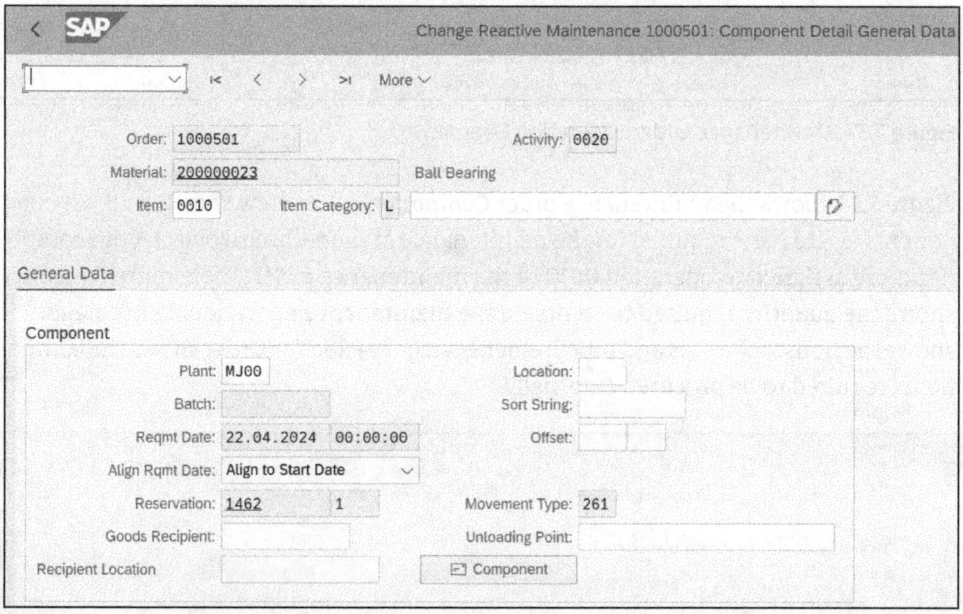

Figure 7.36 Components Details: Material Reservation

The reservation will have the requested quantity and plant details. Movement type **261** indicates the goods issue posting for the maintenance order. During the goods movement from the inventory, this type references the maintenance order.

7.3.5   Procurement of Spare Parts and Services

Procurement is directly integrated with plant maintenance for material planning. Processing the maintenance order triggers the creation of a purchase requisition with reference to maintenance order to procure nonstock materials from external vendors. As shown in Figure 7.36, during maintenance order processing, the system automatically creates material reservations for the stock items (components) and purchase requisitions for out-of-stock items (components). Let's dive deep into how the external procurement of spare parts and other external services required for plant maintenance is planned in plant maintenance.

Figure 7.37 shows the maintenance order **Operations** overview, showing one of the planned operations for the maintenance of a technical object. The maintenance order is in the released state (**REL** status code), and the system has carried out the stock availability check of all components (spare parts).

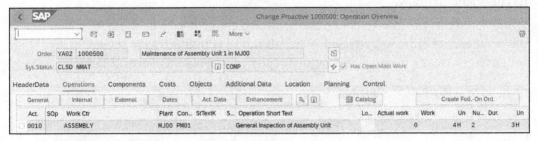

Figure 7.37 Maintenance Order: Operation Overview

Figure 7.38 shows the maintenance order **Components** overview showing all the components/spare parts required for the maintenance of the technical object. Both components are assigned to operation **0010** of the maintenance order. The required quantity shows the quantity required to complete the maintenance operation. Item category **N** shows the nonstock items and procurement category **PReq for Order** shows the components required to be procured externally.

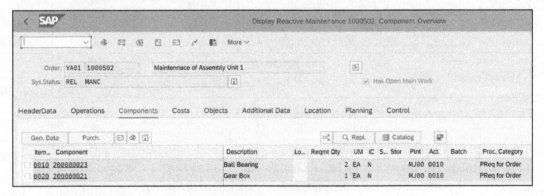

Figure 7.38 Maintenance Order: Components Overview

Next, maintain the purchasing data for components to be procured externally. Select a component item and click the **Purch.** button at the top of the **Component Overview** screen to navigate to the **Component Detail Purchasing Data** view. Figure 7.39 shows the **Purchasing Data** tab of a component in detail. If you maintain the PIR, the purchasing data is copied over to the maintenance order for external procurement. Best practice in purchasing is to create a PIR for all materials procured externally.

Upon the creation of a maintenance order, the system automatically generates a purchase requisition for all components that are not available in the inventory and procured externally from a supplier. You can see purchase requisition **10000979** and item number **10** of the purchase requisition item generated for spare part **200000023**.

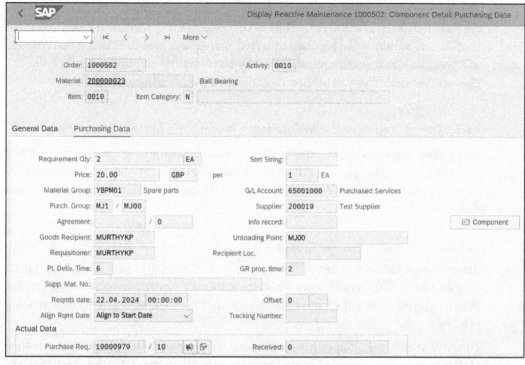

Figure 7.39 Maintenance Order: Purchasing Data of a Component

Figure 7.40 shows the purchase requisition generated from the maintenance order with account assignment category **F** (order). Under the **Account Assignment** tab, you can see the maintenance order reference.

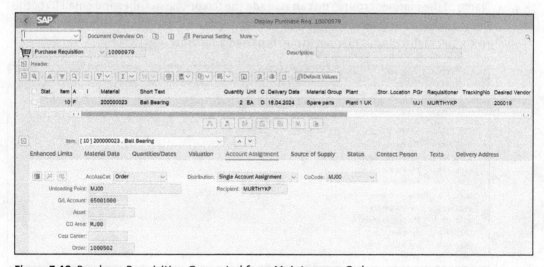

Figure 7.40 Purchase Requisition Generated from Maintenance Order

The purchaser/buyer will convert the purchase requisition into a purchase order. Once the purchase order is approved, a copy of the purchase order will be transmitted to the vendor. The vendor will ship the requested spare parts, and the buying organization will post the goods receipt upon physically receiving the shipment from the vendor. The received materials will be directly consumed by the maintenance order.

7.3.6 Configuration of the Maintenance Process and Material Planning

This section explains the key configuration settings for the maintenance process and material planning for plant maintenance in SAP S/4HANA. Let's dive deep into some of the key configuration settings and understand the configuration settings that control these functions.

Define Maintenance Planner Groups

In this configuration, you can define maintenance planner groups responsible for planning and executing maintenance activities within a maintenance planning plant. The planner group responsible for maintenance planning will be added to the maintenance plan.

To arrive at this configuration step, execute Transaction SPRO and navigate to **Plant Maintenance and Customer Service • Maintenance Plans, Work Centers, Task Lists and PRTs • Basic Settings • Define Maintenance Planner Groups.**

Figure 7.41 shows the maintenance planner groups defined for the respective plant and their description. Click **New Entries** to define a maintenance planner group for a plant. Specify the plant, a three-digit alphanumeric key for the planner group (**PG**), and the **Name** of the planner group. You can also add the telephone number and email ID of the planner group if the maintenance planner is a separate department. In larger organizations, the maintenance planner group is part of a separate department.

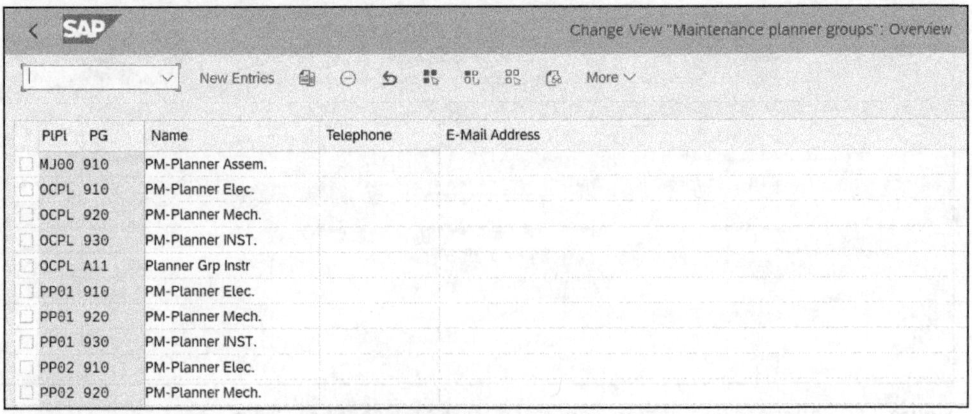
Change View "Maintenance planner groups": Overview
New Entries
More
PlPl PG Name Telephone E-Mail Address
MJ00 910 PM-Planner Assem.
OCPL 910 PM-Planner Elec.
OCPL 920 PM-Planner Mech.
OCPL 930 PM-Planner INST.
OCPL A11 Planner Grp Instr
PP01 910 PM-Planner Elec.
PP01 920 PM-Planner Mech.
PP01 930 PM-Planner INST.
PP02 910 PM-Planner Elec.
PP02 920 PM-Planner Mech.

Figure 7.41 Define Maintenance Planner Groups

Set Maintenance Plan Categories

Maintenance plan categories control the different business functions in plant maintenance. These categories are used in the creation of a maintenance plan, and assigning a maintenance plan category is a required entry.

To arrive at this configuration step, execute Transaction SPRO and navigate to **Plant Maintenance and Customer Service • Maintenance Plans, Work Centers, Task Lists and PRTs • Maintenance Plans • Set Maintenance Plan Categories**.

Click **New Entries** to define maintenance plan categories. Specify a two-digit alphanumeric key and description to define a new define maintenance plan category. Figure 7.42 shows the maintenance plan category defined in configuration. Let's explore the key fields:

- **Call object**
  A call object for the maintenance plan controls the type of object a maintenance call creates upon scheduling the maintenance plan. You can set a specific call object per maintenance plan category. You can choose from the following call objects for the maintenance plan:
  - Maintenance order
  - Service entry sheet
  - Notification
  - Inspection lot
  - Service order

- **Ref. object**
  A reference object controls the subscreen in the maintenance plan. For example, if you're planning to maintain equipment, functional location, and assembly, assign reference object **O100** in this field so that you can assign these technical objects to the maintenance plan. Table 7.4 shows the different reference objects defined in the standard system.

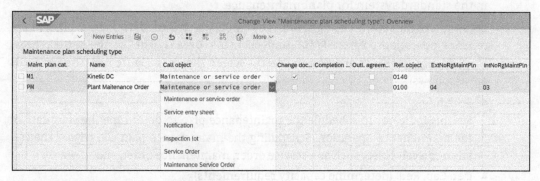

Maint. plan cat.	Name	Call object	Change doc...	Completion ...	Outl. agreem...	Ref. object	ExtNoRgMaintPln	IntNoRgMaintPln
M1	Kinetic DC	Maintenance or service order	✓			O140		
PM	Plant Maitence Order	Maintenance or service order				O100	04	03
		Maintenance or service order						
		Service entry sheet						
		Notification						
		Inspection lot						
		Service Order						
		Maintenance Service Order						

Figure 7.42 Set Maintenance Plan Categories

Reference Object	Description
O100	Functional location + equipment + assembly
O110	Equipment + assembly
O120	Functional location (30) + equipment + assembly
O130	Serial number + material number + device ID
O140	Without reference object
O150	Equipment only
O160	Functional location only
O170	Equipment + serial number + material number
O180	Functional location 1:1 + equipment + assembly
O190	Material sample

Table 7.4 Reference Object for Screen Control

Maintain Control Keys

Control keys are used to define operations in task lists, routings, and maintenance orders. They determine a specific business process within SAP S/4HANA and provide control parameters to drive that process. They control scheduling to determine the start date of the maintenance activity, determination of capacity requirements, and external processing. To arrive at this configuration step, execute Transaction SPRO and navigate to **Plant Maintenance and Customer Service • Maintenance Plans, Work Centers, Task Lists and PRTs • Task Lists • Operation Data • Maintain Control Keys.**

Figure 7.43 shows the initial screen to maintain control keys for operations. SAP S/4HANA provides standard control keys for plant maintenance, production planning, quality inspection, and so on. **PM01**, **PM02**, **PM03**, **PM05**, and **PM07** are defined in the standard system for plant maintenance.

Let's explore the settings using control key **PM02**, which is used in external plant maintenance operations. Select **PM02** and click the **Details** icon at the top or press `Ctrl`+`Shift`+`F2` to arrive at the details screen shown in Figure 7.44. The key fields are as follows:

- **Scheduling**
  Set this indicator to schedule the maintenance plan to determine the start date of the maintenance operation. Scheduling the maintenance plan will trigger the creation of a call object such as a service order, maintenance order, and so on.

- **Det. Cap. Req. (determine capacity requirements)**
  Always set this indicator when the scheduling indicator is set. The system determines the capacity requirements based on scheduling (start date of the maintenance operation).

- **Cost**
  Set this indicator if you want to include the plant maintenance operation with this control key in costing.

- **Confirmation**
  A confirmation is used to monitor maintenance orders. Table 7.5 shows the possible values you can assign to this field for plant maintenance. If you set the confirmation for the maintenance operation as required or optional, you can confirm if the maintenance activity is completed, who performed the maintenance activity, and so on.

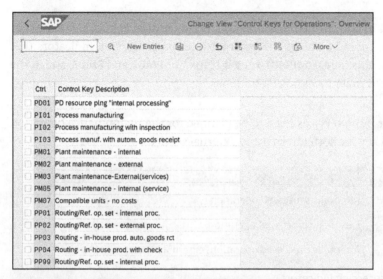

Figure 7.43 Control Keys for Operation: Initial Screen

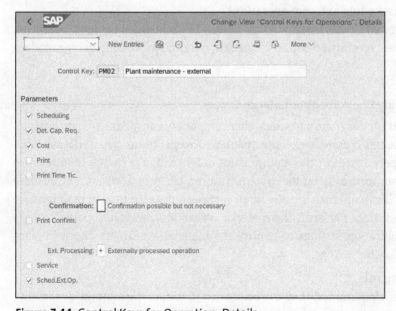

Figure 7.44 Control Keys for Operation: Details

Confirmation	Description
2	Confirmation required
3	Confirmation not possible
Blank	Confirmation possible but not necessary

Table 7.5 Confirmation Keys

- **Ext. Processing**
  This field is used to determine if the maintenance operation is internally or externally performed. Table 7.6 shows the external/internal processing indicators that you can assign to this field. For **PM01**, keep it blank; for **PM02** and **PM03**, assign the value **+** to plan external processing operations.

- **Service**
  Set this indicator if the services need to be planned for the maintenance operation with this control key for both internal and external services.

External Processing	Description
Blank	Internally processed operation
+	Externally processed operation
X	Internally processed operation; external processing possible

Table 7.6 External Processing Indicators

- **Sched.Ext.Op. (schedule external operation)**
  Set this indicator if the external operation needs to be scheduled using standard values. The purchase requisition will be created for external processing using standard values.

Material Availability Check for Maintenance Orders

Required materials such as spare parts and other components are planned in the maintenance orders. Once this is done to execute maintenance operations, a material availability check can be performed for the maintenance order manually by the maintenance planners and/or automatically by the system. You can perform a material availability check manually if the maintenance order is not yet released. At the time of maintenance order release, the system performs the material availability check automatically. Let's explore the configuration settings to control the material availability check in plant maintenance. Follow these steps:

1. **Define checking rule**
   A two-character alphanumeric key will be created to trigger the material availability check when the maintenance order is released. Checking rule **PM** is defined in the

standard system for plant maintenance. You can create a new checking rule to control the material availability check.

To arrive at this configuration step, execute Transaction SPRO and navigate to **Plant Maintenance and Customer Service • Maintenance and Service Processing • Maintenance and Service Orders • Functions and Settings for Order Types • Availability Check for Material, PRTs, and Capacities • Define Checking Rules**.

Figure 7.45 shows the standard checking rule **PM** defined for plant maintenance.

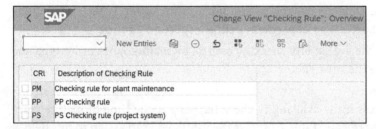

Figure 7.45 Define Checking Rule

2. **Define scope of check**
   In this configuration step, we define the availability checking procedure based on the availability checking group and the checking rule defined in the previous step.

   To arrive at this configuration step, execute Transaction SPRO and navigate to **Plant Maintenance and Customer Service • Maintenance and Service Processing • Maintenance and Service Orders • Functions and Settings for Order Types • Availability Check for Material, PRTs, and Capacities • Define Scope of Check**. This definition will be configured for the combination of checking rule and the checking group of the availability check defined in the previous steps. Figure 7.46 shows the scope of the check defined for availability checking group **02** and checking rule **PM**.

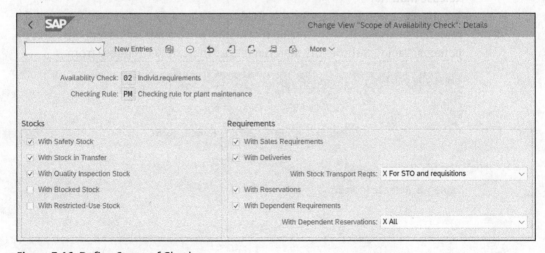

Figure 7.46 Define Scope of Check

Refer to Chapter 4, Section 4.1.5 for detailed information about how to set up the scope of the check.

3. **PM order control—material availability**

   In this configuration step, we can assign the availability checking rule to the maintenance order type by plant.

   To arrive at this configuration step, execute Transaction SPRO and navigate to **Plant Maintenance and Customer Service • Maintenance and Service Processing • Maintenance and Service Orders • Functions and Settings for Order Types • Availability Check for Material, PRTs, and Capacities • Define Inspection Control.**

   Figure 7.47 shows the availability checking control by maintenance order type and plant. Let's explore the key fields:

   – **Availability Check**
     This field controls if the availability check happens at the time of maintenance order creation or release. Table 7.7 shows the values you can assign to this field.

Business Function	Description
1	Check availability during order creation
2	Check availability during order release

Table 7.7 Business Function for Availability Check

   – **Checking Rule**
     Assign the checking rule defined for plant maintenance in this field. By assigning the checking rule, the system automatically determines the scope of the check and performs the material availability check.

   – **Release material**
     Specify a control indicator for this field to take an appropriate action as per the material release indicator selected if a maintenance order is missing certain components. Table 7.8 shows the material release indicators to choose from.

Material Release	Description
1	User decides on release if parts are missing
2	Release permitted despite missing parts
3	No release if parts are missing

Table 7.8 Release Indicators

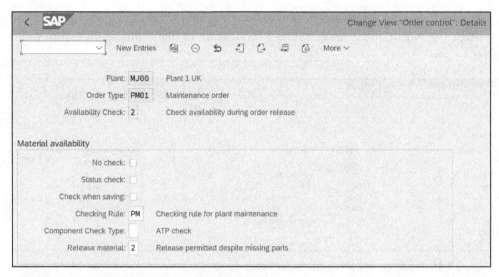

Figure 7.47 PM Order Control: Material Availability

Procurement Configuration for Maintenance Order

The procurement function is directly integrated with plant maintenance for material planning and executing maintenance activities/operations externally based on control keys. Let's explore the configuration settings that control the integration between plant maintenance and procurement functions.

Activate Enhanced Procurement Mode

In this configuration setting, you can assign a procurement mode for the combination of maintenance order type and plant. There are two procurement modes available in the standard system:

- **Enhanced procurement mode**

    The enhanced procurement mode supports lean services. The advantage of lean services in procurement is that they support specialized services for the maintenance operation that can be carried out by internal or external suppliers. Lean services can be maintained in the material master as service products using material type SERV. Lean services can be added to an operation as components and trigger the generation of purchase requisitions for services from the maintenance order.

- **Compatibility procurement mode**

    Compatibility procurement mode doesn't support lean services, but it supports the other types of services procurement.

To arrive at this configuration step, execute Transaction SPRO and navigate to **Plant Maintenance and Customer Service • Maintenance and Service Processing • Maintenance and Service Orders • Functions and Settings for Order Types • Procurement •**

Activate Enhanced Procurement Mode. On the screen shown in Figure 7.48, you can assign the **Procurement Mode** to the combination of plant and maintenance order type.

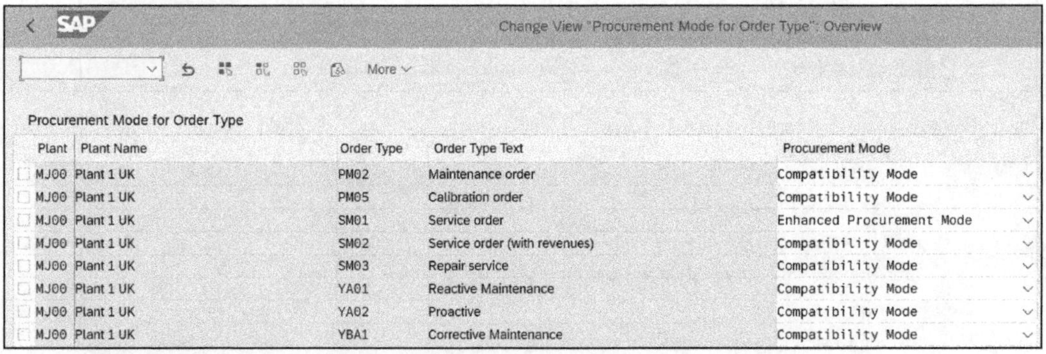

Plant	Plant Name	Order Type	Order Type Text	Procurement Mode
MJ00	Plant 1 UK	PM02	Maintenance order	Compatibility Mode
MJ00	Plant 1 UK	PM05	Calibration order	Compatibility Mode
MJ00	Plant 1 UK	SM01	Service order	Enhanced Procurement Mode
MJ00	Plant 1 UK	SM02	Service order (with revenues)	Compatibility Mode
MJ00	Plant 1 UK	SM03	Repair service	Compatibility Mode
MJ00	Plant 1 UK	YA01	Reactive Maintenance	Compatibility Mode
MJ00	Plant 1 UK	YA02	Proactive	Compatibility Mode
MJ00	Plant 1 UK	YBA1	Corrective Maintenance	Compatibility Mode

Figure 7.48 Procurement Mode for Order Types

By default, **Compatibility Mode** is set irrespective of this configuration. You need this configuration only if you need enhanced procurement mode.

> **Note**
>
> If you activate enhanced procurement mode, it can only support the operations with control key PM02. The external service operation-related control key PM03 is not supported.

Define Collective Purchase Requisition and MRP Relevance

In this configuration setup, you can control the collective purchase requisition creation and the timing of a reservation and purchase requisition creation during maintenance order processing.

To arrive at this configuration step, execute Transaction SPRO and navigate to **Plant Maintenance and Customer Service • Maintenance and Service Processing • Maintenance and Service Orders • Functions and Settings for Order Types • Procurement • Define Collective Purchase Requisition and MRP Relevance.**

You'll arrive directly at the screen shown in Figure 7.49, where the example shows the settings for plant **MJ00** and a maintenance order type combination. Let's explore the key fields:

- **CollReqstn (collective requisition)**
  Set this indicator if you want to create one collective purchase requisition per maintenance order for all externally processed operations and nonstock materials. Collective requisition is useful when there are multiple nonstock materials to be procured externally to perform a maintenance operation.

- **Res/PurRq (reservation relevance/generation of purchase requisition)**
  This field controls when the reservations are relevant for material planning and when the purchase requisitions will be generated for the maintenance order. You can choose from the following values:
  - **Never**
  - **From Release**: Upon release of the maintenance order
  - **Immediately**: Upon creation of the maintenance order
  - **Immediately with blocking indicator**: Upon creation of the maintenance order

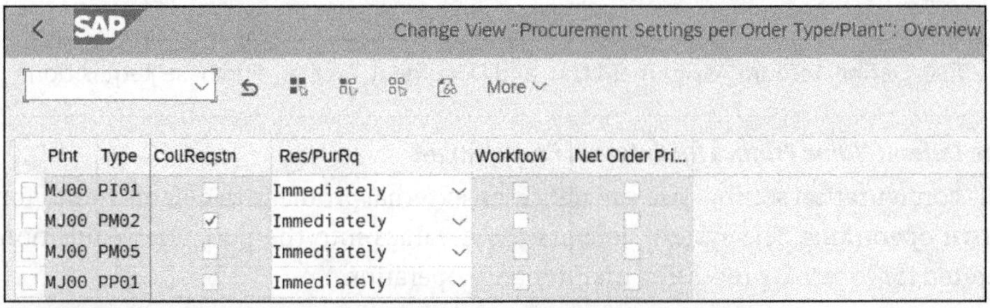

Figure 7.49 Define Collective Purchase Requisition and MRP Relevance

Define Account Assignment Category and Document Type for Purchase Requisitions

In this configuration setup, you can define the purchase requisition document type and account assignment categories for the maintenance order category.

To arrive at this configuration step, execute Transaction SPRO and navigate to **Plant Maintenance and Customer Service • Maintenance and Service Processing • Maintenance and Service Orders • Functions and Settings for Order Types • Procurement • Define Account Assignment Cat. and Document Type for Purchase Requisitions**.

You'll arrive directly at the screen shown in Figure 7.50, where the example shows the settings for order category **30** (maintenance order) defined in the standard system. Let's explore the key fields:

- **Document type**
  Assign the relevant document type for the generation of purchase requisitions if you want to change the standard purchase requisition document type.

- **Account Assignment Categories**
  If you have a specific account assignment category configured with some required entries, you can assign it in these fields. For example, if you have created an account assignment category that requires a maintenance order reference, assign it in one of the fields.

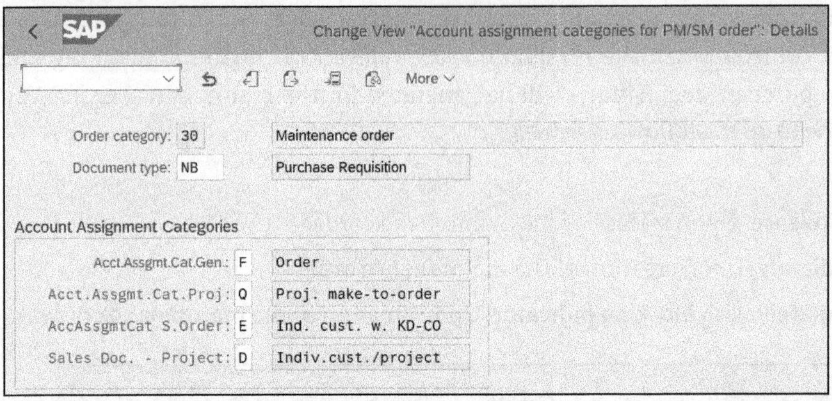

Figure 7.50 Define Account Assignment Cat. and Document Type for Purchase Requisitions

Create Default Value Profiles for External Procurement

In this configuration setting, you can define an external profile with default values for external operations. The system defaults these values into the purchase requisition generated for external processing maintenance operation.

To arrive at this configuration step, execute Transaction SPRO and navigate to **Plant Maintenance and Customer Service • Maintenance and Service Processing • Maintenance and Service Orders • Functions and Settings for Order Types • Procurement • Create Default Value Profiles for External Procurement**.

From the initial screen, click **New Entries** to create a new profile for external procurement. Specify a seven-digit numeric key and description to define a new profile. To edit an existing profile, select the profile and click the **Details** button at the top or press Ctrl + Shift + F2 . Figure 7.51 shows the default values assigned for the external profile. You can assign a default cost element, purchasing organization, purchasing group, and material group.

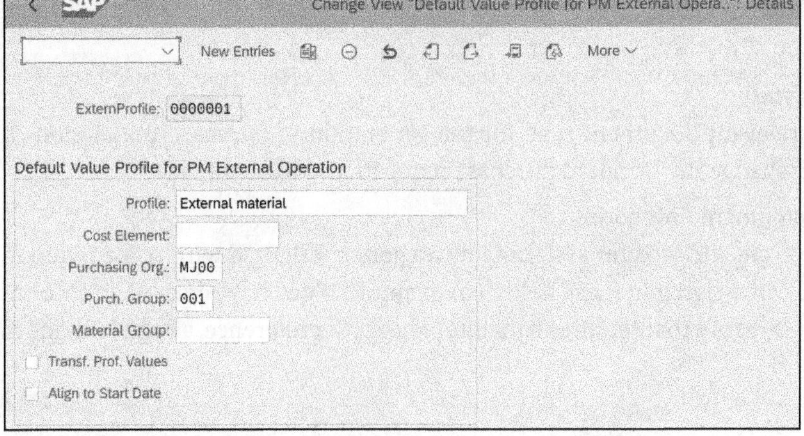

Figure 7.51 Create Default Value Profiles for External Procurement

7.4   Inventory Management

Plant maintenance and inventory management functions are directly integrated in SAP S/4HANA. Inventory management ensures required spare parts and components required for plant maintenance operations are stored in the inventory, issued with reference to plant maintenance orders for maintenance process, and tracked. Inventory management provides visibility into the on-hand inventory and location information where the materials are stored in the inventory.

During material planning, the system generates procurement proposals for any shortage of inventory for plant maintenance based on the stock availability check for spare parts and components required to maintain technical objects or a company's assets. The inventory management function tracks all material movements within the production process that include goods issue of spare parts for maintenance operations and external operations and goods receipt of spare parts and other components. Maintenance planners have direct access to the inventory levels of spare parts and components in real time within the maintenance planning process.

Another important advantage of this integration is that when maintenance orders are created, the system automatically performs the availability check for spare parts and reserves required in-stock materials automatically. This helps to prevent stockouts and ensures the required spare parts and components are available for the respective maintenance order.

Overall, inventory management helps to enhance the operational efficiency of company's assets, reduces maintenance costs, and reduces equipment downtime. Let's dive deep into the integration between the inventory management and plant maintenance functions.

7.4.1   Goods Receipt and Goods Issue

The goods receipt process for plant maintenance involves receiving spare parts and other components for maintenance operations and receiving refurbished materials from external suppliers. Goods receipts will be performed with reference to purchase orders and integrated with maintenance orders to ensure planned maintenance activities/operations receive the required spare parts and components. The goods receipts for spare parts posted without reference to maintenance orders will update the inventory levels so that the stock availability check can be performed and inventory levels can be tracked and valuated. For plant maintenance, the following are goods receipt processes:

1. **Goods receipt of spare parts/components from suppliers**
   Material planning for maintenance orders generates purchase requisitions for spare parts and other components if not available in the inventory. The purchase requisition will be converted into a purchase order to procure required materials for maintenance operations from an external vendor. Once the shipment from the suppliers

arrives at the receiving location within the plant, the goods receipt will be posted manually in SAP S/4HANA with reference to the purchase order using Transaction MIGO and movement type 101. The received materials will be directly consumed by maintenance if the maintenance order is assigned to the purchase order item in the accounting tab.

2. **Asset maintenance through subcontracting from external vendor**
The subcontracting process is a special procurement process in SAP S/4HANA where certain maintenance operations will be outsourced by the organizations to external suppliers and the required components will be provided to the external supplier. For asset maintenance, the technical object or equipment to be maintained or repaired will be planned with an external operation with subcontracting in the maintenance order and the equipment will be added as a component to the maintenance operation. A subcontracting purchase requisition will be generated from the maintenance order process, and it will be converted into a subcontracting order. The asset/equipment to be maintained/repaired will be sent to the service provider with reference to the subcontracting order. The service provider will perform the maintenance operation on behalf of the organization. The subcontractor will return the company's asset or component back to the organization and charge a service fee to the organization. Once the shipment from the subcontractor arrives at the receiving location of the plant, goods receipt will be posted manually with reference to the subcontract order using Transaction MIGO and movement type 101.

Goods issue for maintenance is the process of removing spare parts and components from the inventory for consumption for maintenance operations. Goods issue posting creates both material document and accounting document in the system (depending on the movement types used) to record the material movement and the valuation of the stock in financial accounting respectively. For plant maintenance, the following are the different goods issue processes:

1. **Goods issue of spare parts/components to plant maintenance**
Raw materials and other required materials/components will be issued to production with reference to production order or process order using Transaction MIGO and using movement type 261. MRP ensures the required materials and components to produce the finished products are available at the right time. Inventory management provides a material reservation feature to reserve materials and components for production in advance. The goods issue ensures the required materials are supplied to production for consumption during manufacturing of finished products.

2. **Goods issue of equipment/components to external maintenance operation**
Equipment and other required materials/components will be issued to the external service provider with reference to a subcontract order using Transaction ME2O. Goods issue can be posted with reference to a (sales and distribution) outbound delivery. Movement types play an important role in determining the type of the inventory movement and posting of goods issue. Movement type 541 is used for

goods issue for outbound delivery to ship the components to the subcontractor. Movement type 541 ensures the components provided to the subcontracting vendor are tracked and the ownership remains with the company requesting the external operation.

7.4.2   Movement Types and Stock Postings

Plant maintenance in SAP S/4HANA governs planning and maintenance of a company's assets such as equipment, functional locations, and facilities. Managing material movements to support the internal and external maintenance processes is crucial for the organization. The availability of required spare parts and other components for the maintenance process is essential to effectively managing the maintenance process. There are specific movement types defined in the standard system for material movements to support maintenance processes, as shown in Table 7.9.

Movement Type	Detailed Description
101	Goods receipt. This movement type is used for standard goods receipts from external vendors and subcontractors to receive spare parts and other components into inventory. It increases the stock quantity and value in the receiving storage location. Stock can be received into unrestricted use stock or quality inspection stock or directly to maintenance.
102	Goods receipt reversal. This movement type is used to cancel the goods receipt posted using movement type 101. Stock will be removed from unrestricted use stock or quality inspection stock upon posting this goods receipt reversal.
122	Return to vendor. This movement type is used to return materials to vendor because of poor quality they can't be used in maintenance. Stock will be removed from inventory to return the goods to vendor.
161	Returns purchase order to vendor. This movement type is used to return materials to vendor with reference to a returns purchase order. Stock will be removed from the inventory to return the materials to vendor.
321	Transfer posting from quality inspection to unrestricted. This movement type is used to transfer materials which are accepted in quality inspection into unrestricted use stock.
201	Goods issue for a cost center. This movement type is used to issue goods from the inventory to a cost center for consumption.
261	Goods issue for a production order. This movement type is used to issue goods from the inventory to a maintenance order for maintenance activities.

Table 7.9 Movement Types Used in Plant Maintenance

Movement Type	Detailed Description
541	Goods issue for subcontract supplier. This movement type is used to issue components to the subcontractor to produce/assemble products on behalf of the company that issued the subcontract order. Stock will be transferred from unrestricted use stock to subcontracting stock within the inventory management.
543	Consumption from subcontracting. This movement type is used to post the consumption of components provided to the subcontractor automatically upon posting the receipt of the final product produced/assembled by the subcontractor. This goods movement is automatically posted during the goods receipt with reference to subcontract order.
551	Goods issue for scrapping. This movement type is used to issue the expired/used/defective spare parts and other components used in plant maintenance for the scrapping process. Scrapping will reduce inventory levels.

Table 7.9 Movement Types Used in Plant Maintenance (Cont.)

7.4.3 Reservations

Processing maintenance orders generates material reservations for the stock items (components) and purchase requisitions for out-of-stock items (components) based on the availability check. If the spare parts are available in the inventory, material reservations are created automatically for spare parts for goods issue for maintenance order on a specific date with required quantities. Material reservations are requests made to inventory management to keep aside required quantities of specific materials for a particular purpose, typically for production or maintenance. Material reservations for plant maintenance will be created using movement type 261 for issue of spare parts against the maintenance order.

Material planning is performed for the maintenance order to add required components to execute maintenance operations. Figure 7.52 shows the stock items planned for a maintenance order, indicated by item category L.

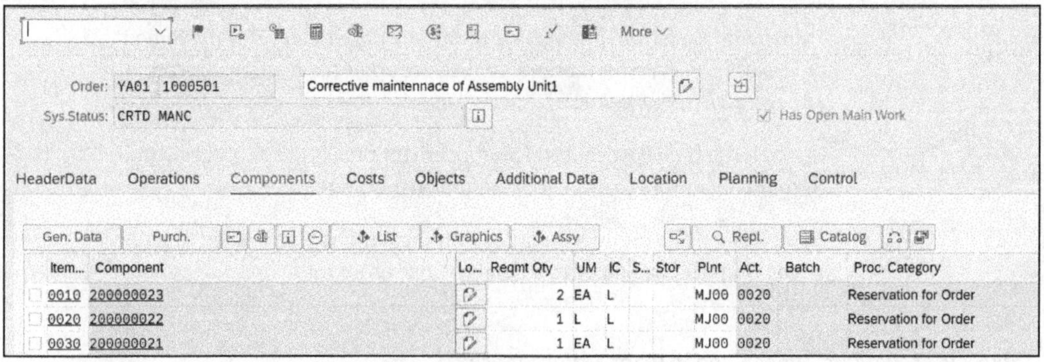

Figure 7.52 Stock Items Planned in Maintenance Order

In configuration, if you have set to trigger the generation of reservations for stock materials immediately, reservations are created upon saving the maintenance order. Refer to our discussion of defining collective purchase requisitions and MRP relevance in Section 7.3.6 for details.

We provided the instructions to create a material reservation in detail in Chapter 6, Section 6.5.4. In this section, we'll examine a material reservation created for plant maintenance to show how the material reservation functionality of materials management is directly integrated with plant maintenance.

Figure 7.53 shows material reservation **1462** created for component **200000023 (Ball Bearing)** in plant **MJ00**. This will ensure the material is available only for the specific maintenance order to perform the maintenance operation.

> **Note**
>
> Reservations generated automatically from production planning and plant maintenance cannot be changed. The system makes sure the requested quantities of the material are reserved for the respective operation.

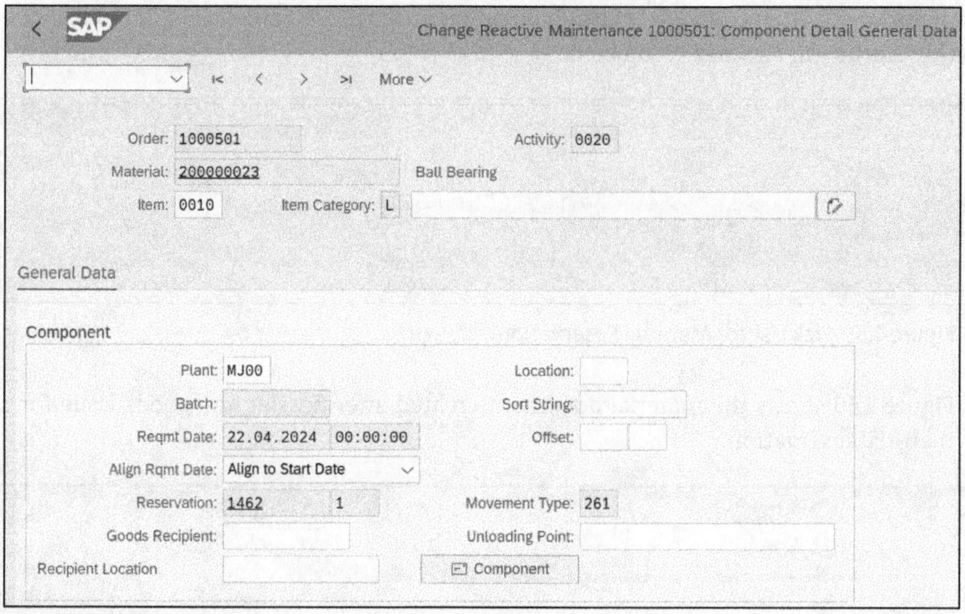

Figure 7.53 Components Details: Material Reservation

Figure 7.54 displays the details of the material reservation created automatically with reference to maintenance order **1000501** for material **200000023 (Ball Bearing)** in plant **MJ00**. A reservation is created upon saving the maintenance order for all stock items listed in the **Components** tab. The system assigns movement type **261** (goods issue for order) to the reservation automatically.

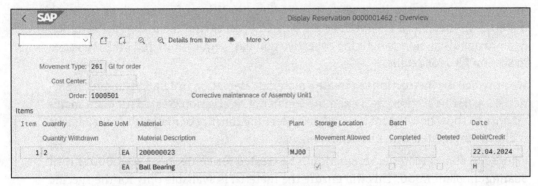

Figure 7.54 Material Reservation for Plant Maintenance

To issue the material for maintenance with reference to the reservation, execute Transaction MB26 (Pick List). In the selection screen of the pick list transaction, you can specify a selection criterion, using the reservation number, plant, material, maintenance order, and so on to refine the pick list to issue materials. After specifying the selection criterion, execute the transaction to list the reservations to be picked for processing. Figure 7.55 shows the pick list for material reservation **1462**. By posting the pick list, material documents and accounting documents are posted and the material reservation status will be set to completed.

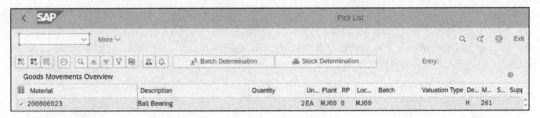

Figure 7.55 Pick List for Material Reservation

Figure 7.56 shows the material document created after posting the goods issue for a material reservation.

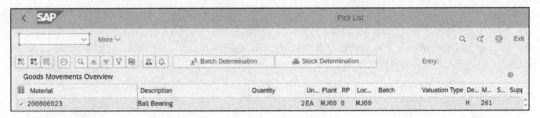

Figure 7.56 Goods Issue for Maintenance with Reference to Reservation

7.4.4 Material Valuation

Material valuation is a crucial aspect of inventory management and financial reporting that helps in determining the price of the material. SAP S/4HANA provides various methods to evaluate materials by standard price and moving average price. The standard price refers to the valuation of inventory at a fixed price over a period, whereas the moving average price of a material is recalculated after each transaction to reflect changes in the cost. The valuation cost of material is stored in the **Accounting 1** view of the material master. Material valuation is typically performed at the plant level, but it can also be performed at the company code level.

In plant maintenance, material valuation is mostly performed based on the condition of the spare parts and other components, such as new, damaged, and refurbished. Similarly, a material produced in house and the same material procured externally can have different prices. Valuation or materials in terms of their condition, procurement type, origin, and so on is called *split valuation*. Split valuation allows you to configure different valuation categories and valuation types within a valuation area (plant). At the time of creation of the material, you can assign a valuation category to the material in the **Accounting 1** view, and then you can define different valuation price for the same material by valuation type in the **Accounting 1** view.

Figure 7.57 shows different valuation categories and valuation types you can define in the configuration at the plant level. The material condition valuation category is highlighted in the dotted lines and is mostly used in plant maintenance to maintain different valuation prices for spare parts.

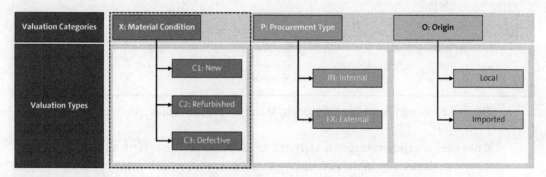

Figure 7.57 Split Valuation in Plant Maintenance

In split valuation, valuation categories and valuation types play a crucial role:

- **Valuation categories**
  This is used to categorize material valuations by different criteria based on business requirements. For example, a material can be evaluated based on its condition (new, damaged, or refurbished) and the material condition can be created as a valuation category to perform split valuation. You can assign the valuation category to the

material master in the **Accounting 1** view to valuate materials at different prices by valuation type.

- **Valuation type**
Valuation types are assigned to valuation categories at the plant level in configuration settings. Valuation types are used to define different valuation prices for the material at the plant level in the **Accounting 1** view. For example, you can define $20.00, $15.00, and $5.00 for new, refurbished, and defective valuation types for the same material.

Figure 7.58 shows valuation category **X** assigned to the material in the **Accounting 1** view at the plant level. This is the first step in defining different valuation prices for the same material, allowing you to define different valuation prices for the material by valuation type. To assign a valuation category to the material, execute Transaction MM02. Specify the **Material**, select the **Accounting 1** view, and specify the **Plant** as the organization unit. Then, press Enter to navigate to the **Accounting 1** view of the material master.

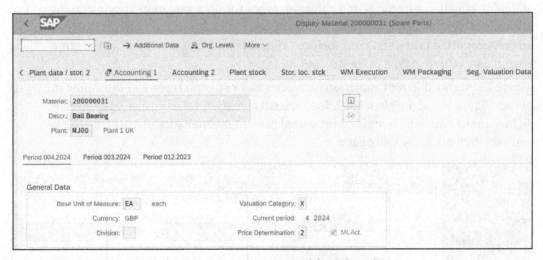

Figure 7.58 Assign Valuation Category to Material for Split Valuation

Once the valuation category is assigned to the material, the system allows the split valuation maintenance in the **Accounting 1** view of the material master by valuation type. You can maintain the split valuation by valuation type for the material. To do that, execute Transaction MM02, specify the material, select the **Accounting 1** view, and specify the plant as the organization unit and the valuation type. Then, press Enter. Repeat this step for every valuation type defined for the valuation category. Figure 7.59 and Figure 7.60 show the valuation price of 20 GBP defined for material **200000031 (Ball Bearing)** for the valuation category **X** and valuation type **C1 (NEW)**.

Figure 7.61 and Figure 7.62 show the valuation price of 15 GBP defined for the same material **200000031 (Ball Bearing)** for the valuation category **X** and valuation type **C2 (REFURBISHED)**.

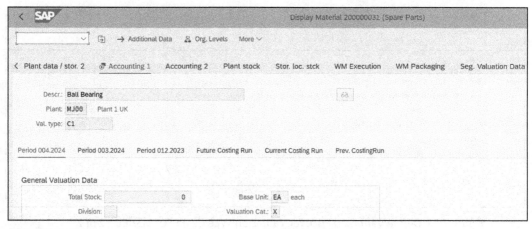

Figure 7.59 Define Valuation Price for Valuation Type C1 (NEW): 1 of 2

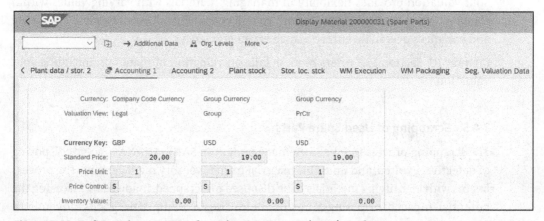

Figure 7.60 Define Valuation Price for Valuation Type C1 (NEW): 2 of 2

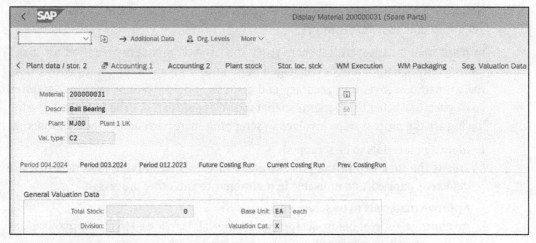

Figure 7.61 Define Valuation Price for Valuation Type C2 (REFURBISHED): 1 of 2

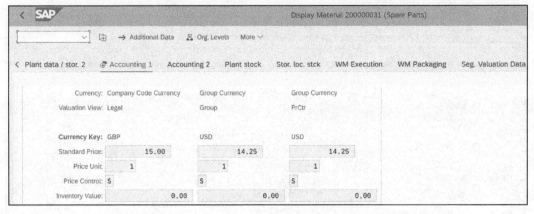

Figure 7.62 Define Valuation Price for Valuation Type C2 (REFURBISHED): 2 of 2

Split valuation provides flexibility in managing materials with varying values within the same plant. It allows organizations to reflect the actual inventory value accurately and enhances overall inventory management efficiency.

Chapter 10, Section 10.3 offers more details on valuation of materials, including split valuation.

7.4.5   Scrapping of Used Spare Parts

The scrapping process in inventory management in SAP S/4HANA involves disposing of defective, expired, and unusable material from inventory management. The process begins with identifying materials to be disposed of/scrapped, followed by assessing the potential impact of disposing of in-stock items and recording of the goods movements. Scrapping is a crucial part of inventory management, and it ensures that nondefective and usable materials and components are available in the inventory for different functions of the organization. It also helps organizations to manage waste effectively to prevent inaccuracies from inventory.

In plant maintenance, critical spare parts and other components required for maintaining critical equipment are managed in inventory management. It is highly essential to maintain inventory accuracy and dispose of defective, unusable, and expired spare parts and other components so that the new spare parts can be restocked. The following are the process steps involved in scrapping spare parts and other components:

1. **Identify materials to be scrapped**
   This is the first step of the process, and it involves identifying materials that are defective, expired, and unusable in maintenance and other processes.

2. **Approve materials to be scrapped**
   An approval process before scrapping of materials is followed in most organizations as the scrapping process reduces inventory levels.

3. **Scrapping of materials**

   Once necessary approvals are received, the defective, unusable, and expired spare parts and other components are withdrawn/issued from inventory to the scrapping process. In SAP S/4HANA, materials can be scrapped from unrestricted use stock, quality inspection stock, and blocked stock. The system records the goods issue to the scrapping process, and a material document and accounting document are created for every goods issue movement to the scrapping process. These documents are useful in the audit process as well. Upon goods issue to scrapping, the inventory levels are reduced, and this helps accurately report and track actual usable quantities of spare parts.

4. **Procure spare parts and other critical components**

   As scrapping of materials reduces inventory levels, maintenance planners can initiate a purchase requisition to procure critical spare parts and other components to bring the inventory levels up.

Goods issue for scrapping can be posted using Transaction MIGO. Table 7.10 shows the movement types used in scrapping of materials in inventory management. You can post the goods issue to scrapping using these movement types based on where the materials to be disposed are stored in the inventory management (i.e., unrestricted use stock, blocked stock, or quality inspection stock). Specify the material and quantity and select the relevant movement type to issue the materials for the scrapping process.

Movement Type	Detailed Description
551	Goods issue/withdrawal for scrapping from unrestricted use stock. This movement type is used to issue the expired/used/defective spare parts and other components used in plant maintenance from unrestricted use stock for scrapping process.
553	Goods issue/withdrawal for scrapping from quality inspection stock. This movement type is used to issue the expired/used/defective spare parts and other components used in plant maintenance from quality inspection stock for scrapping process.
555	Goods issue/withdrawal for scrapping from blocked stock. This movement type is used to issue the expired/used/defective spare parts and other components used in plant maintenance from blocked stock for scrapping process.

Table 7.10 Movement Types Used in Scrapping Process

7.4.6   Configuration of Movement Types

Refer to Chapter 4, Section 4.3.5 for the details of inventory movement type configuration. In this section, we'll explore how the plant maintenance function determines the movement types to generate a material reservation during maintenance order processing. Refer to Section 7.4.3 for more details about the material reservation process for maintenance orders.

To define movement types for material reservations for maintenance orders, execute Transaction SPRO and navigate to **Plant Maintenance and Customer Service • Maintenance and Service Processing • Maintenance and Service Orders • General Data • Define Movement Types for Material Reservations.**

Figure 7.63 shows the movement types for a maintenance order. Let's explore the relevant fields:

- **Package**
  Package **IW01** has been defined in the standard system for plant maintenance. It is an ABAP Workbench object, and similar objects are grouped together in a package.

- **Movement Types**
  For plant maintenance, only movement types **261** and **262** are relevant for material reservation. But do not delete or alter other movement types as these movement types are used in production planning.

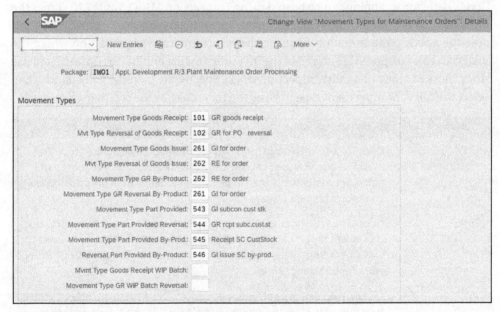

Figure 7.63 Define Movement Types for Material Reservation

7.5   Refurbishment

The refurbishment process involves restoring defective material to its original or usable condition, typically by performing repair operation or activity. This process is commonly used in plant maintenance to restore a piece of equipment and spare parts from a defective condition to their original condition. In SAP S/4HANA, a refurbishment order is created to initiate the refurbishment process for the defective material, followed by goods issue of the defective material to the refurbishment process. The maintenance technicians perform the repair operation and restore the material to its usable condition. This process may involve replacing certain components of the material or a piece of equipment, reassembly, and quality inspection. After the refurbishment process is successfully executed, the material will be received back into inventory. With the use of split valuation, refurbished materials can be valuated at a different price from when they were in new condition. Split valuation plays a vital role in the refurbishment process by providing accurate valuation of defective materials, refurbished materials, and new materials.

Let's explore the internal and external refurbishment processes in detail.

7.5.1 Internal Refurbishment Process

Internal refurbishment involves the refurbishment of faulty/defective material within the same organizational entity, typically within the same plant where the faulty material is located. Defective materials are withdrawn from the inventory for refurbishment and then returned back to the inventory after the refurbishment is completed. Internal refurbishment process starts with the creation of a refurbishment order with an internal operation.

Table 7.11 shows all the key subprocesses impacted in plant maintenance and materials management for internal refurbishment process.

Area	Processes Impacted	Description
Plant maintenance	Create and maintain refurbishment material/equipment	This process supports creation and maintenance of a refurbishment material/equipment with split valuation.
	Create refurbishment order with an internal operation	This process supports the creation of a refurbishment order with internal operation to refurbish a faulty/defective material.
	Planning and executing refurbishment internal	This process supports the execution of a refurbishment operation internally.
	Close refurbishment order	This process supports the settlement of a refurbishment order and business completion of the order.
	Asset disposal process	This process supports the disposal (scrapping) of the faulty/unusable/nonrefurbished materials.
Materials management	Withdrawal of faulty equipment/spare parts from inventory	This process supports the withdrawal of faulty equipment/spare parts from inventory for the internal refurbishment process.
	Return refurbished equipment/spare parts to inventory	This process supports the return of refurbished equipment/spare parts to inventory

Table 7.11 Process Impact: Internal Refurbishment

The prerequisite for the internal refurbishment process is to create a material master for equipment/spare parts with split valuation for new, refurbished, and defective valuation types. See Section 7.4.4 for more details.

Figure 7.64 shows the internal refurbishment process. Let's walk through the process steps:

1. **Create and process internal refurbishment order**
   The process starts with creation of a refurbishment order for the faulty/defective material to be refurbished. You can create the refurbishment order using Transaction IW81 and using the order type defined for refurbishment. Figure 7.65 shows refurbishment order **4000241** created for refurbishment of 1 quantity of material **200000031**.

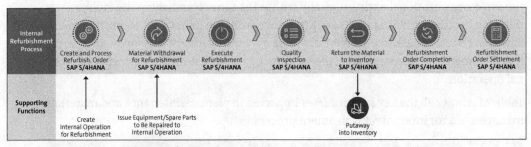

Figure 7.64 Internal Refurbishment Process

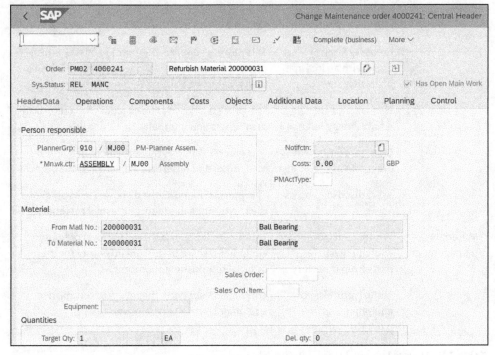

Figure 7.65 Internal Refurbishment Order: Header

Navigate to the **Operations** tab to arrive at the screen shown in Figure 7.66. You can see the internal operation for the refurbishment order with control key **PM01**. Control keys are defaulted based on the order types, but you can change them manually.

Next, let's review the **Components** tab to arrive at the screen shown in Figure 7.67. You can see component **200000031** with batch **C3**, which is the defective material. Material **200000031** was defaulted from the **HeaderData** tab of the order.

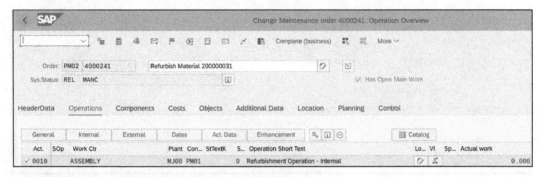

Figure 7.66 Internal Refurbishment Order: Operation

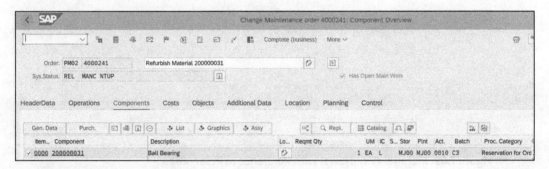

Figure 7.67 Internal Refurbishment Order: Component

2. **Material withdrawal for refurbishment**

   After the refurbishment order is created and released, you can find the reservation created automatically for the defective material. Figure 7.68 shows our example material **200000031** with valuation type **C3** (defective). You can see the material reservation number **1503** was generated automatically from the system for internal refurbishment with reference to order **4000241**.

   The next step is to issue the defective material to the internal refurbishment operation with reference to material reservation. You can use Transaction MB26 (Pick List) for posting goods issue against the material reservation, as shown in Figure 7.69.

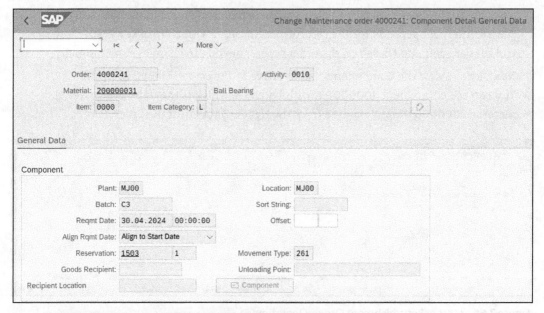

Figure 7.68 Internal Refurbishment Order: Material Reservation

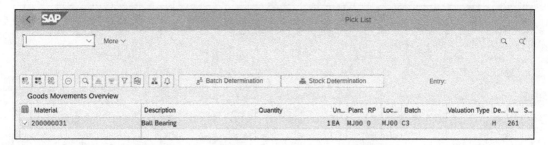

Figure 7.69 Material Withdrawal with Reference to Reservation

3. **Execute refurbishment process**
   Once the defective material is withdrawn from the inventory with reference to the reservation, internal technicians perform the refurbishment operation, during which the defective material will be refurbished.

4. **Quality inspection of refurbish material**
   The next step in the process is to perform quality inspection of refurbished material before returning it back to the inventory. During this process, material characteristics are measured, tested, and inspected using predefined methods and procedures to ensure the refurbished material meets required quality standards.

5. **Return the refurbished material to inventory**
   After the refurbishment process, the material is returned to inventory. The returned material will be valuated at a refurbished price using valuation type C2 (refurbish). To return the material back to inventory, use Transaction MIGO and movement type 262.

6. **Refurbish order completion**

   After the successful execution of a refurbishment process internally and after the refurbished material is returned to inventory, the refurbishment order is set to technically completed as well as business completion. You can set these statuses using Transaction IW32.

7. **Refurbishment order settlement**

   This is the final step of the refurbishment process, and the cost incurred during the refurbishment process will be settled. The system calculates the labor cost and material movement cost automatically. Depending on the valuation types of the refurbish material (C1, C2, and C3), and price control (S: standard price, V: moving average price) of the material set in the **Accounting 1** view of the material master, the system performs the settlement of the refurbishment order. If the price control is set as a moving average price, the settlement will be posted to material; if the price control is set to a standard price, the settlement will be posted to price difference account.

7.5.2   External Refurbishment Process

The external refurbishment process involves sending/issuing the faulty/defective material from inventory to the external vendor or service provider for refurbishment and receiving the refurbished material back into the inventory from the service provider after the refurbishment is completed. The external refurbishment process starts with the creation of a refurbishment order with an external operation.

Table 7.12 shows all the key subprocesses impacted from plant maintenance and materials management for the external refurbishment process.

Area	Processes Impacted	Description
Plant maintenance	Create and maintain refurbishment material/equipment	This process supports creation and maintenance of refurbishment material/equipment with split valuation.
	Create refurbishment order with an external operation	This process supports the creation of a refurbishment order with external operation to get a faulty/defective material refurbished from an external vendor.
	Close refurbishment order	This process supports the settlement of a refurbishment order and business completion of the order.
	Asset disposal process	This process supports the disposal (scrapping) of the faulty/unusable/nonrefurbished materials.

Table 7.12 Process Impact: External Refurbishment

Area	Processes Impacted	Description
Materials management	Sourcing materials and services	This process supports the sourcing for subcontracting vendor (service provider) for refurbishment of faulty equipment/spare parts.
	Create and maintain purchasing master data	This process creation and maintenance of purchasing master data such as subcontracting PIRs to support the external refurbishment process.
	Create and maintain subcontracting purchase requisition	This process supports the creation, generation, and maintenance of subcontracting purchase requisitions originating from the refurbishment order.
	Create and maintain subcontract order	This process supports the creation and maintenance of a subcontracting purchase order for external refurbishment of faulty equipment/spare parts.
	Provide equipment/spare parts to service provider	This process supports the issue of components to subcontractor for external operations.
	Post goods receipt	This process supports the goods receipt of a refurbished material from the service provider after the external refurbishment operation.
	Post component consumption	This process supports the automatic consumption of components (equipment/spare parts) provided to the service provider for external refurbishment operations.
	Supplier invoice management	This process supports the processing of incoming invoice from the service provider (supplier) for external refurbishment operations.

Table 7.12 Process Impact: External Refurbishment (Cont.)

Prerequisites for external refurbishment process are as follows:

1. Create a material master for equipment/spare parts with split valuation for new, refurbish, and defective valuation types. See Section 7.4.4.

2. Business partner (vendor) master data is required for the external service provider. See Chapter 3, Section 3.2.

3. Create and maintain purchasing master data for the service provider such as a subcontracting PIR. See Chapter 3, Section 3.3.

Figure 7.70 shows the external refurbishment process. Let's walk through the process steps:

1. **Create and process a refurbishment order with external operation**
   The process starts with creation of a refurbishment order for the faulty/defective material to be refurbished. You can create the refurbishment order using Transaction

IW81 and using the order type defined for refurbishment. The creation of refurbishment order is like internal refurbishment process explained in Section 7.5.1, but the control key **PM02** is used for the maintenance operation.

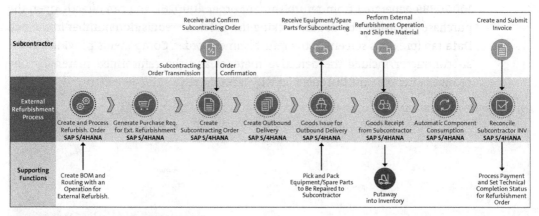

Figure 7.70 External Refurbishment Process

Figure 7.71 shows the external refurbishment operation. This is the only operation set up in the refurbishment order to refurbish defective material.

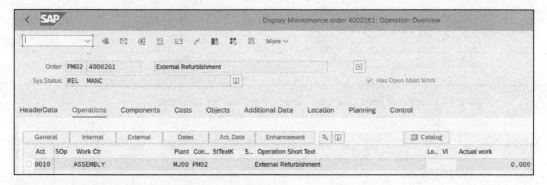

Figure 7.71 External Refurbishment Order: Operation

To generate a subcontracting purchase requisition for the external refurbishment operation, external processing parameters such as the subcontracting PIR are required to be maintained in the refurbishment order. Figure 7.72 shows purchase requisition **10000989** created for the operation with control key **PM02**. You can arrive at this screen by selecting the external refurbishment activity and clicking the **Act. Data** button from the **Operations** tab.

2. **Generation of subcontracting purchase requisition for external refurbishment**
   The system automatically generates subcontracting purchase requisitions for the refurbishment operation from the refurbishment order with account assignment category **F** (order) and item category **L** (subcontracting). The system assigns a subcontractor (supplier) as the source of supply from the external processing parameters. The

components will be added to the **Material Data** tab of the subcontracting purchase requisition automatically from the components assigned to the external operation from the refurbishment order. Figure 7.73 shows subcontracting purchase requisition **10000989** generated from refurbishment order **4000261**. You can directly open the purchase requisition by double-clicking the purchase requisition number in the **Act. Data** tab (previous screen) of the refurbishment order. Components provided to the subcontractor include the defective material and the refurbished material to be received after the refurbishment activity is on the requisition line item.

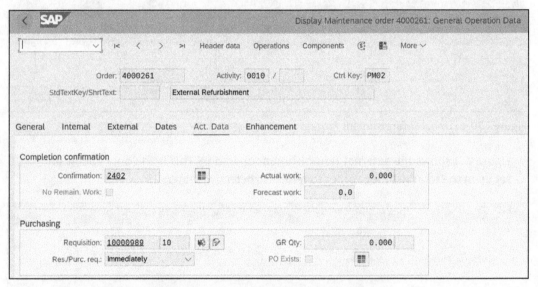

Figure 7.72 External Refurbishment Order: Purchase Requisition

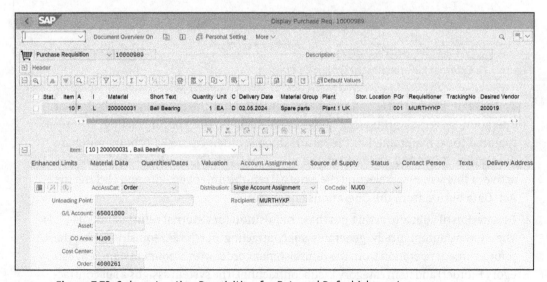

Figure 7.73 Subcontracting Requisition for External Refurbishment

3. **Convert subcontracting purchase requisitions into subcontracting order**

    The subcontracting purchase requisition will be converted into a subcontract order manually by the purchaser using Transaction ME21N or automatically using Transaction ME59. The line item details including components are defaulted into the purchase order. The service fee for the subcontracting operation is defaulted from the subcontracting PIR. The subcontract order will be transmitted to the subcontractor based on the output method: email, EDI, or electronic transmission to the supplier on SAP Business Network.

4. **Create outbound delivery with reference to subcontracting order**

    The next step in the process is to create an outbound delivery to issue the refurbishment material to the subcontractor/service provider. Outbound deliveries will be created using Transaction ME2O with reference to the subcontract order. Inventory and warehouse management activities such as picking of materials and packing activities are performed with reference to outbound deliveries.

5. **Post goods issue for outbound delivery**

    Goods issue will be posted with reference to the outbound delivery after the inventory management functions such as picking and packing of materials are completed. Movement types play an important role in determining the type of inventory movement and posting of goods issue. Movement type 541 is used for goods issue for outbound delivery to ship the components to the subcontractor. Movement type 541 ensures the defective material (component) provided to the subcontracting vendor is tracked and the ownership remains with the company requesting the external operation.

6. **Post goods receipt purchase order**

    The subcontractor/service provider receives the defective material and performs the refurbishment operation. The subcontractor ships the refurbished material. Once the shipment from the subcontractor arrives, the goods receipt will be posted manually in SAP S/4HANA with reference to the subcontract order using Transaction MIGO and movement type 101.

    Upon posting goods receipt, consumption of the components provided to the subcontractor will be posted using movement type 543 automatically. This goods movement will reduce the quantities of the defective material provided to the subcontractor from the inventory.

7. **Post logistics invoice verification**

    The subcontractor sends an invoice for the external refurbishment operation via email (paper invoice), via EDI, or electronically via SAP Business Network. The supplier invoice document will be posted by the accounts payable personnel manually in SAP S/4HANA using Transaction MIRO in the case of a paper invoice. Automatic posting and reconciliation of the invoice happens if an electronic invoice is received from the supplier. The next scheduled payment run will pick up this fully reconciled invoice and post the payment to the supplier based on the preferred payment method of the supplier. Accounts payable personnel can manually process the payment against the fully reconciled invoice as well.

7.5.3   Configuration of Refurbishment Process

In this section, let's explore the configuration settings for the refurbishment process within plant maintenance and materials management in SAP S/4HANA. Let's dive deep into the key configuration settings and understand the configuration settings that control the refurbishment process.

The procurement configuration settings defined in Section 7.3.6 are applicable to the refurbishment process as well. However, there are certain specific configurations applicable to the refurbishment process in addition to those settings.

Indicate Order Types for Refurbishment Processing

In this configuration, you can indicate the maintenance order types for refurbishment process. To create the refurbishment order, you must indicate one of the order types for refurbishment process and you can create a refurbishment order with that order type only.

To arrive at this configuration step, execute Transaction SPRO and navigate to **Plant Maintenance and Customer Service • Maintenance and Service Processing • Maintenance and Service Orders • Functions and Settings for Order Types • Indicate Order Types for Refurbishment Processing**.

Figure 7.74 shows the maintenance order types, and for order type **PM02**, the **Refurbishment order** indicator is set. Set the **Refurbishment order** indicator for the order types you want to use to create refurbishment orders.

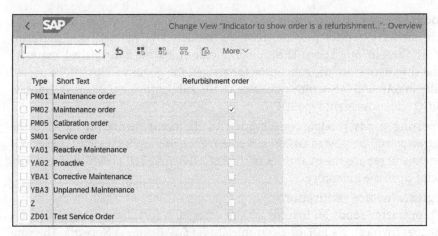

Figure 7.74 Indicate Order Types for Refurbishment Processing

Define Material Provision Indicators

The material provision indicator is assigned to the components in the BOM and refurbishment order. You must assign a material provision indicator to the component of the refurbishment order to provide the defective material to the external service provider (vendor). The system identifies the respective components for material provision.

To arrive at this configuration step, execute Transaction SPRO and navigate to **Plant Maintenance and Customer Service • Master Data in Plant Maintenance and Customer Service • Bills of Material • Item Data • Define Material Provision Indicators**.

Figure 7.75 shows the material provision indicators defined in the standard system. You can define a new one by clicking the **New Entries** button or copy an existing material provision indicator. Let's explore the field level settings:

- **MPCust (material to be provided by customer)**
  Set this indicator if the component is provided by the customer. For example, if you assign a material provision indicator with this indicator to the component of the refurbishment order, the required component to process the refurbishment order is provided by the customer.

- **MPVend (material to be provided by vendor)**
  Set this indicator if the component is provided by the vendor. For example, if you assign a material provision indicator with this indicator to the component of the refurbishment order, the required component to process the refurbishment order is provided by the vendor.

- **A&D MP (A&D-specific material provision)**
  Set this indicator to process refurbishment externally (subcontract) to support aerospace and defense functionalities.

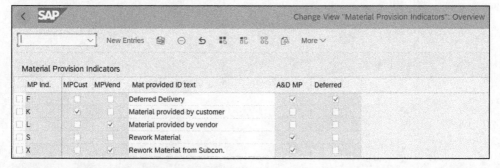

Figure 7.75 Define Material Provision Indicators

Refer to Chapter 6, Section 6.4.2 for additional configuration settings that support the external refurbishment process.

7.6 Summary

In this chapter, we introduced the integration of plant maintenance with materials management with best practice examples and processes. In Section 7.1, we explained the integrated plant maintenance process. In Section 7.2, we explained the master data in plant maintenance with step-by-step procedures to maintain plant maintenance-specific master in SAP S/4HANA. In Section 7.3, we explained the maintenance planning process including maintenance strategies, maintenance task lists, maintenance order processing, material planning for maintenance orders, and procurement of spare parts and other components required for maintenance activities. In Section 7.4, we explained the inventory management processes for plant maintenance and the material valuation process. Finally, in Section 7.5, we explained the internal refurbishment and external refurbishment processes for repair and rework of faulty materials.

In the next chapter, we'll discuss embedded extended warehouse management (EWM) and its integration with materials management in SAP S/4HANA. We'll explain how embedded EWM helps organizations to optimize inventory operations, improve stock visibility, and ensure efficient material handling of inbound material movements, outbound material movements, and internal warehouse movements with materials management integration.

Chapter 8
Extended Warehouse Management

Extended warehouse management (EWM) is an advanced warehouse management system that has evolved to meet today's warehousing needs. It provides a robust variety of tools and functionalities to manage inbound, complex warehouse operations and outbound processes. EWM and materials management are interconnected to ensure an effective flow of materials through the supply chain.

SAP Extended Warehouse Management (SAP EWM) has been embedded in SAP S/4HANA since 2016 and is now referred to as *embedded EWM*. The basic functions of a warehouse are to manage the receiving, storing, and issuing of goods. Today's warehousing requirements are complex and involve a high volume of goods going through distribution centers, aiming to ship goods to customers much faster than before. Embedded EWM is designed to manage highly complex warehouse operations, providing advanced capabilities to manage inbound and outbound processes. This new generation of SAP warehouse management offers flexibility and automation to manage various goods movements. It can handle complex supply chain networks that involve multiple warehouses and distribution centers with a high volume of materials that differ in size, weight, production and expiration dates, serial numbers, batches, country of origin, and so on.

Note

SAP also offers another deployment option for EWM, called *decentralized EWM*. It runs independently from the SAP S/4HANA (ERP) system and is deployed on a separate server. SAP S/4HANA (ERP) and decentralized EWM systems are integrated through standard interfaces. This option is a good fit for organizations with complex warehouse structures.

Materials management and embedded EWM integration ensures achieving greater supply chain efficiencies and control, which leads to reduced operational costs. This integration confirms deeper coordination between procurement, inventory management, and embedded EWM. Organizations will benefit from the real-time stock synchronization between embedded EWM and materials management after inbound processing, outbound processing, consumption postings, and physical inventory. This integration ensures real-time visibility of inventory and smooth material flow through the supply chain.

In this chapter, we'll dive into the cross-functional setup of embedded EWM and materials management in SAP S/4HANA. We'll start with the master data and then move on to the basics of setting up embedded EWM. We'll explore core processes like inbound and outbound delivery processing, including configuration instructions, and provide insight into internal warehouse movements and batch management.

8.1   Master Data in Extended Warehouse Management

Material master data is the central repository related to materials that contain all material information related to purchasing, sales, materials requirement planning, accounting, storage, quality management, and EWM. Refer to Chapter 3, Section 3.1 for more details. This section covers master data specific to embedded EWM. Master data plays a vital role in executing and warehouse operations flexibly, ensuring accurate inventory management and seamless integration with materials management. Let's walk through some of the key master data elements used in embedded EWM.

8.1.1   Supply Chain Units

A supply chain unit is the physical location that represents the physical address of the warehouse. A supply chain unit in embedded EWM can be a shipping office or a goods receiving office or both. Transaction /SCMB/SCUMAIN is used to create, change, and display supply chain units.

Figure 8.1 shows the initial screen of a supply chain unit. Enter a unique alphanumeric code of up to 20 digits for a supply chain unit. The **Type** field distinguishes different types of supply chain units. Production plant, distribution center, shipping point, warehouse, and so on are some of the supply chain unit types defined in the standard system. Type **1008** is defined for a warehouse.

Figure 8.1 Supply Chain Unit: Initial Screen

After entering the supply chain unit type, click the **Create** button to create a new supply chain unit. You can also choose to display or change the supply chain unit with the appropriate buttons.

Figure 8.2 shows the header data and **General** tab of the supply chain unit. Enter the **Description** of the supply chain unit in the header.

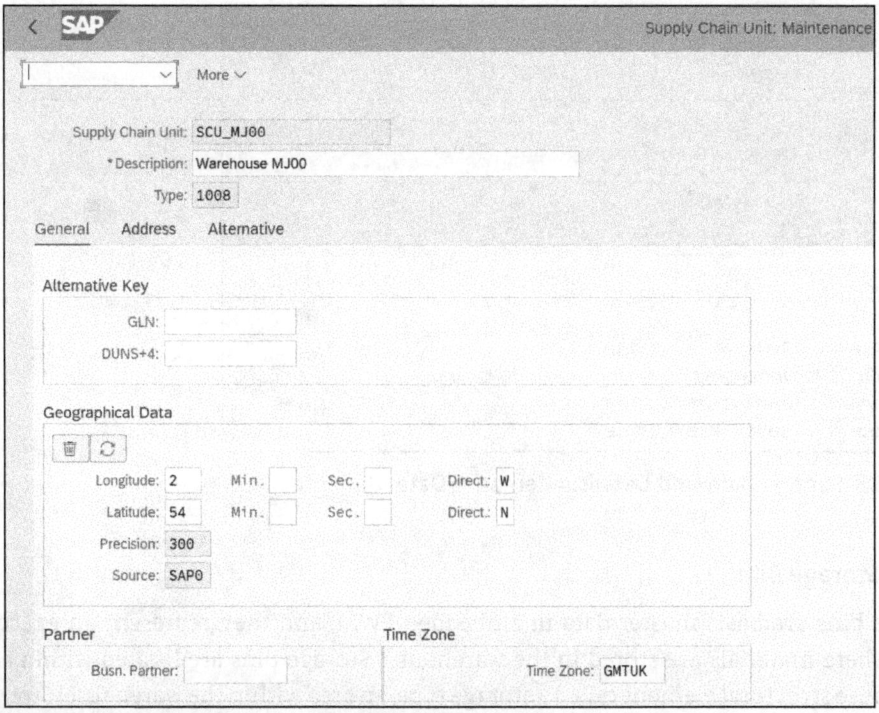

Figure 8.2 Supply Chain Unit Details: General Data

The **Alternative Key** section of the **General** tab has the following fields:

- **GLN**
  A global location number (GLN), also called an operational point, is a 13-digit global code used to identify locations both electronically and physically. Assign a GLN for the location of the supply chain unit.

- **DUNS+4**
  This is an alphanumeric key used to uniquely identify a physical location. Here it is used to uniquely identify the physical location of the warehouse. The Data Universal Numbering System (DUNS) number is a nine-digit number, and a four-digit number is added as suffix to identify an alternative location at the same physical address.

In the **Geographical Data** section, the system automatically selects **SAP0** in the **Source** field, which automatically derives the **Latitude** and **Longitude** information. Hence it is recommended to enter the supply chain unit address in the **Address** tab.

Figure 8.3 shows the **Alternative** tab of the supply chain unit. One or more business attributes are assigned to the supply chain unit in this tab.

Business attribute roles (**Bus.Attri.**) are predefined in the standard system. These are used to define the physical location or an organizational unit such as **INV** (warehouse), **RO** (goods receiving office), **SO** (shipping office), and so on.

Figure 8.3 Supply Chain Unit Details: Alternative Data

8.1.2   Storage Bins

Storage bins are basic master data in embedded EWM, and they represent an exact place where materials are stored in the warehouse. Storage bins are located within a warehouse structural element called a storage type, an area within the warehouse. Storage types control how the materials are placed into the storage bins and removed from the storage bins. The quantities of materials stored in a storage bin are called quants. Before defining storage bins for storage types, storage bin types, access types, bin structures, and so on must be defined in the configuration settings. We'll discuss storage bins in more detail in Section 8.2.5.

8.1.3   Warehouse Product Master

The warehouse product master is important master data in embedded EWM, and it is a replica of the material master from SAP S/4HANA. Replication of the material master as the warehouse product master in embedded EWM is required to perform warehouse activities using that product. After the replication, the warehouse product master is enriched with certain embedded EWM warehouse-related data. Material master data is created for plants assigned to the warehouse in SAP S/4HANA, and the same warehouse is mapped with an embedded EWM warehouse number. Replication of the material master as a warehouse product master in embedded EWM is required to perform warehouse activities. After the materials are replicated as products in embedded EWM, warehouse-dependent data can be updated to perform warehouse activities smoothly.

To create the warehouse product, execute Transaction /SCWM/MAT1. Figure 8.4 shows the initial screen for the creation/replication of a material as a warehouse product into embedded EWM. You can create, maintain, and display warehouse products in the same transaction.

The initial data has the following input fields:

- **Product Number**
  In this field, enter the material number from SAP S/4HANA created for the plant and storage locations that are mapped to the warehouse number integrated with the embedded EWM warehouse.
- **Warehouse No.**
  In this field, enter the embedded EWM warehouse number in which you want to create the warehouse product.
- **Party Entitled to Dispose**
  In this field, enter the business partner of the plant created as the default party entitled to dispose.

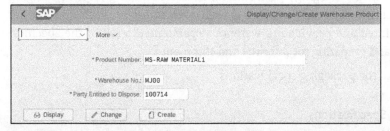

Figure 8.4 Create, Change, and Display Warehouse Product: Initial Screen

Click the **Create** button to arrive at Figure 8.5, which shows the warehouse product maintenance screen for product **MS-RAW MATERIAL1**. The warehouse product in embedded EWM is created with the same number as the material master from SAP S/4HANA.

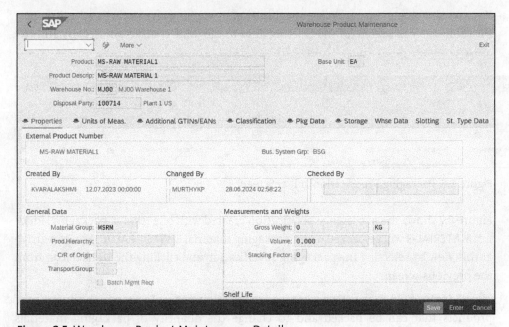

Figure 8.5 Warehouse Product Maintenance: Details

You can enrich the warehouse product by updating the packaging data, storage data, warehouse data, slotting data, and storage type data.

8.1.4   Packaging Specifications

A packaging specification is vital master data in embedded EWM that defines the different packaging levels for a product to putaway into the storage bin and for shipping the products out of the warehouse. It contains the detailed specification of a product quantity that can be packaged using certain packaging material. A product can be packed at multiple levels using different packaging materials at each level. A packaging specification contains information about the products that can be packed (you can specify multiple products in a packaging specification), all the packaging levels, packaging materials and quantities that can be packed at each level, and so on. Packaging specifications can be printed and provided to warehouse personnel to follow the specifications for packing and preparing for putaway and shipment.

Let's dive deep into the packaging specifications.

Create Packaging Specifications

To create a packaging specification, execute Transaction /SCWM/PACKSPEC. Figure 8.6 shows the initial screen for the creation of the packaging specification. Enter the packaging specification group (**PS Group**) and **Description** to initiate the creation of the packaging specification. Upon clicking the **Create** icon after specifying the **PS Group**, the system automatically assigns an internal number from the number range.

You can also create the packaging specification by copying an existing packaging specification. You can search for the existing ones in the same transaction to copy and create a new one.

Figure 8.6 Packaging Specification: Initial Screen

Figure 8.7 shows the packaging specification created with one level of packaging. Product **MATERIAL-3** will be packed using packaging material **EWMS4-PAL00**. You can arrive at this view by selecting the packaging specification and clicking the **Details** icon from the previous screen.

The **PS Status** field shows the packaging specification status. To change the quantity of products that can be packed and to change the quantity of packaging materials required, double-click the product and packaging material from the structural display on the left-hand side. Assign the supply chain unit (**SC Unit**) and production supply area (**PSA**) belonging to the embedded EWM warehouse in the **Determination** tab.

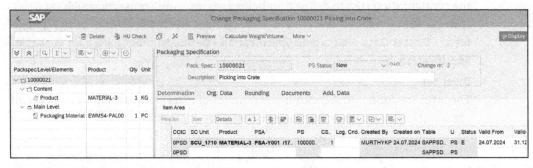

Figure 8.7 Packaging Specification: Details

Once all the updates are done, activate the packaging specification by clicking the **Activate** button in the menu bar.

Define Packaging Specification Group

In this configuration step, we'll define a packaging specification group for creating packaging specifications.

To arrive at the configuration step to define a packaging specification group, execute Transaction SPRO and navigate to **SCM Extended Warehouse Management • Extended Warehouse Management • Master Data • Packaging Specification • Define Packaging Specification Group**.

Figure 8.8 shows standard packaging specification group **SAP**. Click **New Entries** to define a new packaging specification group. Specify a four-digit alphanumeric key and description to define a new packaging specification group.

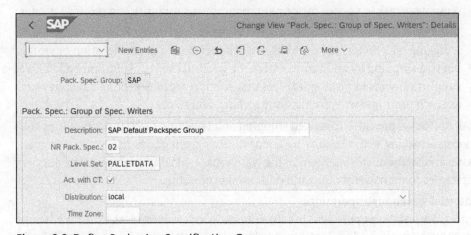

Figure 8.8 Define Packaging Specification Group

The following are the key fields of the packaging specification group:

- **NR Pack. Spec.**
  Assign a number range defined for packaging specifications in this field. To arrive at the configuration step to define number ranges, execute Transaction SPRO and navigate to **SCM Extended Warehouse Management • Extended Warehouse Management • Master Data • Packaging Specification • Define Number Range for Packaging Specification**.

- **Level Set**
  The level set defines and controls the number of level structures for the packaging specifications. You can define the level set in the configuration.

8.2   Basics of Extended Warehouse Management

It is always challenging to keep the optimum inventory levels at all times while handling a high volume of inventory transactions. Embedded EWM provides advanced capabilities to effectively manage and optimize inventory levels. The following are the key functionalities and features of embedded EWM, as illustrated in Figure 8.9:

- **Warehouse structure**
  Complex warehouse structures can be defined in embedded EWM to manage inbound and outbound logistics. The warehouse structure consists of storage types, storage sections, and storage bins. These structural elements are organized in a hierarchical manner to support inbound, outbound, and inventory control processes. It also supports the creation of both physical and logical locations within the warehouse.

- **Inbound and outbound processing**
  Embedded EWM manages goods receipt and goods issue processes effectively and creates the following tasks to manage inbound and outbound processing:
  - A putaway task is a goods movement that moves a quantity of a received product from a source location to a destination putaway bin based on putaway strategies.
  - A picking task directs the removal of a product from a source location or storage bin to a shipping location from where the goods issue is posted. Embedded EWM supports advanced picking features that support various picking strategies such as batch picking, wave picking, zone picking, and so on.

  SAP S/4HANA provides advanced shipping and receiving (ASR) functionalities such as cross-docking—that is, moving goods directly from goods receipt area to goods issue area without storing them in the warehouse—picking strategies, and putaway strategies to manage inbound and outbound processing.

- **Internal warehouse operations**
  Internal warehouse operations involve internal goods movements, transfer postings, inventory counts, and replenishment. Embedded EWM provides functionalities for

real-time inventory tracking and inventory control activities. Internal warehouse movements involve movement of stock from one storage bin to another. Physical inventory and cycle counting are performed in the warehouse to maintain inventory accuracy.

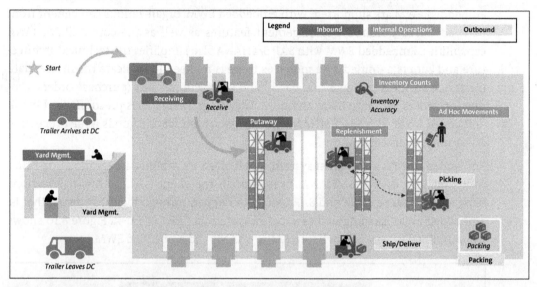

Figure 8.9 Extended Warehouse Management: Key Functionalities

- **Value-added services**
  Embedded EWM supports a cross-docking process that involves moving goods directly from the goods receipt area to goods issue area without storing them in the warehouse. This will save effort and accelerate the order fulfillment process. It supports value-added services such as kitting, labeling, and assembly.

- **Resource management**
  Embedded EWM manages human resources (labor), equipment, and other resources. Task assignment to resources is supported in embedded EWM and ensures optimized resource utilization. Tasks can be either manually assigned to resources or can be automated in embedded EWM. This provides tools and functionalities to monitor task execution by resource. Resource management maximizes the overall efficiency of the resources.

- **Integration with other functions**
  Embedded EWM integrates with materials management, which is the focus of this chapter. In addition, it integrates with embedded transportation management (TM), quality management, production planning, plant maintenance, and so on.

In the following sections, we'll walk through the step-by-step instructions to get embedded EWM up and running. We'll perform setup activities in both the SAP S/4HANA (ERP) system and embedded EWM, and then set up core structural elements for the warehouse.

8.2.1   Integration of Embedded EWM with SAP S/4HANA

One of the main differentiators of SAP S/4HANA is that the platform is equipped with embedded applications such as EWM to improve supply chain efficiencies and to enhance user experience. In SAP S/4HANA, both basic warehouse management and embedded EWM are supported. With embedded EWM, organizations can benefit from both advanced warehouse management features as well as the core SAP S/4HANA capabilities. Embedded EWM with SAP S/4HANA also simplifies the technical architecture and adopts a unified data model as there is no need to replicate the master data (material, business partners, and batch) and transactional data (purchase orders, production orders, etc.) into EWM anymore. The integration layer is preconfigured in the embedded EWM with SAP S/4HANA, and so goods receipt and goods issue processing are seamlessly integrated.

Embedded EWM is a one-client system, and it offers a simplified integration into other functionalities of SAP S/4HANA. Even though the client is the same, transactions between embedded EWM and SAP S/4HANA happen through logical systems; that is, embedded EWM is configured as a logical system in SAP S/4HANA. Figure 8.10 shows the simplified integration between SAP S/4HANA and embedded EWM.

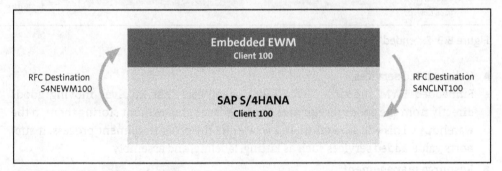

Figure 8.10 Embedded EWM and SAP S/4HANA Integration

Because of this simplified integration using logical systems, a warehouse number must be created in SAP S/4HANA as an organization unit and in embedded EWM configurations inside SAP S/4HANA.

Let's explore the basic settings required for the integration between SAP S/4HANA and embedded EWM, including the logical system assignment and RFC destination setup.

Name Logical System

In this configuration step, you define the logical systems for embedded EWM integration with SAP S/4HANA. For the distribution model to work, it is required to define logical systems for both SAP S/4HANA and embedded EWM in the same client.

To arrive at this configuration step, execute Transaction SPRO and navigate to **Integration with Other SAP Components • Extended Warehouse Management • Basic Settings for Setting Up the System Landscape • Name Logical System**.

Figure 8.11 shows the logical systems **S4NCLNT100** and **S4NEWM100** defined for SAP S/4HANA and embedded EWM respectively for client 100. You can define new logical systems by clicking **New Entries**. Enter a logical system ID and name to define a logical system.

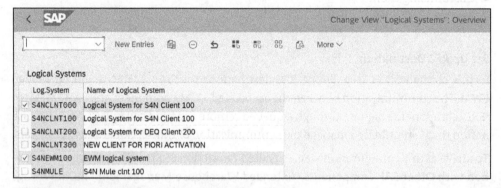

Figure 8.11 Logical Systems for SAP S/4HANA and EWM

Assign Logical System to a Client

In this configuration step, assign the logical system defined for SAP S/4HANA to the client where the business processes are to be executed.

To arrive at this configuration step, execute Transaction SPRO and navigate to **Integration with Other SAP Components • Extended Warehouse Management • Basic Settings for Setting Up the System Landscape • Assign Logical System to a Client**.

As shown in Figure 8.12, enter a **Logical system** (**S4NCLNT100** in this example) to assign to client **100**. The standard currency of the client is also required, which you can enter in the **Currency** field.

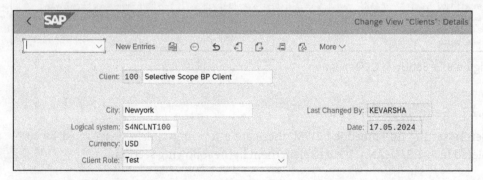

Figure 8.12 Assign Logical System to Client

The **Client Role** defines the environment in which the client is running. The following are the standard roles available for the client:

- Production client
- Test client
- Training client
- Demo client
- Customizing client
- SAP reference client

Set Up RFC Destination

In this configuration step, define RFC destinations for SAP S/4HANA and embedded EWM. The communication between the embedded EWM component and SAP S/4HANA (ERP) components happen through a queued remote function call (qRFC). An RFC destination thus is required to manage the communication of transactional and master data.

To arrive at this configuration step, execute Transaction SPRO and navigate to **Integration with Other SAP Components • Extended Warehouse Management • Basic Settings for Setting Up the System Landscape • Set Up RFC Destination**. To set up a new RFC destination, click the **Create** icon.

Figure 8.13 shows ABAP connections **S4NCLNT100** and **S4NEWM100** defined as RFC destinations.

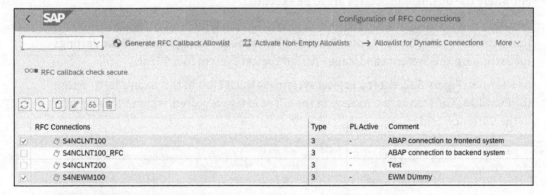

Figure 8.13 Set Up RFC Destination

8.2.2 Setup in SAP S/4HANA

Before setting up embedded EWM, there are a few prerequisites that must be completed in SAP S/4HANA. We'll explain them in the following sections.

Warehouse Number in SAP S/4HANA

A warehouse number is required to be defined in the SAP S/4HANA (ERP) system as an organizational unit. On the embedded EWM side, the SAP S/4HANA warehouse number and the corresponding embedded EWM warehouse numbers are mapped to establish the linkage so that you can perform transactions between SAP S/4HANA's core ERP system and embedded EWM. We'll explain the warehouse number setup for embedded EWM in Section 8.2.3.

Definition

To arrive at the configuration step to define a warehouse number in SAP S/4HANA, execute Transaction SPRO and navigate to **Enterprise Structure • Definition • Logistics Execution • Define Warehouse Number**.

Figure 8.14 displays warehouse number **MJ0**. Click **New Entries** to maintain a new warehouse number or copy an existing warehouse number to create a new one. Define a three-character unique alphanumeric or numeric key to define a warehouse number.

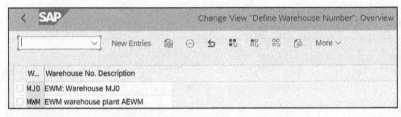

Figure 8.14 Define Warehouse Number in SAP S/4HANA.

Assignment

Warehouses are the physical locations within a plant. A plant can have multiple warehouses, and a warehouse number is assigned to the combination of plant and storage location in SAP S/4HANA.

To assign a warehouse number to the combination of plant and storage location, execute Transaction SPRO and navigate to **Enterprise Structure • Assignment • Logistics Execution • Assign Warehouse Number to Plant and Storage Location**.

To assign a warehouse number to the combination of plant and storage location, click **New Entries** and specify the plant, storage location, and warehouse number. Figure 8.15 shows that warehouse number **MJ0** is assigned to the combination of plant **MJ00** and storage location **MJ00**.

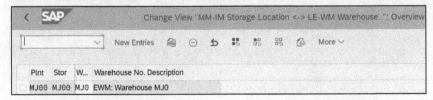

Figure 8.15 Assign Warehouse Number to Plant and Storage Location

Define Shipping Point and Assign it to Plant in SAP S/4HANA

A shipping point is a location within the plant where goods are loaded to the transportation vehicles to ship the goods to internal and external customers. It represents a final stage in outbound logistics from where the goods issue movements are posted in SAP S/4HANA after the picking and packing operations. It is the highest-level organizational unit in shipping. Refer to Chapter 2, Section 2.6.2 for more details on how to create a shipping point, the attributes of a shipping point, and assignment of a shipping point to a plant.

Figure 8.16 shows that shipping point **MJ00** was assigned to plant **MJ00**. Assign a shipping point to the same plant assigned to the warehouse in SAP S/4HANA.

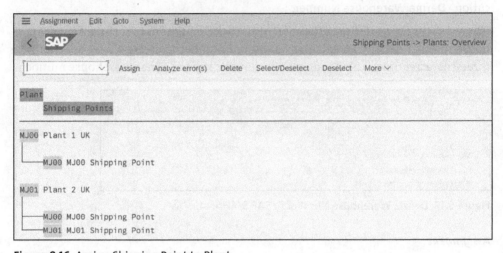

Figure 8.16 Assign Shipping Point to Plant

A goods receiving point is also represented by shipping point. In this configuration step, assign the shipping point defined for a goods receiving point to the combination of a plant and storage location. To arrive at this configuration step, execute Transaction SPRO and navigate to **Logistics Execution • Shipping • Basic Shipping Functions • Assign Goods Receiving Points for Inbound Deliveries**.

To assign the shipping point as a goods receiving point, click **New Entries** and specify the plant, storage location, and shipping point. Figure 8.17 shows that shipping point **MJ00** was assigned to the combination of plant **MJ00** and storage location **MJ00**.

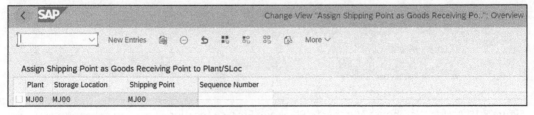

Figure 8.17 Assign Shipping Point as Goods Receiving Point

8.2.3 Setup in Embedded EWM

A warehouse number is the highest-level organizational unit in embedded EWM and manages inbound inventory, outbound inventory, and internal warehouse activities. Warehouses play a vital role in supply chain management, and important operations such as receiving of goods, storing, picking, packing, and shipping of goods are managed in the warehouse. Each warehouse is identified by a unique warehouse number in embedded EWM. Within a warehouse, there are storage areas, staging areas for inbound processing, and staging areas for outbound processing of goods. There must be a link between the embedded EWM warehouse number and the corresponding SAP S/4HANA warehouse number.

In the following sections, we'll walk through the instructions to set up the warehouse number in embedded EWM.

Define Warehouse Numbers

To define a warehouse number in embedded EWM, execute Transaction SPRO and navigate to **SCM Extended Warehouse Management • Extended Warehouse Management • Master Data • Define Warehouse Numbers**.

Figure 8.18 shows warehouse number **MJ00** created as an embedded EWM warehouse number. Click **New Entries** to create a warehouse number. Enter a four-digit alphanumeric key to define a warehouse number and enter a description for it.

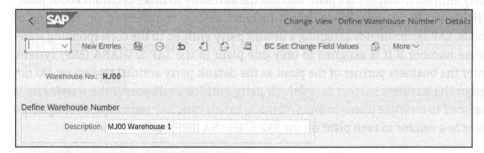

Figure 8.18 Define Warehouse Number in EWM

Assign Warehouse Numbers

In this configuration step, the embedded EWM warehouse numbers are assigned to the supply chain units and a custodian is created as a business partner and default party entitled to dispose.

To assign supply chain units and business partners to the embedded EWM warehouse number, execute Transaction SPRO and navigate to **SCM Extended Warehouse Management • Extended Warehouse Management • Master Data • Assign Warehouse Numbers**.

Figure 8.22 shows that SAP S/4HANA warehouse number **MJO** has been enabled for embedded EWM. When you create a warehouse number in SAP S/4HANA (ERP), it will automatically create an entry in the embedded EWM system overview configuration table. Here you need to change the **Ext. WM** dropdown value to **EWM (Extended Warehouse Management)**.

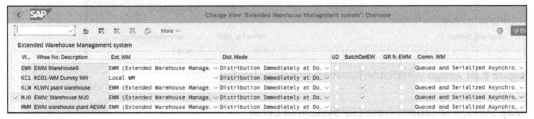

Figure 8.22 Configure EWM-Specific Parameters for SAP S/4HANA Warehouse

The following are the other key fields of this configuration step. These settings are available only if you have set the value for the **Ext. WM** field to **EWM (Extended Warehouse Management)**:

- **UD (unchecked delivery)**
  If you set this indicator, unchecked deliveries are distributed from the SAP S/4HANA (ERP) system for planning purposes in embedded EWM. An unchecked delivery is a delivery with a reduced scope of availability check. Unchecked deliveries are triggered by purchase orders at the time of creation and changing of purchase orders. No ATP check happens for unchecked deliveries. Unchecked deliveries are converted into deliveries when all the necessary availability checks are performed.

- **BatchDetEWM (batch determination in EWM)**
  If you set this indicator, the batch determination criteria will be sent to embedded EWM from the SAP S/4HANA (ERP) system so that batch determination for deliveries happens in embedded EWM system while taking the warehouse layout into account.

- **GR fr. EWM (goods receipt from EWM only)**
  If you set this indicator, the inbound deliveries for goods receipt for process orders and production orders can be created in the embedded EWM system only. But if you don't set this indicator, the inbound deliveries for goods receipt for process orders and production orders can be created in both the SAP S/4HANA (ERP) and embedded EWM systems at the same time in parallel, and this could result in creating overdeliveries for an order.

Leave the other field values set to the default values.

Define Warehouse Number Control

In this configuration step, you can define various controlling parameters and basic data for the embedded EWM warehouse number. To define basic data and control parameters for the embedded EWM warehouse number, execute Transaction SPRO and navigate to **SCM Extended Warehouse Management** • **Extended Warehouse Management** • **Master Data** • **Define Warehouse Number Control**.

The system automatically displays the all the embedded EWM warehouse numbers in the overview screen. From the overview screen, double-click the warehouse number for which you want to define basic data and control parameters, or select the embedded EWM warehouse number and click the **Details** icon, or press Ctrl + Shift + F2 .

Figure 8.23 shows of the first part of the warehouse number control data for embedded EWM warehouse number **MJ00**. The **Define Warehouse Number Control** section has the following fields:

- **Weight Unit**
  Weight unit is required to control the maximum weight allowed in a storage bin that belongs to this warehouse. You can enter the weight unit as **KG** (kilograms), **LB** (pounds), **G** (grams), **TON** (tons), and so on.
- **Volume Unit**
  A volume unit is required to control the maximum volume allowed in a storage bin that belongs to this warehouse. You can enter volume unit as **L** (liters), **M3** (cubic meters), **GAL** (gallons), **YD3** (cubic yards), and so on.
- **Time Unit**
  A time unit is required to calculate the time for various warehouse operations within the warehouse. You can enter time unit as **MIN** (minutes), **D** (days), **HR** (hours), **S** (seconds), and so on.

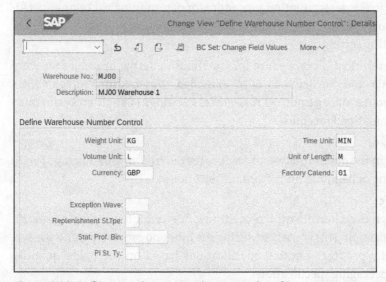

Figure 8.23 Define Warehouse Number Control: 1 of 2

- **Unit of Length**
  A length unit is required to calculate distances within the warehouse. You can enter unit of length as **M** (meters), **MI** (miles), **YD** (yards), **FT** (feet), and so on.

- **Currency**
  The currency of the warehouse is required to calculate resource costs: costs for full-time employees operating the warehouse, tolerance value checks during physical inventory, and so on. The currency of the warehouse will come in handy when the embedded EWM warehouse manages the stock from plants belonging to multiple countries and company codes. Enter a desired currency code in this field.

- **Factory Calend.**
  The factory calendar controls the warehouse operating dates, holidays, and so on. Enter the factory calendar for the warehouse.

- **Replenishment St.Tpe (replenishment stock type)**
  Usually, replenishment of a material happens from unrestricted use stock. In this field, enter the stock types defined for the embedded EWM warehouse number in configuration for replenishment purposes.

- **PI St. Ty. (physical inventory stock type)**
  The physical inventory stock type is a default stock type used during the physical inventory process. You can define one default stock type per inventory control area.

Figure 8.24 shows the second part of the warehouse number control data for embedded EWM warehouse number **MJ00**. The **Determination Procedures for Condition Technique** section has the following fields:

- **Proced. Whse-Int. Proc.**
  In this field, assign the determination procedure for determining packaging specifications for internal warehouse operations. The determination procedure is defined in the configuration settings.

  To define the determination procedure for warehouse-internal processes, execute Transaction SPRO and navigate to **SCM Extended Warehouse Management • Extended Warehouse Management • Cross-Process Settings • Condition Technique • Maintain Determination Procedure**.

- **Pick-HU Procedure**
  In this field, assign the determination procedure for determining packaging specifications for picking of handling units from the warehouse.

- **Palletization Proc.**
  In this field, assign the determination procedure for determining packaging specifications for palletization. This is enabled during the inbound processing, and the system automatically proposes packing specifications based on the pallet quantity specified in the packaging specifications.

- **PhysInv Print. Proc.**
  In this field, assign the determination procedure for determining packaging specifications for physical inventory. This is enabled during the inbound processing when printing a physical inventory document.

- **Receive HUs Proc.**
  In this field, assign the determination procedure for determining packaging specifications for received handling units during inbound processing.

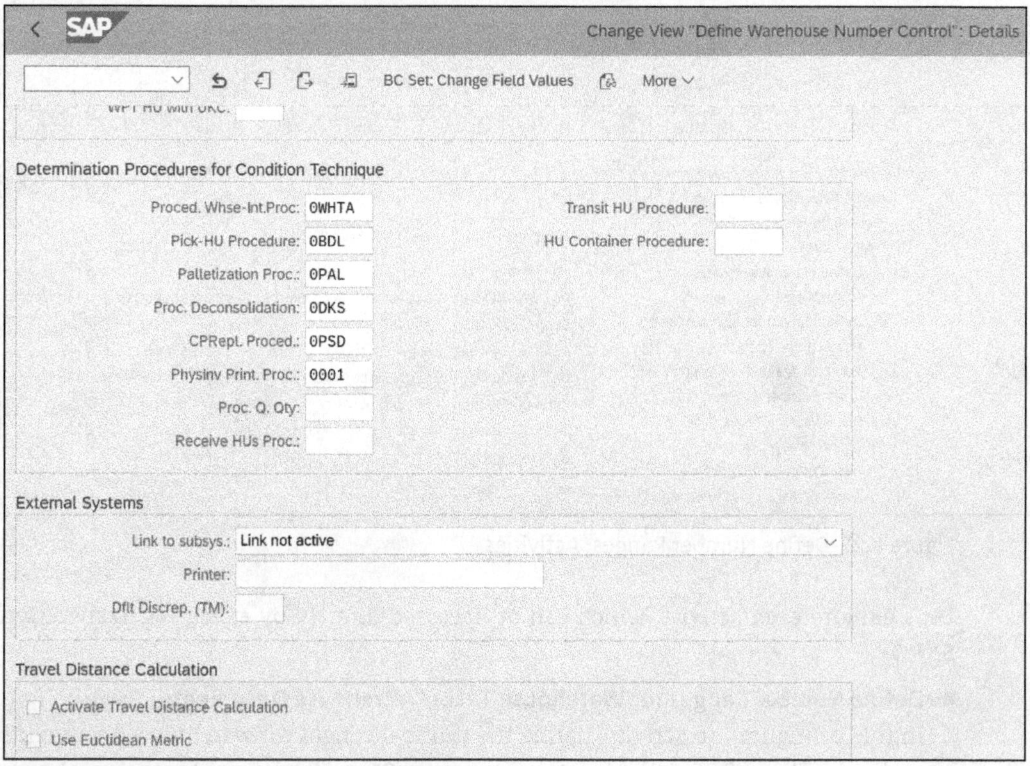

Figure 8.24 Define Warehouse Number Control: 2 of 2

Define Number Ranges and Assign Them to EWM Warehouse Number

In this configuration step, you can define number range intervals for the following:

- Warehouse tasks and warehouse documents
- Waves
- Warehouse orders
- Physical inventory documents
- Value-added services

To define number ranges and assign number range intervals to an embedded EWM warehouse number, execute Transaction SPRO and navigate to **SCM Extended Warehouse Management • Extended Warehouse Management • Master Data • Define Number Ranges**.

Figure 8.25 shows the activities for defining number ranges for the embedded EWM warehouse number. Double-click each activity to define the number ranges for the embedded EWM warehouse number.

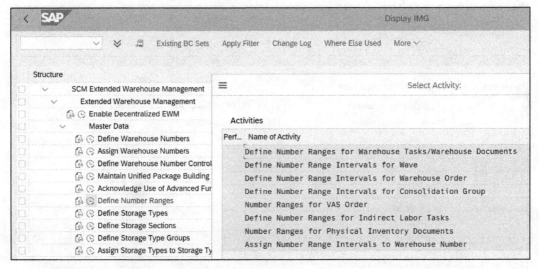

Figure 8.25 Define Number Ranges: Activities

Let's explore each activity, which can be accessed directly by executing Transaction SNRO:

- **Define Number Ranges for Warehouse Tasks/Warehouse Documents**
  In this configuration activity, define the number ranges for warehouse documents and tasks. Figure 8.26 shows the number ranges **01** and **02** defined for the warehouse number **MJ00**.

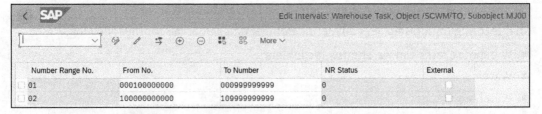

Figure 8.26 Define Number Range for Warehouse Tasks and Warehouse Documents

In this configuration step, both internal and external number ranges can be defined with reference to number range object /SCWM/TO.

- **Define Number Range Intervals for Waves**
  In this configuration activity, define the number ranges for waves for the embedded EWM warehouse number.

  In this configuration step, both internal and external number ranges can be defined with reference to number range object /SCWM/WAVE.

- **Define Number Range Intervals for Warehouse Order**
  In this configuration activity, define the number ranges for warehouse order for the embedded EWM warehouse number.

  In this configuration step, both internal and external number ranges can be defined with reference to number range object /SCWM/WHO.

- **Number Ranges for VAS Order**
  In this configuration activity, define the number ranges for value added services for the embedded EWM warehouse number.

  In this configuration step, both internal and external number ranges can be defined with reference to number range object /SCWM/VASO.

- **Number Ranges for Physical Inventory Documents**
  In this configuration activity, define the number ranges for physical inventory documents for the embedded EWM warehouse number.

  In this configuration step, both internal and external number ranges can be defined with reference to number range object /SCWM/PIDO.

- **Assign Number Range Intervals to Warehouse Number**
  In this configuration activity, you can assign the number range intervals to the embedded EWM warehouse number for warehouse tasks, warehouse documents, waves, warehouse orders, and so on.

Figure 8.27 shows that number range intervals were assigned to embedded EWM warehouse number **MJ00** for different documents.

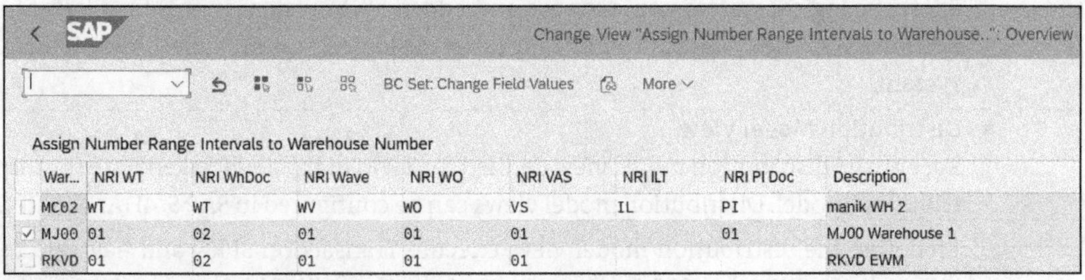

War...	NRI WT	NRI WhDoc	NRI Wave	NRI WO	NRI VAS	NRI ILT	NRI PI Doc	Description
MC02	WT	WT	WV	WO	VS	IL	PI	manik WH 2
MJ00	01	02	01	01	01		01	MJ00 Warehouse 1
RKVD	01	02	01	01	01			RKVD EWM

Figure 8.27 Assign Number Range Intervals to Warehouse Number

8.2.4 Generate Distribution Model for Transaction Data Transfer to Embedded EWM

To enable transaction and master data flow from SAP S/4HANA (ERP) to embedded EWM, an Application Link Enabling (ALE) distribution model is required. Inbound deliveries and outbound deliveries are created in the SAP S/4HANA system, and warehouse tasks for putaway (inbound processing) and picking (outbound processing) happen in the embedded EWM warehouse. Hence the inbound deliveries and outbound deliveries are required to be distributed to embedded EWM. Without the generation of a distribution model from SAP S/4HANA to embedded EWM, distribution of deliveries is not possible. Figure 8.28 illustrates the distribution of deliveries from SAP S/4HANA to embedded EWM using the distribution model.

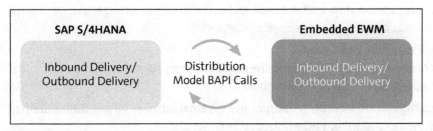

Figure 8.28 Distribution Model from SAP S/4HANA to Embedded EWM

To generate a distribution model from SAP S/4HANA to embedded EWM, execute Transaction SPRO and navigate to **Integration with Other SAP Components • Extended Warehouse Management • Basic Settings for EWM Linkage • Generate Distribution Model for Transaction Data Transfer to SAP EWM**.

Figure 8.29 shows the initial screen. Enter the following details and execute to generate the distribution model:

- **Warehouse Number**
  Enter the warehouse number from the SAP S/4HANA (ERP) system that will be integrated with embedded EWM in this field.

- **Logical System of SAP EWM**
  Enter the logical system name for the embedded EWM system in this field (refer to Section 8.2.1). The system automatically takes the logical system of the SAP S/4HANA system.

- **Distribution Model View**
  Specify the distribution model view defined for EWM in this field to generate the distribution model. Distribution model views can be configured in SAP S/4HANA (ERP).

  To define the distribution model view, execute Transaction SPRO and navigate to **Integration with Other SAP Components • Integration with SAP Cloud for Customer • Communication Setup • Manually Adjust Integration Settings for Data Exchange • Maintain Distribution Model**.

- **Objects**
  Select the objects to be distributed via the distribution model in this field. You can select **Inbound Delivery**, **Outbound Delivery**, **Production Material Request**, or **All**.

Next, click the **Execute** button at the bottom of the screen to generate the distribution model. Figure 8.30 shows the distribution model generated for SAP S/4HANA (ERP) warehouse **MJ0** to distribute the selected objects from SAP S/4HANA (logical system name **S4NCLNT100**) to embedded EWM (logical system name **S4NEWM100**).

Generate Distribution Model from SAP ERP to SAP EWM
Save as Variant...
More
Warehouse Number: MJ0
Logical System of SAP EWM: S4NEWM100
Distribution Model View: EWM
Objects
Inbound Delivery:
Outbound Delivery:
Production Material Request:
All:
Action
Create Entries:
Check Entries:

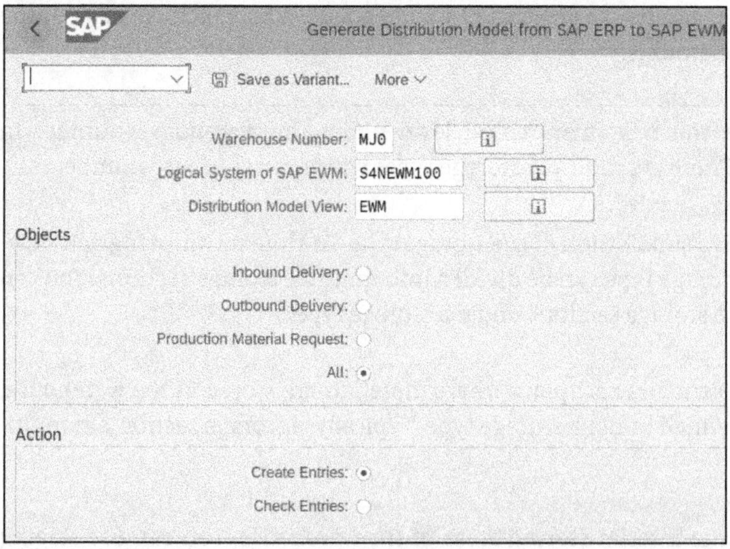

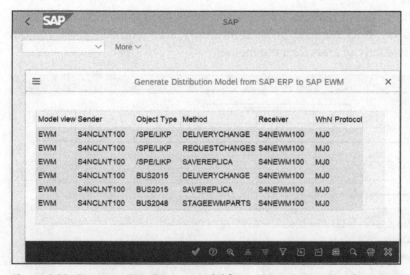

Figure 8.29 Generate Distribution Model from ERP to EWM: Initial Screen

SAP
More
Generate Distribution Model from SAP ERP to SAP EWM
Model view Sender Object Type Method Receiver WhN Protocol
EWM S4NCLNT100 /SPE/LIKP DELIVERYCHANGE S4NEWM100 MJ0
EWM S4NCLNT100 /SPE/LIKP REQUESTCHANGES S4NEWM100 MJ0
EWM S4NCLNT100 /SPE/LIKP SAVEREPLICA S4NEWM100 MJ0
EWM S4NCLNT100 BUS2015 DELIVERYCHANGE S4NEWM100 MJ0
EWM S4NCLNT100 BUS2015 SAVEREPLICA S4NEWM100 MJ0
EWM S4NCLNT100 BUS2048 STAGEEWMPARTS S4NEWM100 MJ0

Figure 8.30 Generate Distribution Model from ERP to EWM: Details

8.2.5 Warehouse Structural Elements

Embedded EWM is structured based on organizational elements or structural elements like any other warehouse management system. In embedded EWM, the setup of warehouse structural elements starts with the warehouse number creation (refer to Section 8.2.3). You can set up one or more warehouse numbers in embedded EWM. Every warehouse number has its own structure.

Figure 8.31 shows the warehouse structure in embedded EWM, which consists of the following structural elements:

- **Storage type**
  Storage types are the next hierarchical level under the warehouse number in embedded EWM. There are multiple storage types within a warehouse number.

- **Storage section**
  A storage section is a subdivision of the storage type. To store fast-moving and slow-moving items, a storage type can be divided into multiple storage sections. You can define one or more storage sections under a storage type.

- **Storage bin**
  Storage bins represent an exact place where materials are stored in the warehouse. Storage bins are defined under a storage type. Typically, a storage section has multiple storage bins.

- **Staging area**
  Staging areas are the interim storing areas of the warehouse. You can use staging areas for inbound movements and/or outbound movements.

- **Activity area**
  Activity areas of the warehouse are the areas where specific warehouse activities such as putaway, stock removal, physical inventory, and so on take place. You can define multiple activity areas for a warehouse.

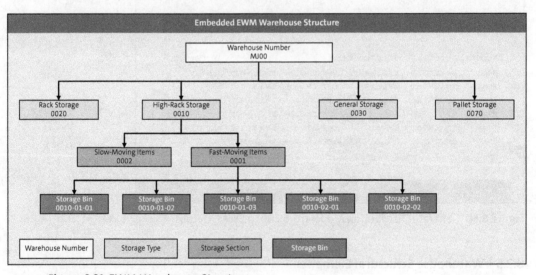

Figure 8.31 EWM Warehouse Structure

Let's dive into every structural element of a warehouse in embedded EWM.

Storage Types

Storage types are the key components of a warehouse. Storage types represent logical or physical areas within the warehouse where materials are logically or physically sorted based on material handling strategies such as putaway strategies and stock removal strategies. Storage types are defined in the warehouse based on storage capacity, storage area requirement, type of materials to be stored, special storage requirements, and so on. Storage types are the next hierarchical level in embedded EWM below the warehouse number. A storage type can have one or more storage sections and storage bins.

Doors and yards are also created as storage types as the materials are stored in that area of the warehouse temporarily; that is, different storage areas within the warehouse are created as storage types, and storage sections and storage bins are defined under storage types.

To define storage types for an embedded EWM warehouse number, execute Transaction SPRO and navigate to **SCM Extended Warehouse Management • Extended Warehouse Management • Master Data • Define Storage Types**.

Figure 8.32 shows the overview of storage types defined for embedded EWM warehouse **MJ00**. Click **New Entries** to define a new storage type. It is recommended to copy an existing similar storage type to create a new one. Specify a four-digit alphanumeric key and description of the storage type. High-rack storage, pallet storage, bulk storage, fixed-bin storage, production supply area, goods issue area, packing area, quality inspection area, returns storage type, and so on are the most common storage types used in the warehouse.

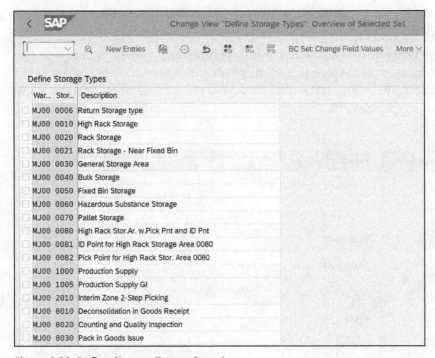

Figure 8.32 Define Storage Types: Overview

Let's dive deep into the details of a storage type. Select the storage type and click the **Details** icon at the top or press [Ctrl]+[Shift]+[F2] to arrive at the screen shown in Figure 8.33, which shows the **General** data of storage type **0010** (**High Rack Storage**) defined for warehouse **MJ00**.

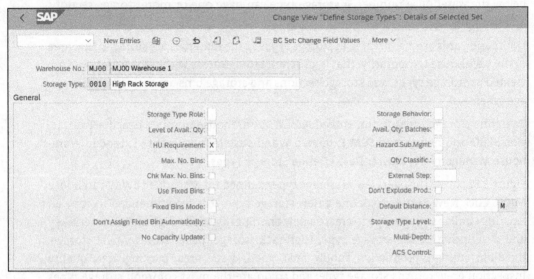

Figure 8.33 Storage Type Details: General Data

The following are the key fields in the **General** data section of the storage type:

- **Storage Type Role**

  A storage type role classifies different storage types. Based on the storage type role, the system controls the direction of the goods movement (inbound or outbound) and accordingly recognizes the door for putaway or stock removal. Table 8.1 shows the storage type roles and descriptions. Assign an appropriate storage type role for the storage type.

Storage Type Role	Description
Blank	Standard storage type
A	Identification point
B	Pick point
C	Identification and pick point
D	Staging area group
E	Work center
F	Doors

Table 8.1 Storage Type Roles

Storage Type Role	Description
G	Yard
H	Automatic storage retrieval (material flow control)
I	Work center in staging areas group
J	Automatic warehouse (controlled by MFS)
K	Production supply

Table 8.1 Storage Type Roles (Cont.)

- **Storage Behavior**

  Storage behavior controls the storage bin determination during putaway (inbound). It defines the structure and attributes of a storage bin created in the storage type. Following are the different storage behaviors:

  - Blank (standard warehouse): This is mostly used in general storage types such as high-rack storage. The storage bins created within the storage type are generic ones.

  - 1 (pallet warehouse): This is used for storage of pallets and similar handling units in the storage bins within the storage type.

  - 2 (bulk storage): This is used to store in bulk within the storage bin. Bulk storage type typically contains only one bulk storage bin. Large quantities of materials are stored in bulk in this area of the warehouse.

  - 3 (flexible storage): This is used to create flexible storage bins. These bins are created as and when needed and this adds flexibility for handling goods within the warehouse.

  Assign an appropriate storage behavior in this field for the storage type.

- **Level of Avail. Qty**

  The level of available quantity controls the stock removal strategies. You can set one of the following levels:

  - 1 (storage bin): If you set the available quantity level as storage bin, during stock removal process system creates the warehouse task without reference to handling unit.

  - Blank (highest-level HU): If you set the available quantity level as the highest-level handling unit, then during stock removal, the system creates the warehouse task with reference to a handling unit.

- **Avail. Qty: Batches**

  This is an indicator that controls the available quantity of the storage bin for batches. This parameter is also needed for stock removal strategies. You can set one of the following values:

- Blank (available quantity): Batch-specific. Maintain this value in this field if you are managing available quantities in the storage bin in batches and creating warehouse tasks for stock removal by mentioning batch-specific details.
- 1 (available quantity): Batch-neutral. Maintain this value in this field if you are not managing available quantities in the storage bin in batches and creating warehouse tasks for stock removal without mentioning batch-specific details.

- **HU-Requirement**
  You can specify in this field whether handling units are required or allowed in the storage bin within the storage type. You can set one of the following values:
  - Blank (HU allowed but not required): This setting allows you to store handling units, but it is not mandatory.
  - **X** (HU required): This setting requires you to store handling units. You cannot store the materials in this storage type without handling units.
  - **Y** (HU not allowed): Storing of handling units in this storage type is not allowed.

- **Hazard. Sub. Mgmt.**
  You can specify the level of hazardous substance check in this field. You can set the following values:
  - Blank (No hazardous substance check): This setting does not check hazardous substance.
  - 1 (hazardous substance check only at storage type level): This setting allows a hazardous substance check only at the storage type level.
  - 2 (hazardous substance check at storage type and storage section levels): This setting allows a hazardous substance check at the storage type level as well as the storage section level.

- **Max. Num. Bins**
  In this field, you can specify the maximum number of storage bins allowed in the storage type. The system will not allow you to create storage bins exceeding this number.

- **Qty. Classific.**
  In this field, you can specify the storage unit for the storage type. The system uses this setting to determine the storage type based on the packaging material coming from packaging specifications for the product. Table 8.2 shows the different values that you can choose from.

Quantity Classification	Description
1	Eaches
2	Cartons
3	Tiers

Table 8.2 Quantity Classification

Quantity Classification	Description
4	Pallets
C	Case quantity
L	Layer quantity
P	Pallet quantity

Table 8.2 Quantity Classification (Cont.)

- **Chk. Max. Num. Bins.**
  If you set this indicator, the system checks the maximum number of storage bins allowed in the storage type during the creation of warehouse tasks.

- **Use Fixed Bins**
  If you set this indicator, fixed storage bins are used in the storage type. The system can automatically assign the products to the storage bins, or you can manually assign the materials to the storage bins. Only those fixed storage bins assigned to the material are allowed to be stored in these storage bins. If there are no fixed bins assigned to the material at the time of putaway, the system creates a fixed bin within the storage type automatically.

- **Fixed Bins Mode**
  This setting is used for putaway of products into the fixed storage bins. This setting controls the putaway of materials into optimum fixed storage bins if the slotting function is active. One of the following values can be set in this field:
  - Blank (putaway to optimum fixed storage bins only)
  - **A** (putaway to optimum fixed storage bins preferred)

- **Default Distance**
  A default average distance in meters to the first storage bin is maintained in this field for planning purposes.

- **Don't Assign Fixed Bin Automatically**
  If you set this indicator, the system doesn't create a fixed storage bin automatically if the use of fixed storage bins for the product is active and the product/material is not assigned to a fixed bin. In such a case, assign a fixed bin to the material manually.

- **Storage Type Level**
  If there are multiple storage levels in the warehouse, you can specify on which level the storage type is located.

Next, scroll down further to view putaway control attributes. Figure 8.34 shows the **Putaway Control** data of storage type **0010** (**High Rack Storage**) defined for warehouse **MJ00**. The following are the key fields:

- **Confirm Putaway**
  If you set this indicator, the warehouse task created for the putaway of materials into the storage type must be confirmed.

- **HU Type Check**
  If you set this indicator, the system checks if the handling unit type mentioned in the warehouse task for putaway is allowed in the storage type.

- **ID Point Active**
  If you set this indicator, an identification point is activated and goods movements for putaway will be received into the identification point to identify products/materials and create appropriate handling units and putaway into the storage bin within this storage type.

- **Don't Put Away HUs**
  If you set this indicator, putaway of handling units into the storage bins of this storage type is not allowed.

- **Check Max. Quantity**
  If you set this indicator, if the putaway quantities exceed the maximum storage type quantity, the system doesn't allow the putaway of the materials.

- **Putaway Rules**
  Putaway rules control the determination of a destination storage bin for putaway. Table 8.3 shows the keys for the putaway rules and their description.

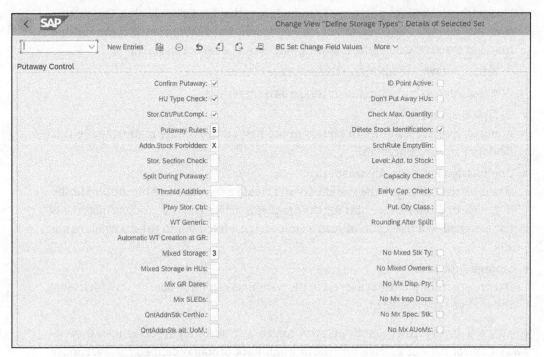

Figure 8.34 Storage Type Details: Putaway Control

Putaway Rule	Description	Impact
2	Addition to existing stock/empty bin	The system searches for a partially filled storage bin. If found, the materials are putaway into the partially empty storage bin. If not found, the materials are putaway into the empty storage bin.
3	Consolidation group	The system searches for an appropriate storage bin where similar handling units with a consolidated group must be stored.
4	General storage area	Addition to existing stock and mixed storage is allowed for the storage type.
5	Empty bin	The system searches for an empty bin for putaway and addition to existing stock is not allowed.
6	Transit warehouse: staging area for door	Used in transit warehouses only.

Table 8.3 Putaway Rules

- **Delete Stock Identification**
  If you set this indicator, stock identification of stock in this storage type is deleted.

- **Addn. Stock Forbidden**
  The additional stock forbidden field controls if a quant of the same material with the same attribute can be putaway into a storage bin which has a similar quant as an addition to the existing stock or not. Table 8.4 shows the values that you can set in this field.

Addition to Stock Forbidden	Description
Blank	Addition to existing stock permitted
X	Addition to existing stock generally not permitted
M	Product putaway profile decides

Table 8.4 Additional Stock Forbidden

- **SrchRule EmptyBin**
  This rule is used to search for empty storage bins for putaway of stocks into those storage bins. Table 8.5 shows the values you can set. If you have set storage bin sorting functionality in embedded EWM, you can set the search rule sorting according to this definition. Bin sorting functionality is explained later in this section. You can set 1 (near to fixed bin) and the system will search for an empty bin near the fixed bin in

the same storage type. Or you can set the product/material to decide the empty storage bin and maintain the master data for it.

Empty Bin Search Rule	Description
Blank	Sorting according to definition
1	Near to fixed bin
2	Product decides

Table 8.5 Empty Bin Search Rule

- **Stor. Section Check**
  You can enable a storage section check for the system to consider a storage section during the putaway into storage. Table 8.6 shows the values that you can set.

Storage Section Check	Description
Blank	No storage section determination or check
X	Storage section determination and check
Y	Storage section determination; no check

Table 8.6 Storage Section Check

- **Level: Add. To Stock**
  This field controls the level at which the stock is added to the existing stock. You can set this at the storage bin or highest handling unit level.
- **Split During Putaway**
  This field controls if the stock split is allowed during putaway if the putaway quantity can't be stored entirely into the destination storage bin in this storage type based on the maximum quantity allowed in the storage bin. Table 8.7 shows the values you can set in this field.

Split During Putaway	Description
Blank	Don't split during putaway
1	Split during putaway
2	Product master decides

Table 8.7 Split During Putaway

- **Capacity Check**
  This field controls if the capacity check is performed for the storage bin. For certain storage bins, a capacity check is required, and the system checks against the

maximum weight and volume allowed for the storage bin. You can turn this off for certain storage bins where capacity doesn't matter. Table 8.8 shows the values you can set in this field.

Capacity Check Method	Description
Blank	No check according to key figure
1	Check according to key figure product
2	Check according to key figure packaging material
3	Check according to key figures product and packaging material
4	No check against key figure, weight, and volume

Table 8.8 Capacity Check Method

- **Early Cap. Check**
  To improve overall efficiency and performance of warehouse operations, a early capacity check is useful. If you set this indicator, the system checks the storage bin capacity during the storage bin search during the warehouse task creation for putaway.

- **Thrshld. Addition**
  If the **Split During Putaway** setting was enabled, you can set a threshold percentage to check against.

- **Ptwy. Stor. Ctrl.**
  This field controls how the putaway of materials into the storage bin is performed. Warehouse tasks can be created with handling unit details or products for putaway depending on the key assigned to this field. Table 8.9 shows the values you can set in this field.

Putaway Storage Control	Description
Blank	Storage control: dynamically evaluated
1	Storage control: putaway with HU warehouse task
2	Storage control: putaway with product warehouse task

Table 8.9 Putaway Storage Control

- **Put. Qty. Class.**
  In this field, you can specify the storage unit for the storage type for putaway. The system uses this setting to determine the storage type based on the packaging material coming from packaging specifications for the product during putaway. Table 8.10 shows the different values that you can choose from.

Quantity Classification	Description
1	Eaches
2	Cartons
3	Tiers
4	Pallets
C	Case quantity
L	Layer quantity
P	Pallet quantity

Table 8.10 Putaway Quantity Classification

- **WT Generic**

  This field controls the destination storage data determination during the creation of a warehouse task. Table 8.11 shows the different values that you can set in this field. Based on the setting, a destination storage type, a storage type, and a storage section; or a storage type, a storage section, and a storage bin, is determined.

Warehouse Task Generic	Description
Blank	Not generic (storage type, storage section, and storage bin)
1	Storage type and storage section
2	Only storage type

Table 8.11 Warehouse Task Generic

- **Automatic WT Creation at GR**

  This field controls the automatic warehouse task creation in embedded EWM during goods receipt in SAP S/4HANA to a storage bin that belongs to this storage type. You can enable this feature only if SAP EWM is embedded in SAP S/4HANA. If this feature is enabled, it syncs with inventory management and warehouse management by creating a warehouse task automatically upon goods receipt. Table 8.12 shows the different values that you can set in this field.

Automatic Warehouse Task Creation at Goods Receipt	Description
Blank	Inactive
1	Active
2	Warehouse process type decides

Table 8.12 Automatic Warehouse Task Creation at Goods Receipt

- **Mixed Storage**

  This field controls how mixed storage is handled in the storage bin that belongs to this storage type. Table 8.13 shows the different values that you can set in this field.

Mixed Storage	Description
Blank	Mixed storage without limitations
1	Several non-mixed HUs with the same product/batch
2	Several HUs with different batches of the same product
3	One HU allowed per bin

Table 8.13 Mixed Storage

- **Mixed Storage in HUs**

  This field controls the mixed storage of different materials and batches in the same handling unit, either in the same storage bin or not. Table 8.14 shows the different values that you can set in this field.

Mixed Storage in HU	Description
1	Mixed storage not allowed
2	Several batches of the same product per HU
Blank	Mixed storage without limitations in HUs

Table 8.14 Mixed Storage in HU

- **Mix GR Dates**

  This field controls if different quants with different goods receipt dates can be put-away into the same storage bin or not. Table 8.15 shows the different values that you can set in this field.

Mixing Goods Receipt Dates	Description
Blank	Allowed—most recent date dominant
2	Allowed—earliest date dominant
1	Not allowed
3	Product putaway profile decides

Table 8.15 Mixing Goods Receipt Dates

- **Mix SLEDs**
  This field controls if different quants with different shelf-life expiration dates (SLEDs) can be putaway into the same storage bin or not. Table 8.16 shows the different values that you can set in this field.

Mixing SLEDs	Description
Blank	Allowed—most recent date dominant
2	Allowed—earliest date dominant
1	Not allowed
3	Product putaway profile decides

Table 8.16 Mixing SLEDs

- **No Mixed Stk Ty**
  Set this indicator if you want the materials stored in the storage type not to be stored in any other storage type.

- **No Mixed Owners**
  Set this indicator if you want the materials stored in the storage type to have only one owner.

- **No Mix Insp Docs**
  Set this indicator if you want to disable storing of multiple inspection lots belong to the same material in the storage type.

- **No Mix Spec Stk**
  Set this indicator if you want to disable storing of multiple special stock types belong to the same material in the storage type.

Figure 8.35 shows the **Stock Removal Control** data, **Goods Movement Control** data, and **Replenishment** data of storage type **0010** (**High Rack Storage**) defined for the warehouse **MJ00**.

The following are the key fields of **Stock Removal Control**:

- **Confirm Removal**
  If you set this indicator, the warehouse task created for the stock removal from the storage type must be confirmed.

- **Pick Pnt Active**
  If you set this indicator, the stock removal from the storage type will happen through a pick point.

- **Stock on Resource**
  If you set this indicator, the stock removal strategy will consider the stock located in resources that belong to the storage type.

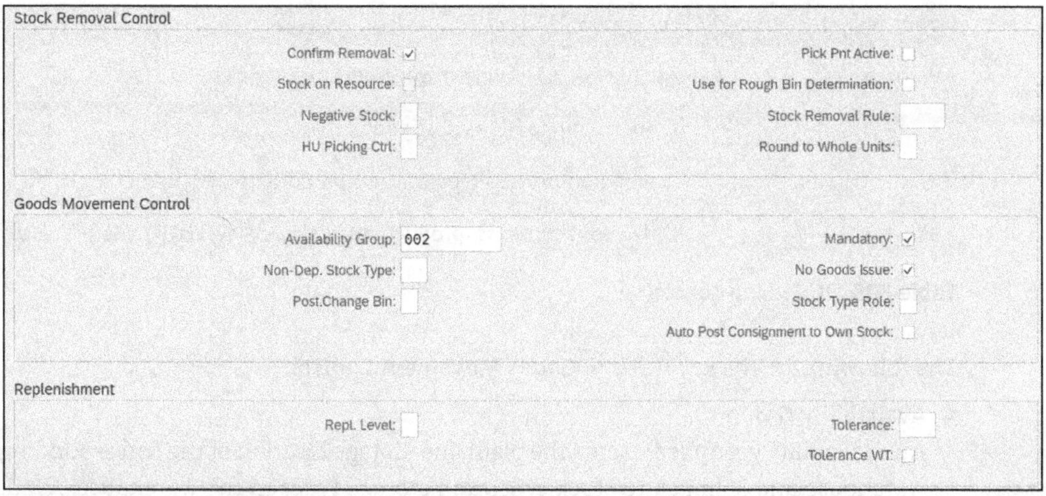

Figure 8.35 Storage Type Details: Stock Removal Control and Other Controls

- **Use for Rough Bin Determination**
  If you set this indicator, the system uses this storage type for rough (approximate) storage determination.

- **Negative Stock**
  This field controls if negative stock is allowed or negative available stock is allowed in the storage bin. Table 8.17 shows the different values that you can set in this field.

Negative Stock	Description
Blank	Don't allow negative stock
A	Allow negative available stock
X	Allow negative stock

Table 8.17 Negative Stock

- **Stock Removal Rule**
  This field controls the sorting of quants in the storage bin for stock removal process. First in, first out (FIFO), last in, first out (LIFO), and so on are the important stock removal rules that can be configured for the warehouse.

- **HU Picking Ctrl**
  This field controls the system behavior during the picking of handling units based on the settings. Table 8.18 shows the various values that you can set in this field.

Control for HU Picking	Description
Blank	Adopt source HU with lower-level HUs into pick-HU
1	Propose source HU as destination HU
2	Warehouse process type controls proposal for destination HU
3	Only adopt contents (production and lower-level HUs) into pick-HU

Table 8.18 HU Picking Control

The following are the key fields of **Goods Movement Control**:

- **Availability Group**
  An availability group represents the plant and storage location of the corresponding warehouse and is linked to stock types and storage types in the warehouse. Availability groups are assigned to storage location in configuration settings. Assigning the availability group to the storage type helps in the transfer posting of stocks in the warehouse. Assign the availability group defined for the warehouse in the configuration settings in this field.

- **Mandatory**
  If you set this indicator, assigning an availability group to the storage type is mandatory.

- **Non-Dep. Stock Type**
  This represents a stock type that is independent of the availability group. Table 8.19 shows the nondependent stock types that you can assign to this field.

Location-Independent Stock Type	Description
BB	Blocked stock
FF	Unrestricted use stock
QQ	Stock in quality inspection
RR	Blocked stock returns

Table 8.19 Location-Independent Stock Type

- **No Goods Issue**
  If you set this indicator, goods issue posting is not allowed using this storage type.

- **Post.Change Bin**
  This field controls the posting changes in the storage bin that belongs to this storage type. Table 8.20 shows the values that you can set in this field to control the posting changes in the storage bin.

Posting Change	Description
Blank	Posting change always in storage bin
1	Posting change according to mixed storage setting
2	Posting change never in storage bin (create warehouse task)

Table 8.20 Storage Type Control for Posting Change Storage Bin

- **Storage Type Role**
  You can set storage type roles for certain storage types, and the system automatically performs a posting change based on the stock type role assigned. Table 8.21 shows the stock type roles that you can assign to this field.

Stock Type Role	Description
C	Customs blocked stock
S	Scrapping stock
N	Normal stock

Table 8.21 Storage Type Role

- **Auto Post Consignment to Own Stock**
  If you set this indicator, the system automatically posts the consignment stock to own stock when a stock movement is posted into the storage bin belonging to this storage type. To learn more about the consignment process, refer to Chapter 1, Section 1.1.4.

In embedded EWM, storage types control key functionalities and hence most of the control parameters are set at the storage type level.

Storage Sections

A storage section is a subdivision of storage type, and it represents a group of storage bins. A storage type can have multiple storage sections defined based on fast-moving items, slow-moving items, and so on. Storage sections play a vital role in determining the storage bin for putaway strategies.

To define storage sections for the storage types for an embedded EWM warehouse number, execute Transaction SPRO and navigate to **SCM Extended Warehouse Management • Extended Warehouse Management • Master Data • Define Storage Sections**.

To create a storage section for a storage type, click **New Entries** and specify an embedded EWM warehouse number, a storage type, and a four-digit alphanumeric key for the storage section in the **Sec** field. Specify the description for the storage section. Figure

8.36 shows the storage sections defined for storage types within embedded EWM warehouse **MJ00**.

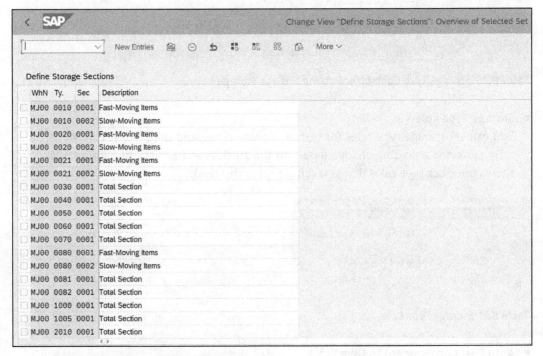

Figure 8.36 Define Storage Sections

For certain storage types, there is only one storage section, defined as **Total Section**, meaning there are no sections within the storage type and the entire storage area of the storage type is the section. Hence, at least one storage section must be defined for the storage type, technically.

Storage Bins

We introduced storage bins, which are the basic master data element in embedded EWM, in Section 8.1.2.

Before defining storage bins for storage types, the storage bin types, access types, bin structure, and so on must be defined in the configuration settings. Let's explore the details.

Define Storage Bin Types

You can define storage bin types based on the size, volume, and type of materials and handling units that can be stored in the storage bins.

To define storage bin types for an embedded EWM warehouse number, execute Transaction SPRO and navigate to **SCM Extended Warehouse Management • Extended Warehouse Management • Master Data • Storage Bins • Define Storage Bin Types**.

Figure 8.37 shows the overview of storage bin types defined for embedded EWM warehouse number **MJ00**. Click **New Entries** to define a new storage bin type. You can copy an existing storage bin type and create a new one. Enter a four-digit alphanumeric key and description to define a storage bin type. Specify the **Max.Depth** for the storage bin type only for multidepth storage types. Multidepth storage types contain storage bins that can store multiple handling units, and the depths of the storage bins are defined based on the handling units that can be stored in the storage bin.

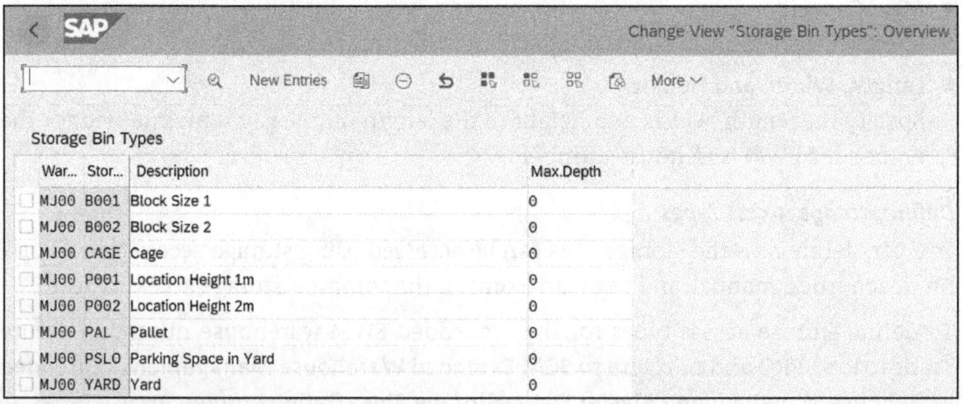

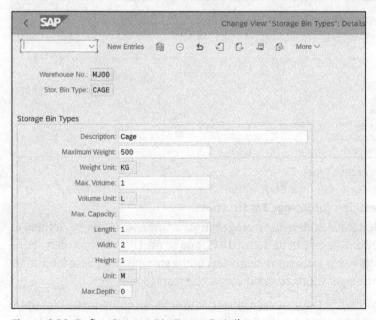

Figure 8.37 Define Storage Bin Types: Overview

To navigate to the storage bin type details, select the bin type and click the **Details** icon at the top or press Ctrl + Shift + F2. Figure 8.38 shows the details of storage bin type **CAGE** defined for embedded EWM warehouse number **MJ00**.

Figure 8.38 Define Storage Bin Types: Details

The following are the important details to be maintained in the storage bin types:

- **Maximum Weight**
  Specify the maximum allowed weight for the storage bin in the weight unit (e.g., KG) maintained for the embedded EWM warehouse number.

- **Max. Volume**
  Specify the maximum allowed volume for the storage bin in the volume unit (e.g., L) maintained for the embedded EWM warehouse number.

- **Max. Capacity**
  Specify the maximum allowed capacity for the storage bin.

- **Length, Width, and Height**
  Specify the length, width, and height in the length unit (e.g., M) maintained for the embedded EWM warehouse number.

Define Storage Access Types

You can define how the storage bins can be accessed using storage access types. Forklift, reach truck, manual, and so on are some of the common storage access types.

To define storage access types for the embedded EWM warehouse number, execute Transaction SPRO and navigate to **SCM Extended Warehouse Management • Extended Warehouse Management • Master Data • Storage Bins • Define Storage Access Types**.

Figure 8.39 shows the overview of storage access types defined for embedded EWM warehouse number **MJ00**. Click **New Entries** to define a new storage access type. You can copy an existing storage access type and create a new one. Enter a four-digit alphanumeric key and description to define a storage access type.

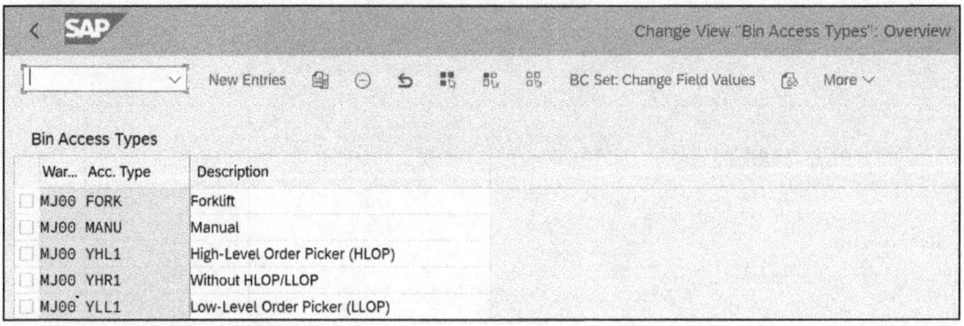

Figure 8.39 Define Bin Access Types

Define Storage Bin Identifiers for Storage Bin Structures

You can define one-letter identifiers for storage bin structures. Aisle can be defined as A, stack can be defined as S, level can be defined as L, and so on. Likewise, you can define the parts of the storage bin as structural identifiers of the storage bin. These identifiers are useful in warehouse operations to easily locate the storage bins.

To define storage bin identifiers for storage bin structures for an embedded EWM warehouse number, execute Transaction SPRO and navigate to **SCM Extended Warehouse**

Management • Extended Warehouse Management • Master Data • Storage Bins • Define Storage Bin Identifiers for Storage Bin Structures.

Figure 8.40 shows the overview of storage bin identifiers defined for embedded EWM warehouse number **MJ00**. Specify a letter to define storage bin identifiers.

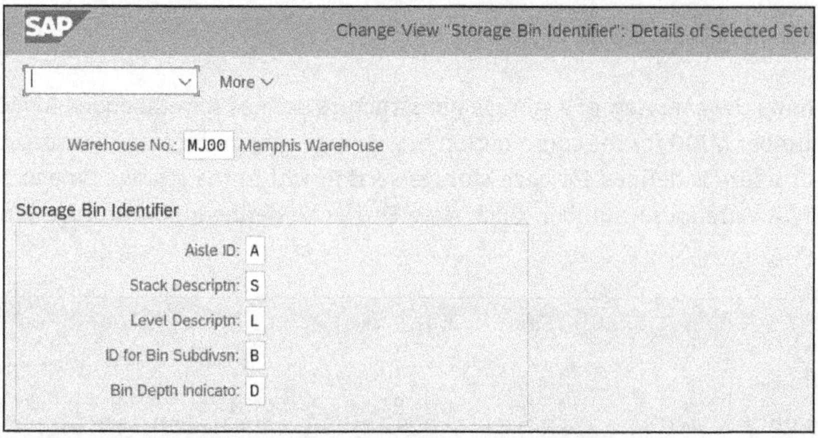

Figure 8.40 Define Storage Bin Identifiers for Storage Bin Structures

> **Note**
>
> Define storage bin identifiers during the initial set up of the embedded EWM warehouse number. Once you start the warehouse operations, it is not possible to define activity areas independently for the storage type.

Define Storage Bin Structure

The storage bin is the smallest spatial unit in a warehouse, and it represents the exact position in the warehouse where products are/can be stored. The address of a storage bin is frequently derived from a coordinate system, so a storage bin is often referred to as a *coordinate*; for example, the coordinate 01-02-03 could be a storage bin in aisle 01, stack 02, and level 03.

Storage bins are defined under storage types within the embedded EWM warehouse number. Storage types control the putaway of materials into storage bins and stock removal/withdrawal from storage bins. Hence the storage bin structure needs to be defined per storage type. In embedded EWM, the storage bin coordinate is 18 characters in length and must be unique within the warehouse. Figure 8.41 shows a basic structure example for a storage bin.

To define a storage bin structure for an embedded EWM warehouse number, execute Transaction SPRO and navigate to **SCM Extended Warehouse Management • Extended Warehouse Management • Master Data • Storage Bins • Define Storage Bin Structure**.

Figure 8.41 Example of Basic Storage Bin Structure

Figure 8.42 shows the overview of a storage bin structure defined for embedded EWM warehouse number **MJ00** for the combination of a storage type and storage section. A storage bin structure is defined for each storage section within the storage type and embedded EWM warehouse number. Click **New Entries** to define a new storage bin structure type.

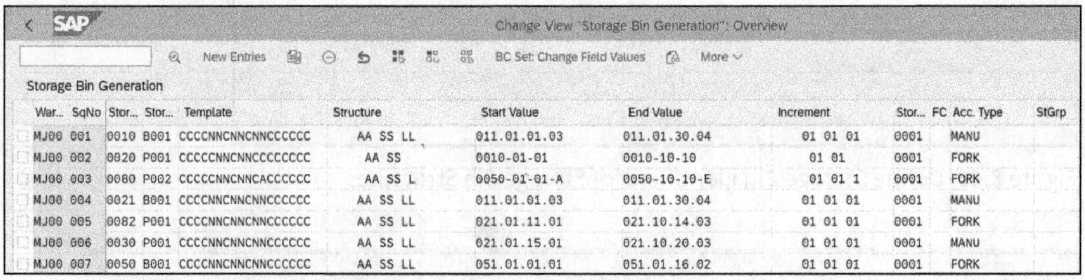

War...	SqNo	Stor...	Stor...	Template	Structure	Start Value	End Value	Increment			Stor...	FC	Acc. Type	StGrp
MJ00	001	0010	B001	CCCCNNCNNCNNCCCCCC	AA SS LL	011.01.01.03	011.01.30.04	01	01	01	0001		MANU	
MJ00	002	0020	P001	CCCCNNCNNCCCCCCCCC	AA SS	0010-01-01	0010-10-10	01	01		0001		FORK	
MJ00	003	0080	P002	CCCCCNNCNNCACCCCCC	AA SS L	0050-01-01-A	0050-10-10-E	01	01	1	0001		FORK	
MJ00	004	0021	B001	CCCCNNCNNCNNCCCCCC	AA SS LL	011.01.01.03	011.01.30.04	01	01	01	0001		MANU	
MJ00	005	0082	P001	CCCCNNCNNCNNCCCCCC	AA SS LL	021.01.11.01	021.10.14.03	01	01	01	0001		FORK	
MJ00	006	0030	P001	CCCCNNCNNCNNCCCCCC	AA SS LL	021.01.15.01	021.10.20.03	01	01	01	0001		MANU	
MJ00	007	0050	B001	CCCCNNCNNCNNCCCCCC	AA SS LL	051.01.01.01	051.01.16.02	01	01	01	0001		FORK	

Figure 8.42 Define Storage Bin Structure: Overview

Figure 8.43 shows the first part of the storage bin structure details. Let's explore the fields within the **Bin Definition**:

- **Template**
  A template is used to generate storage bins automatically. There are three letters used to define the template for the storage bin structure:
  - **C:** This letter represents a character common to all storage bins: a space, a dash, and so on.
  - **N:** This letter represents a numeric value: 0, 1, 2, and so on.
  - **A:** This letter represents a letter: A, B, C, and so on.

  For template CCCCNNCNNCNNCCCCCC, you can generate the storage bins as 01-01-01, 01-01-02, 01-01-03, 01-02-01, 01-02-02, and so on.

- **Structure**
  The structure controls how the system increments storage bin numbering during auto generation of storage bins. If you define the structure as AA SS LL or AA SS, the increments are independent of each other. If you define the structure as AA, multidigit increments would happen.

- **Start Value**
  In this field, you specify the start value for the automatic generation of storage bins.

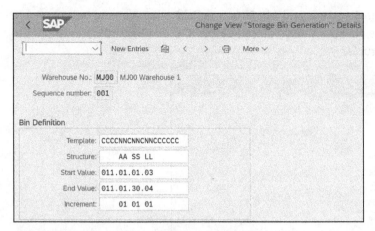

Figure 8.43 Define Storage Bin Structure: Details: 1 of 2

- **End Value**
  In this field, you specify the end value for the automatic generation of storage bins.

- **Increment**
  In this field, you specify the increment for the generation of storage bins automatically.

Figure 8.44 shows the second part of the storage bin structure details. Let's explore the fields within the **XYZ-Coordinates** section; these coordinates are used to locate the exact position of the storage bin in the warehouse:

- **X-Start**
  An X coordinate is used to measure the distance of the storage bin in the x direction in terms of the unit of length defined in the warehouse number.

- **Y-Start**
  A Y coordinate is used to measure the distance of the storage bin in the y direction in terms of the unit of length defined in the warehouse number.

- **Z-Start**
  A Z coordinate is used to measure the distance of the storage bin in the z direction in terms of the unit of length defined in the warehouse number.

- **X Increment**
  An X increment is the increment in the x direction in terms of the unit of length defined in the warehouse number.

- **Y Increment**
  A Y increment is the increment in the y direction in terms of the unit of length defined in the warehouse number.

- **Z Increment**
  A Z increment is the increment in the z direction in terms of the unit of length defined in the warehouse number.

The **X in Structure**, **Y in Structure**, and **Z in Structure** field values depend on storage bin identifiers for storage bin structures. You can define X, Y, and Z coordinates to refer to the storage bin structures.

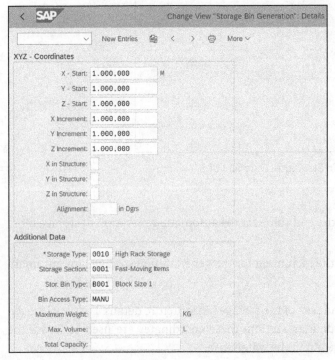

Figure 8.44 Define Storage Bin Structure: Details: 2 of 2

Finally, let's explore the key fields within the **Additional Data** section:

- **Storage Type**
  The storage type is mandatory. The storage bin structure must be created for a storage type. Assign the storage type in which the storage bin structure needs to be defined.

- **Storage Section**
  If the storage type has multiple storage sections, you can define the storage bin structure for every storage section. Assign the storage section in which the storage bin structure needs to be defined.

- **Stor. Bin Type**
  Assign the relevant storage bin type defined for the embedded EWM warehouse number in this field.

- **Bin Access Type**
  Assign the relevant storage access type defined for the embedded EWM warehouse number in this field.

You can also define the **Max. Volume**, **Maximum Weight**, and **Total Capacity** in the **Additional Data** section.

Create Storage Bin

There are multiple ways of creating a storage bin for the embedded EWM warehouse number. The following are three different ways of creating the storage bin:

- **Manual storage bin creation using Transaction /N/SCWM/LS01**
Use this method to create storage bins when there are only a few storage bins to be created for the storage type.

- **Generate storage bins in bulk using the storage bin structure template using Transaction /N/SCWM/LS10**
Use this method when identical storage bins are to be created in bulk as defined in the storage bin structure within the same storage type and storage section.

- **Upload storage bins using upload utility using Transaction /N/SCWM/SBUP**
Use this method when nonidentical storage bins are to be created in bulk and the dimension and structure of the storage bins varies, and storage bins are to be created for multiple storage types and storage sections.

To create a storage bin manually, specify the embedded EWM warehouse number in the **Warehouse No.** field and specify the storage bin as per the storage bin structure defined in the previous step. Then, press Enter to add the details of the storage bin. Figure 8.45 shows that storage bin **0010-01-01** was created manually using Transaction /N/SCWM/LS01 for the embedded EWM warehouse **MJ00**, storage type **0020**, and storage section **0001**. Specify the storage bin you want to create manually. You can update other details of the storage bin as necessary.

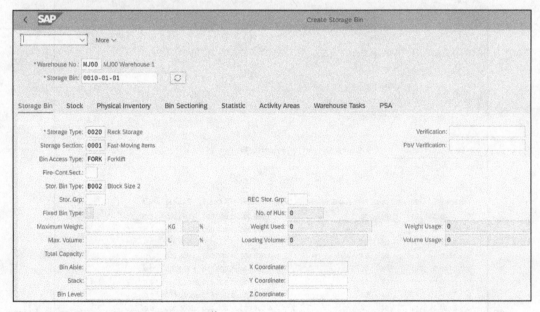

Figure 8.45 Create Storage Bin Manually

Figure 8.46 shows the mass creation of storage bins using a storage bin structure via Transaction /N/SCWM/LS10 for embedded EWM warehouse **MJ00**, storage type **0010**, and storage section **0001**.

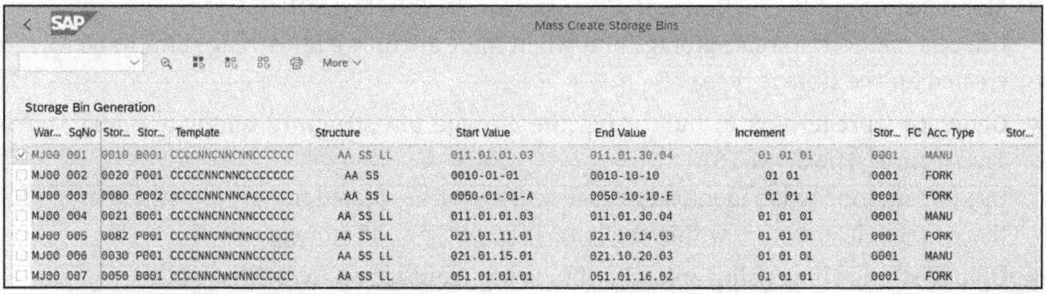

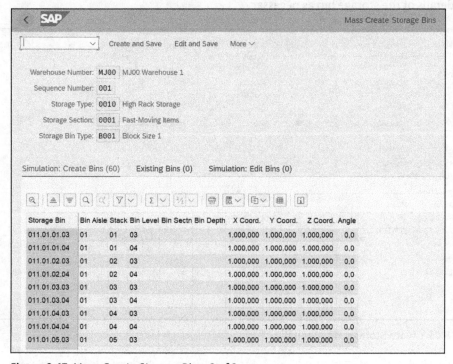

Figure 8.46 Mass Create Storage Bins: 1 of 2

The system displays all the storage bin structures defined in the embedded EWM warehouse number. Select the storage structure for which you want to mass-create storage bins. The system automatically simulates the storage bins as per the coordinates defined in the storage bin structure.

Double-click the selected storage bin structure to display the simulation of storage bins as shown in Figure 8.47. From this screen, you can click **Create and Save** to mass-create storage bins.

Figure 8.47 Mass Create Storage Bins: 2 of 2

Define Warehouse Door

In this configuration step, define warehouse doors and activities to be performed through the door such as inbound movements, outbound movements, or both.

To define warehouse doors for an embedded EWM warehouse number, execute Transaction SPRO and navigate to **SCM Extended Warehouse Management • Extended Warehouse Management • Master Data • Warehouse Door • Define Warehouse Door.**

Figure 8.48 shows the overview of warehouse doors defined for embedded EWM warehouse number **MJ00**. Click **New Entries** to define a new warehouse door for the embedded EWM warehouse. Specify a four-digit alphanumeric or numeric key and a description to define a new warehouse door. The key fields are as follows:

- **Load.Dir.**
  The loading direction of the door defines the possible activities through the warehouse door, such as inbound, outbound, or both.

- **DfStgArGrp (default staging area group)**
  A staging area group is used to group together multiple staging areas. Assign a default staging area group for the warehouse door.

- **DfStgAre (default staging area)**
  A staging area is an interim storing area of the warehouse. Assign the staging area for the warehouse door.

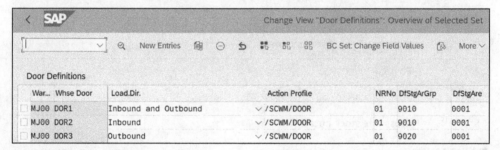

War...	Whse Door	Load.Dir.	Action Profile	NRNo	DfStgArGrp	DfStgAre
MJ00	DOR1	Inbound and Outbound	∨ /SCWM/DOOR	01	9010	0001
MJ00	DOR2	Inbound	∨ /SCWM/DOOR	01	9010	0001
MJ00	DOR3	Outbound	∨ /SCWM/DOOR	01	9020	0001

Figure 8.48 Define Warehouse Door

Define Activities for Activity Areas

Various goods movements within the warehouse such as putaway of stock, stock removal, and internal warehouse movements are represented by activities. In this configuration step, define activities for activity areas within the embedded EWM warehouse.

To define activities for activity areas for the embedded EWM warehouse number, execute Transaction SPRO and navigate to **SCM Extended Warehouse Management • Extended Warehouse Management • Master Data • Activity Areas • Activities • Define Activities for Activity Areas.**

Figure 8.49 shows the overview of activities defined for embedded EWM warehouse number **MJ00**. Click **New Entries** to define a new activity for the embedded EWM warehouse. Specify a four-digit alphanumeric key and description to define a new activity.

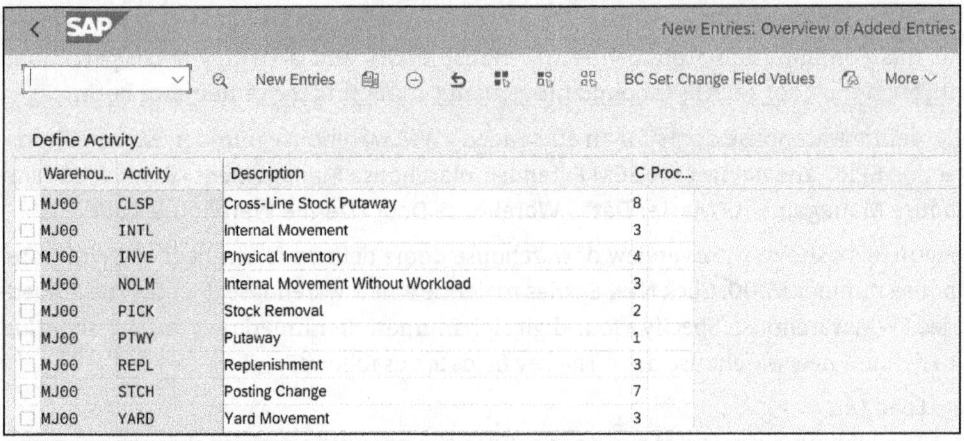

Warehou...	Activity	Description	C Proc...
MJ00	CLSP	Cross-Line Stock Putaway	8
MJ00	INTL	Internal Movement	3
MJ00	INVE	Physical Inventory	4
MJ00	NOLM	Internal Movement Without Workload	3
MJ00	PICK	Stock Removal	2
MJ00	PTWY	Putaway	1
MJ00	REPL	Replenishment	3
MJ00	STCH	Posting Change	7
MJ00	YARD	Yard Movement	3

Figure 8.49 Define Activities for Activity Areas

Assign the relevant warehouse process category (**C**) for each activity. This distinguishes different goods movements happening within the warehouse. Table 8.22 shows the warehouse process categories.

Category	Description
1	Putaway
2	Stock removal
3	Internal warehouse movement
4	Physical inventory
7	Posting change
8	Cross-line stock putaway

Table 8.22 Warehouse Process Categories

Define Activity Areas

Activity areas of the warehouse are where specific warehouse activities such as putaway, stock removal, physical inventory, and so on take place.

To define activity areas for the embedded EWM warehouse number, execute Transaction SPRO and navigate to **SCM Extended Warehouse Management • Extended Warehouse Management • Master Data • Activity Areas • Define Activity Areas.**

Figure 8.50 shows the overview of activity areas defined for embedded EWM warehouse number **MJ00**. Click **New Entries** to define a new activity area for the embedded EWM warehouse. Specify a four-digit alphanumeric or numeric key and a description to define a new activity area. Best practice recommendation is to create an activity area for each storage type and also to keep the four-digit activity area key the same as the storage type.

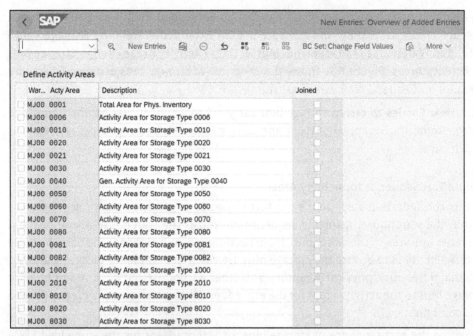

Figure 8.50 Define Activity Areas

Assign Storage Bins to Activity Areas

In this configuration step, assign the storage bins to activity areas for the embedded EWM warehouse. It is recommended to assign the storage type to activity areas and the system will consider the storage bins assigned to the storage type.

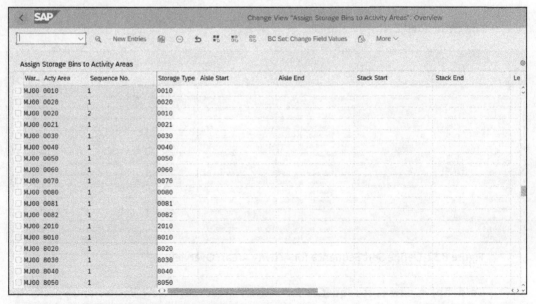

Figure 8.51 Assign Storage Bins to Activity Areas

To assign storage bins to activity areas for the embedded EWM warehouse number, execute Transaction SPRO and navigate to **SCM Extended Warehouse Management • Extended Warehouse Management • Master Data • Activity Areas • Assign Storage Bins to Activity Areas**. Figure 8.51 shows the overview of storage bins assigned to activity areas for embedded EWM warehouse number **MJ00**.

Click **New Entries** to assign storage bins assigned to activity areas for the embedded EWM warehouse. Best practice recommendation is to assign storage types to a relevant activity area.

Define Sort Sequence for Activity Areas

In this configuration step, define the sort sequence of storage bins for activity areas within the warehouse. A sorting sequence optimizes the warehouse operations and prevents one area of the warehouse from becoming heavily loaded. The basic purpose of this activity is to ensure the storage bins are easily accessible for picking, putaway, internal warehouse, physical inventory, and other warehouse activities. Assignment of storage bins to the activity areas for the embedded EWM warehouse is a prerequisite to this configuration.

To define the sort sequence of storage bins for activity areas for the embedded EWM warehouse number, execute Transaction SPRO and navigate to **SCM Extended Warehouse Management • Extended Warehouse Management • Master Data • Activity Areas • Define Sort Sequence for Activity Areas.**

Figure 8.52 shows the overview of a sort sequence defined for activity areas for embedded EWM warehouse number **MJ00**. Click **New Entries** to define the sort sequence of storage bins for activity areas for the embedded EWM warehouse number. Assign a sort sequence, storage type, and so on for every activity and activity area combination for the embedded EWM warehouse.

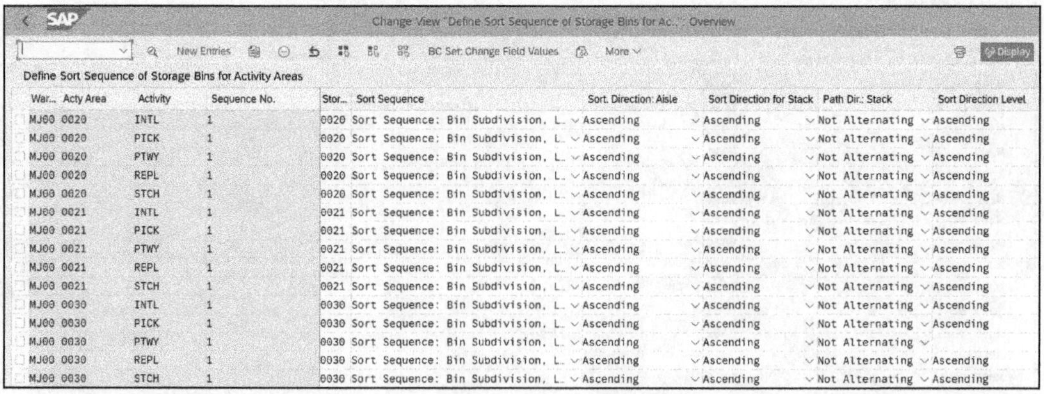

War...	Acty Area	Activity	Sequence No.	Stor...	Sort Sequence	Sort. Direction: Aisle	Sort Direction for Stack	Path Dir.: Stack	Sort Direction Level
MJ00	0020	INTL	1	0020	Sort Sequence: Bin Subdivision, L...	Ascending	Ascending	Not Alternating	Ascending
MJ00	0020	PICK	1	0020	Sort Sequence: Bin Subdivision, L...	Ascending	Ascending	Not Alternating	Ascending
MJ00	0020	PTWY	1	0020	Sort Sequence: Bin Subdivision, L...	Ascending	Ascending	Not Alternating	Ascending
MJ00	0020	REPL	1	0020	Sort Sequence: Bin Subdivision, L...	Ascending	Ascending	Not Alternating	Ascending
MJ00	0020	STCH	1	0020	Sort Sequence: Bin Subdivision, L...	Ascending	Ascending	Not Alternating	Ascending
MJ00	0021	INTL	1	0021	Sort Sequence: Bin Subdivision, L...	Ascending	Ascending	Not Alternating	Ascending
MJ00	0021	PICK	1	0021	Sort Sequence: Bin Subdivision, L...	Ascending	Ascending	Not Alternating	Ascending
MJ00	0021	PTWY	1	0021	Sort Sequence: Bin Subdivision, L...	Ascending	Ascending	Not Alternating	Ascending
MJ00	0021	REPL	1	0021	Sort Sequence: Bin Subdivision, L...	Ascending	Ascending	Not Alternating	Ascending
MJ00	0021	STCH	1	0021	Sort Sequence: Bin Subdivision, L...	Ascending	Ascending	Not Alternating	Ascending
MJ00	0030	INTL	1	0030	Sort Sequence: Bin Subdivision, L...	Ascending	Ascending	Not Alternating	Ascending
MJ00	0030	PICK	1	0030	Sort Sequence: Bin Subdivision, L...	Ascending	Ascending	Not Alternating	Ascending
MJ00	0030	PTWY	1	0030	Sort Sequence: Bin Subdivision, L...	Ascending	Ascending	Not Alternating	
MJ00	0030	REPL	1	0030	Sort Sequence: Bin Subdivision, L...	Ascending	Ascending	Not Alternating	Ascending
MJ00	0030	STCH	1	0030	Sort Sequence: Bin Subdivision, L...	Ascending	Ascending	Not Alternating	Ascending

Figure 8.52 Define Sort Sequence for Activity Areas: Overview

To define a new sort sequence for activity areas, click **New Entries** and specify the embedded EWM warehouse number, specify the activity area for which you're defining the sort sequence, select the activity from the dropdown list, and specify the sequence number. Figure 8.53 shows one of the entry details of the sort sequence of storage bins for activity area.

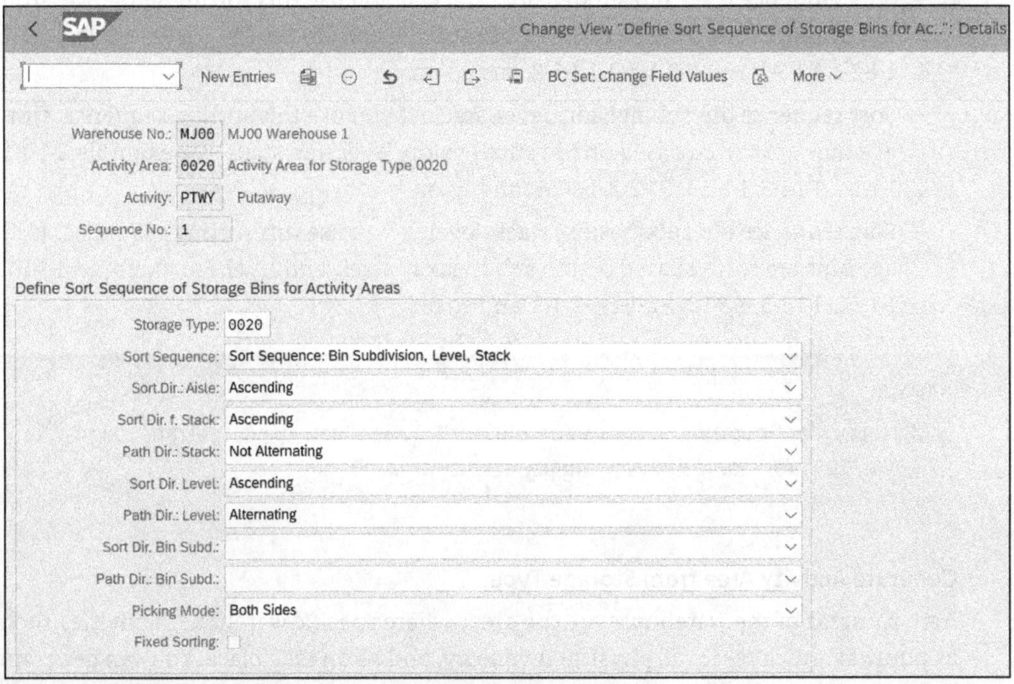

Figure 8.53 Define Sort Sequence for Activity Areas: Details

This example is defined for warehouse **MJ00**, activity area **0020**, and activity **PTWY**. Let's explore the details:

- **Storage Type**
  Assign the storage type to which the activity area belongs to. Storage bins are created in the storage type and a sorting sequence is defined at this level.

- **Sort Sequence**
  In this field, define the sort sequence of storage bins for the activity area of the warehouse. Sorting is based on aisle (A), stack (S), level (L), and bin (B) subdivisions. The following are the sort sequences defined in the standard system. A typical storage bin structure is A-S-L-B. Let's understand the sorting of storage bin based on sort sequence:

  - Sort sequence stack, level, bin subdivision: If you use this sorting sequence, storage bins are sorted based on stack, level, and bin subdivision. For example, 1-1-1-1, 1-2-1-1, 1-3-1-1, 1-1-2-1, 1-2-2-1, 1-3-2-1, and so on.

  - Sort sequence stack, bin subdivision, level: If you use this sorting sequence, storage bins are sorted based on stack, bin subdivision, and level. For example, 1-1-1-1, 1-2-1-1, 1-3-1-1, 1-1-1-2, 1-2-1-2, 1-3-1-2, and so on.

- Sort sequence level, stack, bin subdivision: If you use this sorting sequence, storage bins are sorted based on level, stack, and bin subdivision. For example, 1-1-1-1, 1-1-2-1, 1-1-3-1, 1-2-1-1, 1-2-2-1, 1-2-3-1, and so on.

- Sort sequence level, bin subdivision, stack: If you use this sorting sequence, storage bins are sorted based on level, bin subdivision, and stack. For example, 1-1-1-1, 1-1-2-1, 1-1-3-1, 1-1-1-2, 1-1-2-2, 1-1-3-2, and so on.

- Sort sequence bin subdivision, level, stack: If you use this sorting sequence, storage bins are sorted based on bin subdivision, level, and stack. For example, 1-1-1-1, 1-1-1-2, 1-1-1-3, 1-1-2-1, 1-1-2-2, 1-1-2-3, and so on.

- Sort sequence bin subdivision, stack, level: If you use this sorting sequence, storage bins are sorted based on bin subdivision, stack, and level. For example, 1-1-1-1, 1-1-1-2, 1-1-1-3, 1-2-1-1, 1-2-1-2, 1-2-1-3, and so on.

> **Note**
>
> After the configuration settings for bin sorting, you can execute Transaction /N/ SCWM/SBST for storage bin sorting per activity area.

Generate Activity Area from Storage Type

Activity areas of the warehouse are the areas where specific warehouse activities such as putaway, stock removal, physical inventory, and so on take place. You can generate an activity area from the storage type for the embedded EWM warehouse.

To generate an activity area from a storage type for the embedded EWM warehouse number, execute Transaction SPRO and navigate to **SCM Extended Warehouse Management • Extended Warehouse Management • Master Data • Activity Areas • Generate Activity Area from Storage Type**.

Figure 8.54 shows the initial screen, where you enter a **Warehouse Number**, **Storage Type**, and **Activity**. In our example, we're generating an activity area from storage type **0010** in warehouse **MJ00**. Specify an activity defined for the embedded EWM warehouse only if you want to generate that specific activity. Leave the **Activity** field blank if you want to generate all activities defined for the respective embedded EWM warehouse number for the storage type.

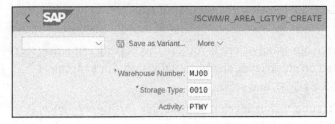

Figure 8.54 Generate Activity Area from Storage Type

Define Staging Areas

Staging areas are the interim storing areas of the warehouse. You can use staging areas for inbound movements and/or outbound movements. Define a staging group prior to defining staging areas for the warehouse.

To define staging areas for the embedded EWM warehouse number, execute Transaction SPRO and navigate to **SCM Extended Warehouse Management • Extended Warehouse Management • Master Data • Staging Areas • Define Staging Areas**.

Figure 8.55 shows the staging area **0001** defined for embedded EWM warehouse **MJ00**. You can define multiple staging areas for different for different warehouse activities. Click **New Entries** to create a new staging area or copy an existing staging area and create a new one. Enter a four-digit alphanumeric key to define a staging area.

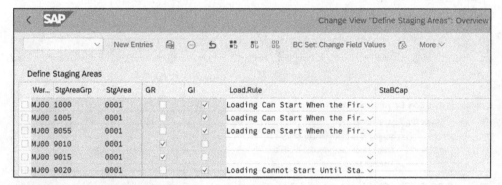

Figure 8.55 Define Staging Areas

Let's explore the key fields in this configuration:

- **StgAreaGrp**
  A staging area group is used to group together multiple staging areas. This field is mandatory to define a staging area for the warehouse.

- **GR**
  Set the goods receipt activities indicator to perform inbound activities in the respective staging area of the warehouse.

- **GI**
  Set the goods issue activities indicator to perform outbound activities in the respective staging area of the warehouse.

- **Load.Rule**
  A loading rule controls the timing of the loading activities in the staging area. If you don't have any such restrictions, leave this field blank. You can set one of the following values other than blank:

  - Loading can start when the first handling unit has arrived.
  - Loading cannot start until the staging has been completed.
  - Loading cannot start until a 24-hour wait time has passed.

Define Staging Area and Door Determination Groups

This setting helps to determine staging area and doors for inbound and outbound transactions.

To define staging area and door determination groups for the embedded EWM warehouse number, execute Transaction SPRO and navigate to **SCM Extended Warehouse Management • Extended Warehouse Management • Master Data • Staging Areas • Define Staging Area and Door Determination Groups**.

Click **New Entries** to define a new staging area and door determination group for the embedded EWM warehouse number. Specify the embedded EWM warehouse number and a four-digit alphanumeric key in the **SA/DDetGrp** field and a description. Figure 8.56 shows the staging area and door determination groups created for warehouse **MJ00**.

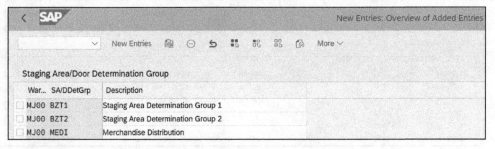

Figure 8.56 Define Staging Area and Door Determination Groups

8.3   Inbound Delivery Processing

Inbound delivery processing in embedded EWM refers to receiving goods from an external supplier, from a supplying plant (stock transfer process), from customer/intercompany returns, from in-house production, and so on.

In the following sections, we'll dive deep into the inbound delivery process in embedded EWM and explore SAP S/4HANA inventory management and embedded EWM integration. We'll cover goods receipt, availability groups, and step-by-step configuration instructions for inbound delivery processing.

8.3.1   Integrated Inbound Deliveries

In embedded EWM, an inbound delivery is a mandatory document that integrates SAP S/4HANA and embedded EWM for inbound processing. The inbound delivery document contains all the relevant information such as item details (material, quantity, delivery date, etc.), batch details, serial number details, receiving location (plant and storage location), and so on to process the goods receipt integrated with SAP S/4HANA and embedded EWM. Inbound delivery for external procurement and stock transfer process is created in SAP S/4HANA using the confirmation control settings in the purchase order. You can receive an advanced shipping notification (ASN) from an external supplier electronically (via EDI or from a B2B application), which will create an inbound delivery in SAP S/4HANA. In case of stock transfer orders, an ASN output can be sent upon goods issue against outbound delivery using ALE IDoc technology to automatically create an inbound delivery in the receiving plant. The confirmation control key that is relevant for inbound delivery creation must be assigned to the purchase order and stock transfer order line items to create the inbound delivery in SAP S/4HANA. You must create an inbound delivery in SAP S/4HANA manually if you don't receive an electronic ASN from the supplier as not all suppliers are able to send the ASN electronically.

Once the inbound delivery is created in SAP S/4HANA, the system checks if the plant and storage location combination assigned to the inbound delivery is assigned to a warehouse number enabled in embedded EWM. If so, the system automatically replicates the inbound delivery from SAP S/4HANA into embedded EWM using qRFC.

The *goods receipt* process with reference to inbound delivery will start when the shipment physically arrives at the yard. The goods are then unloaded for inbound delivery processing. Goods receipt is processed against the inbound delivery in embedded EWM, and the system automatically updates the inbound delivery in SAP S/4HANA using qRFC. The goods receipt posting is automatically performed against the inbound delivery using movement type 101 in SAP S/4HANA, and a material document and an accounting document are created for quantity and value update in inventory management and financial accounting respectively.

After the goods receipt processing in embedded EWM, goods are putaway into one or more storage bins within the warehouse using putaway strategies. Putaway strategies are configured in embedded EWM to determine destination storage bins for inbound processing (putaway).

> **Note**
>
> The warehouse task is a vital document in embedded EWM to process goods movements within the warehouse. For the putaway process, warehouse task is created automatically using the post processing framework (PPF) with reference to the replicated inbound delivery. You can create a warehouse task for putaway manually as well.

Materials management functions in SAP S/4HANA (ERP) and warehouse activities in embedded EWM are tightly integrated to process inbound deliveries. Table 8.23 shows the key process impacts between SAP S/4HANA materials management and embedded EWM. For inbound delivery processing, purchasing and inventory management processes of SAP S/4HANA materials management are deeply impacted.

Area	Processes Impacted	Description
Materials management	Sourcing materials and services	This process supports the sourcing of a best source of supply to procure materials and products from a supplier.
	Create and maintain purchasing master data	This process supports the creation and maintenance of purchasing master data such as source list and purchasing info record (PIR) to support the purchasing of materials/products from a supplier.
	Create and manage outline agreements	This process supports creation and maintenance of outline agreements such as contracts or scheduling agreements with a supplier to procure materials and products.
	Create and maintain purchasing documents	This process focuses on creation of purchasing documents such as purchase requisitions and purchase orders to procure materials/products from a supplier.
	Create and maintain stock transfer orders	This process supports the creation and maintenance of stock transfer order for both intracompany and intercompany stock transfer processes.
	Receive and process ASN	This process supports the creation of inbound delivery automatically from the ASN, which is received into SAP S/4HANA automatically.
	Create inbound delivery manually	This process supports the creation of inbound delivery manually in SAP S/4HANA if electronic ASN is not received from the supplier.
	Post goods receipt	This process supports the posting of goods receipt for inbound delivery to receive materials into stock in the inventory management.
	Process incoming invoice	This process supports creation and reconciliation of incoming invoice received from the supplier.
Extended warehouse management (EWM)	Monitor inbound deliveries	This process supports the monitoring of inbound deliveries in embedded EWM.

Table 8.23 Process Impact

Area	Processes Impacted	Description
Extended warehouse management (EWM) (Cont.)	Perform goods receipt	This process supports the goods receipt with reference to inbound delivery or handling unit in embedded EWM and update the inbound delivery in SAP S/4HANA.
	Perform unloading	This process supports unloading of inbound shipments and creating and confirming warehouse tasks for unloading.
	Determine and perform quality management requirements	This process supports identifying products for quality inspection and performing quality inspection of incoming products/materials in the warehouse.
	Determine putaway requirements and perform putaway	This process supports the determination of an appropriate storage type, storage section, and storage bin as per putaway strategy, considering capacity requirements of storage bins, and moving the stock to the selected storage bin.
	Manage serialization	This process supports the creation of serial numbers for the material/product relevant for serialization management during the inbound delivery processing.
	Manage batch management	This process supports the creation of batches for the material/product which is managed in batches during the inbound delivery processing.

Table 8.23 Process Impact (Cont.)

8.3.2 Inbound Delivery Process

The inbound delivery process is mainly applicable for external procurement, internal stock transfer, in-house production receipt, and customer returns. We'll walk through inbound delivery for external procurement and internal stock transfer in the following sections.

Inbound Delivery Process for External Procurement

Figure 8.57 shows the inbound delivery process for external procurement, which involves steps in both SAP S/4HANA and embedded EWM. Let's have a closer look at the process steps:

1. **Create purchase requisition in SAP S/4HANA**
   Purchase requisitions are generated automatically from material requirements planning (MRP) for stock items or created manually using Transaction ME21N in

SAP S/4HANA. The system assigns a source of supply for raw materials from purchasing master data.

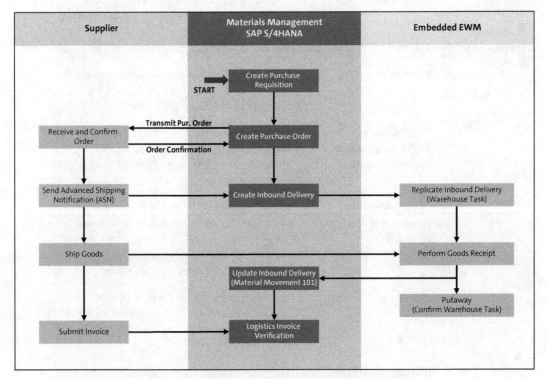

Figure 8.57 EWM Inbound Delivery Processing: External Procurement

2. **Convert purchase requisitions into purchase order**
   The purchase requisition is converted into a purchase order manually by the purchaser using Transaction ME21N or automatically in SAP S/4HANA to procure materials from external supplier. The line item details are defaulted into the purchase order that includes the source of supply, unit cost, and delivery address. A purchase order will be transmitted to the third-party supplier based on the output method: email, EDI, or electronic transmission to the supplier on SAP Business Network. Output determination configuration in SAP S/4HANA drives the purchase order output mechanism. Depending on the confirmation control key, order confirmation from the supplier may be required and the confirmation control key gets defaulted from the PIR.

3. **Create inbound delivery**
   Inbound delivery is the required document in embedded EWM. The conformation control key assigned to the purchase order line item controls the creation of inbound delivery. The supplier can send the ASN via EDI or electronic transmission to the supplier on SAP Business Network or another B2B platform. The ASN received from the supplier is created as an inbound delivery in SAP S/4HANA. An inbound

delivery can be created manually with reference to a purchase order using Transaction VL31N. Inbound deliveries contain shipment information such as materials, quantities, delivery dates, batches, ship-to location, serial numbers, packaging instructions, transportation details, and so on.

4. **Replicate inbound delivery into embedded EWM (automated)**
This is an automated step. The system reads the plant and storage location details from the inbound delivery items and checks if the warehouse number assigned to the plant and storage location combination is an embedded EWM-enabled warehouse. If yes, the system automatically replicates the inbound delivery from SAP S/4HANA into embedded EWM using qRFC. The replicated inbound delivery becomes warehouse request in embedded EWM for goods receipt and putaway.

The warehouse request in embedded EWM enables the processing of warehouse activities such as putaway, picking, physical inventory, and so on. From the inbound delivery warehouse request warehouse tasks can be created for putaway of goods into destination storage bins.

5. **Perform goods receipt in embedded EWM**
When the goods (shipment) arrive at the warehouse, goods receipt can be posted against the inbound delivery in embedded EWM. Transaction /SCWM/PRDI is used to process goods receipt in embedded EWM. In the same transaction, you can unload the inbound delivery, and you can reject the inbound delivery. If an inbound delivery is rejected, it updates the zero quantities in the inbound delivery on the SAP S/4HANA side.

If you post goods receipt for the inbound delivery in embedded EWM, the inbound delivery on the SAP S/4HANA side will be updated using qRFC function module and goods movement 101 will be posted. A material document and an accounting document will be created for quantity and value update in inventory management and financial transaction recording in financial accounting.

6. **Putaway**
Putaway is the process of moving the received products/materials from yard to the destination storage bins within the warehouse. Embedded EWM automatically determines the path for putaway by finding the storage type, storage section and storage bins using putaway strategy. At the inbound delivery item level, embedded EWM stores the putaway status.

Upon the creation of warehouse request (inbound delivery replication), embedded EWM automatically creates warehouse tasks for putaway activity. The warehouse task contains the destination storage bin details and the task details.

7. **Logistics invoice verification**
Goods receipt posted in embedded EWM allows for a three-way match (logistics invoice verification) process for an incoming invoice from the supplier in SAP S/4HANA. The supplier submits an invoice for the goods supplied via email (paper invoice), EDI, or electronically via SAP Business Network. The supplier invoice

document will be posted by accounts payable manually in SAP S/4HANA using Transaction MIRO in the case of a paper invoice. Automatic posting and reconciliation of the invoice happens if an electronic invoice is received from the supplier. Invoice reconciliation is a process of matching purchase order, goods receipt, and supplier invoice to verify any discrepancies with purchase order item price and purchase order item quantity and received quantity. Taxes will be reconciled against the vendor charged tax and the SAP S/4HANA internal tax engine or from the external tax engine calculated tax. Once the invoice is fully reconciled, it will be free for payment processing in the system. The next scheduled payment run will pick up this fully reconciled invoice and post the payment to the supplier based on the preferred payment method of the supplier. Accounts payable personnel can manually process the payment against the fully reconciled invoice as well.

Inbound Delivery Process for Stock Transfer

Figure 8.58 shows the inbound delivery process for intercompany stock transfer, which involves steps in both SAP S/4HANA and embedded EWM. The stock transfer process involves both outbound delivery and inbound delivery processing in embedded EWM and SAP S/4HANA. This is one of the most highly integrated and used processes. Let's take a closer look at the process steps:

1. **Create stock transfer purchase requisitions**
   The MRP run generates stock transfer purchase requisitions in the receiving plant automatically based on stock availability at the supplying plant. The stock transfer purchase requisition can also be created manually using Transaction ME51N. Intracompany stock transfer purchase requisitions will have the supplying plant ID as the source of supply, whereas intercompany stock transfer purchase requisitions will have the business partner of the supplying plant as the source of supply. The valuation price of the stock transfer purchase requisition in most cases will be the standard price or moving average price of the material maintained in the supplying plant. Item category of the requisition line item indicates the type of stock transfer purchase requisition, item category U indicates the intracompany stock transfer purchase requisition and the standard item category indicates intercompany stock transfer purchase requisition.

2. **Create stock transfer order**
   The stock transfer order is the main driver of the stock transfer process, and this document is the combination of sales order and purchase order in the same document. With reference to the stock transfer order, an outbound delivery and goods issue and intercompany billing document will be created on the sales and distribution side. On the purchasing side, the inbound delivery, goods receipt, and intercompany invoice verification and settlement will be processed. The stock transfer order can be created manually or with reference to stock transfer purchase requisition(s) using Transaction ME21N. The item category and source of supply details of

the stock transfer order are pretty much the same as explained in the stock transfer purchase requisition process step. Unlike the other purchase order types, output of the stock transfer order to the supplying plan is not required as it is an internal document and SAP S/4HANA system automatically creates a stock/requirements list that can be displayed using Transaction MD04.

The transfer price is the internal purchase unit price used in the intercompany stock transfer process. The receiving company will be billed by the supplying company at this unit price. An intercompany stock transfer order will be created by the receiving plant using the transfer price. The transfer price for the material can be maintained in the PIRs.

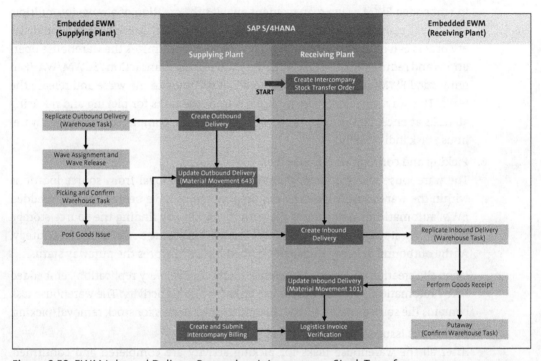

Figure 8.58 EWM Inbound Delivery Processing: Intercompany Stock Transfer

3. **Create replenishment delivery (outbound delivery)**
   Shipping data from the line item level of the stock transfer order is used to create a replenishment delivery using Transaction VL10B. SAP S/4HANA performs availability check before the outbound delivery can be created. Outbound delivery facilitates the movement of materials from the supplying plant to the receiving plant. Picking and packing of materials will be performed within the inventory management against the outbound delivery. It helps to manage the inventory movements efficiently.

4. **Replicate outbound delivery into embedded EWM (automated)**
   This is an automated step. The system reads the plant and storage location details

from the outbound delivery items and checks if the warehouse number assigned to the plant and storage location combination is an embedded EWM-enabled warehouse. If yes, the system automatically replicates the outbound delivery from SAP S/4HANA into embedded EWM using qRFC. The replicated outbound delivery becomes an outbound delivery order in embedded EWM for picking, packing, and shipping of goods from the warehouse. You can display the replicated outbound delivery order in embedded EWM using Transaction /SCWM/PRDO.

Warehouse tasks are created with reference to outbound delivery orders to execute the warehouse activities.

5. **Wave assignment and wave release in embedded EWM**
   In embedded EWM, wave management enables the creation of a wave for multiple warehouse tasks. For picking, a wave can be created, and multiple outbound delivery orders can be assigned to it. Wave management optimizes the warehouse operations and reduces effort. You can create a wave using Transaction /SCWM/WAVE in embedded EWM. Assign the outbound delivery orders to the wave and release the wave. This way, you can create multiple warehouse tasks for picking and releasing all tasks at once. This is an optional step, and you can complete picking for a warehouse task individually.

6. **Picking and confirm warehouse task**
   The warehouse task for picking directs the stock removal from source locations within the warehouse where the required materials have been stored. Embedded EWM automatically determines the path for picking by finding the source storage type, source storage section, and source storage bins using a stock removal strategy. At the outbound delivery item level, embedded EWM stores the putaway status.

   Upon the creation of warehouse request (outbound delivery replication), embedded EWM automatically creates warehouse tasks for picking activity. The warehouse task contains the source storage bin details and the task details for stock removal/picking.

7. **Post goods issue in embedded EWM**
   After all the warehouse tasks for picking activity are completed and confirmed, goods issue can be posted for the outbound delivery in embedded EWM using Transaction /SCWM/PRDO. In the same transaction, you can load the outbound delivery.

   If you post goods issue for the outbound delivery in embedded EWM, the corresponding outbound delivery on the SAP S/4HANA will be updated using qRFC, and goods movement 643 (goods issue for intercompany stock transfer) will be posted automatically. A material document and an accounting document will be created for quantity and value update in inventory management and financial transaction recording in financial accounting.

8. **Create inbound delivery in SAP S/4HANA**
   An inbound delivery is required in embedded EWM. It's created with reference to the stock transfer order on the receiving plant side. A conformation control key assigned to the stock transfer order line item controls the creation of inbound

delivery in SAP S/4HANA. It can be created manually with reference to purchase order using Transaction VL31N. You can also send an ASN output to trigger an IDoc from the outbound delivery during the goods issue posting in SAP S/4HANA. Inbound deliveries contain the shipment information such as materials, quantities, delivery date, batches, ship-to location, serial numbers, packaging instructions, transportation details, and so on.

9. **Replicate inbound delivery into embedded EWM (automated)**
This is an automated step. The system reads the plant and storage location details from the inbound delivery items and checks if the warehouse number assigned to the plant and storage location combination is an embedded EWM-enabled warehouse. If yes, the system automatically replicates the inbound delivery from SAP S/4HANA into embedded EWM using qRFC. The replicated inbound delivery becomes warehouse request in embedded EWM for goods receipt and putaway.

The warehouse request in embedded EWM enables the processing of warehouse activities such as putaway, picking, physical inventory, and so on. From the inbound delivery warehouse request, warehouse tasks can be created for putaway of goods into destination storage bins.

10. **Perform goods receipt in embedded EWM**
When the goods (shipment) arrive at the warehouse, goods receipt can be posted against the inbound delivery in embedded EWM. Transaction /SCWM/PRDI is used to process goods receipt in embedded EWM. In the same transaction, you can unload the inbound delivery, and you can reject the inbound delivery. If an inbound delivery is rejected, it updates the zero quantities in the inbound delivery on the SAP S/4HANA side.

If you post goods receipt for the inbound delivery in embedded EWM, the inbound delivery in SAP S/4HANA will be updated using qRFC, and goods movement 101 will be posted. A material document and an accounting document will be created for quantity and value update in inventory management and financial transaction recording in financial accounting.

11. **Putaway**
Putaway is the process of moving the received products/materials from yard to the destination storage bins within the warehouse. Embedded EWM automatically determines the path for putaway by finding the storage type, storage section and storage bins using putaway strategy. At the inbound delivery item level, embedded EWM stores the putaway status.

Upon the creation of a warehouse request (inbound delivery replication), embedded EWM automatically creates warehouse tasks for putaway activity. The warehouse task contains the destination storage bin details and the task details.

12. **Create intercompany billing document (supplying side)**
The intercompany billing document is required only for the intercompany stock transfer process involving two different legal entities. After the goods issue is

posted from the supplying plant, SAP S/4HANA allows you to create the intercompany billing document with reference to the outbound delivery using Transaction VFO1 on the sales (supplying plant) side. This document will reference both the outbound delivery and stock transfer order. The transfer price from the stock transfer order will be copied over to the billing document. Best practice is to send the billing document to the business partner of the receiving plant. An EDI message can be triggered as an output so that a logistics invoice verification document can be posted automatically on the receiving (purchasing) side.

13. **Post and reconcile intercompany invoice (receiving side)**
    This step is required only for the intercompany stock transfer process involving two different legal entities. It is also called the logistics invoice verification document, created with reference to the intercompany stock transfer order using Transaction MIRO in SAP S/4HANA. The system matches the invoice quantity and price with the received quantities against the stock transfer order and the transfer price and quantities of the stock transfer order. If there are any discrepancies during the invoice reconciliation (three-way match), the system blocks the invoice from payment processing. Those discrepancies will be handled by the exception handlers (accounts payable and buyer/planner). A fully reconciled invoice with no exceptions will be cleared for payment (intercompany settlements).

8.3.3   Goods Receipt Integration with Materials Management

Materials management functions in SAP S/4HANA and embedded EWM functions are deeply integrated. As previously explained, the purchasing function creates purchase orders for external procurement and purchase order contains the necessary confirmation control to create an inbound delivery. The purchase order must be assigned a confirmation control key, which allows inbound delivery creation. The inbound delivery is the bridge between SAP S/4HANA and embedded EWM.

In this section, we'll dive deep into the integration point between SAP S/4HANA materials management and embedded EWM.

Define Storage Bin for Goods Receipt Zone

Goods are first received into a generic storage area or goods receiving area for goods receipt and putaway into warehouse. A goods receipt zone with a storage type, storage section, and storage bin must be created in embedded EWM to support the goods receipt process.

To do so, create a storage bin using Transaction /SCWM/LS01 (refer to our discussion of creating storage bins in Section 8.2.5 for more details). Figure 8.59 shows that storage bin **GR-ZONE** has been created for the storage type **9010** (**Provide in Goods Receipt**), and storage section **0001** (**Total Section**) in warehouse number **MJ00**.

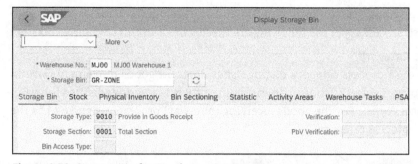

Figure 8.59 Storage Bin for Goods Receipt Zone

Warehouse Process Types for Putaway

Embedded EWM uses warehouse process types to control the creation of warehouse tasks for putaway (inbound processing), picking (outbound processing), posting changes (internal warehouse movements), and so on. Warehouse process types are configured for warehouse numbers in embedded EWM. To define warehouse process types for an embedded EWM warehouse number, execute Transaction SPRO and navigate to **SCM Extended Warehouse Management • Extended Warehouse Management • Cross-Process Settings • Warehouse Task • Define Warehouse Process Type**. To define a warehouse process type, click **New Entries** from the initial screen and specify the embedded EWM warehouse number, a four-digit alphanumeric key in the **Whse Proc. Type** (warehouse process type) field, and a description. Figure 8.60 shows warehouse process type **1010** (putaway) defined for warehouse number **MJ00** for the putaway process.

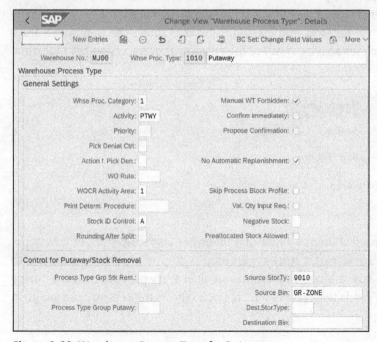

Figure 8.60 Warehouse Process Type for Putaway

The goods receiving area (storage type **9010** and storage bin **GR-ZONE**) is assigned to the warehouse process type. The following are the key fields of this configuration step:

- **Whse Proc. Category**
  The warehouse process category distinguishes different goods movements within the warehouse. Table 8.24 shows the warehouse process categories defined for warehouse process types. For the goods receiving (putaway) process, assign key **1** as the warehouse process category.

Category	Description
1	Putaway
2	Stock removal
3	Internal warehouse movement
7	Posting change

Table 8.24 Warehouse Process Categories for Warehouse Process Types

- **Activity**
  Activities are defined for the warehouse number in embedded EWM. Refer to our discussion of defining activities for activity areas in Section 8.2.5 for details. For the goods receiving activity, assign **PTWY** as the activity.

- **WOCR Activity Area**
  A warehouse order creation rule (WOCR) defines how the warehouse tasks should be grouped together for putaway and picking. Table 8.25 shows the warehouse order creation rule search for activity area. For the putaway process, assign key **1** (destination) in this field.

WOCR Activity Area	Description
Blank	Source
1	Destination

Table 8.25 WOCR Activity Area

- **Stock ID Control**
  Stock identification is a unique value that identifies a specific stock of a material using attributes such as batch, stock type, quantity, and so on. The key assigned in this field controls how the system identifies stock in every product/material during goods receipt and goods issue processes. Table 8.26 shows the keys you can assign in this field.

Stock Identification Control	Description
Blank	No stock identification
A	Stock identification only if externally predetermined
B	Create stock identification if none exists
C	Always assign stock identification anew

Table 8.26 Stock Identification Control

The **Control for Putaway/Stock Removal** section has the source storage type **9010** and storage bin **GR-ZONE** assigned to the source storage type and source storage bin fields, respectively.

Determine Warehouse Process Type for Inbound Process

In this configuration step, define the automatic determination of a warehouse process type in embedded EWM. The system automatically creates a warehouse request when the inbound delivery is replicated from SAP S/4HANA to embedded EWM, and in the warehouse request, the warehouse process type will be automatically determined based on this configuration setup.

To determine a warehouse process type for an inbound process for an embedded EWM warehouse number, execute Transaction SPRO and navigate to **SCM Extended Warehouse Management** • **Extended Warehouse Management** • **Cross-Process Settings** • **Warehouse Task** • **Determine Warehouse Process Type**.

To define determination criteria for the warehouse process type, click **New Entries** and specify the embedded EWM warehouse number in the **War...** field, select a document type from the dropdown list in the **Doc...** field, select the delivery priority from the dropdown list in the **Del.Prio.** field, select the process type indicator from the dropdown list in the **ProTypeDet** field, select the process indicator from the dropdown list in the **Process Ind.** field, and select the process type from the dropdown list in the **Proc...** field. Figure 8.61 shows that warehouse process type **1010** (putaway) has been assigned to the combination of warehouse number **MJ00** and document type **INB**.

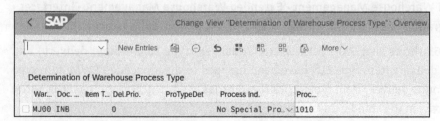

Figure 8.61 Determine Warehouse Process Type for Inbound Process

A document type (**Doc…**) classifies different types of documents in embedded EWM. Document type **INB** is a standard embedded EWM document type for inbound delivery.

Map Document Types from ERP System to EWM for Inbound

This interface setup is one of the main integration settings for the inbound process. In this configuration step, map the delivery type from SAP S/4HANA for the inbound delivery to the embedded EWM document type for inbound delivery creation.

To map the SAP S/4HANA delivery type to embedded EWM document type for inbound delivery, execute Transaction SPRO and navigate to **SCM Extended Warehouse Management • Extended Warehouse Management • Interfaces • ERP Integration • Delivery Processing • Map Document Types from ERP System to EWM**.

Figure 8.62 shows that SAP S/4HANA (ERP) delivery type **EL** (inbound delivery) has been mapped to the embedded EWM document type **INB** (inbound delivery). Click **New Entries** to add a new mapping if this setup doesn't exist in the standard system.

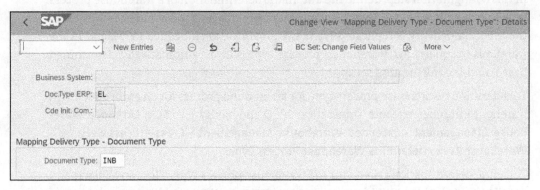

Figure 8.62 Map Document Types from ERP to EWM

Map Item Types from ERP System to EWM for Inbound

In this configuration step, map the inbound delivery item type from SAP S/4HANA to the embedded EWM item type for inbound delivery creation.

To arrive at this configuration step, execute Transaction SPRO and navigate to **SCM Extended Warehouse Management • Extended Warehouse Management • Interfaces • ERP Integration • Delivery Processing • Map Item Types from ERP System to EWM**.

Figure 8.63 shows that SAP S/4HANA (ERP) delivery type **EL** (inbound delivery) and standard delivery item type **ELN** have been mapped to the embedded EWM item type **IDLV** (inbound delivery item type). Click **New Entries** to add a new mapping if this setup doesn't exist in the standard system.

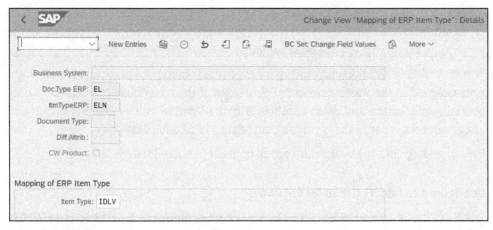

Figure 8.63 Map Item Types from ERP to EWM

Define Number Ranges for LE Deliveries

In this configuration step, define a delivery document number interval for the SAP S/4HANA (ERP) system in embedded EWM.

To arrive at this configuration step, execute Transaction SPRO and navigate to **SCM Extended Warehouse Management • Extended Warehouse Management • Interfaces • ERP Integration • Delivery Processing • Define Number Ranges for LE Deliveries**. You can arrive at this configuration step directly by executing Transaction SNRO.

In this configuration step, both internal and external number ranges can be defined with reference to number range object /SCWM/DLNO. Figure 8.64 shows number range **01** defined for delivery document replication from SAP S/4HANA.

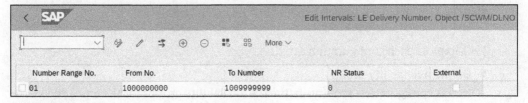

Figure 8.64 Define Number Ranges for LE Deliveries

8.3.4   Availability Groups for Inventory Management

Availability groups are logical keys defined in embedded EWM to determine the storage location and stock type of the SAP S/4HANA system. Availability groups are defined in configuration settings in embedded EWM. They play a vital role in the inbound process as they are used to set up an interface between SAP S/4HANA and embedded EWM. Storage locations and stock types such as unrestricted use stock, quality stock, and blocked stock are part of inventory management in SAP S/4HANA. Stock of materials in inventory management is stored at a storage location within a plant in one of the stock types. Embedded EWM doesn't have a storage location and stock type concept, so the availability group concept has been provided.

You must define an availability group for every combination of plant and storage location assigned to the warehouse in SAP S/4HANA. If the warehouse is assigned to only one storage location and plant combination, then define one availability group and assign all stock types in the configuration setup in embedded EWM.

Let's dive deep into the availability group concept in embedded EWM.

Configure Availability Group for Putaway

There are four activities in this configuration step, as shown in Figure 8.65. To arrive at this configuration step, execute Transaction SPRO and navigate to **SCM Extended Warehouse Management** • **Extended Warehouse Management** • **Goods Receipt Process** • **Configure Availability Group for Putaway**.

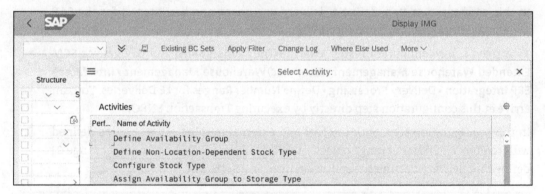

Figure 8.65 Configure Availability Group for Putaway: Activities

Let's explore each configuration activity:

- **Define Availability Group**

  In this configuration activity, define the availability groups for the embedded EWM warehouse number. Define at least one availability group per storage location assigned to the warehouse in SAP S/4HANA (ERP).

  Click **New Entries** to define a new availability group and specify an embedded EWM warehouse number, an alphanumeric key of up to 10 digits for the availability group, and a description. Figure 8.66 shows availability groups **001** (**Goods in Putaway (ROD)**) and **002** (**Goods Completely Available (AFS)**) defined for embedded EWM warehouse number **MJ00**. ROD stands for *ready on dock* and AFS stands for *available for sale*.

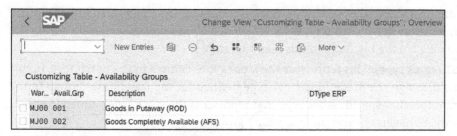

Figure 8.66 Activity 1: Define Availability Group

> **Note**
> If you do not have separate storage locations in inventory management and have only one storage location to manage receiving and shipping processes, define only one availability group for the warehouse.

- **Define Non-Location-Dependent Stock Type**
  In this configuration activity, define the non-location-specific stock types for the embedded EWM warehouse number. Define the stock types used in different processes to manage them in the warehouse. Non-location-specific stock types represent stock types defined in inventory management in SAP S/4HANA.

  Click **New Entries** to define a new non-location-specific stock type and specify the embedded EWM warehouse number, and a two-digit alphanumeric key for the non-location-specific stock type, and a description. Figure 8.67 shows non-location-specific stock types **BB** (blocked stock), **FF** (unrestricted-use stock), **QQ** (quality inspection stock), and **RR** (blocked stock returns) for embedded EWM warehouse number **MJ00**.

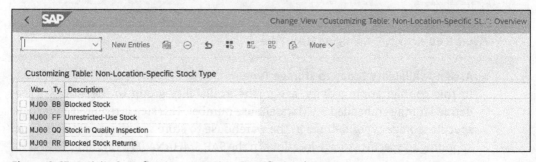

Figure 8.67 Activity 2: Define Non-Location-Specific Stock Type

- **Configure Stock Type**
  In this configuration activity, map the stock types for embedded EWM warehouse to the non-location-specific stock types which represent stock types defined in inventory management and availability groups.

Figure 8.68 shows the warehouse stock types mapped to an availability group and non-location-specific stock types for embedded EWM warehouse number **MJ00**. The following are the key fields of this mapping step:

- **ST (stock type):** This represents the stock types defined in the warehouse. In this field, assign the predefined stock type defined for the warehouse to map to the non-location-specific stock type that represents the stock type defined in inventory management and the availability group.

- **Avail.Grp (availability group):** In this field, assign the availability group for the stock type. For example, if you have a storage location in which you store blocked stocks in inventory management in SAP S/4HANA and you have defined an availability group defined for that storage location in embedded EWM, assign the availability group to the stock type defined for blocked stock in embedded EWM.

- **Ty. (non-location-specific stock type):** In this field, assign the non-location-specific stock type defined for the embedded EWM warehouse.

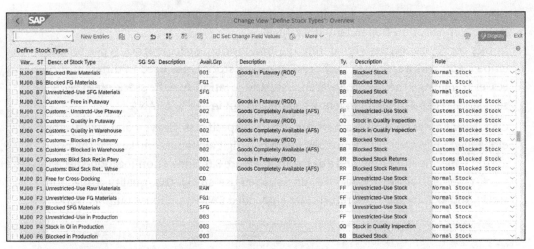

Figure 8.68 Activity 3: Configure Stock Type

- **Assign Availability Group to Storage Type**

  In this configuration activity, assign the availability group to the storage types defined for the embedded EWM warehouse number. This step is required if you have specific storage types defined in the warehouse to store specific stock types corresponding to specific storage locations in the SAP S/4HANA (ERP) system.

  Refer to our discussion of storage types in Section 8.2.5 for more details on how to assign the availability group to storage type.

Map Storage Locations from ERP System to EWM

In this configuration step, map the availability groups defined in the embedded EWM warehouse to the storage locations in SAP S/4HANA. This is one of the crucial steps for the integration setup between SAP S/4HANA and embedded EWM.

To arrive at this configuration step, execute Transaction SPRO and navigate to **SCM Extended Warehouse Management • Extended Warehouse Management • Interfaces • ERP Integration • Goods Movements • Map Storage Locations from ERP System to EWM**.

Click **New Entries** to map storage location from the SAP S/4HANA (ERP) system to embedded EWM. Specify the plant, a storage location, and the logical system of the SAP S/4HANA (ERP) system. Then, for this combination, assign the embedded EWM warehouse number and the availability group. Figure 8.69 shows that availability group **001** defined for embedded EWM warehouse number **MJ00** was assigned to the combination of plant **MJ00** and storage location **MJ00**.

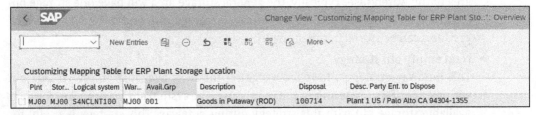

Figure 8.69 Map Storage Locations from ERP System to EWM

8.3.5   Configuration for Inbound Delivery Processing

Configuration of embedded EWM for inbound processing involves defining a putaway strategy. A *putaway* is a task for goods movement of received products from a receiving location to a destination storage bin based on putaway strategies. An effective putaway strategy not only optimizes warehouse activities but also optimizes resource utilization while improving overall efficiency.

The main purpose of putaway strategies is to move received products and materials into the right storage bins while utilizing the capacity and storage area of the storage bins. The system considers the type of material (hazardous goods, liquids, chemical, etc.), size and volume of materials, sequence of storage type search, capacity of storage bins, whether an addition to the existing storage bin stock is allowed or not, and so on for putaway. You can define a storage type search in such a way for the system to search for destination storage bins from one end of the warehouse to the other so that warehouse space can be used effectively. This helps with stock removal/picking for outbound deliveries, replenishment of stock to production, and so on.

The standard system provides the following putaway strategies:

- **Manually assigning storage bins**
  This putaway strategy allows the selection of storage bins at the time of putaway of stock in the warehouse. The system does not automatically propose a storage type and storage bin in the warehouse task for putaway; instead, the warehouse personnel searches for an empty storage bin and manually assigns the storage bin to the warehouse task and confirms putaway of the materials into the storage bin. This strategy is used for low-risk items.

- **Fixed storage bins**

  This putaway strategy allows for assigning a specific storage bin to the product/material. During inbound processing, the system automatically proposes the storage type, storage section, and storage bin in the warehouse task created for putaway of the product. This strategy is used for high-risk items that may require specific storage requirements. This putaway strategy affects the stock removal/picking process as the picking of products stored in the fixed storage bins must be done manually.

- **General storage strategy**

  This putaway strategy allows for assigning a storage bin from the general storage area of the warehouse. Typically, the general storage area will have one storage bin created per storage section. This is a mixed storage bin, and different materials/products can be stored in this storage bin.

- **Next empty bin strategy**

  This putaway strategy allows for assigning the next available empty storage bin within a specific storage area in the warehouse. The system always proposes the next available storage bin with this strategy during putaway. This strategy is useful for products with irregular sizes and is mostly used for high-rack and shelf storage types.

- **Addition to existing stock**

  This putaway strategy allows for assigning the storage bin with the same product to optimize the storage bin capacity. This strategy is used for non-batch-managed materials with similar specifications. The capacity of the storage bin must be defined, and the capacity must exist for adding to the existing stock. In case of limited capacity in the storage bin for additions to existing stock, the system finds an empty storage bin for putaway.

- **Near picking bin strategy**

  This putaway strategy allows for assigning the storage bins near the picking or stock removal area to allow faster processing of outbound deliveries. This strategy is used for fast-moving items that require high-frequency picking operations to minimize picking time.

- **Pallet storage strategy**

  This putaway strategy allows for assigning storage bins to store handling units (pallets, industrial pallets, etc.). The system checks the handling unit type and determines the storage bin for putaway.

- **Bulk storage strategy**

  This putaway strategy allows for assigning the storage bins for products that come in large quantities for storage in a bulk storage area. (e.g., consumer goods such as bottled drinks). Multiple handling unit types can be stored in a bulk storage area, and this area allows mixed storage.

Before configuring putaway strategies, the warehouse structure including storage types, storage sections, and storage bins must be configured in embedded EWM. Let's explore the configuration settings for putaway strategies.

Define Warehouse Number Parameters for Putaway

In this configuration step, define the control parameters for storage bin determination for putaway of stock during the inbound processing (goods receipt).

To define control parameters for the embedded EWM warehouse number, execute Transaction SPRO and navigate to **SCM Extended Warehouse Management • Extended Warehouse Management • Goods Receipt Process • Strategies • Define Warehouse Number Parameters for Putaway**.

All embedded EWM warehouse numbers defined in EWM are displayed in this configuration setting, and you can directly define the warehouse number parameters for putaway for any embedded EWM warehouse number. Figure 8.70 shows the control parameters set for embedded EWM warehouse number **MJ00**. Let's explore the control parameters:

- **Prio. StTp (priority: storage type)**
  This parameter controls the priority of the storage type search for determining a destination storage bin during the putaway process. If the priority is set to **High**, which is recommended for this field, the system prioritizes the storage type and alternative storage types are not prioritized.

- **Prio.Sec. (priority: storage section)**
  This parameter controls the priority of the storage section search for determining a destination storage bin during the putaway process. If the priority is set to **Medium**, which is recommended for this field, the system searches for an alternative storage bin type first if the **Priority: Storage Bin Type** field is set to **Low** and then it searches for an alternative storage section.

- **Prio.Type (priority: storage bin type)**
  This parameter controls the priority of the storage bin type search for determining a destination storage bin during the putaway process. If the priority is set to **Low**, which is recommended for this field, the system prioritizes the alternate storage bin type before the alternate storage section search and storage type search.

- **No BinDet. w/o Slot. (no bin determination without slotting)**
  If you set this indicator, the storage bin search works for the products that are previously slotted for the warehouse.

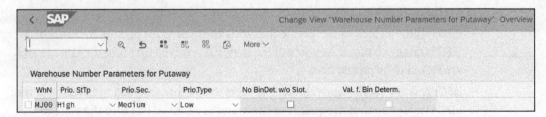

Figure 8.70 Define Warehouse Number Parameters for Putaway

Storage Type Search

This is the first step in the determination of the destination storage bin. The system searches for the destination storage bin when the warehouse task is created for putaway, and it uses the putaway strategy for the destination storage bin search. Under this configuration node, there are various key settings that will enable the storage type search for putaway. Let's explore the key settings.

Define Storage Type Groups

Storage type groups are defined for the embedded EWM warehouse number and specify stock removal and putaway rules for every group.

To define storage type groups for an embedded EWM warehouse number, execute Transaction SPRO and navigate to **SCM Extended Warehouse Management • Extended Warehouse Management • Goods Receipt Process • Strategies • Storage Type Search • Definition of Groups • Define Storage Type Groups.**

Figure 8.71 shows storage type groups **M001**, **M002**, and **M003** defined for embedded EWM warehouse number **MJ00**.

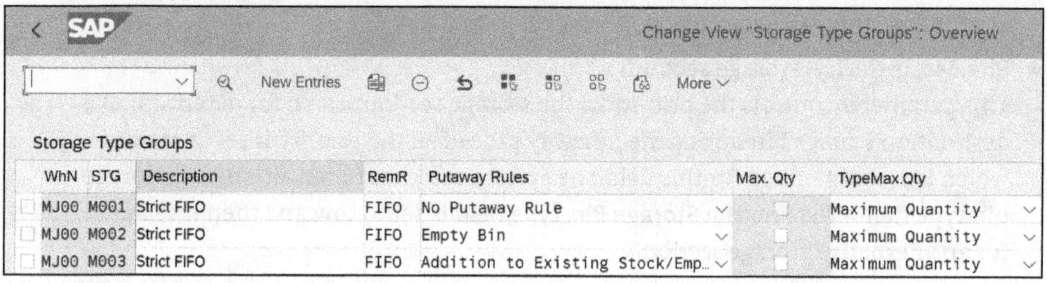
Change View "Storage Type Groups": Overview
New Entries
More
Storage Type Groups
WhN STG Description RemR Putaway Rules Max. Qty TypeMax.Qty
MJ00 M001 Strict FIFO FIFO No Putaway Rule Maximum Quantity
MJ00 M002 Strict FIFO FIFO Empty Bin Maximum Quantity
MJ00 M003 Strict FIFO FIFO Addition to Existing Stock/Emp... Maximum Quantity

Figure 8.71 Define Storage Type Groups

Click **New Entries** to define a new storage type group. You can copy an existing storage type group and create a new one. Create a four-digit alphanumeric key and description to define a new storage type group. The following are the key fields of this configuration:

- **RemR (stock removal rule)**
  Stock removal rules define the logic to search the source storage bin for picking. One of the following stock removal rules can be assigned to the storage type group:
  - **FIFO**: This is a stock removal rule that sorts the quants for picking in terms of the FIFO strategy so that the quants of the material received first into the storage area are sorted to be picked first.
  - **LIFO**: This is a stock removal rule that sorts the quants for picking in terms of the LIFO strategy so that the quants of the material received last into the storage area are sorted to be picked first.

- **Putaway Rules**

  A putaway rule defines the logic to search the destination storage bin for putaway. One of the following putaway rules can be assigned to the storage type group:

  - **No Putaway Rule**: If you assign this rule, the system will not search for a destination storage bin for the respective storage type group for putaway.

  - **Empty Bin**: If you assign this rule, the system will always search for an empty destination storage bin for the respective storage type group for putaway.

  - **Addition to Existing Stock/Empty Bin**: If you assign this rule, the system checks if adding to an existing storage bin stock is possible. If so, it proposes that storage bin; otherwise, it proposes an empty storage bin.

  - **General Storage Area**: If you assign this rule, the system searches for a storage bin in the general storage area that allows mixed storage. General storage area will have one storage bin created per storage section.

Assign Storage Types to Storage Type Groups

In this configuration step, assign storage types to storage type groups to control the storage type search for putaway.

To assign the storage types to storage type groups for an embedded EWM warehouse number, execute Transaction SPRO and navigate to **SCM Extended Warehouse Management • Extended Warehouse Management • Goods Receipt Process • Strategies • Storage Type Search • Definition of Groups • Assign Storage Types to Storage Type Groups**.

Figure 8.72 shows that storage types of warehouse number **MJ00** have been assigned to storage type groups. Click **New Entries** to assign a storage type to the combination of an embedded EWM warehouse number (**WhN**) and storage type group (**STG**). You can copy an existing assignment and create a new one.

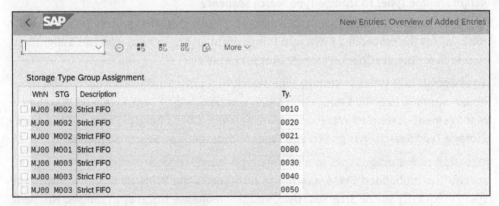

WhN	STG	Description	Ty.
MJ00	M002	Strict FIFO	0010
MJ00	M002	Strict FIFO	0020
MJ00	M002	Strict FIFO	0021
MJ00	M001	Strict FIFO	0080
MJ00	M003	Strict FIFO	0030
MJ00	M003	Strict FIFO	0040
MJ00	M003	Strict FIFO	0050

Figure 8.72 Assign Storage Types to Storage Type Groups

Define Storage Type Search Sequence for Putaway

In this configuration step, define the storage type search sequence, a four-digit alphanumeric key for putaway for the embedded EWM warehouse number. The system uses the storage type sequence maintained here to search for the destination storage bins for putaway. If a destination storage bin is identified in a storage type, the system assigns the storage bin to the warehouse task created for putaway.

To define the storage type search sequence for the embedded EWM warehouse number, execute Transaction SPRO and navigate to **SCM Extended Warehouse Management • Extended Warehouse Management • Goods Receipt Process • Strategies • Storage Type Search • Define Storage Type Search Sequence for Putaway**.

Figure 8.73 shows the storage type search sequence defined for embedded EWM warehouse number **MJ00** for putaway. Click **New Entries** to define a new storage type search sequence. You can copy an existing definition and create a new one. Define a four-digit alphanumeric key and description to define the storage type search sequence.

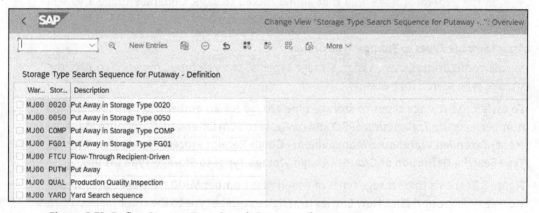

Figure 8.73 Define Storage Type Search Sequence for Putaway

Assign Storage Types to Storage Type Search Sequence

In this configuration step, assign the storage type to a storage type search sequence for putaway for the embedded EWM warehouse number. The system uses this assignment to search for the destination storage bins for putaway.

To assign storage types to storage type search sequence for the embedded EWM warehouse number, execute Transaction SPRO and navigate to **SCM Extended Warehouse Management • Extended Warehouse Management • Goods Receipt Process • Strategies • Storage Type Search • Assign Storage Types to Storage Type Search Sequence**.

To assign new storage types to a storage type search sequence, click **New Entries** and specify the embedded EWM warehouse number in the **WhN** field, the storage type search sequence in the **Srch Seq.** field, and the sequence number in the **Seq No.** field, and then assign the storage type. Figure 8.74 shows the storage type search sequence **PUTW** defined for putaway process have been assigned to the storage types **0050**, **0020**, **0010**, and **0080** in a sequence for warehouse number **MJ00**.

The **Seq. No.** field defines the sequence in which the system must search the storage types for determining a destination storage bin for putaway.

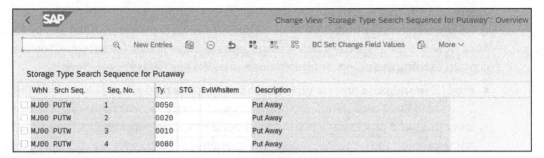

Figure 8.74 Assign Storage Types to Storage Type Search Sequence

Define Putaway Control Indicator

Putaway control indicators control the putaway of certain specific materials into a specific storage area (storage type) within the warehouse.

To define the putaway control indicators for an embedded EWM warehouse number, execute Transaction SPRO and navigate to **SCM Extended Warehouse Management • Extended Warehouse Management • Goods Receipt Process • Strategies • Storage Type Search • Define Putaway Control Indicator**.

Figure 8.75 shows putaway control indicators **0020** and **0050** defined for embedded EWM warehouse number **MJ00** for putaway. Click **New Entries** to define a new putaway indicator. You can copy an existing definition and create a new one. Specify a four-digit alphanumeric key and description to define the putaway control indicator.

Figure 8.75 Define Putaway Control Indicator for Putaway

Specify Storage Type Search Sequence for Putaway

In this configuration step, specify a storage type search sequence for putaway using putaway control indicators and other criteria. As previously explained, the storage type search sequence is assigned to the storage type, which the system uses to determine destination storage bins for putaway.

To specify a storage type search sequence for the embedded EWM warehouse number, execute Transaction SPRO and navigate to **SCM Extended Warehouse Management • Extended Warehouse Management • Goods Receipt Process • Strategies • Storage Type Search • Specify Storage Type Search Sequence for Putaway**.

Figure 8.76 shows the storage type search sequence for putaway specified for embedded EWM warehouse number **MJ00** using putaway control indicators **0020** and **0050**. Click **New Entries** to specify a new storage type search sequence for putaway. You can copy an existing definition and create a new one. The key fields are as follows:

- **Proc./… (warehouse process type/warehouse process type group)**
  Embedded EWM uses warehouse process types for warehouse activities such as putaway (inbound processing), picking (outbound processing), posting changes (internal warehouse movements), and so on. Warehouse process types are configured for warehouse numbers in embedded EWM. Assign the warehouse process type defined for putaway for warehouse number **MJ00** in this field. (See our discussion of warehouse process types in Section 8.3.3 for more information.)

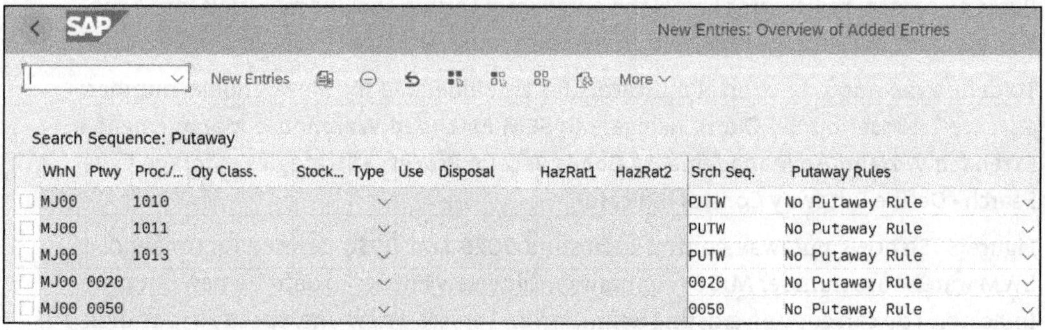

Figure 8.76 Specify Storage Type Search Sequence for Putaway

- **Qty Class. (quantity classification)**
  You can use quantity classification as a criterion for storage type search. The quantity classification comes from the packaging materials. You can assign one of the quantity classifications defined in Table 8.27 if you want the storage type search criteria to include quantity classification.

Quantity Classification	Description
1	Eaches
2	Cartons
3	Tiers
4	Pallets
C	Case quantity
L	Layer quantity
P	Pallet quantity

Table 8.27 Quantity Classification

- **Stock... (stock type/group)**
  You can use the stock type defined for the respective warehouse number in embedded EWM as a search criterion for storage type search. Refer to our discussion of configuring stock types in Section 8.3.4 for more details.

- **HazRat1/HazRat2 (hazard rating)**
  Hazard ratings are assigned to hazardous materials. You can specify the hazard ratings in this field for the system to determine an appropriate storage type. You can specify two hazard ratings for the storage type search per sequence.

- **Srch Seq. (storage type search sequence)**
  In this field, assign the storage type search sequence defined for the respective embedded EWM warehouse number. The system uses the storage type sequence maintained here to search for destination storage bins for putaway.

- **Putaway Rules**
  A putaway rule defines the logic to search the destination storage bin for putaway. The same putaway rules we discussed earlier can be assigned here (**No Putaway Rule**, **Empty Bin**, **Addition to Existing Stock/Empty Bin**, and **General Storage Area**).

Optimize Access Strategy for Storage Type Search: Putaway

Embedded EWM provides various control parameters for storage type search for putaway such as putaway control indicators, warehouse process types, quantity classifications, stock types, stock categories, usage indicators, hazard ratings, and so on. To optimize the storage type search, embedded EWM provides an approach that can be configured in a sequential order.

To optimize a storage type search for the embedded EWM warehouse number, execute Transaction SPRO and navigate to **SCM Extended Warehouse Management • Extended Warehouse Management • Goods Receipt Process • Strategies • Storage Type Search • Optimize Access Strategy for Storage Type Search: Putaway**.

Figure 8.77 shows the storage type search optimization for putaway defined for embedded EWM warehouse number **MJOO**.

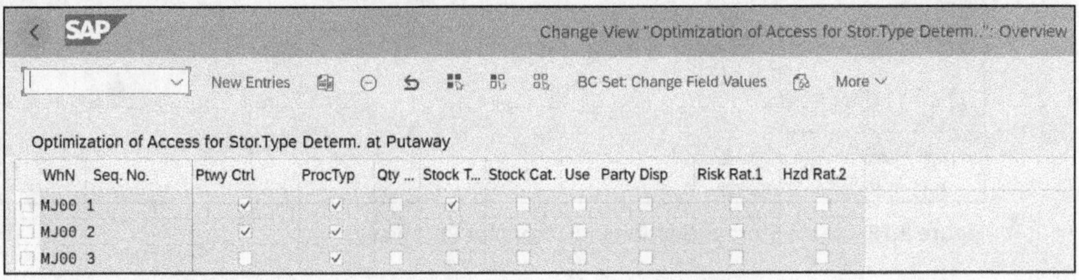

Figure 8.77 Optimize Access Strategy for Storage Type Search: Putaway

Click **New Entries** to define a new sequence. Let's walk through the sequences defined for embedded warehouse number **MJOO**:

- Sequence number **1** considers the putaway control indicator, warehouse process type, and stock type to optimize the storage type search.
- Sequence number **2** considers the putaway control indicator and warehouse process type to optimize the storage type search.
- Sequence number **3** considers only the warehouse process type to optimize the storage type search.

Storage Section Search

This is the second step in the determination of destination storage bin. The system searches for the destination storage bin at the time of creation of warehouse task for putaway and it uses the putaway strategy for the destination storage bin search. Under this configuration node, there are certain key settings that will enable the storage section to search for putaway. We'll discuss them in the following sections.

Create Storage Section Indicators

Storage section indicators are used to define certain attributes of products to search the storage sections after the system searches for a destination storage type within the embedded EWM warehouse number. Storage section indicators can be defined manually in the warehouse product master record or can be determined automatically through the slotting process.

To define storage section indicators for the embedded EWM warehouse number, execute Transaction SPRO and navigate to **SCM Extended Warehouse Management • Extended Warehouse Management • Goods Receipt Process • Strategies • Storage Section Search • Create Storage Section Indicators**.

Click **New Entries** to create a new storage search indicator and specify a four-digit numeric key to define the storage search indicator and enter the description. Figure 8.78 shows storage section indicators **0001** (fast-moving items) and **0002** (slow-moving items) defined for embedded EWM warehouse number **MJ00**.

Figure 8.78 Create Storage Section Indicators for Putaway

Maintain Storage Section Search Sequence

In this configuration step, maintain the storage section search sequence for every storage type and warehouse number combination. You can refine the storage section search using the storage section indicators, hazard ratings, and a sequence number.

To define storage section indicators for the embedded EWM warehouse number, exe-cute Transaction SPRO and navigate to **SCM Extended Warehouse Management** • **Extended Warehouse Management** • **Goods Receipt Process** • **Strategies** • **Storage Section Search** • **Maintain Storage Section Search Sequence**.

To maintain a new storage section search sequence, click **New Entries** and specify the embedded EWM warehouse number, the storage type of the warehouse in which the storage section is to be searched, a storage section indicator, and a sequence number for the search sequence. Then, to this combination, assign the storage section defined in the same embedded EWM warehouse and storage type. Figure 8.79 shows the storage section search criteria defined for warehouse number **MJ00** and for different storage types. A storage type can have multiple storage sections, so it is vital to automatically search for the relevant storage section during putaway. Do not use the storage section indicator if the storage type has only one storage section defined (general storage area, bulk storage, etc.).

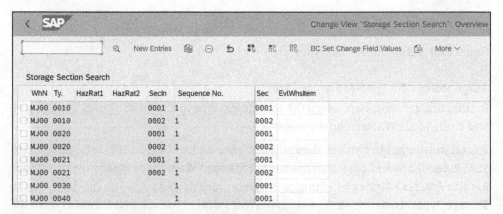

Figure 8.79 Maintain Storage Section Search Sequence for Putaway

Storage Bin Determination

This is the third step in the determination of a destination storage bin. The system searches for the destination storage bin at the time of creation of a warehouse task for putaway, and it uses the putaway strategy for the destination storage bin search. Under this configuration node, there are certain key settings that will enable the storage bin determination for putaway. We'll explore them in the following sections.

Define Storage Bin Types

You can define storage bin types based on the size, volume, and type of materials and handling units that can be stored in the storage bins.

To define storage bin types for the embedded EWM warehouse number, execute Transaction SPRO and navigate to **SCM Extended Warehouse Management** • **Extended Warehouse Management** • **Goods Receipt Process** • **Strategies** • **Storage Bin Determination** • **Define Storage Bin Types**.

Figure 8.80 shows the overview of storage bin types defined for embedded EWM warehouse number **MJ00**. Click **New Entries** to define a new storage bin type. You can copy an existing storage bin type and create a new one. Enter a four-digit alphanumeric key and description to define a storage bin type. Refer to our discussion of defining storage bin types in Section 8.2.5 for more details.

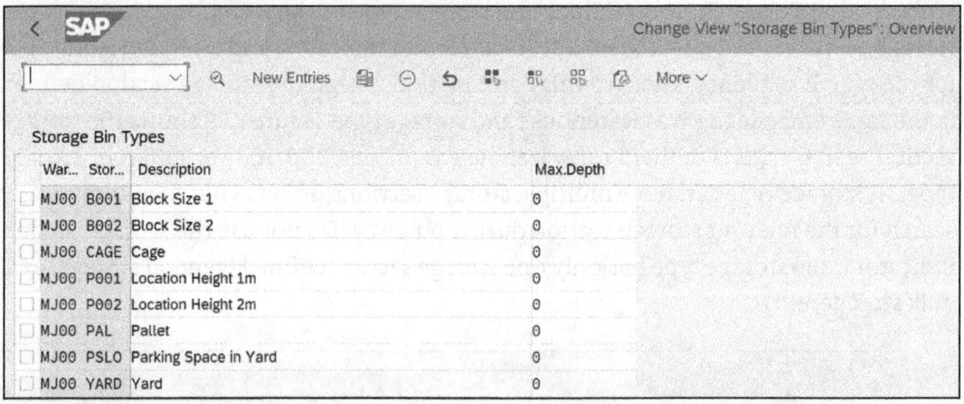

War...	Stor...	Description	Max.Depth
MJ00	B001	Block Size 1	0
MJ00	B002	Block Size 2	0
MJ00	CAGE	Cage	0
MJ00	P001	Location Height 1m	0
MJ00	P002	Location Height 2m	0
MJ00	PAL	Pallet	0
MJ00	PSLO	Parking Space in Yard	0
MJ00	YARD	Yard	0

Figure 8.80 Define Storage Bin Types

Assign Storage Bin Types to Storage Types

In this configuration step, assign the storage bin type to a combination of storage type and embedded EWM warehouse number.

To assign storage bin types to storage types, execute Transaction SPRO and navigate to **SCM Extended Warehouse Management • Extended Warehouse Management • Goods Receipt Process • Strategies • Storage Bin Determination • Assign Storage Bin Types to Storage Types**. To define storage bin types to a storage type, click **New Entries** from the initial screen and specify the embedded EWM warehouse number and storage type. To this combination, assign the storage bin type.

Figure 8.81 shows the storage bin types that are assigned to the storage types defined in the warehouse **MJ00**.

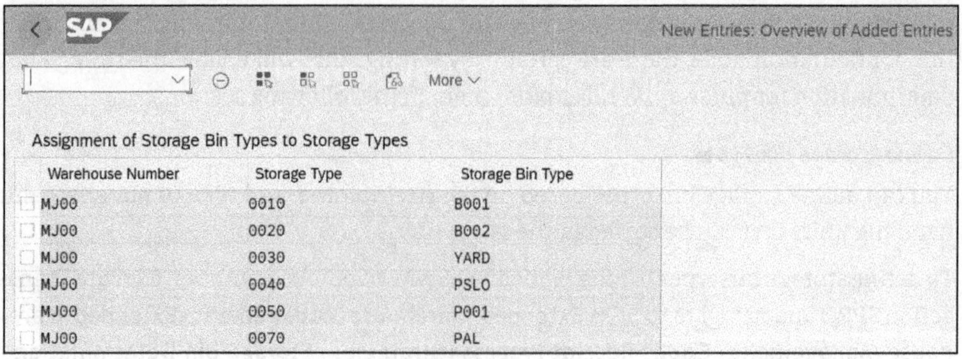

Warehouse Number	Storage Type	Storage Bin Type
MJ00	0010	B001
MJ00	0020	B002
MJ00	0030	YARD
MJ00	0040	PSLO
MJ00	0050	P001
MJ00	0070	PAL

Figure 8.81 Assign Storage Bin Types to Storage Types

Warehouse Product Maintenance for Putaway

As we introduced in Section 8.1.3, the warehouse product master is important master data in embedded EWM, and it is a replica of the material master from SAP S/4HANA. Replication of the material master as a warehouse product master into embedded EWM is required to perform warehouse activities using that product. After the replication, the warehouse product master is enriched with certain embedded EWM warehouse-related data. To create the warehouse product, execute Transaction /SCWM/MAT1.

Figure 8.82 shows the warehouse data maintained for putaway of product **MS-RAW MATERIAL1** into embedded EWM warehouse number **MJ00**. Note the following:

- Putaway control indicator **0020** has been maintained under the **Putaway** section of the warehouse data for the product. Putaway control indicators control the storage type search.

- Storage section indicator **0001** has been maintained under the **Putaway** section of the warehouse data for the product. Storage section indicators control the storage section search.

- Storage bin type **B002** has been maintained under the **Putaway** section of the warehouse data for the product. Storage bin types control the storage bin search.

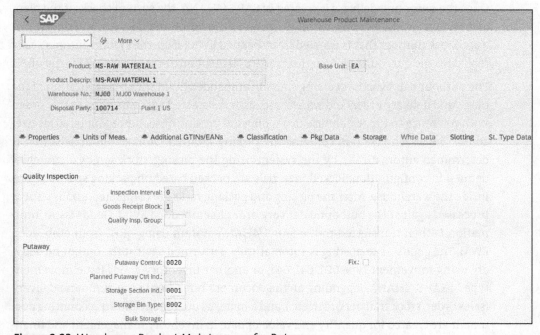

Figure 8.82 Warehouse Product Maintenance for Putaway

8.4 Outbound Delivery Processing

Outbound delivery processing is used to ship goods to customers (sales order process) or receiving plants (stock transfer process). In embedded EWM, *outbound delivery processing* refers to picking of requested materials from the warehouse, packing of picked materials, and goods issue/shipping of requested materials to a customer with reference to sales order or receiving plant (stock transfer process) from the warehouse.

In the following sections, we'll dive deep into the outbound delivery process in embedded EWM and explore SAP S/4HANA inventory management and embedded EWM integration, including goods issue with step-by-step configuration instructions.

8.4.1 Integrated Outbound Deliveries

Outbound deliveries are mandatory documents in embedded EWM that integrate SAP S/4HANA and embedded EWM for outbound processing. The outbound delivery document contains all the relevant information such as item details (material, quantity, delivery date, etc.) and customer details such as delivery address to process the goods issue integrated with SAP S/4HANA and embedded EWM. The outbound delivery is an internal document created with reference to a sales order or stock transfer order in SAP S/4HANA. Creation of outbound deliveries can be automated using background processing in SAP S/4HANA.

Once the outbound delivery is created in SAP S/4HANA, the system checks if the plant and storage location combination assigned to the outbound delivery is assigned to a warehouse number that is enabled in embedded EWM. If so, the system automatically replicates the outbound delivery from SAP S/4HANA into embedded EWM using qRFC.

The outbound delivery processing starts in embedded EWM with the replication of the outbound delivery order, and warehouse tasks are created with reference to this order. With reference to the warehouse tasks, physical picking of requested materials is executed. The warehouse task contains the picking location (storage bin details) that is determined automatically by the system using the picking/stock removal strategies defined in configuration. Picked materials are packed based on packing specifications inside the warehouse. After the picking and packing tasks are completed, goods issue is processed against the outbound delivery order in embedded EWM. Goods issue information is then transferred to the SAP S/4HANA system using qRFC from embedded EWM. The goods issue posting is automatically performed against the outbound delivery using movement type 601, 641, 643, or another inventory goods issue movement type in SAP S/4HANA depending on the document reference in the outbound delivery (sales order, stock transfer order, etc.), and a material document and an accounting document are created for quantity and value update in inventory management and financial accounting, respectively.

Materials management functions in SAP S/4HANA (ERP) and warehouse activities in embedded EWM are tightly integrated to process outbound deliveries. Table 8.28 shows the key process impacts between SAP S/4HANA materials management and embedded EWM. For outbound delivery processing, inventory management processes of SAP S/4HANA materials management are deeply impacted.

Area	Processes Impacted	Description
Materials management	Create and maintain purchasing master data	This process supports the creation and maintenance of purchasing master data such as source list and PIR to support the stock transfer process.
	Create and maintain purchasing documents	This process focuses on creation of purchasing documents such as stock transfer requisitions and stock transfer orders to procure materials/products from another plant.
	Post goods issue	This process supports the posting of goods issue for outbound delivery to ship the goods to the receiving plant or customer depending on the outbound delivery details.
Extended warehouse management (EWM)	Manage picking and packing	This process supports determination of an appropriate storage type, storage section, and storage bin as per the picking strategy in embedded EWM as well as confirms picking tasks and packing of picked quantity if required.
	Manage exception handling during outbound process	This process supports the exception handling of full/partial bin denial during outbound process.
	Monitor loading	This process supports the determination of criteria and relevance for loading of picked and packed materials and performing loading activities.
	Perform goods issue	This process supports the goods issue with reference to outbound delivery order in embedded EWM and updates the outbound delivery in SAP S/4HANA.
	Display stock overview	This process supports checking stock levels in the embedded EWM warehouse.

Table 8.28 Process Impact

8.4.2   Outbound Delivery Process

The outbound delivery process is applicable for customer sales orders and stock transfer orders. For the stock transfer process, refer to Section 8.3.2 for details. In this section, let's dive deep into outbound delivery processing in embedded EWM using the sales order process.

Figure 8.83 shows the outbound delivery process for a sales order, which involves steps in both SAP S/4HANA and embedded EWM. Let's take a closer look at the process steps:

1. **Creation of sales order**

   Customer orders requesting goods or services will be converted into sales orders in SAP S/4HANA using Transaction VA01. Creation of a sales order can be fully automated via an EDI technique. The sales order can also be created with reference to the following documents:

   - An inquiry helps to keep track of customer interactions.
   - A quotation helps keep consistency with pricing and other terms.
   - Creating a sales order with reference to another sales order helps to copy the details over and maintain consistency with pricing and other details.
   - A common practice is to have a customer contract created for the long term. Creating the sales order with reference to a contract helps to keep consistency with pricing and other terms while ensuring contractual agreements.
   - Scheduling agreements ensure recurring deliveries to customers over a period. Creating the sales order with reference to a scheduling agreement helps to keep compliance with delivery schedules and improve overall efficiency.
   - A sales order with reference to a billing document helps in scenarios where customers have special billing agreements and payment terms.

   The system allows you to enter the business partner ID of the customer created in the SAP S/4HANA system, materials, requested quantity, requested delivery date, and pricing details during the creation of a sales order.

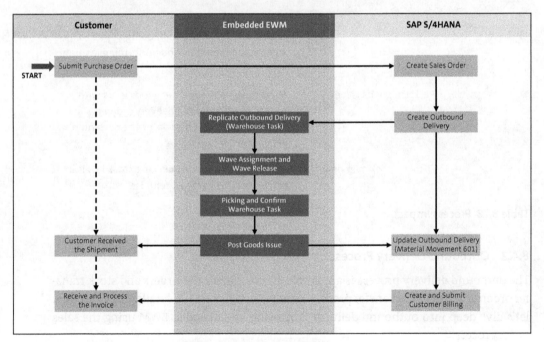

Figure 8.83 EWM Outbound Delivery Processing: Sales Order

2. **Availability check**

   The availability check is the most important functionality of the sales order process that ensures enough of the requested materials are available in the stock to fulfill the sales order by the requested delivery date. The availability check will be triggered automatically during the creation of a sales order using the checking rule. Performing an availability check ensures timely delivery of products to the customers and optimizes inventory management while improving customer satisfaction. After a successful availability check, the system confirms the sales order and automatically allocates the requested quantities to the sales order to deliver the requested quantities of the material on the requested date.

3. **Create outbound delivery**

   The next step in the sales order processing after the requested quantities are available to fulfill the customer order is to create an outbound delivery. Outbound deliveries will be created using Transaction VL01N with reference to a sales order. Outbound deliveries help organizations to prepare for the shipments of goods to the customer and contain specific quantities of the materials to be shipped and delivery dates for scheduling shipments. Inventory and warehouse management activities such as picking of materials and packing activities are performed with reference to outbound deliveries.

4. **Replicate outbound delivery into embedded EWM (automated)**

   This is an automated step. The system reads the plant and storage location details from the outbound delivery items and checks if the warehouse number assigned to the plant and storage location combination is an embedded EWM-enabled warehouse. If yes, the system automatically replicates the outbound delivery from SAP S/4HANA into embedded EWM using qRFC. The replicated outbound delivery becomes an outbound delivery order in embedded EWM for picking, packing, and shipping of goods from the warehouse. You can display the replicated outbound delivery order in embedded EWM using Transaction /SCWM/PRDO.

   Warehouse tasks are created with reference to outbound delivery orders to execute the warehouse activities. The warehouse task contains the picking location (storage type, storage section, and storage bin) details that are determined automatically based on the picking strategy defined in embedded EWM.

5. **Wave assignment and wave release in embedded EWM**

   In embedded EWM, wave management enables the creation of a wave for multiple warehouse tasks. For picking, a wave can be created, and multiple outbound delivery orders can be assigned to it. Wave management optimizes the warehouse operations and reduces efforts. You can create a wave using Transaction /SCWM/WAVE in embedded EWM. Assign the outbound delivery orders to the wave and release the wave. This way, you can create multiple warehouse tasks for picking and releasing all tasks at once. This is an optional step, and you can complete picking for a warehouse task individually.

6. **Picking and confirm warehouse task**

   The warehouse task for picking directs the stock removal from source locations within the warehouse where the required materials have been stored. Embedded EWM automatically determines the path for picking by finding the source storage type, source storage section, and source storage bins using a stock removal strategy. At the outbound delivery item level, embedded EWM stores the putaway status.

   Upon the creation of a warehouse request (outbound delivery replication), embedded EWM automatically creates warehouse tasks for picking activity. The warehouse task contains the source storage bin details and the task details for stock removal/picking.

7. **Post goods issue in embedded EWM**

   After all the warehouse tasks for picking activity are completed and confirmed, goods issue can be posted for the outbound delivery in embedded EWM using Transaction /SCWM/PRDO. In the same transaction, you can load the outbound delivery.

   If you post goods issue for the outbound delivery in embedded EWM, the corresponding outbound delivery on the SAP S/4HANA side will be updated using qRFC, and goods movement 601 (goods issue for customer sales order) will be posted automatically. A material document and an accounting document will be created for quantity and value update in inventory management and financial transaction recording in financial accounting.

8. **Create and submit customer billing document**

   A customer billing/invoice document can be created with reference to a sales order or outbound delivery document. Best practice is to create the billing document after the shipment (goods issues for outbound delivery) is completed. The customer billing document can be created using Transaction VF01. The system calculates the output tax and determines an output tax code based on the country of origin, shipping address, and products/services delivered. Output tax will be added to the billing document and collected from the customer. The seller will pay the taxes on behalf of the customer to the local government authority of the shipping country and state. The billing document will be transmitted to the customer via email or EDI.

9. **Process incoming payment**

   Incoming payments or payments received from the customers against the outstanding billing documents will be processed in SAP S/4HANA as part of the accounts receivable function. Payment receipts are entered into the system using Transaction F-28. A reference document (check, eCheck, etc.), payment amount, payment date, and payment method will be recorded in the payment receipt document. This process can be automated as well.

8.4.3   Goods Issue Integration with Materials Management

Goods issue integration between SAP S/4HANA materials management and embedded EWM starts from the replication of outbound deliveries from SAP S/4HANA to embedded

EWM. For the replicated outbound delivery, in embedded EWM, an outbound delivery order (warehouse request) is created, against which warehouse tasks for picking and packing are created. Upon posting of goods issue from embedded EWM, the outbound delivery is updated in SAP S/4HANA with the goods issue posting. The outbound delivery in SAP S/4HANA can be created with reference to stock transfer orders (both intracompany and intercompany) and sales orders. Outbound deliveries, once created in SAP S/4HANA, are automatically replicated to embedded EWM and create warehouse tasks for picking the requested materials from the warehouse. The source storage type, storage section, and storage bins are automatically determined and assigned to the warehouse task for picking using picking strategies. After the warehouse personnel complete the picking from the source storage bins and confirm the warehouse tasks, the goods issue is posted in embedded EWM and the corresponding outbound delivery in SAP S/4HANA is updated.

Let's explore how outbound deliveries are replicated and linked between SAP S/4HANA and embedded EWM.

Define Storage Bin for Goods Issue Zone

Goods are moved to the goods issue area within the warehouse after stock removal from the picking locations (storage bins) and from where the goods issue is posted for outbound deliveries. A goods issue zone with a storage type, storage section, and storage bin must be created in embedded EWM to support the goods issue process.

To do so, create a storage bin using Transaction /SCWM/LS01 (refer to our discussion of creating storage bins in Section 8.2.5 for more details). Figure 8.84 shows that storage bin **GI-ZONE** has been created for the storage type **9020** (provide in goods issue) and storage section **0001** (total storage) in warehouse number **MJ00**.

Figure 8.84 Storage Bin for Goods Issue Zone

Warehouse Process Types for Picking

Embedded EWM uses warehouse process types for warehouse activities such as putaway (inbound processing), picking (outbound processing), posting changes (internal warehouse movements), and so on. Warehouse process types control the creation of warehouse tasks for putaway, picking, packing, and posting changes. Warehouse process types are configured for the warehouse number in embedded EWM.

To define warehouse process types for an embedded EWM warehouse number, execute Transaction SPRO and navigate to **SCM Extended Warehouse Management • Extended Warehouse Management • Cross-Process Settings • Warehouse Task • Define Warehouse Process Type**.

To define a warehouse process type for picking, click **New Entries** from the initial screen and specify the embedded EWM warehouse number, a four-digit alphanumeric key in the **Whse Proc. Type** (warehouse process type) field, and a description. Figure 8.85 shows warehouse process type **2010** (**Stock Removal**) defined for warehouse number **MJ00** for the picking process. The goods issue area (destination storage type **9020** and destination storage bin **GI-ZONE**) is assigned to the warehouse process type.

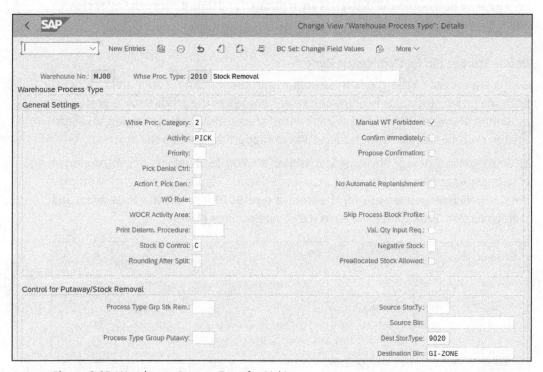

Figure 8.85 Warehouse Process Type for Picking

The following are the key fields of this configuration step:

- **Whse Proc. Category**
  The warehouse process category distinguishes different goods movements within the warehouse. For the goods issue (picking) process, assign key **2** as the warehouse process category.

- **Activity**
  Activities are defined for the warehouse number in embedded EWM. Refer to our discussion of defining activities for activity areas in Section 8.2.5 for details. For the goods issue activity, assign **PICK** as the activity.

- **Stock ID Control**
  Stock identification is a unique value that identifies a specific stock of a material using attributes such as batch, stock type, quantity, and so on. The key assigned in this field controls how the system identifies stock in every product/material during goods receipt and goods issue processes.

The **Control for Putaway/Stock Removal** section has **9020** and **GI-ZONE** assigned to the **Dest.Stor.Type** and **Destination Bin** fields, respectively.

Determine Warehouse Process Type for Outbound Process

In this configuration step, define the automatic determination of a warehouse process type in embedded EWM. The system automatically creates a warehouse request when the outbound delivery is replicated from SAP S/4HANA to embedded EWM, and in the warehouse request, the warehouse process type will be automatically determined based on this configuration setup.

To determine the warehouse process type for the outbound process for an embedded EWM warehouse number, execute Transaction SPRO and navigate to **SCM Extended Warehouse Management • Extended Warehouse Management • Cross-Process Settings • Warehouse Task • Determine Warehouse Process Type**.

To define determination criteria for warehouse process type, click **New Entries** and specify the embedded EWM warehouse number in the **War...** field, select a document type from the dropdown list relevant for the outbound process in the **Doc...** field, select the delivery priority from the dropdown list in the **Del.Prio.** field, select the process type indicator from the dropdown list in the **ProTypeDet** field, select a process indicator from the dropdown list in the **Process Ind.** field, and select the process type from the dropdown list in the field **Proc Type**. Figure 8.86 shows that warehouse process type **2010** (stock removal) has been assigned to the combination of warehouse number **MJ00** and document type **OUTB**.

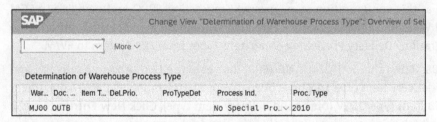

Figure 8.86 Determine Warehouse Process Type for Outbound Process

The document type (**Doc…**) classifies different types of documents in embedded EWM. Document type **OUTB** is a standard embedded EWM document type for outbound delivery.

Map Document Types from ERP System to EWM for Outbound

This is an interface setup and one of the main integration settings for the outbound process. In this configuration step, map the delivery type from SAP S/4HANA for the outbound delivery to the embedded EWM document type for outbound delivery creation.

To map the SAP S/4HANA delivery type to an embedded EWM document type for outbound delivery, execute Transaction SPRO and navigate to **SCM Extended Warehouse Management • Extended Warehouse Management • Interfaces • ERP Integration • Delivery Processing • Map Document Types from ERP System to EWM**.

Figure 8.87 shows that SAP S/4HANA (ERP) delivery type **LF** (outbound delivery) has been mapped to the embedded EWM document type **OUTB** (outbound delivery). Click **New Entries** to add a new mapping if this setup doesn't exist in the standard system.

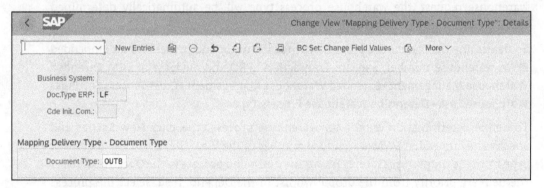

Figure 8.87 Map Document Types from ERP System to EWM for Outbound

Map Item Types from ERP System to EWM

In this configuration step, map the outbound delivery item type from SAP S/4HANA to the embedded EWM item type for outbound delivery creation.

To arrive at this configuration step, execute Transaction SPRO and navigate to **SCM Extended Warehouse Management • Extended Warehouse Management • Interfaces • ERP Integration • Delivery Processing • Map Item Types from ERP System to EWM**.

Figure 8.88 shows that SAP S/4HANA (ERP) delivery type **LF** (outbound delivery) and standard delivery item type **TAN** have been mapped to the embedded EWM document type **OUTB** item type **ODLV** (outbound delivery item type). Click **New Entries** to add a new mapping if this setup doesn't exist in the standard system.

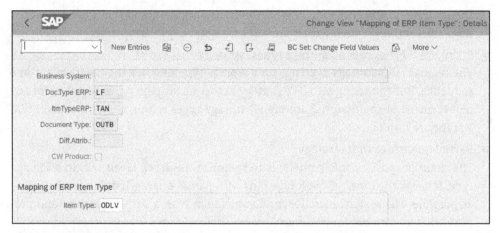

Figure 8.88 Map Item Types from ERP System to EWM for Outbound

8.4.4   Configuration for Outbound Delivery Processing

Defining stock removal strategies is a vital configuration in embedded EWM for outbound delivery process. Picking/stock removal strategies are used to determine the picking location (source storage type, source storage section, and source storage bin) and assign it to the warehouse task created for picking for an outbound delivery order. An effective picking strategy not only optimizes the picking process but also optimizes resource utilization while improving overall efficiency. The main purpose of stock removal strategies is searching for an optimal storage bin and sorting of required materials within the warehouse to move the required products and materials from a source location into a goods issue area (**GI-ZONE**). With stock of requested materials being available in multiple storage bins, stock removal strategies prioritize certain quants and/or handling units over others based on the stock removal strategies defined for the embedded EWM warehouse number.

The standard system provides the following stock removal strategies:

- **FIFO strategy**
  This stock removal strategy considers the age of the quant of the requested material. The oldest quant is searched using this strategy based on the goods receipt date stored in the quant. The system automatically determines the source storage bin that stores the oldest quant of the material in the embedded EWM warehouse number, and that quant is picked first. This strategy is used with materials that expire over time. The storage types are searched in sequence in this strategy, and if the oldest quant is found in the first storage type, the system proposes that storage bin.

- **LIFO strategy**
  This stock removal strategy considers the age of the quant of the requested material. The last quant that is placed in the storage bin is searched using this strategy. The system automatically determines the source storage bin that stores the last quant of

the material in the embedded EWM warehouse number, and that quant is picked first. This strategy is used with materials that won't expire over time.

- **Stringent FIFO (across all storage types) strategy**
  The regular FIFO strategy searches for a storage type that has the oldest quant in a sequence. But the stringent FIFO strategy across all storage types searches for the oldest quant of the material among all storage types within the embedded EWM warehouse number.

- **Partial quantities first strategy**
  The main purpose of this strategy is to optimize the stock levels within a storage type. It keeps the number of storage bins with partial quants/handling units as low as possible. The system searches for a storage bin with a partial quantity and proposes that storage bin for picking.

- **Stock removal suggestion according to quantity**
  This strategy is used when the warehouse has multiple storage types with storage bins that store different quantities of the same material, such as rack storage bins that store small quantities and high-rack storage bins that store large quantities. The system searches for a source storage bin for picking according to the quantities requested.

- **SLED strategy**
  This stock removal strategy considers the SLED of the quant of the requested material. The quant with the oldest SLED is searched using this strategy. The system automatically determines the source storage bin that stores the quant with the oldest SLED of the material in the embedded EWM warehouse number, and that quant is picked first. This strategy is used with materials that expire over time.

- **Fixed storage bins strategy**
  This stock removal strategy allows for determining a specific source storage bin for the requested product/material. During outbound processing, the system automatically proposes the storage type, storage section, and storage bin in the warehouse task created for picking of the product. This strategy is used with high-risk items that may require specific storage requirements.

Before configuring stock removal strategies, the warehouse structure including storage types, storage sections, and storage bins must be configured in embedded EWM. Let's explore the configuration settings for stock removal strategies.

Specify Stock Removal Rules

In this configuration step, specify rules for stock removal. You can define the sequence for every stock removal rule in which the quants found are sorted. Stock removal rules are used further in storage type search and other configuration settings related to stock removal strategies.

To specify stock removal rules for an embedded EWM warehouse number, execute Transaction SPRO and navigate to **SCM Extended Warehouse Management • Extended Warehouse Management • Goods Issue Process • Strategies • Specify Stock Removal Rules.**

Figure 8.89 shows that FIFO and LIFO stock removal rules have been defined for embedded EWM warehouse number **MJ00**. Let's explore the control parameters. Click **New Entries** to define a new stock removal rule. Enter a four-digit alphanumeric key and description to define a new stock removal rule, and then press $\boxed{\text{Enter}}$.

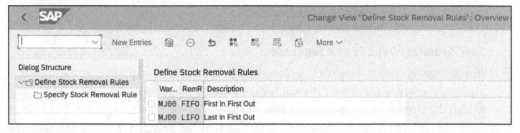

Figure 8.89 Define Stock Removal Rules: Overview

In the **Dialog Structure**, navigate to **Specify Stock Removal Rule**. Figure 8.90 shows the details of the FIFO stock removal rule defined for warehouse number **MJ00**. You should maintain at least one sequence for sorting of the quants found during the search for picking. The key fields are as follows:

- **Sequence No.**
  In this field, maintain the sequence number for sorting of quants found during the search using the stock removal rule. You can create multiple sort sequences for a stock removal rule.

- **Sort Field**
  In this field, specify the fields used for sorting of quants to find the right quant according to the stock removal strategy. For LIFO and FIFO strategies, the goods receipt date is used as a field for sorting. Hence, specify the **WDATU Date and Time of Goods Receipt** field for sorting. The following are some of the other fields that can be used for sorting:
  - **QUAN**: Available quantity
  - **VFDAT**: Shelf-life expiration date
  - **CHARG**: Batch

- **Descending**
  If you set this indicator, quants that are found for the requested material for stock removal are sorted in descending order, and the stock is then removed in descending order according to the requested quantity. Set this indicator for the LIFO strategy.

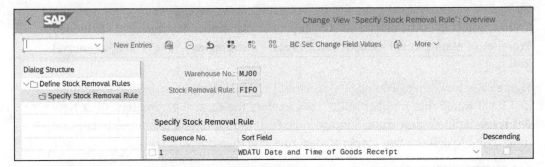

Figure 8.90 Specify Stock Removal Rules: FIFO

Specify Storage Type Search Sequence

In this configuration step, define the storage type search sequence, set a four-digit alphanumeric key for stock removal for the embedded EWM warehouse number, and assign the storage types in a sequence to the storage type search sequence. The system uses the storage type sequence maintained here to search for the quants of the requested material and the source storage bin for stock removal. If a source storage bin is identified in a storage type, the system assigns the storage bin to the warehouse task created for picking.

To define the storage type search sequence for an embedded EWM warehouse number, execute Transaction SPRO and navigate to **SCM Extended Warehouse Management** • **Extended Warehouse Management** • **Goods Issue Process** • **Strategies** • **Specify Storage Type Search Sequence**. Figure 8.91 shows storage type search sequences **ALOC**, **PICK**, **COMP**, and so on defined for embedded EWM warehouse number **MJ00**. Click **New Entries** to define a new storage type search sequence. Enter a four-digit alphanumeric key and description to define it.

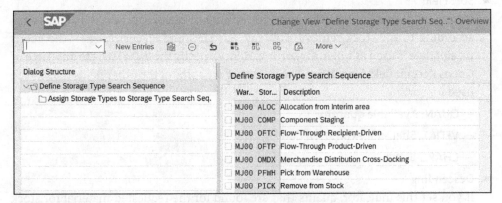

Figure 8.91 Define Storage Type Search Sequence: Overview

To assign the storage types to the storage type search sequence, select a storage type search sequence definition and click **Assign Storage Types to Storage Type Search Seq.**

This opens a new screen where you can assign storage types in a sequence to the storage type search sequence. Click **New Entries** and specify the sequence number in the **Sequence No.** field and the **Storage Type**. For the next storage type assignment for the same storage type search sequence, specify the next sequence number and storage type. Press ⌷Enter⌷ after this assignment. Figure 8.92 shows that storage types **0020**, **0010**, and **0050** have been assigned in a sequence to the storage type search sequence **PFWH** (pick from warehouse) in EWM warehouse number **MJ00**.

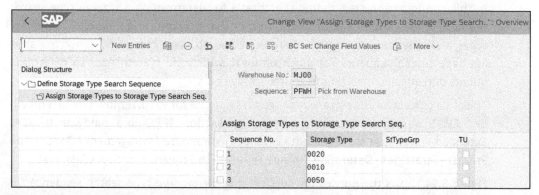

Figure 8.92 Assign Storage Types to Storage Type Search Sequence

Define Stock Removal Control Indicator

Stock removal control indicators control the picking of certain specific materials from specific storage areas (storage types) within the warehouse. Control indicators for stock removal are assigned to specific warehouse products defined in embedded EWM for the warehouse number.

To define the putaway control indicators for an embedded EWM warehouse number, execute Transaction SPRO and navigate to **SCM Extended Warehouse Management • Extended Warehouse Management • Goods Issue Process • Strategies • Define Stock Removal Control Indicator**.

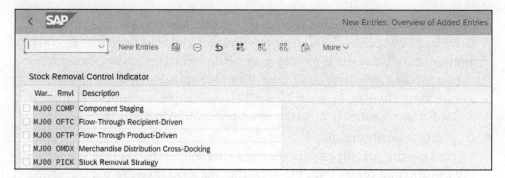

Figure 8.93 Define Stock Removal Control Indicator

Figure 8.93 shows stock removal control indicators **COMP**, **PICK**, **OFTC**, **OMDX**, and **OFTP** defined for embedded EWM warehouse number **MJ00** for stock removal. Click **New Entries** to define a new stock removal control indicator. You can copy an existing definition and create a new one. Specify a four-digit alphanumeric key and description to define a stock removal control indicator.

Determine Storage Type Search Sequence for Stock Removal

In this configuration step, define the criteria for determining a storage type search sequence for stock removal using stock removal control indicators, warehouse process types, and other criteria. As previously explained, storage types are assigned to a storage type search sequence for stock removal, which the system uses to determine quants of requested materials in source storage bins.

To define criteria for the storage type search sequence for stock removal for an embedded EWM warehouse number, execute Transaction SPRO and navigate to **SCM Extended Warehouse Management • Extended Warehouse Management • Goods Issue Process • Strategies • Determine Storage Type Search Sequence for Stock Removal**.

Figure 8.94 shows the entries for determining the storage type search sequence for stock removal for embedded EWM warehouse number **MJ00** using stock removal control indicators **COMP** and **PICK** and warehouse process types **2010**, **Y320**, and so on. Click **New Entries** to enter a new storage type search sequence determination criterion for stock removal. You can copy an existing entry and create a new one.

Let's explore the field level details:

- **2 (two-step picking)**
  Set this indicator if you are adding the storage type search sequence determination criterion for the two-step stock removal process. In the two-step picking process, stock is removed from the source storage bins into an intermediate storage area (storage type), and then from that storage area, stock is moved to goods issue area of the warehouse. There are two warehouse tasks created for the two-step picking process.

- **Whse Process Typ (warehouse process type/warehouse process type group)**
  Embedded EWM uses warehouse process types for warehouse activities such as putaway (inbound processing), picking (outbound processing), posting changes (internal warehouse movements), and so on. Warehouse process types are configured for a warehouse number in embedded EWM. Assign the warehouse process type defined for stock removal for warehouse number **MJ00** in this field.

- **Q (quantity classification)**
  You can use the quantity classification as a criterion for storage type search. Quantity classification comes from the packaging materials. Refer to our discussion of specifying storage types for search sequence for putaway in Section 8.3.5 for more details.

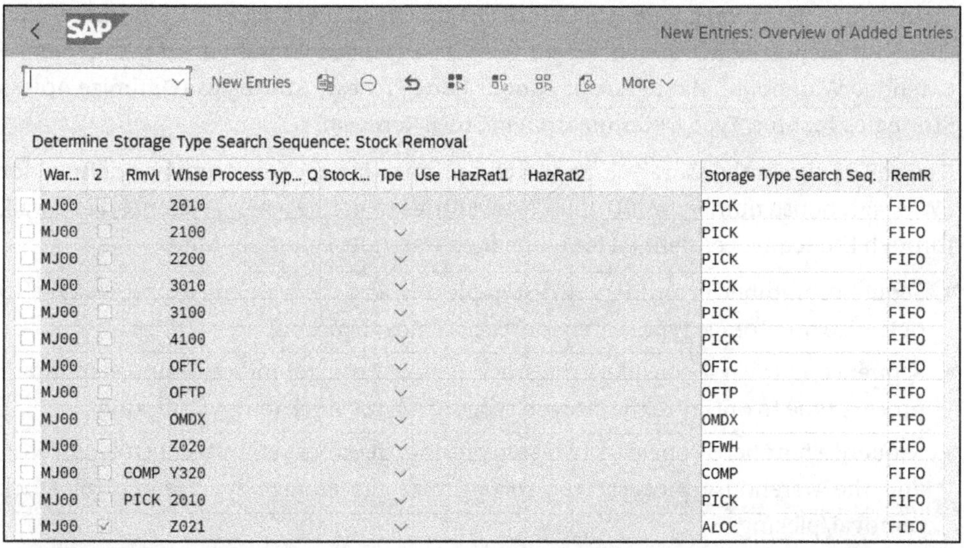

War...	2	Rmvl	Whse Process Typ...	Q	Stock...	Tpe	Use	HazRat1	HazRat2	Storage Type Search Seq.	RemR
MJ00			2010							PICK	FIFO
MJ00			2100							PICK	FIFO
MJ00			2200							PICK	FIFO
MJ00			3010							PICK	FIFO
MJ00			3100							PICK	FIFO
MJ00			4100							PICK	FIFO
MJ00			OFTC							OFTC	FIFO
MJ00			OFTP							OFTP	FIFO
MJ00			OMDX							OMDX	FIFO
MJ00			Z020							PFWH	FIFO
MJ00		COMP	Y320							COMP	FIFO
MJ00		PICK	2010							PICK	FIFO
MJ00	✓		Z021							ALOC	FIFO

Figure 8.94 Determine Storage Type Search Sequence for Stock Removal

- **Stock... (stock type/group)**
  You can use stock type defined for the respective warehouse number in embedded EWM as a search criterion for the storage type search. Refer to our discussion of configuring stock types in Section 8.3.4 for more details about stock types.
- **HazRat1/HazRat2 (hazard rating)**
  Hazard ratings are assigned to hazardous materials. You can specify the hazard ratings in this field for the system to determine the appropriate storage type. You can specify two hazard ratings in the **Hazard Rating 1** and **Hazard Rating 2** fields for the storage type search per sequence.
- **Storage Type Search Seq.**
  In this field, assign the storage type search sequence defined for the respective embedded EWM warehouse number. The system uses the storage type sequence maintained here to search for the storage types for stock removal.
- **RemR (stock removal rule)**
  Stock removal rules are used further in the storage type search and other configuration settings related to stock removal strategies. Assign one of the stock removal rules defined previously.

Optimize Access Strategies for Storage Type Determination in Stock Removal

Embedded EWM provides various control parameters for storage type search for stock removal such as stock removal control indicators, warehouse process types, quantity classifications, stock types, stock categories, usage indicators, hazard ratings, and so on. To optimize the storage type search, embedded EWM provides an approach that can be configured in a sequential order.

To optimize the storage type search for the embedded EWM warehouse number, execute Transaction SPRO and navigate to **SCM Extended Warehouse Management • Extended Warehouse Management • Goods Issue Process • Strategies • Optimize Access Strategies for Stor. Type Determination in Stock Removal**.

Figure 8.95 shows the storage type search optimization for stock removal defined for EWM warehouse number **MJ00**. Click **New Entries** to define a new sequence. Let's walk through the sequences defined for embedded warehouse number **MJ00**:

- Sequence number **1** considers two-step picking and the warehouse process type to optimize the storage type search for stock removal/picking.
- Sequence number **2** considers the stock removal control indicator and warehouse process type to optimize the storage type search for stock removal/picking.
- Sequence number **3** considers two-step picking, the stock removal control indicator, and the warehouse process type to optimize the storage type search for stock removal/picking.
- Sequence number **4** considers only the warehouse process type to optimize the storage type search for stock removal/picking.

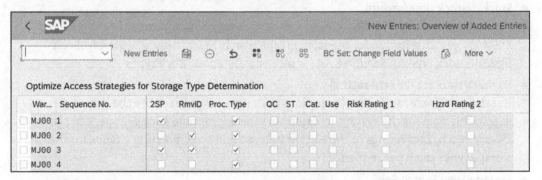

Figure 8.95 Optimize Access Strategies for Storage Type Determination in Stock Removal

Warehouse Product Maintenance for Stock Removal

Let's revisit the warehouse product master, which we introduced in Section 8.1.3 and discussed in the context of inbound deliveries in Section 8.3.5. This time, we'll examine the settings for outbound deliveries. To create the warehouse product, execute Transaction /SCWM/MAT1.

Figure 8.96 shows the warehouse data maintained for stock removal of product **MS-RAW MATERIAL1** into EWM warehouse number **MJ00**. Stock removal control indicator **PICK** has been maintained under the **Stock Removal** section of the warehouse data for the product. Stock removal control indicators control the storage type search, as explained earlier. The system uses this assignment in the warehouse product to search for the storage type for stock removal.

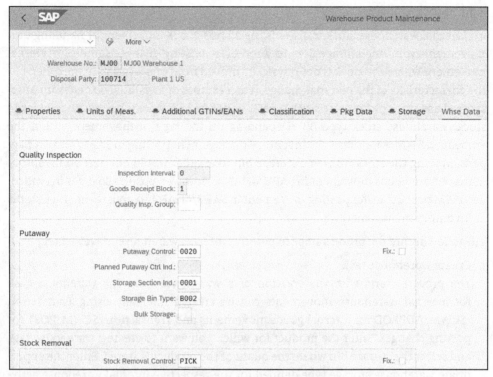

Figure 8.96 Warehouse Product Maintenance for Stock Removal

8.5   Internal Warehouse Movements

Internal warehouse movements involve ad hoc stock transfers within a single warehouse, replenishment, physical inventory, and posting changes. These movements are necessary to optimize the storage space, reduce putaway and stock removal time, and ensure efficient warehouse operations. To optimize the storage space, stock in storage bins is transferred to different storage bins within the same storage type or across storage types. Posting changes deal with stock types, stock categories, and so on.

The replenishment activity is vital within a warehouse, and it supports production, plant maintenance, and other processes. During replenishment, stock of a material is moved from a reserved storage area to the picking location within the warehouse.

Finally, physical inventory is an important operation within the warehouse that helps to keep accurate inventory records in embedded EWM, and it involves counting of materials stored in the warehouse physically and comparing them with the system records to identify discrepancies, if any.

These internal warehouse movements in embedded EWM will update inventory management on the SAP S/4HANA side, and it helps to keep the inventory management and warehouse management in sync. Let's explore these processes in detail, including configuration instructions.

8.5.1   Stock Transfers

An internal warehouse stock transfer is the ad hoc goods movement or handling unit movement from one storage area to another. A posting change is another internal movement where the stock is not physically moved from one location to the other, but the characteristic of the material changes, such as stock of a material posted from unrestricted use stock (warehouse stock type FF as defined for warehouse MJOO) to blocked stock (warehouse stock type BB). Depending on the type of movement within the embedded EWM warehouse, the inventory management posting in SAP S/4HANA happens. Moving the stock from one storage bin to another would not post an inventory management goods movement in SAP S/4HANA. However, posting changes in embedded EWM post a transfer posting movement in SAP S/4HANA to update the stock status in inventory management.

The following are the process steps to perform internal warehouse movements:

1. **Create warehouse task**
   The process starts with the creation of a warehouse task. The warehouse task for internal warehouse movements can be created manually using Transaction /SCWM/ADPROD for internal goods movements and Transaction /SCWM/POST for posting changes. Enter the product for which you want to create a warehouse task and select the storage bin where the quant of the product is stored. Enter the appropriate warehouse storage type defined for the respective internal warehouse activity. The appropriate warehouse process type is selected for internal stock transfers.

2. **Confirm warehouse task**
   The next step of the process is to complete the warehouse activity and confirm the warehouse task created for internal warehouse movements. This step can be performed manually or automatically. For certain warehouse movements like posting changes, the creation of a warehouse task and confirming of the task can happen at the same time.

3. **Update inventory management**
   Certain internal warehouse movements are updated automatically into the SAP S/4HANA inventory management function. Like goods issue and goods receipt posting, transfer posting movements are posted automatically in inventory management using qRFC. Material documents will be created for the respective goods movements.

Let's discuss the key configuration settings for internal warehouse movements involving product stock.

Warehouse Process Types for Internal Stock Movements
Like warehouse process types for inbound delivery processing and outbound delivery processing, embedded EWM uses warehouse process types for internal warehouse movements such as warehouse supervision (goods movement between storage bins)

and posting changes. Warehouse process types control the creation of warehouse tasks for putaway, picking, packing, and posting changes. When the warehouse tasks are created for internal warehouse movements, enter the appropriate warehouse process type.

Warehouse process types are configured for a warehouse number in embedded EWM. To define warehouse process types for an embedded EWM warehouse number, execute Transaction SPRO and navigate to **SCM Extended Warehouse Management • Extended Warehouse Management • Cross-Process Settings • Warehouse Task • Define Warehouse Process Type**.

To define warehouse process types for internal stock movements, click **New Entries** from the initial screen and specify the embedded EWM warehouse number, a four-digit alphanumeric key in the **Whse Proc. Type** (warehouse process type) field, and a description. Posting changes and goods movements in embedded EWM are two different processes, so a different warehouse process type is required for each activity:

- Figure 8.97 shows warehouse process type **9999** (**Warehouse Supervision**) defined for warehouse number **MJ00** for stock movements between storage bins.
- Figure 8.98 shows warehouse process type **4010** (**Transfer Posting**) defined for warehouse number **MJ00** for posting changes.

Let's explore the field-level settings for these two warehouse process types:

- **Whse Proc. Category**
  The warehouse process category distinguishes different goods movements within the warehouse. Table 8.29 shows the warehouse process categories defined for the warehouse process type. For the warehouse supervision process, assign key **3** as the warehouse process category, and for posting change activity, assign key **7** as the warehouse process category.

Category	Description
1	Putaway
2	Stock removal
3	Internal warehouse movement
7	Posting change

Table 8.29 Warehouse Process Types

- **Activity**
  Activities are defined for the warehouse number in embedded EWM. Refer to our discussion of defining activities for activity areas in Section 8.2.5 for details. For the warehouse supervision process, assign key **INTL** as the activity, and for posting change activity, assign key **STCH** as the activity.

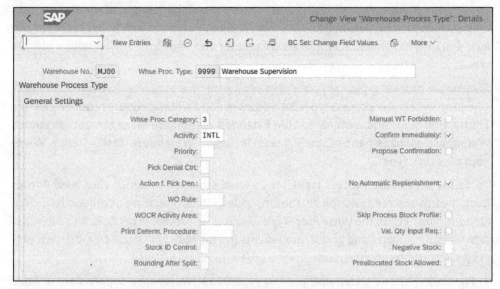

Figure 8.97 Warehouse Process Type for Warehouse Supervision

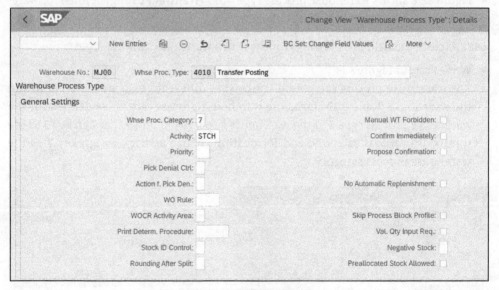

Figure 8.98 Warehouse Process Type for Posting Changes

Activate Synchronous Goods Movement Posting

When you post goods movements such as goods receipt, goods issue, and posting changes in embedded EWM, the corresponding goods movements in the inventory management in SAP S/4HANA are posted through qRFC.

To activate synchronous goods movement for an embedded EWM warehouse number, execute Transaction SPRO and navigate to **SCM Extended Warehouse Management** •

Extended Warehouse Management • Interfaces • ERP Integration • Goods Movements • Activate Synchronous Goods Movement Posting.

Set the **Active** indicator to activate the synchronous goods movement posting for the embedded EWM warehouse number. Figure 8.99 shows that the synchronous goods movement posting is active for warehouse number **MJ00** in embedded EWM.

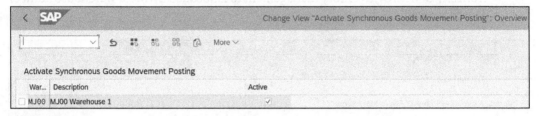

Figure 8.99 Activate Synchronous Goods Movement Posting between EWM and SAP S/4HANA

Note

Synchronous goods movements can happen in embedded EWM only; this is not supported in decentralized EWM systems.

Settings for Synchronous Goods Movements

By activating this feature and setting it up for the embedded EWM warehouse, when you post goods movements such as goods receipt, goods issue, and posting changes in SAP S/4HANA inventory management function using Transaction MIGO, the corresponding goods movements in the warehouse management in embedded EWM are posted through qRFC. There are two settings to activate this feature in embedded EWM.

Note

This feature is available in the embedded EWM system only; it is not supported in the decentralized EWM system.

Make Settings for Synchronous Goods Movement Posting

In this configuration step, make settings for synchronous goods movement postings initiated from SAP S/4HANA inventory management using Transaction MIGO to embedded EWM. These settings are applicable for goods receipts with reference to a purchase order (movement type 101) as well as goods receipt with reference to a production order (movement type 101).

To make settings for synchronous goods movement posting for an embedded EWM warehouse number, execute Transaction SPRO and navigate to **SCM Extended**

Warehouse Management • Extended Warehouse Management • Interfaces • ERP Integration • Goods Movements • Settings for Synchronous Goods Movements (MIGO) • Make Settings for Synchronous Goods Movement Posting.

You can make these settings directly in the input fields for the embedded warehouse number. Figure 8.100 shows the settings for synchronous goods movement posting from SAP S/4HANA inventory management to embedded EWM for warehouse number **MJ00** in embedded EWM.

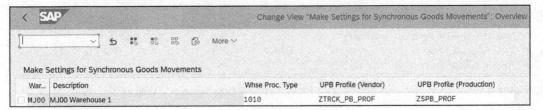

Figure 8.100 Make Settings for Synchronous Goods Movement Posting

The following key settings can be made in this step:

- **Whse Proc. Type (warehouse process type)**
  Embedded EWM uses warehouse process types for warehouse activities such as putaway (inbound processing) and the like. Warehouse process types are configured for the warehouse number in embedded EWM. Assign the warehouse process type defined for putaway for warehouse number **MJ00** in this field. The system uses this warehouse process type during goods receipt movement sync from SAP S/4HANA to embedded EWM only if the putaway strategy to find the destination storage bin doesn't work.

- **USB Profile (Vendor)**
  In this field, assign a predefined unified package building profile for goods receipt from an external supplier. This profile helps to create a packaging proposal when the goods movement sync between SAP S/4HANA and embedded EWM happens.

- **USB Profile (Production)**
  In this field, assign a predefined unified package building profile for goods receipt from production. This profile helps to create a packaging proposal when the goods movement sync between SAP S/4HANA and embedded EWM happens.

Activate Synchronous Goods Movements in the Warehouse

In this configuration step, activate synchronous goods movements for the embedded EWM warehouse number for synchronizing goods movements from SAP S/4HANA inventory management to embedded EWM for goods receipts, goods issues, and/or transfer postings. Then link the corresponding inventory management movement types to the warehouse for inbound, outbound, and transfer postings.

To activate synchronous goods movements in the warehouse, execute Transaction SPRO and navigate to **SCM Extended Warehouse Management • Extended Warehouse Management • Interfaces • ERP Integration • Goods Movements • Settings for Synchronous Goods Movements (MIGO) • Activate Synchronous Goods Movements in the Warehouse**.

Directly set the **Transfer**, **Receipt**, and **Gds Issue** (goods issue) indicators to activate the synchronous goods movement posting in the embedded EWM warehouse number. Figure 8.101 shows the synchronous goods movements for transfer, receipt, and goods issue activated for warehouse number **MJ00** in embedded EWM.

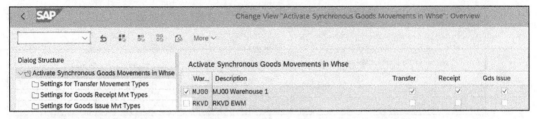

Figure 8.101 Activate Synchronous Goods Movements in Warehouse: Overview

You can activate the synchronous goods movements in the warehouse for all goods movements or a few specific movements that are posted in inventory management based on movement types. Use the indicators to activate (check the checkbox) or deactivate (uncheck the checkbox) synchronous goods movements in the warehouse for transfer posting (**Transfer**), goods receipt (**Receipt**), and goods issue (**Gds Issue**).

Embedded EWM provides additional settings to further refine this feature by activating and deactivating the synchronous goods movements in the warehouse at the movement type level. In the **Dialog Structure**, navigate to **Settings for Transfer Movement Types**, where you can select the checkboxes under **Post. Chg** to activate posting changes. Figure 8.102 shows that inventory management movement types **301**, **311**, and **310** are activated for posting changes in the warehouse.

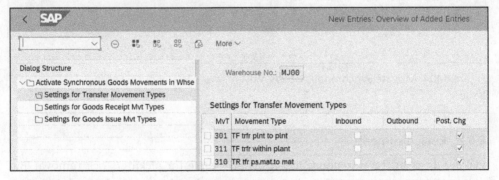

Figure 8.102 Activate Synchronous Goods Movements in Warehouse: Settings for Transfer Postings

Next, navigate to **Settings for Goods Receipt Mvt Types**, where you can activate movement types for goods receipt by selecting the **Inbound** checkboxes. Figure 8.103 shows that inventory management movement types **501**, **451**, and **561** are activated for goods receipt in the warehouse. Movement type 101 for goods receipt is not listed here as the previous configuration step covers movement type 101 for goods receipt from external suppliers and from production.

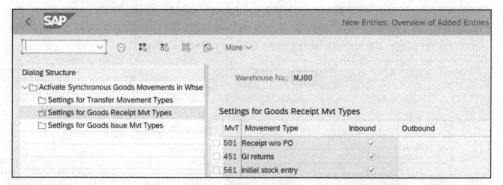

Figure 8.103 Activate Synchronous Goods Movements in Warehouse: Settings for Goods Receipt

Finally, navigate to **Settings for Goods Issue Mvt Types**, where you can activate movement types for goods issue by selecting the **Outbound** checkboxes. Figure 8.104 shows that inventory management movement types **201**, **221**, and **261** are activated for goods issue in the warehouse.

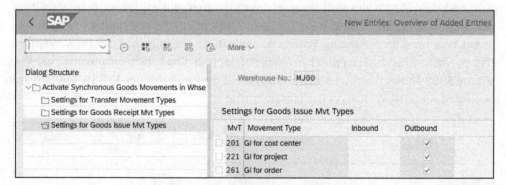

Figure 8.104 Activate Synchronous Goods Movements in Warehouse: Settings for Goods Issue

8.5.2   Replenishment

Picking areas in the warehouse are easily accessible for picking/stock removal. But often picking areas go out of stock and must be replenished regularly. Replenishment is a warehouse activity that involves moving the stock of products from a reserved storage area to the picking area. The picking area can have a fixed bin or a storage type, and embedded EWM provides strategies to replenish the picking area.

The following are the various replenishment strategies in embedded EWM:

- **Planned replenishment**

  A planned replenishment strategy follows a min-max quantity approach to replenish the picking areas. The replenishment quantity is calculated based on a minimum quantity and maximum quantity maintained in the warehouse product under the storage type view of the warehouse product. If the stock of the product in the picking area falls below the minimum quantity maintained, a warehouse task will be created from a scheduled task that can be scheduled using Transaction /SCWM/REPL to replenish the picking area to bring up the stock to the maximum quantity maintained in the warehouse product data.

- **Order-related replenishment**

  An order-related replenishment strategy is targeted to fulfill open quantities of outbound delivery orders. The replenishment is initiated when the required quantity falls below the minimum quantity level. The system calculates the replenishment quantity automatically while considering all open outbound delivery orders to be fulfilled.

- **Crate part replenishment**

  A crate part replenishment strategy is targeted for production supply and to replenish the production supply areas with fixed quantities of products required for production. The replenishment will be triggered when the production supply areas fall below the minimum quantities. The system considers the replenishment quantity maintained for the product in the production supply area.

- **Automatic replenishment**

  An automatic replenishment strategy creates a warehouse task for replenishment automatically. This strategy follows a min-max quantity approach to replenish the picking areas. The replenishment quantity is calculated based on minimum quantity and maximum quantity maintained in the warehouse product under the storage type view of the warehouse product. If the stock of the product in the picking area falls below the minimum quantity maintained, a warehouse task will be created automatically to replenish the picking area to bring up the stock to the maximum quantity maintained in the warehouse product data.

- **Direct replenishment**

  A direct replenishment strategy is aimed at addressing pick denials due to insufficient stock at the picking areas. The replenishment is initiated by an exception code, and it creates a warehouse task for replenishment automatically. This strategy follows the min-max quantity approach to replenish the picking areas. The replenishment quantity is calculated based on minimum quantity and maximum quantity maintained in the warehouse product under the storage type view of the warehouse product.

Stock removal strategies are used to identify storage types for replenishment. The system determines the source storage types, source storage sections, and source storage bins using stock removal strategies and updates the warehouse tasks created for replenishment. The warehouse process type also plays an important role in the replenishment. For more details on stock removal strategies, refer to Section 8.4.4.

Let's dive deep into the configuration settings for the replenishment process in embedded EWM.

Activate Replenishment Strategies in Storage Types

In this configuration step, activate the replenishment strategy by storage type. The storage types activated for replenishment are the picking areas of the warehouse. This setting is required for replenishment of storage types/picking areas.

To activate replenishment strategies for storage types for an embedded EWM warehouse number, execute Transaction SPRO and navigate to **SCM Extended Warehouse Management • Extended Warehouse Management • Internal Warehouse Processes • Replenishment Control • Activate Replenishment Strategies in Storage Types**.

To activate replenishment control for storage types, click **New Entries** and specify the embedded EWM warehouse number and the storage type, and then select an appropriate **Repl. Strat.** (replenishment strategy) from the dropdown list. Figure 8.105 shows the overview of storage types assigned to replenishment strategies for warehouse number **MJ00**.

Warehouse Numb...	Storage Ty...	Repl. Strat.
MJ00	0020	Automatic Replenishment
MJ00	0050	Planned Replenishment
MJ00	0050	Automatic Replenishment
MJ00	0050	Order-Related Replenishmen...
MJ00	1000	Crate Part Replenishment

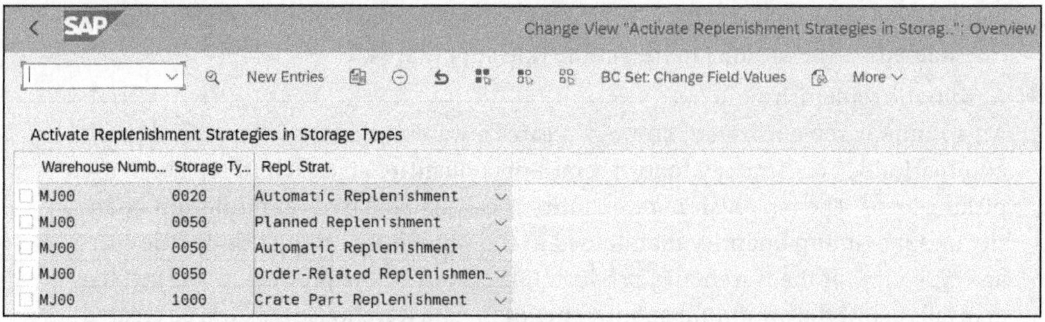

Figure 8.105 Activate Replenishment Strategies in Storage Types: Overview

On this screen, you must assign the warehouse process type defined for replenishment for warehouse number **MJ00** in the **Whse Proc. Category** (warehouse process category) field. Warehouse types play a vital role in the source storage type search for replenishment. Refer to our discussion of determining the storage type search sequence for stock removal in Section 8.4.4 for more details. Warehouse process category **3** is used for internal warehouse movement and activity **REPL** is used for replenishment.

Select the first line for storage type **0020** in warehouse number **MJOO** and click the **Details** button at the top or press Ctrl+Shift+F2 to reach the details screen shown in Figure 8.106.

Figure 8.106 shows warehouse process type **3010** defined for warehouse number **MJOO**. Let's explore the key fields:

- **Qty Type Used**

  In this field, specify the quantity type used for replenishment. Specify one of the following values:

  - **Physical Quantity**
  - **Available Quantity**

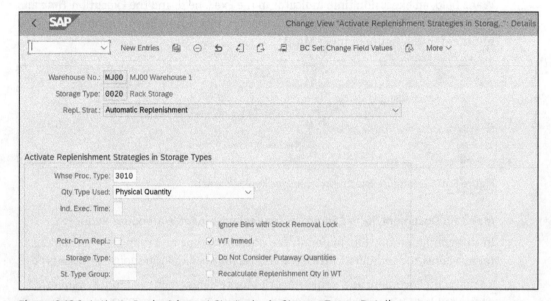

Figure 8.106 Activate Replenishment Strategies in Storage Types: Details

- **WT Immed.**

  If you set the *warehouse task immediately* indicator, the system creates the warehouse task for the replenishment. If you don't set this indicator, the system creates a warehouse request for replenishment and the warehouse task must be created manually then.

- **Storage Type**

  Enter the source storage type for the replenishment. If you enter a storage type in this field, the replenishment always happens from this storage type.

- **St. Type Group**

  Enter the source storage type group for the replenishment. If you enter a storage type group in this field, the replenishment always happens from this storage type group.

Configure Execution Times for Replenishment

In this configuration step, enter the execution time in hours or minutes for the replenishment activity. The system uses this time to calculate the planned completion in the warehouse task.

To configure the execution time for replenishment for an embedded EWM warehouse number, execute Transaction SPRO and navigate to **SCM Extended Warehouse Management • Extended Warehouse Management • Internal Warehouse Processes • Replenishment Control • Configure Execution Times for Replenishment**.

To configure execution times for replenishment for the embedded EWM warehouse number, click **New Entries** and specify an embedded EWM warehouse number in the **War...** field, an execution time indicator in the **Exe...** field, and the **Execution Time** and the **Time Unit**. Figure 8.107 shows the execution time for replenishment activity is **10** minutes for warehouse number **MJ00**.

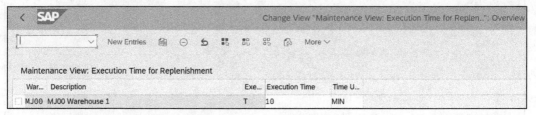

Figure 8.107 Configure Execution Times for Replenishment

Maintain Document/Item Categories for Replenishment Warehouse Request

In this configuration step, maintain the document type and item type for replenishment against the combination of a warehouse number and warehouse process type.

To maintain document and item categories for replenishment warehouse requests for an embedded EWM warehouse number, execute Transaction SPRO and navigate to **SCM Extended Warehouse Management • Extended Warehouse Management • Internal Warehouse Processes • Replenishment Control • Maintain Document/Item Categories for Replenishment Warehouse Request**.

Directly specify the document type and item type for the combination of embedded EWM warehouse number and warehouse process type for replenishment. Figure 8.108 shows that document type **SRPL** and item type **SRPL** are maintained for replenishment for the combination of warehouse number **MJ00** and warehouse process type **3010** (**Replenishment**).

The warehouse document type and item type are used to create a warehouse request for warehouse activities. **SRPL** is a standard document type for replenishment and **SRPL** is also created as a replenishment item type in the standard system. This assignment is vital for the creation of a warehouse task for replenishment.

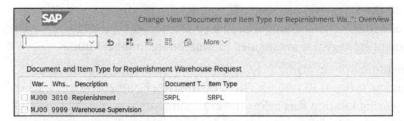

Figure 8.108 Maintain Document and Item Categories for Replenishment Warehouse Request

Warehouse Product Maintenance for Replenishment

8

The warehouse product master in embedded EWM stores the picking location (storage type) and replenishment quantity, minimum quantity, and maximum quantity information. The system uses this data for replenishment based on the replenishment strategy enabled for the storage type. To create the warehouse product, execute Transaction /SCWM/MAT1.

In this transaction, replenishment details and the storage type for replenishment for a product can be defined. Click **Adopt Data** after maintaining such details for the product. Figure 8.109 shows the storage type data maintained for replenishment for a warehouse product in embedded EWM warehouse number **MJ00**.

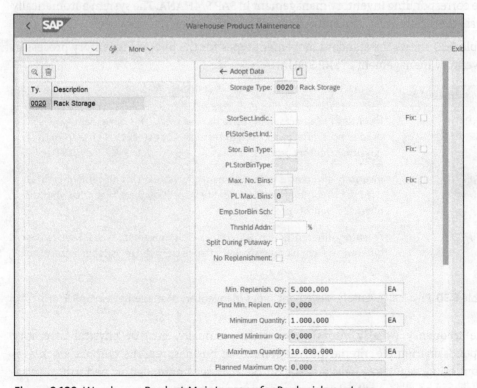

Figure 8.109 Warehouse Product Maintenance for Replenishment

Storage type **0020** (**Rack Storage**) has been maintained as a picking area for the warehouse product, and in the earlier configuration step for storage type **0020**, the automatic replenishment strategy was maintained for warehouse number **MJ00**.

The replenishment quantity is maintained as 5,000 EA, minimum quantity as 1,000 EA, and maximum quantity as 10,000 EA. Based on the replenishment strategy, when the stock at the picking location falls below the minimum quantity, a warehouse task will be created to replenish the picking area to increase the stock to the maximum quantity, and the system calculates the quantity to be replenished in multiples of the replenishment quantity.

8.5.3   Physical Inventory

Businesses largely benefit from physical inventory as it improves the overall efficiency of warehouse operations while reducing costs. This is an internal warehouse process where warehouse products stored in the storage bin are counted periodically and compared with the system stock record for the same product. Physical inventory can be performed on all stock types (unrestricted use stock, quality inspection stock, and blocked stock). Any difference found between the physical stock and system stock record will be posted to adjust the stock levels in the system. This difference can be positive or negative, and the difference posting in embedded EWM automatically updates the corresponding inventory management in SAP S/4HANA. The system automatically posts the differences in inventory management based on the stock type.

Table 8.30 shows the standard movement types for the physical inventory process in inventory management in SAP S/4HANA.

Movement Type	Detailed Description
701	Inventory differences in unrestricted use stock. This movement type is used to post differences from unrestricted use stock that arise from the physical inventory process.
703	Inventory differences in quality inspection stock. This movement type is used to post differences from quality inspection stock that arise from the physical inventory process.
707	Inventory differences in blocked stock. This movement type is used to post differences from blocked stock that arise from the physical inventory process.

Table 8.30 Physical Inventory Movement Types in Inventory Management in SAP S/4HANA

The frequency (weekly, monthly, quarterly, annually, etc.) of physical inventory depends on multiple criteria: the type of material, business reasons, purpose, stock level monitoring, regulatory reasons, and so on. It is recommended to perform physical inventory on all less critical products and materials annually.

Figure 8.110 shows the physical inventory process in embedded EWM and its integration with inventory management in SAP S/4HANA. Let's explore the process steps:

1. **Create physical inventory document**
   The process starts with the creation of a physical inventory document in embedded EWM. This document serves as a basis for physical inventory in the warehouse and it contains the warehouse product, storage bin, stock type, batch, and activity area details. Use Transaction /SCWM/PI_CREATE to create the physical inventory document. The initial status of the physical inventory document must be set to **Active** to start the process.

2. **Counting**
   Counting in the warehouse starts after the physical inventory document is created. Warehouse personnel perform this activity based on the physical inventory document details. After the counting is performed at the designated storage bin, count results such as the actual quantity, counting date and time, and so on are entered manually using Transaction /SCWM/PI_PROCESS. Recounting can be initiated from the same transaction in certain situations.

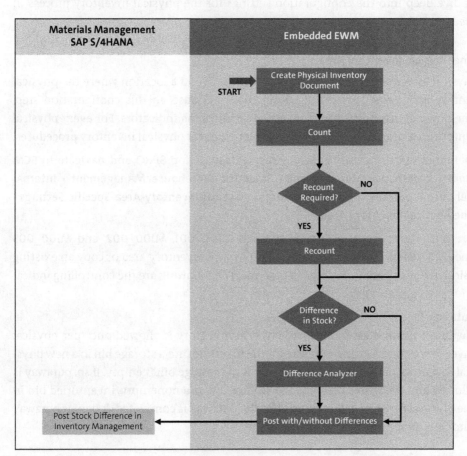

Figure 8.110 Physical Inventory Process in EWM

3. **Post physical inventory**

   After the actual quantity from the counting results is documented, activate the document and save it using Transaction /SCWM/PI_PROCESS. If there are any differences between the actual quantity and the system record quantity, the differences are posted and the system record in embedded EWM is adjusted.

4. **Difference analyzer**

   This step is required if there are any differences found between actual quantity and the system record quantity. Use Transaction /SCWM/DIFF_ANALYZER to analyze the differences found. After the analysis, you can post the differences in the same transaction. Posting differences will create a warehouse material document.

5. **Post differences in inventory management in SAP S/4HANA (automatic)**

   Posting of differences in embedded EWM will automatically update the inventory management in SAP S/4HANA. The system posts a material document using movement type 701, 703, or 707 based on the stock type. A corresponding accounting document will be created to post the differences in financial accounting.

Let's dive deep into the configuration settings for the physical inventory process in embedded EWM.

Define Physical Inventory Area

A physical inventory area is an organizational unit and a location where the physical inventory activity is carried out within the warehouse. In this configuration step, define physical inventory areas and specify controlling indicators. For every physical inventory area, assign the relevant document types for physical inventory procedures.

To define physical inventory areas, execute Transaction SPRO and navigate to **SCM Extended Warehouse Management • Extended Warehouse Management • Internal Warehouse Processes • Physical Inventory • Physical-Inventory-Area-Specific Settings • Define Physical Inventory Area**.

Figure 8.111 shows physical inventory areas **MJ00_001**, **MJ00_002**, and **MJ00_003** defined. Click **New Entries** to define a new physical inventory area or copy an existing physical inventory area and create a new one. The following are the controlling indicators for this configuration:

- **Putaway PI**

  If you set this indicator, putaway physical inventory is allowed once per physical inventory period. During putaway for the first time into a storage bin in a new physical inventory period, the system checks if the storage bin is empty. If so, putaway is allowed and the system records the storage bin as a nonempty/inventoried bin in the physical inventory document after the putaway is completed. It denies putaway into the storage bin if the bin is not empty.

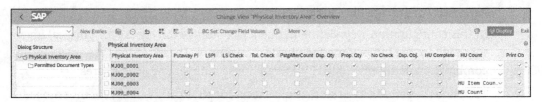

Figure 8.111 Physical Inventory Area: Overview

- **LSPI**

  If you set this indicator, a low stock check with physical inventory is allowed once per physical inventory period. During physical inventory, the system checks if the storage bin has low stock. If so, the storage bin is documented in the physical inventory document as a nonempty/inventoried bin.

- **LS Check**

  If you set this indicator, a low stock check without physical inventory is activated. The system doesn't create a physical inventory document, and the storage bin doesn't become designated as a nonempty/inventoried bin.

- **Tol. Check**

  If you set this indicator, a tolerance check for low stock check will be activated.

- **PstgAfterCount**

  If you set this indicator, the physical inventory document is posted immediately after the count results are entered in the physical inventory document.

In the **Dialog Structure**, navigate to **Permitted Document Types**. Figure 8.112 shows the allowed physical inventory document types for physical inventory areas. These document types provide the controlling procedures for the physical inventory activity. You can specify multiple physical inventory document types per physical inventory area. The following are the most common physical inventory document types:

- Annual physical inventory (product specific)
- Ad hoc physical inventory (storage bin specific)
- Ad hoc physical inventory (product specific)

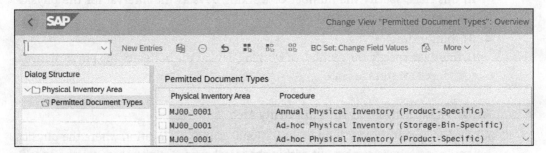

Figure 8.112 Permitted Document Types for Physical Inventory Area: Overview

Periodicity of Storage Bin Check

In this configuration step, define frequency of storage bin checks for physical inventory and set the buffer time for physical inventory. These settings are made for the combination of the embedded EWM warehouse number and activity area.

To define the periodicity of storage bin checks for physical inventory for the embedded EWM warehouse number, execute Transaction SPRO and navigate to **SCM Extended Warehouse Management • Extended Warehouse Management • Internal Warehouse Processes • Physical Inventory • Physical-Inventory-Area-Specific Settings • Periodicity of Storage Bin Check**.

Figure 8.113 shows the physical inventory interval for storage bin checks and the buffer time set for embedded EWM warehouse number **MJ00** and activity area **0001**. Click **New Entries** to define a new entry for the physical inventory interval for storage bin checks and the buffer time for the combination of the embedded EWM warehouse number and activity area.

Figure 8.113 Periodicity of Storage Bin Check per Activity Area

The following are the key fields of this configuration:

- **Acty Area (activity area)**
  Activity areas are the specific areas within the warehouse where specific warehouse activities such as putaway, stock removal, physical inventory, and so on take place. For this configuration step, assign the activity area in which the physical inventory takes place.

- **Int. (physical inventory interval)**
  In this field, specify the number of working days as the interval for the physical inventory check for storage bins.

- **BT (buffer time)**
  In this field, specify the number of working days as the buffer for the physical inventory check for storage bins.

Assign Physical Inventory Area to Activity Area

A physical inventory area is an organizational unit and a location where the physical inventory activity is carried out within the warehouse. Activity areas are the specific areas within the warehouse where specific warehouse activities such as putaway, stock removal, physical inventory, and so on take place. Storage types are assigned to activity areas as explained in Section 8.2.5. In this configuration step, define physical inventory areas of the embedded EWM warehouse that are assigned to the activity areas for the embedded EWM warehouse number.

To assign physical inventory areas to activity areas for the embedded EWM warehouse number, execute Transaction SPRO and navigate to **SCM Extended Warehouse Management • Extended Warehouse Management • Internal Warehouse Processes • Physical Inventory • Warehouse-Number-Specific Settings • Assign Physical Inventory Area to Activity Area**.

Figure 8.114 shows that physical inventory area **MJ00_001** was assigned to activity area **0001** of warehouse number **MJ00**. Click **New Entries** to assign a new physical inventory area to an activity area for the embedded EWM warehouse number.

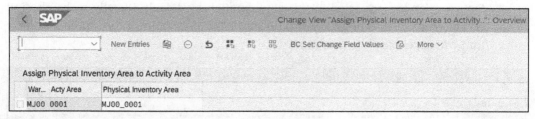

Figure 8.114 Assign Physical Inventory Area to Activity Area

Define Number Range for Physical Inventory Documents

In this configuration step, define a number range for the physical inventory documents.

To define number ranges, execute Transaction SPRO and navigate to **SCM Extended Warehouse Management • Extended Warehouse Management • Internal Warehouse Processes • Physical Inventory • Warehouse-Number-Specific Settings • Define Number Range for Physical Inventory Documents**. You can arrive at this configuration step directly by executing Transaction SNRO.

In this configuration step, both internal and external number ranges can be defined with reference to number range object /SCWM/PIDO.

Figure 8.115 shows that number range **01** was defined for physical inventory documents for embedded EWM warehouse number **MJ00**.

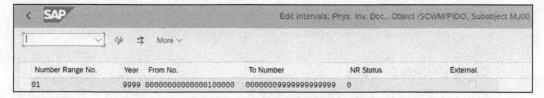

Figure 8.115 Define Number Range for Physical Inventory Documents

Specify Physical Inventory-Specific Settings in the Warehouse

In this configuration step, make embedded EWM warehouse-specific settings for physical inventory. This configuration view displays the embedded EWM warehouse number, currency key of the warehouse, and factory calendar by default. You can specify the fiscal year variance, number range, and tolerance check details for the embedded EWM warehouse number.

To specify physical inventory-specific settings in the warehouse for the embedded EWM warehouse number, execute Transaction SPRO and navigate to **SCM Extended Warehouse Management • Extended Warehouse Management • Internal Warehouse Processes • Physical Inventory • Warehouse-Number-Specific Settings • Specify Physical-Inventory-Specific Settings in the Warehouse**.

All embedded EWM warehouse numbers and their corresponding currency codes and factory calendars are listed in this configuration by default. Directly make physical inventory-specific settings. Figure 8.116 shows the physical inventory-specific settings for EWM warehouse number **MJ00**.

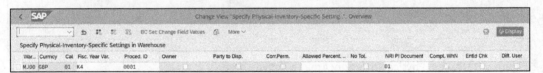

Figure 8.116 Specify Physical Inventory-Specific Settings in Warehouse

The following are the key fields of this configuration:

- **Fisc. Year Var. (fiscal year variant)**
  Fiscal year variants establish the relationship between financial accounting and the calendar year. In this field, specify the fiscal year variant for physical inventory.

- **Allowed Precent. (allowed percentage difference in counting)**
  In this field, specify the allowed percentage difference for counting. This is a tolerance setup and the system will automatically suggest recounting if the difference between actual quantity and the system-recorded quantity is more than the tolerance that has been set.

- **No Tol. (no tolerance)**
  If you set this indicator, there will be no tolerance check during counting.

- **NRI PI Document**
  In this field, assign the number range for the physical inventory document defined for the embedded EWM warehouse in the previous configuration step.

Configure Cycle Counting

In this configuration step, configure the cycle counting indicator for the embedded EWM warehouse number and specify the physical inventory interval in days and a buffer for physical inventory in days. The cycle counting indicator is assigned to the warehouse product, and the system uses that data to determine the physical inventory interval (days) and buffer (days) for cycle counting of the warehouse product.

To configure cycle counting for the embedded EWM warehouse number, execute Transaction SPRO and navigate to **SCM Extended Warehouse Management • Extended Warehouse Management • Internal Warehouse Processes • Physical Inventory • Warehouse-Number-Specific Settings • Configure Cycle Counting**.

Figure 8.117 shows that cycle counting indicators **A**, **B**, and **C** are configured for embedded EWM warehouse number **MJ00**. Cycle counting intervals of **30** days, **60** days, and **90** days are maintained respectively for cycle counting indicators **A**, **B**, and **C**.

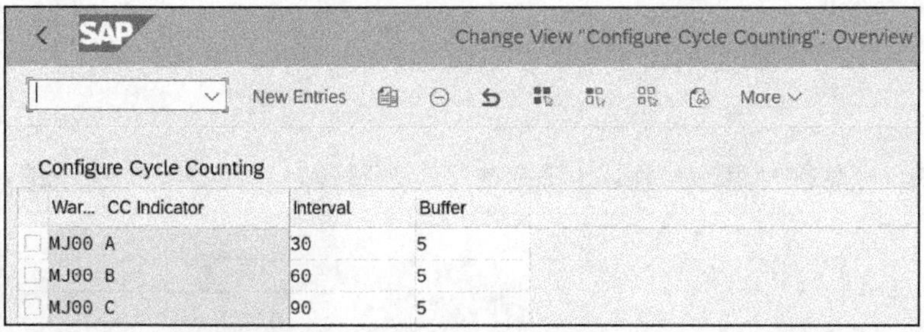

Figure 8.117 Configure Cycle Counting

To create a new cycle counting indicator for the embedded EWM warehouse number, click **New Entries** and maintain the following key fields:

- **CC Indicator**
  In this field, specify a cycle counting indicator, a single-digit alphabetic or numeric character for cycle counting of products. This key is used to define different physical inventory intervals and buffers and assign different warehouse products.

- **Interval**
  In this field, specify the number of working days as an interval for physical inventory checks for storage bins.

- **Buffer**
  In this field, specify the number of working days as a buffer for physical inventory checks for storage bins.

Warehouse Product Maintenance for Physical Inventory

The warehouse product master in embedded EWM stores physical inventory data, and the cycle counting indicator is assigned to the warehouse product. The system uses this data to determine the physical inventory interval (days) and buffer (days) for cycle counting of a warehouse product. To create/change/display the warehouse product, execute Transaction /SCWM/MAT1.

After executing the transaction, specify the warehouse product for which you need to maintain the physical inventory data. Press ⌑Enter⌑ to display the details of the product. Then, in the **Whse Data** (warehouse data) tab of the product details, specify the **Cycle Counting Indicator**. Figure 8.118 shows cycle counting indicator **A** maintained in the **Whse Data** tab for warehouse product **MS-RAW MATERIAL1** in embedded EWM warehouse number **MJ00**.

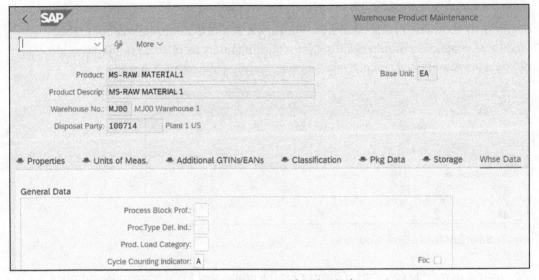

Figure 8.118 Warehouse Product Maintenance for Physical Inventory

8.6   Batch Management

Batches are used to enhance product characteristics; to define quality attributes of a product; and to trace products by production date, expiration date, status, specific characteristics, and so on. Batch management is a regulation and compliance process in certain industries. Quality control is an essential process, and it is made easier with batch management. Defect management is another area where batch management plays a vital role. A defective product is easily tracked in inventory management and warehouse management if the product is batch managed. Batches are created through inbound processing and changes to batch master data can be made in embedded EWM. Changes are updated in the batch master record in SAP S/4HANA through qRFC. Material management, sales and distribution, and warehouse management functions track materials with unique batch master records. Materials management creates, tracks, and stores batch stock of materials. Meanwhile, the warehouse management function is used to search for specific stock of materials in the warehouse.

Let's dive deep into the batch management process in embedded EWM and its integration with SAP S/4HANA materials management.

8.6.1   Batch Management Process

Batch management offers precision in tracing a particular lot of products. A lot can be produced in house or procured externally; batch management makes it easier to manage stocks in warehouse management and inventory management. A batch is always linked to a specific lot of a specific product.

As a prerequisite, the material master record in SAP S/4HANA must have batch management active to manage materials in batches. If any batch characteristics change, the system creates a new batch number for that lot. A batch is defined by the following main attributes:

- Product number (material master data)
- Batch number
- Country of origin
- Batch status
- Production date
- Expiration date

You can define additional attributes for batches in terms of characteristics. SAP provides a class for batch management, and you can configure batch characteristics using the batch class based on the business requirements. Refer to Chapter 6, Section 6.5.5 for the detailed batch management process in SAP S/4HANA.

In embedded EWM, the batch management process starts with the creation of batches. Batches are created during the inbound processing automatically based on settings in embedded EWM. Batches can be changed in embedded EWM. Batches are created either during inbound delivery processing or using Transaction MSC1N (Create Batch). Batch data attributes such as production date, expiration date, country of origin, batch status (restricted/non restricted), supplier's batch number or inhouse production batch number, and other characteristics are captured and adopted from the inbound delivery. SAP S/4HANA transfers these attributes to embedded EWM via inbound delivery. Once the inbound delivery is replicated to EWM, batch master record creation can be initiated manually from the inbound delivery using Transaction /SCWM/PRDI, and you can add additional batch details while creating the batch. The system can automatically create the batch using inbound delivery item characteristics. All the attributes available in the inbound delivery item are copied over to the batch record.

On the outbound delivery processing side, the system will initiate picking of the right batch of the right products as per the delivery request. The outbound delivery is created in SAP S/4HANA with reference to a stock transfer order, sales order, and so on. The outbound delivery will have the attributes for picking, packing, and goods issue of products from the warehouse. Batch determination criteria are part of the outbound deliveries. Once the outbound deliveries are replicated from SAP S/4HANA to embedded EWM, an outbound delivery order (warehouse request) is created, and the batch determination criteria are replicated along with the outbound delivery. Based on the batch determination criteria, a relevant batch of the product is determined for picking/ stock removal. You can directly enter batches in the outbound delivery, and the system will search for the batch in the warehouse. The warehouse task is created with the storage type, storage section, and storage bin information to pick the batches. The warehouse personnel pick the batches and confirm the warehouse task for picking.

The following are the prerequisites and basic configuration settings in SAP S/4HANA and embedded EWM systems for batch management.

Activation of Batch Level in SAP S/4HANA

In SAP S/4HANA, batches can be activated at the client level, plant level, and material level as a prerequisite. Best practice is to create and maintain the batch at the material level. For embedded EWM, the batch level must be activated at either the material level or client level. Activating batching at the plant level is not supported in embedded EWM. During the goods receipt of the materials for the first time into inventory, a batch number will be created with the production date, expiration date, and other attributes. You can change the batches in embedded EWM.

To arrive at this SAP S/4HANA configuration step to activate the batch level, execute Transaction SPRO and navigate to **Logistics—General** • **Batch Management** • **Specify Batch Level and Activate Status Management**.

Figure 8.119 shows that the batch level has been activated at the material level.

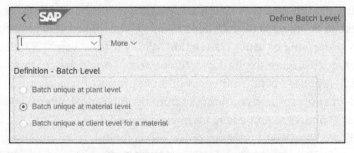

Figure 8.119 Activation of Batch Level in SAP S/4HANA

Activate Internal Batch Number Assignment

This is another prerequisite step and an SAP S/4HANA configuration step. In this configuration, you can activate the internal number range for batches.

To arrive at this SAP S/4HANA configuration step, execute Transaction SPRO and navigate to **Logistics—General** • **Batch Management** • **Batch Number** • **Activate Internal Batch Number Assignment**.

Figure 8.120 shows that the internal batch number assignment is activated for batch master creation.

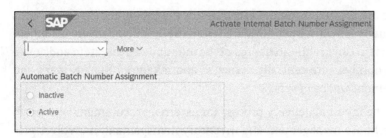

Figure 8.120 Activate Internal Batch Number Assignment

Configure Embedded EWM-Specific Parameters

This is an integration configuration in SAP S/4HANA. In this configuration step, enable the batch determination feature in embedded EWM for the SAP S/4HANA-defined warehouse number that is linked to the embedded EWM warehouse number. When you enable this feature, the batch determination criteria will be passed over to embedded EWM through outbound delivery replication. Embedded EWM uses the search criteria to determine the relevant batch of the requested material to fulfill outbound delivery.

To arrive at this SAP S/4HANA configuration step to activate batch determination in embedded EWM, execute Transaction SPRO and navigate to **Integration with Other SAP Components • Extended Warehouse Management • Basic Settings for EWM Linkage • Configure SAP EWM-Specific Parameters**.

All embedded EWM warehouse numbers are listed in this configuration by default. Directly make the batch management settings here. Figure 8.121 shows that SAP S/4HANA warehouse number **MJO** has been enabled for embedded EWM, and the **BatchDetEW** (batch determination in EWM) indicator was set.

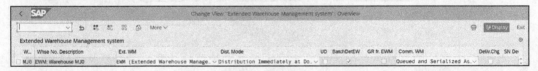

Figure 8.121 Configure EWM-Specific Parameters for Batch Management in SAP S/4HANA

With this setup, there will be no batch determination happening in SAP S/4HANA. Instead, it happens in embedded EWM.

8.6.2 Inbound Batch Management

Inbound batch management involves the creation of batch master data using a supplier's (external) batch number or an internal batch number if the supplier is not sending batch details. The inbound process in embedded EWM starts after the inbound delivery is replicated and a warehouse request for inbound processing is created. Batches for the products received from an external supplier or from an in-house production process are created using the attributes of the inbound delivery. A batch number can be created manually, can be initiated from the inbound delivery in embedded EWM, or the system can use the attributes of the inbound delivery item and create a batch. The batch numbers are centrally managed and inventory management and warehouse management will be in sync.

Figure 8.122 shows the inbound delivery process for external procurement with batch movements in warehouse management and inventory management. It involves process steps in SAP S/4HANA and embedded EWM. Inbound delivery processing steps have been explained previously; refer to Section 8.3.2 for more details.

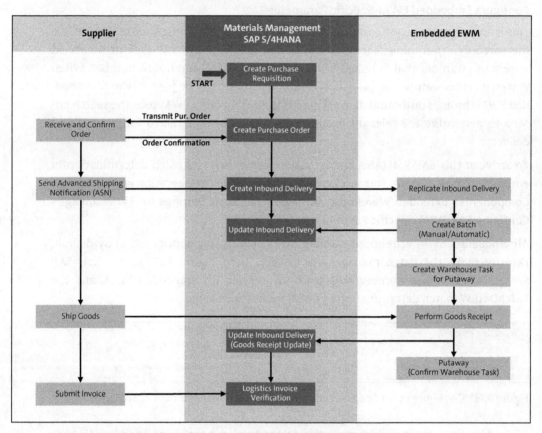

Figure 8.122 EWM Inbound Delivery Processing: Batch Movements

The creation of a batch happens in embedded EWM during the inbound delivery processing, and it automatically updates the batch details in the inbound delivery if no external batch number exists in the inbound delivery. If a supplier sends an external batch number, you can create the same batch number during inbound delivery processing. The batch number will have all the attributes that are part of the inbound delivery, and you can enrich the batch by adding additional characteristics.

Let's walk through the settings in embedded EWM that are necessary for batch management in inbound deliveries.

In this configuration step, define the batch management and shelf-life expiration data check for inbound delivery and outbound delivery processing. This configuration settings for batch management are made with reference to a combination of document category, embedded EWM warehouse number, document type, and item type in embedded EWM for the warehouse request.

To define settings for delivery, execute Transaction SPRO and navigate to **SCM Extended Warehouse Management • Extended Warehouse Management • Cross-Process Settings • Batch Management • Define Setting for Delivery—Warehouse Request**.

Figure 8.123 shows the SLED check and batch management settings for the inbound delivery warehouse request for embedded EWM warehouse number **MJ00**, document category **PDI** (inbound delivery), item type **IDLV** (standard inbound delivery item), and document type **INB** (standard inbound delivery). The following are the key fields of this configuration:

- **Sel. Criteria**
  If you set this field to **Check**, it checks the selection characteristics of the inbound delivery item against the batch characteristics. It is recommended to set this field to **No Check** for inbound delivery warehouse requests.

- **GdsMvt OK Rstd Batch**
  If you set this indicator, goods movement with restricted batches is allowed. You can set this for inbound delivery warehouse requests, but it is not recommended for outbound delivery warehouse requests.

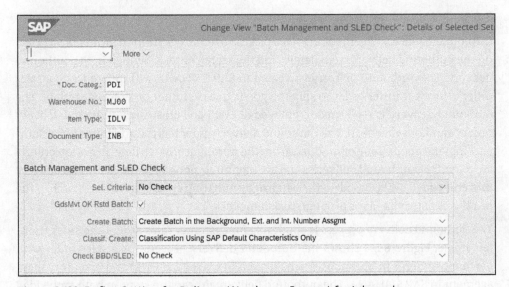

Figure 8.123 Define Setting for Delivery: Warehouse Request for Inbound

- **Create Batch**

  This field is only enabled for the inbound delivery category, and in this field, you can set criteria for batch creation during inbound delivery processing in the warehouse. One of the following values can be set:

  - Manual batch creation allowed.
  - Batch creation not allowed.
  - Create batch in the background with external number assignment only.
  - Create batch in the background with external or internal number assignment.

- **Classif. Create**

  This field is only enabled for the inbound delivery category, and it controls how the batch characteristics are valuated during the creation of a batch during inbound delivery processing. One of the following values can be set:

  - Classification using SAP default characteristics only.
  - Classification in the background when creating a batch.

- **Check BBD/SLED**

  This field is only enabled for the inbound delivery category, and it controls how the SLED or best-before date is checked during inbound delivery processing. One of the following values can be set:

  - No check
  - Lock item, manual release allowed.
  - Lock item, manual release not allowed.

8.6.3   Outbound Batch Management

Outbound batch management involves picking batch stock from the warehouse based on the outbound delivery item details and the search criteria defined in the outbound delivery. The outbound delivery is created in SAP S/4HANA with reference to a sales order or stock transfer order to ship the goods to a customer or a receiving plant. The outbound delivery is replicated to embedded EWM and creates an outbound delivery order (warehouse request). The outbound delivery order has the attributes for picking, packing, and goods issue of products from the warehouse. Batch determination criteria are part of the outbound deliveries and are replicated along with the outbound delivery into embedded EWM. Based on the batch determination criteria, a relevant batch of the product is determined for picking/stock removal.

The outbound delivery process is applicable for customer sales orders and stock transfer orders. Figure 8.124 shows the outbound delivery process for a sales order with batch movements in warehouse management in embedded EWM and inventory management in SAP S/4HANA. The outbound delivery processing steps in both embedded EWM and SAP S/4HANA have been explained previously; refer to Section 8.4.2 for more details.

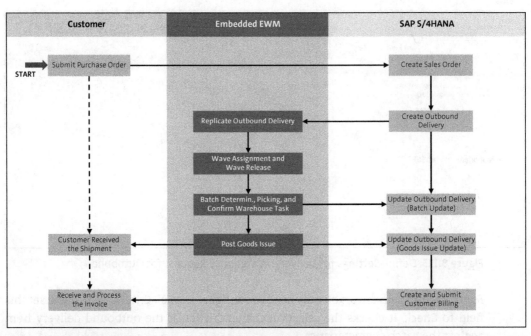

Figure 8.124 EWM Outbound Delivery Processing: Batch Movements

Unlike inbound delivery processing, there is no batch creation in the outbound delivery processing. The warehouse task created for picking/stock removal contains the source storage area details, and the source storage type, storage section, and storage bins are determined based on the batch determination criteria. Once the batch stock is picked and the warehouse task is confirmed, there will be an automated update to the outbound delivery in SAP S/4HANA through qRFC. The outbound delivery in SAP S/4HANA is updated with the batch number details based on warehouse task confirmation.

Let's walk through the settings in embedded EWM that are necessary for batch management in outbound deliveries.

In this configuration step, define the batch management and SLED check for outbound delivery processing. This configuration settings for batch management are made with reference to a combination of document category, embedded EWM warehouse number, document type, and item type in embedded EWM for the warehouse request.

To define settings for delivery, execute Transaction SPRO and navigate to **SCM Extended Warehouse Management • Extended Warehouse Management • Cross-Process Settings • Batch Management • Define Setting for Delivery—Warehouse Request**.

Figure 8.125 shows the SLED check and batch management settings for an outbound delivery warehouse request for embedded EWM warehouse number **MJ00**, document category **PDO** (outbound delivery), item type **ODLV** (standard outbound delivery item), and document type **OUTB** (standard outbound delivery).

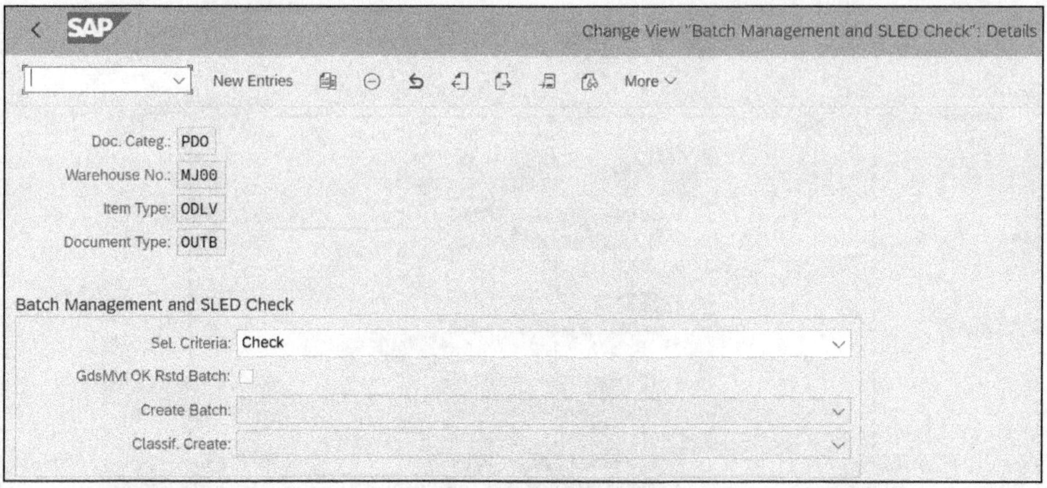

Figure 8.125 Define Settings for Delivery: Warehouse Request for Outbound

For outbound deliveries, you only need to configure the **Sel. Criteria** field. If you set this field to **Check**, it checks the selection characteristics of the outbound delivery item against the batch characteristics.

8.7   Summary

In this chapter, we introduced the integration of embedded extended warehouse management with materials management with best practice examples and processes. In Section 8.1, we explained the master data in extended warehouse management with step-by-step procedures to maintain extended warehouse management master data. In Section 8.2, we explained the basics of EWM and the warehouse structural elements definitions. In Section 8.3, we explained the inbound delivery processing in EWM with best practice processes, configuration, and integration with SAP S/4HANA (ERP). In Section 8.4, we explained the outbound delivery processing in EWM with best practice processes, configuration, and integration with SAP S/4HANA (ERP). In Section 8.5, we explained the internal warehouse processes in EWM and integration with SAP S/4HANA (ERP). Finally, in Section 8.6, we explained the batch management process in EWM.

In the next chapter, we'll discuss the integration between transportation management and materials management for optimizing logistics and ensuring smooth freight execution and settlement. We'll also explain the order management, freight planning, freight execution, and freight settlement processes for purchasing with best practice examples, master data requirements, and configuration settings.

Chapter 9
Transportation Management

Transportation management in SAP S/4HANA takes care of all physical transfers of goods between locations. It provides features to automate transportation planning, mode, determination, and carrier selection. It manages freight rates and provides features like freight charge calculation and carrier collaboration. Transportation management is tightly integrated with materials management for inbound and outbound shipment processing as well as freight order settlement.

Transportation management is one of the pillars of the supply chain, and it is a vital function in both inbound and outbound logistics. Transportation management is about not only transferring goods from one location to another but also meeting a customer's delivery requirements such as the delivery date and ship-to location considering the capacity of resources. It deals with point-to-point transfer of goods where point A is the ship-from (source) location and point B is the ship-to (destination) location. Means of transport and mode of transport play an important role in transporting goods from a source location to a destination location. In simple terms, transportation management transfers goods or cargo from a source location to a destination location using a specific means and mode of transport.

Transportation management operates with the following business partners:

- **Shippers**
  Also called consigners, these are the owners/suppliers of the commodity being shipped, and the transportation is initiated from a shipping point of the shipper. A shipper can be a customer or a supplier.

- **Business partner (customer-related, such as ship-to party, sold-to party, etc.)**
  Customers are business partners who receive goods from a shipper. Goods are transported from a shipping point (source) to a customer's ship-to location (destination) using transportation management.

- **Business partner (supplier-related, such as goods supplier, vendor, etc.)**
  For inbound transportation, suppliers are business partners who supply the goods that they own, and transportation is managed by either the supplier or the buyer. Goods are transported from the supplier location (source) to the receiving location (destination) using transportation management.

- **Logistics service providers (LSPs)**
  Also called brokers, these are the third-party service providers who manage transportation of goods on behalf of shippers by arranging carriers.

- **Carriers**
  Carriers are also called forwarding agents; they are the actual transporters of goods by various means from a source location to a destination location, and they work with shippers. LSPs can also act as carriers.

Movement of goods from a vendor location to a receiving location using transportation management is called *inbound freight*, whereas movement of goods from a shipping point to a customer's location using transportation management is called *outbound freight*. SAP S/4HANA offers embedded transportation management (TM) as a deployment option, with SAP Transportation Management (SAP TM) requiring additional licensing. Like embedded extended warehouse management (EWM), which we discussed in Chapter 8, TM is embedded in SAP S/4HANA with centralized master data. Embedded TM can be integrated with embedded EWM as well.

In this chapter, we'll walk through the integrated TM and materials management functionalities in SAP S/4HANA, including the key TM master data and configuration instructions for cross-functional TM and materials management processes.

9.1   Integrated Transportation Management Processes

Materials management in SAP S/4HANA and embedded TM functions are tightly integrated. All major functions of materials management such as purchasing, inventory management, logistics invoice verification, and valuation and account determination are integrated with SAP transportation processes. Figure 9.1 shows the end-to-end transportation management process with SAP S/4HANA and embedded TM integration.

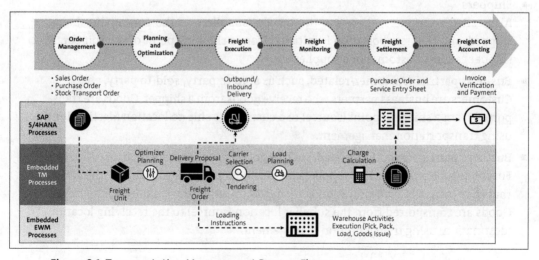

Figure 9.1 Transportation Management Process Flow

Let's explore the functions of embedded TM:

1. **Order management**

   The transportation process starts with order management in embedded TM. When a purchase order or sales order is created in SAP S/4HANA, an order-based transportation requirement (OTR) is created in embedded TM. Order management controls the type of purchase order (third-party order, standard purchase order, consignment order, subcontracting order, etc.), and sales orders are managed in transportation management. Delivery-based transportation requirements (DTRs) can also be created in embedded TM. The inbound deliveries and outbound deliveries create a demand/requirement in embedded TM. Embedded TM provides tools and capabilities to combine inbound and outbound deliveries. Delivery documents in SAP S/4HANA have a **TM Status** tab, which indicates the deliveries are managed in embedded TM. Using order management, changes in demand from the procure-to-pay process and order-to-cash process can be managed in embedded TM.

2. **Transportation planning and optimization**

   Transportation planning and optimization begin in embedded TM with the creation of freight units. Freight units are planning documents in transportation management that represent the goods to be transported and how and where they should be transported. A freight unit is automatically created in transportation management from OTRs and/or DTRs. It contains the item details from different purchase orders, sales orders, inbound deliveries, and outbound deliveries including quantities, source location, destination location, and so on. Once the freight unit is created in embedded TM, the subsequent transportation management process will begin.

   Embedded TM provides a manual transportation planning option as well as an automatic planning option aiming to improve transportation efficiency while reducing transportation costs. Embedded TM provides a planning optimizer engine to automate transportation planning. Transportation networks, resource availability and capacity, carrier selection and types of carriers, product-specific vehicle requirements (old storage, hazardous material handling, etc.), cost, and distance are the key considerations for transportation planning and optimization. The features of the planning optimizer include the following:

   - For route optimization, the system considers the minimum distance or minimum cost based on business requirements.
   - For schedule optimization, the system considers the shipping date and time, the date and time of the shipment delivery, the number of stops required, and the number of transshipments required.
   - For resource utilization and vehicle selection, the system considers the available resources, capacities, product-specific vehicle requirements, choice of route, and so on.
   - For load optimization, the system considers the vehicle size and capacity, package weight and volume, product-specific vehicle requirements, customer-specific requirements, and so on.

3. **Freight execution**

Once the transportation planning is completed and a freight order is created, a freight order or freight booking document is created based on the mode of transportation. Freight orders and freight bookings are called *freight execution documents* as the physical movement of goods starts once these documents are created. Freight orders and freight bookings are similar documents, and the differentiator is the mode of transport. A freight order is created if the mode of transport is road or rail, whereas a freight booking is created if the mode of transport is air or ocean. Because a carrier is assigned to the freight execution document to execute transportation, it is also called the *subcontracting document*. The freight execution document displays the status of a shipment as it captures all major events or milestones during the execution process, such as pick-up date, delivery date, and so on.

Embedded TM can be integrated with embedded EWM to support the freight execution process. During freight execution, load planning is performed. Based on load planning, picking, packing, and loading can be executed from embedded EWM.

4. **Freight monitoring**

Tracking and tracing of shipments is an important process in transportation management. Shipments are continuously monitored as soon as the transportation begins from the source location until it is delivered and unloaded from the vehicle at the destination location. The freight execution process can be tracked and traced in real time using the freight monitoring feature of embedded TM. Users can get end-to-end process visibility and monitor the progress of every freight execution process.

5. **Freight settlement**

A freight settlement document is created in embedded TM to settle the carrier costs with reference to a freight order and freight booking documents. The freight costs calculated in embedded TM for freight orders and freight bookings and the financial accounting postings and settlements happen in SAP S/4HANA. Freight costs are recorded as accruals in financial accounting, and the carriers will be paid in SAP S/4HANA to clear accruals. The initiated or completed status of freight execution documents triggers the posting of freight cost in financial accounting by creating a service purchase order and service entry sheet. A carrier can submit an invoice for services via electronic data interchange (EDI) or by sending a copy of the invoice by email. If the invoice is received via EDI, the system automatically posts the invoice in SAP S/4HANA; otherwise, accounts payable personnel post the invoice manually. Evaluated receipt settlement (ERS) can be leveraged to automate the invoice posting.

If an LSP is involved in the freight planning and execution process, a forwarding settlement document instead of freight settlement document is created. The service purchase order and the rest of the settlement process in SAP S/4HANA is like the freight settlement process. The LSP will be paid in the forwarding settlement process.

6. **Freight cost accounting**

Financial account postings happen in SAP S/4HANA. The freight settlement document or forwarding settlement document created in embedded TM calculates the freight cost or service charges only, but the financial accounting postings happen in SAP S/4HANA with reference to the service purchase order, service entry sheet, and incoming invoice. After the invoice posting, payment processing is done in SAP S/4HANA to settle the costs with the carrier or LSP (forwarding agent).

Embedded TM provides the following benefits:

- Integrated transportation management across supply chain to improve responsiveness to changes in demand
- Optimized planning for multiple modes of transport regardless of direction or geography
- Reduced transportation spending by leveraging economies of scale
- Synchronized transportation planning and execution
- Supports business-to-business integration with the carriers and LSPs for effective collaboration and improves operational efficiency
- Better alignment with the organization's strategic goals and objectives
- Cost distribution of freight to achieve profitability and material valuation
- Quick point of issue identification with enhanced and real-time tracking and tracing of shipments
- Automated mode and carrier selection and route optimization

Seamless integration between SAP S/4HANA materials management and embedded TM ensures operational efficiency. Table 9.1 shows all the key processes impacted.

Area	Processes Impacted	Description
Transportation management (TM)	Manage master data in transportation management	This process supports the creation and maintenance of transportation specific master data such as resources, transportation network, transportation mode, and so on.
	Perform transportation order management	This process supports the evaluation of SAP S/4HANA requirements for transportation management and creation and maintenance of freight units for inland transportation, rail transportation, air transportation, and ocean transportation.

Table 9.1 Process Impact: Transportation Management

Area	Processes Impacted	Description
Transportation management (TM) (Cont.)	Manage transportation planning	This process supports the manual and automated transportation planning that includes air freight planning, ocean freight planning, rail freight planning, road/truck planning, and so on, plus maintenance of a freight order, freight booking, or forwarding order.
	Process tendering/subcontracting	This process supports the tendering process to select appropriate carriers, sharing freight order/freight booking with the carrier and receiving confirmation from carriers.
	Perform load planning	This process supports the management of the transportation cockpit, performing load optimization, managing the load plan, managing load distribution elements, and managing load planning exceptions.
	Manage transportation execution and monitoring	This process supports the execution of physical transfer of goods, customs and compliance management, updating status of freight documents, monitoring of freight document status, and so on.
	Manage freight settlement	This process supports the calculation of freight charges, creation and maintenance of freight settlement documents, transferring of freight settlement documents to SAP S/4HANA, and receiving confirmation from SAP S/4HANA.
	Manage strategic freight procurement	This process supports the strategic sourcing of carriers and forwarding agents (LSPs) through a request for quotation (RFQ) process as well as creation and maintenance of freight agreements.
Materials management	Create and maintain purchasing documents	This process supports the creation of service purchase order in SAP S/4HANA from the freight settlement document from embedded TM.
	Inventory management	This process supports picking, packing, goods issue, and loading of goods for shipment (freight documents) in inventory management.
	Supplier invoice management	This process supports the processing of an incoming invoice from the carrier or forwarding agents.

Table 9.1 Process Impact: Transportation Management (Cont.)

9.2   Master Data in Transportation Management

During order management, transportation-related requirements from orders such as sales orders, purchase orders, and stock transfer orders are transferred from SAP S/4HANA to embedded TM. Order management tells embedded TM the source location, destination location, and the products to be transported using embedded TM functionality. The material master is the core master data for almost all functions of materials management. Material master data is created and maintained in SAP S/4HANA and transferred to embedded TM as products. All transportable products and packaging materials are transferred from SAP S/4HANA to embedded TM. The unit of measure and conversion factors, transportation group, loading group, storage data, packaging data, and so on are the key product attributes required in embedded TM. Refer to Chapter 3, Section 3.1 for more details.

Another core master data element used in transportation management and maintained centrally in SAP S/4HANA is the business partner. The following are the business partner roles required in embedded TM:

- **Carrier**
  Business partner created with the partner role, responsible for execution of freight order or freight booking

- **Forwarding agent**
  Business partner created with the partner role, responsible for execution of forwarding order

- **Ship-to party**
  Customer-specific business partner role that represents the destination location for transportation of goods for outbound transportation

- **Sold-to party**
  Customer-specific business partner role that represents the party to which the goods are sold

- **Vendor**
  Business partner created with the supplier role that acts as a shipper for inbound transportation

Refer to Chapter 3, Section 3.2, for more details about business partners.

This section covers the master data created in embedded TM. Let's dive in.

9.2.1   Resource

Transportation resources are vital master data elements in embedded TM. Resources are used in planning and execution of transportation. Resource capacity and availability are major considerations in transportation planning and optimization. Resources are required for all transportation activities from loading to handling, from transporting to unloading, and so on. The following are the different resource types available in the standard system:

- **Calendar resource**
  Calendar resources operate in shifts and may have downtimes. These resources are used in goods issue and goods receipt processes for loading, handling, and unloading purposes. They are assigned to shipping and receiving locations.

- **Vehicle resource**
  Vehicle resources are specific means of transport used to transport goods from point A to point B. Trucks, ships, railcars, trailers, cargo ships, and so on are examples of vehicle resources. A means of transport is assigned to vehicle resource master data.

- **Handling resource**
  Handling resources are used in handling activities during goods issue and goods receipt processing. These are specific tools used at receiving and shipping locations. Forklifts are handling resources.

- **Driver**
  A driver is a person who operates a vehicle resource and is defined as a separate resource in embedded TM. Drivers will have specific qualifications to operate a vehicle resource (e.g., a truck). Availability and capacity of drivers are important criteria in transportation planning.

- **Transportation units**
  Transportation units are used to carry products during transportation. Containers and the like are transportation units.

To explore the details of a transportation resource, let's create a vehicle resource. You can use Transaction /SCMTMS/RES01 to create resources manually. Figure 9.2 shows the initial screen for defining resource master data, where you fill out the following fields:

- **Resource**
  Enter a resource name as the unique identifier to create a resource master data. Specify a resource name of up to 40 characters in this field.

- **Location**
  A location represents a source, destination, or transit location in a transportation network. Handling resources are assigned to a particular location for loading, unloading, or material handling activities. Specify a location if you are creating a handling resource.

- **Organization Unit**
  Organization units in embedded TM are independent organization units that support transportation management. The embedded TM organization units manage orders/requirements, purchasing, transportation planning, and execution. Specify the organization unit in this field that the resource belongs to.

■ **Resource Type**

This field distinguishes different resource types. Choose the resource type from the dropdown menu. Table 9.2 shows the resource types available in the standard system.

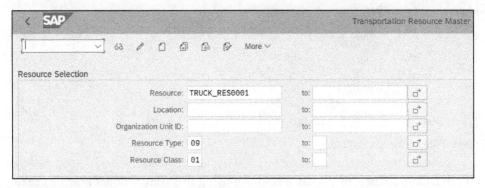

Figure 9.2 Define Resource Master Data: Initial Screen

Resource Type	Description
09	Vehicle resource
10	Calendar
12	Transportation unit
13	Handling resource

Table 9.2 Resource Type

■ **Resource Class**

A resource class is used to group identical resources. Table 9.3 shows the resource classes available in the standard system to choose from.

Resource Class	Description
01	Truck
02	Trailer
03	Locomotive
04	Railcar
05	Vessel
06	Airplane
10	Container

Table 9.3 Resource Class

Resource Class	Description
11	Unit load device (ULD)
12	Box
20	Doors
21	Dolly
22	Forklift
30	Driver

Table 9.3 Resource Class (Cont.)

You can create, change, or display a resource using this transaction. Click the **Create** icon to create a new resource for embedded TM. You can also create a resource via the copy function. Figure 9.3 shows the resource creation using the **Copy from Template** feature. Specify the resource name and specify the template to copy from in the **Means of Transport** field to create the resource.

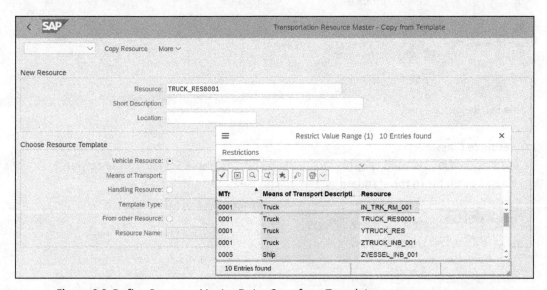

Figure 9.3 Define Resource Master Data: Copy from Template

The system allows you to change/edit the resource details after you create it using the **Copy from Template** icon. After specifying the fields in the initial screen, click the **Create Resources** icon at the top or press F5 to arrive at the header data of the resource. Figure 9.4 shows the header data of the vehicle resource. There are four tabs in the header: **Vehicle**, **Calendar**, **TranspUnit**, and **Handling.** If you're creating a vehicle resource, update the details under the **Vehicle** tab. There are various data tabs to be updated to create the resource; we'll walk through the most important ones.

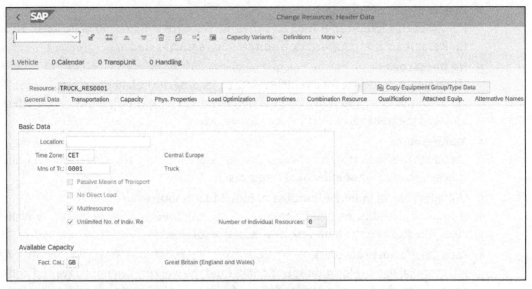

Figure 9.4 Define Resource Master Data: General Data

The **General Data** tab has the following key fields:

- **Time Zone**
  This is a basic data element of the resource. Specify a time zone where the resource is located.

- **Mns of Tr. (means of transport)**
  The means of transport is the vehicle resource or the vehicle classification. It's either a single vehicle resource or a group of identical vehicle resources. Specify **0001** as the means of transport for the truck resource in this field. Table 9.4 shows the means of transport defined in the standard system.

Means of Transport	Description
0001	Truck
0002	Rail
0003	Airplane
0004	Courier, express, and delivery company
0005	Ship
0006	Car
0007	Trunk (subcontracting)
0008	Railcar

Table 9.4 Means of Transport

- **Passive Means of Transport**
  If you set this indicator, it indicates that the resource will not move by itself. This indicator is set for transportation unit resources such as trailers.

- **No Direct Load**
  If you set this indicator for a vehicle resource, loading the freight directly to the vehicle is not possible; you must load the freight into a transportation unit (e.g., trailer) and load the transportation unit into the vehicle.

- **Multiresource**
  If you set this indicator, it indicates that there are multiple resources, and you can specify the number of individual resources.

- **Unlimited No. of Indiv. Re (number of individual resources)**
  If you set this indicator, it indicates there is an unlimited number of resources available; you must set the **Multiresource** indicator for this.

- **Fact. Cal. (factory calendar)**
  A factory calendar is location specific. It defines the working calendar days and holidays. Assign the factory calendar defined for the specific location for resource planning purposes.

Figure 9.5 shows the header data and the **Transportation** tab of the vehicle resource. Under the **Resource Validity** section, you can enter valid from and valid to dates.

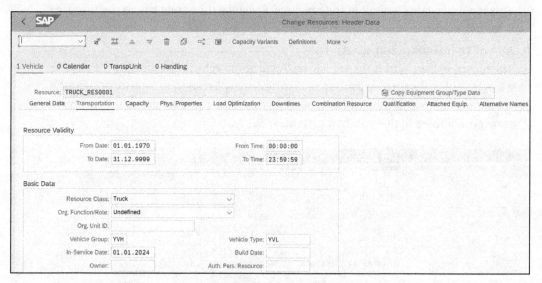

Figure 9.5 Define Resource Master Data: Transportation Data

Under the **Basic Data** section of the transportation data, **Resource Class** and **Org. Unit ID** are copied from the initial screen. The following are the other key fields:

- **Org. Function/Role**
  The organization function/role defines the function/role of the organization unit assigned to the resource. This could be a sales group, purchasing group, execution

and planning group, and so on. **Undefined** will be assigned to this field by default, and you can select the organization function from the dropdown menu.

- **Vehicle Group**
  A vehicle group is used in conjunction with a vehicle type to group identical vehicle resource groups, handling resource groups, and transportation unit resource groups. You can define vehicle groups in configuration settings and assign the relevant vehicle group for the resource.

- **Vehicle Type**
  A vehicle type is used for different vehicle resource types, handling resource types, and transportation unit resource types. You can define vehicle types in the configuration settings and assign the relevant vehicle type for the resource.

- **Owner**
  Assign the business partner of the owner (if any) in this field. You can define the vehicle owners as business partners.

- **Registration Number**
  In this field (not shown; you can scroll down to view it), specify the registration number of the vehicle as provided by the local authority in this field. In the **Registration Number Country/Region** field, assign the country/region in which the vehicle was registered.

Figure 9.6 shows the header data and the **Capacity** tab of the vehicle resource. On this tab, the **Dimensions** of the vehicle resource can be defined in terms of mass, volume, or length. Select the dimension to define the capacity of the resource. Specify the capacity and unit of measure based on the dimension you selected; for example, setting **Capacity** to **22** and **Unit** to **TON** indicates that the maximum capacity of the vehicle resource is 22 tons.

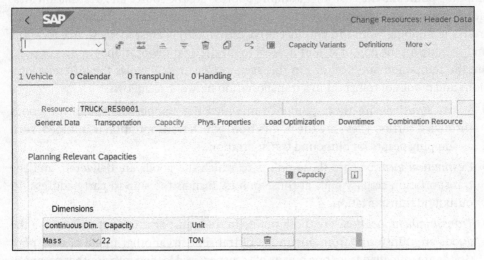

Figure 9.6 Define Resource Master Data: Capacity Data

In the **Phys. Properties** tab, you can define the gross weight, tare weight, tare volume, maximum length, maximum width, and maximum height of the resource. Likewise, in other data tabs, specific information for the resource master data is maintained.

9.2.2   Transportation Network

The transportation network is the most important master data element in embedded TM for transportation management to operate. Using a transportation network, the transportation planner can choose the best route for transferring the goods or requirements arising from order management based on delivery date requested by the customer, means of transport, transportation distance/cost, and so on. If you use a transportation planning optimizer to automate the planning process, the transportation network must be maintained accurately. The transportation network defines the zones, locations, transshipment points, transportation lanes, carrier profile, and so on to transport the goods from point A to point B, from point B to point C, and so on. It consists of multiple locations, customer sites, transshipment locations, transportation lanes, and transportation zones.

Let's explore the important attributes to define a transportation network.

Locations

Locations are the nodes of the transportation network. They represent the source, destination, and transit locations in a transportation network. Let's look at an example. Imagine a customer has requested some finished products to be delivered to Seattle, and the finished products are produced and shipped from Birmingham in the UK. In this scenario, depending on the standard delivery or expedited delivery, road, ocean (for standard delivery), and air (for expedited delivery) are the means of transport. The shipping point and the customer's ship-to address are the source and destination locations, respectively. This route requires two transit locations, one to transfer the goods from the shipping point to a port (ocean) or airport (air) nearest to the source location, and another for the port or airport nearest to the destination location. Airports or ports are the transshipment locations in this scenario. Figure 9.7 shows the different locations and modes of transport in a transportation network, as follows:

- *Source locations* are those points from which the shipper transports the goods through a carrier. These are the points from which transportation is initiated, such as shipping points for outbound transportation.
- *Destination locations* are those points to which the goods are delivered, and the transportation activity ends at those points, such as the ship-to party address for outbound transportation.
- *Transshipment locations* are those points in the transportation network where the goods are transferred from one mode of transport to another mode of transport. Goods are unloaded from one means of transport and loaded into another means of transport at transshipment locations.

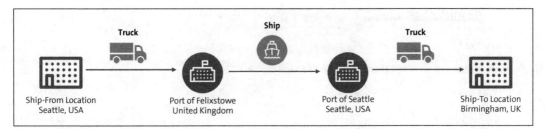

Figure 9.7 Transportation Locations

The number of freight documents (freight order or freight booking) is dependent on the number of transshipment locations. In the preceding example, three freight documents are created (two freight orders and one freight booking). Transportation stages are derived from transshipment locations as well. There are three stages in the example. Transportation stages are calculated using the following formula:

Transportation stages = Number of locations in the selected route – 1

Locations can be created automatically via background processing or manually. Business partner locations, shipping points, and so on are automatically created in embedded TM during the integration of orders (purchase orders, sales orders, and stock transfer orders). You can also use Transaction /SCMTMS/LOC3 to create location master data manually. Figure 9.8 shows the initial screen for creating location master data, where you can fill out the following fields:

- **Location**
  Enter a location name as the unique identifier to create location master data. It is recommended to create the location with the same reference object ID; for example, if you are creating a location for a business partner (supplier), specify the location as the number of the business partner (supplier).

- **Location Type**
  The location type distinguishes different location types in transportation management. The standard system provides sufficient location types, shown in Table 9.5. A location type is mandatory to create location master data. Select the relevant location type; for example, if you are creating location master data for a shipping point, select **1003** as the location type.

Location Type	Description
1001	Production plant
1002	Distribution center
1003	Shipping point
1010	Customer

Table 9.5 Location Types

Location Type	Description
1011	Vendor
1020	Transportation service provider
1021	Business partner
1030	Terminal
1200	Loading point
1100	Port
1110	Airport
1120	Railway station
1130	Container freight station
1140	Hub
1170	Warehouse
1180	Carrier warehouse
1190	Rail junction
1191	Border crossing point

Table 9.5 Location Types (Cont.)

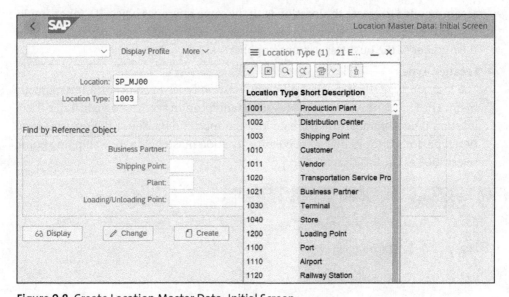

Figure 9.8 Create Location Master Data: Initial Screen

- **Find by Reference Object**
  The fields under **Find by Reference Object** are search fields. If you want to display or change location master data that is created with reference to a **Business Partner**, **Plant**, **Shipping Point**, or **Loading/Unloading Point**, specify the reference object and click **Display** or **Change**.

After specifying the fields in the initial screen, click the **Create** button at the bottom. Figure 9.9 shows the header details and the **General** tab of the location master data. Enter the description of the location. The **General** tab contains the global identifiers, reference objects, and geographical data of the location master data. Let's explore the details:

- **Identifier**
  The following fields can be used to specify a unique identifier that is the globally accepted standard for the location master data. These unique identifiers are used in EDI and used by airlines, vessels, and so on to uniquely identify a location for transportation management:

  - **GLN** (global location number): A 13-digit code that uniquely identifies a location master data. GLN is used in EDI to collaborate with business partners.
  - **DUNS+4**: A Data Universal Numbering System (DUNS) number plus a four-character suffix assigned to a business to uniquely identify a business to facilitate business transactions.
  - **UN/LOCODE**: The United Nations Location Code for Trade and Transportation.
  - **IATA Code**: An International Air Transport Association (IATA) code. These codes are used by airlines and airports to uniquely identify a location globally: SEA (Seattle), BLR (Bangalore), LAX (Los Angeles), and so on.

- **Reference Object**
  A reference object is automatically populated based on the location type and the reference object used to create the location. It includes the following fields:

  - **Object**: This field is populated with the object reference ID, such as shipping point, plant, business partner, and so on.
  - **Object Type**: This field is populated with the type of the reference object that is referenced to create the location master data.

- **Geographical Data**
  Geographical data is automatically populated from the address of the location master data. It contains longitude, latitude, and time zone information.

In the **Address** tab, detailed address information is populated from the reference object. You can enter the complete physical address of the location.

Figure 9.10 shows the **TM** data tab for the location. Let's explore the attributes to be maintained.

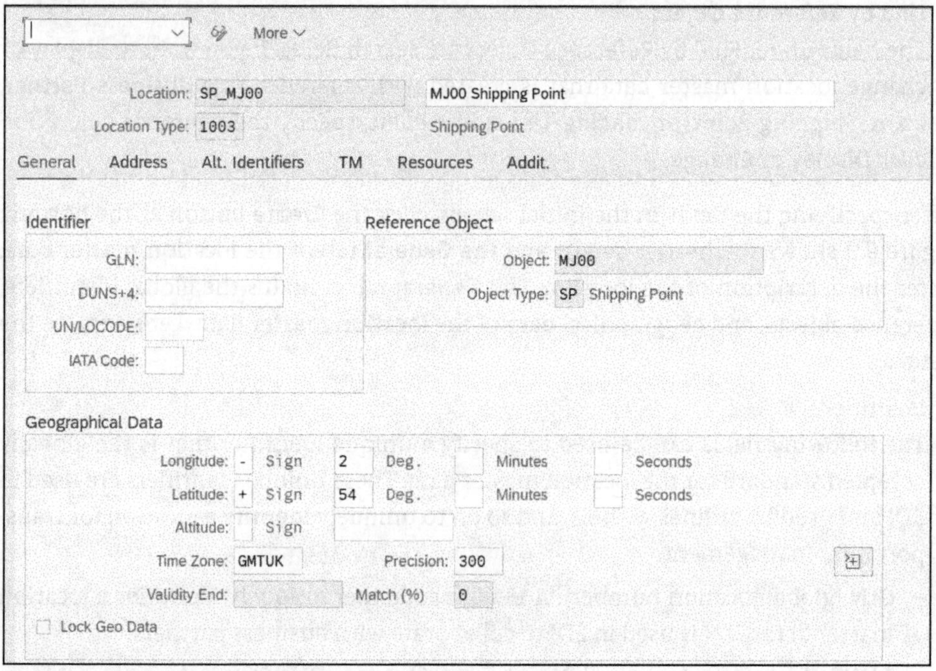

Figure 9.9 Create Location Master Data: General Data

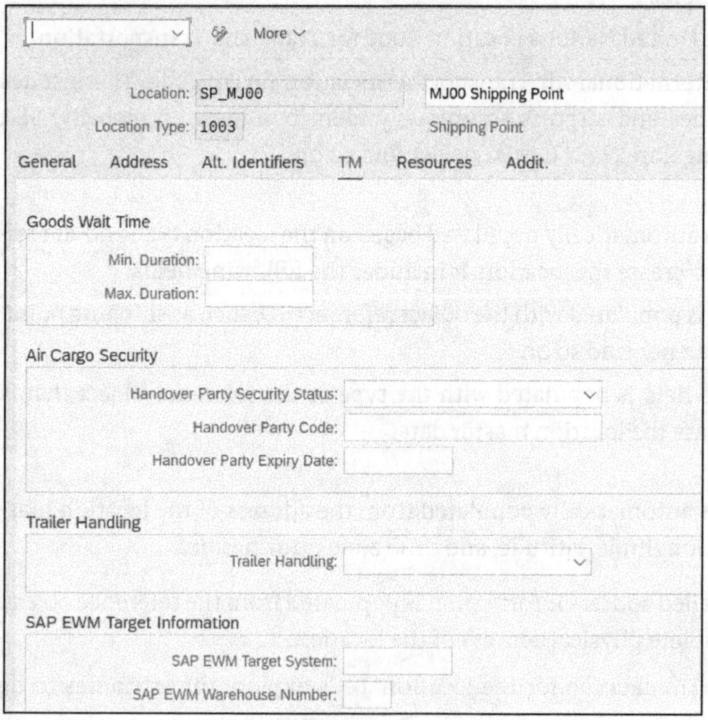

Figure 9.10 Create Location Master Data: TM Data

Goods wait times are required for the planning optimizer to calculate the best route to transport the goods. In the **Goods Wait Time** section, specify the minimum duration and maximum duration for the goods at this location. Typically, these times are specified for border crossing points, customs checking locations, unloading and reloading points, and so on.

The **Air Cargo Security** section includes the following fields, which are to be maintained if the location is a cargo facility at the airport:

- **Handover Party Security Status**
  For the security of the cargo being transported, whether the shipper or the agent who will manage the shipment at the cargo facility is known or unknown is specified in this field. The goods to be transported from this location will be handed over to the shipper/agent at this location. The standard system provides the following security statuses of the goods handover party:
  - **Known shipper**
  - **Unknown shipper**
  - **Account shipper**
  - **Regulated agent**
- **Handover Party Code**
  A unique code is issued to the goods handover party by the relevant local authority of the location. Specify the handover party code in this field.
- **Handover Party Expiry Date**
  Specify the expiration date of the security status of the handover party in this field.

The **Trailer Handling** section should be maintained if coupling and uncoupling of trailers at this location is possible. One of the following values can be maintained based on the recoupling strategy of trailers at this location:

- **Recoupling not possible**
  If you maintain this strategy, coupling and uncoupling of trailers are not allowed at this location.
- **Recoupling possible**
  If you maintain this strategy, coupling and uncoupling of trailers are allowed at this location.
- **Recoupling possible on pickup and delivery only**
  If you maintain this strategy, coupling and uncoupling of trailers are allowed at this location for pickup and delivery only.
- **Trailer swap possible only**
  If you maintain this strategy, a trailer from one truck can be uncoupled and then recoupled into another truck at this location.

The **SAP EWM Target Information** section should be maintained if embedded EWM and embedded TM are integrated as per business requirements:

- **SAP EWM Target System**
  Specify a logical system name created for the embedded EWM system in SAP S/4HANA for the RFC connection between embedded EWM and TM.

- **SAP EWM Warehouse Number**
  Specify the corresponding embedded EWM warehouse number that is associated with the location master data for integration.

Figure 9.11 shows the **Resources** data of the location for transportation management activities. The resources are assigned to the loading/unloading locations within the plant/warehouse.

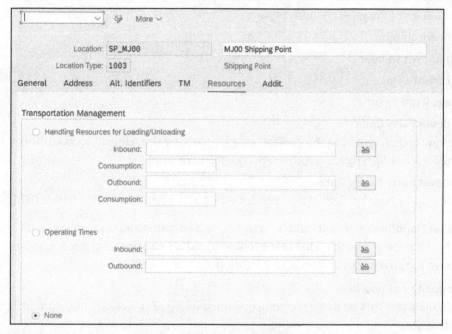

Figure 9.11 Create Location Master Data: Resource Data

You can set one of the following radio buttons and maintain the corresponding attributes. As a prerequisite, resource master data must be maintained before assigning the resources in this tab.

- **Handling Resources for Loading/Unloading**
  If you set this radio button, maintain the resources for loading or unloading at the location during inbound and outbound delivery processing.

- **Consumption**
  The capacity of the resources for inbound and outbound delivery processing must be maintained separately. You can maintain the consumption capacity of the

resources in dimensional (weight, volume, etc.) or nondimensional (number of resources required for inbound and outbound delivery processing) units.

- **Operating Times**
  If you select this radio button, maintain a calendar resource for inbound and outbound delivery processing at this location. The resource can be assigned to a calendar or shift to determine goods issue and goods receipt processing times.

- **None**
  If you select this radio button, no resources are maintained to handle transportation management activities at this location. By default, this radio button will be set for all external locations such as port, airport, customer, vendor, business partner location, and so on.

Transportation Zone

A transportation zone is a grouping of different locations, different regions, or different postal codes used to determine the route for transportation. A transportation zone is part of the transportation network. The US East zone, US West zone, Northern Europe zone, and South India zone are some examples of transportation zones. Transportation zones are used to determine the best routes based on departure zones and receiving zones.

The following are the four different types of transportation zones that can be created in embedded TM:

- **Direct zone**
  Direct zones are created by grouping different locations, including shipping locations, delivery locations, transshipment locations, and others.

- **Postal code zone**
  Postal code zones are created by grouping a range of portal codes of a country/region.

- **Region zone**
  Region zones are created by grouping countries and regions; for example, the US West zone is specified by grouping different states/regions such as California, Washington, Oregon, and so on within the US, whereas the Europe West zone is specified by grouping different countries such as France, Netherlands, the United Kingdom, Germany, and so on.

- **Mixed zone**
  Mixed zones are created by grouping different locations, different portal codes, and/or different countries/regions. You can create mixed zones with a combination of any of two attributes from locations, postal codes, and countries/regions.

You can use Transaction /SCMTMS/ZONE to maintain transportation zones manually. Use **Create**, **Change**, **Display**, and other icons, as shown in Figure 9.12, to perform a particular action. To create a new transportation zone, click the **Create** icon. The **Maintain**

Zone popup screen will be displayed, as shown in Figure 9.12. Specify a 20-digit alphanumeric name (unique identifier) for the transportation zone in the **Zone** field.

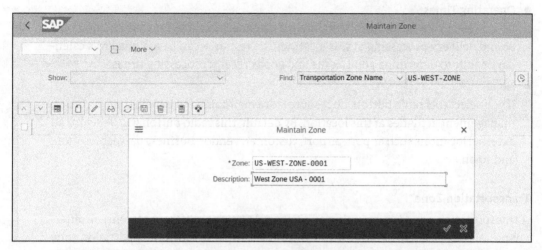

Figure 9.12 Maintain Transportation Zone: Initial Screen

After specifying the unique name and description for the transportation zone, click the green checkmark and scroll down to define a direct zone, postal code zone, region zone, or mixed zone.

For better understanding of transportation zones, let's create a mixed transportation zone. Figure 9.13 shows the **Zone—Location** tab for creating a direct zone or mixed zone.

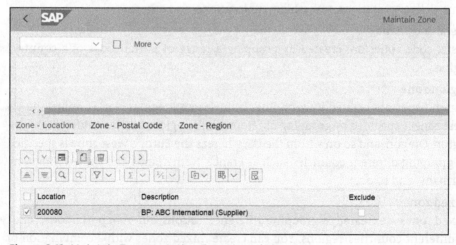

Figure 9.13 Maintain Transportation Zone: Locations

Click the **Create** icon to assign relevant locations to the transportation zone. Group locations that will fall under the zone you are creating. If you maintain only the **Zone— Location** tab, a direct zone will be created. In our example, location **200080**, a location of business partner ABC International (supplier), has been assigned to the transportation zone **US-WEST-ZONE-0001**.

Figure 9.14 shows the **Zone—Postal Code** tab for maintaining the transportation zone. Click the **Create** icon to assign relevant postal codes of a country/region to the transportation zone. If you maintain only the **Zone—Postal Code** tab, a postal code zone will be created. In our example, a postal code range from **98001** to **98045** has been assigned to transportation zone **US-WEST-ZONE-0001**.

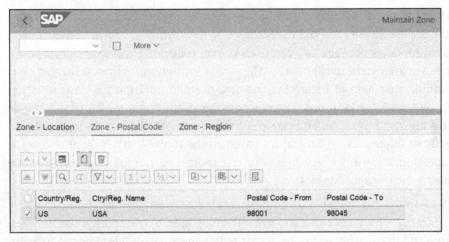

Figure 9.14 Maintain Transportation Zone: Postal Codes

Figure 9.15 shows the **Zone—Region** tab for maintaining a transportation zone. Click the **Create** icon to assign a relevant country/region to the transportation zone. If you maintain only the **Zone—Region** tab, a region zone will be created. In our example, country **US** and region **WA** (Washington) have been assigned to transportation zone **US-WEST-ZONE-0001**.

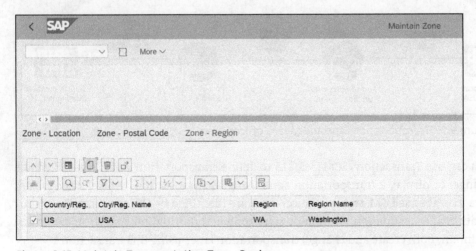

Figure 9.15 Maintain Transportation Zone: Regions

Transportation Lane

A transportation lane is a direct route between two transportation zones or locations, one being the departure zone and the other being the destination zone. You can create transportation lanes between two locations or two transportation zones. Transportation lanes are important master data within a transportation network for transportation planning that help to determine the distance and duration for transportation between two zones or locations. To transport the goods from one location/zone to another, a means of transport and carrier for the means of transport are required. You can specify multiple means of transport based on the source location/zone and destination location/zone. For every means of transport, you can assign a carrier while maintaining the transportation lane.

Figure 9.16 shows the transportation lane created between two locations. Multiple means of transport such as truck and ship are being used to transport goods between the source and destination locations. A carrier (business partner) is assigned to every means of transport. Transportation distance, duration, and costs are maintained for every means of transport, which helps in transportation planning.

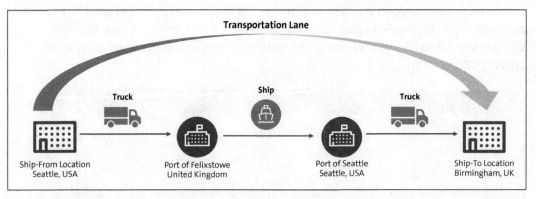

Figure 9.16 Transportation Lane and Means of Transport

You can use Transaction /SCMTMS/TL5 to define transportation lanes. You can create, change, or display a transportation zone using this transaction. Unlike other master data in embedded TM, transportation lanes are identified by the combination of start location and destination location or start zone and destination zone. You can mass-maintain (both create and change) transportation lanes using this transaction.

Figure 9.17 shows the initial screen for defining the transportation lane for start zone **US-WEST-ZONE-0001** and destination zone **DEP-ZONE-UK001**. After filling the **Start Location/Zone** and **Dest. Loc./Zone** fields, click **Create**.

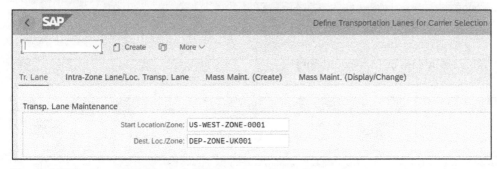

Figure 9.17 Transportation Lane: Initial Screen

Means of transport and carrier details are updated in the next screen for defining the transportation lanes. Means of transport are added first, and then the carrier information that corresponds to the means of transport is added. Figure 9.18 shows the **Means of Transport** details added to the transportation lane, which you can open by clicking the **Create** icon at the top if you want to add new means of transport details to the transportation lane. You can change the existing means of transport by selecting the respective means of transport and clicking the **Details** icon. You can delete an existing means of transport by selecting the respective means of transport and clicking the **Delete** icon. Let's explore the key fields in the means of transport details screen.

The **Validity** section of the means of transport details has the following fields:

- **Means of Trans.**
  A means of transport represents a vehicle resource or vehicle classification. It's either a single vehicle resource or a group of identical vehicle resources. Specify the means of transport to be maintained in the transportation lane in this field. You can define multiple means of transport for a transportation lane. Table 9.6 shows the means of transport and descriptions defined in the standard system.

Means of Transport	Description	Transportation Mode
0001	Truck	ROAD
0002	Rail	RAIL
0003	Airplane	AIR
0004	Courier, express, and delivery company	MAIL
0005	Ship	SEA

Table 9.6 Means of Transport

Means of Transport	Description	Transportation Mode
0006	Car	ROAD
0007	Truck (subcontracting)	ROAD
0008	Railcar	RAIL

Table 9.6 Means of Transport (Cont.)

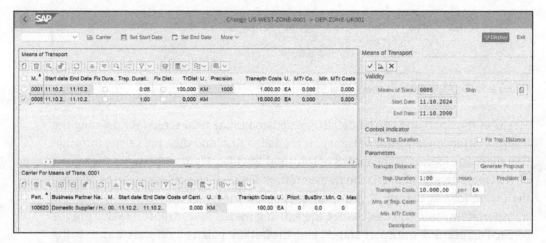

Figure 9.18 Transportation Lane: Means of Transport

- **Start Date**
  Specify the start date to define the validity period for the means of transport.

- **End Date**
  Specify the end date to define the validity period for the means of transport.

The **Control Indicator** section of the means of transport details has the following fields:

- **Fix Trsp. Duration**
  If you set this indicator, the transportation duration will not be overwritten by the system calculation. The transportation duration is dependent on the distance and the speed of the vehicle. But if you set this indicator, the duration is fixed irrespective of the means of transport, speed of the vehicle, and transportation distance.

- **Fix Trsp. Distance**
  If you set this indicator, the transportation distance will not be overwritten by the system calculation. The transportation distance is calculated by the system based on aerial (in a straight line) distance between the source and destination locations. But if you set this indicator, the distance is fixed irrespective of the aerial distance.

The **Parameters** section of the means of transport details has the following fields:

- **Transptn. Distance**
  Specify the transport distance in kilometers to be travelled by the means of transport from point A to point B. **Precision** indicates if the distance is entered manually or determined by the geographical information system or line of flight distance.

- **Trsp. Duration**
  Specify the transport duration in HHH:MM format for the time the means of transport takes to transport goods from point A to point B. **Precision** indicates if the duration is entered manually or determined by the geographical information system or line of flight distance.

- **Transportn Costs**
  Specify the variable transportation costs for the means of transport to transport the entire shipment between the transportation zones/locations.

- **Mns of Trsp. Costs**
  Specify the per-kilometer cost for the means of transport in this field.

- **Min. MTr. Costs**
  Specify the minimum means of transportation cost, which is distance-dependent, in this field.

After the means of transport is added, carrier information for the respective means of transport is added in the transportation lane details. Figure 9.19 shows the **Carrier** details added to the transportation lane for means of transport **0005**, which you navigate to by selecting **0005** at the top and then clicking the **Create** icon under the carrier.

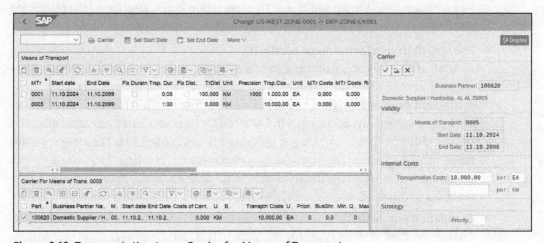

Figure 9.19 Transportation Lane: Carrier for Means of Transport

Let's explore the key fields in the carrier details screen:

- **Business Partner**
  Specify the business partner created with the carrier partner role in this field that is responsible for execution of the freight order or freight booking. As a prerequisite, a

business partner must have been created in the system for the carrier that corresponds to the means of transport.

- **Transportation Costs**
  Specify the variable transportation costs for the carrier to transport the entire shipment between the transportation zones/locations.

- **Priority**
  Specify the priority of the transportation service provider (carrier) as an integer value (0 to 10).

Carrier Profile

Carriers are created as business partners with the carrier business partner role. The carrier profile is defined in embedded TM to maintain additional attributes and capabilities of the carrier for transportation management. A business partner for each carrier with the carrier business partner role is a prerequisite to create the corresponding carrier profile in embedded TM. Carrier profiles are used in carrier selection. A carrier profile contains the freight group of products, transportation groups, freight code sets, transportation lane, fixed cost of transportation, and other attributes.

Figure 9.20 shows that business partner **300041 (Test Carrier)** was created with the carrier business partner role (for details on creating a business partner, see Chapter 3, Section 3.2). It also shows the purchasing data of the business partner (carrier) created in purchasing organization **MJ00**. The purchasing data of the business partner (carrier) is like a purchasing vendor. An additional **Transportation Purch. Org. Data** tab appears for the business partner (carrier) where you can maintain the following information:

- **FSD. Prf. ID (freight settlement profile ID)**
  As part of the freight settlement, settlement documents are posted in materials management to settle the transportation costs incurred with the carrier. The freight settlement profile is assigned to the business partner (carrier) for the system to create freight settlement documents in SAP S/4HANA materials management after the freight order or freight booking is processed in embedded TM. The freight settlement profile contains the controlling parameters for creating freight settlement documents. The freight settlement profile ID defined in configuration settings is assigned to this field.

- **Calculation Profile ID**
  Charge profiles contain controlling parameters for freight charge calculation. Freight charges are calculated using transportation lanes (between locations). Freight charges can be calculated from freight orders or freight bookings, forwarding orders, and freight settlement documents. The calculation profile ID defined in the configuration settings is assigned to this field.

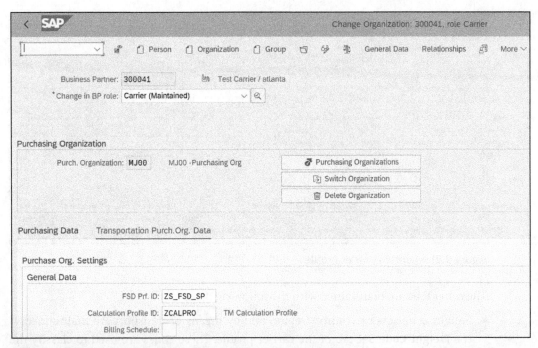

Figure 9.20 Business Partner with Carrier Business Partner Role

After creating a business partner with a carrier business partner role, you can maintain the carrier profile in embedded TM using Transaction /SCMTMS/TSPP. The carrier profile is maintained for the business partner (carrier), and in the initial screen, you need to specify the business partner with the carrier role for which you are maintaining the carrier profile. Click the **Create** icon in the initial screen and then in the **Business Partner** popup, specify the business partner and press ⎡Enter⎤. Carrier details are added to the carrier section (first half) of the screen. Figure 9.21 shows the carrier profile maintained for business partner (carrier) **300041**.

In the **Transportation** tab of the carrier profile, you can define multiple combinations of the following:

- Source location/zone
- Destination location/zone
- Means of transport
- Validity period

Click the **Create** icon under the **Transportation** tab to add transportation details.

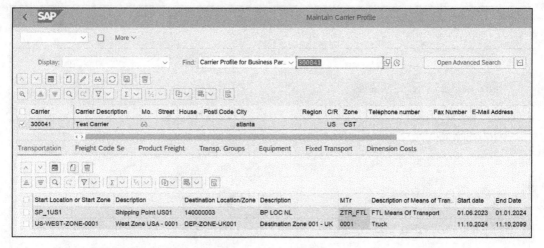

Figure 9.21 Maintain Carrier Profile

The other tabs are maintained with the following data:

- Freight code sets by country/region and by means of transport are maintained in the **Freight Code Set** tab of the carrier profile. These codes are used to classify the products to be transported using the carrier.

- Product freight groups are maintained in the **Product Freight** tab of the carrier profile. Product freight groups are used to classify products by freight group and freight class.

- Transport groups are maintained in the **Transp. Groups** tab of the carrier profile. Transport groups are used to group products with similar transportation requirements: in container, on pallet, on trailer, and so on.

- Means of transport and fixed transportation costs are maintained in the **Fixed Transport** tab of the carrier profile.

- The means of transport; dimensions for cost determination such as by duration, by distance, or by quantity; and the transportation cost by dimension are maintained in the **Dimension Costs** tab of the carrier profile.

9.2.3   Incoterms

Incoterms and Incoterm locations play a vital role in transportation planning. Incoterms and Incoterm locations are defined in purchase orders, stock transfer orders, sales orders, and so on. Incoterms are the sets of trading terms/rules for suppliers and buyers for the sale/purchase of goods. Incoterm locations indicate from where the goods are picked up for transportation. When the demand is generated in embedded TM from the purchase orders, sales orders, and so on, Incoterms and Incoterm locations are considered in freight unit creation. Hence the mapping of an Incoterm location to a transportation location is required as the Incoterm locations defined in the purchasing documents are free text values that can't be considered as is in embedded TM for building freight units.

Incoterms are defined in the configuration settings. You can define Incoterms centrally if you have deployed embedded TM as an internal component of SAP S/4HANA. If you are using a standalone (decentralized) TM system, define Incoterms in both SAP S/4HANA (ERP) and the standalone TM component of SAP S/4HANA.

To define Incoterms in transportation management, execute Transaction SPRO and navigate to **Transportation Management • Basic Functions • General Settings • Incoterms • Define Incoterms.**

Figure 9.22 shows the incoterms defined in the standard system. Click **New Entries** to define new Incoterms and enter the description for the new Incoterm you are defining.

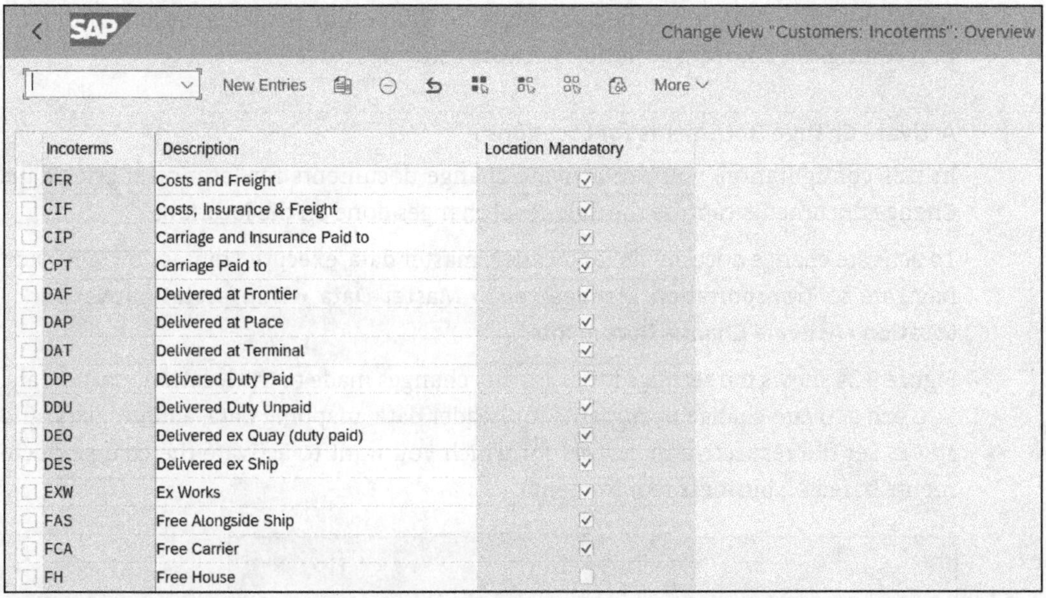

Incoterms	Description	Location Mandatory
CFR	Costs and Freight	✓
CIF	Costs, Insurance & Freight	✓
CIP	Carriage and Insurance Paid to	✓
CPT	Carriage Paid to	✓
DAF	Delivered at Frontier	✓
DAP	Delivered at Place	✓
DAT	Delivered at Terminal	✓
DDP	Delivered Duty Paid	✓
DDU	Delivered Duty Unpaid	✓
DEQ	Delivered ex Quay (duty paid)	✓
DES	Delivered ex Ship	✓
EXW	Ex Works	✓
FAS	Free Alongside Ship	✓
FCA	Free Carrier	✓
FH	Free House	

Figure 9.22 Define Incoterms

If you set the **Location Mandatory** indicator, specifying the Incoterm location is mandatory whenever the respective Incoterm is used in any document. The Incoterm location is used in transportation planning and these locations are mapped to transportation locations.

To map the predefined transportation locations to Incoterm locations, execute Transaction /SCMTMS/INCLOC_MAP. Figure 9.23 shows that Incoterm locations **SEATTLE** and **LONDON** have been mapped to transportation locations **DE_01_09** and **PT_LONDON**.

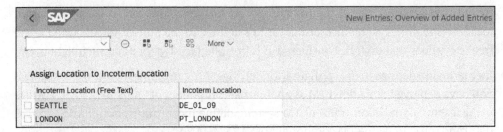

Figure 9.23 Incoterm Location to Transportation Location Mapping

9.2.4   Configuration Settings for Master Data

Let's explore the key configuration settings for master data in embedded TM. Configuration settings for business partners in embedded TM such as driver, carrier, ship-to party, shipper, bidder, consignee, and so on are configured like the business partner configuration settings explained in Chapter 3. This section explains the key configuration settings for the transportation network setup, resources, and so on.

Activate Change Documents for Locations

In this configuration, you can activate change documents for location master data. Change documents capture the field-level changes done on the locations.

To activate change documents for location master data, execute Transaction SPRO and navigate to **Transportation Management • Master Data • Transportation Network • Location • Activate Change Documents**.

Figure 9.24 shows the settings for capturing changes made to the location master data. You can activate change documents for header data, mapping data, address data, and so on. Set the respective indicators for which you want to activate the change documents based on business requirements.

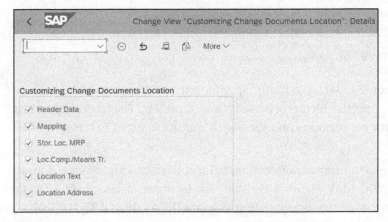

Figure 9.24 Activate Change Documents for Location Master Data

Define IATA Location Codes

IATA codes are used by airlines and airports to uniquely identify a location globally: SEA (Seattle), BLR (Bangalore), LAX (Los Angeles), and so on.

To define IATA location codes required to create location master data, execute Transaction SPRO and navigate to **Transportation Management • Master Data • Transportation Network • Location • Define IATA Location Codes**. To define new IATA location codes, click **New Entries** and specify the three-character alphanumeric IATA key in the **IATA: Location** field and a description. Then specify the **C/R** (country/region key), **Re…**(region), **IATA: City**, and **Time Zone** field values. Figure 9.25 shows three IATA codes, **SEA**, **BLR**, and **BHX**, defined to uniquely identify the Seattle, Bangalore, and Birmingham airports. Enter the three-digit airport code defined for the respective airport, a description of the airport, country and region of the airport, and the time zone to define IATA location codes.

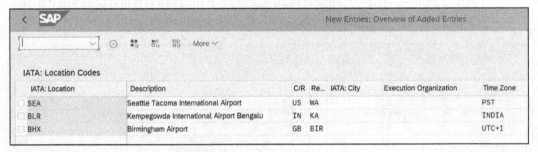

Figure 9.25 Define IATA Location Codes

Define UN/LOCODE

The United Nations Location Code for Trade and Transportation (UN/LOCODE) is a geocoding scheme developed and maintained by the United Nations by country.

To define the UN/LOCODE to create location master data, execute Transaction SPRO and navigate to **Transportation Management • Master Data • Transportation Network • Location • Define UN/LOCODE**. To define a new UN/LOCODE, click **New Entries** and specify a five-character alphanumeric key and description. Figure 9.26 shows the UN/LOCODEs **SEA** and **BHM** defined to uniquely identify geographic locations Seattle and Birmingham, as defined and maintained by the United Nations. Enter the UN/LOCODE defined by the United Nations and its description.

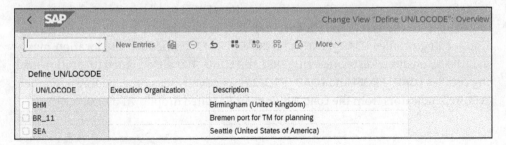

Figure 9.26 Define UN/LOCODE

Define Loading Direction Profile

The loading direction profile defines the direction from which a location or a resource can be loaded or unloaded.

To define a loading direction profile for location master data, execute Transaction SPRO and navigate to **Transportation Management • Master Data • Transportation Network • Location • Define Loading Direction Profile**.

To define a new loading direction profile, click **New Entries** and specify an alphanumeric key of up to 20 characters and a description. Figure 9.27 shows loading direction profiles **LOAD_LEFT**, **LOAD_RIGHT**, **LOAD_TOP**, **LOAD_BACK**, and **NO_PREF**.

Set the relevant direction indicators for the loading direction profile and select the loading direction preference from the **LoDirPref** dropdown menu. For the direction profiles for loading from left and right, select both the **Load Left** and **Load Right** indicators and set the appropriate loading direction preference.

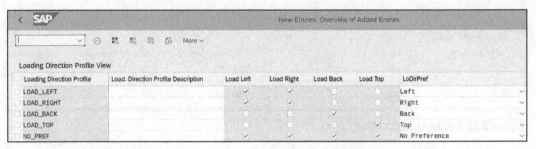

Figure 9.27 Define Loading Direction Profile

Define Location Loading Profile

The loading profile controls the loading pattern at the location. In this configuration step, a loading location profile is defined, and the loading direction profile will be assigned to the loading profile. It also defines the loading equipment requirement at the location for loading and unloading.

To define a location loading profile for location master data, execute Transaction SPRO and navigate to **Transportation Management • Master Data • Transportation Network • Location • Define Location Loading Profile**.

To define a new location loading profile, click **New Entries** and specify an alphanumeric key of up to 20 characters and the **Loc. Load. Profile Description**. Figure 9.28 shows location loading profiles **LOAD_PROF1** and **LOAD_PROF2** defined for the location master data. Location direction profiles **NO_PREF** and **LOAD_TOP** are assigned to location loading profiles **LOAD_PROF1** and **LOAD_PROF2** respectively. Select the loading equipment required indicators from the **Load. Equipment Required** dropdown menu.

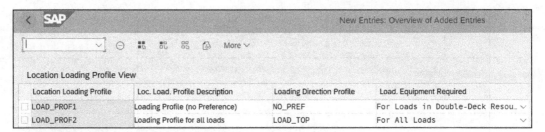

Figure 9.28 Define Location Loading Profiles

Activate Change Documents for Transportation Lane

In this configuration, you can activate change documents for transportation lanes. Change documents capture the field-level changes made in the transportation lane.

To activate change documents for the transportation lane, execute Transaction SPRO and navigate to **Transportation Management • Master Data • Transportation Network • Transportation Lane • Activate Change Documents**.

Figure 9.29 shows the settings for capturing changes made to the transportation lane. Set the respective indicators for which you want to activate the change documents for transportation lane-related master data changes based on business requirements:

- **Transportation Lane**
  If you set this indicator, change documents will be created for changes to the transportation lane.

- **Pr-Spec. Trans. Lane**
  If you set this indicator, change documents will be created for changes to the product-specific transportation lane.

- **Version dep. data**
  If you set this indicator, change documents will be created for changes to the version-dependent fields of a product-specific transportation lane.

- **Means of Transport**
  If you set this indicator, change documents will be created for changes to the means of transport.

- **Product-Specific MTr**
  If you set this indicator, change documents will be created for changes to the product-specific means of transport.

- **TSP**
  If you set this indicator, change documents will be created for changes to the transportation service provider.

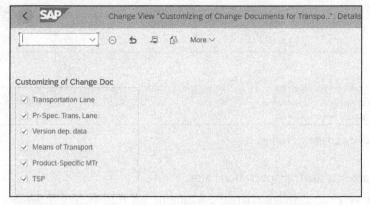

Figure 9.29 Activate Change Documents for Transportation Lane

Define Transportation Mode

In this configuration, define the transportation mode and assign a relevant means of transport.

To define a transportation mode, execute Transaction SPRO and navigate to **Transportation Management • Master Data • Transportation Network • Transportation Lane • Define Transportation Mode**. Figure 9.30 shows transportation modes **01 (Road)**, **02 (Rail)**, **03 (Sea)**, **04 (Inland Waterway)**, **05 (Air)**, and **06 (Postal Service)** defined in the system. Relevant mode of transportation categories are assigned to the respective mode of transportation. These categories are defined in the standard system to recognize any user-defined transportation modes. Assign the relevant means of transport in the **MTr** column for every transportation mode.

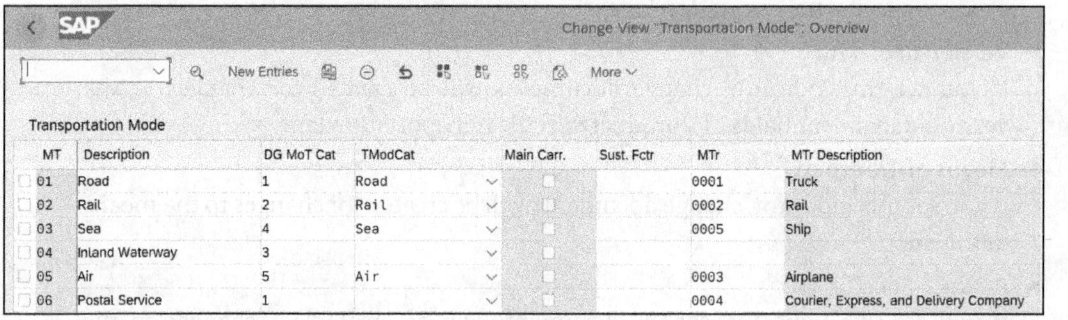

MT	Description	DG MoT Cat	TModCat	Main Carr.	Sust. Fctr	MTr	MTr Description
01	Road	1	Road	☐		0001	Truck
02	Rail	2	Rail	☐		0002	Rail
03	Sea	4	Sea	☐		0005	Ship
04	Inland Waterway	3		☐			
05	Air	5	Air	☐		0003	Airplane
06	Postal Service	1		☐		0004	Courier, Express, and Delivery Company

Figure 9.30 Define Transportation Mode

Activate Change Documents for Resource

In this configuration, you can activate change documents for resource master data. Change documents capture the field-level changes made to a resource.

To activate change documents for resource master data, execute Transaction SPRO and navigate to **Transportation Management • Master Data • Resources • General Settings • Activate Change Documents for Resource**.

Figure 9.31 shows the settings for capturing changes made to the resource. Set the **Activate CD Resource** indicator to activate the change documents for a resource.

Figure 9.31 Activate Change Documents for Resource

Define Resource Class

Resource classes are used to group resources with identical classifications/attributes. Resources such as vehicles, handling resources, and so on are defined based on resource classes. In this configuration, you can define a resource class that can be used to define resource master data.

To define resource class, execute Transaction SPRO and navigate to **Transportation Management • Master Data • Resources • General Settings • Define Resource Class**.

To define a new resource class, click **New Entries** and specify a two-digit numeric key and description. Then select a **Res. Type** (resource type) option from the dropdown list to assign it to the resource type, which identifies the usage of the resource. Figure 9.32 shows the resource classes defined in embedded TM. The following resource types are defined in the standard system:

- **Vehicle Resource**
- **Transportation Unit Resource**
- **Handling Resource**
- **Driver**

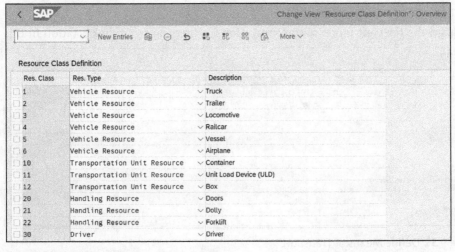

Figure 9.32 Define Resource Class

Define Means of Transport

A means of transportation groups identical vehicle resources. While the mode of transportation defines how the goods are transported (via sea, air, road, etc.), the means of transport defines the group of vehicles that can be used to carry the goods using the mode of transport.

For example, sea is the *mode* of transport, and cargo ship and container ship are the *means* of transport. In this configuration, define the means of transportation and assign the relevant mode of transportation.

To define means of transportation, execute Transaction SPRO and navigate to **Transportation Management • Master Data • Resources • Means of Transport and Compartment • Define Means of Transport**.

Figure 9.33 shows the means of transportation defined in the standard system. Click **New Entries** to define a new means of transportation. You can copy an existing means of transportation and create a new one. Specify a 10-digit alphanumeric key to define a means of transport and its description.

The following are the key fields of this configuration:

- **Std Code (standard code)**
  Specify the United Nations standard code for the means of transport for EDI purposes to exchange transportation documents with carriers electronically. The following are the standard codes used in the UN/EDIFACT standard:
  - **031**: Truck
  - **006**: Aircraft
  - **072**: Rail
  - **011**: Ship

- **MT (mode of transportation)**
  Assign the relevant mode of transportation to the means of transportation in this field.

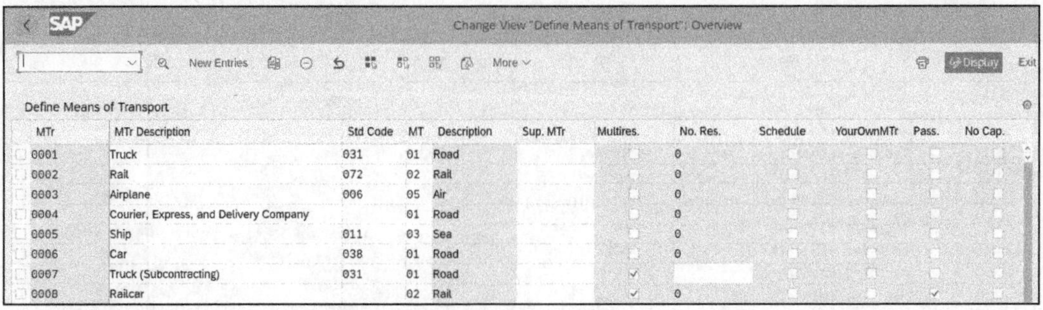

MTr	MTr Description	Std Code	MT	Description	Sup. MTr	Multires.	No. Res.	Schedule	YourOwnMTr	Pass.	No Cap.
0001	Truck	031	01	Road			0				
0002	Rail	072	02	Rail			0				
0003	Airplane	006	05	Air			0				
0004	Courier, Express, and Delivery Company		01	Road			0				
0005	Ship	011	03	Sea			0				
0006	Car	038	01	Road			0				
0007	Truck (Subcontracting)	031	01	Road		✓					
0008	Railcar		02	Rail		✓	0			✓	

Figure 9.33 Define Means of Transport

- **Multires. (multiresource)**
  If you set this indicator, it indicates that multiple resources are assigned to the means of transport. Specify the number of individual resources (a numeric value) when you set this indicator.

- **YourOwnMTr. (your own means of transport)**
  If you set this indicator, the vehicle resource for the respective means of transport is owned by the company itself, and an external service from a carrier (business partner) is not required.

- **Res. Class (resource class)**
  A resource class is used to group resources with identical classifications/attributes. Specify the resource class for the means of transport.

- **UnltdNoRes (unlimited number of induvial resources)**
  If you set this indicator, it indicates that there is an unlimited number of resources available, and you must set the **Multires.** indicator with this as a prerequisite.

- **Default MTr (default means of transport)**
  If you set this indicator, the means of transport will be proposed by default in the freight documents.

9.3   Procurement Integration

There are two integration modes possible for integrating embedded TM and SAP S/4HANA. As previously stated, embedded TM is embedded in SAP S/4HANA. But you have an option to run embedded TM transactions in a standalone mode, where SAP S/4HANA (ERP) is deployed in one box and embedded TM in a different box as a standalone system. With the latest SAP Supply Chain Management (SAP SCM) edition for SAP S/4HANA, embedded TM can be integrated with SAP S/4HANA (ERP) using the following integration modes:

- **Internal TM component integration**
  If you deploy this mode, leverage the internal TM component embedded in SAP S/4HANA for transportation management. ERP components and TM components will be embedded within SAP S/4HANA.

- **External TM component integration**
  If you deploy this mode, embedded TM will be a standalone system in the SAP SCM edition for SAP S/4HANA. The SAP S/4HANA (ERP) components will be part of another SAP S/4HANA (ERP) system.

Let's dive deep into the procurement integration with embedded TM. We'll then walk through the embedded TM functions and provide step-by-step instructions for cross-functional configuration.

SAP Business Client

Transactional data in embedded TM is displayed and managed in SAP Business Client. SAP Business Client integrates with SAP GUI; if you are accessing SAP S/4HANA in SAP GUI, you can navigate to SAP Business Client using Transaction NWBC. The following are some of the important apps available in SAP Business Client for embedded TM:

- Transportation Cockpit: For manual transportation planning
- Manage Fright Units: To display and manage freight units
- Manage Freight Orders: To display and manage freight orders
- Manage Air/Ocean Freight Bookings: To display and manage freight bookings
- Manage Rate Tables: To display and manage rate tables
- Manage Freight Agreements: To display and manage freight agreements
- Monitor Accrual Postings—Freight Documents: To monitor freight documents to post accruals

We'll explore the features of these apps throughout this section.

9.3.1   Integrated Procurement and Transportation Management Process

Figure 9.34 shows the data flow between SAP S/4HANA (ERP) components and embedded TM components of SAP S/4HANA, and the carrier who executes transportation between the source location and destination location. Integration of embedded TM and SAP S/4HANA procurement functions of materials management supports the inbound transportation process. Purchase orders and schedule lines for scheduling agreements are created in SAP S/4HANA for external procurement of materials and products, and stock transfer orders are created for internal procurement from the same legal entity or another legal entity.

OTRs arising from purchase orders, schedule lines, and stock transfer orders that are created in SAP S/4HANA (ERP) can be used in transportation planning in embedded TM. SAP S/4HANA sends these orders to the embedded TM component as soon as they are created and released. The OTR is automatically created in the embedded TM component with reference to the received purchase order/schedule line/stock transfer order from SAP S/4HANA (ERP).

The order management function of embedded TM converts the transportation demands arising from purchase orders, scheduling agreements (schedule line releases), and stock transfer orders into shippable units known as freight units in embedded TM. Freight units are created based on the rules and incompatibilities defined for freight unit building. As a result, purchase orders, stock transfer orders, and schedule lines are consolidated into one or more freight units. Freight units are the basis for transportation planning in embedded TM.

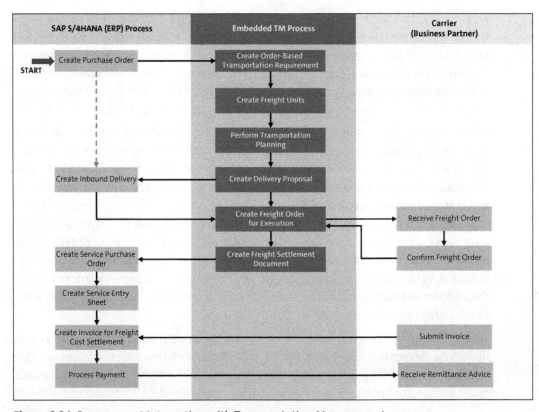

Figure 9.34 Procurement Integration with Transportation Management

Automatic creation of freight units from transportation requirements is based on freight unit building rules, source and destination locations, incompatibilities, business requirements, and packaging information. Freight units can be edited, a freight unit can be split into multiple freight units, and multiple freight units can be merged into one freight unit. For manual transportation planning, the transportation cockpit can be used. The Transportation Cockpit app provides tools and capabilities to perform transportation planning. In manual planning, freight units are assigned to vehicle resources to create transportation plans. Transportation plans include pickup dates and delivery dates, resources (vehicle assignments), and routes to transport goods between source location and destination location.

The embedded TM component provides automatic planning capabilities using vehicle scheduling and routing (VSR) optimization, load optimization, and load planning features. The VSR optimizer generates transportation plans from freight units and creates planned freight orders. The main objective of the VSR optimizer is to assign the best resource (e.g., vehicle) based on capacities at the best possible cost and to determine the best route and finalize the pickup and delivery dates. You can run VSR optimizer even after performing manual transportation planning to optimize the resources and minimize the transportation cost.

After the transportation planning is executed, embedded TM creates a delivery proposal from the transportation planning results. The delivery proposal will be sent to the SAP S/4HANA (ERP) system for the creation of an inbound delivery with reference to the purchase order or stock transfer order. The inbound delivery is created automatically in SAP S/4HANA from the delivery proposal. Immediately, SAP S/4HANA (ERP) transfers the inbound delivery to the embedded TM component for the freight order execution.

To execute transportation, freight orders need to be created with reference to freight units during transportation planning using the Manage Freight Order app. Freight orders are created for transportation via road (mode of transport) using trucks (means of transport), whereas freight bookings are created for ocean and air transportation. In the Transportation Cockpit app, freight orders/freight bookings can be manually created and the freight orders/freight bookings that are created from transportation planning can be edited. You can assign and remove freight units from freight orders/freight bookings. You can also manually update the carrier in freight orders/freight bookings. Carrier selection and transmitting freight orders/freight bookings to carriers is called the subcontracting process as the transportation activities are subcontracted to a carrier (business partner).

Once the freight orders and freight bookings are transmitted to carriers (business partners), you can set the relevant status for every transportation stage of transportation execution for tracking and tracing shipments. Table 9.7 shows the standard freight order and freight booking execution statuses in transportation management.

Execution Status	Description
Ready for execution	This is the initial execution status of the freight order or freight booking. After the carrier selection, the freight document is transmitted to the carrier and the shipment is scheduled and ready for execution.
Execution started	This status can be set when the freight order or freight booking execution is started. At the source location, deliveries are picked and packed, the goods are loaded to the truck or other means of transport, and the vehicle is ready to depart.
In transit	After the vehicle departs, the execution status of freight documents can be set to in transit.
Arrived at the intermediate location	When the shipment arrives at transshipment locations, the execution status of the freight documents can be set as arrived at the intermediate location.
Arrived at destination	When the shipment arrives at the destination location, the execution status of freight documents can be set as arrived at destination.
Executed	When all transportation activities for the shipment are completed, the final status of the freight document can be set to executed.

Table 9.7 Fright Document Execution Status

Freight order or freight booking documents that are ready for settlement can be displayed in the Freight Orders Worklist app. For every freight order or freight booking, you can create a settlement document from the Freight Order Worklist app. Once the freight settlement document is created, documents can be edited, and freight charge calculations can be triggered. Freight charges are calculated automatically from the charge calculation template or from the agreement if a contract agreement with the carrier is used in the freight order or freight booking document.

After the freight charges are calculated, the freight settlement document is posted. Upon posting the settlement document from embedded TM, a service purchase order and service entry sheet are created in the materials management function of SAP S/4HANA automatically. A service purchase order is created for the carrier as a business partner (vendor) and with a service item that references the service master record. The freight charges are added as the net price in the service purchase order.

For the carrier (business partner), if ERS, also called self-billing, is active, the system posts the invoice automatically using a background job. If ERS is not enabled for the carrier, the carrier (business partner) will submit an incoming invoice electronically either via EDI or via email. Accounts payable processes the incoming invoice. Payment processing will be performed against the reconciled invoice and a remittance advice will be sent to the carrier (vendor).

9.3.2   Flow of Purchasing Documents into Transportation Management

Procurement integration with TM begins with the generation of transportation requirements from purchasing documents such as purchase orders, stock transfer orders, and inbound deliveries. The flow of purchase orders created in SAP S/4HANA (ERP) into embedded TM can be controlled by configuring the filters using purchase order document types, purchasing organizations, and purchasing groups. Only those purchasing documents created using specific document types, purchasing organizations, and purchasing groups will flow into embedded TM. This will help organizations to manage transportation for those purchasing documents that they are responsible for. The transportation of goods for the remaining purchasing documents is performed by the suppliers themselves using their transportation management systems.

Sending purchasing documents into embedded TM prior to inbound delivery creation allows companies to have ample time to perform better transportation planning and optimization. The order management feature of transportation management converts the demand arising from the flow of purchasing documents into shippable units called freight units. A freight unit document is automatically created in embedded TM based on the freight unit building rule configured/defined in transportation management. Embedded TM displays the document flow through the lifecycle of the freight unit document that includes planning, tendering, and execution documents. Freight units that are created from purchasing documents contain general data, item data, subcontracting data, transportation execution data, reference documents data, statuses, and dangerous goods data. Purchase orders can be broken down into multiple freight units considering the capacity of resources and other key factors. Multiple purchase orders can be combined into one freight unit as well.

In the following sections, we'll dive deep into the integration setup of orders and deliveries from SAP S/4HANA (ERP) into embedded TM to generate the transportation demands/requirements.

Purchase Order Document Flow

Figure 9.35 shows the standard purchase order created in SAP S/4HANA. The **TM Status** tab is added automatically in the purchase order header details when the purchase order flows into embedded TM. It displays various statuses of the purchase order in embedded TM.

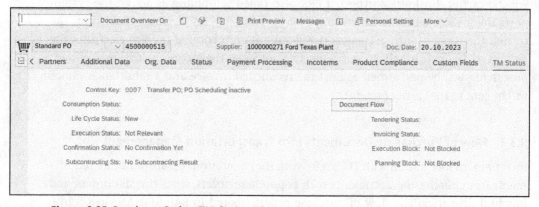

Figure 9.35 Purchase Order: TM Status View

From the **TM Status** tab, when you click **Document Flow**, you can display the embedded TM document flow for the purchase order, as shown in Figure 9.36.

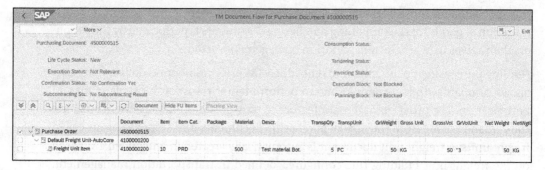

Figure 9.36 TM Document Flow for Purchase Order

The sales orders, stock transfer orders, and schedule lines for scheduling agreements (for standalone embedded TM scenario only) are replicated to the embedded TM function as demand signals, and the system creates a freight unit with reference to those documents. Embedded TM provides the flexibility to transfer deliveries only to embedded TM and provides the features to propose the delivery creation (both inbound and outbound deliveries) in SAP S/4HANA after the transportation planning is performed. If your business requirement is not to replicate orders (purchase orders, sales orders, etc.), inbound and outbound deliveries can be replicated from SAP S/4HANA to embedded TM. Freights units are created with reference to the demand generated from inbound and outbound deliveries.

Define Control Keys for Document Integration

Let's move on to configuration settings for integrating orders and deliveries into an internal TM component, starting with control keys. Control keys in embedded TM control the documents that will flow/integrate into embedded TM for transportation planning and execution. The order management function of embedded TM converts the transportation demand arising from purchase orders, stock transfer orders, sales orders, inbound deliveries, and outbound deliveries into shippable units known as freight units in embedded TM. Control keys can be configured to control the flow/integration of these documents into embedded TM from SAP S/4HANA.

To define control keys for document integration, execute Transaction SPRO and navigate to **Integration with Other SAP Components • Transportation Management • Logistics Integration • Define Control Keys for Document Integration**.

Figure 9.37 shows the control keys for document integration defined in the standard system. Click **New Entries** to define a new control key for document integration or copy an existing control key and create a new one. Specify a four-character alphanumeric key to define a new control key and enter the description.

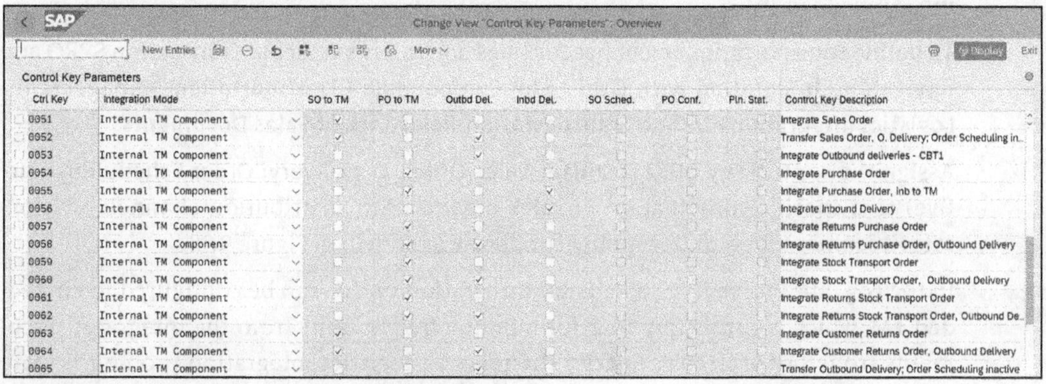

Ctrl Key	Integration Mode	SO to TM	PO to TM	Outbd Del.	Inbd Del.	SO Sched.	PO Conf.	Pln. Stat.	Control Key Description
0051	Internal TM Component	✓							Integrate Sales Order
0052	Internal TM Component	✓		✓					Transfer Sales Order, O. Delivery; Order Scheduling in...
0053	Internal TM Component			✓					Integrate Outbound deliveries - CBT1
0054	Internal TM Component		✓						Integrate Purchase Order
0055	Internal TM Component		✓		✓				Integrate Purchase Order, Inb to TM
0056	Internal TM Component				✓				Integrate Inbound Delivery
0057	Internal TM Component		✓						Integrate Returns Purchase Order
0058	Internal TM Component		✓	✓					Integrate Returns Purchase Order, Outbound Delivery
0059	Internal TM Component		✓						Integrate Stock Transport Order
0060	Internal TM Component		✓	✓					Integrate Stock Transport Order, Outbound Delivery
0061	Internal TM Component		✓						Integrate Returns Stock Transport Order
0062	Internal TM Component		✓	✓					Integrate Returns Stock Transport Order, Outbound De...
0063	Internal TM Component	✓							Integrate Customer Returns Order
0064	Internal TM Component	✓		✓					Integrate Customer Returns Order, Outbound Delivery
0065	Internal TM Component			✓					Transfer Outbound Delivery; Order Scheduling inactive

Figure 9.37 Define Control Keys for Document Integration

The following are the key fields of this configuration:

- **Integration Mode**

  The integration mode indicates if the documents are integrated into an external embedded TM system or an internal embedded TM system that is a component of SAP S/4HANA. For embedded TM in SAP S/4HANA, select **Internal TM Component** as the integration mode.

- **SO to TM**

  If you set this indicator for the control key, sales orders are transferred/integrated into embedded TM from SAP S/4HANA.

- **PO to TM**

  If you set this indicator for the control key, purchase orders are transferred/integrated into embedded TM from SAP S/4HANA.

- **Outbd to TM**

  If you set this indicator for the control key, outbound deliveries are transferred/integrated into embedded TM from SAP S/4HANA.

- **Inbd to TM**

  If you set this indicator for the control key, inbound deliveries are transferred/integrated into embedded TM from SAP S/4HANA.

- **PO Conf.**

  If you set this indicator for the control key, purchase order confirmations (from the suppliers) are transferred/integrated into embedded TM from SAP S/4HANA.

Define Transportation Relevance of Sales Documents

This configuration setup controls only the relevant sales documents from integrating into embedded TM from SAP S/4HANA. You can restrict the sales documents integration to certain combinations based on sales organization, distribution channel, division, sales document type, and shipping condition. Only those sales documents with the specific combinations maintained in this configuration setup will be integrated into embedded TM.

To define transportation relevance for sales documents, execute Transaction SPRO and navigate to **Integration with Other SAP Components • Transportation Management • Logistics Integration • Define Transportation-Relevance of Sales Documents**.

Assign the control key **0052 (Transfer Sales Order, O. Delivery; Order Scheduling inactive)** to a specific combination of sales organization, distribution channel, division, sales document type, and shipping condition, as shown in Figure 9.38.

In the **Log. Int. Profile** field, a logistics integration profile can be configured in embedded TM, which controls the rules for building freight units from the integrated documents from SAP S/4HANA. Specify the relevant logistics integration profile to create freight units for sales documents in this field. The logistics integration profile is defined in the configuration settings.

To define a logistics integration profile, execute Transaction SPRO and navigate to **Transportation Management • Integration • Logistics Integration • Internal TM Component Integration • Define Logistics Integration Profile.**

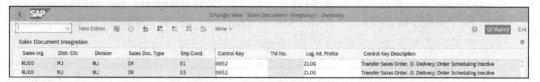

Figure 9.38 Define Transportation Relevance of Sales Documents

Define Transportation Relevance of Purchasing Documents

This configuration setup controls only the relevant purchasing documents from integrating into embedded TM from SAP S/4HANA. You can restrict the purchasing documents integration to certain combinations based on purchasing organization, purchasing group, and purchasing document type. Only those purchasing documents with the specific combinations maintained in this configuration setup will be integrated into embedded TM.

To define transportation relevance for purchasing documents, execute Transaction SPRO and navigate to **Integration with Other SAP Components • Transportation Management • Logistics Integration • Define Transportation-Relevance of Purchasing Documents.**

Assign the control key **0055 (Integrate Purchase Order, inb to TM)** to a specific combination of purchasing organization, purchasing group, and purchasing document type, as shown in Figure 9.39.

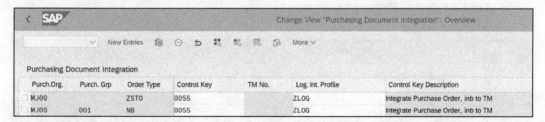

Figure 9.39 Define Transportation Relevance of Purchasing Documents

Define Transportation Relevance of Delivery Documents

This configuration setup controls only the relevant delivery documents from integrating into embedded TM from SAP S/4HANA. You can restrict the delivery document integration to certain combinations based on shipping point/receiving plant, delivery type, and shipping condition. Only those delivery documents with the specific combinations maintained in this configuration setup will be integrated into embedded TM.

To define transportation relevance for delivery documents, execute Transaction SPRO and navigate to **Integration with Other SAP Components • Transportation Management • Logistics Integration • Define Transportation-Relevance of Delivery Documents**.

As shown in Figure 9.40, assign the control key **0056 (Integrate Inbound Delivery)** and **0053 (Integrate Outbound Deliveries)** to delivery types **EL** (inbound delivery) and **LF** (outbound delivery) with a specific shipping condition.

Shipping conditions are defined in the configuration settings based on business requirements. To define shipping conditions, execute Transaction SPRO and navigate to **Shipping • Basic Shipping Functions • Shipping Point and Goods Receiving Point Determination • Define Shipping Conditions**.

Table 9.8 shows the typical shipping conditions in transportation management.

Shipping Condition	Description
1	Standard
2	Pickup
3	Ground
4	Express saver
5	Air—next day
6	Air—2nd day
7	Air—3rd day

Table 9.8 Typical Shipping Conditions

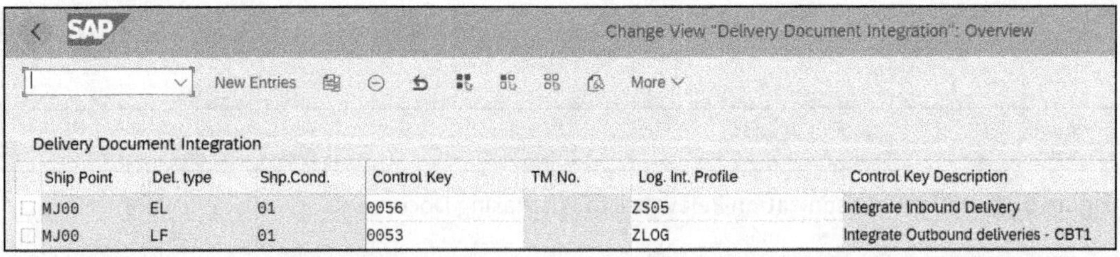

Figure 9.40 Define Transportation Relevance of Delivery Documents

Define Settings for Materials Management Scheduling Agreements Integration

This configuration setup controls only the relevant schedule lines for the scheduling agreement created in materials management from integrating into embedded TM from SAP S/4HANA. You can restrict the schedule lines for scheduling agreement document integration to certain combinations based on purchasing organization, purchasing group, plant, and scheduling agreement document type. Only those schedule lines for scheduling agreement with the specific combinations maintained in this configuration setup will be integrated into embedded TM.

To define transportation relevance for scheduling agreements created in materials management in SAP S/4HANA, execute Transaction SPRO and navigate to **Integration with Other SAP Components** • **Transportation Management** • **Logistics Integration** • **Define Settings for MM Scheduling Agreements Integration**.

As shown in Figure 9.41, maintain the scheduling agreement document types **LP** (scheduling agreement without release) and **LPA** (scheduling agreement with release) for purchasing organization **MJ00**, purchasing group **001**, and plant **MJ00**.

Specify the horizon in **Days** from the current date for the system to integrate the delivery schedules from scheduling agreements to be integrated into embedded TM.

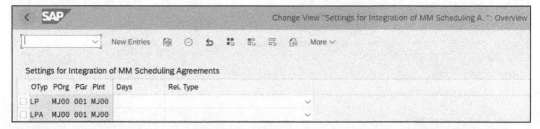

Figure 9.41 Define Settings for Materials Management Scheduling Agreements Integration

---

Note

Integration of the scheduling agreement into embedded TM is possible only if you are running embedded TM as an external system (standalone TM). If you're running the embedded TM component within SAP S/4HANA (ERP) as an internal component, replication of a scheduling agreement (delivery schedule) is not possible. Only the inbound deliveries can be replicated in that scenario.

Configuration for External Transportation Management System

Let's explore how to integrate orders and deliveries into an external (standalone) TM component deployed in another SAP S/4HANA system. The following are the key configuration settings for integrating orders and deliveries into an external TM component.

Define Order-Based Transportation Requirement Types

Orders are transferred to the external TM system using web services enabled between SAP S/4HANA (ERP) and the external TM system. Requirements generated from the purchase orders, stock transfer orders, sales orders, and so on are controlled by defining OTR types. Based on the controlling parameters set in the OTR type, the system processes the business requirements transferred from the SAP S/4HANA (ERP) system. You can define in the OTR type if the freight unit is automatically created when the business requirements are transferred into the embedded TM system from SAP S/4HANA.

To define OTR types, execute Transaction SPRO and navigate to **Transportation Management • Integration • Logistics Integration • External TM System Integration • Order-Based Transportation Requirement • Define Order-Based Transportation Requirement Types**.

Figure 9.42 shows that OTR type **OTR1** was defined to capture the order-based requirements coming from SAP S/4HANA (ERP) in the external TM system. Click **New Entries** to define a new OTR type or copy an existing OTR type and create a new one.

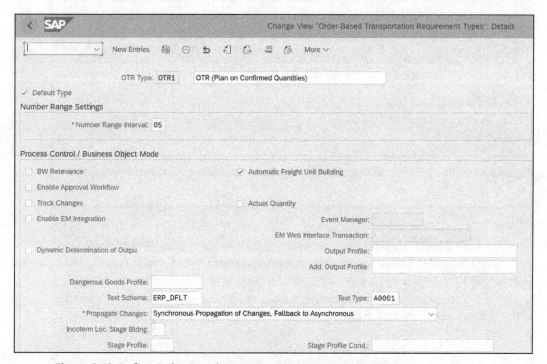

Figure 9.42 Define Order-Based Transportation Requirement Types

The following are the key fields of this configuration setup:

- **Default Type**
  If you set this indicator, the respective OTR type is used by the transportation management system as the default whenever the business requirements (orders) are transferred to the embedded TM system from SAP S/4HANA.

- **Automatic Freight Unit Building**
  If you set this indicator, freight unit building is executed automatically by the system.

- **Enable Approval Workflow**

  If you set this indicator, an approval workflow is required for further processing of orders (requirements) coming from SAP S/4HANA (ERP).

- **Dangerous Goods Profile**

  A dangerous goods profile is defined in the configuration settings in transportation management to control how the system behaves when orders (purchase order, sales order, stock transfer order, etc.) containing dangerous goods are transferred to embedded TM as business requirements. Select the relevant dangerous goods profile from the dropdown menu.

Assign SD/MM Text Types to TM Text Types for SO/PO

The text types maintained in the orders (purchase orders, stock transfer orders, sales orders, etc.) in SAP S/4HANA (ERP) are required to be mapped to the corresponding text types in the external SAP transportation management system. This configuration is required only if an external TM component is deployed in another system. The text in purchase orders, stock transfer orders, sales orders, and so on contains special instructions and business process-relevant information. The text types relevant for transportation management are displayed in the **Notes** section of the transportation documents (freight order, forwarding order, etc.).

To arrive at this configuration step, execute Transaction SPRO and navigate to **Transportation Management • Integration • Logistics Integration • External TM System Integration • Order-Based Transportation Requirement • Assign SD/MM Text Types to TM Text Types for SO/PO**.

Figure 9.43 shows the purchase order text types mapped to the corresponding transportation management text types. If you have defined a custom text type for the purchase orders in SAP S/4HANA (ERP), click **New Entries** to map that text type to the corresponding transportation management text type.

If you set the **Not TM-Rel** indicator, the mapped purchase order text type is not relevant for transportation management. The text that exists in the respective text type in the purchase order will not be copied over to the freight order or forwarding order when the purchase order document is transferred to embedded TM for processing.

> **Note**
>
> In this configuration setup, you can also map the sales order text types to the corresponding text types in embedded TM. Sales order text type 0001 is mapped to A0001, 0002 is mapped to A0002, and so on.

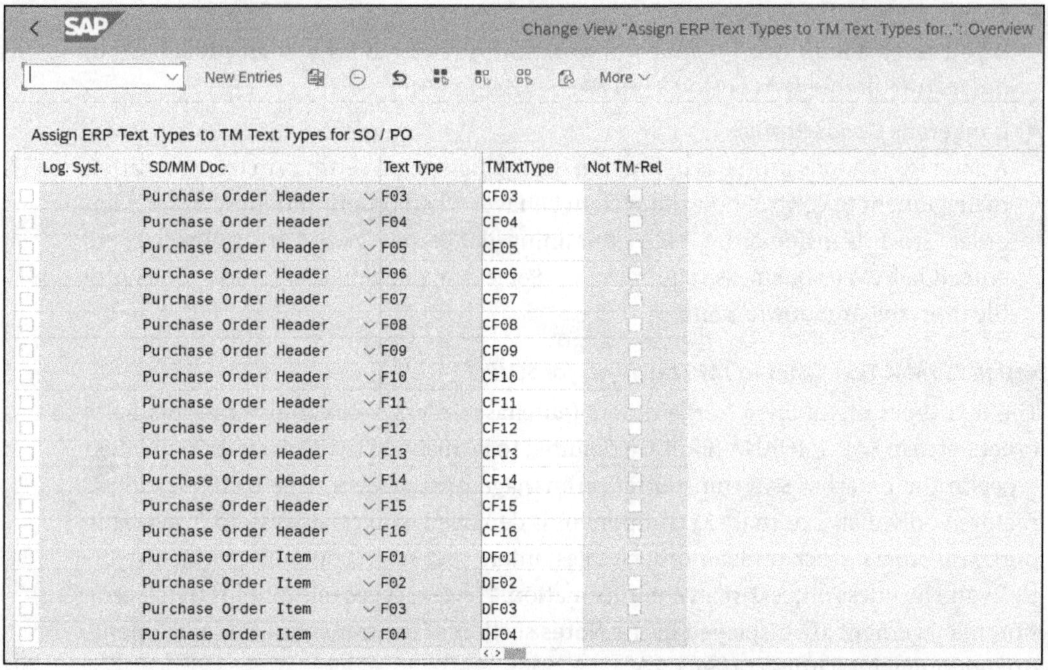

Log. Syst.	SD/MM Doc.	Text Type	TMTxtType	Not TM-Rel
	Purchase Order Header	F03	CF03	
	Purchase Order Header	F04	CF04	
	Purchase Order Header	F05	CF05	
	Purchase Order Header	F06	CF06	
	Purchase Order Header	F07	CF07	
	Purchase Order Header	F08	CF08	
	Purchase Order Header	F09	CF09	
	Purchase Order Header	F10	CF10	
	Purchase Order Header	F11	CF11	
	Purchase Order Header	F12	CF12	
	Purchase Order Header	F13	CF13	
	Purchase Order Header	F14	CF14	
	Purchase Order Header	F15	CF15	
	Purchase Order Header	F16	CF16	
	Purchase Order Item	F01	DF01	
	Purchase Order Item	F02	DF02	
	Purchase Order Item	F03	DF03	
	Purchase Order Item	F04	DF04	

Figure 9.43 Assign Sales and Distribution and Materials Management Text Types for TM Text Types for Sales Orders and Purchase Orders

Define Delivery-Based Transportation Requirement Types

Deliveries are transferred to the external TM system using web services enabled between SAP S/4HANA (ERP) and the external TM system. Requirements generated from the inbound deliveries and outbound deliveries are controlled by defining DTR types. Based on the controlling parameters set in the DTR type, the system processes the business requirements transferred from the SAP S/4HANA (ERP) system. You can define in the DTR type if the freight unit is automatically created when the business requirements (deliveries) are transferred into the embedded TM system from SAP S/4HANA.

To define DTR types, execute Transaction SPRO and navigate to **Transportation Management • Integration • Logistics Integration • External TM System Integration • Delivery-Based Transportation Requirement • Define Delivery-Based Transportation Requirement Types.**

Figure 9.44 shows that DTR type **DTR1** was defined to capture the delivery-based requirements (inbound deliveries and outbound deliveries) coming from SAP S/4HANA (ERP) in the external TM system. Click **New Entries** to define a new DTR type or copy an existing DTR type and create a new one.

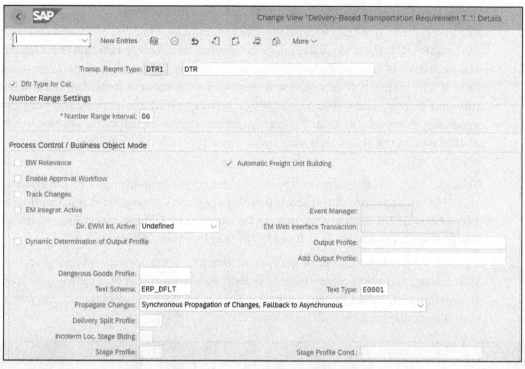

Figure 9.44 Define Delivery-Based Transportation Requirement Types

The following are the key fields of this configuration setup:

- **Dflt Type for Cat. (default type for category)**
  If you set this indicator, the respective DTR type is used by the transportation management system as default whenever the business requirements (inbound deliveries and outbound deliveries) are transferred to embedded TM system from SAP S/4HANA.

- **Automatic Freight Unit Building**
  If you set this indicator, freight unit building is executed automatically by the system.

- **Enable Approval Workflow**
  If you set this indicator, an approval workflow is required for further processing of deliveries (requirements) coming from SAP S/4HANA (ERP).

- **Dangerous Goods Profile**
  A dangerous goods profile is defined in the configuration settings in transportation management to control how the system behaves when deliveries containing dangerous goods are transferred to embedded TM as business requirements. Select the relevant dangerous goods profile from the dropdown menu.

Assign Logistics Execution Text Types to TM Text Types for Deliveries

The text types maintained in the deliveries (inbound delivery and outbound delivery) in the SAP S/4HANA (ERP) system must be mapped to the corresponding text types in the external SAP transportation management system. This configuration is required only if an external TM component is deployed in another system. The text in the inbound delivery and outbound delivery contains special instructions and business process relevant information. The text types relevant for transportation management is displayed in the **Notes** section of the transportation documents (freight order, forwarding order, etc.).

To assign logistics execution text types to TM text types for deliveries, execute Transaction SPRO and navigate to **Transportation Management • Integration • Logistics Integration • External TM System Integration • Delivery-Based Transportation Requirement • Assign Logistics Execution Text Types to TM Text Types for Deliveries.**

Figure 9.45 shows the delivery text types are mapped to the corresponding transportation management text types. If you have defined a custom text type for the deliveries in SAP S/4HANA (ERP) system, click **New Entries** to map that text type to the corresponding transportation management text type.

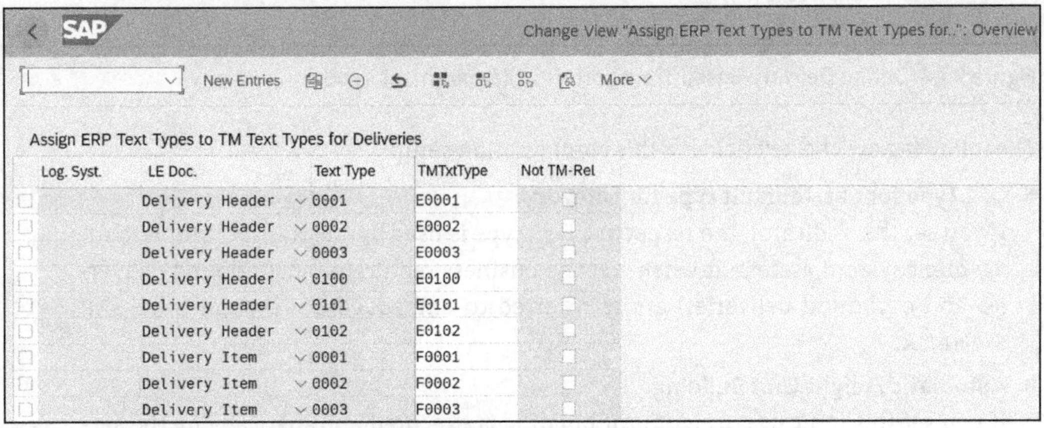

Log. Syst.	LE Doc.	Text Type	TMTxtType	Not TM-Rel
	Delivery Header	0001	E0001	
	Delivery Header	0002	E0002	
	Delivery Header	0003	E0003	
	Delivery Header	0100	E0100	
	Delivery Header	0101	E0101	
	Delivery Header	0102	E0102	
	Delivery Item	0001	F0001	
	Delivery Item	0002	F0002	
	Delivery Item	0003	F0003	

Figure 9.45 Assign Logistics Execution Text Types to TM Text Types for Deliveries

If you set the **Not TM-Rel** indicator, the mapped purchase order text type is not relevant for transportation management. The text that exists in the respective text type in the purchase order will not be copied over to the freight order or forwarding order when the purchase order document is transferred to embedded TM for processing.

9.3.3 Transportation Planning

Transportation planning is the vital process of transportation management aimed at reducing transportation costs, optimizing the usage of resources, and reducing transportation time. Properly designing and managing transportation networks; route planning; maintaining supply networks, distribution networks, and LSPs; and so on supports the transportation planning process.

In the following sections, we'll walk through the key configuration settings required for transportation planning. But first, let's take a closer look at the data flow.

Transportation Planning Data Flow

Figure 9.46 shows the data flow from a demand document for planning through the capacity documents. Freight units are the basis for transportation planning in embedded TM. Freight units are created from business requirements generated from purchase orders, stock transfer orders, sales orders, inbound deliveries, and outbound deliveries in embedded TM. Capacities are assigned to the freight units during planning by considering various factors such as requested delivery dates, source location, destination location, and so on. Transportation units and freight orders are created from the transportation planning process.

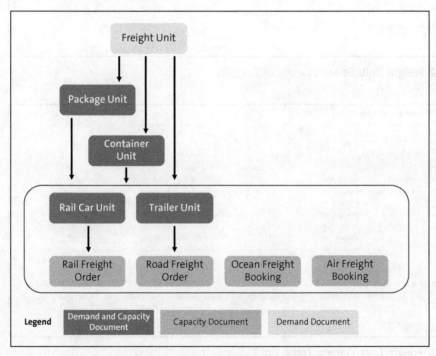

Figure 9.46 Transportation Planning with Embedded TM

Freight units are created from OTRs, DTRs, and forwarding orders. Freight unit building rules control the splitting and merging of freight units. Figure 9.47 and Figure 9.48 show the **General Data** views of freight unit **4100000200** created automatically from an OTR. You can arrive at the freight unit display view directly from the document flow of the purchase order, as shown previously in Figure 9.36. You can also arrive at the freight unit details from the Transportation Cockpit app.

The **General Data** view of the freight unit document (also called a business document) contains generic information such as the freeing unit type, freight unit building rule, transportation mode, and so on, plus the transportation data such as total distance, organizational data, Incoterms, and more. Other important data in the **General Data** view is the location information. Source location and destination location are automatically determined in the freight unit document.

Figure 9.47 Freight Unit: General Data: 1 of 2

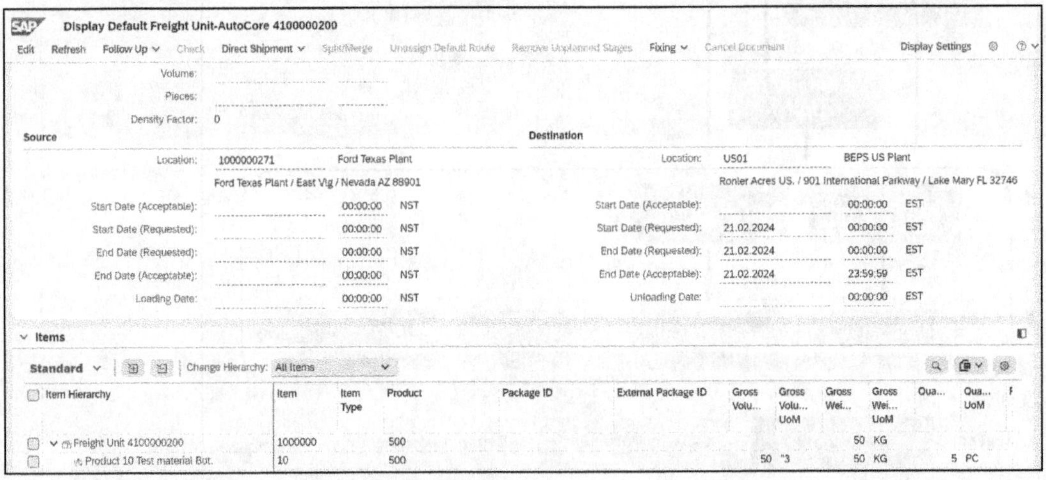

Figure 9.48 Freight Unit: General Data: 2 of 2

Figure 9.49 shows the **Business Partner** data view of freight unit document. This displays partner functions such as shipper, ship-to party, carrier, shipper, and invoicing party. The carrier (business partner) can be updated during the transportation planning process.

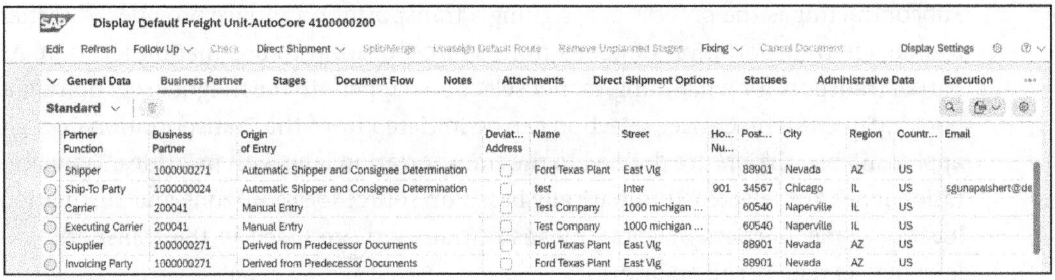

Figure 9.49 Freight Unit: Business Partner Data

The transportation unit consolidates several freight units, and it is used in the transportation of goods. Transportation units such as containers, trailers, railcar units, package (pallet, carton) units, and so on represent transportation requirements as well as capacity documents. TUs can be created manually using the Transportation Cockpit app. Transportation unit types play a vital role in the creation of a transportation unit. Transportation units can be created automatically from forwarding orders and via the VSR optimizer for transporting via truck (only).

Freight orders are created with reference to the carrier or LSP to execute transportation. Freight orders are created for transportation units, and you can create freight orders directly from freight units. The following are the different types of freight orders that you can create:

- **Road freight order**
  A road freight order is a transportation document created from transportation planning to carry out the transportation of goods via the road (mode of transport) using trucks. Trailers and package units (pallets, cartons, etc.) containing goods to be transported are loaded onto the truck.

- **Rail freight order**
  A rail freight order is a transportation document created from transportation planning to carry out the transportation of goods via the rail (mode of transport) using trains. Railcar units are loaded into trains for transportation.

- **Ocean freight booking**
  An ocean freight booking is a transportation document created from transportation planning to carry out the transportation of goods via the ocean (mode of transport) using ships. Freight bookings reserve space in the ships for transportation of freight. Container units are loaded onto the ship for transportation.

- **Air freight booking**
  An air freight booking is a transportation document created from transportation planning to carry out the transportation of goods via the air (mode of transport) using aircrafts. Freight bookings reserve space in aircrafts for transportation of freight. Package units are loaded into the aircraft for transportation.

Subcontracting is the process of assigning a transportation service provider or carrier to manage transportation of goods from source location to destination location. As part of transportation planning, carrier selection happens in the freight order or freight booking document. Carrier selection can be initiated from the Transportation Cockpit app. Available carriers are defined in the transportation lane with priorities. Transportation lanes are selected automatically based on source location/zone and destination location/zone. Carriers by every means of transport are listed in the transportation lane master data based on priorities.

Define Item Types for Freight Order Management

Let's dive into the key configuration settings required for transportation planning, starting with item types. Item types are defined for transportation management business documents such as freight order, freight booking, transportation unit, and so on. Item types are created with reference to item categories. Allowed item types are assigned to the freight order type, transportation unit type, freight booking type, and so on. For service item types, the allowed service types are assigned in the configuration settings.

To define item types for freight order management, execute Transaction SPRO and navigate to **Transportation Management** • **Freight Order Management** • **Define Item Types for Freight Order Management**.

Figure 9.50 shows the overview of item types defined for freight order management. Freight order items are created using item types. Item types represent a vehicle such as truck, trailer, railcar, and so on for transportation of packages/goods. Click **New Entries** to define a new item type for freight order management. Specify a four-digit alphanumeric key to define an item type for freight management and specify the item type description.

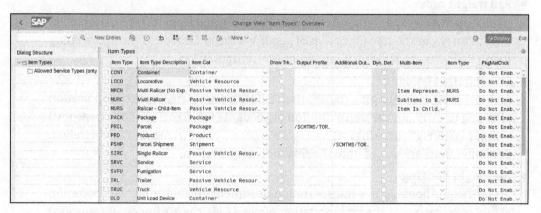

Figure 9.50 Define Item Types for Freight Order Management

The following are the key fields in this configuration:

- **Item Cat**
  The item category is used to categorize the item types. Select a relevant item category for every item type from the dropdown menu. Item categories are predefined in the standard system.

- **Multi-item**
  Multi-items are multiple items of the same category. Multi-items can be created for passive vehicle resources, containers, products, and packages.

In the **Dialog Structure**, navigate to **Allowed Service Types** to get to the screen shown in Figure 9.51, which shows the allowed service types for the **SRVC** (service) item type. Service orders can be created for freight order items and freight booking items in embedded TM for service activities to be performed such as insurance, cleaning, customs clearance, and so on. Select allowed service types for the service item types by clicking the **Srvc. Type** field; an icon will appear. Click the icon to display the list of service types in the dropdown menu.

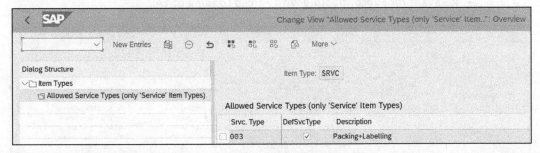

Figure 9.51 Allowed Service Types for Service Item Types (Only)

Define Freight Order Types

Freight order types are the document type for freight order documents that control their behavior. The attributes defined in the freight order types control the freight order processing in embedded TM. Freight order types control if the transportation activities can be subcontracted and how the shipper, consignee, and so on are determined during the freight order processing. This controls the freight order execution and event management processes.

To define freight order types, execute Transaction SPRO and navigate to **Transportation Management • Freight Order Management • Freight Order • Define Freight Order Types**.

Figure 9.52 shows the overview of freight order types defined for freight order processing. Click **New Entries** to define a new freight order type for freight order management. Specify a four-character alphanumeric key to define a new freight order type and specify an item type description.

The following are the key fields in this configuration:

- **NR**

  Define the number range for the freight orders. To define number ranges for freight orders, execute Transaction SPRO and navigate to **Transportation Management** • **Freight Order Management** • **Define Number Range Intervals for Freight Order Management**. You can arrive at this configuration step directly by executing Transaction SNRO and defining the number ranges for freight orders against number range object /SCMTMS/TO. Specify the number range interval for freight order creation in this field.

- **Subcontr. Relevance**

  Subcontracting relevance indicates if the freight order is executed by a subcontractor. If the subcontracting relevance is not set, freight execution is not performed by the subcontractor. The following values can be set in this field:

  - **Relevant for Subcontracting**
  - **Not Relevant for Subcontracting**
  - **Relevant for Subcontracting with Forced Loading of TAL**

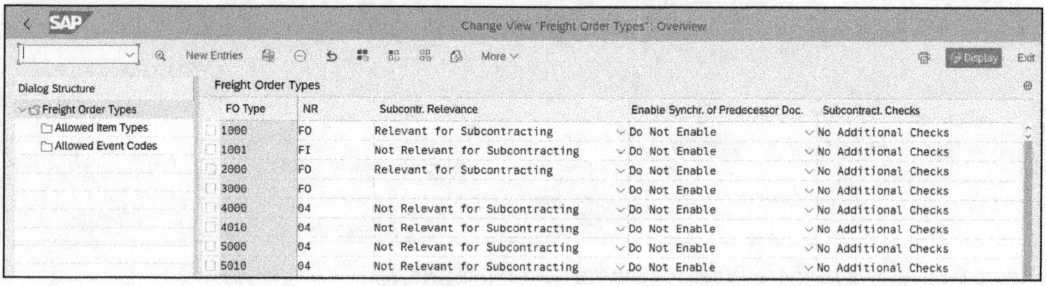

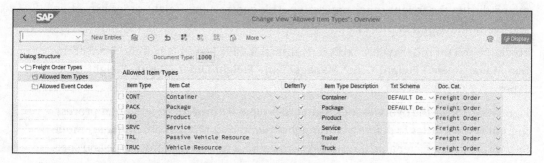

Figure 9.52 Define Freight Order Types: Overview

Navigate to **Allowed Item Types** in the **Dialog Structure** and click **New Entries** to assign item types to the freight order type. Click the **Item Type** field and an icon will appear; click the icon to display the list of item types in the dropdown menu. Figure 9.53 shows the allowed item types for freight order document type **1000**.

Figure 9.53 Allowed Item Types for Freight Order Types

Assign the allowed item types defined in the configuration step in the previous section for every freight order document type. Set the default item type (**DefItmTy**) indicator for at least one item type allowed for the freight order document type.

Finally, navigate to **Allowed Event Codes** in the **Dialog Structure**. Figure 9.54 shows the event codes assigned to freight order document type **1000**. During the freight order execution, you can set the status of freight order processing using these event codes. Assign the allowed event codes for every freight order document type to monitor the status of freight order execution. To assign the allowed event codes to the freight order type, click **New Entries**. To select the event codes, click the **Event** field and an icon will appear; click the icon to display the list of event codes in the dropdown menu.

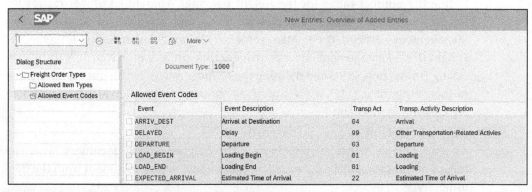

Figure 9.54 Allowed Event Codes for Freight Order Types

Define Freight Booking Types

Freight booking types are the document types for freight booking documents that control their behavior. Freight booking documents are used to reserve freight space in a ship or an aircraft for freight transportation. The attributes defined in the freight booking types control the freight booking processing in embedded TM. The freight booking type controls how the shipper, consignee, and so on are determined during the freight booking processing. This controls the freight booking execution and event management processes. You can assign default means of transport, transportation mode, and so on in the freight booking type definition.

To define freight order types, execute Transaction SPRO and navigate to **Transportation Management • Freight Order Management • Freight Booking • Define Freight Booking Types**.

Figure 9.55 shows the overview of freight booking types defined for freight booking execution. Click **New Entries** to define a new freight booking type for freight order management. Specify a four-character alphanumeric key to define a new freight booking type and specify the item type description.

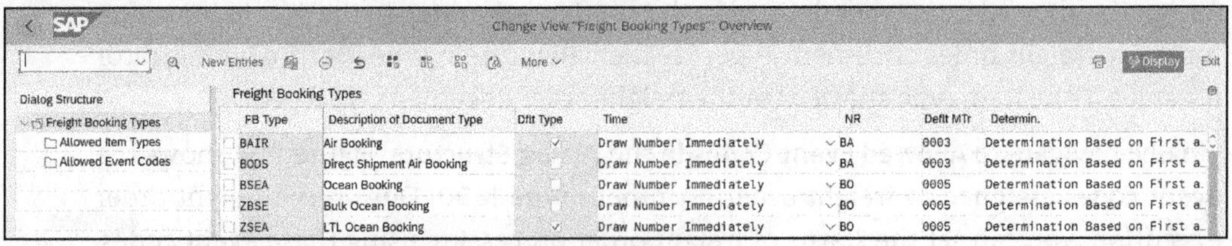

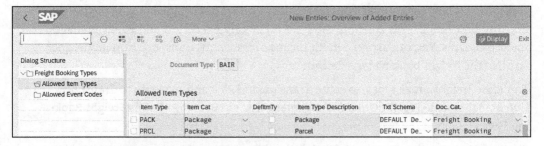

Figure 9.55 Define Freight Booking Types: Overview

The following are the key fields in this configuration:

- **NR**
  Define the number range for the freight bookings. To define number ranges for freight bookings, execute Transaction SPRO and navigate to **Transportation Management • Freight Order Management • Define Number Range Intervals for Freight Order Management**. You can arrive at this configuration step directly by executing Transaction SNRO and defining the number ranges for freight orders against number range object /SCMTMS/TO. Specify the number range interval for freight booking creation in this field.

- **Deflt MTr. (default means of transport)**
  Assign a default means of transport for the freight booking document type. The means of transport will be defaulted to the freight booking document from the document type.

Navigate to **Allowed Item Types** in the **Dialog Structure** to reach the screen shown in Figure 9.56, which shows the allowed item types for freight booking document type **BAIR**. Assign the allowed item types defined in the earlier configuration step for every freight booking document type. To assign item types to the freight booking types, click **New**. To select the item types, click the **Item Type** field; an icon will appear. Click the icon to display the list of item types in the dropdown menu.

Figure 9.56 Allowed Item Types for Freight Booking Types

Move on to **Allowed Event Codes** in the **Dialog Structure**. Figure 9.57 shows the event codes assigned to freight booking document type **BAIR**. During the freight booking execution, you can set the status of freight booking processing using these event codes. Assign the allowed event codes for every freight booking document type to monitor the status of freight booking execution. To assign the allowed event codes to the freight booking type, click **New Entries**. To select the event codes, click the **Event** field and an icon will appear; click the icon to display the list of event codes in the dropdown menu.

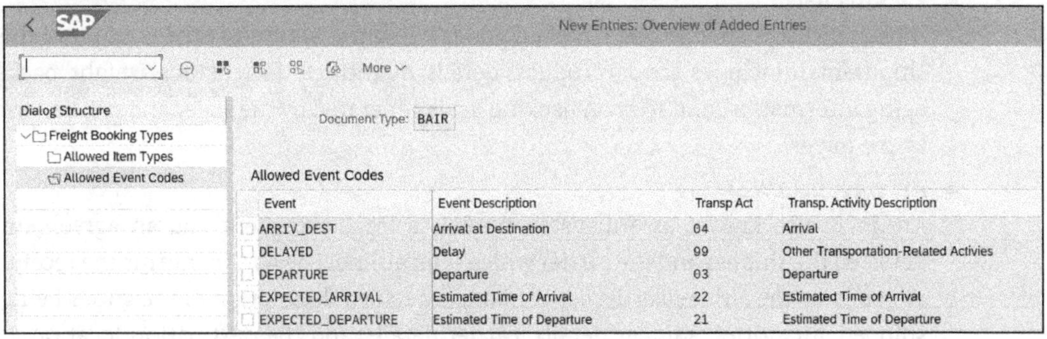

Figure 9.57 Allowed Event Codes for Freight Booking Types

9.3.4   Transportation Execution

Freight execution or transportation execution starts when the freight execution documents such as freight order and freight booking are created, and the vehicle and carrier are determined and assigned to the freight execution documents. Freight orders and freight booking documents are transmitted to the carrier. The carrier confirms the freight order and begins the transportation preparation activities based on the schedule. The freight order or freight booking creation, carrier selection, and assignment steps are part of the transportation planning. The vehicle is ready to pick up the freight units or transportation units; from this moment onwards the transportation execution starts. During the freight document execution, required freight documentation depending on the type of shipment is generated in embedded TM. The following are some of the important documentation requirements to execute transportation:

- **Bill of lading**
  A bill of lading serves as a legal document and it's an agreement between the shipper and the carrier which acts as a receipt for the transportation of goods to be carried out by the carrier. It contains the type of goods being shipped, quantities, name of the shipper, carrier details, and the destination location. It also acts as the title document, and it is evidence of the ownership of the goods being transported indicating the shipper has the title of the goods. It also serves as an acknowledgement that the carrier has received the goods for transportation.

- **Commercial invoice**
  Commercial invoices are used in international trade, and they serve as legal documents for custom clearance purpose. It's a type of bill from the seller to the buyer and it contains detailed description of goods, quantity of goods, unit price of each material/product, international commercial terms, and so on. A commercial invoice is used by the customs department to calculate tariffs/custom duties to be paid.

Commercial invoices may avoid custom clearance delays; hence it is important documentation for freight execution. Commercial invoices are not required to execute transportation within the European Union.

- **Packing list**
  A packing list is an itemized list of all goods being transported or the content of the shipment. It includes goods/products details, quantities, dimensions, weight, packaging information, and so on. A packing list is generated by the seller and will be sent to the buyer.

- **Air waybill (AWB)**
  An AWB, also known as a dispatch note, is a legal document and an agreement between the shipper and the carrier which is an airline. The AWB confirms the receipt of goods by the airline (carrier). It contains a detailed description of the goods being shipped, quantities, shipper details, carrier details, and the destination location. It also acts as the title document, and it is evidence of the ownership of the goods being transported indicating the shipper has the title of the goods. It also serves as an acknowledgement that the carrier has received the goods for transportation.

Further documentation is required for freight document execution, such as customs declaration, dangerous goods declaration, certificate of origin, and so on.

At the shipping point or source location, loading of deliveries into the vehicle happens as part of shipping activities. Picking and packing of deliveries are performed at the source location and the goods are staged for the vehicle's arrival. Loading the deliveries into the vehicle is performed with reference to the freight order. Loading and unloading points are assigned to the freight order to facilitate the loading process.

Once the freight execution starts, the execution status, process status, and event status of freight order and freight booking documents can be monitored at every stage of transportation, including the status at the transshipment locations. The Track Freight Movement app can be used to monitor and track statuses. The execution statuses can be set manually in embedded TM. The status codes are maintained in the configuration settings for freight order types and freight booking types. Refer to the configuration settings in Section 9.3.3 for more details. If embedded TM is integrated with SAP Event Management, the execution status can be set automatically. Table 9.9 shows the execution status of freight documents.

Execution Status	Description
Ready for execution	This is the initial execution status of freight order or freight booking. After the carrier selection, the freight document is transmitted to the carrier, and the shipment is scheduled, the shipment is ready for execution.

Table 9.9 Execution Status of Freight Document

Execution Status	Description
Execution started	This status can be set when the freight order or freight booking execution is started. At the source location, deliveries are picked & packed, the goods are loaded to the truck or other means of transport and the vehicle is ready to be departed.
In-transit	After the vehicle is departed, the execution status of freight documents can be set to in-transit.
Arrived at the intermediate location	When the shipment arrived at the transshipment locations, the execution status of freight documents can be set as arrived at the intermediate location.
Arrived at destination	When the shipment arrived at the destination location, the execution status of freight documents can be set as arrived at destination.
Executed	When all transportation activities for the shipment are completed, the final status of the freight document can be set to executed.

Table 9.9 Execution Status of Freight Document (Cont.)

Embedded TM displays the business process status of freight documents based on the most recent event that occurred. Table 9.10 shows the process statuses for freight documents.

Process Status	Description
As planned	This process status is set when there are no delays occurred during the freight order/booking execution. This status indicates that the process is being executed as per the schedule created during transportation planning.
Early	This status is set when a particular event occurred during the business process earlier than planned.
Late	This status is set when a particular event occurred during the business process later than planned.
Overdue	This status is set when a particular event during the business process occurs outside of the expected window. By default, if an event occurs after 10 mins from the planned time, the process status will be set to overdue.
Delayed	This status is set when a delay event is reported.

Table 9.10 Process Status of Freight Document

Freight execution and monitoring are the most important processes in transportation management. The physical freight movement from source location to destination location happens during the freight execution process. Carrier management, freight execution, and freight monitoring are the main functions of transportation execution and monitoring process.

> **Note**
>
> The freight order and freight booking documents are used to execute freight in transportation management. A forwarding order is a contract with the forwarding agent (freight forwarder) who performs the service of moving goods from the shipper's location to the destination location according to the agreed-upon terms, whereas freight orders and freight bookings are executed by a carrier. Freight orders can propose delivery creation in the SAP S/4HANA (ERP) system, and they can be tracked and traced for all events during freight execution.
>
> You can create forwarding orders manually. The forwarding order is used as a basis for transportation planning, and freight units can be created with reference to a forwarding order, whereas a freight order is created after planning and during the freight execution. Freight orders are issued to carriers, and they contain planned departure data, freight units to be loaded, and freight execution data.

9.3.5   Freight Charge Calculation

Freight charge management is another vital function of embedded TM. Actual freight charges as well as estimations can be calculated in embedded TM. The system considers the freight agreement, a long-term contract with the carrier that provides transportation services, for freight charge calculation.

The transportation service details and the rates are defined in the freight agreement. Freight agreements are created within embedded TM as part of strategic freight procurement. Strategic freight procurement is the core functionality of embedded TM for prebidding planning and RFQ creation. A historical spend analysis and prioritization of carriers are prerequisites for the creation of an RFQ to source the best carriers and forwarding agents. Selected carriers are invited for the bidding process while considering their historical performance. Bidding responses are automatically compared in embedded TM via the optimizer, and the carriers are ranked based on business rules and key considerations. A carrier is awarded the transportation business based on the best rate and service quoted during the RFQ process. A contract agreement with the awarded carrier can be created with a single click in embedded TM, or an existing contract can be extended. This core functionality of strategic freight procurement helps to explore the best services and rates for transporting goods between source and destination transportation zones and between source and destination locations.

Although the strategic freight procurement functionality of embedded TM doesn't integrate directly with the procurement function of materials management, the core concept is influenced by the strategic procurement (RFQ process) function of materials management in SAP S/4HANA. Figure 9.58 shows how the system calculates freight charges using freight agreements, transportation charge calculation sheets (TCCSs), rate tables, and scales.

The system also considers the TCCS as the basis for transportation charge calculation. The freight agreement includes a TCCS, which can be created as a master data element and included in freight agreements. The TCCS includes various freight charge types for charge calculation. A freight agreement can have multiple transportation charge calculation sheets depending on the modes of transport and stages. Basic freight charges, fuel charges, surcharges, and so on are added to the TCCS in a sequence using charge types. Charge types are configured in embedded TM and can be selected during the transportation charge calculation sheet creation in master data, or you can define the charges for charge types in a sequence directly in the freight agreement. TCCS provides a calculation procedure (i.e., how to calculate different transportation charges in a sequence) to embedded TM.

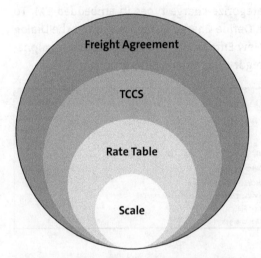

Figure 9.58 Freight Charge Calculation in Embedded TM

TCCSs consist of one or more rate tables. Rate tables are used to group charges for transportation services. The rate table contains the general data, calculation base, and scales. The general data of the rate table contains general information for the system to calculate freight charges. Calculation basis is mandatory, and it controls if the freight charge calculations are amount based or not amount based. The rates are listed by validity period in the rate table. Scales are used to define parameters that in turn define rates.

The charges calculated using freight agreements, TCCSs, rate tables, and scales are displayed and managed in the freight order document, which serves as the execution document in embedded TM. Use the Manage Freight Order app to display and manage freight orders.

Let's dive deep into the key configuration settings required for transportation planning.

Define Transportation Charges

Freight charge calculation for different types of freight charges happens in the freight order, forwarding order, freight settlement, forwarding settlement, carrier selection, and other documents in embedded TM. To integrate freight settlement documents with SAP S/4HANA, charge categories, subcategories, and charge types need to be defined in embedded TM. Once defined, a combination of charge category, charge subcategory, and charge type is assigned to an account assignment category defined in SAP S/4HANA (e.g., cost center).

To define transportation changes, execute Transaction SPRO and navigate to **Integration with Other SAP Components • Transportation Management • Invoice Integration • Invoicing • Definition of Transportation Charges**.

You can define charge categories, charge subcategories, and charge types in this configuration setup. Figure 9.59 shows the charge categories defined in the standard system for basic freight, transportation and additional charges, and miscellaneous and other charges. Charge categories are used to categorize charge types in embedded TM. To define new charge categories, double-click **Define Charge Categories** under the **Dialog Structure** on the left-hand side and click **New Entries**. Specify a three-character alphanumeric key for defining a new charge category and enter a description.

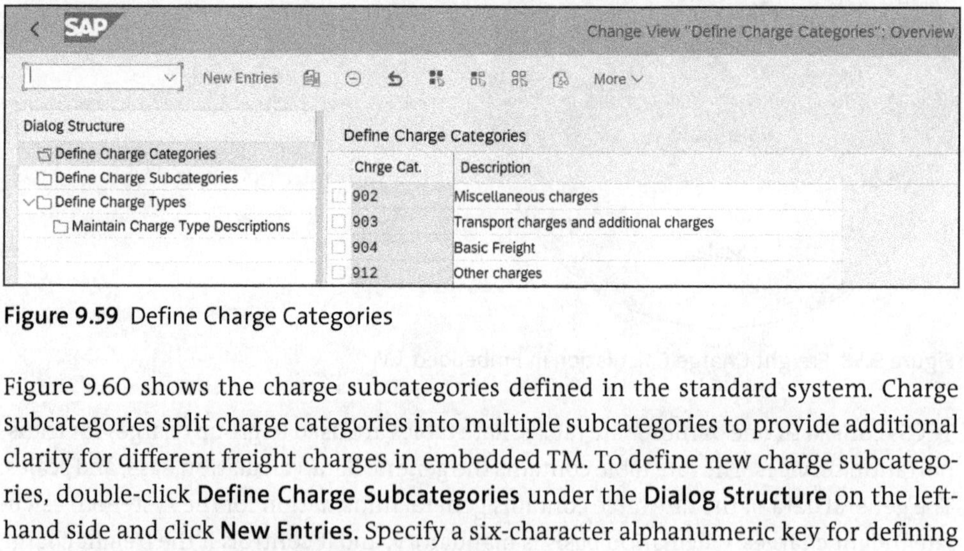

Figure 9.59 Define Charge Categories

Figure 9.60 shows the charge subcategories defined in the standard system. Charge subcategories split charge categories into multiple subcategories to provide additional clarity for different freight charges in embedded TM. To define new charge subcategories, double-click **Define Charge Subcategories** under the **Dialog Structure** on the left-hand side and click **New Entries**. Specify a six-character alphanumeric key for defining a new charge subcategory and enter a description.

Figure 9.61 shows the charge types defined in embedded TM. Relevant charge categories and subcategories are assigned to the charge types. Charge types are in turn used in calculation sheets (calculation procedure) as a basis for freight charge calculation including scales. To define new charge types, double-click **Define Charge Types** from the **Dialog Structure** on the left-hand side and click **New Entries**. Specify a 15-character alphanumeric key for defining a new charge type and enter the description. Assign the charge category (**Chrge Cat.**) and subcategory (**Ch. Subcat**) to the new charge type.

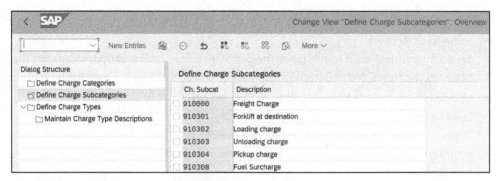

Figure 9.60 Define Charge Subcategories

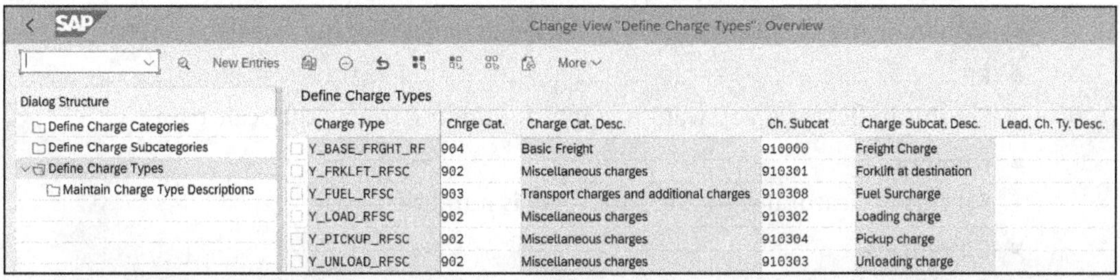

Charge Type	Chrge Cat.	Charge Cat. Desc.	Ch. Subcat	Charge Subcat. Desc.	Lead. Ch. Ty. Desc.
Y_BASE_FRGHT_RF	904	Basic Freight	910000	Freight Charge	
Y_FRKLFT_RFSC	902	Miscellaneous charges	910301	Forklift at destination	
Y_FUEL_RFSC	903	Transport charges and additional charges	910308	Fuel Surcharge	
Y_LOAD_RFSC	902	Miscellaneous charges	910302	Loading charge	
Y_PICKUP_RFSC	902	Miscellaneous charges	910304	Pickup charge	
Y_UNLOAD_RFSC	902	Miscellaneous charges	910303	Unloading charge	

Figure 9.61 Define Charge Types

Define Scale Bases

The scale base is a parameter used as a basis for transportation charge calculation. A scale base can be a product, amount, distance, business partner, country, and so on. To define scale bases, execute Transaction SPRO and navigate to **Transportation Management • Basic Functions • Charge Calculation • Data Source Binding • Define Scale Bases**.

Figure 9.62 shows the scale bases defined in the standard system for freight charge calculation. Click **New Entries** to define a new scale base. Specify a six-character alphanumeric key to define a new scale base and its description.

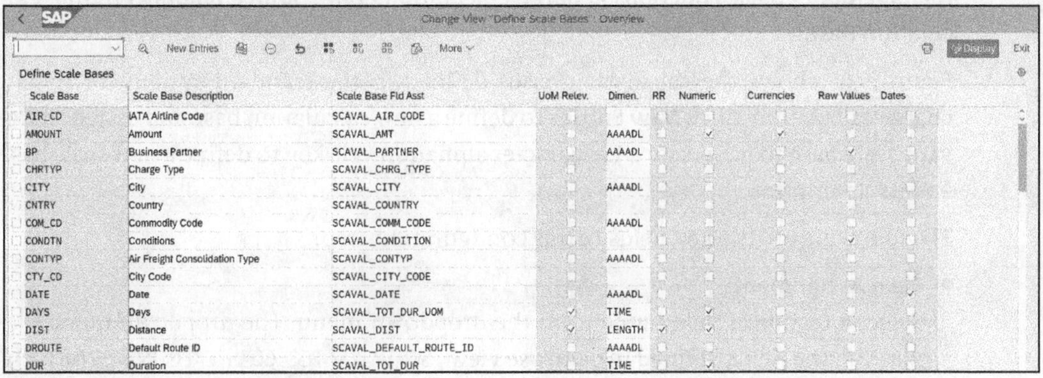

Scale Base	Scale Base Description	Scale Base Fld Asst	UoM Relev.	Dimen.	RR	Numeric	Currencies	Raw Values	Dates
AIR_CD	IATA Airline Code	SCAVAL_AIR_CODE							
AMOUNT	Amount	SCAVAL_AMT		AAAADL		✓	✓		
BP	Business Partner	SCAVAL_PARTNER		AAAADL				✓	
CHRTYP	Charge Type	SCAVAL_CHRG_TYPE							
CITY	City	SCAVAL_CITY		AAAADL					
CNTRY	Country	SCAVAL_COUNTRY							
COM_CD	Commodity Code	SCAVAL_COMM_CODE		AAAADL					
CONDTN	Conditions	SCAVAL_CONDITION						✓	
CONTYP	Air Freight Consolidation Type	SCAVAL_CONTYP		AAAADL					
CTY_CD	City Code	SCAVAL_CITY_CODE							
DATE	Date	SCAVAL_DATE		AAAADL					✓
DAYS	Days	SCAVAL_TOT_DUR_UOM	✓	TIME		✓			
DIST	Distance	SCAVAL_DIST	✓	LENGTH	✓	✓			
DROUTE	Default Route ID	SCAVAL_DEFAULT_ROUTE_ID		AAAADL					
DUR	Duration	SCAVAL_TOT_DUR		TIME		✓			

Figure 9.62 Define Scale Bases

The following are the key fields of this configuration step:

- **Scale Base Fld Asst**

  Select a technical field name from the dropdown menu. The dropdown menu displays all the fields defined in database view /SCMTMS/S_TCSCALE_ITEM_DB. The technical field name is used to store the scale base information in the database.

- **UoM Relev.**

  If you set the unit of measure relevance indicator, it indicates that the scale base requires a unit of measure when maintained as a basis for charge calculation in a rate table.

- **Dimen.**

  Select the relevant dimension from the dropdown list for the scale base. The dimension indicates the technical abbreviation as defined in the standard system.

- **RR**

  Set this indicator if you want to apply a rounding rule to the scale base.

- **Numeric**

  Set this indicator if the scale base value is numeric.

- **Currencies**

  Set this indicator if the scale base value requires currencies (e.g., amount).

- **Dates**

  Set this indicator if the scale base is a calendar date.

Define Calculation Bases

A calculation base defines an actual basis for transportation charge calculation. It is assigned to the calculation base to provide factors for charge calculation. A calculation base can be net weight, gross weight, source location details, destination location details, business partner, vehicle gross weight, date, duration, and so on.

To define calculation bases, execute Transaction SPRO and navigate to **Transportation Management • Basic Functions • Charge Calculation • Data Source Binding • Define Calculation Bases**.

Figure 9.63 shows the calculation bases defined in the standard system for freight charge calculation. Click **New Entries** to define a new calculation base and assign a relevant scale base to it. Specify a 15-character alphanumeric key to define a new scale base and its description.

The following are the key fields of this configuration step:

- **Field Assignment**

  Select a technical field name from the dropdown menu. The dropdown menu displays all the fields defined in database view /SCMTMS/S_TCC_COMM_CALC_BASE. The technical field name is used to store the calculation base information in the database.

- **Currency Field Asst**
  Select a technical field name for the currency from the dropdown menu for amount-based calculation bases (e.g., insurable value). The dropdown menu displays all the fields defined in database view /SCMTMS/S_TCC_COMM_CALC_BASE. The technical field name is used to store the calculation base information in the database.
- **Unit Field Assgmt**
  Select a technical field name for currency from the dropdown menu for unit-based calculation bases (e.g., gross volume, gross weight). The dropdown menu displays all the fields defined in database view /SCMTMS/S_TCC_COMM_CALC_BASE. The technical field name is used to store the calculation base information in the database.

Calculation Base	Scale Base	Description	Field Assignment	Currency Field Asst	Unit Field Assgmt	Helper Class Assgmt
ACTUAL_DATE	DATE	Actual Execution Date	ACTUAL_DATE			/SCMTMS/CL_TCC_CB_CONV_TS
ACTUAL_DIST	DIST	Actual Distance	ACTUAL_DIST_VAL		ACTUAL_DIST_UNI	/SCMTMS/CL_TCC_CB_ACTUAL_DIST
ADD_AGR_PARTY	BP	Additional Party				
AIR_CD	AIR_CD	IATA Airline Code	TSP_IATA_CODE			
BILL_FROM_PARTY	BP	Bill-from Party	BILL_FROM_PARTY_UUIC			/SCMTMS/CL_TCC_CB_PARTY_ID
BILL_TO_PARTY	BP	Bill-to Party	BILL_TO_PARTY_UUID			/SCMTMS/CL_TCC_CB_PARTY_ID
CALC_DATE	DATE	Calculation Date				/SCMTMS/CL_TCC_CB_CALC_DATE
CASH_ON_DLVRY	IND	Cash on Delivery	COD_INDICATOR			
CHRG_TYP	CHRTYP	Charge Type				
CHRG_WEIGHT	WEIGHT	Chargeable Weight	CHRG_WEI_VAL		CHRG_WEI_UNI	/SCMTMS/CL_TCC_CB_CHRG_WEIGHT
CHRG_WEIGHT_NET	WEIGHT	Chargeable Weight Based ...	CHRG_WEI_VAL_NET		CHRG_WEI_UNI_NET	/SCMTMS/CL_TCC_CB_CHRG_WT_NET
COMMODITY_CODE	COM_CD	Commodity Code	COMMODITY_CODE			/SCMTMS/CL_TCC_CB_COMM_CODE
COMM_CODE_IATA	COM_CD	IATA Commodity Code	COMMODITY_CODE_IATA			
COMM_CODE_MUL	COM_CD	Commodity Code Multiple	COMMODITY_CODE			/SCMTMS/CL_TCC_CB_COMM_CODE

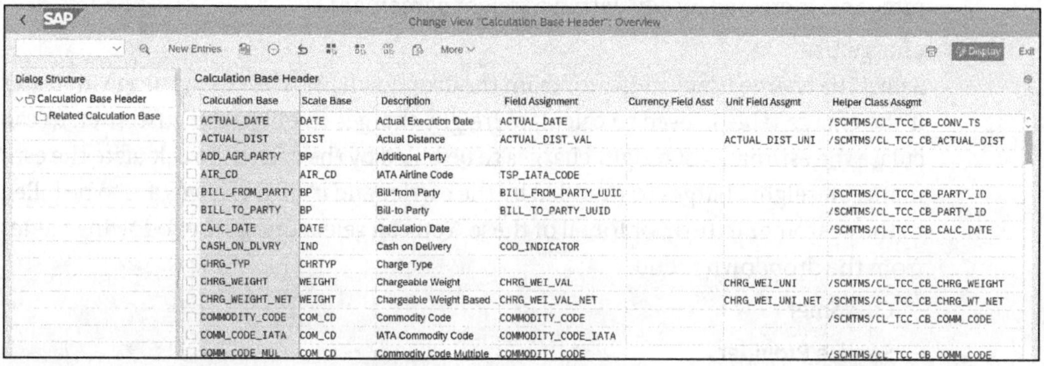

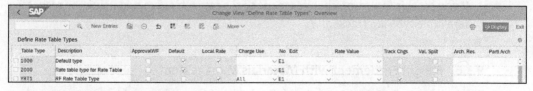

Figure 9.63 Define Calculation Bases

Define Rate Table Types

In this configuration step, define the attributes for rate table creation for freight charge calculation. Attributes or controlling parameters are defined in the rate table types which are used to create rate tables and local rate tables. TCCSs consist of one or more rate tables. Rate tables are used to group charges for transportation services. The rate table contains the general data, calculation base, and scales.

To define rate table types, execute Transaction SPRO and navigate to **Transportation Management • Master Data • Rate Tables • Define Rate Table Types.**

Figure 9.64 shows an overview of rate table types. Click **New Entries** to define a new rate table type. Specify a four-character alphanumeric key to define a new rate table type and its description.

Table Type	Description	ApprovalWF	Default	Local Rate	Charge Use	No Edit	Rate Value	Track Chgs	Val. Split	Arch. Res.	Partl Arch
1000	Default type		✓	✓		E1					
2000	Rate table type for Rate Table		✓			E1					
YRT1	RF Rate Table Type			✓	All	E1		✓			

Figure 9.64 Define Rate Table Types

The following are the key fields of this configuration step:

- **ApprovalWF**
  If you set this indicator, a rate table created using the respective rate table type will require approval.

- **Default**
  If you set this indicator, the respective rate table type is used as a default whenever a rate table type is not specified to create a rate table.

- **Local Rate**
  If you set this indicator, the respective rate table type is used to create a local rate table. Rate tables are used to define and maintain rates for charge calculation. A local rate table is created locally within a freight agreement.

- **Charge Use**
  Select the charge usage indicator from the dropdown list. The charge usage indicates if the charge table is used to calculate freight charges to bill to the customer, to calculate the estimates of freight charges to be billed by the carrier, to calculate the estimates of freight charges to be billed by the carrier and if such charges are to be billed to the customer in turn, or for all of these. You can select one of the following values from the dropdown menu:
  - **Customer**
  - **Service Provider**
  - **Customer & Service Provider**
  - **Internal**
  - **All**
  - **Dispute**

- **Edit**
  If you leave this field blank, the rate table can be edited if it is awaiting approval. If you do not want to allow the rate table to be edited, select the **Not Editable** value from the dropdown menu.

- **Rate Value**
  If you leave this field blank, the rate values can be edited if the rate table is awaiting approval. If you do not want to allow the rate values to be edited, select the **Not Editable** value from the dropdown menu.

- **Track Chgs**
  If you set this indicator, changes made to the rate tables created using the respective rate table type are tracked.

9.3.6 Freight Settlement with Materials Management

The freight settlement process is the final step of the end-to-end transportation management process, and it integrates with SAP S/4HANA materials management for freight settlement with carriers and forwarding agents. Integration of embedded TM with materials management happens at the initial step (to replicate orders and deliveries into embedded TM) and the final step of the transportation management process.

The charges calculated using freight agreement, TCCSs, rate tables, and scales are displayed and managed in the freight order document, which serves as the execution document in embedded TM. Use the Manage Freight Order app to display and manage freight orders. After the freight charges are calculated for the freight order, a freight settlement document is created. The freight settlement document is created for every freight order, forwarding order, and service order. It serves as a basis for settling the freight charges in the materials management function of SAP S/4HANA (ERP). The freight settlement document is created with reference to a freight order or forwarding order, and most of the data is copied from the freight order and forwarding order to the freight settlement document.

The status of the freight execution documents (freight order, freight booking, etc.) must be *ready for confirmation* to trigger the settlement process and the integration with materials management. Typically, the transportation buyer is responsible for confirming the freight order and posting the financial postings. Upon posting the settlement document from transportation management, a service purchase order and service entry sheet are created in the materials management function of SAP S/4HANA automatically. A service purchase order is created for the carrier as a business partner (vendor) and with a service item that references a service master record. The freight charges are added as the net price in the service purchase order.

For the carrier (business partner), if ERS, or self-billing, is active, the system posts the invoice automatically using a background job. If ERS is not enabled for the carrier, the carrier (business partner) will submit an incoming invoice electronically either via electronic data interchange (EDI) or via email. Accounts payable processes the incoming invoice. Payment processing will be performed against the reconciled invoice, and a remittance advice will be sent to the carrier (vendor).

Let's explore the key configuration settings in transportation planning for integrating/posting freight settlement documents in materials management in SAP S/4HANA (ERP).

Assign Service Master Record and Account Assignment Category

In this configuration setup, the account assignment category and activity number (service master record) are assigned to a combination of a charge category, charge subcategory, and charge type. This assignment is necessary to create a service purchase order with the service master data and account assignment category for freight settlement postings in SAP S/4HANA when the freight cost settlement document is integrated.

To assign a service master record and account assignment category to a charge type, execute Transaction SPRO and navigate to **Integration with Other SAP Components** •

Transportation Management • Invoice Integration • Invoicing • Assignment of Transportation Charge Types • Assign Service Master Record and Account Assignment Category.

Figure 9.65 shows different service master records, with account assignment category K (cost center) assigned to different combinations of change category, charge subcategory, and change type. Click **New Entries** to assign a service master record and account assignment category to a new combination of change category, charge subcategory, and change type.

Map Transportation Charge Type to Service Master Record

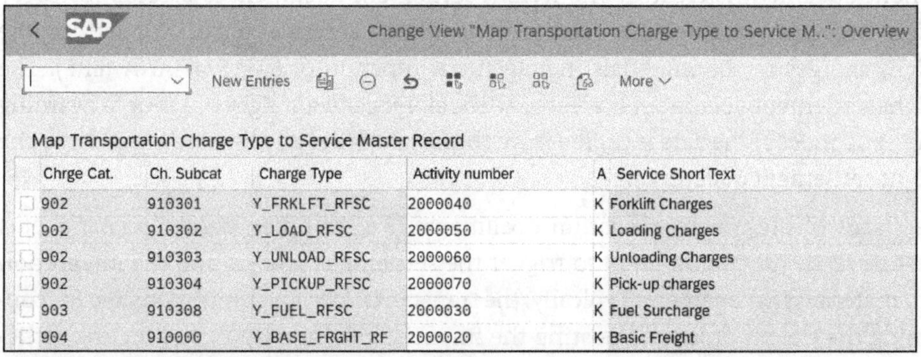

Chrge Cat.	Ch. Subcat	Charge Type	Activity number	A	Service Short Text
902	910301	Y_FRKLFT_RFSC	2000040	K	Forklift Charges
902	910302	Y_LOAD_RFSC	2000050	K	Loading Charges
902	910303	Y_UNLOAD_RFSC	2000060	K	Unloading Charges
902	910304	Y_PICKUP_RFSC	2000070	K	Pick-up charges
903	910308	Y_FUEL_RFSC	2000030	K	Fuel Surcharge
904	910000	Y_BASE_FRGHT_RF	2000020	K	Basic Freight

Figure 9.65 Assign Service Master Record and Account Assignment Category

The following are the key fields of this configuration setup:

- **A (account assignment category)**
  The account assignment category associates purchasing documents such as purchase orders with cost objects or accounts. Cost objects are controlling objects such as cost centers, internal orders, sales orders, maintenance orders, network orders, and so on. During goods receipt and invoice postings with reference to account-assigned purchase orders, the costs related to procurement are posted to the respective financial accounts. By assigning the account assignment category as **K** (cost center), a service purchase order is created with account assignment category **K**, and freight charges are settled to a cost center in SAP S/4HANA.

- **Activity number (service master record)**
  An activity number is an internal number assignment for the service master data. The service master record is created in the external service management function of materials management in SAP S/4HANA. A service purchase order is created with reference to a service master record, followed by a service entry sheet and invoice receipt process. By creating and assigning a service master record in this configuration setup, you ensure the freight charges are settled using the service purchase order with the service master record assigned to the line items of the service purchase order.

Service master records for external service management are created using Transaction AC01 in SAP S/4HANA. Figure 9.66 shows the service master record. The valuation class and material group assigned to the service master record ensure automatic general

ledger account determination and postings during goods receipt and invoice receipt
with reference to the service purchase order.

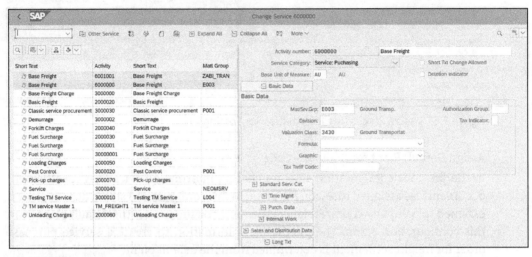

Figure 9.66 Service Master Data for Transportation Management

Assign Purchasing Information for Posting

Mapping of SAP S/4HANA organizational units and other attributes is required for
freight settlement document integration to create a service purchase order in materials management. Service purchase order creation needs organizational units such as a
plant and purchasing group, document type for service purchase order, and material
group. In this configuration setup, a purchase order document type, material group,
plant, and purchasing group are assigned to a freight settlement document type and
freight settlement document category.

Prerequisites for this configuration setup include the following:

- Configure the organization/enterprise structure in SAP S/4HANA.

- Configure the purchase order document type for the creation of a service purchase
  order. NB is the standard purchase order document type.

- Configure the freight settlement document types to assign purchasing information
  in embedded TM.

To assign purchasing information for posting, execute Transaction SPRO and navigate
to **Integration with Other SAP Components • Transportation Management • Invoice
Integration • Invoicing • Mapping of Organizational Units • Assign Purchasing Information for Posting**.

Figure 9.67 shows that purchase order document type **NB**, material group **P001**, plant
MJ00, and purchasing group **01** are assigned to freight settlement document type **001**.
To assign the purchasing information to a new freight settlement document type, click

727

New Entries, specify the freight settlement document type (**FSD Type**), and assign the purchasing information to it.

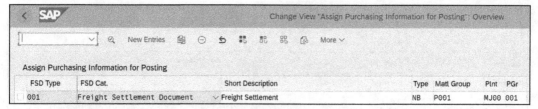

Figure 9.67 Assign Purchasing Information for Posting

Assign Default Plant for Company Code

In the previous configuration step, purchasing information for a freight settlement document was assigned based on the plant. In this configuration step, a default plant is assigned to company codes. Freight settlement documents get the default plant from this configuration setup. The purchasing information to create a service purchase order for freight settlement is determined from the default plant.

To assign a default plant for a company code, execute Transaction SPRO and navigate to **Integration with Other SAP Components • Transportation Management • Invoice Integration • Invoicing • Mapping of Organizational Units • Assign Default Plant for Company Code**.

Figure 9.68 shows that plants **MJ00** and **MJ01** are assigned as the defaults for company codes **MJ00** and **MJ01**. Specify the default plant for every company code in which freight settlement documents are being created.

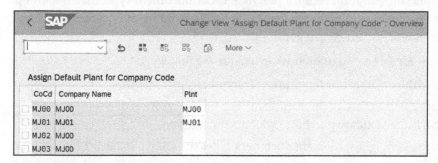

Figure 9.68 Assign Default Plant for Company Code

Assign Transportation Charge Types to Condition Types

To distribute the charges to a purchase order pricing condition, a combination of a charge category, charge subcategory, and charge type is assigned to a purchase condition type.

As a prerequisite for this step, define a pricing condition type in materials management in SAP S/4HANA with the condition category set as freight. See Chapter 10, Section 10.2.3 for details on how to create pricing condition types.

To assign a transportation charge type to a purchase condition type, execute Transaction SPRO and navigate to **Integration with Other SAP Components • Transportation Management • Invoice Integration • Mapping for Cost Distribution • Assign Transportation Charge Types to Condition Types**.

Figure 9.69 shows that purchase condition type **WTF1** (proposed distribution of freight charges) was assigned to different combinations of charge category, charge subcategory, and charge type. Click **New Entries** to assign a new pricing condition type defined in materials management in SAP S/4HANA to a combination of charge category, charge subcategory, and charge type.

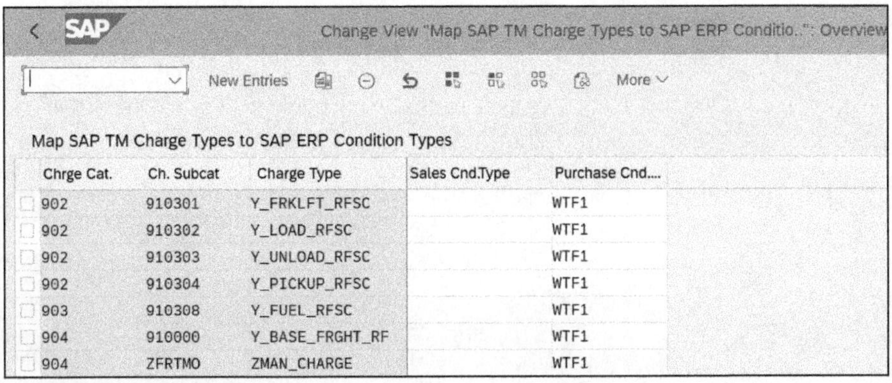

Chrge Cat.	Ch. Subcat	Charge Type	Sales Cnd.Type	Purchase Cnd....
902	910301	Y_FRKLFT_RFSC		WTF1
902	910302	Y_LOAD_RFSC		WTF1
902	910303	Y_UNLOAD_RFSC		WTF1
902	910304	Y_PICKUP_RFSC		WTF1
903	910308	Y_FUEL_RFSC		WTF1
904	910000	Y_BASE_FRGHT_RF		WTF1
904	ZFRTMO	ZMAN_CHARGE		WTF1

Figure 9.69 Assign Transportation Charge Types to Condition Types

9.4 Summary

In this chapter, we introduced the integration of transportation management with materials management and offered best practice examples and processes. In Section 9.1, we explained the integrated transportation management process. In Section 9.2, we explained the master data in transportation management with step-by-step procedures to maintain transportation management-specific master data. Finally, in Section 9.3, we explained the procurement and logistics integration with transportation management to process purchase orders, stock transfer orders, and sales orders in transportation management, with all the integration points and configuration settings that support this process.

In the next chapter, we'll discuss the integration between finance and materials management, ensuring smooth business processes, accurate financial reporting, and a smooth flow of goods through the supply chain. We'll also explain the process and setup of account determination in inventory management, accounts assigned to purchasing documents, and supplier invoice processing concepts with best practice examples, master data requirements, and configuration settings.

Figure 5.29 Assign Transportation Charge Types to Condition Types

5.4 Summary

Chapter 10
Finance

Finance in SAP S/4HANA is responsible for recording financial transactions, providing insights into a company's financial performance, and generating financial statements. It plays a vital role in managing finance processes and ensuring compliance. Finance and materials management functions are tightly integrated to facilitate seamless data flow among procurement, inventory management, and financial accounting processes.

Recording and reporting financial transactions is crucial for organizations. The general ledger is the central component of the financial accounting function, which provides a comprehensive view of the financial status of an organization by collecting financial data from different processes. The following are the different functions of financial accounting in SAP S/4HANA:

- *Accounts payable* is a subprocess in finance that manages all outgoing payments to suppliers after incoming vendor invoice processing and reconciliation.
- *Accounts receivable* is a subprocess in finance that manages all incoming payments from customers after invoice/billing documents are issued to customers.
- *Asset accounting* is a subprocess in finance that manages a company's fixed assets through their lifecycle.
- The *financial reporting* function of finance provides comprehensive tools for generating financial reports to help monitor the financial performance of an organization.

This chapter covers all the integration points between finance and materials management in SAP S/4HSANA. Procurement, inventory management, and logistics invoice verification (LIV) processes within materials management are greatly integrated with finance processes.

Procurement integration with finance starts with the purchase requisition process itself, which is the first step of the procure-to-pay process. The purchase requisition will have an account assignment at the line item level for indirect procurement and a material and its valuation price at the line item level for direct procurement. When the purchase requisition is converted into a purchase order, account assignment details are copied over, and the price determination happens for a direct purchase order using the price determination functionality. Upon goods receipt, quantity and value updates happen in materials management, and the financial transaction is recorded in financial accounting.

Inventory management is responsible for managing inventory and valuation. Valuation and automatic account determination configuration within materials management has financial accounting settings that ensure the flow of inventory valuation data during goods movements into finance for the accurate reporting of on-hand inventory value.

The LIV process within materials management ensures reconciliation/matching of an incoming supplier invoice with a purchase order and goods receipt. Upon posting the supplier invoice, the system generates an accounting document, and the financial transaction is recorded in financial accounting. Payments to the suppliers will be processed in accounts payable based on the posted invoice verification document.

We'll walk through each of these core integration areas in this chapter, including step-by-step instructions for cross-functional configuration. But first, let's dive deep into the master data in financial accounting.

10.1   Master Data in Finance

Master data in finance plays a vital role in collecting financial transactions for reporting purposes. These are critical for financial data accuracy and financial operational efficiency. It is necessary to maintain the financial master data accurately to ensure the integrity of financial processes and reliability of financial information. In the following sections, we'll explore the key master data elements used in SAP S/4HANA financial accounting.

10.1.1   General Ledger Accounts

General ledger accounts are the central components of financial accounting, and they serve as a central repository for all financial transactions. General ledger accounts are used to record financial transactions accurately and to generate financial reports to help monitor the financial performance of the organization. They are integrated with the purchasing, inventory management, and invoicing processes of materials management.

In SAP S/4HANA, the general ledger account is uniquely identified by an account number. There are different types of general ledger accounts:

- **Balance sheet accounts**
  Balance sheet accounts are used to record assets, liabilities, and equity. Balance sheet accounts provide the financial position of an organization. There are three subcategories of balance sheet accounts:

- Asset accounts: Asset accounts represent a company's resources such as equipment, plant, infrastructure, and so on that are owned by the organization. Asset accounts are further subcategorized as follows:
  - Current assets: Current assets include cash accounts (bank balance, short-term investment accounts, etc.), inventory (on-hand inventory of all valuated materials, materials in production, consignment stock at customer location, etc.), prepaid expenses (downpayment made to suppliers), and accounts receivable (amounts due from customers).
  - Fixed assets: These are noncurrent accounts that are long-term resources. These accounts include property, plant, and equipment (PP&E), intangible assets (nonphysical assets such as patents, trademarks, etc.), long-term investments, and other noncurrent assets.
- Liability accounts: Liabilities are obligations that the organization owes to external parties. Liability accounts are further subcategorized as follows:
  - Current liabilities: Current liabilities include accounts payable (amounts owed to suppliers for the procured goods and services), short-term loans, accrued expenses (tax accruals, wages to employees, utility charges), deferred revenues (downpayment/advance payment received from customers), and other short-term liabilities.
  - Noncurrent liabilities: These are long-term liabilities/obligations, typically longer than one year. These accounts include long-term loans and other long-term liabilities.
- Equity accounts: Equity is finance and represents the value of all the company's assets minus the liabilities. Equity accounts are further subcategorized as follows:
  - Share capital: These accounts record the funds raised by issuing the company's stocks.
  - Treasury stock: These accounts record the stock that is repurchased by the company and held in the treasury.
  - Other equity accounts: These include additional paid-in capital.

- **Profit and loss accounts**
  Revenues and expenses are recorded in profit and loss (P&L) accounts. There are three subcategories of P&L accounts:
  - Revenue accounts: Sales revenue and other income are recorded in these accounts. Sales revenue comes from the sale of goods and services.
  - Expense accounts: These accounts record the cost of goods sold (COGS; the total cost of manufacturing of sellable goods including raw materials and other costs) and operating expenses (rent, expenses from indirect procurement, utility charges, and other expenses).
  - Nonoperating expenses: These accounts record the interest owed, loss from sold assets, and so on.

- **Reconciliation accounts**
  These are special types of general ledger accounts that integrate the subledgers such as accounts payable and accounts receivable with the general ledger accounts. They ensure all financial transactions recorded in the subledgers are reflected automatically in the general ledger accounts. These accounts are configured in SAP S/4HANA to maintain the integrity of financial data. There are three subcategories of reconciliation accounts:
  - Customer reconciliation accounts: All financial transactions related to customers are recorded in the customer reconciliation accounts. During the customer invoice posting, SAP S/4HANA updates the accounts receivable subledger with the outstanding amount due from the customer and automatically posts the entry to the customer reconciliation account in the general ledger.
  - Vendor reconciliation accounts: All financial transactions related to suppliers are recorded in the vendor reconciliation accounts. During the supplier invoice posting, SAP S/4HANA updates the accounts payable subledger with the outstanding amount owed to the supplier and automatically posts the entry to the vendor reconciliation account in the general ledger.
  - Asset reconciliation accounts: All financial transactions related to fixed assets are recorded in the customer reconciliation accounts. During the asset acquisition process, SAP S/4HANA updates the asset subledger with the value of the fixed asset and automatically posts the entry to the asset reconciliation account in the general ledger.

Let's discuss the various attributes of general ledger accounts. Execute Transaction FS00 to create, edit, or display a general ledger account centrally.

You'll arrive at an initial screen, where you enter the following fields and click the **Create** icon to create the general ledger account:

- **G/L Account**
  Enter a unique general ledger account number in this field. The allowed number range is defined in the configuration settings. Refer to Section 10.1.6 for details about the number range. After entering the new general ledger account number, you can click the **Create** icon or select **G/L Account • Create** from the menu bar to create the general ledger account.
- **Company Code**
  Enter the company code in which the general ledger account is to be created. The system determines the chart of accounts from the company code.

Figure 10.1 shows the header details and the **Type/Description** tab of general ledger account **63001000** created in company code **MJ00**.

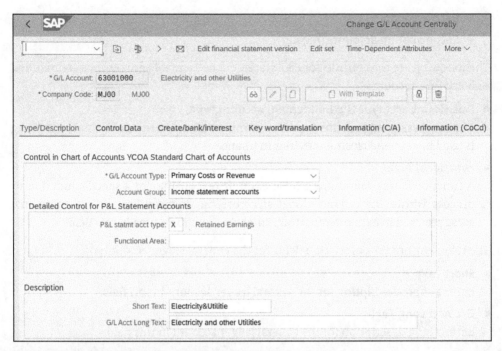

Figure 10.1 General Ledger Account: Type/Description

The following are the controls in the chart of accounts. Chart of accounts **YCOA** was automatically assigned to the general ledger account from the configuration settings. Refer to Section 10.1.6 for details. Let's explore these fields:

- **G/L Account Type**
  The general ledger account type distinguishes different general ledger accounts. Select the general ledger account type from the dropdown; the following are some common selections:
  - **Balance Sheet Account**
  - **Nonoperating Expense or Income**
  - **Primary Costs or Revenue**
  - **Secondary Costs**
  - **Cash Account**

- **Account Group**
  An account group defines the criteria for general ledger account creation. It controls the number range and screen layout for the general ledger account. Select the predefined general ledger account group from the dropdown menu; the following are some common selections:
  - **Fixed asset accounts**
  - **Income statement accounts**

- Cash accounts
- Reconciliation accounts (AR/AP)

The following are the controls for P&L statement accounts. These settings are required only for P&L accounts:

- **P&L statmt acct type (P&L statement account type)**
  You must assign a predefined retained earning account in this field. This field setting is used in year-end closing activities in finance.

- **Functional Area**
  A functional area represents a business unit responsible for a specific function or process within the organizational structure. In financial accounting, functional areas are used in expense reporting. Enter the functional area in this field.

The following are the description fields for the general ledger account:

- **Short Text**
  Enter a short description for the general ledger account in this field.

- **G/L Acct Long Text**
  Enter a long description for the general ledger account in this field.

Let's move on to the **Control Data** tab, as shown in Figure 10.2. The following fields control the company code-related data of the general ledger account and are relevant for financial accounting:

- **Account Currency**
  Assign a currency code to this field to post values into the general ledger account. If the company code currency is assigned, you can post the values to this account in any currency. If a currency other than the company code currency is assigned to this field, you can post values to this general ledger account only in that currency.

- **Balances in Local Crcy Only**
  It is recommended to set this indicator for balance sheet accounts so that the balances are updated in the local currency (company code currency).

- **Tax Category**
  If you want to use this general ledger account for tax-relevant postings, set the appropriate indicator; otherwise, leave this field blank. Typically, expense accounts used in indirect procurement will have this field set to either – or *. The following are the different values that you can set in this field:
  - -: Only input tax allowed
  - *: All tax types allowed
  - +: Only output tax allowed

- **Posting without tax allowed**
  If you want to post taxable and nontaxable expenses into this general ledger account, set this indicator.

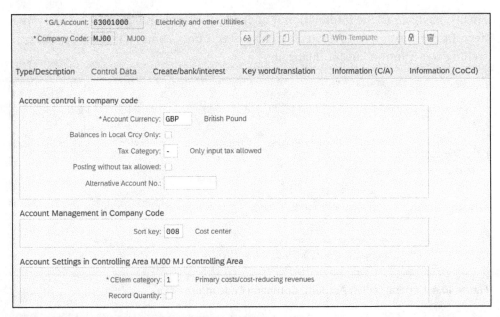

Figure 10.2 General Ledger Account: Control Data

Next, navigate to the **Create/bank/interest** tab, as shown in Figure 10.3. The following fields control the document creation (e.g., invoice document) in a company code:

- **Field status group**
  A field status group controls the screen layout of a document that uses this general ledger account. Field status groups are defined in configuration settings (Section 10.1.6). Select the predefined field status group from the dropdown menu.

- **Post Automatically Only**
  Set this indicator for all general ledger accounts used in automatic account determination settings. If you set this indicator, the system posts values into this general ledger account automatically.

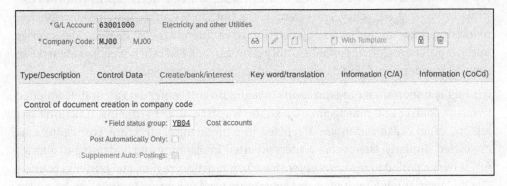

Figure 10.3 General Ledger Account: Create/Bank/Interest

Finally, navigate to the **Information (CoCd)** tab, as shown in Figure 10.4. The system defaults the creation date, created by, chart of accounts, country/region key, and controlling area information automatically.

Figure 10.4 General Ledger Account: Company Code Information

Refer to Section 10.1.6 for the configuration settings for general ledger account creation and maintenance.

10.1.2   Cost Center

A cost center is an organizational unit that is used extensively as master data in financial accounting to track and manage costs incurred by various functions and departments of the organization. A cost center can be used to track the cost incurred by a specific geographical region, a specific activity, or a business area of the organization. Each cost center represents a specific department, function, region, or activity. A cost center will have a manager, a user who's responsible for managing the department; an activity; a region; or a functional area. The main purpose of the cost center is to track, report, and analyze costs incurred and to make strategic decisions to control costs while improving efficiency. For example, the human resources (HR), production, and marketing departments of an organization have their own cost centers. The salaries, indirect purchases (office supplies, utility charges, etc.), and other expenses incurred by each department are tracked and reported using these cost centers.

It is highly important for organizations to maintain cost center data accurately for effective cost control and management. Like the way that all departments, functions, and regions of an organization are organized in a hierarchical structure, cost centers are organized similarly. Hence cost centers are used in planning and budgeting to allocate funds over a period to the cost centers based on function, region, and business reasons. The budget or funds allocated to cost centers are used to procure indirect goods and services for a department/function/activity belonging to the cost center. The cost center tracks spending and supports budget management by not allowing departments and the like not to overspend over a period beyond the budget limit set for the respective cost centers.

In the following sections, we'll explain how to set up cost centers and cost center budgeting.

Create Cost Centers

To create cost centers, navigate SAP Easy Access menu path **Accounting • Controlling • Cost Center Accounting • Master Data • Cost Center • Individual Processing • KS01— Create**. You can arrive at this directly by executing Transaction KS01.

Figure 10.5 shows the initial screen for the creation of a cost center. You can copy from an existing cost center or create it from scratch.

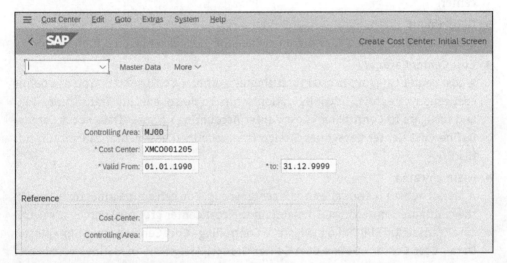

Figure 10.5 Cost Center: Initial Screen

On the initial screen, fill out the following fields and press $\boxed{\text{Enter}}$ to navigate to the **Basic data** view to create a cost center:

- **Controlling Area**
  The controlling area is the highest organizational unit in controlling and is used for internal controlling and cost accounting purposes. A cost center is defined within a controlling area. Enter the controlling area that belongs to the company code in this field.

- **Cost Center**
  Enter a unique 10-character alphanumeric code to define a cost center.

- **Valid From/to**
  These values indicate the validity period of the cost center. Enter the valid from date and valid to date for the cost center.

Figure 10.6 shows the general (header) data and basic data of the cost center. The **Basic data** tab of the cost center has the following fields:

- **Name**
  Enter a short name for the cost center in this field.

- **Description**
  Enter a detailed description of the cost center in this field.

- **User Responsible**
  Enter the user ID of a user stored in SAP S/4HANA in this field. This user is the owner of the cost center.

- **Person Responsible**
  Enter the name of the user responsible in this field. This user is the owner of the cost center.

- **Department**
  Enter the name of the department to which the cost center belongs in this field.

- **Cost Center Category**
  A cost center category is used to distinguish different cost centers. You can define cost center categories in configuration setup. To do so, execute Transaction SPRO and navigate to **Controlling • Cost Center Accounting • Master Data • Cost Center • Define Cost Center Categories**. Assign the predefined cost center category in this field.

- **Hierarchy area**
  A hierarchy area is also called a cost center group. You can define a hierarchy level in the configuration settings. To maintain the cost center group/hierarchy area, execute Transaction SPRO and navigate to **Controlling • Cost Center Accounting • Master Data • Cost Center • Define Cost Center Groups**. Assign the predefined cost center group in this field.

- **Company Code**
  Assign the company code to which the cost center belongs. The company code must be part of the controlling area of the cost center.

- **Business Area**
  Assign the business area to which the cost center belongs.

- **Functional Area**
  Assign the functional area to which the cost center belongs.

- **Currency**
  The currency key is defaulted automatically from the controlling area.

- **Profit Center**
  Assign the associated profit center to this field (not shown; you can scroll down to view this field). You can assign the same profit center to multiple cost centers.

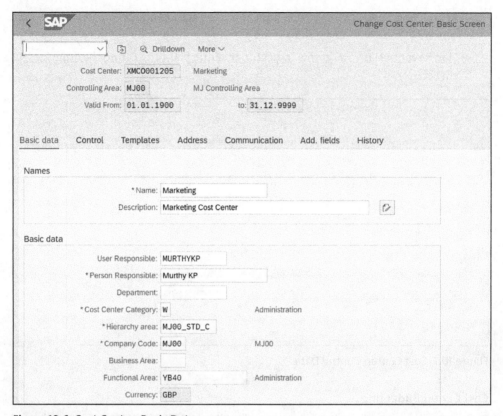

Figure 10.6 Cost Center: Basic Data

Switch to the **Control** tab, as shown in Figure 10.7. The following are the control parameters that you can set in the cost center master data:

- **Record Quantity**
  To record certain overhead costs that depend on units of measure, set this indicator.
- **Lock**
  You can lock the cost center for certain cost postings:
  - **Actual primary costs**: If you set this indicator, no primary costs, such as external expenses (indirect purchases, utilities, etc.), can be posted to this cost center.
  - **Plan primary costs**: If you set this indicator, no primary costs, such as external expenses, can be planned for this cost center.
  - **Act. secondary costs**: If you set this indicator, no secondary costs, such as internally incurred costs (e.g., labor cost from production or maintenance operations), can be posted to this cost center.
  - **Plan Secondary Costs**: If you set this indicator, no secondary costs, such as internally incurred costs, can be planned for this cost center.

- **Actual Revenues**: If you set this indicator, revenues cannot be posted to this cost center.
- **Plan Revenues**: If you set this indicator, revenue planning cannot be done for this cost center.

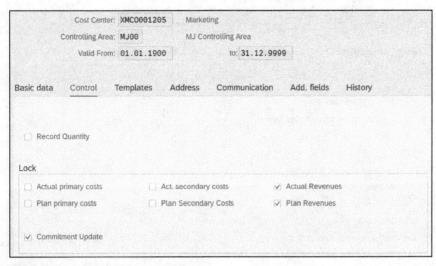

Figure 10.7 Cost Center: Control Data

Cost Center Budgeting

Budget management is an important functionality in SAP S/4HANA, and it tracks and controls the costs incurred by the cost center. During the consumption postings for the cost center, the system checks if sufficient budget is available for an expense/consumption posting. If sufficient budget is available, it allows the posting; otherwise, it throws an appropriate error message and halts the transaction posting.

The budget availability check is integrated with purchasing. During the creation of a purchase requisition and the purchase order itself, a real-time budget availability check is performed. Only if a sufficient budget is available does the system allow the creation of a purchase requisition and purchase order.

You can display and change the budget allocated to the cost center using Transaction KPZ2. Enter the controlling area of the cost center to arrive at the initial screen shown in Figure 10.8.

Enter the following values to start budget planning for a cost center:

- **Profile**
  Enter the planning profile, also called the budgeting profile, defined in the configuration settings (Section 10.1.6) in this field. This profile contains the control parameters for budgeting.

- **Cost Center**
  Enter the cost center for which the budget is to be planned in this field.

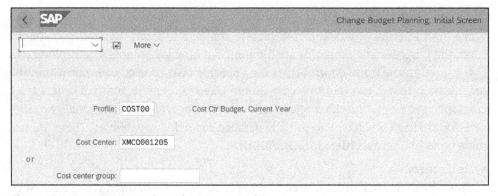

Figure 10.8 Cost Center Budget Planning: Initial Screen

Press Enter from the initial screen or click the **Overview** screen icon at the top of the initial screen to navigate to the budget planning period overview screen. Figure 10.9 shows the budget planning period overview for cost center **XMCO001205** in controlling area **MJ00**. The system displays the posting period automatically based on the posting period variant definition and assignment to a company code in the configuration settings.

Enter the budget or allocate funds for each posting period for fiscal year 2024, as shown in Figure 10.9.

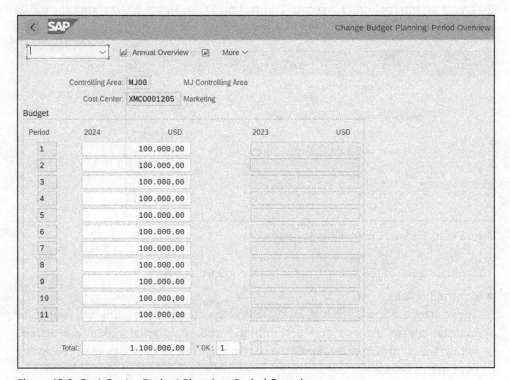

Figure 10.9 Cost Center Budget Planning: Period Overview

10.1.3   Internal Orders

While both internal orders and cost centers are used to track, manage, and control costs, the purposes and scenarios are different. An internal order is a controlling tool that is used to track and control costs for a specific task, project, or event within the organization that is not tied to a cost center. Like cost centers, internal orders track costs incurred from a specific project, task, or event; provide the functionality to allocate and manage budgets; and provide detailed reporting capabilities. There are two main types of internal orders in SAP S/4HANA:

- **Real orders**
  This type of internal order is used to settle collected costs to other cost objects such as cost centers, fixed asset accounts, and so on.

- **Statistical orders**
  This type of internal order is used for statistical purposes only; these orders can't be used to settle costs with other cost objects.

Internal orders in SAP S/4HANA are a highly flexible tool to track, control, and report costs incurred for a specific purpose. We'll walk through creating them and setting up internal order budgeting in the following sections.

Create Internal Orders

To create an internal order, execute Transaction KO01. Figure 10.10 shows the initial screen for the creation of an internal order.

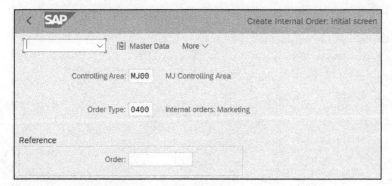

Figure 10.10 Internal Order: Initial Screen

The initial screen of the internal order has the following fields:

- **Controlling Area**
  The controlling area is the highest organizational unit in controlling and is used for internal controlling and cost accounting purposes. Internal orders are defined within the controlling area. Enter the controlling area that belongs to the company code in this field.

- **Order Type**

  The order type distinguishes different types of internal orders and controls the number range of the internal order and how the internal order is costed. Enter the order type defined in the configuration settings in this field.

- **Reference Order**

  If you want to create an internal order with reference to another order, enter the order number in this field. The system copies the field values from the reference internal order.

Press ⌷Enter⌷ from the initial screen or click **Master Data** at the top of the initial screen to navigate to the assignments view to create an internal order. Figure 10.11 shows the general (header) data and assignment details of the internal order. Enter the **Description** for the internal order. Then, the **Assignments** tab of the internal order has the following fields:

- **Company Code**

  Assign the company code to which the internal order belongs. The company code must be part of the controlling area of the internal order.

- **Business Area**

  A business area represents a specific area of responsibility within the organizational structure. Assign the business area to which the internal order belongs.

- **Plant**

  Assign the plant to which the internal order belongs.

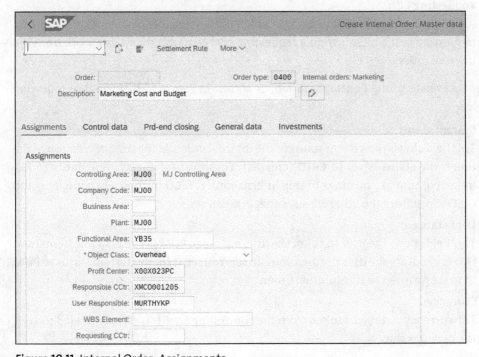

Figure 10.11 Internal Order: Assignments

- **Functional Area**
  A functional area represents a business unit responsible for a specific function or process within the organizational structure. Assign the functional area to which the internal order belongs.

- **Object Class**
  An object class categorizes an internal order according to the business functions for controlling and managing costs. One of the following values can be assigned:
  - **Overhead**
  - **Production**
  - **Investment**
  - **Earnings, Sales**

- **Profit Center**
  Assign a profit center in this field. The system posts costs into the profit center along with the internal order.

- **Responsible CCtr**
  Assign a responsible cost center for the order. This cost center may or may not be the one to which internal order costs are settled.

- **User Responsible**
  Enter the user ID of the user stored in SAP S/4HANA in this field. This user is the owner of the internal order.

- **Requesting CCtr**
  If you don't assign a requesting cost center, the responsible cost center can also be a requesting cost center. Assign a requesting cost center that requests work from the internal order.

Next, navigate to the **Control data** tab, as shown in Figure 10.12. It has the following fields:

- **System status**
  This field shows the system status of the internal order. At the time of creation of the order, the status is set to **CRTD** (created). You must set the system status to **REL** (released) to make postings to this internal order. **TECO** (technically complete) and **CLSD** (closed) are the other statuses that you can set.

- **User status**
  This field shows the user status of the internal order. **PLIM** (write plan line items) and **LKD** (locked) are the user statuses you can set. You must set the system status to **PLIM** to make postings to this internal order.

- **Currency**
  The currency code is defaulted from the controlling area. It is recommended to keep the default value.

- **Statistical order**
  If you set this indicator, the internal order will be used for statistical purposes only. You must assign a value to **Actual posted CCtr** when you set the statistical order indicator.

- **Plan-integrated order**
  If you set this indicator, the internal order will be considered in integrated planning. Plan-integrated orders are used to plan cost elements, and activity inputs are integrated with cost centers and business processes. The system updates the plan allocations directly to the cost centers or business processes so that the plan-integrated internal orders can be settled directly to cost centers or business processes.

- **Revenue postings**
  This indicator is controlled by the order type. You can set this indicator only for relevant order types that allow revenue postings into the internal orders.

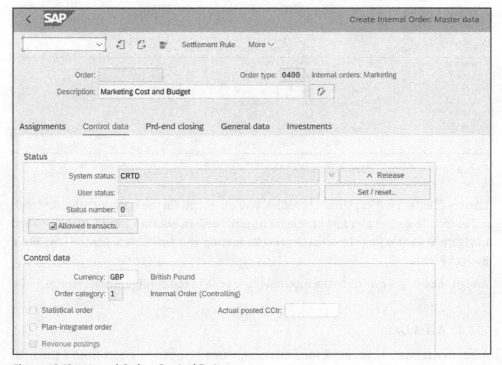

Figure 10.12 Internal Order: Control Data

Internal Order Budgeting

Internal orders are more commonly used in budgeting. Budget management is an important functionality in SAP S/4HANA, and it tracks and controls the costs incurred by the internal order. During the consumption postings against the internal order, the system checks if sufficient budget is available for the expense/consumption posting. If sufficient budget is available, it allows the posting; otherwise, it throws an appropriate error message and halts the transaction posting.

The budget availability check is integrated with purchasing. During the creation of a purchase requisition and the purchase order itself, a real-time budget availability check is performed. Only if sufficient budget is available for the internal order does the system allow the creation of a purchase requisition and purchase order.

You can display and change the budget allocated to the internal order using Transaction KO22. Enter the internal order number and press Enter to arrive at the screen shown in Figure 10.13.

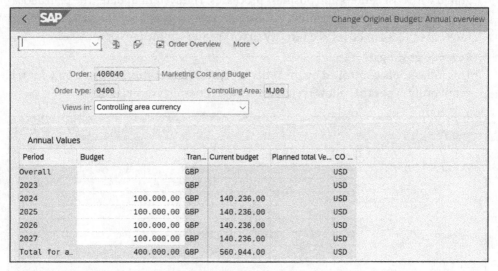

Figure 10.13 Internal Order Budget Planning

Like cost center budget planning profiles, you can maintain the budget profile for internal orders. To define a budget profile for internal orders, execute Transaction SPRO and navigate to **Controlling • Internal Orders • Budgeting and Availability Control • Maintain Budget Profile**.

Assign the budget profile to the internal order type in the configuration settings.

10.1.4   Asset Data

Asset master data holds basic information about a company's assets from a financial accounting perspective. This data is vital for asset procurement, asset management, and asset accounting. Asset master data is used to record the transactions for procuring/constructing fixed assets, sale of fixed assets, and to report asset accounting data. Creation of asset master data is mandatory before purchasing the asset. Asset master data contains general information such as the type of asset, location details of the asset, equipment assignments data, origin data of the asset, and depreciation areas to valuate the asset.

To create asset master data, execute Transaction AS01. Figure 10.14 shows the initial screen for the creation of asset master data, where you fill out the following fields:

- **Asset Class**
  The asset class is used to distinguish different types of assets. Asset classes are defined in configuration settings (Section 10.1.6) that control the account assignment, screen layout, number ranges, and other parameters. Select an asset class from the dropdown menu to assign it to this field.

- **Company Code**
  Assign the company code to which the asset master data belongs.

- **Number of Similar Assets**
  This field controls the number of similar asset master data elements to be created. For example, if there are 10 similar equipment assets acquired by the company, then you can create 10 asset master data records at once by entering 10 in this field.

- **Reference**
  If you want to create a new asset with reference to another asset, enter the **Asset** number, **Subnumber**, and **Company Code** of the existing asset.

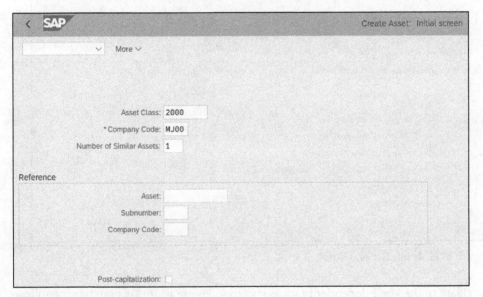

Figure 10.14 Asset: Initial Screen

Press ⌷Enter⌷ from the initial screen to navigate to the **General** tab to create an asset master record. Figure 10.15 shows the header details and the **General** tab of the new asset creation screen. The system assigns an internal number to the new asset in the **Asset** field until the creation process is completed. A new number will be generated upon saving the asset master record based on the number ranges assigned in the asset class configuration. Let's explore the fields in the **General** tab:

- **Description**
  Provide a short description of the asset in this field. Use the **Create Long Text** icon next to the description field to provide detailed information about the asset.

- **Account Determ.**
  The account determination is defaulted from the asset class into this field. This determines the asset reconciliation account from financial accounting. Changes to this field cannot be made directly.

- **Serial number**
  If you have acquired this asset from an external manufacturer, assign the manufacturer's serial number in this field.

- **Inventory Number**
  This is a unique number assigned to every asset in the organization for tracking and tracing purpose. Assign the inventory number of the asset in this field.

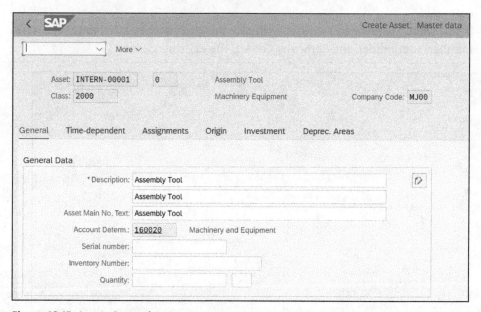

Figure 10.15 Asset: General Data

Navigate to the **Time-dependent** tab, as shown in Figure 10.16. This tab has the following fields:

- **Cost Center**
  Assign the cost center responsible for recording the profit/loss and depreciation value of the asset.

- **Internal Order**
  If you want to post the depreciation value of the asset into an internal order directly, assign the internal order to this field.

- **Plant**
  Assign the plant in which the asset is located to perform plant-specific analysis in asset accounting.

- **Location**
  Enter the location of the asset. Locations are located within the plant and are defined in configuration settings as organizational units. To define locations for a plant, execute Transaction SPRO and navigate to **Enterprise Structure • Definition • Logistics — General • Define Location**.

- **Functional Area**
  Assign a functional area for the asset. A functional area represents a business unit responsible for a specific function or process within the organizational structure. In financial accounting, functional areas are used in expense reporting. See Chapter 2, Section 2.2.5 for more details.

- **WBS Element**
  Assign a work breakdown structure (WBS) element in this field to post the gain/loss and depreciation value of the asset. See Chapter 11, Section 11.2.2 for more details.

- **Profit Center**
  Assign a profit center in this field. The system posts costs into the profit center along with asset data.

- **Segment**
  Assign a segment to which the asset belongs to. A segment is an organizational unit within financial accounting. It represents an area within the organization such as sales, marketing, and so on, or a geography, or a product that generates revenue and incurs expenses. Financial statements can be created against segments.

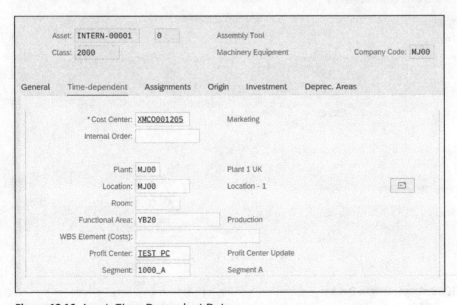

Figure 10.16 Asset: Time-Dependent Data

Next, go to the **Assignments** tab, as shown in Figure 10.17. You can assign the equipment master record to integrate an asset with a piece of equipment. Refer to Chapter 7, Section 7.2.2 for details about equipment master data. Click the **Equipment number** field and then click the small icon that appears at the end of the field to search for an equipment master data record to assign to the asset.

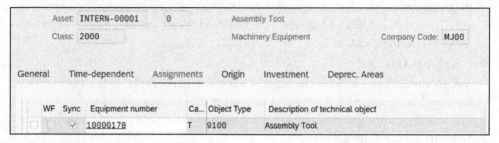

Figure 10.17 Asset: Assignments Data

Finally, navigate to the **Origin** tab, as shown in Figure 10.18. The following are the key fields to be maintained in this tab:

- **Vendor**
  Assign the vendor ID in this field for the vendor from which the asset was procured.

- **Manufacturer**
  Enter the name of the manufacturer of the asset in this field. Set the **Asset purch. new** indicator if the asset was purchased in its new condition; otherwise, set the **Purchased used** indicator.

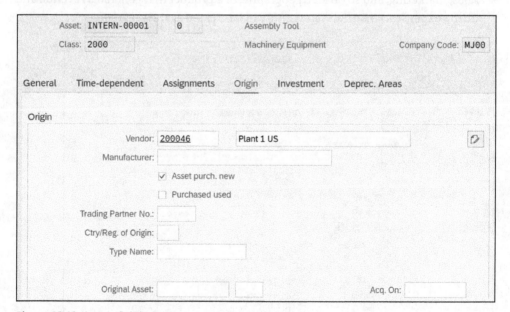

Figure 10.18 Asset: Origin Data

10.1.5   Tax Data

Tax data is a crucial part of sales and procurement. SAP S/4HANA provides tools and procedures to manage tax data such as tax codes, condition records, and tax determination procedures. Tax calculation, tax reporting, and paying accurate taxes on sales and purchases is important for organizations to stay compliant with regulatory requirements. Typically, taxes on purchases (input tax) are paid to the suppliers during supplier invoice processing as part of accounts payable. Taxes on sales are collected from the customers (output tax) with reference to the customer billing document as part of accounts receivable.

Taxes in materials management are based on procurement of goods and services, and tax on purchases is called *input tax*. Tax reconciliation is an important process during supplier invoice reconciliation that compares the supplier-submitted tax in the invoice with the system-calculated tax. It is important for both businesses (the buyer's and the seller's organizations) to accurately calculate the tax and make payments to the local tax authority. There are two important tax categories for taxes on purchases:

- **Indirect taxes**
  Indirect taxes are imposed on purchasing of goods and services from an external vendor. The supplier submits an invoice for the goods and services sold, with tax details added to the total invoice amount. The buying organization pays the invoice amount with the tax amount to the supplier. The supplier pays the taxes to the local tax authority (government). Indirect taxes are common across all regions and countries. Indirect taxes are named differently in different regions/countries of the world. For example, indirect taxes in the US are called sales and use taxes; in European and South American countries, they are called value-added taxes; in Japan, they are called consumption taxes; in India, they are called goods and service taxes; and so on. The tax rates are different in different countries, and they depend on the location, goods, services, import/export scenario, and whether the purchase is for consumption, research and development, or manufacturing. Hence it is important to maintain the tax data accurately in SAP S/4HANA based on regulatory requirements by country.

- **Withholding taxes**
  The buying organization in certain countries deducts a portion of taxes for certain goods and services during the supplier invoice processing and then pays the deducted taxes directly to the local tax authority (government). These taxes are called withholding taxes. Withholding taxes are used in certain countries such as South American countries, Italy, Spain, India, Japan, the US for certain foreign transactions, and so on. Withholding taxes are maintained in the vendor master record in SAP S/4HANA at the company code level. During the invoice processing, the accounts payable personnel apply the withholding taxes to the invoices based on local regulations (localizations). The taxes are deducted based on the withholding

10

tax code and the base amount applied to the invoice. This portion of the withheld taxes is not paid to the supplier during payment processing, but the collected withholding taxes are paid directly to the government at regular intervals during the fiscal year.

Tax authorities audit the taxes paid on purchases, and the organizations maintain accurate tax reports. SAP S/4HANA supports tax data maintenance by country to comply with regulatory requirements. There are two main tax components:

- **Tax procedure**
  The tax procedure supports the determination of tax codes in materials management. Standard tax determination procedures provided by SAP S/4HANA are enough to determine tax codes. Necessary master data needed for tax code determination such as condition records, purchasing info record (PIR), and so on are maintained to determine taxes during purchasing and invoicing processes. The system provides standard tax procedures by country, and this can be further configured in financial accounting global settings to maintain a condition technique, access sequence, condition type, and tax procedure.

- **Tax codes**
  Tax codes are maintained mainly by country and by jurisdiction in certain countries such as the United States, Canada, and so on. The tax code contains the tax rate for tax calculations and tax types such as value-added tax (VAT), goods and services tax (GST), sales tax, and so on based on the localization requirements.

Transaction FTXP is used to create tax codes. Tax codes are created by country, and a jurisdiction code is required if you create a tax code for the United States; for the UK, jurisdiction codes are not required. Figure 10.19 shows the initial screen for maintaining a tax code for the US, and the tax code for the US is maintained by country and jurisdiction code. Figure 10.20 shows the initial screen for maintaining a tax code for country GB (the UK), and the tax code for GB is maintained by country only.

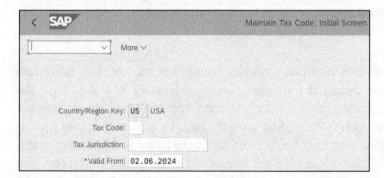

Figure 10.19 Maintain Tax Code for United States

Figure 10.20 Maintain Tax Code for United Kingdom

Let's dive deep into the tax code details by creating one for the UK. Execute Transaction FTXP and enter "GB" (the code for the United Kingdom) in the **Country/Region Key** field to arrive at the screen shown in Figure 10.21. Enter a two-digit key for the tax code and the description for it.

The **Tax Type** field indicates if the tax code belongs to input tax or output tax. Select the **V** indicator for input tax and **A** for output tax.

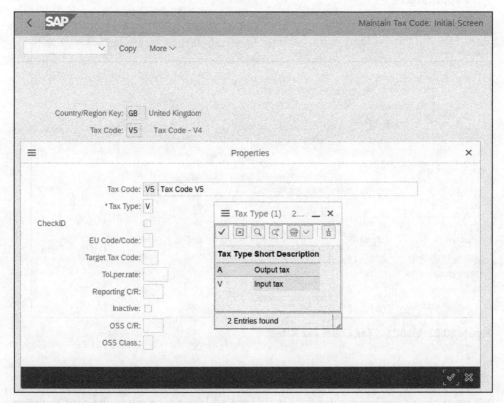

Figure 10.21 Maintain Tax Code: Initial Screen

From the initial screen, click the green checkmark to navigate to the tax rate maintenance screen. Figure 10.22 shows the tax rate maintenance screen for the tax code. Tax procedure **TAXGB** is defaulted from the country key **GB** automatically. The condition type/tax type, account key, sequence of the condition types, level, and from level are copied from tax procedure **TAXGB**. Let's take a closer look at these fields:

- **Cond. Type**
  Condition types are used in tax procedures, pricing procedures, and others. In a tax procedure, a condition type is used to differentiate different tax types to calculate the tax rate for the tax code. Condition types are configured in a specific sequence in the tax procedure. In Transaction FTXP to maintain the tax code, you cannot remove or add tax condition types.

- **Acct Key**
  Account keys are used to determine accounts for financial postings. In a tax procedure, account keys are assigned for condition types. In Transaction FTXP to maintain the tax code, you cannot remove or add tax account keys.

- **Tax Percent. Rate**
  These are input fields for every tax condition type. Enter the tax rate as a percentage as applicable.

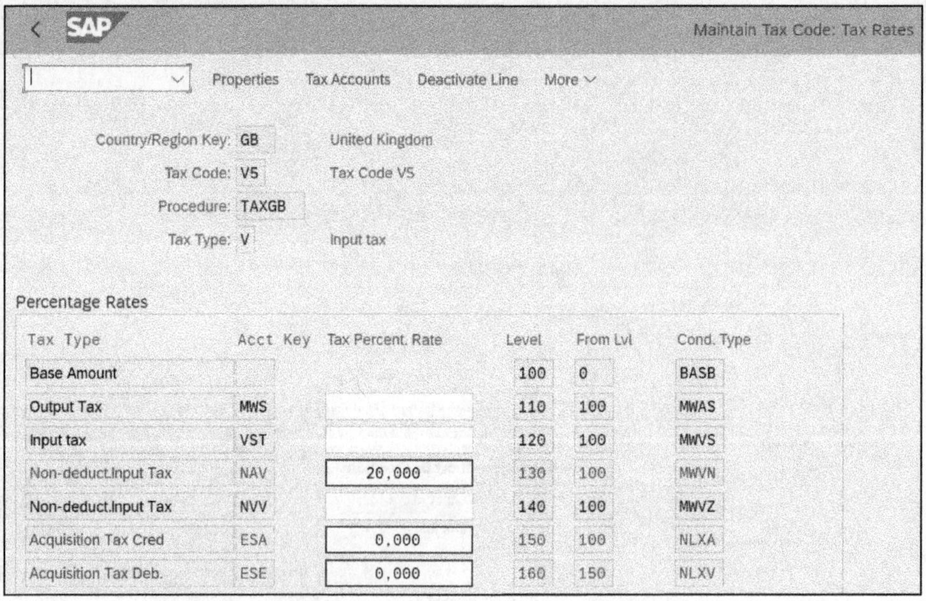

Figure 10.22 Maintain Tax Code: Tax Rates

10.1.6   Configuration Settings for Master Data

We've explored the key master data in finance, so let's now explore the configuration settings that control the master data in SAP S/4HANA. We'll provide step-by-step instructions in the following sections.

Edit Chart of Accounts List

The chart of accounts is the list of accounts listed under different account groups, which an organization uses to record financial transactions in the general ledger. In this configuration step, you can maintain the chart of accounts. It is recommended to maintain the global chart of accounts—that is, the same chart of accounts for all company codes. However, the chart of accounts is defined based on business and legal requirements. There are three types of charts of accounts:

- **Operational chart of accounts**
  These are the lists of accounts used to record financial transactions on a day-to-day basis.

- **Group chart of accounts**
  These are the lists of accounts used by the entire corporate group for consolidation purposes.

- **Country-specific chart of accounts**
  These are the lists of accounts used by a specific country for financial transactions on a day-to-day basis. The requirements for a country-specific chart of accounts are mainly driven by legal/localization requirements.

  To maintain a chart of accounts, execute Transaction SPRO and navigate to **Financial Accounting • General Ledger Accounting • Master Data • G/L Accounts • Preparations • Edit Chart of Accounts List**.

Figure 10.23 displays predefined chart of accounts **YCOA**. Click **New Entries** to maintain a new chart of accounts or copy an existing chart of accounts to create a new one. Define a four-character unique alphanumeric or numeric key to define a chart of accounts and enter a description for it.

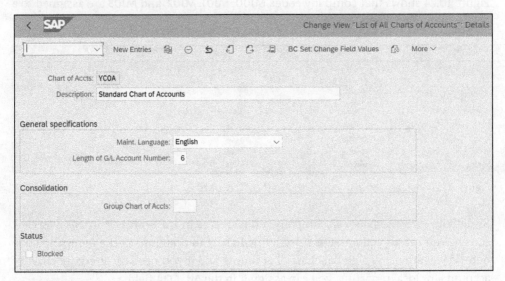

Figure 10.23 Chart of Accounts

The following are the general specifications of the chart of accounts:

- **Maint. Language**
  Assign a maintenance language for the chart of accounts in this field. All accounts created with this chart of accounts will be displayed and maintained in this language.

- **Length of G/L Account Number**
  Specify the length of the general ledger account numbers in this field. You can specify up to 10 digits in length because the standard field length of a general ledger account number is 10 digits in SAP S/4HANA.

Under the **Consolidation** section, you can enter the group chart of accounts if you want to maintain a group account number in the general ledger account as additional information.

Set the **Blocked** indicator if you do not want to create any general ledger account using this chart of accounts. You can create general ledger accounts under a chart of accounts only if it is in released status.

Assign Company Code to Chart of Accounts

General ledger accounts are created at the company code level. To create a list of general ledger accounts using the defined chart of accounts, the company code must be assigned to the chart of accounts in SAP S/4HANA.

To assign company codes to chart of accounts, execute Transaction SPRO and navigate to **Financial Accounting** • **General Ledger Accounting** • **Master Data** • **G/L Accounts** • **Preparations** • **Assign Company Code to Chart of Accounts**.

Figure 10.24 shows that company codes **MJ00**, **MJ01**, **MJ02**, and **MJ03** are assigned to a single chart of accounts, **YCOA**.

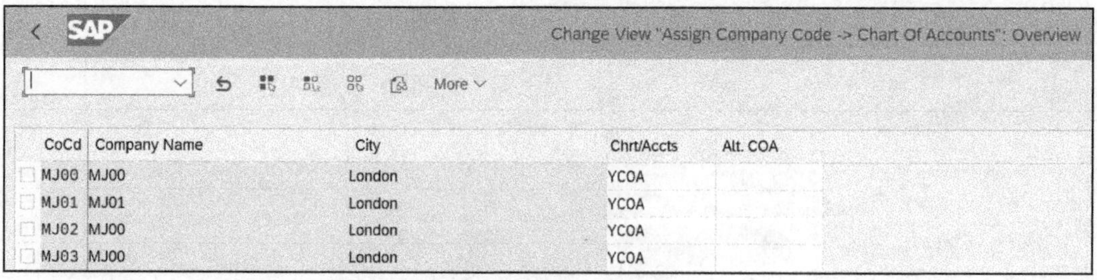

Figure 10.24 Assign Company Code to Chart of Accounts

By default, the system lists all company codes created in SAP S/4HANA in this overview screen. You can directly assign a predefined chart of accounts to the company code here. If you are using a global chart of accounts and if you need an alternate chart of accounts for local reporting, you can assign it in the **Alt. COA** field.

Define Account Group

For better management of accounts, it is highly important to group similar accounts together using account groups. The structure of different general ledger account types is defined using account groups.

To define account groups, execute Transaction SPRO and navigate to **Financial Accounting • General Ledger Accounting • Master Data • G/L Accounts • Preparations • Define Account Groups**.

Figure 10.25 shows different account groups and the structure of the accounts defined for chart of accounts **YCOA**.

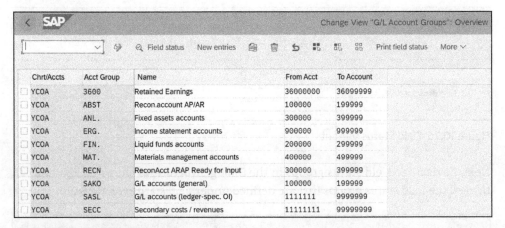

Chrt/Accts	Acct Group	Name	From Acct	To Account
YCOA	3600	Retained Earnings	36000000	36099999
YCOA	ABST	Recon.account AP/AR	100000	199999
YCOA	ANL.	Fixed assets accounts	300000	399999
YCOA	ERG.	Income statement accounts	900000	999999
YCOA	FIN.	Liquid funds accounts	200000	299999
YCOA	MAT.	Materials management accounts	400000	499999
YCOA	RECN	ReconAcct ARAP Ready for Input	300000	399999
YCOA	SAKO	G/L accounts (general)	100000	199999
YCOA	SASL	G/L accounts (ledger-spec. OI)	1111111	9999999
YCOA	SECC	Secondary costs / revenues	11111111	99999999

Figure 10.25 Account Groups

Click **New Entries** to define a new account group for the chart of accounts or copy an existing account group to create a new one. The fields are as follows:

- **Acct Group**
  Define a four-character alphanumeric value in this field to define an account group for the chart of accounts.

- **From Acct**
  To define the structure of the general ledger account for the account group, enter a lower limit for the general ledger account number interval.

- **To Account**
  To define the structure of the general ledger account for the account group, enter the upper limit of the general ledger account number interval.

Define Field Status Variants

Field status variants are used to group the field status groups together. Field status groups are assigned to the general ledger accounts to control the screen layout and define the status of the fields, such as optional entry, required entry, and suppressed. The field status groups that are defined under field status variants control which fields of a document (invoice, accounting document, etc.) are optional for input, mandatory for input required, or suppressed.

To define field status variants, execute Transaction SPRO and navigate to **Financial Accounting • Financial Accounting Global Settings • Ledgers • Fields • Define Field Status Variants**.

Figure 10.26 shows field status variants **0010**, **0123**, and **1000**. Click the **New Entries** icon or press $\boxed{\text{F5}}$ on your keyboard to define new field status variants. It is recommended to copy an existing field status variant and create a new one. Fill in the **Field Status Name** field to name the field status variant.

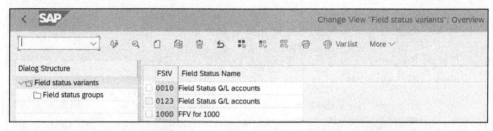

Figure 10.26 Field Status Variants

Next, navigate to **Field status groups** in the **Dialog Structure**, as shown in Figure 10.27, to view the field status groups that are defined under field status variant **0010**.

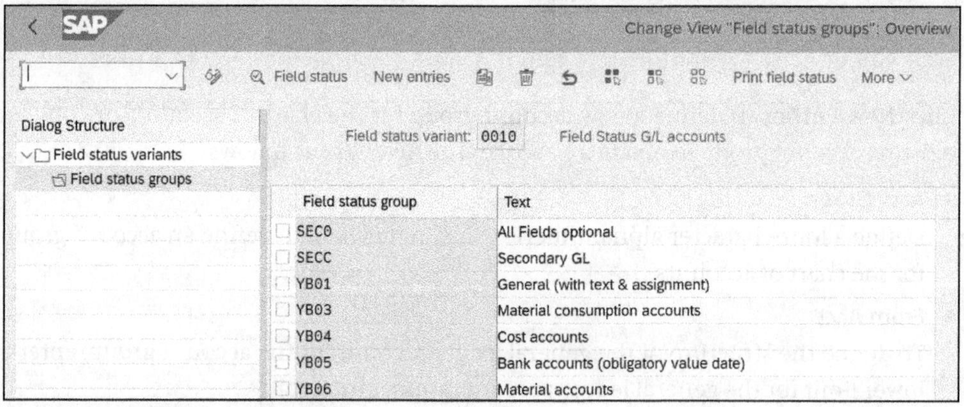

Figure 10.27 Field Status Groups

Click **New Entries** to define a new field status group. It is recommended to copy an existing field status group and create a new one. Enter a short text for the field status group.

To define the status of each field or to define the layout of a document, click the **Field status** button from the field status group screen or press $\boxed{\text{Ctrl}}$+$\boxed{\text{F3}}$. Figure 10.28 displays all subgroups under field status variant **0010** and field status group **YB04**. Double-click each subgroup to define the screen layout or the field statuses.

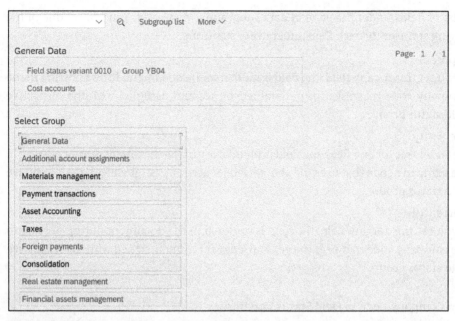

Figure 10.28 Subgroups under Field Status Group

Figure 10.29 shows the maintenance view of the field status group for the general data subgroup.

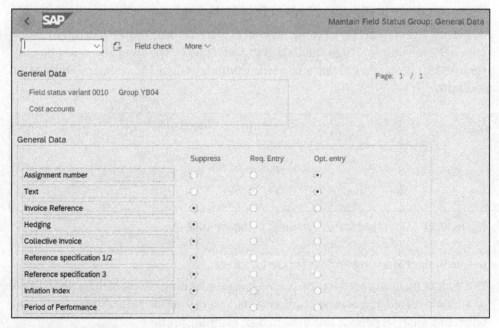

Figure 10.29 Maintain Field Status Group

The list of fields under the general data subgroup is displayed. You can set one of the following statuses for each field under every subgroup:

- **Suppress**
  If you set this for any field, the field is hidden in the document during posting in the company code that uses the general ledger account assigned with the respective field status group.

- **Req. Entry**
  If you set this for any field, the field is mandatory in the document during posting in the company code that uses the general ledger account assigned with the respective field status group.

- **Opt. entry**
  If you set this for any field, the field is optional in the document during posting in the company code that uses the general ledger account assigned with the respective field status group.

Assign Company Code to Field Status Variants

To apply the screen layout defined in the field status groups, you must assign a company code to a field status variant.

To assign a company code to a field status variant, execute Transaction SPRO and navigate to **Financial Accounting • General Ledger Accounting • Master Data • G/L Accounts • Preparations • Assign Company Code to Field Status Variants**.

By default, the system lists all company codes created in SAP S/4HANA in this overview screen. Here you can assign the field status variants to the company codes directly. Figure 10.30 shows that field status variant **0010** is assigned to company codes **MJ00** and **MJ01**.

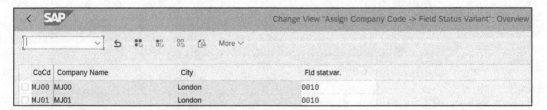

Figure 10.30 Assign Field Status Variant to Company Code

Define Budget Planning Profiles for Cost Centers

The budget planning profile controls the budget planning of cost centers periodically or annually based on business requirements. You can define past, current, and future budgeting analysis periods using the budget planning profiles.

To define budget planning profiles for cost centers, execute Transaction SPRO and navigate to **Controlling • Cost Center Accounting • Budget Management • Define Budget Planning Profiles**.

Figure 10.31 shows the initial screen for maintaining the cost center budget planning profiles. Click **New Entries** to define a new budget planning profile. It is recommended to copy an existing budget planning profile and create a new one. Enter a key for the new budget planning profile (**Profile**) and its description (**Text**).

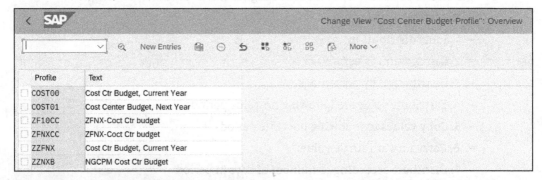

Figure 10.31 Cost Center Budget Planning Profile: Initial Screen

Select a cost center budget profile and click the **Details** icon at the top or press [Ctrl]+[Shift]+[F2] to arrive at the details screen. Figure 10.32 shows the details screen of the budget planning profile settings. The **Time Frame** settings are as follows:

- **Past**
  This field controls if you can plan the budget for the cost centers for past years from the start year. Leave this field blank if you do not want to plan the budget for the past years from the start year.

- **Future**
  This field controls if you can plan the budget for the cost centers for the future years from the start year. If you enter "3" in this field, you can plan the cost center budget for three years into the future from the start year.

- **Start**
  This field controls the start year for budget planning of cost centers. If you enter "2" in this field, the budget planning will start after two years from the current fiscal year.

- **Annual Values**
  If you set this indicator, the system allows you to plan the cost center budget annually.

- **Period Values**

  If you set this indicator, the system allows you to plan the cost center budget periodically. The annual budget will be distributed to periods of the fiscal year. The **Distribution key** field value will control how the annual budget value can be distributed to periods.

- **Distribution Key**

  Using a distribution key, you can control how the budget planning values are distributed to periods. One of the following standard values can be assigned:

  - **0**: Manual distribution

  - **1**: Equal distribution

  - **2**: Distribution as before

  - **3**: Distribution by percentage

  - **4**: Distribute values to following no-value periods

  - **5**: Copy values to following no-value periods

  - **6**: Carry forward single value

  - **7**: Distribute according to number of days in period

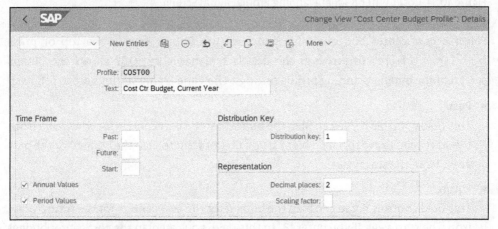

Figure 10.32 Cost Center Budget Planning Profile: Details

Define Internal Order Types

Order types contain default values and various control parameters. An order type is mandatory to create an internal order in SAP S/4HANA.

To define internal order types, execute Transaction SPRO and navigate to **Controlling • Internal Orders • Order Master Data • Define Order Types**.

Figure 10.33 shows the initial screen for defining internal order types. Click **New Entries** or press F5 on your keyboard to define a new order type based on business requirements. Enter a key for the new order type and its description.

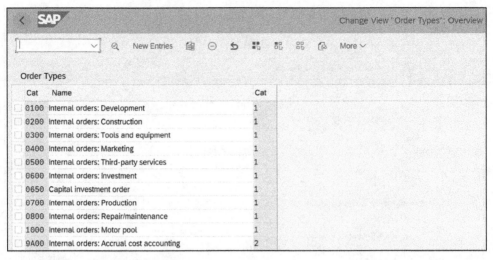

Figure 10.33 Internal Order Type: Overview

Select an order type and click the **Details** icon at the top or press `Ctrl`+`Shift`+`F2` to navigate to the details screen. Figure 10.34 shows the details of the internal order type. Let's dive deep into the field-level settings, starting with the header details:

- **Order category**
  The order category distinguishes different types of orders in SAP S/4HANA, such as maintenance orders, production orders, internal orders, process orders, network orders, and so on. Each order category controls the structure and functionality of the order. For internal order types, the system automatically defaults the internal order (controlling) as the order category.

- **Number range interval**
  Assign a predefined number range interval to this field.

The **General parameters** section settings for the internal order type are as follows:

- **Settlement prof.**
  A settlement profile determines how and where the accumulated costs in the internal order are settled. Select a predefined settlement profile from the dropdown menu.

- **Planning Profile**
  A planning profile determines how the cost planning is performed for the internal order. Select a predefined planning profile from the dropdown menu.

- **Budget Profile**
  A budget profile determines how the budget planning of internal orders is performed periodically or annually based on the business requirements. Select a predefined budget profile from the dropdown menu.

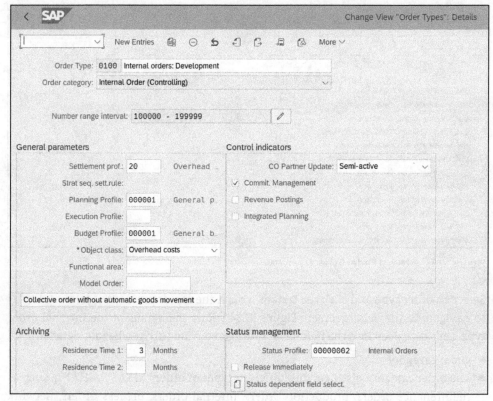

Figure 10.34 Internal Order Type: Details

- **Object class**
  An object class is used to categorize internal orders by their business function to control the cost flow within controlling. Assign one of the following object classes in this field:
  - Overhead
  - Production
  - Investment
  - Profitability analysis and sales
- **Model Order**
  A model order references an internal order number. If you assign an internal order to this field, it acts as a template to create new internal orders.

Define Asset Classes

An asset class is a vital component of asset accounting that defines the rules, default values, and controlling parameters to manage fixed assets. An asset class is mandatory to create an asset master record in SAP S/4HANA.

To define asset classes, execute Transaction SPRO and navigate to **Financial Accounting • Organizational Structures • Asset Classes • Define Asset Classes**.

Figure 10.35 shows the details screen for asset class **2000** (**Machinery and Equipment**). Click **New Entries** or press F5 to define a new asset class based on business requirements.

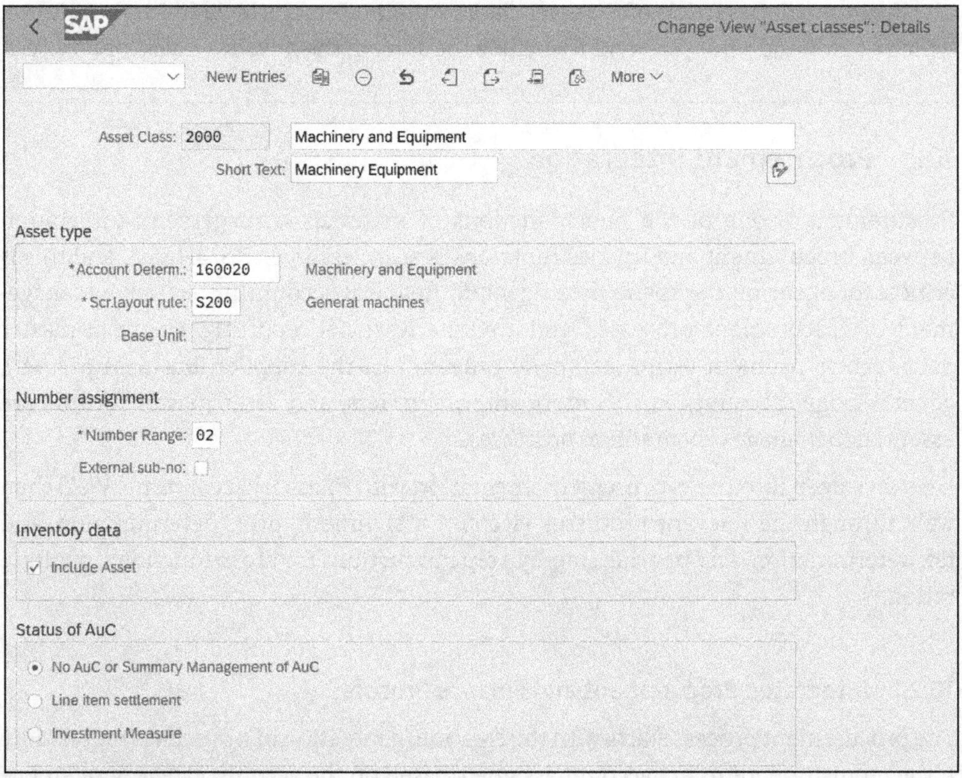

Figure 10.35 Asset Class Details

The **Asset type** section of the asset class settings has the following fields:

- **Account Determ.**
  Account determination is the bridge between the asset master record and the general ledger. It determines the asset reconciliation account from financial accounting. Assign a predefined account determination key from the dropdown menu.

- **Scr.layout rule**
  The screen layout rule controls the mandatory, optional, and suppressed fields, plus other details of the asset master record. Assign a predefined screen layout rule from the dropdown menu.

The **Number assignment** section of the asset class settings has the following fields:

- **Number Range**
  Assign the number range for the asset master record. The number assignment is determined from this number range.

- **External sub-no**

  Set this indicator if you want the external subnumber assignment for the asset. The external subasset number is used to identify a part of the fixed asset.

The **Inventory data** section of the asset class settings offers the **Include Asset** indicator, which you should set if you want to include the asset in the physical inventory process.

10.2   Procurement Integration

Procurement is one of the main functions of materials management. Integration between procurement and finance functions is seamless in SAP S/4HANA, and this is critical for ensuring the spend data flow into financial accounting, effective management of procurement processes, and accurate financial reporting. Financial master data such as payment terms, company code data of the supplier (business partner), general ledger accounts, cost centers, internal orders, and asset master records are essential to manage procurement processes.

Let's dive deep into the procurement integration with financial accounting. We'll then walk through the relevant processes (account assignment, price determination, and tax determination) and provide step-by-step instructions for cross-functional configuration.

10.2.1   Integrated Procurement and Finance Process

The procurement process starts with the creation/generation of a purchase requisition. If the requisition items are account assigned, based on the account category reference maintained in the requisition item, the financial master data (such as general ledger account, cost center, internal order, and asset master record) are assigned in the **Account Assignment** tab at the item level. Account assignment data from the purchase requisition is copied into the purchase order when the purchase requisition is converted to a purchase order. Price determination and tax determination happens during the creation of a purchase order. During the goods receipt, the system generates material documents and accounting documents automatically. The material documents represent the inventory increases and accounting documents represent the accurate recording of the financial postings during goods receipt.

The next step of the process is to match the incoming invoice (supplier-submitted invoice) with the purchase order and goods receipt: this is called three-way matching and invoice reconciliation. In certain scenarios, a two-way match will be performed where an incoming invoice is matched with a purchase order only, followed by the invoice approval process. During invoice reconciliation, taxes are also reconciled, and supplier-submitted tax is compared with the system-calculated tax amount. The final step of the process is to process the payment against invoices that are free for payment and without any matching exceptions/variances.

Table 10.1 shows all the key processes impacted in both the finance and materials management functions.

Area	Processes Impacted	Description
Finance	Maintain financial master data	This process supports creation and maintenance of financial master data such as general ledger account and cost center, internal order, asset master record, and so on.
	Manage withholding tax determination	This process supports the maintenance of withholding tax data, managing withholding taxes on invoices, and managing withholding tax payments.
	Manage indirect tax determination—accounts payables	This process supports the maintenance of tax codes, tax condition records, tax master data, and tax exemption certificates, and managing indirect tax on purchase order and invoice.
	Process vendor payment (manual/automatic)	This process supports the processing of manual and automatic payments to suppliers to clear vendor open items and issue manual checks or post electronic payments to suppliers.
	Enable accounts payable reporting	This process supports the display of various financial reports such as financial data of suppliers (business partners), goods receipt or invoice receipt clearing accounts, payment lists, list of vendor line items, payment run reports, and so on.
Materials management	Maintain purchasing master data	This process supports creation and maintenance of purchasing master data such as source lists and PIRs.
	Create and maintain purchase requisition	This process supports the creation and maintenance of purchase requisitions.
	Create and maintain purchase order	This process supports the creation and maintenance of a purchase order.
	Post goods receipt	This process supports the processing of a goods receipt for the purchase order.
	Receive and process incoming invoice	This process supports the processing of an incoming invoice from the supplier.

Table 10.1 Process Impact: Procure-to-Pay Process

Figure 10.36 shows the process flow of the end-to-end procurement process with all integration points with finance. Let's take a closer look at the process steps:

1. **Creation of purchase requisition**

   Creation of a purchase requisition in SAP S/4HANA can be either manual or automatic. To create the purchase requisition manually, execute Transaction ME51N. Purchase requisitions are automatically generated from a sales order (third-party process) to procure trading goods from a third-party vendor, from a production order, from a maintenance order, from Project System, and from MRP. Purchase requisitions may contain account assignment data at the line item level based on the account category reference assigned to the purchase requisition item. Hence, financial master data is essential to create purchase requisitions with account assignments. Account-assigned requisitions are created for direct consumption to a cost center, internal order, or other order.

2. **Creation of purchase order**

   A purchase requisition is converted to a purchase order manually by the purchaser using Transaction ME21N or automatically in SAP S/4HANA. The line item details are defaulted into the purchase order, including the account assignment, source of supply, unit cost, and delivery address. During the creation of a purchase order, price determination and tax code determination happens, and the pricing information and tax details are updated for every purchase order line item. The purchase order will then be transmitted to the supplier based on the output method: email, electronic data interchange (EDI), or electronic transmission to the supplier on SAP Business Network. Output determination configuration in SAP S/4HANA drives the purchase order output mechanism. Depending on the confirmation control key, order confirmation from the supplier may be required, and the confirmation control key is defaulted from the PIR.

3. **Post goods receipt purchase order**

   The supplier ships the requested materials to the buyer, and once the shipment from the supplier arrives, goods receipt is posted manually in SAP S/4HANA with reference to the purchase order or inbound delivery depending on the confirmation control key assigned to the purchase order. Transaction MIGO and movement type 101 are used to post goods receipt for a purchase order. Upon goods receipt, a material document and an accounting document are generated automatically. The material document represents the inventory increases, and the accounting document represents the accurate recording of the financial postings during goods receipt. The accounting document consists of the following:

   - One or more debit line items for an inventory account if the purchase order line is not account assigned, or an expense account if the purchase order line is account assigned
   - A credit line item for a goods receipt/invoice receipt clearing account

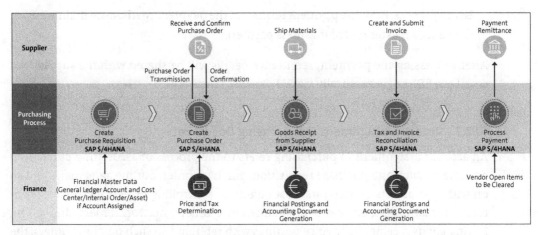

Figure 10.36 Procurement and Finance Integration

4. **Process and reconcile supplier invoice**

   The supplier sends an invoice to the buying organization against the fulfilled purchase order via email (paper invoice), via EDI, or electronically via SAP Business Network. The supplier invoice document will be posted by accounts payable manually in SAP S/4HANA using Transaction MIRO for a paper invoice. Automatic posting and reconciliation of the invoice happens if an electronic invoice is received from the supplier either via EDI or from SAP Business Network. Invoice reconciliation is a process of matching the purchase order, goods receipt, and supplier invoice to verify any discrepancies with the purchase order item price and purchase order item quantity and received quantity. Taxes will be reconciled against the vendor-charged tax and the SAP S/4HANA internal tax engine or from the external tax engine calculated tax. Once the invoice is fully reconciled, it will be free for payment processing in the system. Upon invoice posting, an accounting document is generated automatically, and it represents the accurate recording of the financial postings during invoice reconciliation. The accounting document consists of the following:

   – A credit line for the supplier account

   – A debit line item for the goods receipt/invoice receipt clearing account

5. **Process vendor payment**

   Posting of a supplier invoice generates open vendor line items to be processed for payments on the finance side. The accounts payable team either schedules the payment run to clear vendor open items or processes the payments manually. Transaction F110 is used to process vendor payments in SAP S/4HANA. Payment terms and the payment method are crucial data for payment processing. Payment terms control the due date of vendor payments and any potential discounts. The payment method controls if the payment to the vendor is made electronically or manually by

10

issuing checks. Both the payment terms and the payment method are maintained in the vendor master record (business partner).

After processing the payment, remittance details will be shared with the supplier via email, via EDI, or electronically via SAP Business Network.

10.2.2   Account Assignment in Purchasing Documents

An account assignment in purchasing refers to the process of associating purchasing documents such as a purchase requisition, purchase order, outline agreement, and so on with cost objects or accounts. Cost objects or controlling objects include cost centers, internal orders, sales orders, maintenance orders, network orders, and so on. During goods receipt and invoice postings with reference to such purchase orders, the costs related to procurement are posted to the respective financial accounts. These financial postings are tracked, analyzed, and reported. Account assignment to a purchasing document is done manually during the creation of a purchasing document. This can happen automatically when the purchase requisitions are generated from the production order, sales order, maintenance order, or Project System. Assigning proper accounting details in the purchasing documents helps organizations to ensure accurate cost allocation, effective financial planning, and tracking, and enables organizations to maintain financial control.

Account assignment categories determine how costs are allocated to the appropriate financial accounts. The following are the most common account assignment categories in the standard system used in procurement:

- **Cost center**
  Cost centers typically represent departments of an organization unit (company code). The cost center is used to allocate expenses from a specific department to financial accounts.

- **Internal order**
  Internal orders are used to track and control costs for a specific task, project, or event within the organization that is not tied to a cost center.

- **Project**
  A project is used to track and control costs for a specific project within the organization.

- **Asset**
  An asset is used to record the transactions from procuring/constructing fixed assets and to report asset accounting data.

Account assignment data drives the purchasing document approval process, especially for indirect procurement. Figure 10.37 shows a purchase order with account assignment

K (cost center) assigned to line item number **10**. The system automatically displays the **Account Assignment** tab in the item details when the purchase order item is account assigned. Because the account assignment is to a cost center, the system displays the **Cost Ctr** (cost center) field for input, and it is a required field. A general ledger account (expense account) is required to complete the purchase order creation.

During the goods receipt and invoice posting, the system uses the account assignment data to post the costs into financial accounting and to generate an accounting document.

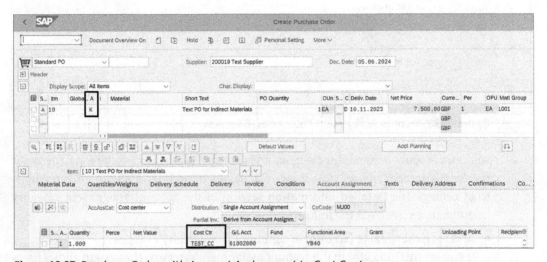

Figure 10.37 Purchase Order with Account Assignment to Cost Center

Configuration settings include setting up account assignment categories, defining account assignment category and item category combinations, and viewing default value settings for asset classes. Let's explore the configuration settings that control this process.

Maintain Account Assignment Categories

In this configuration step, you can maintain the account assignment categories for purchasing documents.

To maintain account assignment categories, execute Transaction SPRO and navigate to **Materials Management • Purchasing • Account Assignment • Maintain Account Assignment Categories**.

Figure 10.38 displays the overview of the account assignment categories defined in the standard system. Click the **New Entries** button to create a new one. You can copy an existing one to create a new one from it. Enter a single-character key and a description to create a new account assignment category.

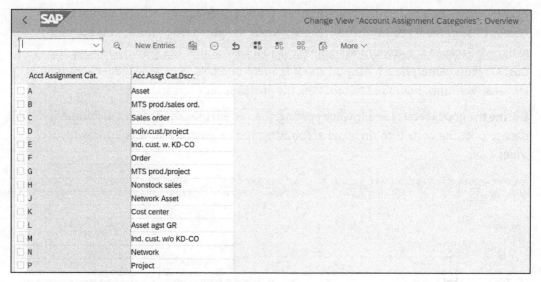

Figure 10.38 Account Assignment Categories: Overview

Select an account assignment category and click the **Details** icon at the top to arrive at the details screen. Figure 10.39 shows the details of account assignment category **K** (cost center). The **Detailed information** section of the account assignment category maintenance screen has the following fields:

- **Acct.assg.changeable (account assignment changeable)**
  If you set this indicator in the account assignment category, the system allows users to change the account assignment data in the follow-on documents, such as goods receipt and invoice receipt.

- **AA Changeable at IR (account assignment changeable at invoice receipt)**
  If you set this indicator in the account assignment category, the system allows users to change the account assignment data only in the invoice receipt.

- **Derive acct. assgt. (derive account assignment)**
  If you set this indicator in the account assignment category, the system derives the preliminary account from the general ledger account.

- **Del.CstsSep. (delivery costs separate)**
  If you set this indicator in the account assignment category, delivery costs for purchase orders are posted separately to another account.

- **Consumption posting**
  The consumption posting keys specify the categories of consumption postings that happen during the goods receipt with reference to purchase order items with the pertinent account assignment category. Table 10.2 shows the consumption posting keys defined in the standard system. For account assignment categories for cost center, order, and network, assign posting indicator **V** (consumption).

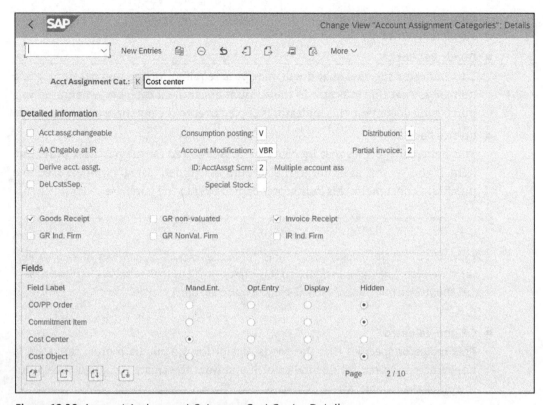

Figure 10.39 Account Assignment Category: Cost Center Details

Consumption Posting	Description
A	Asset
V	Consumption
E	Accounting via sales order
U	Unknown
P	Accounting via project

Table 10.2 Consumption Posting Indicators

- **Account Modification**
  An account modifier helps with account determination during automatic goods movement postings. Assign the account modifier based on the automatic account determination settings. Refer to Section 10.3.2 for more details.

- **Special Stock**
  If you assign a special stock indicator in this field, the stock during the goods receipt

for a purchase order with this account assignment category is posted to the special stock in inventory.

- **Goods Receipt**
  This indicator specifies that goods receipt is required for the purchase order line item. If you set this indicator in the account assignment category, when used in a purchasing document, this indicator is copied over to the purchase order line item.

- **Invoice Receipt**
  This indicator specifies that invoice receipt is required for the purchase order line item. If you set this indicator in the account assignment category, when used in a purchasing document, this indicator is copied over to the purchase order line item.

> **Note**
>
> If you do not set the invoice receipt indicator in the account assignment category, the purchase order line items assigned to this category will have the free of charge indicator set automatically, and there will be no invoice posting possible.

- **GR non-valuated**
  This indicator specifies that the goods receipt for the purchase order line item is nonvaluated. Do not set this indicator if you want to valuate the goods receipt for the purchase order item.

The **Fields** section of the account assignment category maintenance screen has the field status controls. You can set account assignment data fields such as cost center, internal order, asset, functional area, and so on as optional, suppressed, or required fields.

Define Combination of Item Categories/Account Assignment Categories

In this configuration step, you can define the account assignment category and item category combinations that are allowed in purchasing documents.

To define the account assignment category and item category combination, execute Transaction SPRO and navigate to **Materials Management • Purchasing • Account Assignment • Define Combination of Item Categories/Account Assignment Categories**.

Click **New Entries** to define a new combination of allowed item categories and account assignment categories in purchasing documents. Then, specify an item category in the **ItCat** field and an account assignment category in the **AAC** field, and press ⏎ Enter . Figure 10.40 shows the definition of allowed account assignment category and item category combinations. For example, for item category **B** (limits), the only allowed account assignment categories are **K** (cost center) and **U** (unknown) in the purchasing documents.

ItCat.	AAC	Text	Acc.Assgt Cat.Dscr.
	K	Standard	Cost center
	L	Standard	Asset agst GR
	M	Standard	Ind. cust. w/o KD-CO
	N	Standard	Network
	P	Standard	Project
	Q	Standard	Proj. make-to-order
	T	Standard	All new aux.acc.ass.
	U	Standard	Unknown
	X	Standard	All aux.acct.assgts.
B	K	Limit	Cost center
B	U	Limit	Unknown
L	F	Subcontracting	Order
L	P	Subcontracting	Project
L	Q	Subcontracting	Proj. make-to-order
L	X	Subcontracting	All aux.acct.assgts.

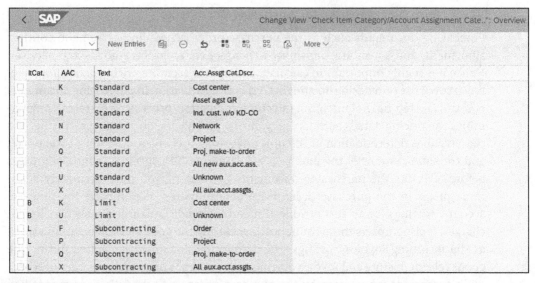

Figure 10.40 Define Item Category and Account Assignment Category Combinations

Assign Default Values for Asset Class

In this configuration step, you can assign a default asset class to the material groups. To assign default values for the asset class, execute Transaction SPRO and navigate to **Materials Management • Purchasing • Account Assignment • Assign Default Values for Asset Class.**

Figure 10.41 shows the asset classes that are assigned to material groups.

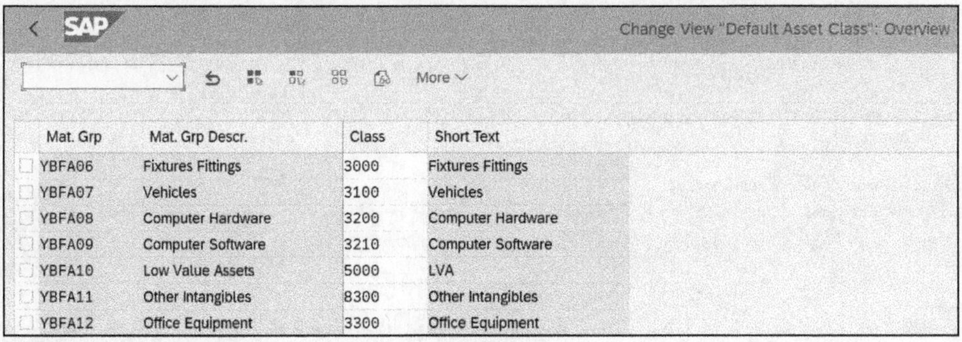

Mat. Grp	Mat. Grp Descr.	Class	Short Text
YBFA06	Fixtures Fittings	3000	Fixtures Fittings
YBFA07	Vehicles	3100	Vehicles
YBFA08	Computer Hardware	3200	Computer Hardware
YBFA09	Computer Software	3210	Computer Software
YBFA10	Low Value Assets	5000	LVA
YBFA11	Other Intangibles	8300	Other Intangibles
YBFA12	Office Equipment	3300	Office Equipment

Figure 10.41 Assign Default Values for Asset Class

For example, when you create a purchase order with account assignment category **A** (asset) and material group **YBFA06**, the system defaults asset class **3000**. The system allows you to assign the asset master record created with asset class **3000** only to the purchasing order item.

10.2.3   Price Determination for Purchasing Documents

Purchasing documents such as the purchase order, scheduling agreement, contract agreement, and so on are transmitted to suppliers to procure goods and services. Hence it is highly important to calculate and update accurate pricing details for every line item of the purchasing documents. Price determination in purchasing documents is a crucial step of the purchasing document creation process. SAP S/4HANA determines the pricing details automatically during purchasing document processing, and that includes determination of the unit price, freight charges, discounts, scale prices, and the total amount of the purchasing documents. To trigger the approval process before releasing the purchasing documents to vendors, you must accurately determine prices in the purchase order. Price determination also impacts the financial account postings for accrual postings for certain condition amounts, such as freight charges. It also impacts the account postings during the goods receipt postings as well as the incoming invoice postings. The accounting document generated during the goods receipt posting and invoice posting with reference to a purchase order or scheduling agreement is impacted by the purchase order and scheduling agreement line item price. Price determination in purchasing documents will also influence the three-way and two-way matching processes, also called invoice reconciliation. Refer to Section 10.4.1 for more details.

Figure 10.42 shows the **Conditions** tab for purchase order line item **10**. During the purchase order creation, the system automatically determined the unit price of **100.00 GBP** for condition type **PB00** and calculated the condition value of **10,000.00 GBP** for the total quantity of **100 EA**.

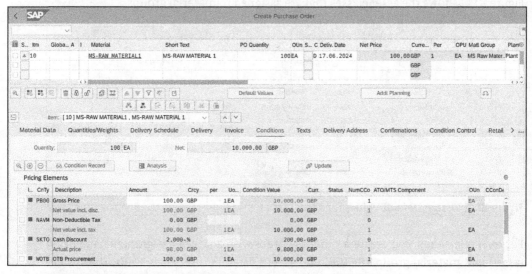

Figure 10.42 Price Determination in Purchasing Documents

It also calculated a discount of **200 GBP** for condition type **SKTO** automatically by applying a 2% discount on the base amount 10,000.00 GBP.

Figure 10.43 shows the condition record for condition type **PB00**. Best practice is to maintain the condition record in the PIR using Transaction ME11 for the combination of a material and a supplier at the purchasing organization level. The system automatically determines the unit price from the condition record in the purchasing documents.

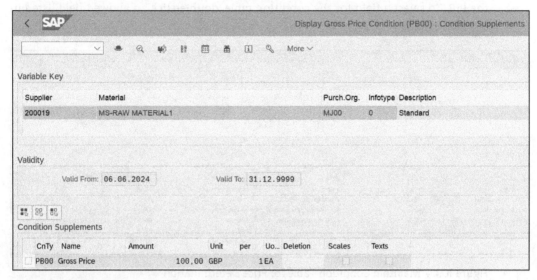

Figure 10.43 Condition Record for Price Determination

Price determination functionality in SAP S/4HANA leverages the condition technique. A pricing procedure is required to be set up for determining condition values in purchasing documents comprising condition types, access sequences, and condition records. Let's explore the configuration settings that drive the price determination in purchasing documents.

Maintain Condition Tables

Condition records for price determination are set up based on the criteria defined in the condition tables. Condition tables are set up using fields from table KOMG, KOMK, or KOMP only. Condition tables are assigned to an access sequence, and the access sequence is assigned to condition types. Condition records are set up based on condition types defined in a price determination procedure. This is called the *condition technique* in SAP S/4HANA.

To maintain condition tables, execute Transaction SPRO and navigate to **Materials Management • Purchasing • Conditions • Define Price Determination Process • Maintain Condition Tables**.

To create a new condition table, click the **Create Condition Table** activity from the initial screen. Specify a three-digit numeric value for the table name. You can also create this with reference to an existing condition table by directly specifying the table number under the **Copy from Condition Table** option in the initial screen. Then, press Enter to navigate to the field overview screen.

Figure 10.44 shows the maintenance screen for the condition table. You can use the fields from the **Field Catalog** on the right-hand side to create the condition table. Condition table **066** is defined in the standard system for maintaining condition records in the PIR. To select a field for the condition table, double-click an allowed field listed in the field catalog.

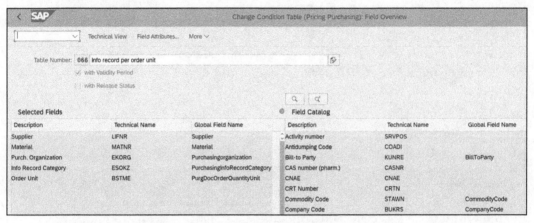

Figure 10.44 Maintain Condition Table for Price Determination

Define Access Sequences for Purchasing

The access sequence is part of the condition technique, and it provides a search strategy with one or more search criteria for price determination for different condition types in SAP S/4HANA. You can also define a sequence of condition tables within the access sequence to define different search criteria for the condition types, with the first search criterion having the highest priority and the last one having the lowest priority. Search criteria is assigned to the access sequence in the condition table.

To arrive at the configuration step to maintain an access sequence for price determination for purchasing, execute Transaction SPRO and navigate to **Materials Management • Purchasing • Conditions • Define Price Determination Process • Define Access Sequences for Purchasing**.

Figure 10.45 shows the access sequence in purchasing. Click **New Entries** to define the access sequence. You can also copy an existing access sequence and create a new one. Enter a four-digit key and a description to define an access sequence for price determination in purchasing documents.

The next step in defining the access sequence s to assign condition tables defined for the search criteria. Select the access sequence and then navigate to **Accesses** in the **Dialog Structure** to assign condition tables to the access sequence. Click **New Entries** to assign new condition tables to the access sequence. Figure 10.46 shows that condition tables **4MB**, **067**, **066**, and so on are assigned to the access sequence **0002**.

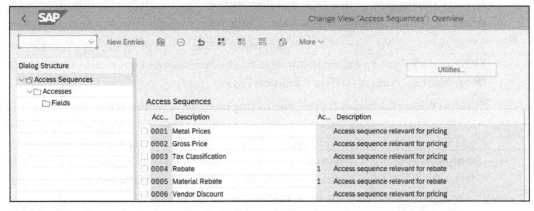

Figure 10.45 Access Sequences in Purchasing: Overview

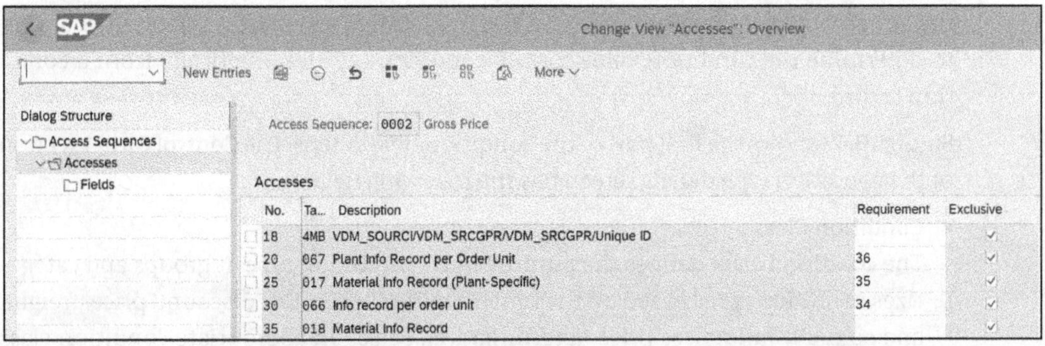

Figure 10.46 Access Sequences in Purchasing: Accesses

Next, navigate to **Fields** in the **Dialog Structure**. Figure 10.47 shows the fields copied from condition table **066** assigned to access sequence **0002**. You can't assign new fields for the search criteria. To modify the fields, you need to change the condition table maintenance per the instructions in the previous section.

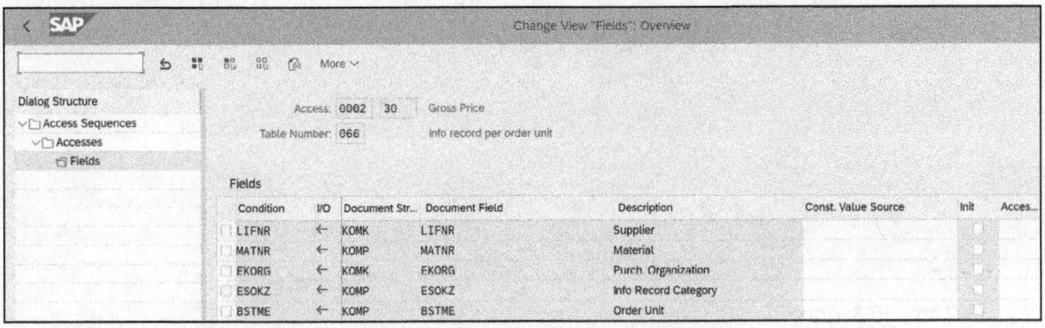

Figure 10.47 Access Sequences in Purchasing: Access Fields

Define Condition Types

Condition types are a vital component of price determination functionality in both materials management and sales and distribution. You can define condition types for prices, discounts, freight charges, taxes, and so on.

To define a condition type for price determination in purchasing, execute Transaction SPRO and navigate to **Materials Management • Purchasing • Conditions • Define Price Determination Process • Define Condition Types**.

Click **Set Pricing Condition Types—Purchasing** from the **Activities** screen to arrive at the condition type overview screen. To define a condition type, click **New Entries**. You can also copy an existing condition type and create a new one. Enter a four-digit key and a description to define a condition type for price determination. Figure 10.48 shows the **Control Data 1**, **Group Condition**, and **Changes which can be made** sections of the condition type details screen for standard condition type **PB00**.

The header details of the condition type display the condition type key and description. The access sequence defined in the configuration settings (see the previous section) is the key attribute of the condition type. You must assign an access sequence if you want to determine the condition value for the condition type automatically from a condition record.

Figure 10.48 shows the first part of the condition type details. The **Control Data 1** section of the condition type details screen has the following fields:

- **Condition Class**
  The condition class defines the purpose of the condition type. It groups and categorizes condition types based on the functionality such as tax, discount, price, freight, and so on. It influences price determination. Select an appropriate condition class from the dropdown menu. Table 10.3 shows the condition classes defined in the standard system.

Condition Class	Description
A	Discount or surcharge
B	Prices
C	Expense reimbursement
D	Taxes
E	Extra pay
F	Fees or differential (only IS-OIL)
G	Tax classification

Table 10.3 Condition Classes

Condition Class	Description
H	Determining sales deal
P	Compare price protection
Q	Totals record for fees (only IS-OIL)
W	Wage withholding tax

Table 10.3 Condition Classes (Cont.)

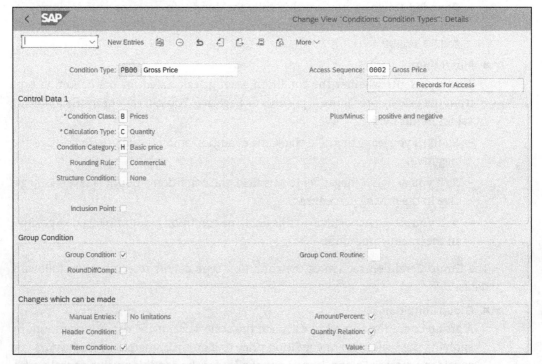

Figure 10.48 Define Pricing Condition Type: 1 of 2

- **Calculation Type**

  A calculation type specifies how the condition amount is calculated in the pricing procedure for purchasing documents. It determines the basis for the condition amount calculation. Select an appropriate calculation type from the dropdown menu. Percentage, fixed amount, quantity, gross weight, volume, and so on can be used as calculation types.

- **Condition Category**

  A condition category further categorizes the condition type based on functionality and business purpose such as tax, freight, discount, and so on. Select an appropriate condition category from the dropdown menu.

- **Rounding Rule**

  The condition amount calculated from the pricing procedure is rounded off based on the rounding rule assigned to the condition type. There are three options available in the standard system:

  - **Commercial**: If you set this rule, the condition amount rounds off according to commercial standards; for example, 100.544 GBP rounds off to 100.54 GBP and 100.546 GBP rounds off to 100.55 GBP.

  - **Rounded Up**: If you set this rule, the condition amount rounds up; for example, 100.544 GBP rounds up to 100.55 GBP and 100.546 GBP rounds up to 100.55 GBP.

  - **Rounded Down**: If you set this rule, the condition amount rounds down; for example, 100.544 GBP rounds down to 100.54 GBP and 100.546 GBP rounds down to 100.54 GBP.

- **Plus/Minus**

  This represents whether the condition amount calculated for the condition type from the pricing procedure is positive or negative. You can set one of the following values in this field:

  - Blank: If you leave this field blank, the condition amount can be both positive and negative.

  - **X**: If you assign **X** (negative) to this field, the condition amount is taken as negative in the pricing procedure.

  - **A**: If you assign **A** (positive) to this field, the condition amount is taken as positive in the pricing procedure.

The **Group Condition** section of the condition type details screen has the following fields:

- **Group Condition**

  A group condition is used to calculate the scale price or the condition amount by applying the scale price or condition price to a group of materials/products. If you consider a material alone, it may not qualify for the condition value or scale price, whereas if you consider a group of materials, the scale price can be easily applied. Set this indicator if you want to apply a group condition based on business requirements.

Note

The scale price is most used in condition records, PIRs, and outline agreements in materials management. It allows us to define prices and discounts on varying quantities. See the following examples:

- 1 to 10 units, $100 per unit
- 11 to 100 units, $95 per unit
- 101 to 1,000 units, $85 per unit

- **RoundDiffComp. (rounding difference comparison)**
  If you set this indicator, the system applies the rounding differences arising from a group condition to the largest value item after comparing the condition amount in the header with that of the item.

- **Group Cond. Routine**
  Assign a group condition routine to calculate the basis for the scale value when a group condition occurs in a pricing procedure. Table 10.4 shows the different routines available in the standard system.

Routine	Description
1	Overall document
2	Across all condition types
3	Material pricing group
8	Commercial rounding

Table 10.4 Group Condition Routine

The **Changes which can be made** section of the condition type details screen has the following fields:

- **Manual Entries**
  You can set the priority in this field for the condition type to determine if a manually entered condition has the priority or an automatically determined condition. Table 10.5 shows the manual entries that can be assigned to this field.

Manual Entries	Description
Blank	No limitations
A	Freely definable
B	Automatic entry has priority
C	Manual entry has priority
D	Not possible to process manually

Table 10.5 Manual Entries

- **Header Condition**
  If you set this indicator, the condition type can be entered in the **Conditions** tab at the header level in the purchasing documents.

- **Item Condition**

  If you set this indicator, the condition type can be entered in the **Conditions** tab at the item level in the purchasing documents.

- **Amount/Percent.**

  If you set this indicator, the system allows you to change the amount/percentage of this condition type during the purchasing documents processing.

- **Quantity Relation**

  If you set this indicator, the system allows you to change the conversion factor of this condition type during the purchasing documents processing.

- **Value**

  If you set this indicator, the system allows you to change the value of this condition type during the purchasing documents processing.

Figure 10.49 shows the second part of the condition type details. The **Master Data** section of the condition type details screen has the following fields:

- **Proposed Valid From**

  The system automatically proposes a valid from date when you create an agreement using this condition type. Table 10.6 shows the available standard values.

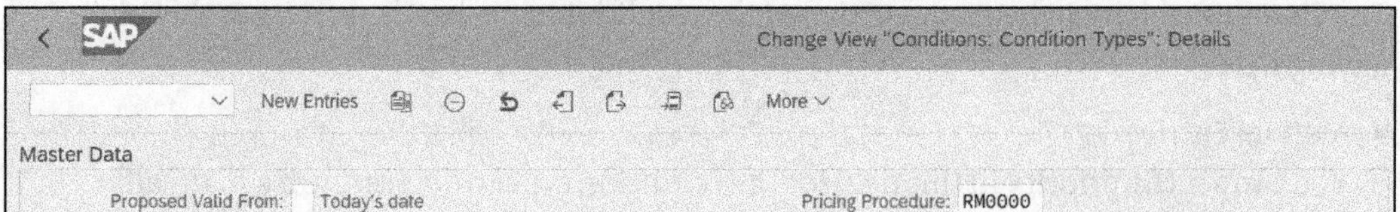

Figure 10.49 Define Pricing Condition Type: 2 of 2

Proposed Valid From	Description
Blank	Today's date
1	First day of the week
2	First day of month
3	First day of year
4	No proposal

Table 10.6 Proposed Valid From

- **Proposed Valid To**
  The system automatically proposes a valid to date when you create an agreement using this condition type. Table 10.7 shows the available standard values.

Proposed Valid To	Description
Blank	31.12.9999
2	End of the current year
1	End of the current month
5	No proposal
0	Today's date

Table 10.7 Proposed Valid To

- **Ref. Condition Type**
  You can assign a reference condition type in this field if the condition type is like another condition type, which means that you don't have to create separate condition records.

- **Pricing Procedure**
  If you want to restrict the use of the condition type to a specific pricing procedure only, assign the pricing procedure in this field.

The **Scales** section of the condition type details screen has the following fields:

- **Scale Base Type**
  Specify a scale base type such as quantity scale, weight scale, volume scale, value scale, and so on for determining the scale price. Select an appropriate value from the dropdown menu.

- **Check Scale**
  In this field, you can specify the scale amount if must be entered in an ascending order or descending order.

- **Scale Routine**
  In this field, you can specify an alternate routine for the determination of scale base value.

- **Scale Unit**
  In this field, you can specify a unit of measure for the system to calculate the scale price when a group condition is determined.

The **Control Data 2** section of the condition type details screen has the following fields:

- **Currency Conversion**
  If you set this indicator, the system controls the currency conversion if the condition amount currency is different from the document currency.

- **Exclusion**
  If you set an exclusion indicator, the system excludes discounts during price determination in purchasing documents. Figure 10.50 shows the exclusion indicators defined in the system that you can assign here. To define a condition type for price determination in purchasing, execute Transaction SPRO and navigate to **Materials Management • Purchasing • Conditions • Define Price Determination Process • Define Exclusion Indicators**.

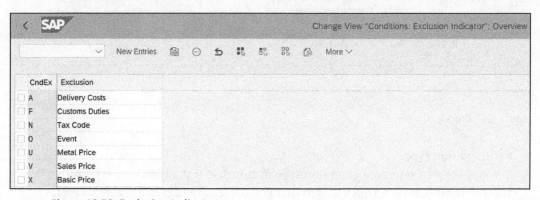

Figure 10.50 Exclusion Indicators

- **Accruals**
  If you set this indicator, the system posts the condition amount to financial accounting as accruals.

- **Used for Var.Config.**
  If you set this indicator, this condition type is used in variant configuration.

- **Rel. for Acct Assigt. (relevant for account assignment)**
  In this field, you can specify how the system performs account assignment for this condition type.

- **Intercomp.Billing**
  If you set this indicator, the system considers this condition type as relevant for intercompany billing.

Set Calculation Schema for Purchasing

The pricing procedure, also called the *calculation schema*, ensures accurate price determination during the purchase order and outline agreement creation process. It consists of control data with different condition types defined in a sequential order to determine pricing details such as base price, freight, discount, and so on in the purchasing documents. SAP S/4HANA provides standard procedures for price determination in purchasing. You can copy the existing procedure and create a new one if any changes are required based on business requirements. To define pricing procedures, execute Transaction SPRO and navigate to **Materials Management • Purchasing • Conditions • Define Price Determination Process • Set Calculation Schema—Purchasing**.

To define a new pricing procedure or a schema, click **New Entries**. You can also copy an existing schema and create a new one.

Figure 10.51 shows the overview of calculation schemas for the price determination procedure. The header details of the overview screen include the following fields:

- **Usage**
  This field contains the area of usage for the defined schemas. Usage code **A** represents pricing.

- **Application**
  This field contains the application area of usage for the defined schemas. Application **M** represents purchasing.

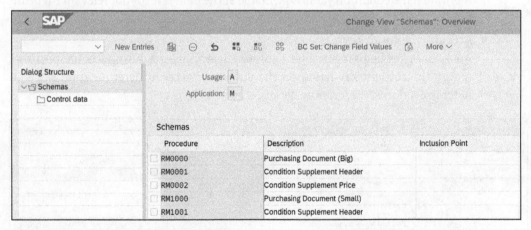

Figure 10.51 Calculation Schema: Overview

Select procedure **RM0000**, then go to **Control data** in the **Dialog Structure** to arrive at the screen shown in Figure 10.52. Let's explore the different columns of the control data:

- **Step**
  A step is a sequential number that defines the sequence of condition types maintained in the procedure. Enter step numbers in intervals of 10 (best practice).

- **Con... (condition type)**
  A condition type is used in price determination procedures to define different pricing categories such as gross price, discounts, freight, surcharge, and so on. Assign the condition types in a sequential order.

- **From and To Step**
  These columns are used in percentage-based condition types. The system calculates the total amount automatically by totaling the amount from **From Step** to **To Step** during the price determination. Define the step numbers accurately to calculate the condition amounts based on a certain base amount.

- **Man... (manual only)**
  If you set this indicator, the condition type is not automatically added to the purchasing documents, but it can be added manually.

- **Re... (required)**
  If you set this indicator, the condition type is mandatory during price determination for purchasing documents.

- **Stati... (statistical)**
  If you set this indicator, the amount calculated is for statistical purposes only and doesn't alter the total amount determined.

- **Print Type**
  This field controls the output of pricing details in purchasing documents. You can mention if the condition value is printable at the item level or not relevant for printing.

- **Account... (account key)**
  An account key helps in automatic account determination during account postings. Assign the account key based on the automatic account determination settings. Refer to Section 10.3.2 for more details.

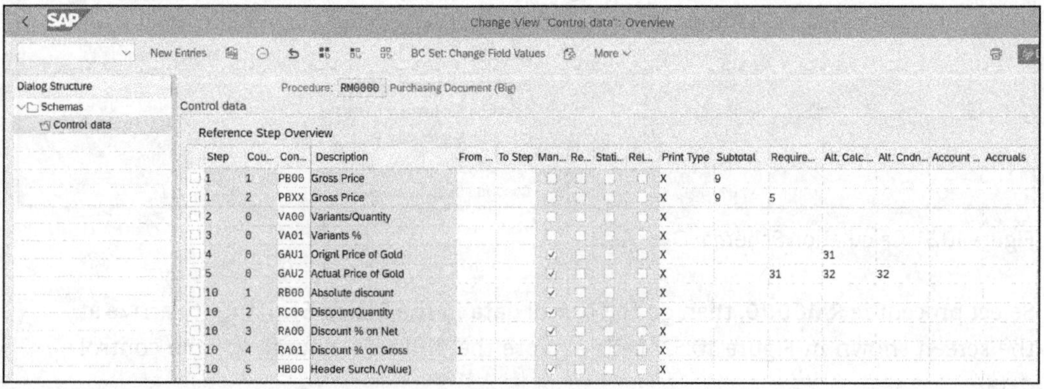

Figure 10.52 Calculation Schema: Control Data

Define Schema Groups

To determine the pricing procedure automatically in the purchasing documents, you must define schema groups for purchasing organizations and suppliers. You can control the determination of different pricing procedures for different suppliers and purchasing organizations using this configuration step. In this configuration step, you can define schema groups for suppliers and schema groups for purchasing organizations and can assign schema groups to purchasing organizations.

To define schema groups for purchasing organizations and suppliers, execute Transaction SPRO and navigate to **Materials Management • Purchasing • Conditions • Define Price Determination Process • Set Calculation Schema—Purchasing**.

Figure 10.53 shows the initial screen with activities for defining schema groups. Double-click each activity to perform the configuration settings.

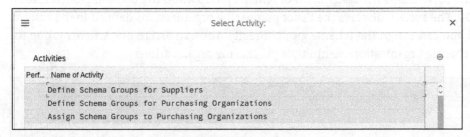

Figure 10.53 Define Schema Groups: Activities

Figure 10.54 shows the creation of schema groups for suppliers. Click **New Entries** and enter a two-digit key and a description to define a schema group for suppliers.

Sch.Grp Supp	Description
01	Default vendor schema group
RM	Purchase order Big
Z0	Non Taxable
Z1	Pricing for Local vendor
ZA	Schema Group for Suppliers
ZE	Schema Grp for ZE Suppliers

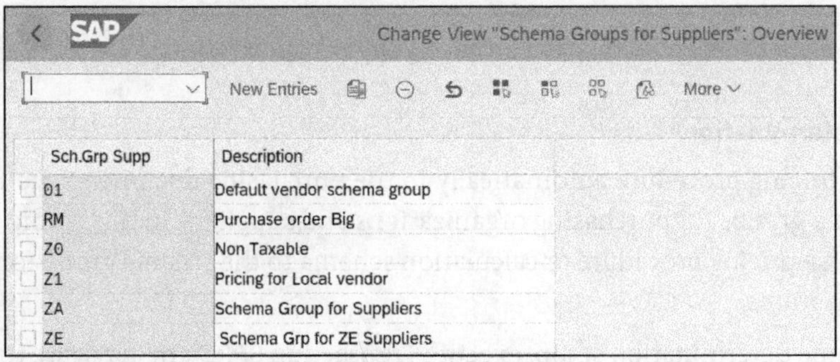

Figure 10.54 Define Schema Groups for Suppliers

Figure 10.55 shows the creation of schema groups for purchasing organizations. Click **New Entries** and enter a two-digit key and a description to define a schema group for purchasing organizations.

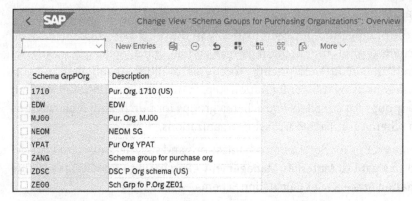

Figure 10.55 Define Schema Groups for Purchasing Organizations

Figure 10.56 shows the assignment of schema group **MJ00** to purchasing organization **MJ00**. The system displays the list of purchasing organizations defined in the system, and you can assign the schema group directly. You can assign one schema group for purchasing organizations to multiple purchasing organizations.

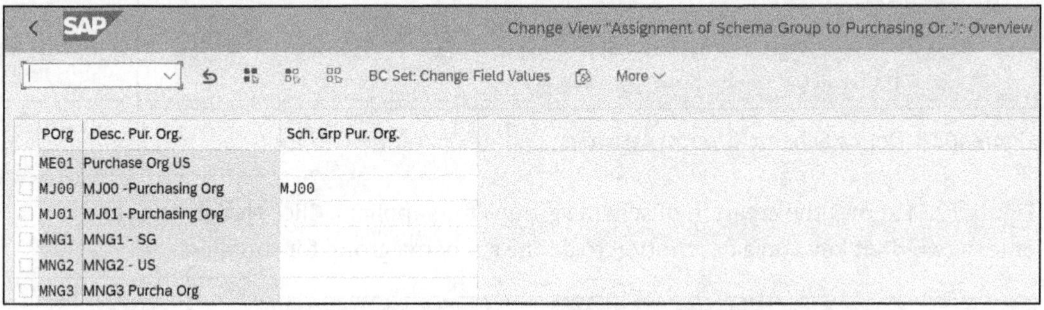

Figure 10.56 Assignment of Scheme Group to Purchasing Organization

Define Schema Determination

To determine the pricing procedure automatically in the purchasing documents, you must define schema groups for purchasing organizations and suppliers. In this configuration step, assign a pricing procedure or calculation schema to the schema group for purchasing organizations.

To define the schema determination setup, execute Transaction SPRO and navigate to **Materials Management • Purchasing • Conditions • Define Price Determination Process • Define Schema Determination**.

In this step, you can determine the calculation schema for standard purchase orders, stock transfer orders, and market price determination. Click the **Determine Calculation Schema for Standard Purchase Orders** activity from the initial screen to navigate to the overview screen. Figure 10.57 shows the overview of the determination of a calculation schema for standard purchase orders.

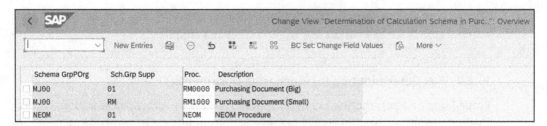

Figure 10.57 Determination of Calculation Schema for Standard Purchase Orders

Click **New Entries** to define a new criterion for calculation schema determination for standard purchase orders. Then specify a purchasing organization in the **Schema Grp-POrg** field and a calculation schema group for a supplier in the **Sch.Grp Supp** field, and to this combination assign the calculation schema (pricing procedure). The schema group for purchasing organization **MJ00** is assigned to two calculation schemas (pricing procedures) **RM0000** and **RM1000** based on schema groups for suppliers **01** and **RM**.

Schema groups for suppliers will be assigned to business partners (suppliers) in the **Purchasing** view at the purchasing organization level. Figure 10.58 shows the schema group **01** defined for suppliers is assigned to the supplier master record (business partner) in the **Purchasing Data** view for the purchasing organization **MJ00**. To assign the calculation schema group (supplier) to the business partner with a supplier role, execute Transaction BP and display the business partner with a supplier role in edit mode, then navigate to the **Purchasing Data** view. Select the purchasing organization to display the purchasing attributes specific to the purchasing organization. Then, assign the calculation schema group (supplier) in the **Schema Group Supplier** field.

You can define multiple schema determination settings for the combination of a schema group for purchasing organizations and a schema group for suppliers. By assigning different schema groups to suppliers, you can control the pricing procedure determination.

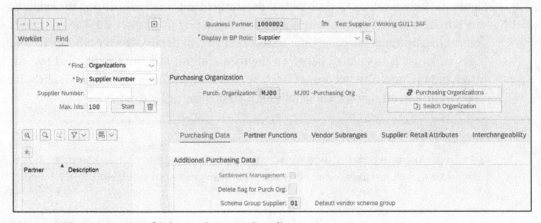

Figure 10.58 Assignment of Schema Group to Supplier

This is how the system automatically determines the pricing procedure based on schema groups for the purchasing organization and supplier.

10.2.4   Tax Determination for Purchasing Documents

Embedding and automating tax determination and tax reconciliation is vital to effectively manage the procure-to-pay process as both indirect and withholding taxes are important aspects of goods and services procurement. Both indirect and withholding tax calculation are integral parts of the procure-to-pay process. Indirect tax includes taxes levied on goods and services purchased and paid to the supplier as part of their invoice, tax that in turn is paid to the government by the supplier. Withholding taxes are not included in the supplier invoice, but the buying organization is responsible for withholding a portion of taxes from the invoice and paying them directly to the government.

Best practice is to calculate indirect taxes on a purchase order in SAP S/4HANA automatically. The taxes calculated on a purchase order are typically an estimated tax. However, the importance of determining taxes during a purchase order is as follows:

- **Compliance**
  In certain countries (e.g., Brazil), tax details are required to be included in purchase orders sent to suppliers. It is also best practice in other countries for buying organizations to include taxes on their purchase orders.

- **Visibility**
  This process provides visibility of expected indirect taxes to spend controllers and buyers.

- **Approvals**
  Including indirect taxes on a purchase order can drive the approval process based on total landed cost.

Figure 10.59 shows that tax code **V4** was determined automatically during the creation of a purchase order for company code **MJ00** belonging to country **GB**. The tax rate is determined from the tax code; refer to Section 10.1.5 for details. The tax amount for the purchase order is calculated based on the base amount (quantity multiplied by unit price) and tax rate. The tax amount can be displayed in the purchase order; click the **Taxes** button to display the tax details for the purchase order item.

Note

Tax in a purchase order is used for statistical purposes. But in certain countries, it is required to include taxes in the purchase order transmitted to suppliers.

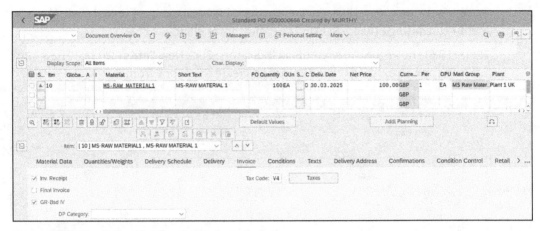

Figure 10.59 Tax Determination in Purchase Order

Figure 10.60 shows the condition record maintained for tax determination. Tax code **V4** is determined in the purchase order from the condition record maintained for condition type **MWVN**. You can create condition records for tax code determination using Transaction MEK1.

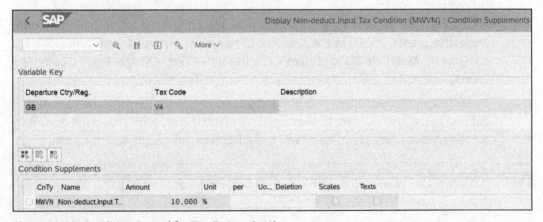

Figure 10.60 Condition Record for Tax Determination

Tax determination in SAP S/4HANA leverages the condition technique. The tax determination procedure is set up to determine taxes in purchasing documents comprising condition types, access sequences, and condition records. Let's explore the configuration settings that drive the tax determination in purchasing documents.

Set Up Tax Indicators

Tax indicators are logical keys used in tax determination. These indicators are used in material, plant, and account assignment categories to set up condition tables and access sequences to drive tax determination.

To set up tax indicators, execute Transaction SPRO and navigate to **Materials Management • Purchasing • Taxes • Set Up Tax Indicator for Plant.**

Figure 10.61 shows the overview of tax indicators. Click **New Entries** to create new ones. You can copy an existing tax indicator and create a new one from it.

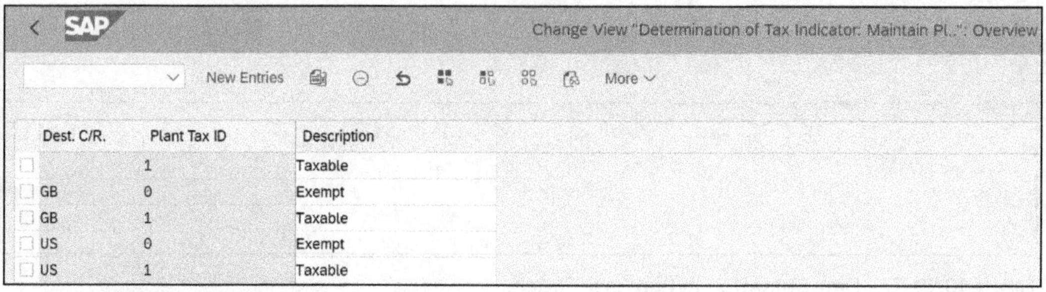
SAP
Change View "Determination of Tax Indicator: Maintain Pl...": Overview
New Entries
More
Dest. C/R.
Plant Tax ID
Description
1 Taxable
GB 0 Exempt
GB 1 Taxable
US 0 Exempt
US 1 Taxable

Figure 10.61 Set Tax Indicator for Plant

The destination country (**Dest. C/R**) is the country where the plant is located. Based on business requirements, if the tax indicators are needed for plants located in multiple countries, set them up for every country. Enter a single-digit key for tax indicators in the **Plant Tax ID** field.

Next, assign the plant tax indicator (**Plant Tax ID**) defined in the previous step to the respective plants. To assign the plant tax ID to plants, execute Transaction SPRO and navigate to **Materials Management • Purchasing • Taxes • Assign Tax Indicators for Plants.**

By default, the system displays all plants. You can directly assign the plant tax indicator defined in the previous step to the respective plants. Figure 10.62 shows the tax indicator assigned to every plant. Only one tax indicator can be assigned to one plant.

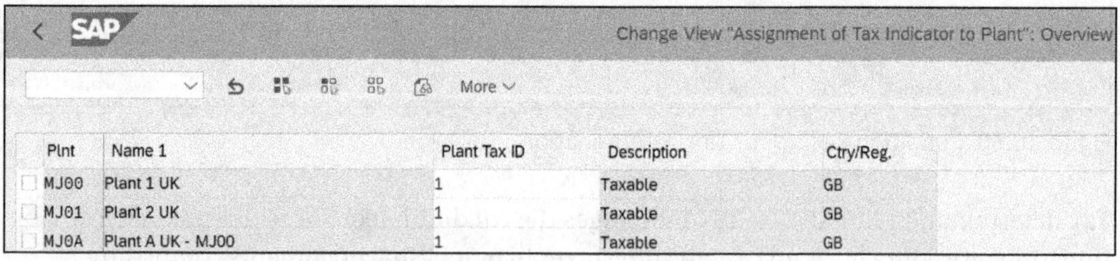
SAP
Change View "Assignment of Tax Indicator to Plant": Overview
More
Plnt
Name 1
Plant Tax ID
Description
Ctry/Reg.
MJ00 Plant 1 UK 1 Taxable GB
MJ01 Plant 2 UK 1 Taxable GB
MJ0A Plant A UK - MJ00 1 Taxable GB

Figure 10.62 Assign Tax Indicators for Plants

Note

Similarly, you can set up tax indicators for material and account assignment categories. Once the required tax indicators are set up, you must assign them to the respective materials and account assignment categories.

Maintain Condition Tables

We discussed condition tables the context of price determination for purchasing documents in Section 10.2.3. Maintaining condition tables is also necessary for tax determination.

To maintain condition tables, execute Transaction SPRO and navigate to **Materials Management • Purchasing • Conditions • Define Price Determination Process • Maintain Condition Tables**.

To create a new condition table, click the **Create Condition Table** activity from the initial screen. Specify a three-digit numeric value for the table name. You can also create it with reference to an existing condition table by directly specifying the table number under the **Copy from Condition Table** option in the initial screen. Then, press Enter to navigate to the field overview screen. Figure 10.63 shows the maintenance screen for the condition table. You can use the fields from **Field Catalog** on the right-hand side to create the condition table. To select a field for the condition table, double-click an allowed field listed in the field catalog.

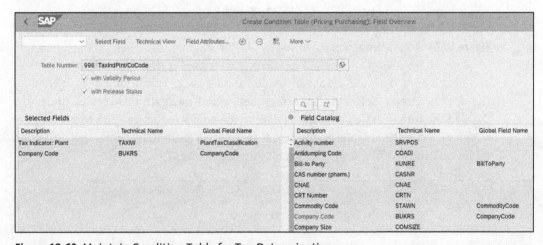

Figure 10.63 Maintain Condition Table for Tax Determination

You can create multiple condition tables to define the tax determination criteria based on business requirements.

Maintain Access Sequences

We discussed access sequences in the context of price determination for purchasing documents in Section 10.2.3. Maintaining an access sequence is also an important step in tax determination.

To maintain an access sequence for tax determination, execute Transaction SPRO and navigate to **Financial Accounting • Financial Accounting Global Settings • Tax on Sales/Purchases • Basic Settings • Check Calculation Procedure**.

Click **Access Sequences** from the **Activity** screen to arrive at the access sequence overview screen, as shown in Figure 10.64. To define a four-digit access sequence key, click **New Entries**. You can also copy an existing access sequence and create a new one. Enter a four-digit key and a description to define an access sequence for tax determination.

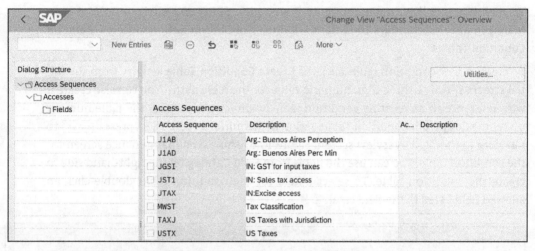

Figure 10.64 Access Sequences: Overview

The next step in defining the access sequence is to assign condition tables defined for the search criteria. Select the access sequence and navigate to **Accesses** under the **Dialog Structure** to assign condition tables to the access sequence. Click **New Entries** to assign new condition tables to the access sequence. Figure 10.65 shows that condition tables **003** and **4AV** are assigned to access sequence **MWST**.

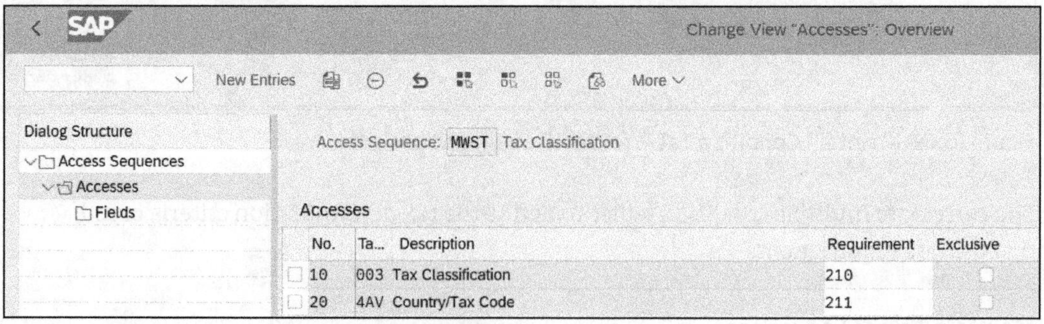

Figure 10.65 Access Sequences: Accesses

By navigating to **Fields** under the **Dialog Structure**, you'll arrive at the screen shown in Figure 10.66, which shows the fields copied from condition table **003** assigned to access sequence **MWST**. You can't assign new fields for the search criteria. To modify the fields, you need to change the condition table maintenance per our instructions in the previous section.

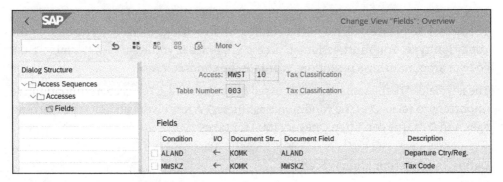

Figure 10.66 Access Sequences: Fields

Define Condition Types

We discussed condition types in the context of price determination for purchasing documents in Section 10.2.3. Defining condition types is also a key step for tax determination. To define a condition type for tax determination, execute Transaction SPRO and navigate to **Financial Accounting • Financial Accounting Global Settings • Tax on Sales/Purchases • Basic Settings • Check Calculation Procedure.**

Click **Define condition types** from the **Activity** screen to arrive at the condition type overview screen. To define a condition type, click **New Entries.** You can also copy an existing condition type and create a new one. Enter a four-digit key and a description to define a condition type for tax determination. Figure 10.67 shows the **Control Data 1**, **Group Condition**, and **Changes which can be made** sections of the condition type details screen.

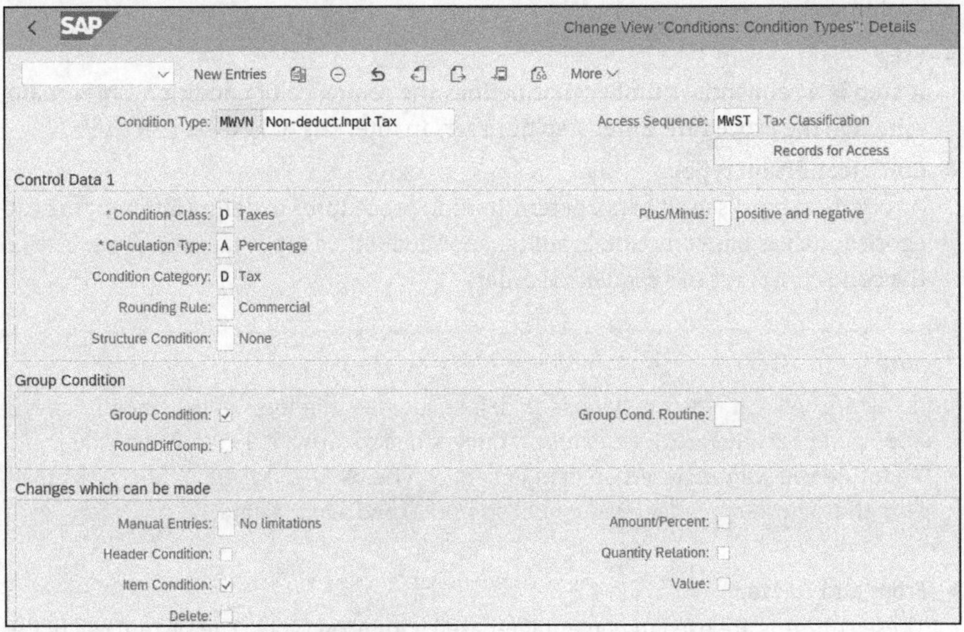

Figure 10.67 Define Tax Condition Type

Header details of the condition type display the condition type key and description. The **Access Sequence** defined in the configuration settings is the key attribute of the condition type. You must assign an access sequence if you want to determine a tax code or amount using a condition record for the condition type.

The key fields are the same as we discussed in Section 10.2.3. For tax determination, it's important to select **D** as the condition class (taxes), **A** for the calculation type (percentage), and **D** for the condition category (tax) for taxes.

Define Tax Determination Procedures

The tax determination procedure ensures the accurate tax determination during the purchase order and outline agreement creation process. It consists of control data with different condition types defined in a sequential order to determine tax details such as tax code, tax rate, and tax amount in the purchasing documents. SAP S/4HANA provides standard procedures for tax determination. You can copy the existing procedure and create a new one if any changes are required based on business requirements.

To define tax determination procedures, execute Transaction SPRO and navigate to **Financial Accounting • Financial Accounting Global Settings • Tax on Sales/Purchases • Basic Settings • Check Calculation Procedure**.

Click **Define procedures** from the **Activity** screen to arrive at the procedures overview screen. To define a procedure, click **New Entries**. You can also copy an existing procedure and create a new one.

Figure 10.68 shows the **Control Data** area for procedure **TAXGB**. Let's explore the different columns:

- **Step**
  A step is a sequential number that defines the sequence of condition types maintained in the procedure. Enter step numbers in intervals of 10 (best practice).

- **Con... (condition type)**
  A condition type is used in tax determination procedures to define different tax categories such as base amount, input tax, nondeductible input tax, and so on. Assign the condition types in a sequential order.

> **Note**
>
> In SAP S/4HANA, it is recommended to define line item numbers of documents (sales order, purchase order, etc.), sequential numbers, and so on in intervals of 10 so that if in the future you want to insert an item in between items 10 and 20, you can, for example, define line item number/sequential number 11 and add the data.

- **From** and **To Step**
  These columns are used in percentage-based condition types. The system calculates the total amount automatically by totaling the amount from **From Step** to **To Step**

during the tax determination. Define the step numbers accurately to calculate the tax amounts based on a certain base amount.

- **Man... (manual only)**
  If you set this indicator, the condition type is not automatically added to the purchasing documents, but it can be added manually.
- **Re... (required)**
  If you set this indicator, the condition type is mandatory during tax determination for purchasing documents.
- **Stati... (statistical)**
  If you set this indicator, the amount calculated is for statistical purposes only and doesn't alter the total amount determined.
- **Print Type**
  This field controls the output of pricing details in purchasing documents. You can mention if the condition value is printable at the item level or not relevant for printing.
- **Account... (account key)**
  An account key helps in automatic account determination during tax postings. Assign the account key based on the automatic account determination settings. Refer to Section 10.3.2 for more details.

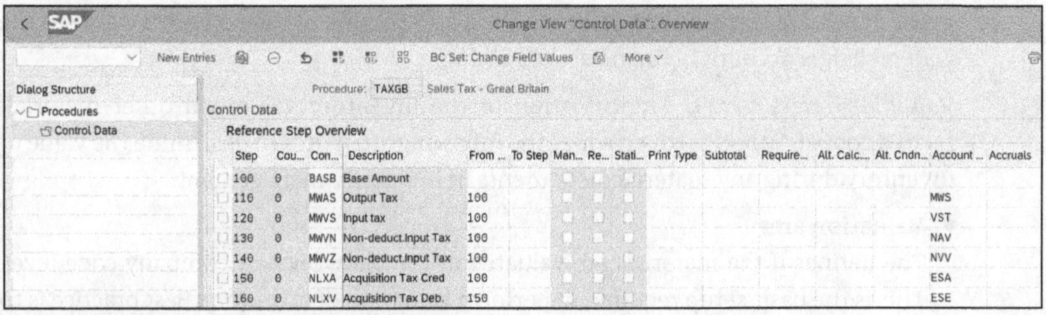
Change View "Control Data": Overview
New Entries
BC Set: Change Field Values
More
Dialog Structure
Procedures
Control Data
Procedure: TAXGB Sales Tax - Great Britain
Control Data
Reference Step Overview
Step Cou... Con... Description From ... To Step Man... Re... Stati... Print Type Subtotal Require... Alt. Calc... Alt. Cndn... Account ... Accruals
100 0 BASB Base Amount
110 0 MWAS Output Tax 100 MWS
120 0 MWVS Input tax 100 VST
130 0 MWVN Non-deduct.Input Tax 100 NAV
140 0 MWVZ Non-deduct.Input Tax 100 NVV
150 0 NLXA Acquisition Tax Cred 100 ESA
160 0 NLXV Acquisition Tax Deb. 150 ESE

Figure 10.68 Define Tax Determination Procedure

Assign Country/Region to Calculation Procedure

The tax determination/calculation procedure must be assigned to the respective country. A predefined tax calculation procedure must be defined before performing this activity. The system determines the procedure automatically during the creation of a purchase order and outline agreement based on the country of the company code in which the transactional document is being created.

To assign tax determination procedures to a country/region, execute Transaction SPRO and navigate to **Financial Accounting • Financial Accounting Global Settings • Tax on Sales/Purchases • Basic Settings • Assign Country/Region to Calculation Procedure.**

By default, the system displays all the country/region codes in the **C/R** column. Directly assign the calculation schema (calculation procedure) defined for tax calculation to the respective country in the **Proc.** field. Figure 10.69 shows procedure **TAXGB** assigned to country **GB**. You can assign the tax determination procedure directly to the country/region in this configuration setting.

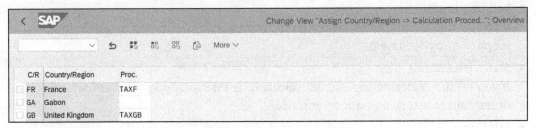

Figure 10.69 Assign Country/Region to Calculation Procedure

10.3   Valuation and Account Determination

Valuation and account assignment are integral parts of materials management and closely integrated with financial accounting. Material movements happening in the inventory management function need to be accurately valued and posted to the right general ledger accounts in financial accounting. This integration ensures the automatic general ledger account determination and posting of the inventory value to the general ledger account in financial accounting.

Valuation refers to the determination of the inventory value and other material-related costs. SAP S/4HANA provides the following attributes to determine the value of inventory during any material movements in inventory management:

- **Valuation area**
  The defines if the materials are valuated at the plant level or company code level. This is the basic setup required to perform valuation of inventory. Best practice is to valuate materials at the plant level.

- **Price control**
  This defines how the materials are valuated. SAP S/4HANA provides two methods for valuing inventory: standard price and moving average price. A price control is assigned to the **Accounting 1** view of material master data at the plant level:

  - Standard price (**S**): The standard price is a static price set for a material, and it does not change with procurement activity. The standard price of the material is fixed until you change it manually in the material master.

  - Moving average price (**V**): The moving average price is the dynamic price that changes with procurement activities such as goods receipt and invoice receipt. The system calculates the moving average price of material based on the purchase price during goods receipt and invoice receipt postings.

Account determination, on the other hand, refers to the determination of the general ledger account automatically during material movement transactions in inventory management. It ensures the inventory value of the material is posted to financial accounting for financial reporting purposes. The following are the key components of this feature:

- **Valuation class**

  The valuation class is a crucial and powerful feature of the inventory valuation functionality. It is the key component in ensuring seamless integration between materials management and financial accounting. The valuation class is assigned to material master data, and it ensures the determination of a general ledger account during material movement transactions in inventory management. The valuation class is defined for material types, and you can define one valuation class for multiple material types or one valuation class per material type depending on the financial accounting and reporting requirements. The valuation class is defaulted from the material type into the **Accounting 1** view of the material master data. The defaulted valuation class can't be changed or deleted in the material master data, and this ensures consistency in financial postings. For example, valuation class 3000 is assigned to material type ROH (raw materials), and valuation class 7920 is assigned to material type FERT (finished goods). In turn, valuation class 3000 is linked to raw material inventory accounts and valuation class 7920 is assigned to finished goods inventory accounts.

- **Valuation category and valuation types**

  These are used to manage and differentiate the material valuation based on different criteria such as origin, status, batch, and quality. For example, a material can be valuated based on its condition/status (new, damaged, or refurbished), and different material statuses can be created as valuation types to perform split valuation. Different valuation types can be defined for valuation categories. You can assign the valuation category to the material master in the **Accounting 1** view to valuate materials at different prices by valuation type. For example, if you define new, damaged, and refurbished valuation types for the *condition/status* valuation category, you can specify a valuation price per valuation type in the **Accounting 1** view of the material master. This feature is called *split valuation*.

Let's dive deep into valuation and account determination.

10.3.1   Price Control for Materials

Price control for materials in SAP S/4HANA manages the valuation price of materials. Price control settings drive the inventory valuation of materials. The price control is automatically updated in the material master data based on the price control settings maintained in the configuration for material types.

Figure 10.70 shows the **Accounting 1** view of the material master data. This view is maintained at the plant level. You can set the price control for materials in this view. For the **MATERIAL-3** material, price control **S** (standard price) has been maintained. Based on the price control settings at the material type level, the price control key is copied over from the configuration settings. But the system allows users to change the price control in the material master data.

Price control settings for materials are configured at the material type level.

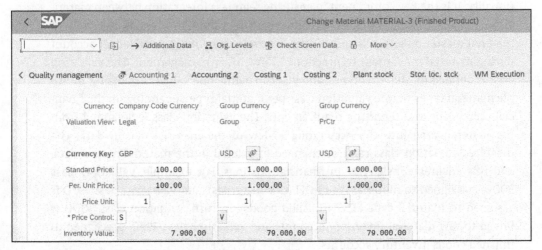

Figure 10.70 Price Control for Materials

In this configuration, we can assign a specific price control for every material type. There are two price controls in SAP S/4HANA: standard price and moving average price, to be assigned to a material type so that the system valuates materials in inventory management accordingly.

To arrive at this configuration step, execute Transaction SPRO and navigate to **Materials Management • Valuation and Account Assignment • Define Price Control for Material Types**.

Figure 10.71 shows the price control indicators assigned to every material type. Let's explore the fields of this configuration:

- **Price Control**
  In this field, assign a price control method (standard price **S** or moving average price **V**) to valuate materials in inventory management. The price control method will be assigned to the material type, and when a new material is created with the respective material type, the price control method will be assigned to the material automatically. There are two important price control methods:
  - **Standard Price (S)**: With this control, materials will be valuated in inventory management at a fixed price. Typically, the standard price control will be assigned to internally manufactured materials or finished goods.

- **Moving Average Price (V):** With this control, materials will be valuated in inventory management based on a weighted average unit price. The system continuously updates the moving average price based on the unit price of the material in the purchase order. Typically, the moving average price control will be assigned to externally procured materials.

- **Price Control Mandatory**
  If this indicator is set in the configuration, the price control method assigned to the material type will become mandatory, and the value can't be changed in the material master record created with reference to the respective material type.

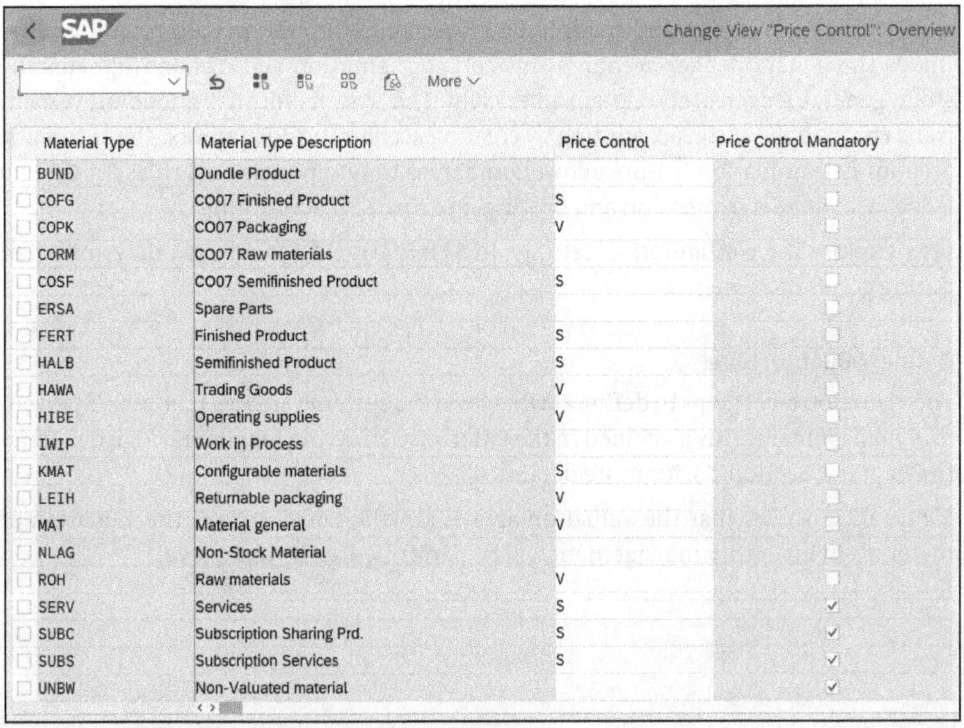

Material Type	Material Type Description	Price Control	Price Control Mandatory
BUND	Oundle Product		
COFG	CO07 Finished Product	S	
COPK	CO07 Packaging	V	
CORM	CO07 Raw materials		
COSF	CO07 Semifinished Product	S	
ERSA	Spare Parts		
FERT	Finished Product	S	
HALB	Semifinished Product	S	
HAWA	Trading Goods	V	
HIBE	Operating supplies	V	
IWIP	Work in Process		
KMAT	Configurable materials	S	
LEIH	Returnable packaging	V	
MAT	Material general	V	
NLAG	Non-Stock Material		
ROH	Raw materials	V	
SERV	Services	S	✓
SUBC	Subscription Sharing Prd.	S	✓
SUBS	Subscription Services	S	✓
UNBW	Non-Valuated material		✓

Figure 10.71 Define Price Control for Material Types

10.3.2 Account Determination in Inventory Management

We introduced automatic account determination in Chapter 4, Section 4.1.6. To recap, account determination is a configurable feature that automatically assigns general ledger accounts to goods movements in inventory management to post the values into financial accounting. The process of account determination considers the following key factors:

- **Material**
  Material type, valuation class, and valuation type of materials

- **Movement type**
  Types of movements such as goods issue, goods receipt, or transfer postings

- **Account assignment categories**
  The account assignment categories such as cost center, asset, order, and so on that are assigned to the purchase orders
- **Plant**
  Typically, the plant is set as the valuation area
- **Valuation grouping**
  Grouping valuation areas (e.g., plants) to simplify account determination logic
- **Chart of accounts**
  The chart of accounts used

During the goods receipt and goods issue movements, the system automatically determines the general ledger accounts through configuration settings and updates the stock general ledger accounts automatically. The system records goods movement transactions in financial accounting by creating accounting documents. The automatic account determination feature allows companies to set up their own rules for general ledger account determination and posting into financial accounting.

Let's explore the configuration settings in SAP S/4HANA for account determination during goods movements.

Define Valuation Level

The valuation level helps to define at what level the system valuates materials in inventory management. This is defined in the enterprise structure for general logistics. Refer to Chapter 2, Section 2.3, for more details.

Figure 10.72 shows that the valuation area is a plant, which means the valuation of materials in inventory management will be performed at the plant level.

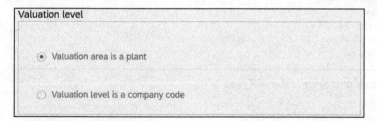

Figure 10.72 Valuation Level

Define Valuation Control

In this step, we can activate the valuation grouping code. By activating the valuation grouping code, valuation areas can be logically grouped using a grouping code to simplify the automatic account determination configuration.

To arrive at this configuration step, execute Transaction SPRO and navigate to **Materials Management • Valuation and Account Assignment • Account Determination • Account Determination Without Wizard • Define Valuation Control**.

As shown in Figure 10.73, select **Valuation grouping code active** to activate the valuation grouping code. In the next configuration steps, we'll discuss how this simplifies the configuration of automatic account determination.

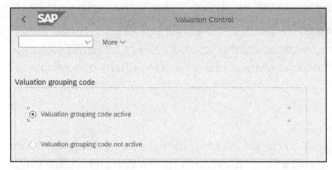

Figure 10.73 Define Valuation Control

Group Together Valuation Areas

In this step, we can group the valuation areas together using a valuation grouping code.

To arrive at this configuration step, execute Transaction SPRO and navigate to **Materials Management • Valuation and Account Assignment • Account Determination • Account Determination Without Wizard • Group Together Valuation Areas**.

You directly assign a valuation grouping code in the **Val.Grpg Code** field. Figure 10.74 shows that valuation areas **MJ00**, **MJ01**, and **MJ0A** have been assigned to valuation account group **MJ00**.

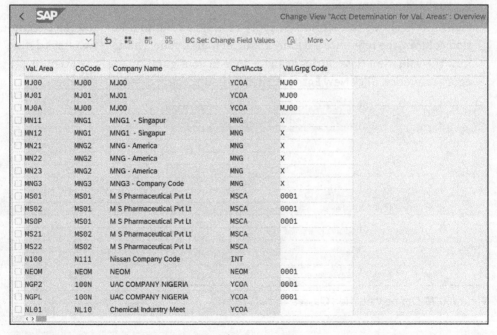

Figure 10.74 Group Valuation Areas Together

This way, we are grouping multiple valuation areas together based on business requirements to determine accounts automatically.

In an earlier configuration step, we defined the valuation level as the plant. Hence, plants will be the valuation areas, and the materials will be valuated at the plant level.

The chart of accounts (**Chrt/Accts**) is a structured list of general ledger accounts used in financial accounting and controls the account types. It provides a framework for organizations to report various financial transactions accurately. Charts of accounts are assigned to company codes.

Define Valuation Class

In this step, we can define valuation classes, which control the general ledger account determination for materials during a goods movement transaction for material valuation. Material types are linked to valuation classes, and the valuation classes are linked to a general ledger account in financial accounting.

To arrive at this configuration step, execute Transaction SPRO and navigate to **Materials Management • Valuation and Account Assignment • Account Determination • Account Determination Without Wizard • Define Valuation Classes**.

Figure 10.75 shows the initial screen of the valuation class configuration. Let's explore the configuration settings:

- **Account category reference**
  The account category reference acts as a bridge between material types and valuation classes. A valuation class is defaulted into the material master data in SAP through the account category reference.

  Figure 10.76 shows the account category reference column (**ARef**) and the **Description** column. the reference is a four-character logical code. The next two steps define how the valuation class settings can be done in configuration using an account category reference. Click **New Entries** to define new account category references.

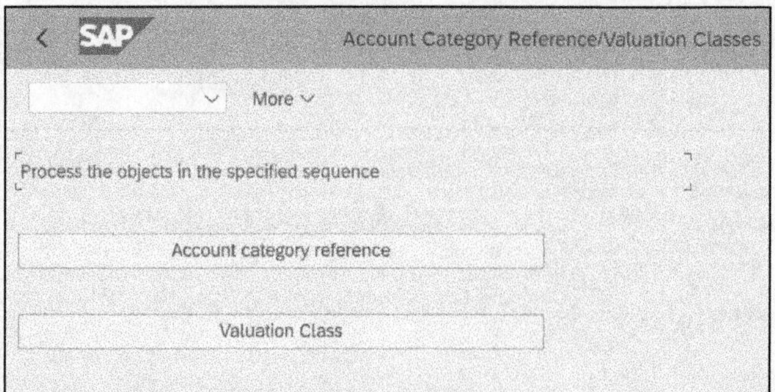

Figure 10.75 Define Valuation Classes: Initial Screen

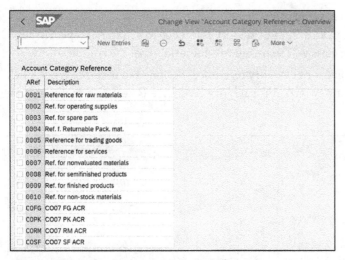

Figure 10.76 Define Account Category References

- **Valuation Class**

  A valuation class controls the determination of general ledger accounts for valuation of material stocks during a goods movement transaction. Using a valuation class, the system determines the same general ledger account for different material types during goods movement or different general ledger accounts for different material types.

  Figure 10.77 shows the defined valuation classes. Typically, valuation classes are defined by type or group of materials. Click **New Entries** to define new valuation classes and assign an account category reference (**ARef**) to the valuation class (**ValCl**).

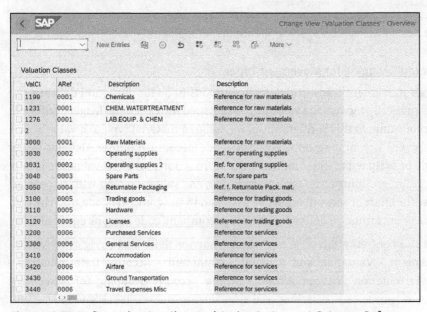

Figure 10.77 Define Valuation Class and Assign to Account Category Reference

- **Material type/account category reference**

  In this step, we can assign an account category reference to a material type. Through this setting, a valuation class will be linked to a material type.

  Figure 10.78 shows the account category references assigned to material types. Click **New Entries** to assign a new account category reference (**ARef**) to a material type (**MTyp**).

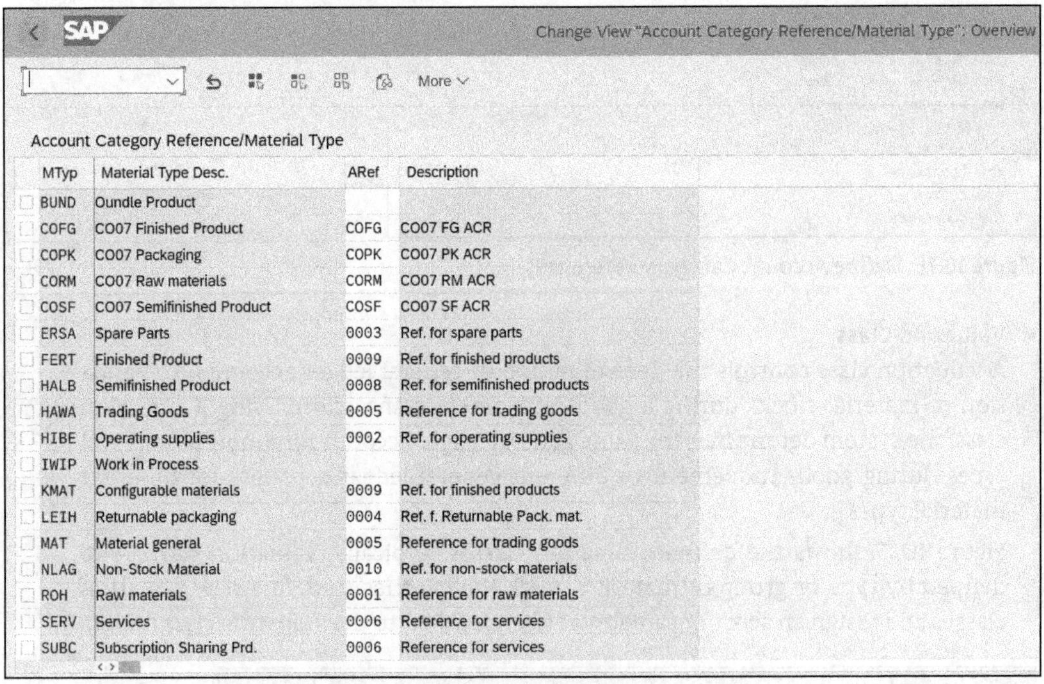

Account Category Reference/Material Type

MTyp	Material Type Desc.	ARef	Description
BUND	Oundle Product		
COFG	CO07 Finished Product	COFG	CO07 FG ACR
COPK	CO07 Packaging	COPK	CO07 PK ACR
CORM	CO07 Raw materials	CORM	CO07 RM ACR
COSF	CO07 Semifinished Product	COSF	CO07 SF ACR
ERSA	Spare Parts	0003	Ref. for spare parts
FERT	Finished Product	0009	Ref. for finished products
HALB	Semifinished Product	0008	Ref. for semifinished products
HAWA	Trading Goods	0005	Reference for trading goods
HIBE	Operating supplies	0002	Ref. for operating supplies
IWIP	Work in Process		
KMAT	Configurable materials	0009	Ref. for finished products
LEIH	Returnable packaging	0004	Ref. f. Returnable Pack. mat.
MAT	Material general	0005	Reference for trading goods
NLAG	Non-Stock Material	0010	Ref. for non-stock materials
ROH	Raw materials	0001	Reference for raw materials
SERV	Services	0006	Reference for services
SUBC	Subscription Sharing Prd.	0006	Reference for services

Figure 10.78 Assign Account Category Reference to Material Type

Define Account Grouping for Movement Types

During goods movement transactions, such as goods issue for outbound delivery, the system automatically records the value of goods issued into a general ledger account in financial accounting. In this configuration, we assign a material type to a value string, transaction event key, and account grouping code/account modifier. General ledger accounts will be assigned to the combination of transaction event key, valuation grouping code, account grouping code/account modifier, and valuation class within the chart of accounts. The chart of accounts must be assigned to the company code, and the general ledger accounts must be defined in the same company code as a prerequisite.

To arrive at this configuration step, execute Transaction SPRO and navigate to **Materials Management • Valuation and Account Assignment • Account Determination • Account Determination Without Wizard • Define Account Grouping for Movement Types**.

Figure 10.79 shows the configuration settings for defining an account grouping for a goods movement transaction.

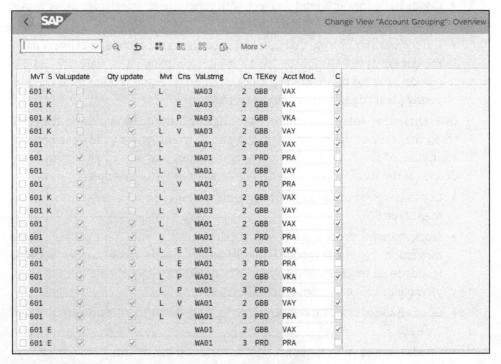

MvT	S	Val.update	Qty update	Mvt	Cns	Val.strng	Cn	TEKey	Acct Mod.	C
601	K		✓	L		WA03	2	GBB	VAX	✓
601	K		✓	L	E	WA03	2	GBB	VKA	✓
601	K		✓	L	P	WA03	2	GBB	VKA	✓
601	K		✓	L	V	WA03	2	GBB	VAY	✓
601		✓		L		WA01	2	GBB	VAX	✓
601		✓		L		WA01	3	PRD	PRA	
601		✓		L	V	WA01	2	GBB	VAY	✓
601		✓		L	V	WA01	3	PRD	PRA	
601	K	✓		L		WA03	2	GBB	VAX	✓
601	K	✓		L	V	WA03	2	GBB	VAY	✓
601		✓	✓	L		WA01	2	GBB	VAX	✓
601		✓	✓	L		WA01	3	PRD	PRA	✓
601		✓	✓	L	E	WA01	2	GBB	VKA	✓
601		✓	✓	L	E	WA01	3	PRD	PRA	
601		✓	✓	L	P	WA01	2	GBB	VKA	✓
601		✓	✓	L	P	WA01	3	PRD	PRA	
601		✓	✓	L	V	WA01	2	GBB	VAY	✓
601		✓	✓	L	V	WA01	3	PRD	PRA	
601	E	✓	✓			WA01	2	GBB	VAX	✓
601	E	✓	✓			WA01	3	PRD	PRA	

Figure 10.79 Define Account Grouping for Movement Types

Let's explore the key fields of this configuration:

- **TEKey (transaction event key)**
  This is an internal key in SAP S/4HANA to identify a specific goods movement transaction. This standard setting is assigned to movement types and controls the determination of general ledger accounts during the goods movement. The following are the standard transaction event keys defined in the system that are relevant for goods movement transactions:

  - **BSX**: This is the transaction event key for all inventory postings. This event will be triggered for all goods movements involving posting of stock into inventory and posting of stock issue from inventory. This transaction increases and reduces the stock value in inventory management depending on goods receipt and goods issue postings. The following are some examples:
    - Initial entry of stock balances into inventory using movement type **561**.
    - Goods receipt of stock into inventory without purchase order using movement type **501**.
    - Goods receipt with reference to purchase order with movement type **101**. The catch here is that if the purchase order is created with an account assignment

such as a cost center, order, or WBS element, the stock will not be posted into inventory. In such cases, transaction event key **BSX** will not be triggered.

- Goods issue for outbound delivery with reference to sales order using movement type **601**.

- A stock transfer process that involves goods issue for outbound delivery using movement types **641** (goods issue for intracompany stock transfer) and **643** (goods issue for intercompany stock transfer). The goods issue from the supplying plant triggers transaction event **BSX**.

- **GBB**: Offsetting entry for inventory postings. This transaction event is triggered when the stock of the material is taken out of inventory for goods issue or consumption of stock for production, a cost center, and so on. This transaction decreases the stock value in inventory management—for example:

  - Goods issue for outbound delivery with reference to sales order using movement type **601**.

  - Stock transfer process involving goods issue for outbound delivery using movement types **641** (goods issue for intracompany stock transfer) and **643** (goods issue for intercompany stock transfer). The goods issue from the supplying plant triggers transaction event **GBB**.

  - Goods issue of stock for consumption for a production order using movement type **261**.

  - Goods issue of stock for consumption for a cost center using movement type **201**.

- **PRD**: Price differences between a material standard price and transactions that differ the standard price—for example:

  - Goods receipt with reference to a purchase order where the material standard price differs from the purchase order unit price.

  - Invoice receipt with reference to a purchase order where the material standard price differs from the unit price in the supplier invoice.

- **WRX**: Goods receipt/invoice receipt clearing account. This transaction event is triggered when a goods receipt is posted with reference to a purchase order and an invoice receipt is posted with refence to a purchase order. The system records the credit and debit entries using the **WRX** transaction event key for goods receipt and invoice receipt postings.

- **BSV**: Stock change account for materials with a standard price. This transaction event is triggered when there is a change to the inventory value during a material movement transaction.

- **VBO**: Consumption of stock provided to a vendor. This transaction event is triggered during the goods receipt from a subcontractor with reference to a subcontracting order. The component consumption that happens during the goods receipt triggers the **VBO** transaction event to record the consumption of components stock provided to the subcontractor.

– **VBR**: Internal goods issue. This transaction event is triggered to record the value of goods issue postings to a production order or a maintenance order. The **VBR** transaction event records the consumption of materials issued from inventory.

– **KBS**: Account-assigned purchase order. This transaction event is triggered during the goods receipt from a vendor with reference to an account-assigned purchase order.

- **Val.strng (value string)**
A value string is also called a posting string, and it contains the rules for every goods movement transaction to determine accounts automatically. The system determines the transaction event key using the value string. It controls the material master data updates and quantity updates through the goods movement posting. The following are the important value strings in the system:

– **WE01**: This value string is assigned for goods receipt into inventory with reference to purchase orders in the standard system.

– **WE01**: This value string is assigned for goods issues.

- **Acct Mod. (account modifier)**
An account modifier or account grouping code further simplifies the automatic account determination when a goods movement transaction is posted in the system. These codes in the standard system are assigned to some of the transaction event keys to refine the account determination further and record the inventory valuation accurately in financial accounting. Account modifiers are assigned to only the following transaction event keys:

– **GBB**
The following account modifiers are assigned based on the goods movement transaction in the standard system for transaction event key **GBB**:

  - **AUF**: For goods receipts for production orders (movement type **101**)
  - **BSA**: For initial entries of stock balances (movement type **561**)
  - **INV**: For expense/revenue from inventory differences (movement type **701**)
  - **VAX**: For goods issues for sales orders without account assignment object (movement type **601**)
  - **VAY**: For goods issues for sales orders with account assignment object (movement type **601**)
  - **VBO**: For consumption from stock of material provided to vendor (movement type **543**)
  - **VBR**: For internal goods issues (movement type **201**, **261**, etc.)
  - **VKA**: For consumption for sales order without delivery (movement type **351**, **411**, etc.)

10

- **VNG**: For scrapping/destruction (movement type **551**)
- **VQP**: For sampling (movement type **333**)
- **ZOB**: For goods receipts without purchase orders (movement type **501**)
- **ZOF**: For goods receipts of a byproduct (movement type **582**)
- **PRD**: The following account modifiers are assigned based on the goods movement transaction in the standard system for transaction event key **PRD**:
  - \<Blank\>: Blank value for goods receipts and invoice receipts for purchase orders
  - **PRF**: For goods receipts for production orders
  - **PRA**: For goods issues and other goods movements
- **KON**: The following account modifiers are assigned based on the goods movement transaction in the standard system for transaction event key **KON**:
  - \<Blank\>: Blank value for consignment liabilities
  - **PIP**: For pipeline liabilities

Configure Automatic Postings

This is an important configuration step where we tie all the settings together to automatically determine accounts during a material movement. In this configuration step, automatic account determination can be activated for a transaction event key by defining an account determination procedure. We use the following settings we configured in the previous steps for automatic account determination:

- Transaction event key
- Chart of accounts
- Valuation grouping code
- Valuation class
- Account modifier

To arrive at this configuration step, execute Transaction SPRO and navigate to **Materials Management • Valuation and Account Assignment • Account Determination • Account Determination Without Wizard • Configure Automatic Postings**.

General ledger accounts need to be created as a prerequisite for this configuration. Figure 10.80 shows the automatic posting procedure activated for every transaction event key in the standard system. Let's discuss how to assign accounts by examining transaction event keys **BSX** and **GBB**. An account modifier or account grouping code is not assigned to transaction event key **BSX**, but account modifiers are assigned to transaction event key **GBB**.

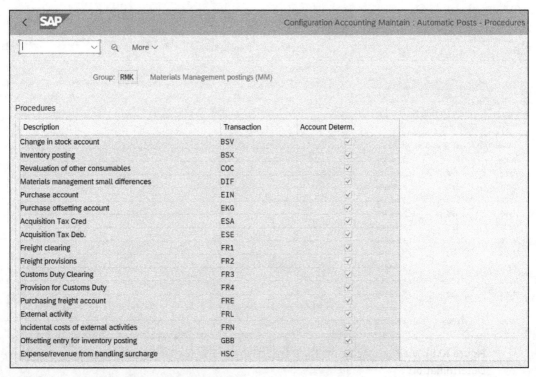

Figure 10.80 Automatic Posting Procedures

Double-click account posting procedure **BSX** to reach the screen shown in Figure 10.81, which is the maintenance view for assigning accounts for automatic account determination for transaction event key **BSX**. You can list the valuation modifier and valuation class directly, and then assign the general ledger account. There are no account modifiers applicable to transaction event key **BSX**; account determination happens based on the valuation grouping code and valuation class. If you have not grouped the valuation area using a valuation grouping code, you can leave the **Valuation mo...** (valuation modifier) column blank. An account must be assigned to a combination of valuation grouping code and valuation class.

For transaction event key **GBB**, because an account grouping code or account modifier exists for it, the account assignment for automatic account determination will be defined using the account modifier in addition to the valuation grouping code and the valuation class. Figure 10.82 shows the maintenance view for assigning accounts for automatic account determination for transaction event key **GBB**. Assign a general ledger account for the combination of valuation grouping code (**Valuation mo...**), account modifier (**General modi...**), and valuation class (**Valuation class**).

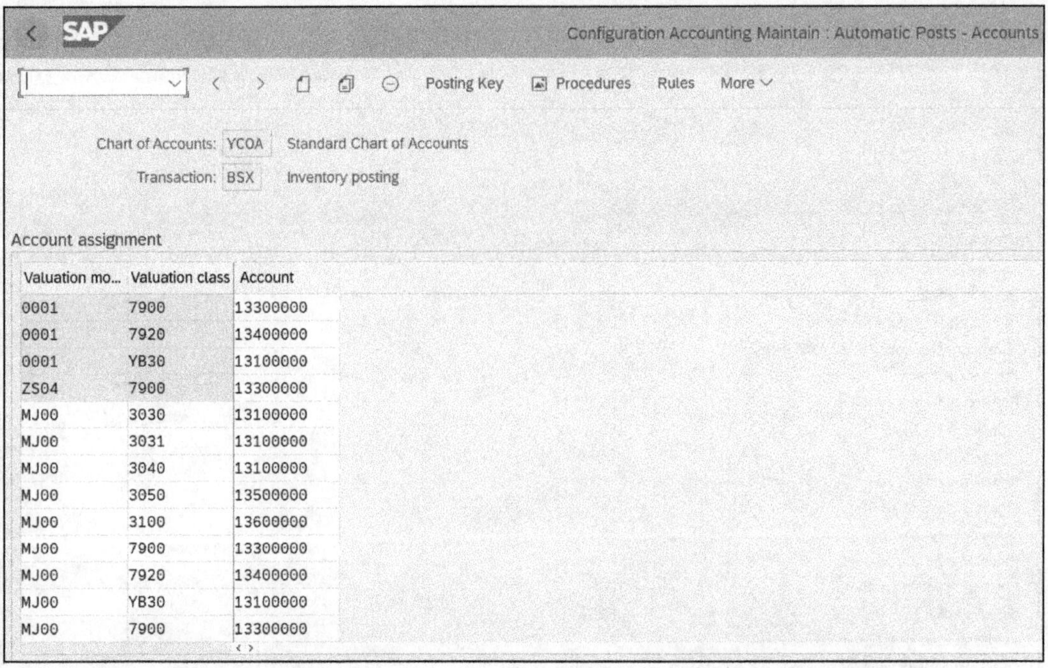

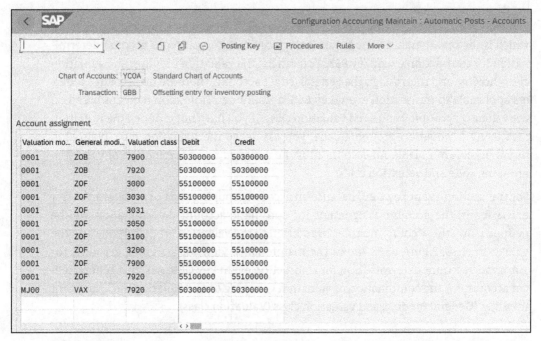

Figure 10.81 Maintain Accounts for Transaction Key BSX for Automatic Account Determination

Figure 10.82 Maintain Accounts for Transaction Key GBB for Automatic Account Determination

Figure 10.83 shows the account postings in financial accounting for the goods issue for outbound delivery involving movement type **601** in plant **MJ00** where the valuation area is the same as the plant. You can see that transaction events **BSX** and **GBB** were posted, and the relevant accounts were determined automatically based on the configuration settings.

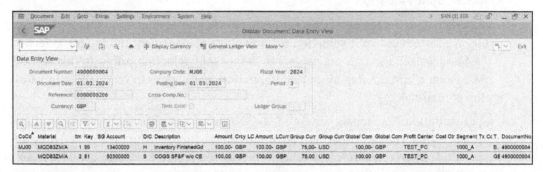

Figure 10.83 Automatic Account Determination for Goods Issue for Outbound Delivery

Figure 10.84 shows the accounting document generated automatically upon posting the goods receipt for a purchase order in plant **MJ00**.

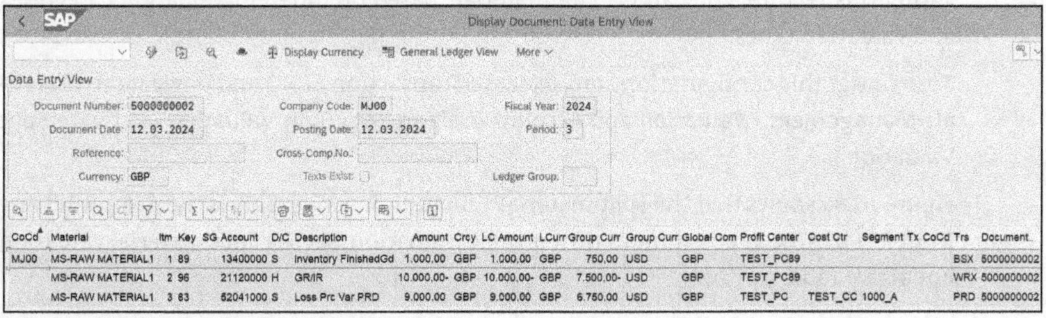

Figure 10.84 Automatic Account Determination for Goods Receipt for Purchase Order

The standard price of the material maintained in the **Accounting 1** view of the material master is 10 GBP, and the unit price of the purchase order is 100 GBP, so there is a difference of 90 GBP to be posted to the price variance account. The following transaction events are triggered to record financial transactions:

- **BSX**
  Inventory accounts are posted at a valuated price of 10.00 GBP per unit and a total value of 1000.00 GBP is debited from inventory account **13400000**.

- **WRX**
  Goods receipt/invoice receipt clearing account **21120000** was credited with 10,000.00 GBP, which is the total purchase order value received.

- **PRD**

  Price variance account **52041000** was debited with the price difference between the valuated price and the purchase order price.

10.3.3   Split Valuation

Material valuation is a crucial aspect of inventory management and financial reporting. We introduced split valuation for inventory management in Chapter 7, Section 7.4.4 and walked through the process to define split valuation prices. In this section, we'll expand into the step-by-step configuration instructions to set up the split valuation of materials.

Split valuation functionality in SAP S/4HANA can be configured based on business requirements. You can either turn on or turn off the functionality, as well as define different valuation categories and valuation types to accurately valuate inventory. Let's dive deep into these settings.

Activate Split Valuation

In this step, you can activate split valuation functionality for an organization. By activating this feature, you can valuate materials based on different valuation categories and valuation types.

To arrive at this configuration step, execute Transaction SPRO and navigate to **Materials Management** • **Valuation and Account Assignment** • **Split Valuation** • **Activate Split Valuation**.

Figure 10.85 shows that the split material valuation is active. This is the default setting in SAP S/4HANA. If you want to turn off split valuation, set the **Split material valuation not active** radio button.

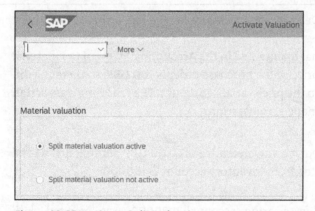

Figure 10.85 Activate Split Valuation

Configure Split Valuation

In this step, you can define global valuation categories and global valuation types and assign the required valuation categories and valuation types to plants to set local definitions based on business requirements. By setting the local definitions, you can select only the required valuation categories and types at the plant level to activate split valuation functionality for an organization. To arrive at this configuration step, execute Transaction SPRO and navigate to **Materials Management • Valuation and Account Assignment • Split Valuation • Configure Split Valuation**.

Figure 10.86 shows the initial screen for split valuation configuration. You can see **Global Types**, **Global Categories**, and **Local Definitions** tabs at the top.

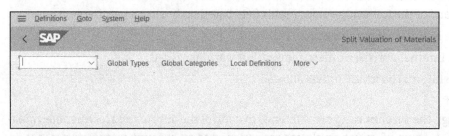

Figure 10.86 Configure Split Valuation: Initial Screen

The first step in this configuration setting is to define global valuation types. Click **Global Types** from the initial screen to arrive at the screen to define global valuation types, as shown in Figure 10.87. To create new valuation types, click the **Create** button. You can also **Change** and **Delete** existing valuation types.

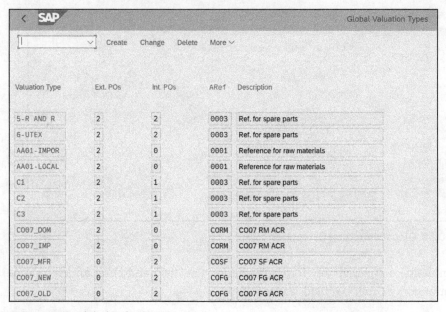

Valuation Type	Ext. POs	Int. POs	ARef	Description
5-R AND R	2	2	0003	Ref. for spare parts
6-UTEX	2	2	0003	Ref. for spare parts
AA01-IMPOR	2	0	0001	Reference for raw materials
AA01-LOCAL	2	0	0001	Reference for raw materials
C1	2	1	0003	Ref. for spare parts
C2	2	1	0003	Ref. for spare parts
C3	2	1	0003	Ref. for spare parts
CO07_DOM	2	0	CORM	CO07 RM ACR
CO07_IMP	2	0	CORM	CO07 RM ACR
CO07_MFR	0	2	COSF	CO07 SF ACR
CO07_NEW	0	2	COFG	CO07 FG ACR
CO07_OLD	0	2	COFG	CO07 FG ACR

Figure 10.87 Global Valuation Types

You can define up to a 10-digit key for **Valuation Type** and enter an appropriate description for it. Other key fields are as follows:

- **Ext. POs**

  You can control if external purchase orders are allowed for this valuation type or not. Set one of the following values in this field based on business requirements:

  - **0**: No external purchase orders allowed
  - **1**: External purchase orders allowed, but warning issued
  - **2**: External purchase orders allowed

- **Int. POs**

  You can control if internal purchase orders are allowed for this valuation type or not. Set one of the following values in this field based on business requirements:

  - **0**: No internal purchase orders allowed
  - **1**: Internal purchase orders allowed, but warning issued
  - **2**: Internal purchase orders allowed

- **ARef**

  Assign the account category reference to control the automatic account determination. A group of valuation classes are assigned to an account category reference. In turn, a valuation class is assigned to an inventory account.

The next step in this configuration setting is to define global valuation categories. Click **Global Categories** from the initial screen to arrive at the screen to define global valuation categories, as shown in Figure 10.88. To create new valuation categories, click the **Create** button. You can also **Change** and **Delete** existing valuation categories.

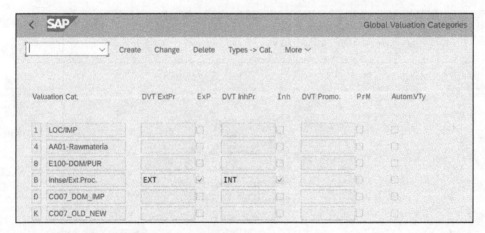

Valuation Cat.		DVT ExtPr	ExP	DVT InhPr	Inh	DVT Promo.	PrM	Autom.VTy
1	LOC/IMP							
4	AA01-Rawmateria							
8	E100-DOM/PUR							
B	Inhse/Ext.Proc.	EXT	✓	INT	✓			
D	CO07_DOM_IMP							
K	CO07_OLD_NEW							

Figure 10.88 Global Valuation Categories

You can define a one-digit key for the **Valuation Cat.** field and enter an appropriate description for it. Other key fields are as follows:

- **DVT ExtPr**

  You can define a default valuation type for external procurement in this field. Select a default valuation type from the dropdown menu.

- **DVT InhPr**

  You can define a default valuation type for in-house production in this field. Select a default valuation type from the dropdown menu.

The next step in this configuration setting is to allocate global valuation types to global valuation categories. Click **Types -> Cat** from the global valuation category definition screen to assign valuation types to a valuation category, as shown in Figure 10.89. Valuation types **C1**, **C2**, and **C3** are allocated to valuation category **X** (batch).

The system displays all valuation types that you can **Activate** or **Deactivate** for a valuation category. Use the function + page up and page down keys on your keyboard to display the full list of valuation types.

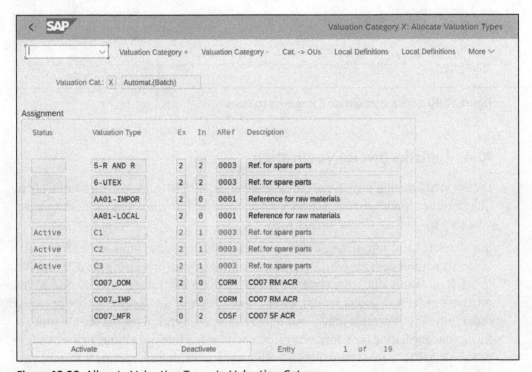

Figure 10.89 Allocate Valuation Types to Valuation Category

The next step in this configuration setting is to define local definitions. Click **Local Definitions** from valuation types to reach the valuation category allocation screen, where you can define local definitions at the plant level. Figure 10.90 shows that valuation category **X** (batch) is allocated to plant **MJ00**. You can activate multiple valuation categories for a plant. The system displays all global valuation categories that you can **Activate** or **Deactivate** for a plant. Use function + page up and page down keys on your keyboard to display the full list of valuation categories.

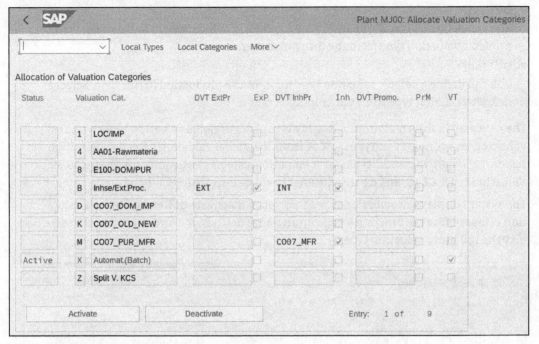

Figure 10.90 Allocate Valuation Categories to Plant

10.4   Logistics Invoice Verification

LIV is a vital function of materials management in SAP S/4HANA. We introduced LIV in Chapter 1, Section 1.3; in this section, we'll dive deeper into the core processes and provide step-by-step configuration instructions.

Standardizing and automating LIV processes across all business units is the focus of the accounts payable department. During the incoming invoice processing in SAP S/4HANA using Transaction MIRO, the system performs a three-way match of an invoice with a purchase order and goods receipt, or two-way match of the invoice with a purchase order only. This process is called *invoice reconciliation*, a critical step of the LIV process. Invoice reconciliation is performed against set rules and tolerance limits in the configuration. Only a fully reconciled invoice without any matching exceptions is free for payment; otherwise, the system blocks the invoice for payment until the matching exceptions are resolved and the invoice is released for payment.

A duplicate check is standard functionality in SAP S/4HANA, and the system automatically performs the duplicate check depending on the criteria set in the configuration settings. You can set the duplicate check criteria based on company code, invoice reference number, and invoice date. The system issues an error message if a duplicate invoice is being entered using Transaction MIGO.

Exception handling to resolve invoice matching exceptions such as price variance, quantity variance, tax variances and so on is an important step of the LIV process. The system sets the payment block if there are any matching exceptions found during invoice reconciliation. Depending on the exception, additional goods receipt may need to be posted, or the purchase order price or quantity may need to be updated to resolve the invoice exception.

> **Note**
>
> SAP S/4HANA provides Transaction MRBR to release blocked invoices manually or automatically via a background job. This is a powerful transaction, and access to it is limited to a few people in the accounts payable department. Best practice is to resolve the exception to release the invoice by fixing the underlying problem: posting goods receipt, updating the purchase order, and so on.

Figure 10.91 shows the invoice-to-pay process. This process integrates materials management and financial accounting functions. While LIV is the process of verifying and reconciling incoming invoices, posting the invoice receipt generates an accounting document to post the financial transaction into financial accounting.

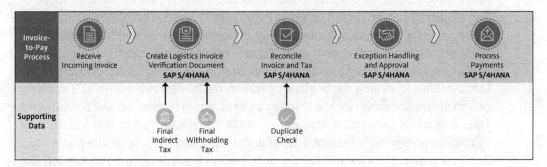

Figure 10.91 Invoice-to-Pay Process

Tax reconciliation is another important step of the LIV process. The vendor-submitted taxes are verified and reconciled. The system will allow you to post the invoice only if the taxes are entered accurately and reconciled fully. The taxes are posted to financial accounting for tracking, auditing, and reporting purposes. Reconciled indirect taxes are paid to the vendor, and in turn the vendor will pay the taxes to the local tax authority. Withholding taxes, if applicable, are collected and withheld by the buying organization and paid directly to the tax authority.

The accounting document generated during the incoming invoice posting displays the vendor account with the amount to be credited, the goods receipt/invoice receipt clearing account with the amount to be debited, and other accounts depending on various factors and business scenarios. The credited vendor account will create vendor open items to be cleared. During payment processing, these vendor open items will be processed depending on the due date of the payment set in the payment terms, and the payment will be made to the supplier based on the payment method.

Let's explore the core LIV processes in detail, including reconciliation, duplicate check, invoice parking and posting, credit memos, tax reconciliation, and payment processing and remittance.

10.4.1   Reconciliation of Incoming Invoice

Invoice reconciliation is the process of identifying and resolving the discrepancies between incoming invoices from the supplier and related documents such as purchase order/scheduling agreements and goods receipt. The main objective of invoice reconciliation is to ensure the financial accounting postings are clean and the entries made in financial books are matched correctly.

The process of invoice reconciliation involves receiving a supplier invoice electronically, via email as a PDF copy, or via fax/mail, posting of the invoice receipt in SAP S/4HANA by the accounts payable personnel, and having the SAP S/4HANA system determine the discrepancies between the invoice receipt and the related documents. Invoice receipts can be posted manually with reference to the incoming invoice document submitted by the supplier or automatically if the supplier submits the incoming invoice electronically via EDI or via SAP Business Network. During the creation of an invoice receipt, the system automatically matches the invoice receipt with the purchase order/scheduling agreement and goods receipt document. If there are any discrepancies found that are outside of the tolerances set, the system blocks the invoice for payment. Depending on the blocking reason, the invoice reconciliation is also a process to determine if the blocked invoice can be accepted or rejected. SAP S/4HANA provides a standard transaction, Transaction MRBR, to release blocked invoices manually by the accounts payable personnel if the discrepancies are within the acceptable range. But certain discrepancies are unacceptable, such as a goods receipt quantity less than the invoice quantity, purchase order quantity less than the invoice quantity, and so on. To resolve such exceptions, collaboration with the supplier is required, and posting of goods receipt, adjusting purchase order quantity and amount, and so on might also be required. In case of any disputes, the invoice document is rejected back to the supplier. In such cases, the supplier must resubmit the corrected invoice.

The invoice reconciliation process in SAP S/4HANA is capable enough to determine a duplicate invoice while posting the invoice receipt in the system and can automatically reject duplicate invoices. Tax reconciliation is another feature of invoice reconciliation, where the taxes charged by the supplier are matched against the system-calculated tax using the internal tax engine of SAP S/4HANA. The invoices can be blocked due to tax discrepancies like invoice, purchase order, and goods receipt matching discrepancies.

In this section, we'll explain the invoice reconciliation process in detail. But before that let's dive deep into the two types of matching process involving incoming invoices, purchase order/scheduling agreements, and the goods receipt documents:

- **Three-way match**

  In a three-way match type of invoice reconciliation, the invoice receipt is matched against the purchase order/scheduling agreement and goods receipt document. If there are any discrepancies found, the invoice will be blocked for payment.

  Figure 10.92 shows the illustration of a three-way match scenario where a supplier has submitted a partial invoice with quantity 50 EA, unit price of $25 per unit, and total amount of $1,250.

  The invoice line item unit price is matched to the respective purchase order line item. This is an exact match as the purchase order line item unit price is $25 per unit.

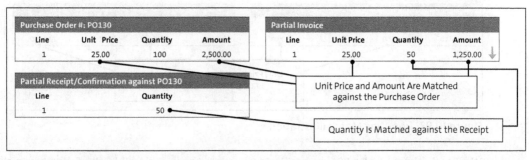

Figure 10.92 Three-Way Match Illustration

  The invoice line item quantity is matched to the purchase order line item quantity and the received quantity of the goods receipt item. This is a match as the invoice quantity is less than or equal to the purchase item quantity and the received quantity. This is how the system performs the three-way match of the incoming invoice.

- **Two-way match**

  In a two-way match type of invoice reconciliation, the invoice receipt is matched against the purchase order only. Certain purchase orders such as framework orders, limit orders created for indirect procurement and for the payment of property rents, marketing/advertising charges, legal charges, tax penalties, and so on do not require goods receipt to be posted. Instead, invoices are approved by the respective departments or requisitioners for payment. During the two-way match, if there are any discrepancies found, the invoice will be blocked for payment.

  Figure 10.93 shows the illustration of a two-way match scenario for both a full invoice and a partial invoice. Scenario 1-A shows the partial invoice scenario, where the purchase order item quantity is 100 EA and unit price is $25 per unit, and the received incoming invoice item quantity is 50 EA and the unit price is $25 per unit. The system passes the two-way match for this scenario as the unit price is an exact match between the purchase order and invoice, and the quantity of the invoice item is less than or equal to the purchase order item quantity.

  Scenario 1-B shows the full invoice scenario, where the purchase order item quantity is 100 EA and unit price is $25 per unit, and the received incoming invoice item quantity is 100 EA and the unit price is $25 per unit. The system passes the two-way match for this scenario as the unit price and quantity exactly match between the purchase order and invoice.

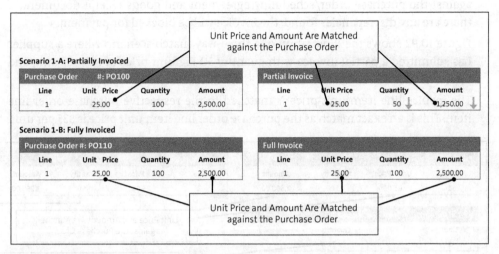

Figure 10.93 Two-Way Match Illustration

We'll walk through the LIV reconciliation process in more detail in the following sections, starting with a simulation of a three-way match and continuing with the step-by-step configuration instructions.

Three-Way Match Simulation

Let's simulate the three-way match functionality by posting a LIV document. Figure 10.94 shows the LIV of incoming invoice **INV671-1** for purchase order **4500000671**.

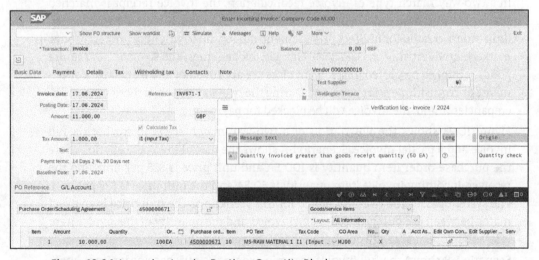

Figure 10.94 Incoming Invoice Posting: Quantity Block

During the invoice verification using Transaction MIRO, the system has performed a three-way match and issued a warning message due to discrepancies with the ordered quantity of 100 EA, received quantity of 50 EA, and invoiced quantity of 100 EA, as shown in Figure 10.95. Because the invoiced quantity of 100 EA is greater than the received quantity of 50 EA, the system is blocking the invoice with a quantity block set at the item level, as shown in Figure 10.94 with the **X** indicator set in the **Qty** column in the line item of the invoice.

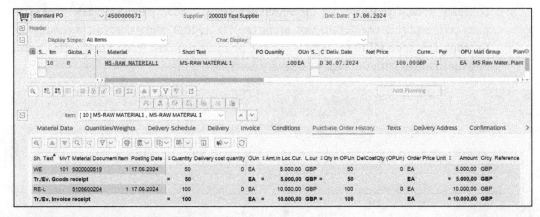

Figure 10.95 Purchase Order History

Due to the quantity discrepancy, the system automatically blocks the invoice for payment. Figure 10.96 shows that a payment block due to invoice verification was set in the **Payment** tab of invoice **5105600204**.

Figure 10.96 Invoice Payment Block

To release the payment block, the remaining quantity of 50 EA must be received using Transaction MIGO (Goods Receipt Posting). Unless the invoice is cleared for payment, you can't process the payment for the invoice.

Maintain Default Values for Tax Codes

In this configuration step, you can maintain default values for tax codes that will be populated during the incoming invoice entry using Transaction MIRO. The system automatically copies the default tax codes set at the company code level into the LIV screen.

To arrive at this configuration step, execute Transaction SPRO and navigate to **Materials Management • Logistics Invoice Verification • Incoming Invoice • Maintain Default Values for Tax Codes**.

The system displays an overview of all existing company codes for which the default tax codes are maintained. Click **New Entries** to maintain default tax codes for a new company code. To edit/display the details of an existing company code, select the company code and click the **Details** icon at the top. Figure 10.97 shows default tax code I1 set for company code **MJ00** for both domestic invoices received from a domestic vendor and for invoices with unplanned delivery costs.

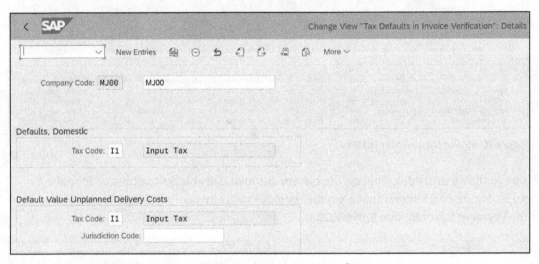

Figure 10.97 Default Values for Tax Codes in Invoice Verification

Configure Supplier-Specific Tolerances

In this configuration step, you can maintain supplier-specific tolerances for each company code using a logical key called a tolerance group. You can define multiple tolerance groups for a company code that have different tolerance values set for invoice verification.

To arrive at this configuration step, execute Transaction SPRO and navigate to **Materials Management • Logistics Invoice Verification • Incoming Invoice • Configure Supplier-Specific Tolerances**.

Figure 10.98 shows two tolerance groups, **0001** and **0002**, created for company code **MJ00**. To define new tolerance groups, click **New Entries** and enter a four-digit numeric key and description for the tolerance group. You can copy an existing tolerance group and create a new one from it (recommended). Once you define the tolerance groups, assign one of them to certain business partners (vendors) and the other one to a different set of business partners (vendors).

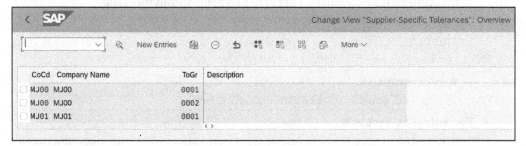

Figure 10.98 Supplier-Specific Tolerances: Overview

To display the details of an existing tolerance group assigned to a company code, select the company code and the tolerance group combination, and then click the **Details** icon at the top. Figure 10.99 shows the details of tolerance group **0001** defined for company code **MJ00**. The header section of the configuration details screen shows currency **GBP** for company code **MJ00** as a default value that can't be changed.

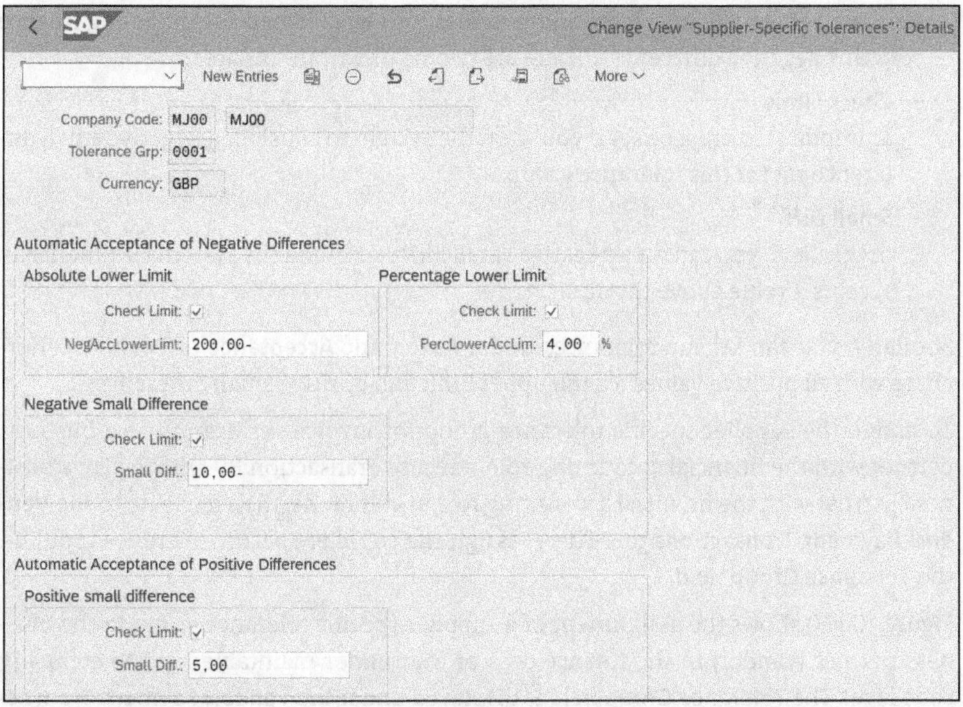

Figure 10.99 Supplier-Specific Tolerances: Details

The **Automatic Acceptance of Negative Differences** section of the configuration step has the following details:

- **Absolute Lower Limit**
  This is an absolute tolerance, and it comes into effect when the incoming invoice contains negative differences. There are two fields within this area:
  - **Check Limit**
    This indicator must be set if you want the system to consider the absolute lower limit.
  - **NegAccLowerLimit**
    In this field, you can set a negative absolute limit.
- **Percentage Lower Limit**
  This is a percentage tolerance, and it comes into effect when the incoming invoice contains negative differences. There are two fields within this area:
  - **Check Limit**
    This indicator must be set if you want the system to consider the percentage lower limit.
  - **PercLowerAccLim**
    In this field, you can set a negative percentage limit.
- **Negative Small Difference**
  This is a percentage tolerance, and it comes into effect when the incoming invoice contains negative differences. There are two fields within this area:
  - **Check Limit**
    This indicator must be set if you want the system to consider a negative small difference set for this tolerance group.
  - **Small Diff.**
    In this field, you can set a negative small difference, and the system automatically accepts a value above this limit.

Similarly, you can set the tolerances under **Automatic Acceptance of Positive Differences** with all positive values for the upper limit and positive small differences.

To assign the supplier-specific tolerance group for invoice verification to a business partner with the financial accounting role, execute Transaction BP and display a business partner with the financial accounting role in edit mode, then navigate to the **Vendor: Payment Transactions** tab. There, assign the supplier-specific tolerance group in the **Tolerance Group** field.

Figure 10.100 shows the assignment of a supplier-specific tolerance group to the business partner (vendor) in the finance data of the vendor maintained at the company code level. The **Tolerance Group** field is displayed under the **Vendor: Payment Transactions** tab. The dropdown shows tolerance groups **0001** and **0002** defined for company code **MJ00.** You can assign one of the tolerance groups in this field.

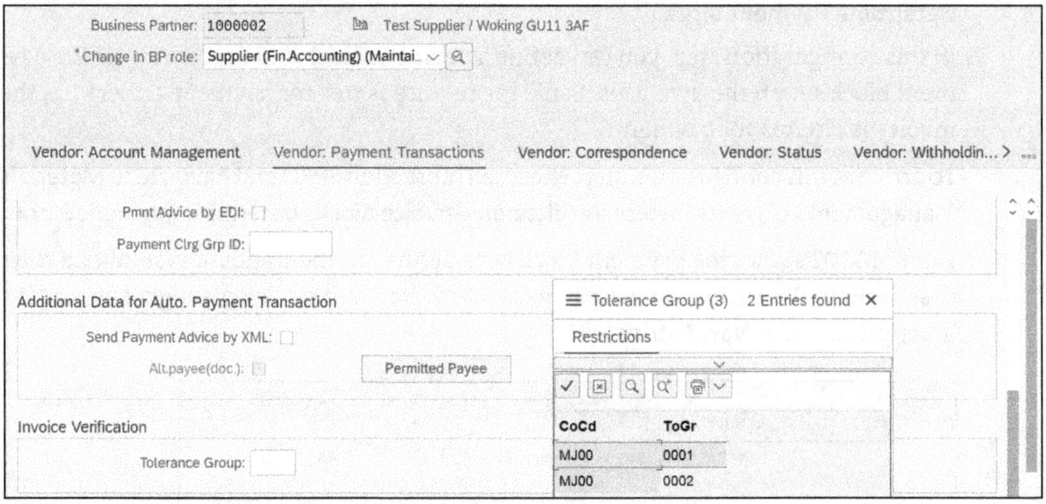

Figure 10.100 Assign Tolerance Group to Supplier Record

Activate Direct Posting to Alternative G/L Accounts

In this configuration step, you can set whether or not direct postings to general ledger accounts, material inventory accounts, and/or asset accounts are possible during incoming invoice posting. During LIV using Transaction MIRO, depending on the configuration settings, additional tabs will appear so that direct postings can be made.

To arrive at this configuration step, execute Transaction SPRO and navigate to **Materials Management • Logistics Invoice Verification • Incoming Invoice • Activate Direct Posting to Alternative G/L Accounts**.

Figure 10.101 shows three indicators: **Dir. posting to G/L acct = active**, **Dir. posting to matl = active**, and **Dir. posting to asset = active**. If you set all the three indicators, then in Transaction MIRO, three additional tabs will appear for the general ledger account, material, and asset. You can make direct postings when you set these indicators.

Figure 10.101 Direct Posting in Logistics Invoice Verification

Determine Payment Block

In this configuration step, you can define a payment block and description. If the payment block set on the invoice is blank, the invoice is free for payment. Otherwise, the invoice is blocked for payment.

To arrive at this configuration step, execute Transaction SPRO and navigate to **Materials Management • Logistics Invoice Verification • Invoice Block • Determine Payment Block**.

Figure 10.102 shows the payment block keys defined in the standard system and their descriptions. To define a new payment block key, enter a single-digit key and its description under **New Entries**.

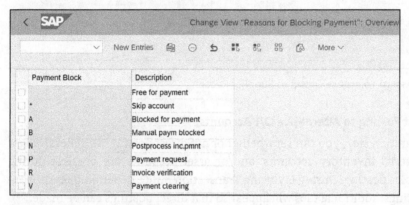

Figure 10.102 Reasons for Blocking Invoice Payment

Set Tolerance Limits for Invoice Postings

In this configuration step, you can define tolerance limits for incoming invoice postings. You can specify tolerance limits for tolerance keys defined for each company code. The tolerance keys are defined for specific checks during LIV.

The following are some of the tolerance keys defined in the standard system:

- **AN**

  This tolerance key checks every invoice line item without order reference against the absolute upper limit defined if you have activated the item amount check in configuration settings.

- **AP**

  This tolerance key checks every invoice line item with order reference against the absolute upper limit defined based on the item amount check setup in configuration settings.

- **BD**

  This tolerance key checks the invoice amount against the absolute upper limit defined, and if the upper limit is not exceeded, it automatically creates a posting line in the accounting document called *expense/income from small differences* upon posting the invoice.

- **BR**

  This tolerance check is triggered when an invoice is posted before the goods receipt. It calculates the percentage variance between the following ratios:

  - Quantity invoiced in order price quantity units to quantity invoiced in order units
  - Quantity ordered in order price quantity units to quantity ordered in order units

  The system compares the variance with the upper and lower percentage tolerance limits.

- **BW**

  This tolerance check is triggered when an invoice is posted after the goods receipt. It calculates the percentage variance between the following ratios:

  - Quantity invoiced in order price quantity units to quantity invoiced in order units
  - Goods receipt quantity in order price quantity units to goods receipt quantity in order units

  The system compares the variance with the upper and lower percentage tolerance limits.

- **PP**

  This tolerance check is based on the variance between the total item amount of the invoice and the quantity invoiced, multiplied by the order unit price based on the tolerance limits defined.

To arrive at this configuration step, execute Transaction SPRO and navigate to **Materials Management • Logistics Invoice Verification • Invoice Block • Set Tolerance Limits**.

Figure 10.103 shows the tolerance keys defined for company code **MJ00**. The tolerance keys are already defined in the standard system. Copy the existing ones to define new keys based on your business requirements.

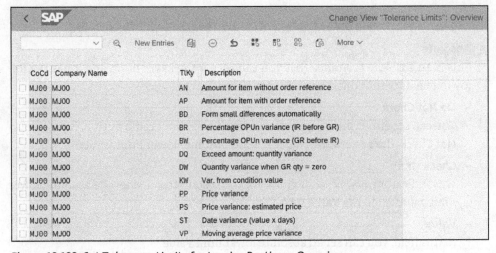

CoCd	Company Name		TlKy	Description
MJ00	MJ00		AN	Amount for item without order reference
MJ00	MJ00		AP	Amount for item with order reference
MJ00	MJ00		BD	Form small differences automatically
MJ00	MJ00		BR	Percentage OPUn variance (IR before GR)
MJ00	MJ00		BW	Percentage OPUn variance (GR before IR)
MJ00	MJ00		DQ	Exceed amount: quantity variance
MJ00	MJ00		DW	Quantity variance when GR qty = zero
MJ00	MJ00		KW	Var. from condition value
MJ00	MJ00		PP	Price variance
MJ00	MJ00		PS	Price variance: estimated price
MJ00	MJ00		ST	Date variance (value x days)
MJ00	MJ00		VP	Moving average price variance

Figure 10.103 Set Tolerance Limits for Invoice Postings: Overview

To maintain the tolerance limits for a tolerance key for a new company code, click **New Entries** and specify the company code and the standard tolerance key. To edit an existing entry, select the entry from the overview screen and click the **Details** icon at the top. Figure 10.104 shows the tolerance limits defined for tolerance key **PP** (price variance) in company code **MJ00**.

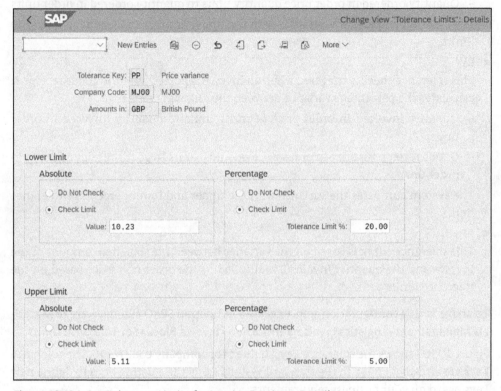

Figure 10.104 Set Tolerance Limits for Invoice Postings: Details

The **Lower Limit** section of the configuration step has the following details:

- **Absolute**
  This is an absolute tolerance that you can set for the lower limit, and it has the following fields within this area:
  - **Do Not Check**
    You can set either this radio button or the **Check Limit** radio button. If you set **Do Not Check**, the system will not check this absolute lower limit value.
  - **Check Limit**
    This indicator must be set if you want the system to consider the absolute lower limit value set in the **Value** field.
  - **Value**
    In this field, you can set an absolute lower limit value.

- **Percentage**
  This is a percentage tolerance that you can set for the lower limit, and it has the following fields within this area:
  - **Do Not Check**
    You can set either this radio button or the **Check Limit** radio button. If you set **Do Not Check**, the system will not check this percentage lower limit value.
  - **Check Limit**
    This indicator must be set if you want the system to consider the percentage lower limit value set in the **Value** field.
  - **Value**
    In this field, you can set a percentage lower limit value.

Similarly, you can set the tolerances under the **Upper Limit** section for tolerance key **PP**.

Activate Item Amount Check

In this configuration step, you can define an invoice item amount check during LIV where the system checks the invoice item amount against certain limits and blocks the invoice for payment if it exceeds those limits.

To arrive at this configuration step, execute Transaction SPRO and navigate to **Materials Management • Logistics Invoice Verification • Invoice Block • Item Amount Check • Activate Item Amount Check**.

As shown in Figure 10.105, activate the item amount check by selecting the **Check item amount** checkboxes for company codes **MJ00**, **MJ01**, and **MJ02**.

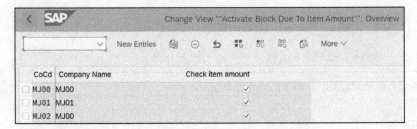

Figure 10.105 Activate Invoice Block Due to Item Amount

Set Item Amount Check

In this configuration step, you can set the item amount check for different item categories for the purchasing documents, such as standard, consignment, subcontracting, limit, service, and so on, at the company code level. You can define multiple entries for the combination of company code and item category.

To arrive at this configuration step, execute Transaction SPRO and navigate to **Materials Management • Logistics Invoice Verification • Invoice Block • Item Amount Check • Set Item Amount Check**.

Figure 10.106 shows that the item amount check has been set for company code **MJ00** and item category **Standard**. You can set the **Goods Receipt** indicator if you want the system to consider goods receipt for the item amount check.

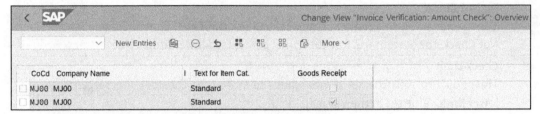

Figure 10.106 Set Item Amount Check for Invoice Verification

Activate Stochastic Block

In this configuration step, you can activate a stochastic block so that the system will block incoming invoices randomly based on the degree of probability set in the configuration.

To arrive at this configuration step, execute Transaction SPRO and navigate to **Materials Management • Logistics Invoice Verification • Invoice Block • Stochastic Block • Activate Stochastic Block**.

Figure 10.107 shows that a stochastic block has been activated for company code **MJ00**. You can directly set the **Stochastic block** indicator to activate it for a company code.

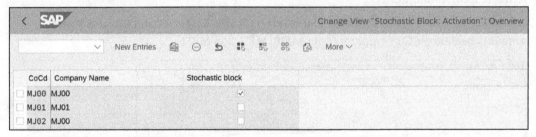

Figure 10.107 Activate Stochastic Block for Invoice Verification

Set Stochastic Block

In this configuration step, you can set the degree of probability for a stochastic block so that the system can block incoming invoices randomly.

To arrive at this configuration step, execute Transaction SPRO and navigate to **Materials Management • Logistics Invoice Verification • Invoice Block • Stochastic Block • Set Stochastic Block**.

To set the stochastic block for a company code, click **New Entries** and specify the company code, threshold value, and percentage. To edit the existing entries, you can make the changes to the **Threshold value** and **Percentage** fields directly in the overview screen. Figure 10.108 shows the threshold value of 6000.00 in the local currency (GBP), and percentage value of 60% is set as the degree of probability for company code **MJ00**.

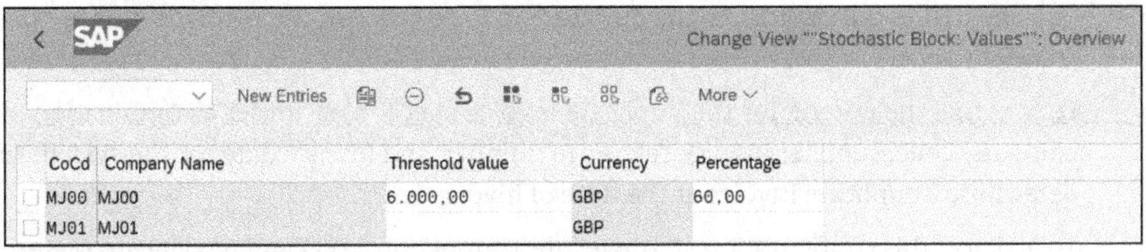

< SAP				Change View ""Stochastic Block: Values"": Overview	
	New Entries	⊖ ↺	More ∨		
CoCd	Company Name	Threshold value	Currency	Percentage	
☐ MJ00	MJ00	6.000,00	GBP	60,00	
☐ MJ01	MJ01		GBP		

Figure 10.108 Set Stochastic Block for Invoice Verification

For example, if the invoice value is 3,000 GBP, the degree of probability = 60 * 3000/ 6000 or 30%.

10.4.2   Duplicate Check

A duplicate check of the incoming invoice is a basic requirement of the accounts payable department. Processing a duplicate invoice is a loss for the organization, so it must be prevented every time a duplicate invoice is received from a supplier. SAP S/4HANA provides a standard functionality to check for duplicate invoices automatically during logistics invoice verification and displays an error message if a duplicate invoice is being entered using Transaction MIRO manually or received automatically via EDI or from a B2B application such as SAP Business Network.

One of the basic settings for the duplicate invoice check is to set the **Check Double Invoice** indicator in the business partner (supplier). Figure 10.109 shows the double invoice check in the **Vendor: Payment Transactions** tab of the vendor master data at the company code level, which is the financial master data of the vendor. During LIV, the system checks if the invoice has already been entered; if so, it prevents a duplicate entry.

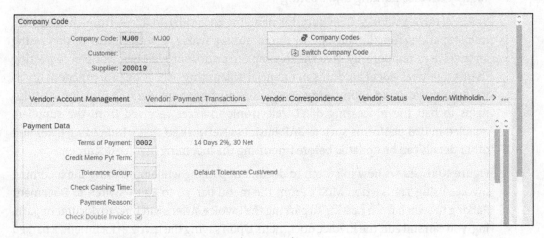

Figure 10.109 Double Invoice Check in Supplier Record

You can define the criteria for the duplicate invoice check in the configuration for the company code.

Let's define the criteria for the duplicate invoice check. You can set a combination of company code, supplier invoice reference number, and invoice date for the system to determine a duplicate invoice at the time of invoice entry.

To arrive at this configuration step, execute Transaction SPRO and navigate to **Materials Management • Logistics Invoice Verification • Incoming Invoice • Set Check for Duplicate Invoices**.

Figure 10.110 shows the duplicate check criteria defined for company code **MJ00**.

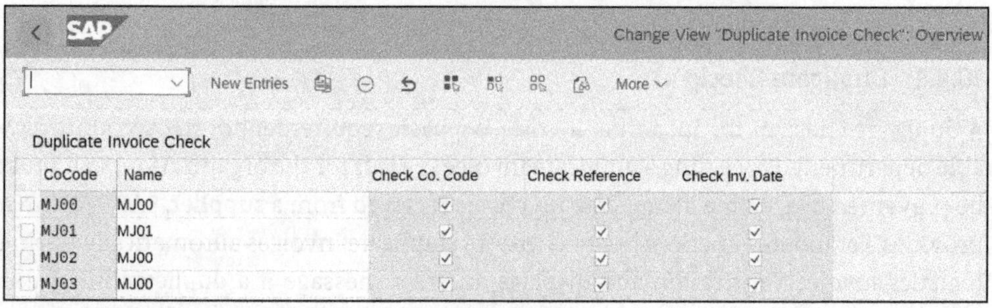

CoCode	Name	Check Co. Code	Check Reference	Check Inv. Date
MJ00	MJ00	✓	✓	✓
MJ01	MJ01	✓	✓	✓
MJ02	MJ00	✓	✓	✓
MJ03	MJ00	✓	✓	✓

Figure 10.110 Set Check for Duplicate Invoice Check

If you want the system to perform the duplicate invoice check with the combination of company code, supplier invoice reference number, and invoice date, set the **Check Co. Code**, **Check Reference**, and **Check Inv. Date** indicators for the respective company code.

10.4.3   Invoice Parking and Posting

SAP S/4HANA provides flexibility for handling incoming invoices from the supplier. Invoice parking functionality allows the accounts payable personnel who are processing an invoice to temporarily save the incomplete or unapproved invoices before posting them to general ledger accounts in financial accounting. This is helpful functionality to validate, update, and approve invoices before posting them. Parking functionality also helps to halt the processing of EDI/electronic invoices received from the suppliers enabled in B2B platforms such as SAP Business Network so that withholding taxes and other details can be updated before importing the incoming invoice document.

Figure 10.111 shows how to switch to document parking while processing the incoming invoice using Transaction MIRO. From the menu bar, go to **Edit • Switch to Document Parking** to switch from posting to parking the invoice. After switching to document parking, you can still edit the invoice document. Upon saving the invoice document, a parked document will be created and an internal SAP S/4HANA invoice number will be saved.

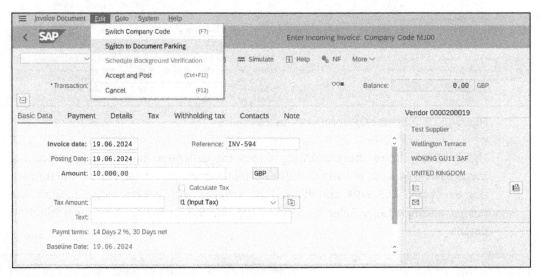

Figure 10.111 Parking of Incoming Invoice

Parked invoices created with reference to a purchase order are updated in the **Purchase Order History** tab of the purchase order at the line item level. You can display the parked invoice documents using Transaction MIR4. Parked invoice documents will not be linked to financial accounting, and the accounting document will not be generated in financial accounting as there are no account postings. Figure 10.112 shows parked invoice **5105600207** updated in the purchase order history.

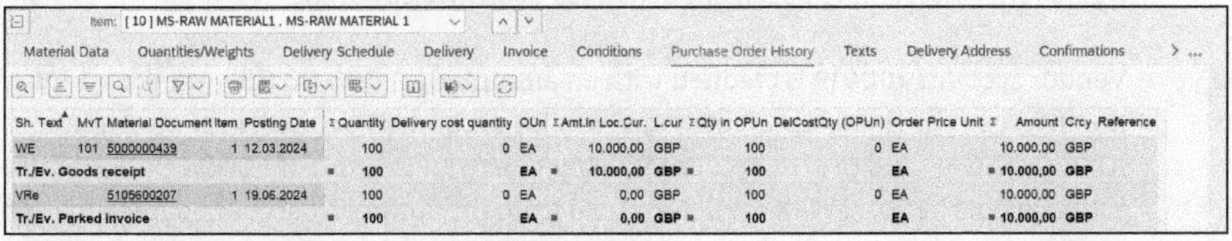

Figure 10.112 Purchase Order History: Parked Invoice

On the other hand, posting an incoming invoice using Transaction MIRO will generate an accounting document in financial accounting. This is one of the integration points between materials management and financial accounting. The posting of the invoice document links the financial accounting with the invoice document. Posted invoices created with reference to a purchase order are updated in the **Purchase Order History** tab of the purchase order at the line item level. You can display the posted invoice documents using Transaction MIR4. Figure 10.113 shows invoice receipt **5105600207** updated in the purchase order history.

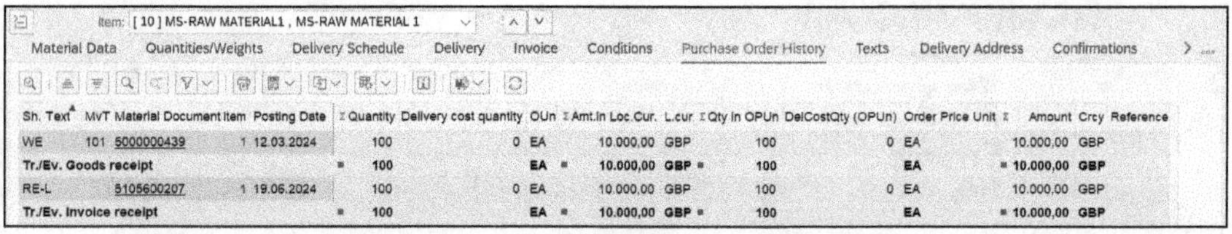

Figure 10.113 Purchase Order History: Posted Invoice

Figure 10.114 shows the accounting document generated upon posting incoming invoice **INV-594** from the supplier. The system automatically determines the transaction event keys **KBS**, **WRX**, and **VST** and the corresponding general ledger accounts are posted automatically. Refer to Section 10.3 for automatic account determination details.

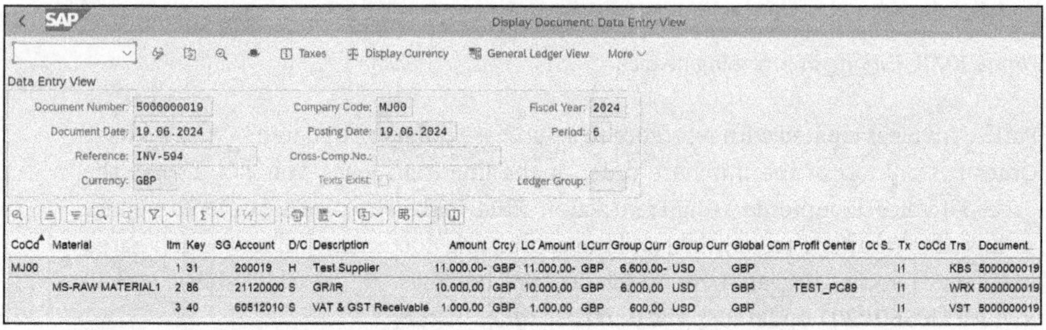

Figure 10.114 Accounting Document Generated upon Invoice Posting

Vendor account **200019** is credited with an amount of 11,000.00 GBP (total amount of the invoice), the goods receipt/invoice receipt account is debited with an amount of 10,000.00 GBP, and the tax account is debited with an amount of 1,000.00 GBP. Transaction events **KBS** and **WRX** are determined from the automatic account determination configuration, and transaction event **VST** is determined from the tax code settings.

The posting to the vendor account generates vendor open items, and after the payment processing these open items will be cleared.

10.4.4   Credit Memo

A *credit memo* is a document that reduces the amount and quantity from a previously posted incoming invoice. A credit memo is submitted by the supplier whenever there is a price reduction due to an overpaid invoice, returns processing, and other quantity/ price adjustment reasons. Technically, in SAP S/4HANA, cancelling an invoice using Transaction MR8M automatically creates a credit memo; that is, it reverses the entire invoice posted previously and creates a credit memo document.

Like an incoming invoice, you can process the incoming credit memo using Transaction MIRO. Select **Credit Memo** for **Transaction Type** to create the credit memo. Figure 10.115 shows the process flow for processing supplier returns and for credit memo processing, which includes the following steps:

1. **Creation of returns purchase order**
   The buyer creates a returns purchase order using Transaction ME21N due to quality issues or if excess materials are ordered and initiates the material returns to the supplier. Shipping data is automatically populated at the item level to facilitate the creation of a returns delivery (outbound).

2. **Creation of returns delivery with reference to returns purchase order**
   Shipping data from the line item level of the returns purchase order is used to create a returns delivery using Transaction VL10B. Best practice is to automate the creation of a returns delivery (outbound) using a batch job (background processing). The returns delivery facilitates the movement of materials from the issuing plant mentioned in the returns purchase order item. Picking and packing of materials will be performed within inventory management with reference to the outbound delivery.

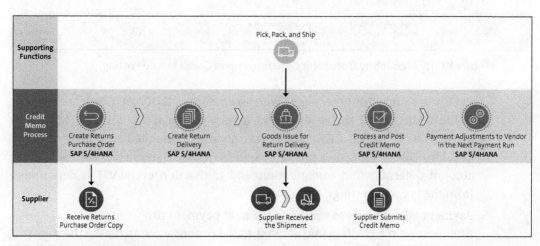

Figure 10.115 Credit Memo Process

3. **Goods issue for return delivery**
   After the picking and packing of returned materials, goods issue will be posted from the issuing plant. Material documents and accounting documents are created for the goods movement while updating the quantity and value of the issued stock in inventory management. Movement types play an important role in determining the type of inventory movement and posting of the goods issue. Movement type 161 is used for goods issue for a vendor returns delivery.

4. **Creation of incoming credit memo**
   Once the supplier receives the returns shipment, a credit memo is issued by their accounts receivable department. The incoming credit memo is processed using

Transaction MIRO with the transaction type set as a credit memo. The posting of the credit memo document links the financial accounting with the credit memo document. A posted credit memo created with reference to a purchase order is updated in the **Purchase Order History** tab of the purchase order at the line item level. You can display the posted credit memo documents using Transaction MIR4.

Figure 10.116 shows the accounting document generated upon posting the incoming credit memo from the supplier. The system automatically determines transaction event keys **KBS**, **WRX**, and **VST** and the corresponding general ledger accounts are posted automatically.

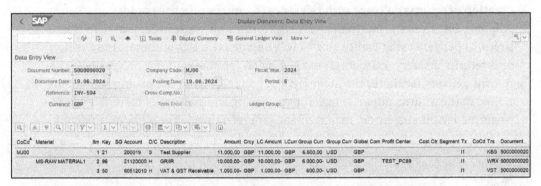

Figure 10.116 Accounting Document Generated upon Credit Memo Posting

Vendor account **200019** is debited with an amount of 11,000.00 GBP (total amount of the credit memo), the goods receipt/invoice receipt account is credited with an amount of 10,000.00 GBP, and the tax account is credited with an amount of 1,000.00 GBP. Transaction events **KBS** and **WRX** are determined from automatic account determination configuration, and transaction event **VST** is determined from the tax code settings.

5. **Payment adjustments to vendor in the next payment run**
   After posting the credit memo amount to the vendor account, during the payment run the system automatically deducts the amount of the credit memo if there are other invoices to be paid to the same supplier.

This is how the system keeps track of every financial transaction in financial accounting. The credit memo process helps to keep the books clean and accurate in financial accounting.

10.4.5   Tax Reconciliation

Indirect taxes charged by the supplier in an incoming invoice need to be verified. Tax reconciliation is part of the invoice reconciliation process and is the process of identifying and managing discrepancies between the tax submitted by the supplier in the incoming invoice and the tax determined within SAP S/4HANA through its native tax solution. The importance of tax determination and reconciliation during invoice management is as follows:

- **Compliance**
  It is important for accounts payable personnel from the buying organization to reconcile taxes and ensure compliance with local and international tax rules for indirect taxes and withholding taxes.

- **Reporting**
  It is important to capture and report reliable tax accounting data to the local tax authority.

- **Accuracy**
  It's best to avoid overpayments, underpayments, and tax penalties by accurately determining and reconciling tax.

The process of tax reconciliation involves verifying and matching the tax amount submitted by the supplier in the invoice against the system-calculated tax. The differences are thoroughly verified before posting the invoice. Upon posting the incoming invoice document, relevant tax accounts are posted based on the tax code settings.

The system uses the internal tax engine of SAP S/4HANA to calculate the tax amount automatically during LIV. Refer to Section 10.1.5 for more details on tax codes. The tax code contains the tax rate and transaction event keys to calculate the tax amount and to determine the relevant general ledger accounts automatically. Invoices can be blocked due to tax amount discrepancies and can trigger an exception handling workflow to route the invoices to the tax department of the organization to resolve these exceptions.

10.4.6   Payment Processing and Remittance

Payment processing is part of the accounts payable process. It involves managing vendor payments with reference to incoming invoices submitted by the suppliers upon providing requested goods and services. LIV generates the accounting documents and posts the vendor accounts with accurate amounts for the invoices to be paid. Posting of the vendor accounts creates vendor open items in financial accounting.

To execute an automatic payment run, use Transaction F110. You can create payment proposals, and the system will process the payments automatically. Figure 10.117 shows the parameters that control the automatic payment transactions.

The payment run can be executed for one or more vendor accounts in one or more company codes. In general, payment processing is a centralized function of the organization and executed periodically based on transaction volume, urgency, and other criteria.

The payment method plays an important role as it controls whether an eCheck, paper check, or direct bank posting must be created upon processing the payment to a vendor account. The vendor master data (business partner) contains the payment method information. During the payment processing, it can be overridden to update a specific payment method.

The supplier's bank account information is maintained in the business partner (vendor master data) along with the payment method, payment terms, withholding tax, and other data to support the payment run.

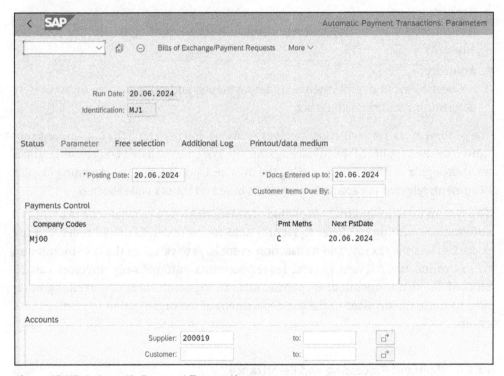

Figure 10.117 Automatic Payment Transaction

Payment remittances are generated as part of the payment run (Transaction F110). *Remittance* refers to the process of sending a detailed statement of payments made to the supplier. This helps suppliers reconcile the payments received with the invoices issued.

Payment remittances to suppliers can be sent via an email notification that can be generated automatically after the payment execution to the supplier's email ID. If the supplier is EDI-enabled or enabled on a B2B application such as SAP Business Network, payment remittances can be sent directly to the supplier's ERP system electronically.

Payment terms are other vital control data elements that define the conditions under which a vendor expects to receive the payment for the goods and services provided. Payment terms are assigned to the vendor master data (business partner) and defaulted into purchasing documents and invoice documents; however, they can be changed in the purchase order, outline agreements, and invoice. Payment terms are used to determine the due date of and the discounts applicable to the invoice.

To maintain payment terms, execute Transaction SPRO and navigate to **Financial Accounting • Accounts Receivable and Accounts Payable • Business Transactions • Incoming Invoices/Credit Memos • Maintain Terms of Payment**.

Figure 10.118 shows the details of payment terms (**Pyt Terms**) type **0002** (**14 Days 2%, 30 Net**).

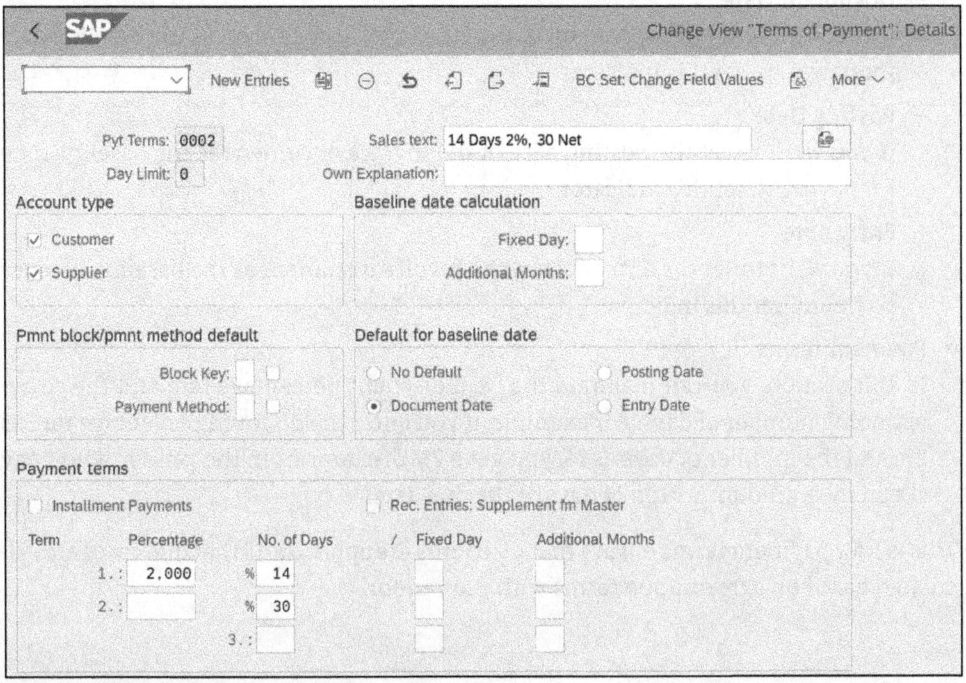

Figure 10.118 Maintain Payment Terms

Let's explore the important details of payment terms:

- **Account type**
  You can define payment terms for customers, suppliers, or both accounts. Set the **Supplier** and **Customer** indicators if the terms belong to both account types.

- **Baseline date calculation**
  The baseline date calculation is controlled by the following fields:

  - **Fixed Day**
    Enter a calendar day for the system to determine that calendar day as the baseline date for payment.

  - **Additional Months**
    Enter additional months as an integer value to calculate the baseline date for payment. The system adds that many months to the calendar month for the baseline date calculation.

845

- **Default for baseline date**

  You can set one of the following values as the default for baseline date determination for payment:

  - **No Default**

    If you want to enter the baseline date for payment manually, set this indicator.

  - **Document Date**

    If you want to set the document date of the invoice document as the baseline date for payment, set this indicator.

  - **Posting Date**

    If you want to set the posting date of the invoice document as the baseline date for payment, set this indicator.

  - **Entry date**

    If you want to set the entry date of the invoice document as the baseline date for payment, set this indicator.

- **Payment terms**

  In this section, you can maintain the cash discount percentage rate and the corresponding number of days. For example, if you enter the discount percentage rate as 2% and the number of days as 14, you get a 2% discount from the vendor if you pay the invoice amount within 14 days of the due date.

This is how payment terms ensure that discounts are applied and that due dates are calculated based on agreed-upon terms with the vendor.

10.5   Summary

In this chapter, we introduced the integration of finance functions with materials management and offered best practice examples and processes. In Section 10.1, we explained the master data in finance with step-by-step procedures to maintain finance-specific master data. In Section 10.2, we explained the integrated procurement and finance process, including account assignment in purchase orders, price determination, and tax determination. In Section 10.3, we explained the valuation and account determination in materials management and its integration with financial accounting. Finally, in Section 10.4, we explained the LIV process, including invoice reconciliation, duplicate invoice checks, credit memo processes, tax reconciliation, and payment processing.

In the next chapter, we'll discuss the integration between Project System and materials management, facilitating the efficient management of project-related procurement and material consumption. We'll also explain the integrated processes, master data in Project System, setup of project procurement, material requirements planning for project procurement, and inventory management.

Chapter 11

Project System

The Project System functionality in SAP S/4HANA provides functionalities to manage projects from planning through execution and completion. It ensures that projects of all sizes are completed on time, within budget, and while meeting the desired quality standards. Integration of Project System and materials management is essential for efficient project execution and control.

A *project* is a set of activities to be performed sequentially within the defined timeline and within the allocated budget to accomplish a specific set of goals—for example, an IT project or construction project. Projects are generally part of an organization's goals over a period and may involve one or more departments of the organization. The set of activities or tasks of the project are executed by the designated team, known as the project team, led by a project manager. The set of tasks and team size are dependent on the complexity of the project. Typically, there are four phases within the project lifecycle: project initiation, project planning, project execution, and project completion. Each phase of the project can be further divided into subphases depending on the complexity. A well-executed project will meet all the project goals within the schedule and budget and be delivered with optimum quality standards.

The integration of Project System with materials management is crucial for execution of the project. The procurement function of materials management helps in procuring required materials and services for the project from external suppliers. A project budget check and control during the procurement process is automatically performed during purchase requisition, purchase order, goods receipt, and invoice receipt transactions. If the budget is exceeded for a work breakdown structure (WBS) element, then purchase requisition, purchase order, goods receipt, and invoice receipt transactions cannot be executed.

The inventory management function of materials management helps in reserving and holding the required material stock for a project with reference to WBS elements. *Project stock* is a special kind of stock in inventory management. Materials are issued to a specific project activity from the project stock with reference to a WBS element.

In this chapter, we'll first look at the Project System structure and master data, and then we'll walk through the core cross-functional process setup: project procurement, material requirements planning (MRP), inventory management, and project budget management.

11.1   Structure of Project System

The Project System functionality enables you to plan and manage all phases and tasks of a project. It facilitates comprehensive project management by integrating with other functions such as materials management, finance and controlling, and so on. It provides tools to monitor and report on the progress of the project in real time. A project is created for a specific company code and controlling area in SAP S/4HANA. The following are the key components of Project System:

- **Project structure**
  Figure 11.1 shows the structure of a project in SAP S/4HANA Project System. It consists of the following:
  - Work breakdown structure: The WBS defines the hierarchical structure of the project, breaking down the project into smaller components to facilitate better control and management. A WBS element is assigned to a specific company code, controlling area, business area, functional area, cost center, and profit center. You can have multiple WBS elements created for a project.
  - Network: The network represents the sequence of project activities, such as internally executed activities, externally executed activities, general cost activities, service activities, material components, and so on. A network is assigned to a specific WBS element. The organizational unit assignments are copied from the WBS element. You can perform forward scheduling and backward scheduling to drive the activity dates. You can have multiple networks created for every WBS element.
  - Activity: An activity represents an internal or external task. Internal activities are performed at a specific work center within the plant, whereas an external activity is performed by a vendor. Activities are defined and executed in a sequence within the network. This is one of the integration points with materials management, and an external activity is supported by the procurement function. You can have multiple activities created for every network.

- **Project planning**
  Successful project execution depends on project planning. Scheduling of the project, resource planning, and cost planning are the three most important aspects of project planning:
  - Cost planning: This includes the estimation of the total project cost including the resource cost, material cost, execution cost, and other overhead. This is done at the WBS level and at the network activity level. Cost planning at the network activity level is done by assigning primary costs with reference to a cost element. Based on the network schedule, the planned costs are distributed by periods. After the detailed planning is done at the network activity level, in SAP S/4HANA it is possible to perform cost planning at a higher level—the WBS level. This is done manually with reference to a cost center with the activity type.

– Date planning: Date planning consists of scheduling project timelines and milestones. Basic dates are entered at the WBS element level manually. More detailed date planning is done at the network activity level. You can derive WBS dates and network activity dates automatically by carrying out the scheduling at the network activity level. Scheduling parameters defined in the configuration settings will determine how the scheduling is carried out. Both forward and backward scheduling are possible for a network activity. The earliest date for the activity is determined for forward scheduling and the latest date for the activity for backward scheduling.

– Resource planning: This involves allocating required resources such as human resources, work centers, equipment, and required materials to the project. These allocations are done for all in-house activities. The system automatically calculates the activity cost based on the human resources costs, capacity of work centers, and actual date of completion. Human resources are assigned to the work center, and in turn a work center is assigned to a network activity.

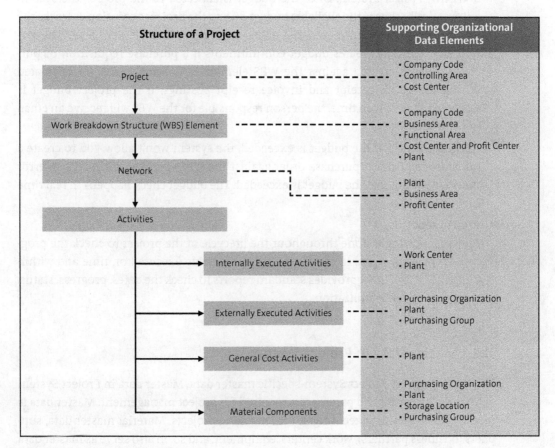

Figure 11.1 Project Structure Overview

- **Project execution**
  After the detailed planning is done at the activity level, each project activity is executed depending on the schedule. SAP S/4HANA Project System provides tools and functionalities to perform the following during project execution:
  - Project progress tracking: This involves monitoring the progress of the project by tracking the activities and updating the status of the project.
  - Time tracking and management: This involves tracking and recording the time spent on each activity. The progress of the project activities is compared against the planned dates.
  - Progress analysis: Progress analysis tools provide planned and actual dates for the project activity progress. This helps in identifying schedule variances and costs to take necessary actions to bring the project back on track.

- **Project controlling**
  Budget allocation and controlling the budget of the project is a vital feature of SAP S/4HANA Project System. Once the budget is allocated to the project, the system automatically checks the available budget during the project execution. It tracks the available funds, commitments, and actual costs at the WBS element level. The system automatically creates budget commitments if a purchase requisition or purchase order is created against the WBS element. The actual costs are calculated during the goods receipt and invoice receipt postings. If the project budget is exceeded at any given time, the person responsible for the WBS will receive an email notification.

  At the WBS level, if the budget is exceeded, the system won't allow you to create a purchase requisition/purchase order against the WBS element. It displays an error message stating that the budget is exceeded. The budget check happens in real time in SAP S/4HANA.

- **Project reporting**
  Project reporting is done throughout the lifecycle of the project to check the progress of the project and to ensure the project is being executed on time and within budget. SAP S/4HANA provides standard reports to check the dates, progress, status, costs, and resource utilization.

11.2   Master Data in Project System

This section covers Project System-specific master data. Master data in Project System defines the structure and parameters required for project management. Master data is required for planning, executing, and controlling projects. Material master data, suppliers (business partners), work centers, equipment, and so on also serve as master data in Project System.

Let's walk through the different master data in Project System.

11.2.1 Project

The project definition is the central component of Project System. It contains the project details, basic data of the project, controlling parameters, and organizational assignment. To create and maintain a project, use Transaction CJ20N (Project Builder).

Figure 11.2 shows the header details and the **Basic Data** tab of project definition **MJ-00-2024**. The header details include the **Project def.** field, where you enter a unique key to create the project definition. Enter the name of the project. The small icon next to the project name is used to enter the long text and provide a detailed description of the project.

The **Basic Data** tab of the project definition contains the project status, responsibilities, and organizational data elements. The **Status** section displays the current system status of the project. The following are the different system statuses:

- **CRTD**
  This is the status of the project at the time of creation. Most business transactions are not permitted for the project if the status is set to **CRTD**.

- **REL**
  This is the released status of the project. You must release the project to allow business transactions. To release the project, use menu path **Edit • Status • Release**.

- **TECO**
  This is the technically completed status, but you can still accrue costs.

- **CLSD**
  This is the closed status of the project. Accounting will be closed for the project.

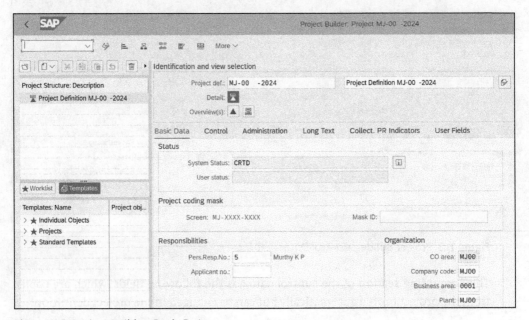

Figure 11.2 Project Builder: Basic Data

The **Responsibilities** section includes the **Pers.Resp.No.** field, where you can assign the project manager defined in the configuration settings as the person responsible.

Finally, the **Organization** section has the following fields:

- **CO area**
  In this field, assign the controlling area responsible for controlling the project and project cost accounting.

- **Company code**
  In this field, assign the company code for the project.

- **Business area**
  In this field, assign the business area for the project.

- **Plant**
  In this field, assign the plant for the project.

Similarly, assign the location, functional area, and profit center for the project. The currency for the project is defaulted from the company code data.

Next, navigate to the **Control** tab of project definition **MJ-00-2024**, as shown in Figure 11.3. The **Control** tab of the project definition includes the **Project Profile** field in the header, where you assign the project profile defined in the configuration settings (Section 10.1.6). Select the project profile from the dropdown menu. The project profile carries control parameters and a planning method for dates and costs.

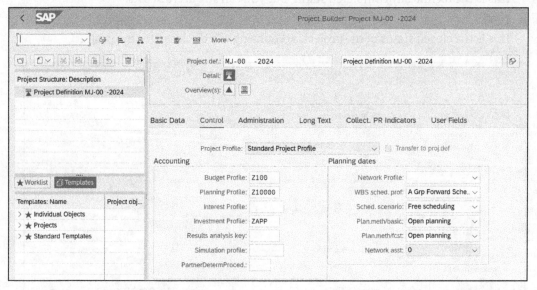

Figure 11.3 Project Builder: Control Data

The **Accounting** section of the control data has the following fields, which are copied from the project profile automatically but can be changed in the project definition:

- **Planning Profile**
  The planning profile controls revenue planning parameters.

- **Budget Profile**
  The budget profile controls the budgeting control parameters.

- **WBS sched. prof.**
  The WBS scheduling profile controls WBS scheduling parameters.

- **Sched. scenario**
  The scheduling scenario controls if the scheduling dates are defined at the top-level hierarchy, lower-level hierarchy, or with free scheduling.

11.2.2   WBS Elements

A WBS in SAP S/4HANA is a hierarchical breakdown of a project into smaller, manageable components. It is vital for managing the planning, execution, monitoring, and controlling of a project. Each WBS element represents tasks and subtasks that are required to complete the project. A WBS is organized in a hierarchical manner under the project definition. WBS elements can be defined at different levels hierarchically; that is, each WBS element can be further broken down into multiple smaller WBS elements, each representing a specific task. The WBS element is integrated with networks and activities.

To create and maintain a WBS element, use Transaction CJ02. The WBS element is created with reference to project definition as the WBS is always created under the project in a hierarchical structure. Figure 11.4 shows the initial screen for the creation of a WBS element with reference to project **MJ-00-2024**.

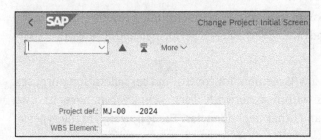

Figure 11.4 Create WBS Element: Initial Screen

After specifying the project definition (**Project def.**) in the initial screen, click the **WBS Element Overview** button at the bottom-right corner of the screen (not shown). Figure 11.5 shows the **WBS Element Overview** screen, where you can define the WBS element and the description. Upon saving the transaction, the system automatically creates a WBS element and links to the project in a hierarchical structure.

The system automatically copies the details of the project definition WBS element into the project definition. **CRTD** will be the status of the WBS element when created.

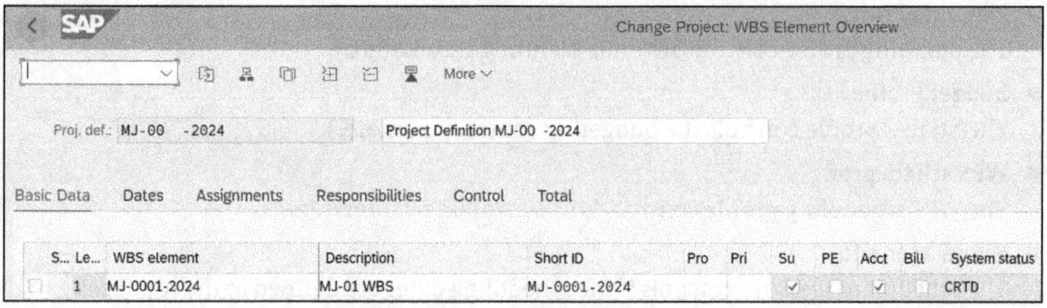

Figure 11.5 Create WBS Element: WBS Overview Screen

To release the WBS element, execute Transaction CJ12 and enter the project definition and WBS element to display the WBS element in change mode. From the menu bar, choose **Edit • Status • Release**.

11.2.3   Network and Activities

Project tasks are controlled, managed, and executed using networks in Project System. A network is a sequence of project activities/tasks, and every project activity is planned, scheduled, and executed sequentially. SAP S/4HANA allows detailed planning of the project tasks at the network level. Network activities are used to plan the following tasks:

- Resource planning such as for human resources, material requirements, and equipment and work center requirements happen at the task level within the network.

- Time planning—for time estimates to complete the activity—is also performed at the task level within the network.

- Cost planning—for estimation of total project costs including resource costs, material costs, execution costs, and other overhead—is performed at the network activity level.

Network activities are linked to a WBS element for control and cost allocation purposes. Activities are individual tasks within a network. Each activity can have its own resources, duration, and costs and other dependencies. The following are the types of activities:

- **Internal processing activity**
  A task or operation is performed internally. Required resources such as materials, human resources, work center, equipment, and so on are planned and provided to execute the activity. Costs associated with internal processing activities are calculated based on labor cost, duration, material cost, and other overhead costs. Materials will be issued from inventory management if available or procured from a vendor for this activity.

- **External processing activity**
  A task or operation is performed externally. External processing activities are

planned with a third-party vendor and the required purchasing master data is updated in the activity. A purchase requisition will be generated automatically upon releasing the network for the external processing activity. The purchasing function converts the purchase requisition to a purchase order and formally submits the order for an external activity with the supplier. The supplier then completes the required task.

- **General cost activity**
  This activity type is used to plan and allocate costs to projects. These activities are not project tasks to be performed internally or externally but are used to plan and control general costs associated with a project such as overhead costs, administrative costs, and other indirect costs.

To create and maintain networks, use Transaction CN21. The network is created at the plant level using a network profile and network type. Figure 11.6 shows the initial screen for creating a network, where you can fill out the following fields:

- **Network Profile**
  In this field, assign a network profile defined in the configuration settings (Section 10.1.6). Select the network profile from the dropdown menu. You can configure default values in the network profile such as plant, unit of measure for duration, unit of measure for work, control keys for external processing, and so on.

- **Network type**
  In this field, assign a key for the network type from the dropdown menu. Like order types, network types are used to distinguish different order categories.

- **Plant**
  Assign the key for the plant in which the activities are processed. This is required to assign the work center, equipment, materials, and other resources to network activities.

- **MRP controller**
  Assign a key for the MRP controller from the same plant assigned to the network. An MRP controller is a person who is responsible for MRP for a group of materials in the plant. This value is required to plan materials required to process the internal activities for the project.

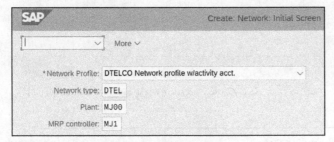

Figure 11.6 Create Network: Initial Screen

After specifying the field values in the initial screen, press ⌷Enter⌷ to arrive at the scheduling data view of the header data of the network, as shown in Figure 11.7. The header data includes the **Scheduling**, **Assignments**, **Control**, and other tabs. The **Scheduling** data tab has different sections, including the **Status** section, which displays the current status of the network and has the following different system statuses:

- **CRTD**
  This is the status of the network at the time of creation. Most business transactions are not permitted for the network if the status is set to **CRTD**.

- **REL**
  This is the released status of the network. You must release the network to allow business transactions. To release the network, follow menu path **Edit • Status • Release**.

- **MANC**
  This status is set when material availability for the network activities is not checked.

- **SETC**
  This status is set when a settlement rule is created for the network.

- **NTUP**
  This status is set when the dates are not updated in the network.

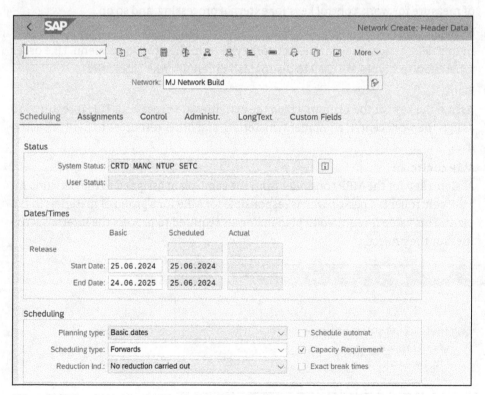

Figure 11.7 Create Network: Scheduling Data

The **Dates/Times** section has the following key fields:

- **Basic Start Date**
  This indicates the basic start date for scheduling the network.

- **Basic End Date**
  This indicates the basic end date for scheduling the network.

- **Scheduled Start Date** and **Scheduled End Date**
  These are determined automatically. The system also determines the **Actual Start Date** and **Actual End Date** based on the processing start and end dates of all activities.

Next, the **Scheduling** section of the **Scheduling** data tab has the following key fields:

- **Planning type**
  A planning type defines a set of dates for network planning. The planning type is defaulted from the project.

- **Scheduling type**
  A scheduling type is used for detailed scheduling of the network. The scheduling type is defaulted from the project. For forward scheduling, the **Basic Start Date** field must be filled, and the **Basic End Date** for backward scheduling.

Move on to the **Assignments** tab of the network header details, as shown in Figure 11.8.

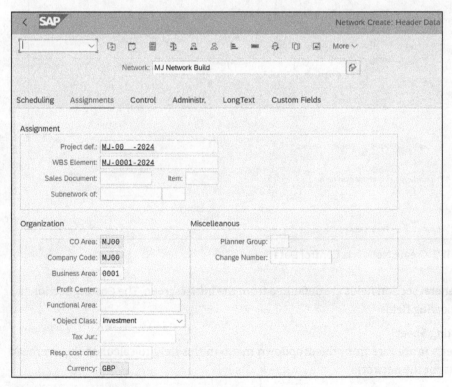

Figure 11.8 Create Network: Assignments Data

The **Assignment** section has the following key fields:

- **Project def.**
  Enter the key for the project definition in this field. This assignment is critical to link the network to the project.

- **WBS Element**
  Enter the key for the WBS element in this field. This assignment is critical to link the network to the WBS element.

The controlling area, company code, business area, functional area, profit center, and currency are copied from the project definition.

Now go to the **Control** tab of network header details, as shown in Figure 11.9.

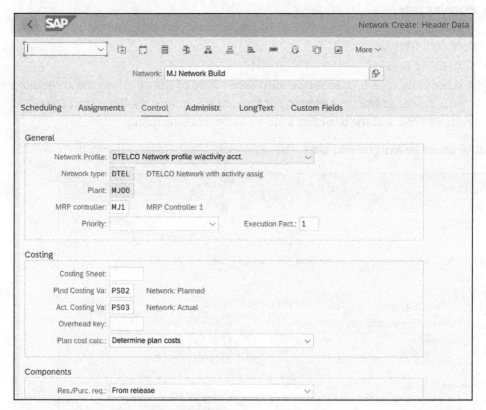

Figure 11.9 Create Network: Control Data

The **General** section fields are defaulted from the initial screen. The **Costing** section has the following fields:

- **Costing Sheet**
  Enter a procedure from the dropdown menu in this field to calculate the overhead costs for the network.

- **Planned Costing Va.**
  The planned costing variant is defaulted from the network profile. However, you can change it manually. This variant is used to determine planned costs.

- **Act. Costing Va.**
  The actual costing variant is defaulted from the network profile. However, you can change it manually. This variant is used to determine actual costs.

- **Plan cost calc.**
  The planned cost calculation procedure controls if the planned costs are calculated and updated upon releasing the network or not.

The **Components** section of the control data includes the **Res./Purc. req** field, which controls whether a reservation and purchase requisition are generated for the network or not. These transactional documents are generated upon saving the new network order or upon releasing the network order.

Next, click **Goto • Activity Overview** in the menu bar at the top or press ⌜F7⌝ from the header data screen to arrive at the **Int. Processing** tab of the **Basic Data Overview (Basic Dates)** screen, where you can maintain internal processing activities and external processing activities. Figure 11.10 shows the internal processing activity defined for the network. The activity will be entered in increments of 10. Enter the description, duration, work center, plant, and other fields to define an internal processing activity. You can define multiple activities based on the nature of the tasks and business requirements.

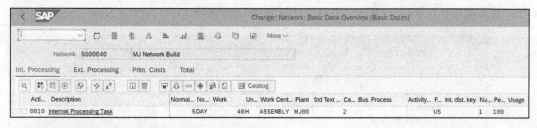

Figure 11.10 Network: Internal Processing Activity

Next, maintain the external processing activities of the network. Click the **Ext. Processing** tab to maintain external processing activities. Figure 11.11 shows the external processing activity defined for the network. The activity will be entered in increments of 10. Enter a description and the purchasing details (purchasing info record [PIR], vendor, cost element, material group, purchasing organization, purchasing group, etc.) to define an external processing activity. You can define multiple activities based on the nature of the tasks and business requirements.

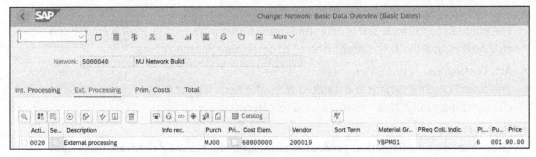

Figure 11.11 Network: External Processing Activity

11.2.4   Configuration Settings for Master Data

Let's explore the configuration settings that control the master data in Project System.

Specify Persons Responsible for WBS Elements

The person assigned is responsible for project management. You can specify the person responsible in this configuration setup. To arrive at this configuration step, execute Transaction SPRO and navigate to **Project System • Structures • Operative Structures • Work Breakdown Structure (WBS) • Specify Persons Responsible for WBS Elements**.

Figure 11.12 shows the person responsible for the projects. To define the person responsible for your project, specify the number of the person responsible in the **Respons.** field (a number value in the incremental order), the name of the responsible person, and the **Office user** (user ID) value of the person defined in the SAP S/4HANA system.

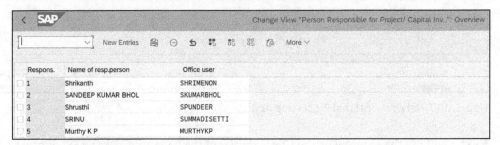

Figure 11.12 Person Responsible for Project

Create Project Profile

The project profile contains default values and controlling parameters for project definition. It contains basic project data, organization data, time scheduling, and cost/revenue/finance data that is defaulted into the project during project definition. From the project definition, it's defaulted into the WBS elements. To arrive at this configuration step, execute Transaction SPRO and navigate to **Project System • Structures • Templates • Standard Work Breakdown Structure • Settings for Standard and Operative WBSs • Create Project Profile**.

Figure 11.13 shows the initial screen for the creation of a project profile. Click **New Entries** to create a new project profile. Enter a seven-digit alphanumeric key for creating the project profile and enter the description. You can copy an existing project profile and create a new one from it.

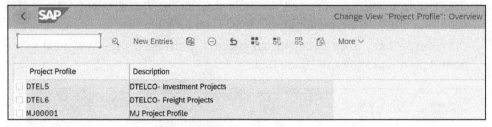

Figure 11.13 Create Project Profile: Initial Screen

To display the details of a project profile, select the project profile and click **Details** button at the top or press $\boxed{\text{Ctrl}}$+$\boxed{\text{Shift}}$+$\boxed{\text{F2}}$. Figure 11.14 shows the **Control** data for project profile **MJ00001 (MJ Project Profile)**. The **Basic Data** section of the control data has the following fields:

- **Project Type**

  A project type uniquely identifies the type of the project. You can assign a project type from the dropdown menu in this field. Investment projects, expense projects, revenue projects, customer projects, and so on are some of the project types.

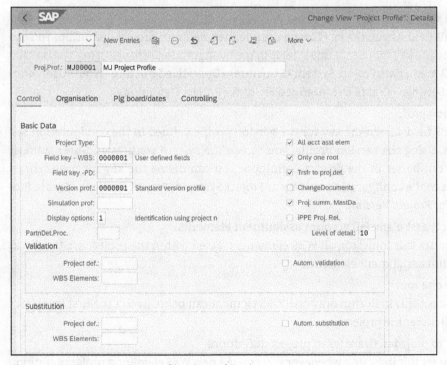

Figure 11.14 Create Project Profile: Control Data

- **Field key—WBS**

  In this field, assign the key for the user-defined fields for WBS elements.

  You can define the key for user-defined fields in the configuration settings using Transaction SPRO menu path **Project System • Structures • Operative Structure (WBS) • User Interface Settings • Create User-Defined Fields for WBS Elements**. Figure 11.15 shows key **0000001** created for user-defined fields for WBS.

  Once the field key is defined, you can assign it to the **Field Key—WBS** field in the project profile, as shown in Figure 11.14.

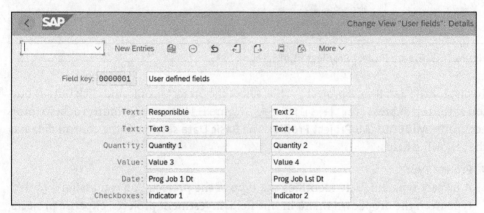

Figure 11.15 Create User-Defined Fields for WBS Elements

- **Field key—PD**

  In this field, assign the key for the user-defined fields for project definition. You can define the key for user defined fields in the configuration settings using Transaction SPRO menu path **Project System • Structures • Operative Structure (WBS) • User Interface Settings • Create User-Defined Fields for Project Definition.**

- **Version prof.**

  In this field, assign the key for the version profile defined in the configuration settings. Using the version profile, you can define system statuses and user statuses that can be set in the project definition. You can define the key for user-defined fields in the configuration settings at **Project Systems • Project Versions • Create Profile for Project Version.**

- **All acct asst elem. (all account assignment elements)**

  If you set this indicator, all WBS elements created within the project are treated as account assignment elements.

- **Only one root**

  If you set this indicator, only one WBS element can be created at the first level (root level) under the project.

- **Trsfr to proj. def. (transfer to project definition)**

  If you set this indicator, whenever you create a new WBS element, a project definition with the same name is created automatically. When you change the WBS element, the corresponding project definition is updated.

- **ChangeDocuments**
  If you set this indicator, every change, including basic data changes, in the WBS will create a change document.

Navigate to the **Organisation** tab next, as shown in Figure 11.16. This tab has the following fields that you can specify so that they are copied over to the project definition:

- **Controlling area**
  In this field, assign the controlling area responsible for controlling the project and for project cost accounting.

- **Company code**
  In this field, assign the company code for the project.

- **Business area**
  In this field, assign the business area for the project.

- **Plant**
  In this field, assign the plant for the project.

Similarly, assign the **Functional Area** and **Profit Center** for the project. The **Project Currency** for the project can be different from the company code currency.

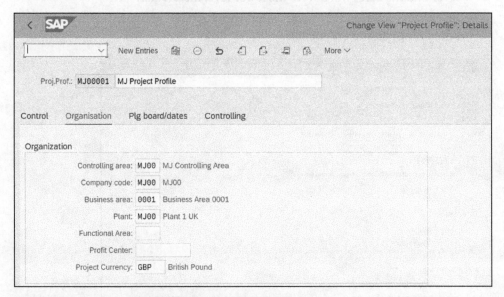

Figure 11.16 Create Project Profile: Organization Data

Next, go to the **Plg board/dates** tab for project profile **MJ00001 (MJ Project Profile)**, as shown in Figure 11.17. The **WBS time scheduling** section has the following fields that you can specify so that they get copied over to the project definition:

- **WBS sched. prof. (WBS schedule profile)**

  In this field, assign the profile for WBS scheduling. You can define the profile in configuration settings using Transaction SPRO menu path **Project System • Dates • Date Planning in WBS • Define Parameters for WBS Scheduling**.

  Figure 11.18 shows the control parameters in the WBS scheduling profile **Z10000000000 (A Grp Forward Scheduling)**. The following are the key fields for defining parameters for WBS scheduling:

  - **Scheduling type**: A scheduling type is used for detailed scheduling of the network. Forward, backward, forward in time, backward in time, current date, and only capacity requirements are the different values that you can select from the dropdown list.

  - **Schedul method (scheduling method)**: The scheduling method controls if the WBS determines dates or the network decides dates for the forward or backward scheduling types. You can select either option in this field.

- **Sched. scenario (scheduling scenario)**

  In this field, assign one of the following based on the business requirements:

  - **1** (bottom-up scenario): If you assign this scenario, the scheduling dates are defined at the lower-level hierarchy and the dates will flow to the top-level hierarchies.

  - **2** (top-down scenario): If you assign this scenario, the scheduling dates are defined at the top-level hierarchy, either at the project definition level or WBS level, and the dates will flow to the lower-level hierarchies.

  - Blank (free scheduling): If you assign this scenario, you are free to choose the scheduling parameters you like.

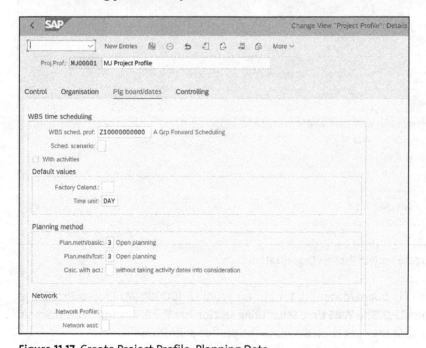

Figure 11.17 Create Project Profile: Planning Data

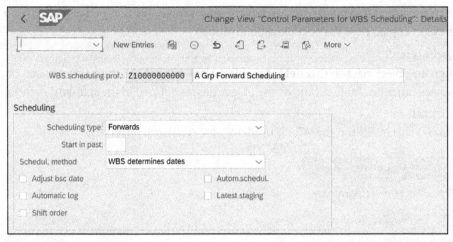

Figure 11.18 WBS Scheduling Profile

Finally, let's move on to the **Controlling** tab for project profile **MJ00001 (MJ Project Profile)**, as shown in Figure 11.19.

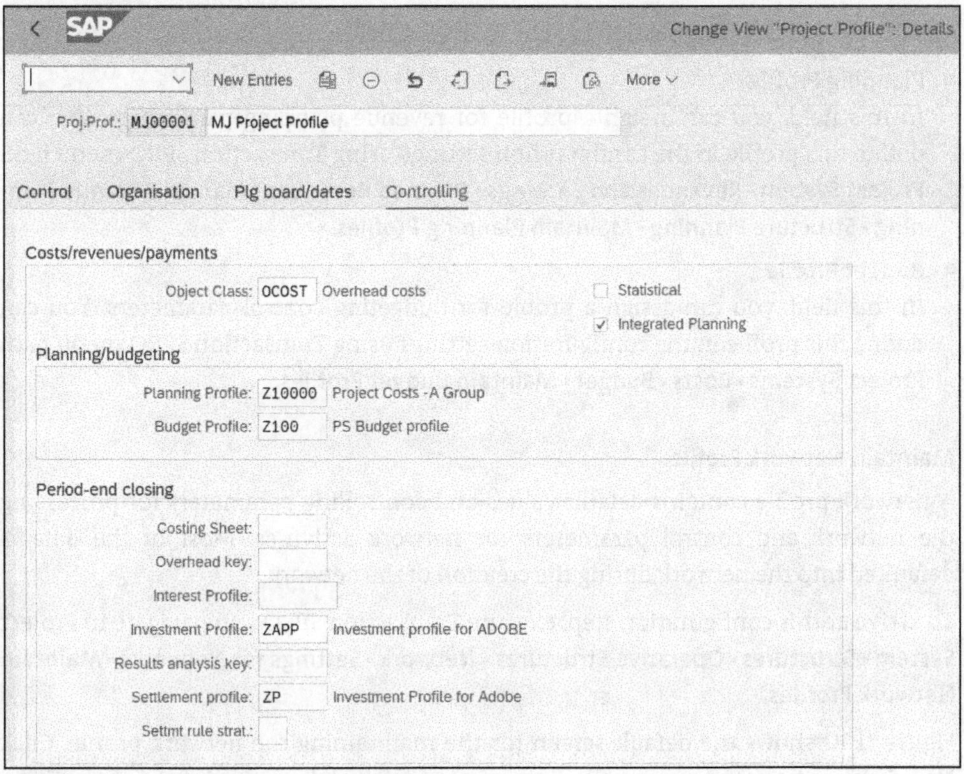

Figure 11.19 Create Project Profile: Controlling Data

The **Costs/revenues/payments** section has the following fields that you can specify so that they get copied over to the project definition:

- **Object Class**
  Assign an object class in this field. An object class categorizes the project based on business function. Table 11.1 shows the object classes defined in the standard system.

- **Statistical**
  If you set this indicator, the WBS is used only for statistical purposes.

Object Class	Description
INVST	Investment
OCOST	Overhead
PROFT	Earnings, sales
PRODT	Production

Table 11.1 Object Class

The **Planning/budgeting** section has the following fields:

- **Planning Profile**
  In this field, you can assign a profile for revenue planning parameters. You can define this profile in the configuration settings using Transaction SPRO menu path **Project System • Revenues and Earnings • Planned Revenues • Manual Revenue Planning • Structure Planning • Maintain Planning Profiles**.

- **Budget Profile**
  In this field, you can assign a profile for budgeting control parameters. You can define this profile in the configuration settings using Transaction SPRO menu path **Project Systems • Costs • Budget • Maintain Budget Profiles**.

Maintain Network Profile

A network profile contains default values and controlling parameters for processing the network and control parameters for network activities. Most of the data is defaulted into the network during the creation of the network.

To arrive at this configuration step, execute Transaction SPRO and navigate to **Project System • Structures • Operative Structures • Network • Settings for Networks • Maintain Network Profiles**.

Figure 11.20 shows the default screen for the maintaining the network profile. Click **New Entries** to create a new profile. Enter a seven-digit alphanumeric key for creating a network profile and enter the description. You can copy an existing network profile and create a new one.

The following are the **Network parameters** that you can define in the **Defaults** tab:

- **Plant**
  Assign the key for the plant in which the activities are processed. You must assign the work center, equipment, materials, and other resources to network activities.

- **Network type**
  In this field, assign a key for the network type from the dropdown menu. Like order types, network types are used to distinguish different network order categories.

- **Planner Group**
  A planner group is a group that is responsible for planning network activities. Assign a planner group that was predefined at the plant level.

- **MRP cont.group (MRP controller group)**
  Assign a key for the MRP controller from the same plant assigned to the network. An MRP controller is a person who is responsible for MRP for a group of materials in the plant. This value is required to plan materials required to process the internal activities for the project.

- **Rel. view (relationship view)**
  Assign a key for the relationship view of the network activities in this field. It controls how the relationships in the activities are displayed. You can select predecessor, successor, mixed (predecessor and successor), or as created.

- **Comp. increment (component increment)**
  Assign an interval for automatic numbering of component items. If you enter "1", the item numbers for components will be created automatically in increments of 1.

- **Op./act. incrmt (operation/activity increment)**
  Assign an interval for automatic numbering of internal processing, external processing, and general cost activities. If you enter "10", the activity numbers will be created automatically in increments of 10.

- **Check. WBS act. (checking WBS activity)**
  Based on the value set in this field, the system checks the activity dates with the basic dates scheduled in the WBS element. The following values can be set:
  - Blank (**No check**): If you set this value, the activity dates are not checked against the WBS basic dates.
  - **W (Exit with warning)**: If you set this value, the activity dates are checked against the WBS basic dates, and if the activity dates do not fall within the WBS basic dates, the system throws a warning message.
  - **E (Error)**: If you set this value, the activity dates are checked against the WBS basic dates, and if the activity dates do not fall within the WBS basic dates, the system throws an error message.

- **Procurement**
  Assign a key for a procurement indicator in this field. This controls the procurement of components required to perform the network activity/task. Table 11.2 shows the procurement indicators that you can assign.

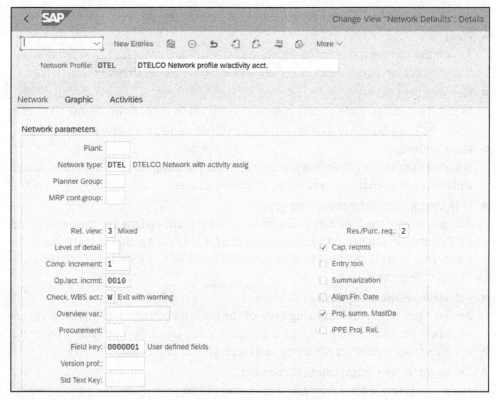

Figure 11.20 Network Profile: Defaults

Procurement Ind.	Description
PEV	Planned independent requirements for WBS
PF	Purchase requisition for WBS
PFS	Third party requisition for WBS
PFV	Preliminary requisition for WBS
WE	Reservation plant stock

Table 11.2 Procurement Indicator

- **Field key**

  Assign a field key for user-defined fields. Field keys are defined in the configuration settings as discussed in the previous section.

- **Res./Purc. req. (reservation/purchase requisition)**

  This field controls if a reservation and purchase requisition are generated for the network or not. These transactional documents are generated upon saving the new network order or upon releasing the network order. You can set one of the following values:

- **1**: Never
- **2**: From release
- **3**: Immediately

- **Cap. reqmts (capacity requirements)**
  Set this indicator if you want the system to calculate capacity requirements upon saving the network.

Next, navigate to the **Graphic** tab of network profile **DTEL**, as shown in Figure 11.21. In this tab, you can set the details for display graphics like graphic profile, color, and so on.

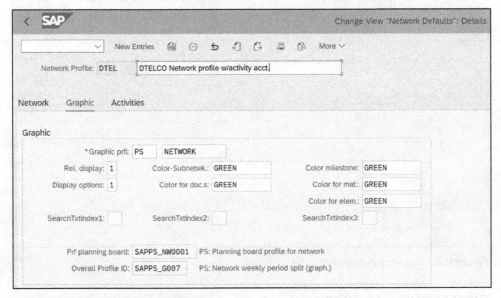

Figure 11.21 Network Profile: Graphic

Let's continue on to the **Activities** tab of network profile **DTEL**, as shown in Figure 11.22. This tab contains the processing keys and control parameters for internal processing activities, external processing activities, general cost activities, and service activities.

Activity parameters for internal processing activities include the **Control key** field. Control keys are used to define activities. This determines a specific business process within SAP S/4HANA Project System and provides control parameters to drive that specific process.

Table 11.3 shows the control keys defined in the standard system for Project System.

Control Key	Description
PS01	Network—internal processing
PS02	Network—external processing
PS03	Network—general costs activity
PS04	Network—internal processing or costing
PS05	Network—replaced by subnetwork

Table 11.3 Control Keys in Project System

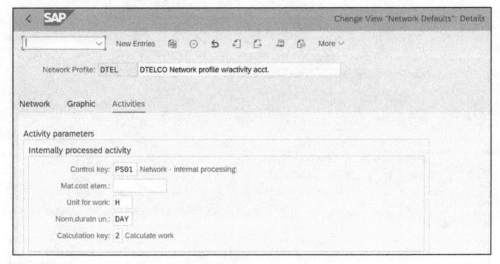

Figure 11.22 Network Profile: Activities

Define Control Key for Activities

Control keys control scheduling to determine the start date of the network activity, determination of capacity requirements, and internal and external processing parameters.

To arrive at this configuration step, execute Transaction SPRO and navigate to **Project System • Structures • Templates • Standard Network • General Settings for Standard and Operative Networks • Define Control Key for Activities**.

Figure 11.23 shows the initial screen for maintaining control keys for network activities. SAP S/4HANA provides standard control keys for plant maintenance, production planning, quality inspection, Project System, and so on. **PS01**, **PS02**, **PS03**, **PS04**, and **PS05** are defined in the standard system for Project System. Click **New Entries** to create new control keys for network activities. It is recommended to copy an existing control key and create a new one.

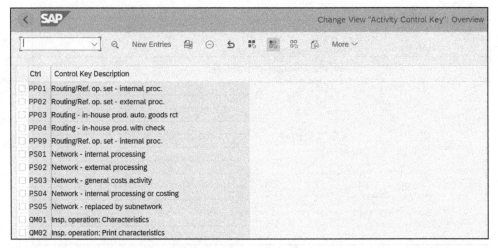

Figure 11.23 Control Keys for Activities: Overview

Let's explore the settings using control key **PS02**, which is used in external processing activities and is also relevant for procurement. To display the details of the control key, select the control key and click the **Details** icon at the top or press `Ctrl`+`Shift`+`F2`. Figure 11.24 shows the details screen for setting up control keys for operations:

- **Scheduling**
  Set this indicator to scheduleg dates for network activity. In Project System, it is used to determine the start date and end date of a network activity.

- **Det. Cap. Req. (determine capacity requirement)**
  Always set this indicator when the scheduling indicator is set. The system determines the capacity requirements based on scheduling for the network activity.

- **Gen. costs act. (general costs activity)**
  Set this indicator for a control key for a general cost activity, such as PS03.

- **Cost**
  Set this indicator if you want to include the network activity in costing.

- **Confirmation**
  A confirmation is used to monitor network activities. Table 11.4 shows the possible values you can assign to this field for Project System.

Confirmation	Description
1	Milestone confirmation (not used in Project System/plant maintenance)
2	Confirmation required
3	Confirmation not possible
Blank	Confirmation possible but not necessary

Table 11.4 Confirmation Keys

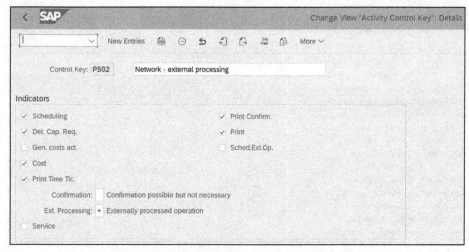

Figure 11.24 Control Keys for Activities: Details

- **Ext. Processing (external processing)**
  This field is used to determine whether the network activity is internal or externally performed. Table 11.5 shows the external/internal processing indicators that you can assign to this field. For **PS01**, keep it blank; for **PS02** and **PM04**, assign the value **+** to plan external processing operations.

External Processing	Description
Blank	Internally processed operation
+	Externally processed operation
X	Internally processed operation/external processing possible

Table 11.5 External Processing Indicators

- **Service**
  Set this indicator if the services need to be planned for the network activity with this control key for both internal and external services.
- **Sched.Ext.Op. (schedule external operation)**
  Set this indicator if the external operation needs to be scheduled using standard values. The purchase requisition will be created for external processing using standard values.

11.3 Project Procurement

Procurement is vital to execute the project, and it helps in procuring materials and services on time. The procurement function is directly integrated with Project System. It ensures the right materials are procured at the right time as required for project execution. Project System tracks all the costs associated with procurement of goods and services. The budget control functionality will ensure the purchasing of goods and services will not exceed the budget allocated for the WBS element. All tasks processed externally by vendors through a procurement function, material procurement, and service procurement are tracked and controlled in Project System. Project System directly generates purchase requisitions for external processing activities, service procurement, and subcontracting processes. The purchasing activities are planned in the network operations/activities. Purchase requisitions are converted into purchase orders and the purchase orders transmitted to vendors. Purchase requisitions and purchase orders are created with reference to a network or WBS element. The procured materials and services are received, and incoming invoices from the vendors are processed.

Let's dive deep into the project procurement process in SAP S/4HANA, including the core cross-functional process, the project structure, and purchase requisition generation from Project System.

11.3.1   Integrated Project Procurement

The procurement function of materials management is directly integrated with Project System. An effective project procurement process is the key to the successful execution of a project, where materials and services are procured in a timely manner. The linking of relevant WBS elements and networks to the purchasing documents ensures accurate cost tracking. SAP S/4HANA provides tools and functionalities to monitor procurement activities and costs to identify variances and issues. This will help the project team to work with procurement to take corrective action.

Table 11.6 shows the key processes impacted within Project System and materials management to support project procurement.

Area	Processes Impacted	Description
Project System	Develop initial project plan	This process supports the initiation of the project that includes developing schedules, creating project baseline, estimating costs, developing a high-level WBS, and so on.
	Prioritize projects	This process supports the prioritization of projects, performing risk analysis, and funding.
	Analyze project requirements	This process supports establishing project performance requirements and criteria by project type, conducting reviews, building a detailed WBS, and creating preliminary project activities.

Table 11.6 Process Impact: Project Procurement

Area	Processes Impacted	Description
Project System (Cont.)	Develop detailed project plan	This process supports the formation of project team, assigning resources to tasks, planning summarized project expenditures by WBS by year, and determining total planned project costs.
	Obtain project approval	This process supports reviewing/modifying a project plan and resubmitting a project for approval, obtaining approval, and approving projects.
	Manage project budgets and funding	This process supports finalizing project budget and funding for the project activities.
	Prepare and schedule project for execution	This process supports receiving final approval and releasing the project.
	Execute project	This process supports completing project staffing, conducting kick-off meetings, and updating WBS details.
	Track and close project	This process supports monitoring the project plan versus the budget versus actual cost.
Materials management	Sourcing materials and services	This process supports the sourcing for vendors for project procurement needs.
	Create and maintain purchasing master data	This process creates and maintains purchasing master data such as source lists and PIRs to support the project procurement needs.
	Create and maintain purchase requisition	This process supports the creation, generation, and maintenance of purchase requisitions created with reference to a WBS, originating from project activities.
	Create and maintain purchase order	This process supports the creation and maintenance of a purchase order for procuring goods and services needed for project execution.
	Post goods receipt	This process supports the goods receipt for a purchase order from the external vendor.
	Supplier invoice management	This process supports the processing of an incoming invoice from the supplier for goods/services provided.

Table 11.6 Process Impact: Project Procurement (Cont.)

Figure 11.25 shows the process of an external processing activity for project procurement. Let's walk through the process steps:

1. **Initiate and prioritize the project**

    This is the first step of the process and of initiating and prioritizing a project. It starts with development of an initial project plan with a timeline, cost estimates, creating a project baseline, and creating a high-level WBS. Project definition, WBS elements, and network creation in SAP S/4HANA happen at this step.

    Project requirements are analyzed, and a detailed project plan is then developed. A project manager is assigned to the project and the core team for the project is formed. Tasks/activities are created in detail and resources are assigned to project tasks. Summarized project costs by WBS and by year are planned and total project costs are determined. Project plans are updated with complete financial analysis, value, risk, and feasibility analysis. The project is then submitted for approval.

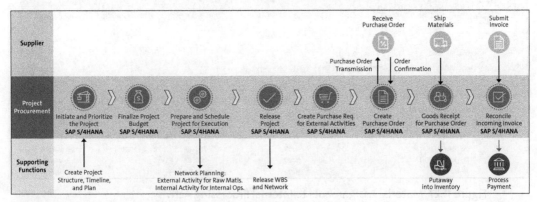

Figure 11.25 Project Procurement with External Processing Activity

2. **Finalize project budget**

    After the development of a detailed project plan and project approval, annually approved budgets are then recorded in the WBS, and budgets are distributed to lower level WBS elements in the project structure and released. Annually approved budgets are then recorded in the WBS. Project scheduling of start and end dates is updated, planned costs updated, and project budgets are maintained to track the budget at lower levels.

3. **Prepare and schedule project for execution**

    At this stage of the process, a final approval can be obtained, and details are finalized. A project kick-off meeting will be conducted. WBS elements are updated. Detailed network activity planning happens here, including internal processing activities and external processing activities. Tasks/activities are released for execution sequentially.

    Figure 11.26 shows the released internal processing activity. These activities are planned within the network using Transaction CN22. Before releasing the activity, resources such as work center, equipment, materials, and so on are planned and assigned to the respective activity. The system status has been set to **REL** after release.

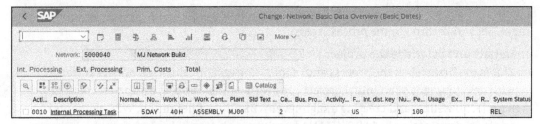

Figure 11.26 Released Network Activity for Internal Processing

Figure 11.27 shows the material assignments for internal processing activity **0010**. Material assignments are made based on the material requirements to complete the task. From the activities screen, click the **Materials** icon to navigate to the material assignments to an activity.

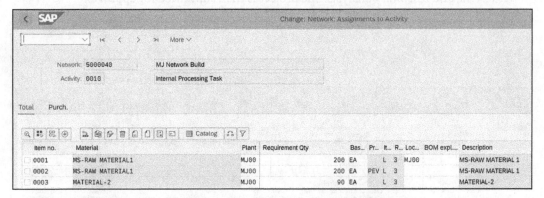

Figure 11.27 Material Assignments to Internal Processing Activity

You can assign stock materials as well as nonstock materials. Materials **MS-RAW MATERIAL1** and **MATERIAL-2** have been assigned to internal processing task **0010** as stock items, indicated by item category **L**. Requirement quantities for each material assigned to the activity are planned as well. For all stock items, a reservation will be created upon releasing the network.

Figure 11.28 shows material reservation **1657** created for material **MS-RAW MATERIAL 1**. The system creates one reservation for all material requirements of the network activity with item category **L** (stock item).

You can plan procurement of nonstock materials to be procured from the vendor directly in the network activity. Figure 11.29 shows that item **0004** is added to procure 100 KG of nonstock material **MATERIAL-3**. This item is planned in the **Purchasing** tab by assigning the PIR, purchasing organization, purchasing group, and vendor to trigger the purchasing process.

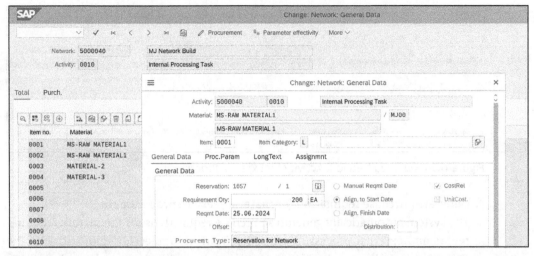

Figure 11.28 Reservation for Material Requirements

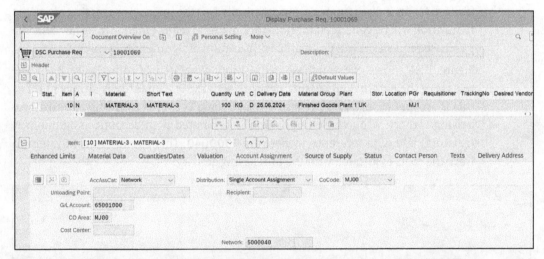

Figure 11.29 Purchase Planning for Nonstock Material Requirements for Internal Activity

Figure 11.30 shows the released external processing activity. This task/operation will be performed by the external vendor, so a purchase order must be submitted for the activity formally. Purchasing master data such as vendor, purchasing organization, purchasing group, PIR or outline agreement, material group, cost element, and so on are added to the external processing task and scheduled sequentially. Upon release of the activity, the system generates a purchase requisition automatically.

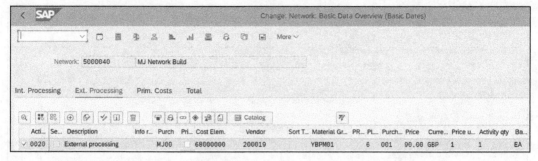

Figure 11.30 Released Network Activity for External Processing

4. **Generate purchase requisition for material/service procurement**
The system automatically generates purchase requisitions for the material require-ments and external processing activities from the Project System function. The external processing activity of the network can be updated to generate subcontract-ing purchase requisition, a service purchase requisition, or a standard purchase req-uisition for an indirect material automatically. For the internal processing activity, a requisition will be generated for nonstock material requirements. In either case, the purchase requisition will be created with account assignment **N** (network).

Figure 11.31 shows the purchase requisition generated automatically from the release of the external processing network activity. The quantity is always 1 unit as this is an operation/task. Purchasing master data is copied from the external pro-cessing activity. A purchase requisition is generated with account assignment **N** (network) and with reference to network **5000040**. This allows Project System to track the costs associated with the purchasing activity.

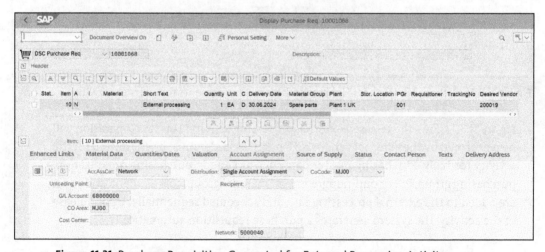

Figure 11.31 Purchase Requisition Generated for External Processing Activity

Figure 11.32 shows the purchase requisition generated for the material requirement for the internal processing activity. Material, requirement quantity, and purchasing-related data are copied from the external purchasing parameters assigned to the material item linked to the internal processing activity. A purchase requisition is generated with account assignment **N** (network) and with reference to network **5000040**. In this way, Project System tracks the costs associated with the purchasing activity.

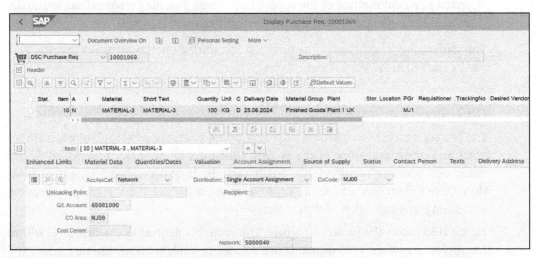

Figure 11.32 Purchase Requisition for Nonstock Material Requirements for Internal Activity

5. **Convert purchase requisitions into purchase order**
   The purchase requisition is converted into a purchase order manually by the purchaser using Transaction ME21N or automatically using Transaction ME59. The line item details including account assignment details are defaulted into the purchase order. The price is defaulted from the PIR. The purchase order will be transmitted to the supplier based on the output method: email, electronic data interchange (EDI), or electronic transmission to the supplier on SAP Business Network.

6. **Post goods receipt for purchase order**
   The supplier ships the material or provides the requested services. Once the shipment from the supplier arrives, the goods receipt will be posted manually in SAP S/4HANA with reference to the purchase order using Transaction MIGO and movement type 101:

   - A service entry sheet will be created for service purchase orders with the service item category.

   - Upon posting goods receipt, the material document and accounting document are created. The transaction is recorded in financial accounting. This can be tracked in Project System.

7. **Post logistics invoice verification**
   The supplier sends an invoice for the service provided or materials supplied via email (paper invoice), via EDI, or electronically via SAP Business Network. The

supplier invoice document will be posted by the accounts payable personnel manually in the SAP S/4HANA system using Transaction MIRO in the case of a paper invoice. Automatic posting and reconciliation of the invoice happens if an electronic invoice is received from the supplier. The next scheduled payment run will pick up this fully reconciled invoice and post the payment to the supplier based on the preferred payment method of the supplier. Accounts payable personnel can manually process the payment against the fully reconciled invoice as well.

The transaction is recorded in the financial accounting upon posting the invoice. This can be tracked in Project System.

11.3.2   Project Structure Setup

The project structure setup starts with developing a detailed project plan that includes goals and objectives, scope of the project, timeline with all major milestones clearly called out, resource requirements, quality standards, and detailed budgeting allocation. Once the project is formally approved, the project team can start the planning, scheduling, and execution of the project.

Figure 11.33 shows the project structure. The project is defined as the first level within the project structure. WBS elements are defined in a hierarchical structure under the project, and they represent smaller components of a project to facilitate better control and management. Networks are defined under WBS elements, and they represent the sequence of project activities such as internally executed activities, externally executed activities, general cost activities, service activities, material components, and so on. A network is assigned to a specific WBS element. Activities represent internal and/or external tasks.

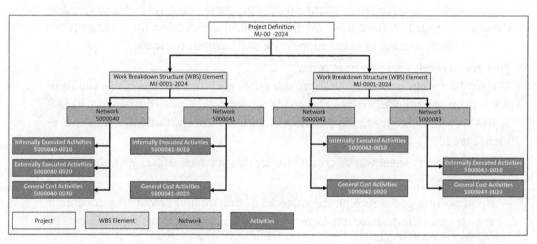

Figure 11.33 Project Structure Details

Internal activities are performed at a specific work center within the plant, whereas an external activity is performed by a vendor. Activities are defined and executed in a sequence within the network.

The project structure setup in SAP S/4HANA starts with the project definition. Transaction CJ20N (Project Builder) is used to define a project. The project definition is the central component of Project System, and it contains the project details, basic data of the project, controlling parameters, and organizational assignments. There is only one project definition per project. See Section 11.2 for details of how to set up the project structure.

The next step is to define WBS elements. Each WBS element represents tasks and subtasks that are required to complete the project. The WBS is organized in a hierarchical manner under the project definition. WBS elements can be defined at different sublevels hierarchically; that is, each WBS element can be further broken down into multiple smaller WBS elements, each representing a specific task. WBS elements are integrated with networks and activities. Transaction CJ02 is used to create WBS elements. You can create multiple WBS elements for a project, and under each WBS element, you can create child-level WBS elements.

The next step in the project structure setup is to define networks under WBS elements. Multiple networks can be created under WBS elements. Networks contain activities that are linked to a WBS for control and cost allocation purposes. Activities are individual tasks within a network. An activity represents the sequence of project activities such as internally executed activities, externally executed activities, general cost activities, service activities, material components and so on. Each activity can have its own resources, duration, costs, and other dependencies.

Setting up a project structure in SAP S/4HANA facilitates detailed planning of tasks, resources, costs, budget allocation, and timelines. A clear hierarchical view of the project structure makes it easier to manage and control complex projects. It makes it easier for the project management team to monitor the execution process and helps organizations to execute projects effectively.

11.3.3   Purchase Requisition Generation from Project System

As part of the project procurement, purchase requisitions are generated automatically upon releasing a network activity. Purchasing provides the necessary master data to support the activity. The network planning group will update the purchasing details for every item to be procured externally. The following purchasing master data is needed for this integrated process, other than the material master data and business partners (vendors):

- **Organization elements**
  Organizational elements such as the purchasing organization, purchasing group, and plant are important to trigger the creation of a purchase requisition automatically.

- **Source list**
  The source list provides the source of supply including the outline agreement, fixed vendor, and preferred vendor details for the material planners to update the purchasing details in the material assignment level within the network activity.

- **Purchasing info record**
  The PIR provides details about the procurement lead time, unit price, and so on. If a standard purchase order needs to be created, a standard PIR must exist for the combination of material and vendor within the plant and purchasing organization levels. If a subcontracting purchase order needs to be created, a subcontracting PIR must exist for the combination of material and vendor within the plant and purchasing organization levels.

- **Outline agreement**
  If there is a contract or a scheduling agreement that exists with a vendor for the nonstock material, it provides contracted terms and pricing information to procure items required for the project activity to be executed. An outline agreement can be entered directly in the external processing activity and at the material assignment level within the network activity.

- **Cost element**
  Cost elements are a special type of general ledger account that belong to cost-relevant accounts within the chart of accounts. This element is mandatory for an indirect procurement activity such as an external processing activity without a material master.

The activity planning group responsible for network planning updates the purchasing data in the respective activity and releases the activities sequentially. The following are the purchasing-related scenarios from Project System that will trigger the creation of purchase requisitions automatically:

- **Nonstock material requirement from internal processing activity**
  During material planning for network activities, you can plan for nonstock materials with item category N. Nonstock materials are not available in the inventory and must be procured from an external party. These materials are consumables, intangible items, and so on. These materials are planned by considering the procurement lead times in the respective network activity.

- **Subcontracting activity**
  The subcontracting process can be initiated as an external processing activity from Project System. The subcontracting process involves sending/issuing the components from inventory to the external subcontractor/vendor/service provider for the subcontracting activity and receiving the finished/semifinished material back into inventory from the subcontractor/vendor.

- **External service activity**
  An external service requisition can be planned for the external processing activity. A service purchase requisition can be triggered from Project System that can be

converted into a service order. The service provider/vendor provides the required service to execute the network activity.

Figure 11.34 shows the external processing parameters added to the external processing activity. These parameters are necessary to trigger the generation of a purchase requisition. Once the external processing parameters are added for procuring a nonstock item for the network activity, a service item for the network activity, or a subcontracting service for the network activity, save the network and release the network activity. A purchase requisition will be generated automatically as explained in Section 11.3.1.

In the activity details view of the external processing activity, you must maintain the vendor, control key, plant, purchasing organization, purchasing group, unit price for the purchase requisition line item, cost element, and so on. Or you can maintain a PIR (**Info record** field) or a contract agreement (**Outl. agreement** field) as the required attributes to create the purchase requisition.

In the **Res./Purc req.** field, if you set the value to **Immediately** it will trigger the generation of the purchase requisition upon the release of the network activity.

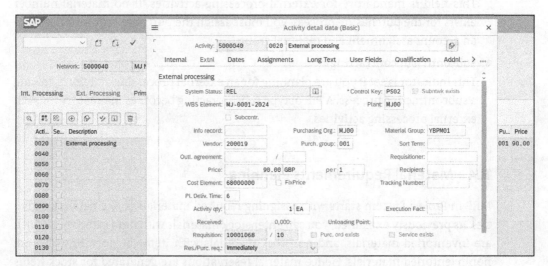

Figure 11.34 Generating Purchase Requisition from Project System

Purchase requisition creation is controlled by network profile settings such as the timing of the purchase requisition creation, whether immediately, from release, or never. If you set the value in the control parameter for the purchase requisition creation as never, the purchase requisition must be created manually. Refer to our discussion of maintaining the network profile in Section 10.1.6 for more details. To maintain the external processing parameters, select the network activity and click the **Details** icon in the **Basic Data Overview** screen of the network to arrive at the activity detail data view. Then, navigate to the **Extnl** tab to maintain the following external processing parameters to generate a purchase requisition:

- **Control Key**
  This plays a vital role in controlling network activities such as internal processing activities and external processing activities. The necessary processing parameters such as procurement parameters required to complete the activity are defined in the control keys.

- **Outl. agreement**
  The outline agreement is optional, but if an outline agreement exists for the material with a vendor, it is recommended to assign it as external processing parameter. The price and terms are copied from the outline agreement to the purchase requisition.

- **Info record**
  This provides details such as the source of supply (vendor), unit price, purchasing lead time, and so on. These are vital for the creation of a purchase requisition to procure materials and services.

- **Material Group**
  This field is mandatory for external processing activities. If no material number exists for the purchasing process, you must assign the purchase requisition item to an account assignment and material group.

- **Subcontr. (subcontracting)**
  This indicator is set if you are planning for an external processing activity involving a subcontracting process. A PIR of type subcontracting becomes mandatory for such external processing activities.

11.4   Material Requirements Planning

MRP in Project System starts with assigning required materials to the network activities. As previously explained, you can enter stock materials with item category L, which are inventoried materials, and nonstock materials with item category N, which are noninventoried materials. Hence material reservations are generated for stock items and purchase requisitions are generated for nonstock items. Nonstock items are directly consumed in the network activity, whereas stock items are issued to the network activity from the inventory management if available. Otherwise, these materials are procured from an external vendor or produced in house (depending on the procurement type assigned to the material master and other attributes) to stock them in inventory management and are later issued to the network activity using the material reservations.

In the following sections, we'll explore MRP for project planning, including the integrated stock monitoring process, and provide step-by-step configuration instructions.

11.4.1   Project Planning with Material Requirements Planning

MRP for a project is performed manually for network activities, or you can transfer the bill of materials (BOM) to the project directly using Transaction CN33. Materials requirements can be included in the network activity manually as previously explained in Section 11.2.3. You can plan both stock items and nonstock items using the material BOM transfer feature or manually. The system creates reservations for stock items with item category L, whereas nonstock items are procured from a vendor and consumed directly to execute a project activity.

Once the required materials are planned for the network activities, MRP controllers run MRP for stock items. The main purpose of MRP in Project System is to provide required materials to execute project activities.

Figure 11.35 shows the MRP process for project activities planned in the networks.

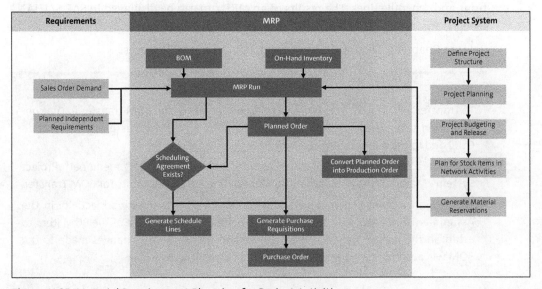

Figure 11.35 Material Requirement Planning for Project Activities

MRP is the main function of an organization that keeps demand and supply in sync. It converts independent requirements arising from sales demand and forecast demand into dependent requirements. It ensures the right quantities of the right material are available for other functions, such as production, Project System, maintenance, and so on to execute their processes at the right time so that organizations always keep optimized inventory levels.

For Project System, the MRP process starts with forecast planning of materials required to execute projects and with material reservations generated from network activities. An MRP run is executed by material planners using Transaction MD01N. The BOM, current inventory levels of the material, reservations, procurement type of materials, procurement lead times, and production lead times are other data components considered as inputs to the MRP run. You can also use the following transactions to run MRP:

- MD01: Total Planning Run, generally used to run MRP for all materials within planning scope
- MD02: Single Item and Multi Level Planning, generally used to run MRP for materials with multilevel BOMs
- MD03: Single Item and Single Level Planning, generally used to run MRP for materials without a BOM

The MRP run considers the on-hand inventory, reservations, and forecasts created for project execution and creates planned orders and procurement proposals depending on the procurement type of the material requirements. Planned orders can further be converted into production orders for in-house manufacturing and purchase requisitions and schedule lines. The results of an MRP run can be displayed in SAP S/4HANA using Transaction MD04 (Stock/Requirements List).

Refer to Chapter 6, Section 6.3 for more details on the MRP process.

Note

To transfer a BOM to a WBS, define a reference point for the BOM transfer in the configuration settings and assign the reference point to BOM items using Transaction CS02 and to the network activity using Transaction CN22.

To create a reference point for the BOM transfer to a WBS, follow menu path **Project System • Material • Bill of Material Transfer • Define Reference Points for BOM Transfer**.

The reference point you configure must be assigned to the network activity in the **Assignments** tab. Also assign the reference point in the basic data of the BOM item to establish the linkage between the network and the BOM. Any changes made to the BOM will automatically be transferred to the network activity.

11.4.2   Stock Monitoring

Stock monitoring for Project System involves tracking and managing inventory levels to ensure the required materials are available for project tasks/activities. This is a continuous process throughout the lifecycle of the project to ensure optimum inventory levels at all times while avoiding stockouts and overstocking. Stock monitoring is integrated with material MRP for better alignment with demand. Stock monitoring involves the following:

- **Stock type monitoring for materials required to execute projects**
  Available stock in unrestricted use stock, stock of materials in quality inspection, and materials in blocked stock in inventory management are monitored regularly. Reserved stock and special stocks such as project stock are also monitored regularly. Only unrestricted use stock is available for project execution, while project stock is

available for the respective WBS element and reserved stock is available for a specific project activity.

- **Stock levels monitoring of materials required to execute projects**
  Stock levels including the maximum stock, minimum stock, and reorder point are continuously monitored for critical materials needed for project execution. The maximum stock level of a material refers to the highest quantity of the material in the inventory. The stock of material should not cross this level to avoid overstocking. The minimum stock level of a material refers to the lowest quantity of the material in the inventory. The stock of material should not fall below this level to avoid stockouts. The *reorder point* is a specific quantity of a material at which a purchase order or production order is placed to bring the stock levels to the maximum level.

The following are the transaction codes used in SAP S/4HANA for stock monitoring:

- **MMBE**
  Displays the stock overview of a material at the plant and storage location levels. This transaction displays stock of material for all stock types and special stock types at the plant and storage location level. If the material is managed in batches, it displays the batch stock as well.

- **MB52**
  Displays warehouse stock of material on hand. This transaction displays the total stock on hand and the total valuation of stock at the plant and storage location levels.

- **MB51**
  Displays material documents for a material based on selection criteria. You can set a combination of material, plant, storage location, batch, vendor, posting date, and fiscal year in the selection criteria to filter the display of material documents.

- **MB5B**
  Displays stock of material for a given posting date. This transaction displays the stock of material, movement type, and material document for a given posting date or date range at the plant and storage location levels.

- **CO24**
  Displays the results of an availability check for materials at the plant and storage location levels.

11.4.3   Configuration Settings for MRP and Material Procurement for Projects

Let's explore the configuration settings for MRP and procurement integrations.

Define Procurement Indicators for Material Components

In this configuration step, you can define the control parameters, priorities, and default item category of materials to be assigned to network activities.

To arrive at this configuration step, execute Transaction SPRO and navigate to **Project System • Material • Procurement • Define Procurement Indicators for Material Components.**

Figure 11.36 shows an overview of procurement indicators defined for the material flow in the network. Click **New Entries** to create a new procurement indicator for material components. You can copy an existing procurement indicator and create a new one. Enter a three-character alphanumeric key and a description to define a new procurement indicator.

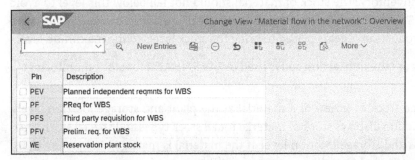

Figure 11.36 Define Procurement Indicators for Material Components: Overview

To display the details of a procurement indicator, select the procurement indicator and click the **Details** icon at the top or press `Ctrl`+`Shift`+`F2`. Figure 11.37 shows the details of procurement indicator **PF** (**PReq for WBS**). Let's explore the details.

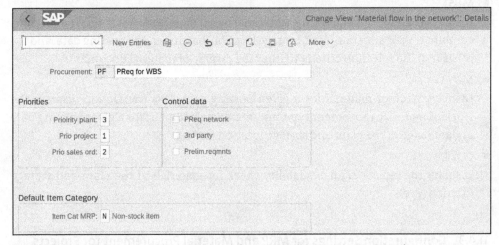

Figure 11.37 Define Procurement Indicators for Material Components: Details

The **Priorities** section of the procurement indicator has the following fields:

- **Priority plant**
  You can enter a numeric value from 1 to 9 to define the priority for a material component to be stored in plant stock, with 1 being the highest priority and 9 being the

lowest priority. If you enter 1 for the priority for plant stock, the material component will be kept in plant stock predominantly.

- **Prio project (priority project)**
  You can enter a numeric value from 1 to 9 to define the priority for material component to be stored in project stock, with 1 being the highest priority and 9 being the lowest priority. If you enter 1, the material component will be kept in project stock predominantly.

- **Prio sales ord. (priority sales order)**
  You can enter a numeric value from 1 to 9 to define the priority for material component to be stored in sales order stock, with 1 being the highest priority and 9 being the lowest priority. If you enter 1, the material component will be kept in sales order stock predominantly.

The **Control data** section of the procurement indicator has the following fields:

- **PReq network (purchase requisition network)**
  Set this indicator along with **Item Cat MRP** set as **L** (stock item) for materials managed in inventory. If you set this indicator, a commitment will be generated immediately upon saving the network in terms of a material reservation. The material reservation will become relevant for MRP.

- **3rd party**
  Set this indicator along with **Item Cat MRP** set as **N** (stock item), for materials not managed in inventory. If you set this indicator, a third-party purchase requisition will be generated for drop shipment to a customer from the vendor directly.

- **Prelim.reqmnts (preliminary requirements)**
  Set this indicator to generate preliminary requirements such as purchase requisitions for components for the network in advance. These preliminary requirements can be adjusted later. Set this indicator along with **Item Cat MRP** set as **N** (stock item), for materials not managed in inventory.

The **Default Item Category** section of the procurement indicator includes the **Item Cat MRP** field, where you can set the item category **L** for stock items and **N** for nonstock items as the default for the procurement parameter. This field goes along with the control data.

Check Account Assignment Categories and Document Types for Purchase Requisitions

In this configuration step, you can define a default document type for a purchase requisition that is generated from a network activity as well as default account assignment categories for different scenarios involving Project System functions.

To arrive at this configuration step, execute Transaction SPRO and navigate to **Project System • Material • Procurement • Check Acct Asst Categories and Document Types for Purc.Reqs.**

You will arrive directly at the screen to maintain account assignment categories and the document type for a purchase requisition for order category **20**. Figure 11.38 shows the account assignment categories for purchase requisition/order generation defined for order category **20** (network). In this configuration, you can set the **Document type**, which is a default document type for purchase requisitions generated from network activities from Project System.

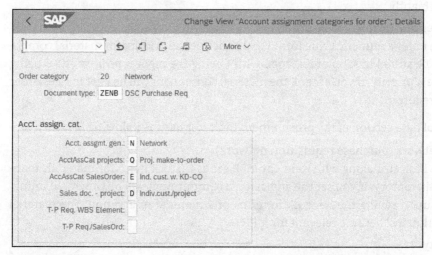

Figure 11.38 Check Account Assignment Categories and Document Types for Purchasing Requisitions

The **Acct. assign. cat.** section of this configuration has the following fields:

- **Acct. assgmt. gen. (account assignment generation)**
  In this field, assign an account assignment category for purchase requisitions and purchase orders generated from network activities for direct consumption in a network activity. You can set **N** (network) as the account assignment category and the received materials for the purchase order will be directly consumed in a network activity.

- **AcctAssCat projects (account assignment category projects)**
  In this field, assign an account assignment category for project stock. Assign account assignment category **Q** for project stock to have the received materials for the purchase order posted to project stock.

- **AccAssCat SalesOrder (account assignment category sales order)**
  In this field, assign an account assignment category for sales order stock. Assign account assignment category **E** for sales order stock and the received materials for the purchase order will be posted to sales order stock.

- **Sales doc. - Project (account assignment category sales document with project settlement)**
  In this field, assign an account assignment category for sales order stock and for the values to be settled to a project.

- **T-P Req. WBS Element (third-party requisition with WBS element)**
  In this field, assign an account assignment category for a third-party purchase requisition with a WBS element as the account assignment.
- **T-P Req./SalesOrd (third-party requisition with sales order)**
  In this field, assign an account assignment category for a third-party purchase requisition with the sales order as the account assignment.

Define Movement Types for Material Movements

In this configuration step, you can maintain movement types for various goods movement transactions within Project System.

To arrive at this configuration step, execute Transaction SPRO and navigate to **Project System • Material • Procurement • Define Movement Types for Material Movements**.

> **Note**
>
> Best practice recommendation is to accept the standard movement type settings in this configuration and not to modify the standard settings.

Figure 11.39 shows the movement types maintained in the standard system for Project System:

- Movement type **281** has been maintained for goods issue of plant stock, project stock, and sales order stock to a network activity.
- Movement type **282** has been maintained for goods issue reversal for plant stock, project stock, and sales order stock to a network activity.

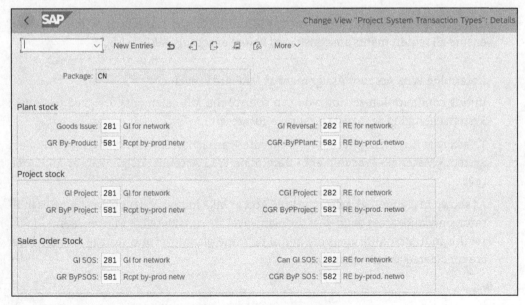

Figure 11.39 Define Movement Types for Material Movements

- Movement type **581** has been maintained for goods receipt of by-products from a network into plant stock, project stock, and sales order stock.

- Movement type **582** has been maintained for reversal of goods receipt of by-products from a network into plant stock, project stock, and sales order stock.

Activate MRP Groups for Requirements Grouping

In this configuration step, you can create MRP groups to manage MRP for multiple projects. This setting is required for materials/components managed in MRP. An MRP group is assigned to materials in the MRP views of the material master record.

To arrive at this configuration step, execute Transaction SPRO and navigate to **Project System • Material • Procurement • Activate MRP Groups for Requirements Grouping**.

Figure 11.40 shows MRP groups **0001**, **0002**, and **0010** created for plant **MJ00**. Click **New Entries** to create a new MRP group for a plant. You can copy an existing MRP group and create a new one for the plant. Enter a four-digit numeric key for the MRP group and enter a short description.

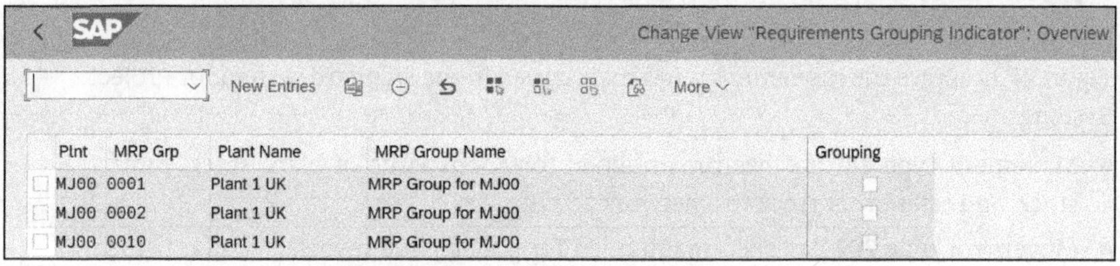

Plnt	MRP Grp	Plant Name	MRP Group Name	Grouping
MJ00	0001	Plant 1 UK	MRP Group for MJ00	
MJ00	0002	Plant 1 UK	MRP Group for MJ00	
MJ00	0010	Plant 1 UK	MRP Group for MJ00	

Figure 11.40 Activate MRP Groups for Requirements Grouping

Using MRP groups, you can manage MRP for multiple projects at a time. This will ensure all requirements are considered before executing MRP.

Determine WBS Account Assignment at MRP Area Level

In this configuration setting, you can control the MRP elements assigned to Project System belonging to storage location MRP areas.

To arrive at this configuration step, execute Transaction SPRO and navigate to **Project System • Material • Procurement • Determine WBS Account Assignment at MRP Area Level**.

As shown in Figure 11.41, select the **WBS Acct at MRP** indicator. If you don't set this indicator, only MRP areas defined at the plant level are considered as MRP elements. If you set this indicator, MRP areas defined at both the plant level and storage location level are considered as MRP elements.

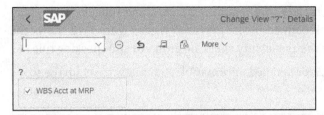

Figure 11.41 Determine WBS Account Assignment at MRP Area Level

Availability Check

Required materials and components are planned in the network activities to execute project tasks. A material availability check can be performed for the network activities manually by project planners and/or automatically by the system during the MRP run. Let's explore the configuration settings that control the material availability check in the network.

Define Checking Rules

A two-character alphanumeric key will be created to trigger the material availability check in networks. Checking rule PS is defined in the standard system for Project System. You can create a new checking rule to control the material availability check.

To arrive at this configuration step, execute Transaction SPRO and navigate to **Project System • Material • Availability Check • Define Checking Rules**.

To create a new checking rule, specify a two-digit alphanumeric key and a description. Checking rule **PS** is defined in the standard system for the availability checking rule for Project System, as shown in Figure 11.42.

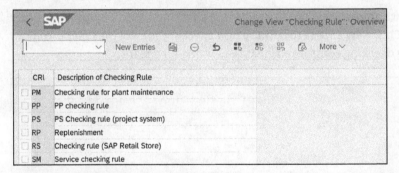

Figure 11.42 Define Checking Rules

Define Checking Scope

In this configuration step, you can define the availability checking procedure based on the availability checking group and the checking rule defined in the previous step.

To arrive at this configuration step, execute Transaction SPRO and navigate to **Project System • Material • Availability Check • Define Checking Scope**.

This definition is configured for the combination of a checking rule and a checking group of the availability check defined in the previous steps. Figure 11.43 shows the scope of the check defined for the availability checking group **01** and checking rule **PS**.

Refer to Chapter 4, Section 4.1.5 for detailed information about how to set up the scope of the check.

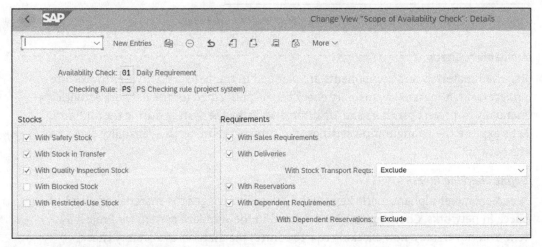

Figure 11.43 Define Checking Scope

Define Checking Control

In this configuration step, we can assign the availability checking rule to the network order type by plant. To arrive at this configuration step, execute Transaction SPRO and navigate to **Project System • Material • Availability Check • Define Checking Control**.

Figure 11.44 shows an overview of the availability checking control by network order type and plant. Click **New Entries** to define a new checking control for a plant. You can copy an existing checking control and create a new one for a plant.

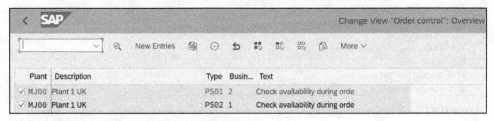

Figure 11.44 Define Checking Control: Overview

To display the details of the checking control for a plant and order type combination, select a specific combination from the overview screen and click the **Details** icon at the top or press Ctrl + Shift + F2. Figure 11.45 shows the details of the checking control defined for network order type **PS02** and plant **MJ00**. The following are the key fields of this configuration:

- **Order Type**
  The order type defines the purpose of the document in different functions such as production planning, plant maintenance, Project System, and so on. The following are the order types defined for network order creation in Project System:
  - **PS01**: Network with header assignment (int. NA)
  - **PS02**: Network with activity assignment(int.NA)
  - **PS03**: Network with header assignment (ext. NA)
  - **PS04**: Networks for sales order (int. NA)
  - **PS05**: Networks for make-to-order (int. NA)

  The availability checking control is defined for the combination of network order type and plant.

- **Availability Check**
  This field controls if the availability check happens at the time of maintenance order creation or release. Table 11.7 shows the values you can assign to this field.

Business Function	Description
1	Check availability during order creation
2	Check availability during order release

Table 11.7 Business Function for Availability Check

- **Checking Rule**
  Assign the checking rule defined for Project System (**PS**) in this field. By assigning the checking rule, the system automatically determines the scope of check and performs the material availability check.

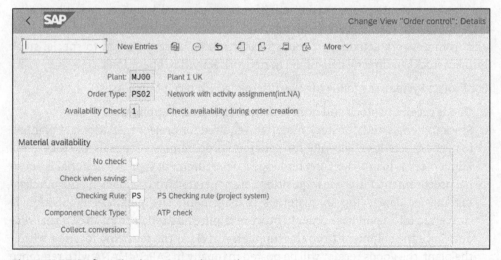

Figure 11.45 Define Checking Control: Details

11.5   Inventory Management

Inventory management is the central function for all material needs for almost every function of the organization. Inventory management ensures the required materials are stored, managed, and issued to business processes. Project System and inventory management are seamlessly integrated in SAP S/4HANA. Inventory management ensures required materials and components required for network tasks/activities are issued and tracked. Inventory management provides visibility into the on-hand inventory and location information where the materials are stored in the inventory and helps the planner groups responsible for project planning to plan and schedule activities accordingly.

Inventory management provides real-time information about stock availability. During MRP, the system generates procurement proposals for insufficient stock. It can then execute project tasks/activities based on the stock availability check for materials and components assigned as required materials to execute network activities. Inventory management tracks all material movements within the project execution process, including goods receipt of project stocks from suppliers, goods issue to project activities, goods issue of components to subcontractors, and other internal inventory movements involving project stocks.

Overall, inventory management ensures projects are effectively executed within the budget and without any delays due to material availability. Let's explore the integration between inventory management and Project System, including both goods receipt and goods issue processes and project stocks.

11.5.1   Goods Receipt for Project System

Goods receipt for Project System involves receiving materials and components required to execute project tasks/activities from external suppliers and subcontractors. Goods receipts are posted with reference to purchase orders originating from MRP and from network activities. A goods receipt can be directly posted to special stock (project stock), or directly consumed in network activities/operations.

For Project System, the following are different goods receipt processes:

1. **Goods receipt of stock and nonstock materials from suppliers**
   Stock materials and nonstock materials required for project execution are planned in network activities. The MRP run generates procurement proposals for stock items if inventory is insufficient and if the special procurement type of a material is external procurement. Purchase requisitions are generated from MRP using the purchasing master data setup in materials management. The purchase requisition is converted into a purchase order to procure required materials from an external vendor. Once the shipment from the suppliers arrives at the receiving location within the plant, the goods receipt will be posted manually in SAP S/4HANA with reference

to the purchase order using Transaction MIGO and movement type 101. The materials received during goods receipt can be posted to unrestricted use stock or project stock (special stock type).

For nonstock materials and external processing activities, purchase requisitions are generated from Project System directly based on external processing parameters and procurement master data assigned to the nonstock materials and external processing activities. Purchase requisitions are created with account assignment N (network) and converted to purchase orders by the buyers manually or automatically via a background job (Transaction ME59 can be scheduled as a background job). Once the materials are received from the vendor, goods receipt will be posted manually in SAP S/4HANA with reference to the purchase order using Transaction MIGO and movement type 101. The received materials are directly consumed by the network activity, and these materials are not managed in the inventory.

Figure 11.46 shows the accounting document generated upon posting the goods receipt for a purchase order line item with account assignment **N** (network). Transaction events **KBS** and **WRX** are triggered to post the costs in financial accounting.

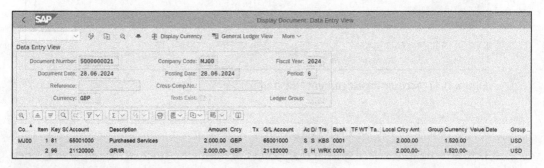

Figure 11.46 Goods Receipt: Accounting Document

Figure 11.47 shows the first line item of the accounting document with transaction event **KBS**, and it is evident the system records the transaction with reference to the network. This way, costs are recorded, tracked, and controlled in Project System.

2. **Goods receipt for subcontracting activity (external processing activity)**
As previously explained, the subcontracting process can be planned in the external processing activity. Subcontracting is a special procurement process where an external supplier performs/executes certain project activities, and the components required for the process will be provided by the buying organization to the subcontracting vendor. Subcontracting purchase requisitions are generated from Project System with reference to the network activity (account assignment **N**). A subcontracting purchase requisition is converted into a subcontracting order. The components required to execute the external processing activity are provided/issued to the subcontracting vendor with reference to the subcontracting order. The subcontractor will perform the external processing activity on behalf

of the buying organization and ship the completed material to the buyer. Once the shipment from the subcontractor arrives at the receiving location of the plant, goods receipt will be posted manually with reference to the subcontract order using Transaction MIGO and movement type 101. The received materials are directly consumed by the network activity, and the external processing activity is marked as completed.

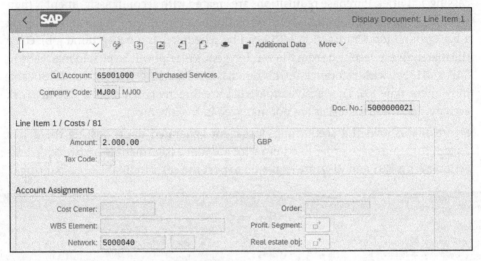

Figure 11.47 Accounting Document Line Item

11.5.2   Goods Issue for Project System

Goods issue for Project System is a process of removing stock items required for network activities, especially for internal processing activities from the inventory for consumption. A goods issue posting creates both a material document and an accounting document in the system to record the material movement and the valuation of the stock in financial accounting. For stock material requirements, Project System generates material reservations. Goods issue from inventory management is processed with reference to reservations using Transaction MB26. You can issue materials from inventory without reference to reservations as well if required.

For Project System, the following are the different goods issue processes:

1. **Goods issue materials to network activities**
   Raw materials and other required materials/components to execute a project are issued to the project with reference to the network using Transaction MIGO and using movement type 281. Materials are issued with reference to material reservations generated automatically from network activities. The goods issue ensures the required materials are supplied to projects for project execution.

   Figure 11.48 shows material reservation **1657** generated for stock material **MS-RAW MATERIAL1** for 200 EA with reference to network activity **5000040 0010**.

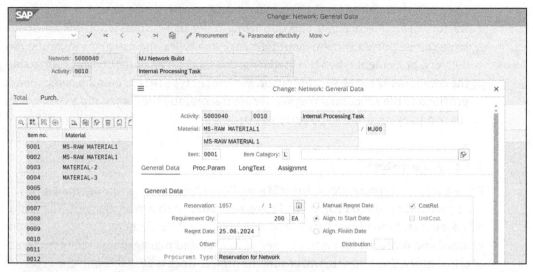

Figure 11.48 Reservation Generated for Material Requirements

Figure 11.49 shows the details of reservation **1657**. The reservation is created with reference to network **5000040** and with movement type **281**. This movement type is used to issue goods from the inventory to a network for consumption. The material, quantity, plant, and so on are copied from the material assignments to the network activity.

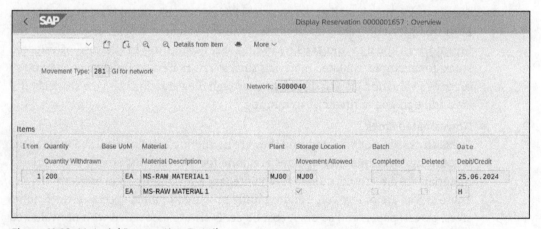

Figure 11.49 Material Reservation Details

2. **Goods issue of components to subcontracting activity**

   The subcontracting process can be planned in the external processing activity. Subcontracting is a special procurement process where an external supplier performs/executes certain project activities, and the components required for the process will be provided by the buying organization to the subcontracting vendor. Materials/

899

components required for the external processing activity are issued to the subcontractor with reference to a subcontracting order using Transaction ME2O. Goods issue can be posted with/without reference to a (sales and distribution) outbound delivery. Movement type 541 is used for goods issue for an outbound delivery to ship the components to the subcontractor. Movement type 541 ensures the components provided to the subcontracting vendor are tracked, and the ownership remains with the company requesting the external operation.

11.5.3   Project Stock

Project stock is a special stock, and it is specifically held in the inventory management for a particular project. Project stocks are associated with a WBS element. The issue/withdrawal from the inventory happens against the WBS element. Project stocks are planned and procured separately as well as tracked and managed separately from regular inventory. The MRP run will ensure the project stock is available only for a specific project as designated. Typically, projects are executed within a tight schedule, and managing and tracking the materials required for the project activities separately ensures projects are executed efficiently while keeping the costs low and making sure the required materials are available on time. Managing and tracking the required materials separately using project stocks is vital for efficiently managing projects. The following are the two types of project stocks:

- **Valuated stock**
  These stocks are valuated at the plant level (if valuation control is set at plant level). Both quantity and value updates are done for valuated stocks. The value of the inventory is updated and tracked in financial accounting in the case of valuated stock; for example, valuated materials such as raw materials, finished goods, and so on have a valuation class assigned, and through the valuation class the stock is valuated and updated in financial accounting.

- **Nonvaluated stock**
  These stocks are not valuated, and they are neither updated nor tracked in financial accounting. Only quantity updates are done for nonvaluated stocks. Nonvaluated materials are directly consumed for maintenance, production, and project activities. These stocks are of low value and typically procured in bulk, such as screws, adhesives, gloves, paper, and so on. A cost object is always associated with nonvaluated stocks.

Table 11.8 shows some of the important standard movement types defined for goods movements (both valuated and nonvaluated inventory) involving project stocks. Reversal of these movements involving project stock are also possible, but those movement types are not listed.

Movement Type	Detailed Description
101	Goods receipt. This movement type is used for standard goods receipts from external vendors. It increases the stock quantity and value in the receiving storage location. Stock can be received into unrestricted use stock or quality inspection stock.
122	Return to vendor. This movement type is used to return materials to the vendor because of poor quality. Stock will be removed from the quality inspection stock to return the goods to the vendor.
161	Returns purchase order to vendor. This movement type is used to return materials to vendor with reference to a returns purchase order. Stock will be removed from the inventory to return the materials to the vendor.
221	Goods issue for a project. This movement type is used to issue goods from the inventory to a project for consumption.
261	Goods issue for a production order. This movement type is used to issue goods from the inventory to a production order for consumption.
281	Goods issue for a network. This movement type is used to issue goods from the inventory to a network for consumption.
301	Transfer posting plant to plant. This movement type is used to transfer materials from one plant to another within the same company code.
309	Transfer posting material to material. This movement type is used to transfer stock of one material to another material within the same plant.
311	Transfer posting from one storage location to another storage location within the same plant. This movement type is used to transfer materials from one storage location to another within the same plant.
321	Transfer posting from quality inspection to unrestricted. This movement type is used to transfer materials from quality inspection to unrestricted use stock.
411	Transfer posting of special stock to unrestricted use stock. This movement type is used to transfer materials from special stock to unrestricted use stock.
501	Goods receipt without a purchase order. This movement type is used to receive materials from a vendor without reference to a purchase order.
551	Goods issue scrapping. This movement type is used to issue discrepant materials from unrestricted use stock to scrapping. This will reduce the inventory of materials.

Table 11.8 Movement Types Defined for Goods Movement Involving Project Stock

Movement Type	Detailed Description
601	Goods issue for delivery. This movement type is used to issue goods from inventory for an outbound delivery to fulfill a customer order.
641	Goods issue for a stock transport order. This movement type is used to issue goods from the inventory for an outbound delivery to fulfill an intra-company stock transfer order.
701	Inventory differences in unrestricted use stock. This movement type is used to post differences from unrestricted use stock that arise from the physical inventory process.
703	Inventory differences in quality inspection stock. This movement type is used to post differences from quality inspection stock that arise from the physical inventory process.
707	Inventory differences in blocked stock. This movement type is used to post differences from blocked stock that arise from the physical inventory process.

Table 11.8 Movement Types Defined for Goods Movement Involving Project Stock (Cont.)

Note

Nonstock materials can't be managed in project stocks in inventory management. Only stock items can be managed in project stock.

11.6 Project Budget Management

Budget management in SAP S/4HANA Project System involves planning, allocating, and controlling project financials using WBS elements. Budget management is a critical functionality, and it tracks and controls the cost incurred by the WBS element. Effective budget management ensures projects are executed within budget and that all the project costs are tracked accurately.

Let's explore the budget management functionality in detail, including instructions for setting up budget allocation.

11.6.1 Integrated Budget Management

The budget availability check is integrated with the purchasing function of materials management. During the creation of a purchase requisition and purchase order, a real-time budget availability check is performed against the available budget in the WBS element. Only if a sufficient budget is available—greater than or equal to the total value of the purchase requisition and purchase order—will the system allow the creation of a purchase requisition and purchase order. For every purchase requisition transaction, the system reduces the available budget of the WBS element.

Figure 11.50 shows the budget check functionality during the purchasing process. Purchase requisitions are created with reference to a WBS element manually, from MRP, or from external procurement systems such as SAP Ariba Buying and Invoicing. At the time of creation of the purchase requisition, the system checks the available budget of the WBS element. If the available budget is insufficient, the system throws an error message and aborts the creation of the purchase requisition.

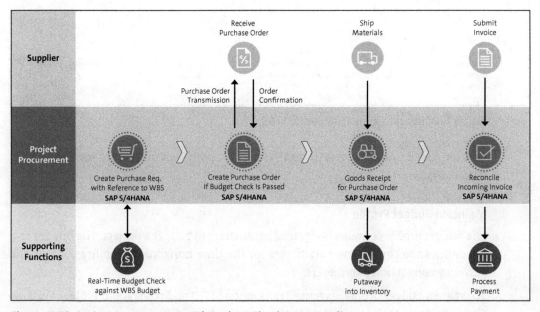

Figure 11.50 Project Procurement with Budget Check Functionality

If the budget check passes, a purchase requisition can be created without any budget issues. Follow-on documents such as the purchase order, goods receipt, and invoice receipt can be created to complete the purchasing process.

11.6.2   Budget Allocation Setup

The budget allocation setup involves creating a budget structure and allocating budget amounts to WBS elements. You can set up tolerances for budget control to ensure real-life scenarios are executed without any hurdles. Budget control ensures that spending doesn't exceed the allocated budget by applying tolerances set. Expenditures are continuously monitored during project execution and reported if there are any variances from planned versus actual expenditures. SAP S/4HANA allows budget allocation adjustments, if necessary, based on scope changes and on resource costs including materials, human resources, and other overhead costs.

To allocate and change budget values for WBS elements, use Transaction CJ30. Our example in Figure 11.51 shows the budget allocation for WBS **MJ-0001-2024**, WBS **MJ-0002-2024**, and WBS **MJ-0003-2024**. The budget values can vary from one WBS element to another based on the tasks to be executed within the WBS elements.

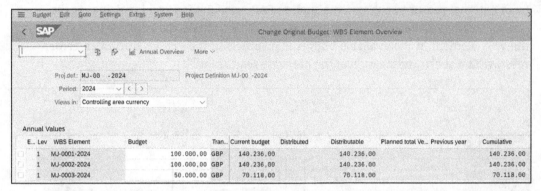

Figure 11.51 Allocate Annual Budget to WBS

Let's explore configuration settings that support budget management functionality in Project System.

Maintain Budget Profile

The budget profile contains budgeting parameters to budget a project. The budget profile defines the controlling parameters for the time horizon, availability control, and currency conversion parameters.

To create a budget profile, execute Transaction SPRO and navigate to **Project System • Costs • Budget • Maintain Budget Profiles**.

Figure 11.52 shows budget profile **Z100 (PS Budget profile)**. Click **New Entries** to create a new budget profile. You can copy an existing budget profile to create a new one. Enter a six-character alphanumeric key and a description to maintain a new budget profile. Let's explore the controlling parameters that can be maintained in the budget profile.

The **Time Frame** section of the budget profile has the following fields:

- **Past**
  In this field, enter a numeric value for the number of years in the past from the current year that you can budget for a project.

- **Future**
  In this field, enter a numeric value for a number of years in the future from the current year that you can budget for a project.

- **Start**
  In this field, enter a numeric value for a number of years from the current year to start budgeting for a project.

- **Total Values**
  Set this indicator if you want to budget for the total/overall values. The system allows you to budget for the total allocated value of the project.

- **Annual Values**
  Set this indicator if you want to budget for the annual values. The system allows you to budget for the annual value allocated to the project.

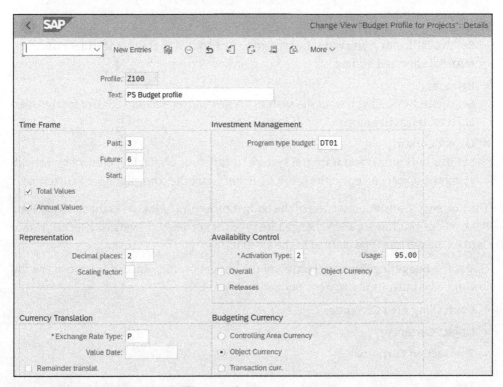

Figure 11.52 Maintain Budget Profile: Details

The **Availability Control** section of the budget profile has the following fields:

- **Activation Type**
  In this field, you can set a key defined for availability control. The following keys are defined in the standard system:
  - **0: Cannot be activated**: Set this key if you do not want availability control for your projects.
  - **1: Automatic activation during budget allocation**: Set this key if you want the system to automatically activate the availability control as soon as the budget is allocated.
  - **2: Background activation**: Set this indicator if you want to activate availability control using a background program whenever the usage of the budget is exceeded.

> **Note**
>
> If you set the availability control key as **2** (background activation when usage exceeded), you must enter the percentage of usage of the budget in the **Usage** field—for example, 95%. As soon as the usage of the budget for a WBS element crosses 95%, availability control is activated through the background process.

- **Overall**
  Set this indicator if you want the system to carry out availability control for the overall/total allocated budget.
- **Releases**
  Set this indicator if you want the system to carry out availability control for the overall/total released budget.
- **Object Currency**
  Set this indicator if you want the system to carry out availability control by considering the object currency (the project currency, not the controlling area currency).

The **Currency Translation** section of the budget profile includes the **Exchange Rate Type** field, where you can set an exchange rate type for currency conversions. Select a standard exchange rate type defined in the system.

Under the **Budgeting Currency** section of the budget profile, you can set one of the following radio buttons to carry out budgeting of projects:

- **Controlling Area Currency**
- **Object Currency**
- **Transaction Currency**

Stipulate Default Budget Profile for Project Definition

This is an important configuration setting where you can assign the budget profile to the project profile. When a new project definition is created using a project profile, the system automatically copies the budget profile assigned to the project profile.

To arrive at this configuration step, execute Transaction SPRO and navigate to **Project System • Costs • Budget • Stipulate Default Budget Profile for Project Definition.**

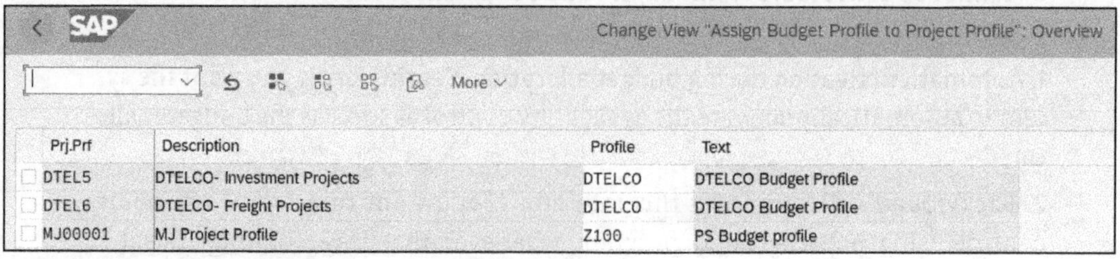
SAP
Change View "Assign Budget Profile to Project Profile": Overview
More
Prj.Prf
Description
Profile
Text
DTEL5
DTELCO- Investment Projects
DTELCO
DTELCO Budget Profile
DTEL6
DTELCO- Freight Projects
DTELCO
DTELCO Budget Profile
MJ00001
MJ Project Profile
Z100
PS Budget profile

Figure 11.53 Assign Default Budget Profile for Project Definition

Figure 11.53 shows that budget profile **Z100** has been assigned to project profile **MJ00001**. The system displays all project profiles defined in the system, and you can directly assign your budget profile. You can assign one budget profile to multiple project profiles.

Define Tolerance Limits

In this configuration setting, you can define tolerance limits for availability control of budgets for your projects. As a prerequisite, you must have activated availability control in the budget profile before setting up tolerance limits.

To arrive at this configuration step, execute Transaction SPRO and navigate to **Project System • Costs • Budget • Define Tolerance Limits.**

Figure 11.54 shows the tolerance limits maintained for controlling area **MJ00** and budget profile **Z100**.

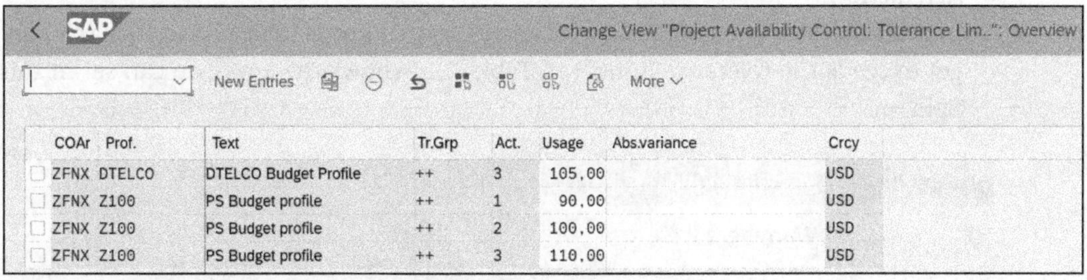

COAr	Prof.	Text	Tr.Grp	Act.	Usage	Abs.variance	Crcy
ZFNX	DTELCO	DTELCO Budget Profile	++	3	105,00		USD
ZFNX	Z100	PS Budget profile	++	1	90,00		USD
ZFNX	Z100	PS Budget profile	++	2	100,00		USD
ZFNX	Z100	PS Budget profile	++	3	110,00		USD

Figure 11.54 Define Tolerance Limits for Budget Availability Control

Click **New Entries** to define tolerance limits for availability control for the combination of controlling area and budget profile. You can copy the existing entries to create a new one. You can define the following values as tolerance limits:

- **Tr.Grp (activity group)**
  In this field, you can assign a key for an activity group for availability control. Table 11.9 lists the values defined in the standard system. If you set the **++** key, the system activates the availability control for all groups.

Activity Group	Description
++	All activity groups
0	Purchase requisition
1	Purchase order
2	Orders for project
3	Goods issue

Table 11.9 Activity Groups for Availability Control

Activity Group	Description
4	Financial accounting document
5	Controlling document
6	Budgeting
7	Funds reservation
8	Fixed prices in project
9	Payroll
10	Travel expenses

Table 11.9 Activity Groups for Availability Control (Cont.)

- **Act. (action)**
  In this field, you can assign a key for an action that needs to be taken when the budget exceeds the tolerance limits set. Table 11.10 shows the keys you can set in this field.

Action Key	Description
1	Warning
2	Warning with mail to person responsible
3	Error message

Table 11.10 Action Keys for Budget Availability Control

- **Usage**
  In this field, you can enter the percentage of usage of the budget as a tolerance limit—for example, 95%. As soon as the usage of the budget for a WBS element crosses 95%, availability control kicks in and takes the action defined in the **Action** field.

- **Abs.variance (permissible variance)**
  In this field, you can enter the maximum absolute value as a tolerance limit. If you set an absolute value of 10,000.00 USD and the total available budget is 200,000.00 USD, for example, the system takes appropriate action when the budget exceeds 210,000.00 USD.

11.7 Summary

In this chapter, we introduced the integration of Project System with materials management and offered best practice examples and processes. In Section 11.1, we explained the structure of Project System as an introduction. In Section 11.2, we explained the master data in Project System with a step-by-step procedure to maintain Project System-specific master data. In Section 11.3, we explained the integrated project procurement process with a simulation of the purchasing process triggered from Project System. In Section 11.4, we explained MRP for Project System and stock monitoring processes. In Section 11.5, we explained the goods receipt and goods issue process for project activities. Finally, in Section 11.6, we explained the budget management in Project System.

In the next chapter, we'll discuss the significance of EDI and external tax engine integration with materials management and offer best practice examples and configuration settings.

11

Chapter 12
Integration with Other External Systems

SAP S/4HANA provides a platform to integrate other external systems and businesses. This chapter focuses on electronic data interchange for the electronic transactional data flow between SAP S/4HANA and business partners, and on the external tax engine integration with SAP S/4HANA.

Standardizing, harmonizing, and automating the business processes in SAP S/4HANA are key objectives of any organization. SAP S/4HANA provides a platform and capabilities to standardize and automate business processes.

The first topic we'll cover in this chapter is electronic data interchange (EDI), which is one of the key capabilities used across businesses to collaborate and integrate ERP systems with SAP S/4HANA. EDI integrates businesses by enabling electronic exchange of transactional data between businesses. In materials management, use of EDI methods starts with sharing the forecast/demand from SAP S/4HANA with the supplier (business partner) before even submitting the purchase order formally. Suppliers largely benefit from that as they can optimize production based on demand. The buying organization can transmit the purchase order once it's released/approved in SAP S/4HANA via EDI. On the other hand, suppliers can create a sales order with reference to the purchase order automatically in their ERP systems. They can submit an order confirmation to the buying organization electronically via EDI. EDI is used in other business processes outside of materials management as well, including quality management, transportation management, and so on.

The second topic we'll cover in this chapter is external tax engine integration with SAP S/4HANA, which has its own benefits over the native or internal tax engine. External tax engines are third-party tax determination systems that calculate indirect taxes for the procure-to-pay process. Indirect taxes include taxes levied on goods and services purchased and paid to the supplier as part of their invoice and taxes that are in turn paid to the government by the supplier. Tax determination rules vary from country to country, so maintaining complex and often changing tax rules manually in SAP S/4HANA requires continuous manual effort. External tax engine integration with SAP S/4HANA fully eliminates manual maintenance of tax details and condition records in SAP S/4HANA. During purchase order creation and logistics invoice processing, SAP S/4HANA automatically calls an external tax engine to calculate indirect taxes by passing the transaction details to it.

We'll dive into integration with these external systems in the following sections, including cross-functional configuration instructions for materials management.

12.1   Electronic Data Interchange

EDI is a widely used methodology for exchanging business transactions (documents) between organizations. EDI uses digitized transaction standards to exchange documents electronically (a paperless process). Through EDI, organizations can collaborate for faster sharing and processing of transactional data while reducing errors. EDI uses specific standards for every transactional data element as these transactions are processed by computers. These standards help organizations, both the senders and receivers of the data, to easily read and understand electronic documents. There are various EDI standards in use today such as EDIFACT (used in Europe), ANSI X12 (used in North America), and more. These standards define the structure of the documents and meet all compliant requirements.

Figure 12.1 shows the various transactions that can be triggered from the buying side (buying organization) and supplying side (supplier organization). These transactions are automatically transmitted in a specific format defined for the specific transactional data. For example, EDI 850 is used to transmit a purchase order to a supplier. The following steps define how the purchase order will be sent to the supplier electronically using EDI:

1. Once the purchase order is created in SAP S/4HANA and released/approved, an output message will be triggered automatically. The output will trigger an IDoc (intermediate document) and be transmitted to an EDI system.

2. The EDI system converts the IDoc into an electronic format based on the specific standard defined for the transactional data. In this case, it is EDI 850.

3. The supplier's ERP system accepts the electronic purchase order received via EDI and converts it into a sales order automatically.

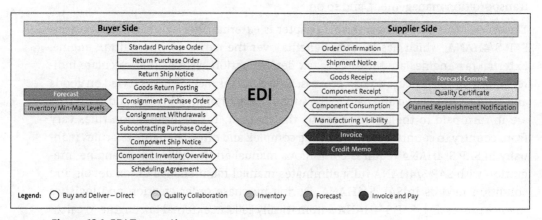

Figure 12.1 EDI Transactions

The following are significant benefits of EDI:

- Real-time exchange of documents between organizations. Electronic documents are triggered, transmitted, and received by the business partners in real time.

- Increased standardization and effective collaboration with business partners, which improves overall process efficiency.

- Paperless transactions. It's easy to process these transactions electronically, and this reduces errors.

- Helps to keep optimum inventory levels on the buying side and optimizes production on the supply side.

- Improves buyer and supplier relationships, reduces costs, and streamlines business processes.

In the following sections, we'll explore EDI integration with materials management and the setup of EDI interfaces.

12.1.1   Integrated EDI with Materials Management

SAP S/4HANA uses IDocs to transmit transactions in real time electronically. IDocs are triggered automatically based on the output condition records defined for a transaction. The main purpose of an IDoc is to transfer transaction data from SAP S/4HANA to another ERP system and receive transactional data into SAP S/4HANA. IDocs are received and processed by EDI systems that convert them into a specific EDI format based on the EDI standard being used. A business partner receives the EDI message and processes it in their ERP system. For inbound documents into SAP S/4HANA, IDocs are generated in the EDI system by converting the EDI message into an IDoc and receiving it into SAP S/4HANA.

Collaborating with suppliers is vital for the organization's objective of standardizing and automating its business processes. Purchasing, inventory management, and logistics invoice verification (LIV) functions of materials management traditionally leverage EDI for collaborating with business partners (suppliers). IDocs are used for inbound and outbound messages in SAP S/4HANA. In a procure-to-pay cycle, purchase orders are transmitted to business partners (suppliers) using outbound IDocs via EDI. Suppliers send an order confirmation and advanced shipping notification (ASN) with reference to the purchase order via EDI, and these messages are received into SAP S/4HANA using inbound IDocs. Upon posting the goods receipt once the shipment arrives, goods receipt confirmation can be sent to the supplier using an outbound IDoc via EDI. After fulfilling the purchase order, suppliers can send an invoice via EDI, and an invoice document is received into SAP S/4HANA using an inbound IDoc.

Figure 12.2 illustrates the use of EDI/IDocs in the procure-to-pay process for collaborating with suppliers. The transactional data flow between the buying side (on SAP S/4HANA) and the business partner (supplier) happens electronically using IDocs.

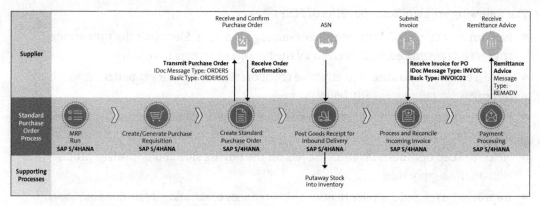

Figure 12.2 Procure-to-Pay Process with EDI

We'll take a closer look at the IDoc components and setup in the following sections.

Message Types for IDocs

Message types distinguish different document types that flow via EDI using IDocs. Every document that flows electronically via EDI is associated with a message type in SAP S/4HANA. Message types are linked to basic IDoc types for generating IDocs for transmission to external systems (business partners) and for receiving IDocs from external systems. For instance, purchase orders, order confirmations, ASNs, invoices, and the like all have their own message types for electronic data transfer. Table 12.1 shows the standard message types used in SAP S/4HANA for transmitting and receiving different documents electronically using IDocs.

Message Type	Description
MATMAS	Material master data
CREMAS	Vendor master data
ORDERS	Purchase order
ORDCHG	Purchase order change
ORDERSP	Purchase order confirmation
DESADV	Advanced shipping notification
INVOIC	Incoming invoice document
MBGMCR	Goods receipt
PURSAG_MAINTAIN	Scheduling agreement delivery schedule
REMADV	Payment advice

Table 12.1 Standard IDoc Message Types

IDoc Types

An IDoc type defines the structure of an IDoc. Every basic type is mapped to a message type in SAP S/4HANA, and message types represent a specific transactional data or master data element.

Figure 12.3 shows the basic types assigned to message types in the standard system. If you create a new custom basic type, you can assign it to a message type (based on the document type) using Transaction WE82. You will arrive directly at the screen shown in Figure 12.3. If you have defined an extension of a basic type to customize an existing basic type of an IDoc, click **New Entries** and assign the basic type and extension to the relevant message type. This assignment is mandatory for electronic data transfer via EDI.

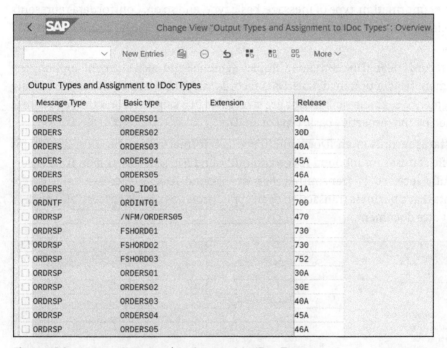

Figure 12.3 Message Types and Assignment to IDoc Types

The standard system provides basic types with standard segments that are structured sequentially. Each segment consists of data fields to carry the transactional data. Each segment contains a group of fields that represent a specific data set: organizational data, taxes, conditions, general data, and so on. IDoc types (basic types) can be customized to add extensions to a standard basic type. These extensions are defined as a custom basic type in SAP S/4HANA. This type contains additional custom segments that can carry custom fields or standard fields that do not exist in the standard IDoc types. IDoc segments can have a parent and child relationship; that is, an IDoc segment can be defined as a parent, and that can have multiple child segments.

Structure of an IDoc

The message types and basic IDoc type of the IDoc represent a specific document and the direction of it, whether inbound into SAP S/4HANA or outbound from SAP S/4HANA. Figure 12.4 shows the outbound IDoc to transmit a purchase order to the business partner (supplier). To arrive at the details of an IDoc, execute Transaction WE02 and specify the selection criteria to display a list of IDocs in another screen. From there, select an IDoc number and click the **Details** icon at the top of the list to arrive at the details of an IDoc, as shown in Figure 12.4. If you know the IDoc number, enter it and execute the transaction to arrive at the details of an IDoc, as shown in Figure 12.4.

There are three main structures of an IDoc:

- The control data of the IDoc consists of technical details such as sender information, receiver information, type of message, basic type, and so on. Control data represents a specific document type such as a purchase order, invoice, order confirmation, and so on.

- Data records of an IDoc contain multiple segments, and each segment contains the document header or item details. Data records carry the business transactional data in segments. Segments of an IDoc are structures in a sequence, and the sequence is defined by the respective basic type of an IDoc.

- The status records of an IDoc contain status information for the IDoc. It contains specific statuses for inbound IDocs and outbound IDocs. It shows if an IDoc is successfully received or transmitted. For an inbound IDoc, a successful status represents that a business transaction of an application was posted successfully, such as an invoice document.

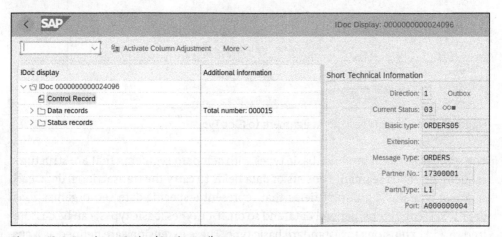

Figure 12.4 Purchase Order (Outbound) IDoc

Every IDoc in the system is identified by a unique number called the IDoc number. You can display an IDoc using the following transactions:

- Transaction WE02 or WE05 to display a list of IDocs by IDoc type, direction, status, and so on. Both transactions are the same.
- Transaction WE09 to display a list of IDocs by selection criteria based on specific business content—for example, a field value of a segment.

The **Short Technical Information** details for an IDoc are derived from the IDoc's control record. Let's explore the details of an IDoc.

Click **Control Record** under the IDoc number from the IDoc display screen to display the control record of the IDoc. Figure 12.5 shows the **Control Record** header details and **Typeinfo** tab. The header details of the control record include the following fields:

- **Direction**
  The direction shows if the IDoc is inbound into SAP S/4HANA or outbound from SAP S/4HANA. This value is set automatically in the control record:
  - **1**: Outbound
  - **2**: Inbound
- **Status**
  This field displays the processing status of an IDoc. These are two-digit codes in the standard system. There are specific status codes defined for outbound IDocs and inbound IDocs. Table 12.2 shows some of the status codes and their descriptions. The following are the two status codes the system sets if an IDoc is successfully processed:
  - **03**, for a successfully processed outbound IDoc
  - **53**, for a successfully processed inbound IDoc

Status Code	Description	Direction
01	IDoc created	Outbound
02	Error passing data to port	Outbound
03	Data passed to port OK	Outbound
04	Error within control information of EDI subsystem	Outbound
51	Application document not posted	Inbound
52	Application document not fully posted	Inbound
53	Application document posted	Inbound
54	Error during formal application check	Inbound
55	Formal application check OK	Inbound
56	IDoc with errors added	Inbound

Table 12.2 IDoc Status Codes

The **Typinfo** tab of the control record has the following fields:

- **Basic type**
  This field is populated automatically, and it defines the IDoc type for the document details that the IDoc is carrying. The IDoc type represents an application document type such as a purchase order, invoice, and so on.

- **Extension**
  SAP S/4HANA provides capabilities to enhance the IDoc types to add custom segments and custom fields. These extensions are assigned to the basic type.

- **Message Type**
  The message type represents a logical definition to identify an application document processed inbound to or outbound from SAP S/4HANA using IDocs. IDoc types are assigned to message types, and the IDoc types are determined from the message types.

- **Test Flag**
  The test flag indicates if the IDoc is dispatched to or received from a test system or production system. If the flag is set, the message originates from the test system.

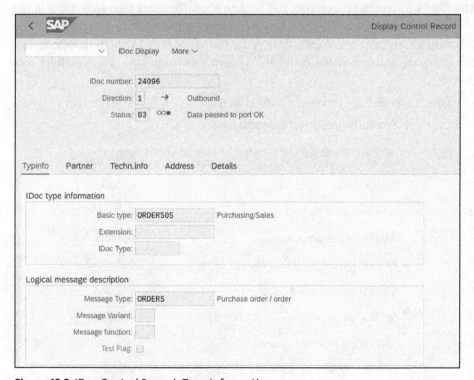

Figure 12.5 IDoc Control Record: Type Information

Figure 12.6 shows the control record header details and the **Partner** tab. The **Partner** tab of the control record has two sections, **Recipient information** and **Sender information**. The **Recipient information** section has the following fields:

- **Port**

  The port is a logical gateway to dispatch and receive IDocs. You must maintain a receiver port and assign it to the partner profile to dispatch IDocs to the external system via EDI.

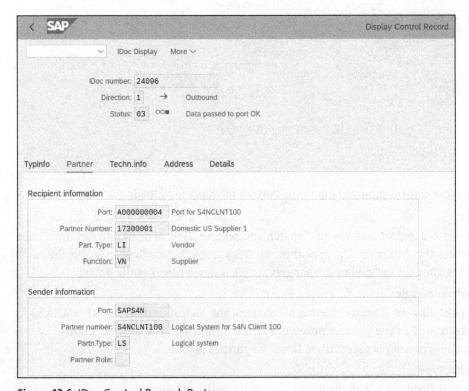

Figure 12.6 IDoc Control Record: Partner

- **Partner Number**

  To process an IDoc in SAP S/4HANA, a partner profile is created for every partner type: logical system, vendor, customer, and so on. While creating a profile, enter the partner number of the business partner (e.g., supplier) defined in the system. The system automatically determines the partner number from the application document (e.g., purchase order) and the partner profile during the IDoc processing.

- **Part. Type**

  The partner type distinguishes different partner types such as customer, vendor, logical system, and so on. The system automatically determines the partner number from the application document (e.g., purchase order) and partner profile during IDoc processing. Table 12.3 shows the different partner types defined in the standard system. You can create a partner profile using partner types for IDoc processing.

- **Function**

  This field defines the function of the recipient of the IDoc: vendor, customer, logical system, and so on.

Partner Type	Description
B	Bank
BP	Benefits provider
GP	Business partner
KU	Customer
LI	Vendor
LS	Logical system
US	User (first 10 characters, no check)

Table 12.3 Partner Type

The **Sender information** section of the **Partner** tab has the following fields:

- **Port**
  A port is a logical gateway to dispatch and receive IDocs. The system automatically derives this information for outbound IDocs from the configuration. For inbound IDocs, the port defined in the partner profile is determined as the port.

- **Partner number**
  For sending an IDoc to an external system, the logical system of SAP S/4HANA is determined as the partner number. For inbound IDocs, the partner number of the business partner is determined from the partner profile.

- **Partn.Type**
  Logical system LS is determined by the standard system for sending IDocs to an external system. For inbound IDocs, the business partner is determined as the partner type.

From the control records screen, go back to the IDoc display screen by clicking the left arrow at the top of the screen. Then click the data records and the individual data segments of the IDoc to display the transactional data that the IDoc is carrying. Figure 12.7 shows the **Data records** list for the IDoc. Data records consist of IDoc segments structured in a hierarchical manner. Some of the IDoc segments contain child/sub segments, such as **EIEDP01** (document item data). Each IDoc segment contains the actual application document data to be sent to or received from an external system/business partner.

Figure 12.7 also displays the actual content of segment **E1EDK01** that belongs to the basic type **ORDERS05** (purchase order). Here, the BSART field contains the document type of the purchase order, the BELNR field contains the purchase order number, the RECIPNT_NO field contains the business partner (supplier) number, and so on.

Figure 12.8 shows the **ORDERS05** basic type IDoc, the segments of the basic type structured in a hierarchical manner, and the descriptions of the segments. You can display the segments and the field names that each segment contains for any basic IDoc type using Transaction WE30.

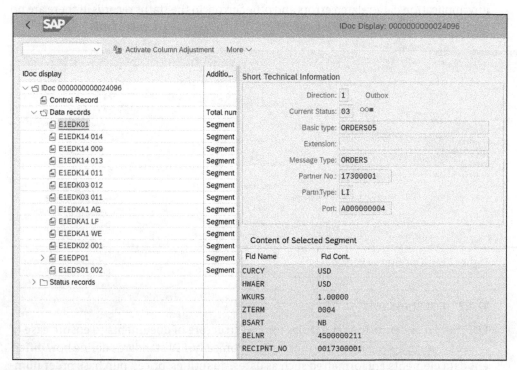

Figure 12.7 IDoc Data Record and Segment E1EDK01 Data

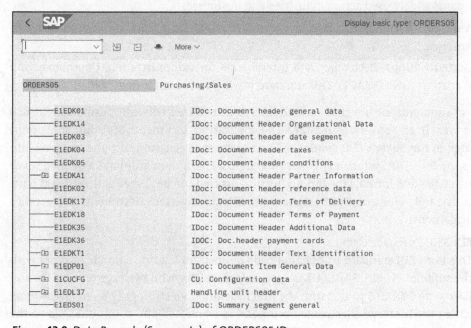

Figure 12.8 Data Records (Segments) of ORDERS05 IDoc

Next, click **Status records** in the left-hand navigation structure in the IDoc display screen to expand the status records of the IDoc, as shown in Figure 12.9. You can display any application data-related errors and other errors in the status records. If there are no errors, the IDoc is processed successfully and displays the success status code.

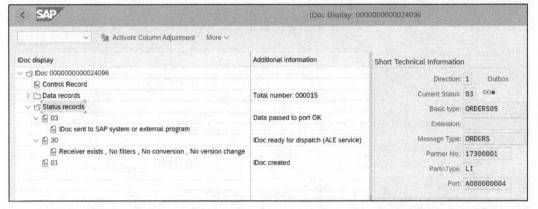

Figure 12.9 IDoc Status Record

12.1.2   EDI Standards

EDI standards define the set of rules for and structure of documents to ensure ease of data exchange between two businesses electronically. EDI standards define how different data elements are formatted such as dates, quantities, prices, purchase order numbers, invoice numbers, and so on within the electronic documents. There are two EDI standards widely used across industries and businesses:

- American National Standard Institute (ANSI) X12: This standard is used in north America.
- United Nations Electronic Data Interchange for Administration, Commerce, and Transport (UN EDIFACT): This standard is widely used in Europe.

These standards ensure that secured data is transferred between businesses (in B2B networks) in an accelerated manner. The application documents (e.g., purchase order) created in the buyer's ERP system will be transmitted electronically and received into the supplier's ERP system using these standards. These standards have a defined set of syntax rules and formats for various documents that can be shared with business partners. The following are the EDI standards defined for various documents in materials management:

- **EDI 850: Purchase Order**
  This is an EDI standard transaction code defined for sharing a purchase order with the supplier via EDI. SAP S/4HANA generates an IDoc with message type ORDERS and IDoc type ORDERS05 to transmit the purchase order electronically to an EDI system, where the message (electronic document) is translated and communicated to the supplier's EDI system.

- **EDI 855: Purchase Order Confirmation**
This is an EDI standard transaction code defined for the supplier sharing a purchase order confirmation with the buyer via EDI. Once the supplier gets a new purchase order electronically via EDI, they confirm the order and send an order confirmation via EDI. This message will be received as a purchase order confirmation message with message type ORDERSP and basic (IDoc) type ORDERS05 into SAP S/4HANA systems that belongs to the buyer.

- **EDI 856: Advanced Shipping Notification**
This is an EDI standard transaction code defined for the supplier sharing an ASN with the buyer via EDI. Once suppliers ship the goods for the purchase order, an ASN with the item details including materials, batch, serial numbers (if any), shipping date, and so on is shared electronically with the buyer via EDI. This message will be received as an inbound delivery with message type DESADV into SAP S/4HANA systems that belong to the buyer.

- **EDI 810: Invoice**
This is an EDI standard transaction code defined for the supplier sharing an invoice for the purchase order with the buyer via EDI. Once the suppliers ship the goods for the purchase order, they generate an invoice with reference to the purchase order and share the invoice document electronically with the buyer via EDI. This message will be received as an incoming invoice with message type INVOIC and basic type INVOIC02 into SAP S/4HANA systems that belong to the buyer.

- **EDI 812: Credit Memo**
This is an EDI standard transaction code defined for the supplier sharing a credit memo with the buyer via EDI. Suppliers create a credit memo for the returns purchase order or for price adjustments for an already submitted invoice.

12.1.3   Configuration of EDI Interfaces

Supplier collaboration is vital in automating, standardizing, and streamlining the procure-to-pay process in materials management. Business documents such as purchase orders, purchase order confirmations, ASNs, invoice documents, and so on are shared between the buyer's and supplier's ERP systems electronically via EDI. Let's explore the setup of these interfaces in SAP S/4HANA, assuming that SAP S/4HANA is the ERP system on the buyer side.

Let's start with the outbound EDI interfaces from SAP S/4HANA.

Outbound EDI Interfaces

The purchase order and the scheduling agreement delivery schedule are two important documents that will be shared by the buyers with supplier business partners. SAP S/4HANA generates IDocs for transmitting purchase orders (message type ORDERS) and scheduling the agreement delivery schedule (message type PURSAG_MAINTAIN) for business partners (suppliers) in real time via EDI.

SAP S/4HANA uses output determination for generating and transmitting outbound IDocs to an external system. Output messages are automatically generated during the creation of a purchase order and transmitted to suppliers after the purchase order is released or approved. NEU is the standard output type used to transmit purchase orders and scheduling agreement delivery schedules.

Let's walk through the main configuration steps for outbound EDI interfaces.

Maintain Output Determination Condition Record

The condition technique is used for output determination for the message output in SAP S/4HANA. Creating an output determination condition record enables the automation of generating and transmitting output messages. To create, change, or display an output condition record, use the following transactions:

- Transaction MN04: Purchase Order Output Condition Record—Create
- Transaction MN05: Purchase Order Output Condition Record—Change
- Transaction MN06: Purchase Order Output Condition Record—Display
- Transaction MN10: Scheduling Agreement Delivery Schedule—Create
- Transaction MN11: Scheduling Agreement Delivery Schedule—Change
- Transaction MN12: Scheduling Agreement Delivery Schedule—Display

Use Transaction MN05 to arrive at the screen shown in Figure 12.10, which is the initial screen for changing an output condition record for output type **NEU**. Specify the output type **NEU** in the entry field and then press ⌷Enter⌷. The **Key Combination** popup will be displayed. Key combinations can be configured based on business requirements to determine the output condition record during the creation of a purchase order and a scheduling agreement delivery schedule. For example, if you maintain an output condition record for purchasing output determination with the key combination set to **Doc.Type/Purch.Org/Vendor**, the output condition record will be determined during the creation of the purchase order automatically when the same combination of a document type, purchasing organization, and vendor is found in the purchase order.

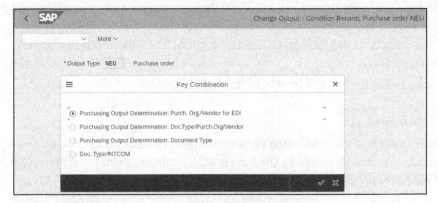

Figure 12.10 Output Condition Record for Output Type NEU: Initial Screen

Press ⌈Enter⌉ after selecting a key combination from the initial screen. Figure 12.11 shows the output condition record details that are created for output type **NEU** and with key combination **Purch.Org/Vendor for EDI**. The following are the key fields of the output condition record:

- **Purch. Organization**
  Specify the purchasing organization in which the output condition record is being created. This field is mandatory as it is one of the fields from the key combination.

- **Supplier**
  Specify the EDI supplier to which the purchase order is to be transmitted electronically via EDI.

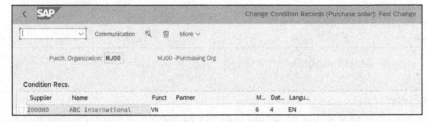

Figure 12.11 Output Condition Record for Output Type NEU: Details

- **Funct (function)**
  Choose a partner function code from the dropdown menu. For a supplier-based condition record, choose **VN** (vendor) as the partner function. For EDI message output, a partner profile must exist for the vendor with partner type **LI** (vendor).

  If you want to send the IDoc output message to a middleware system, choose **LS** (logical system) as the partner function. In such cases, for EDI message output, a partner profile must exist for the logical system with partner type **LS**.

- **M... (medium)**
  Choose a message transmission medium for outbound messages. The system creates messages based on the medium and triggers automatic message outputs such as fax, email (external send), print output, EDI, and so on. Table 12.4 shows the key output mediums in the standard system. For EDI output, select **6** (EDI) as the output medium.

Output Medium	Description
1	Print output
2	Fax
4	Telex
5	External send
6	EDI

Table 12.4 Output Medium

Figure 12.12 shows that output record with output type **NEU** was generated automatically to trigger an EDI output that transmitted a purchase order to supplier **200080**.

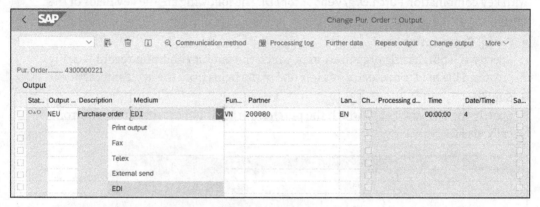

Figure 12.12 Purchase Order Output Determination

Ports in IDoc Processing

Ports are used in IDoc processing to define the communication channel for transmitting outbound IDocs from SAP S/4HANA to external systems and business partners. The following are the different types of ports in SAP S/4HANA:

- **Transactional RFC port**
  These ports are used for communicating the messages (IDocs) via Application Link Enabling (ALE) for establishing communication between two logical systems.

- **File port**
  File ports transmit IDocs in file format to a specific location in a system that can be accessed to retrieve the file in order to transmit it to the target system. File ports are used in EDI.

- **ABAP PI port**
  These ports are used to transfer messages using an ABAP interface. It provides the ability to customize the interface for transmitting and receiving messages (IDocs).

- **XML file port**
  These ports are used for communicating messages (IDocs) as XML files.

Ports are defined in SAP S/4HANA using Transaction WE21. A prerequisite for creating a port for IDoc processing is to define a remote function call (RFC) destination using Transaction SM59. The RFC destination contains the logical address and the authentication of the destination (target) system.

Figure 12.13 shows information for transactional RFC port **A000000004**. You can see the **RFC destination** field, where you specify an RFC destination to communicate the IDoc messages to the target system. The RFC destination contains the logical address and authentication details for the data transfer through IDocs.

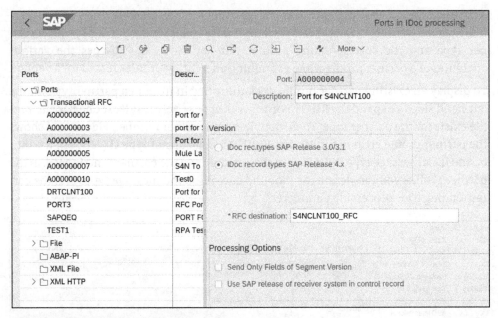

Figure 12.13 Port in IDoc Processing

Create Partner Profile for Outbound IDocs

The partner profile consists of logical attributes to send and receive application documents (messages) electronically using IDocs. For every logical system, vendor, customer, other business partner, and so on, a partner profile must exist in SAP S/4HANA to process inbound and outbound IDocs.

Partner profiles are created in SAP S/4HANA using Transaction WE20 for the partner types shown in Table 12.5.

Partner Type	Description
AD	Address
B	Bank
BP	Benefits provider
GP	Business partner
KU	Customer
LI	Vendor
LS	Logical system
US	User (first 10 characters, no check)

Table 12.5 Partner Types

A partner type is a vital attribute to create a partner profile. To create a new partner profile, execute Transaction WE20 and click the **Create** icon at the top. Then specify a partner type and the corresponding partner number. Now, let's discuss the various attributes of a partner profile while examining an existing profile created for a vendor.

To display an existing partner profile, click the arrow in front of a partner type on the left-hand side to expand the list. This arrow will appear only if at least one partner profile exists for the partner type. Then, double-click a partner number. Figure 12.14 shows the partner profile created for vendor **200080** using partner type **LI** (vendor), the outbound IDoc message type **ORDERS** (purchase order), and the partner function **VN**. In the partner profile, you can define the inbound and outbound message types and the corresponding IDoc processing parameters.

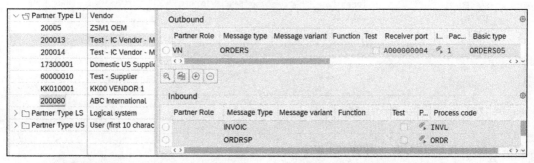

Figure 12.14 Partner Profile for Business Partner with Partner Type LI (Vendor)

Let's dive deep into the settings required in the partner profile for processing outbound IDocs:

- **Partner No.**
  Specify the partner number of the partner for which you are defining the partner profile. You can specify the supplier number of the business partner if you're creating the partner profile for partner type **LI** (vendor).

- **Partn.Type**
  A partner type distinguishes different types of partners for communication. Specify the partner type for transmitting the IDocs. A partner profile can be defined for the logical system (**LS**), vendor (**LI**), customer (**KU**), and other partner types. To transmit a purchase order to an external system, define the partner profile for partner type **LS**, and to transmit a purchase order to a vendor, define the partner profile for partner type **LI**.

- **Message Type**
  Message types distinguish different document types that flow via EDI using IDocs. Every document that flows electronically via EDI is associated with a message type in SAP S/4HANA. For purchase order transmission, specify **ORDERS** (the standard message type for a purchase order) in this field.

Then, under the **Post Processing: Valid Processors** tab, you can define a user who gets the error messages if any during the IDoc processing. Specify "US" in the **Ty.** (user type) field, the user ID of the user who gets error messages in the **Agent** field, and the desired language in the **Lang** field as the postprocessing parameters.

Next, maintain the outbound options. Click the **+** icon under the **Outbound** section of the partner profile initial screen. Figure 12.15 shows the outbound parameters of the partner profile for transmitting a purchase order to an external system via EDI using IDocs.

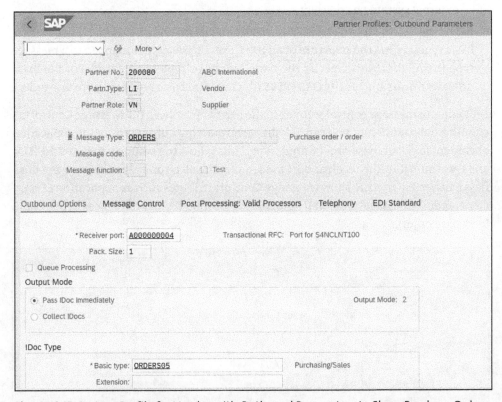

Figure 12.15 Partner Profile for Vendor with Outbound Parameters to Share Purchase Orders

The **Outbound Options** tab has the following key fields:

- **Receiver port**
  Ports are used in IDoc processing to define the communication channel for transmitting outbound IDocs from SAP S/4HANA to external systems and business partners. Specify a receiver port to transmit the IDoc to the EDI system that will send it to the vendor's ERP system.

- **Pack. Size**
  Specify the number of IDocs that can be sent per RFC.

- **Output Mode**

  The output mode defines how the IDocs are transmitted to the external system/ business partner. There are two modes:

  - **Pass IDoc immediately**

    If you select this mode, generated IDocs will be sent to the EDI system/external system immediately.

  - **Collect IDocs**

    If you select this mode, IDocs are collected and transmitted to the EDI system/ external system periodically.

- **Basic type**

  Basic types define the structure of an IDoc. Every basic type is mapped to a message type in SAP S/4HANA. Specify the basic type/IDoc type in this field. For purchase order transmission, specify **ORDERS05** (the standard basic type for a purchase order).

Let's explore message control settings in the partner profile. In the **Message Control** tab of outbound parameters, you can link the corresponding output type of the application document (e.g., purchase order output type). This is how the system generates the IDoc and transmits it to the receiver port for sending the electronic document to the business partner. Figure 12.16 shows **Message Control** attributes such as application **EF** (purchase order), message type **NEU** (purchasing output type), and process code **ME10** (purchase order).

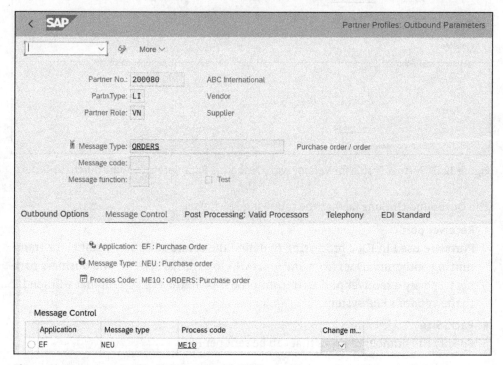

Figure 12.16 Partner Profile for Vendor with Outbound Parameters: Message Control

If the NEU output message is triggered with medium 6 (EDI) from purchase order creation, the system checks the corresponding partner profile for the partner function of the output message using message type NEU and the partner ID of the vendor.

Process code ME10 is defined for the purchase order transmission in the standard system. It contains a function module to process the purchase order IDoc.

Figure 12.17 shows the **EDI Standard** attributes of the partner profile for the purchase order transmission. Message type **850** represents the EDI standard for the purchase order transmission.

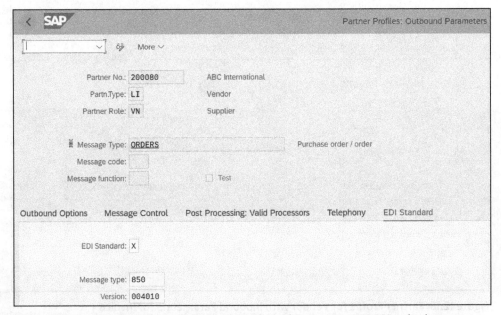

Figure 12.17 Partner Profile for Vendor with Outbound Parameters: EDI Standard

Inbound EDI Interfaces

The purchase order confirmation, ASN, and invoice are the most important documents received from the supplier. To receive them via EDI, IDocs with specific message types and basic types must be generated in the EDI system or external system and sent to SAP S/4HANA. A partner profile for the respective partner and partner type is mandatory to process the inbound IDocs.

Let's walk through the main configuration steps for inbound EDI interfaces.

Partner Profile Settings for Incoming Invoice

To maintain inbound options for a partner (e.g., vendor), go back to the partner profile initial screen. Click the **+** icon under the **Inbound** section of the partner profile initial screen to arrive at the inbound options screen. Figure 12.18 shows the partner profile for partner number **200080** and partner type **LI** (vendor). Inbound parameters of the partner profile are defined for receiving an invoice document from the vendor via EDI using IDocs. The following are the key fields:

- **Message Type**
  Message types distinguish different document types that flow via EDI using IDocs. Every document that flows electronically via EDI is associated with a message type in SAP S/4HANA. For an incoming invoice document, specify **INVOIC** (the standard message type for incoming invoice documents) in this field.

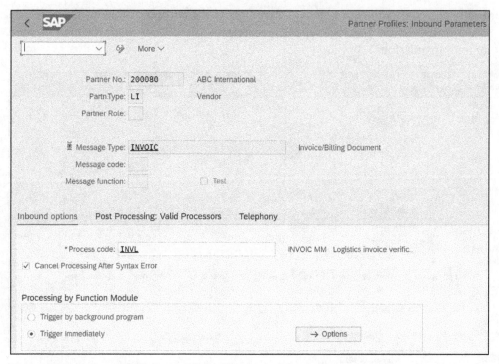

Figure 12.18 Partner Profile for Vendor with Inbound Parameters for Invoice

- **Process code**
  A process code represents the type of inbound transaction data for processing the IDoc. Specify **INVL** (the standard logistics invoice verification process code) for incoming logistics invoices.

- **Cancel Processing After Syntax Error**
  If you enable this indicator, the system cancels IDoc processing if a syntax error is found in the IDoc.

- **Processing by Function Module**
  There are two options to process the inbound IDocs:

  - **Trigger by background program**: If you select this processing mode, inbound IDocs are collected and processed by a background program periodically.

  - **Trigger Immediately**: If you select this processing mode, inbound IDocs are processed immediately.

Partner Profile Settings for Purchase Order Confirmation

Figure 12.19 shows the partner profile for partner **200080** and partner type **LI** (vendor). Inbound parameters of the partner profile are defined for receiving an order confirmation from the vendor via EDI using IDocs. The following are the key fields:

- **Message Type**

  Message types distinguish different document types that flow via EDI using IDocs. Every document that flows electronically via EDI is associated with a message type in SAP S/4HANA. For an incoming invoice document, specify **ORDRSP** (the standard message type for purchase order confirmations) in this field.

- **Process code**

  A process code represents the type of inbound transaction data for processing the IDoc. Specify **ORDR** (the purchase order confirmation process code) for order confirmation.

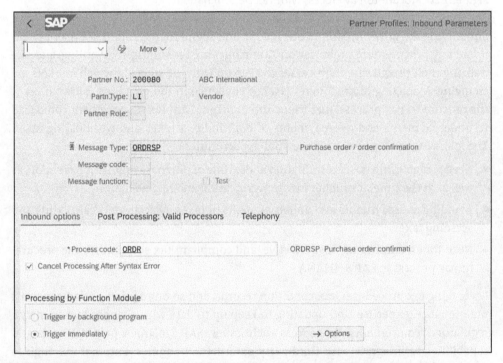

Figure 12.19 Partner Profile for Vendor with Inbound Parameters for Purchase Order Confirmation

12.2   External Tax Engine Integration

Indirect tax calculation is a key part of the procure-to-pay process. Estimated tax is calculated on the purchase order at the line item level, and the actual tax calculation happens at the time of incoming invoice processing. During the incoming invoice processing, tax reconciliation is performed to match the supplier-charged tax and system-calculated tax. Embedding and automating tax determination and reconciliation into the procure-to-pay process is key to successful spend management in SAP S/4HANA. Tax calculation, tax reporting, and paying accurate taxes on purchases are important for organizations to stay compliant with regulatory requirements. Figure 12.20 shows the procure-to-pay process with tax determination steps.

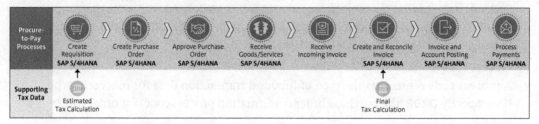

Figure 12.20 Procure-to-Pay Process with Tax Determination

Simplifying the growing complexities of tax and spend management is a major priority for any procurement organization. Companies are redefining, reinventing, and digitalizing their procurement processes to keep up with the speed at which markets are changing. Managing indirect taxes is a key challenge in the digital transformation of the procure-to-pay process. Inefficient and noncompliant tax management could lead to underpayments and overpayments of both indirect taxes and withholding taxes. The following are the key challenges and opportunities:

- Need for real-time tax determination at the time of purchase requisition creation, as well as at the time of supplier invoice verification and reconciliation
- Avoiding overpayments and underpayments of taxes by accurately calculating and accruing taxes
- Need for an automated, cost-effective, and compliant tax solution for the procure-to-pay process in SAP S/4HANA

Leveraging tax procedures, tax condition records, and so on requires periodic manual maintenance, expertise, and updating to keep up to date with changes in a country's regulatory requirements. To address such issues, SAP S/4HANA provides tools and capabilities to integrate with third-party external tax engines. External tax engines provide tax determination capabilities for most countries across regions such as the Americas, EMEA, and Asia-Pacific by leveraging complex tax calculation rules based on regulatory requirements by country. Leveraging an external tax engine reduces setup and periodic maintenance tasks in SAP S/4HANA.

Let's explore the difference between the internal tax engine of SAP S/4HANA and external tax engines.

12.2.1   External Tax Engine versus Internal Tax Engine

SAP S/4HANA provides tools and capabilities to maintain tax calculation procedures to determine taxes, and if the tax is calculated internally within the system using the tax setup, it is from the internal tax engine. SAP S/4HANA supports tax data maintenance by country to comply with regulatory requirements.

An external tax engine is a third-party system that can be integrated with SAP S/4HANA to determine taxes. SAP S/4HANA sends a request payload to the external tax engine to determine taxes; the third-party external tax engine calculates the taxes in turn based on the request and sends a response payload to SAP S/4HANA with the tax results for the respective country. There is no middleware system required to integrate certain common external tax engines with SAP S/4HANA. The request payload contains standard attributes including the document details (e.g., purchase order header and line item details) and business-specific custom attributes to calculate taxes. The response payload contains the tax code, tax type (e.g., VAT, GST, PST, HST, sales tax), tax rate (measured as a percentage), tax amount, jurisdiction code (in certain countries), and so on in addition to the attributes received in the request payload.

Let's explore the concepts of an external tax engine and an internal tax engine.

Internal Tax Engine in SAP S/4HANA

Tax determination in SAP S/4HANA leverages the condition technique. The tax determination procedure is set up to determine taxes in purchasing documents comprising condition types, access sequence, and condition records. Let's explore the condition technique details for tax determination in SAP S/4HANA:

- **Tax procedure**

  The tax procedure supports the determination of tax codes in materials management. Standard tax determination procedures provided by SAP S/4HANA are enough to determine tax codes. Necessary master data records needed for tax code determination such as condition records, purchasing info record (PIR), and so on are maintained to determine taxes during purchasing and invoicing processes. The system provides standard tax procedures by country, and this can be further configured in financial accounting global settings to maintain the condition technique, access sequence, condition types, and tax procedure.

  To define tax determination procedures, execute Transaction SPRO and navigate to **Financial Accounting • Financial Accounting Global Settings • Tax on Sales/Purchases • Basic Settings • Check Calculation Procedure**. Click **Define procedures** from the activity screen to arrive at the procedures overview screen. To define a procedure, click **New Entries**. You can also copy an existing procedure and create a new one from it.

  Figure 12.21 shows the control data of procedure **TAXGB** for the internal tax engine. Tax condition types are maintained in sequence.

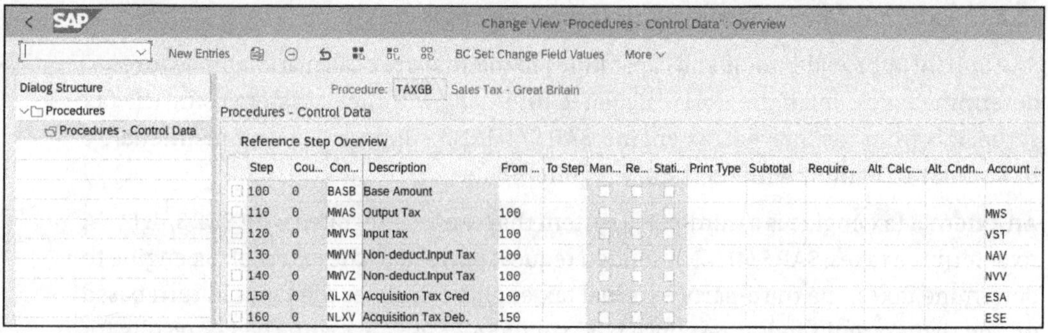

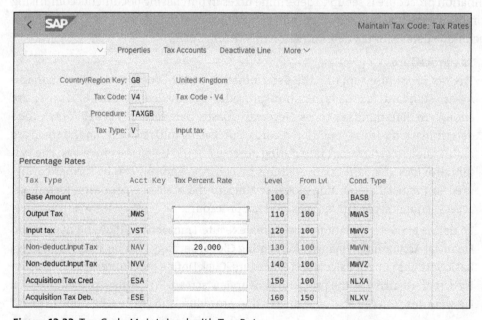

Figure 12.21 Tax Procedure for Internal Tax Engine

- **Maintain tax codes**

  Tax codes are maintained mainly by country and by jurisdiction in certain countries, such as the United States, Canada, and so on. The tax code contains the tax rate for tax calculations and tax types such as sales tax, value-added tax (VAT), goods and services tax (GST), and so on based on the localization requirements.

  Transaction FTXP is used to create tax codes. Figure 12.22 shows tax code **V4** maintained for country **GB** in tax procedure **TAXGB**. A tax rate of 20% is maintained for tax condition type **MWVN**. Tax codes are created by country. A jurisdiction code is required if you create a tax code for the United States, but for Great Britain, jurisdiction codes are not required.

Figure 12.22 Tax Code Maintained with Tax Rates

- **Assign country/region to calculation procedure**
  A tax determination/calculation procedure must be assigned to each country. A predefined tax calculation procedure must be defined before this activity. The system determines the procedure automatically during the creation of a purchase order and an outline agreement based on the ship-to country defined in the **Delivery Address** tab of the purchasing documents.

  To define tax determination procedures, execute Transaction SPRO and navigate to **Financial Accounting • Financial Accounting Global Settings • Tax on Sales/Purchases • Basic Settings • Assign Country/Region to Calculation Procedure**.

  Figure 12.23 shows procedure **TAXGB** assigned to country **GB**. You can assign the tax determination procedure directly to the country/region in this configuration setting.

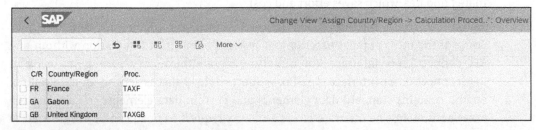

Figure 12.23 Assign Country/Region to Calculation Procedure

- **Tax code condition records**
  SAP S/4HANA uses the condition technique to determine tax codes for sales and purchases. Tax codes are determined automatically during the creation of sales and purchasing documents. You can create condition records for tax code determination using Transaction FV11. Figure 12.24 shows the condition record maintained for tax determination. Tax code **V4** is determined in the purchase order from the condition record maintained for condition type **MWVN**.

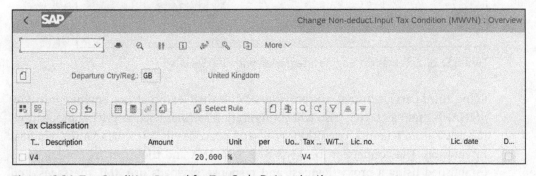

Figure 12.24 Tax Condition Record for Tax Code Determination

Refer to Chapter 10, Section 10.2.4 for more details on internal tax calculation processes in SAP S/4HANA.

Maintaining complex tax determination rules, tax codes, and tax condition records via SAP S/4HANA's internal tax engine requires continuous manual effort. In contrast, let's explore the external tax engines integrated with SAP S/4HANA.

External Tax Engine Integrated with SAP S/4HANA

An external tax engine is maintained by a third party. All regulatory and tax compliance rules are maintained by the third-party vendor periodically to make sure the tax determination rules are compliant with the tax authority rules of each country. In addition, third-party external tax engines allow you to configure customer-specific business rules for tax determination. SAP S/4HANA facilitates the data exchange from a third-party external tax engine. Any changes in tax laws, tax rates, and so on from the respective tax authority are updated frequently to ensure that customers get automated updates, which saves effort and cost.

Figure 12.25 shows the integration between external tax engines and SAP S/4HANA. Some of the most popular external tax systems offer direct integration without any RFC connection or interface. You can also integrate third-party tax engines using an RFC connection or interface. The tax request payload that is sent to an external tax engine contains standard data elements and custom data elements to calculate the taxes accurately. The interface to send the tax request payload can be customized to include custom data elements. External tax engines recognize foreign transactions and domestic transactions, as well as changes in ship-to addresses in case of drop-ship transactions. Taxes on purchase orders, accounts payable invoices, and accounts receivable invoices are calculated accurately.

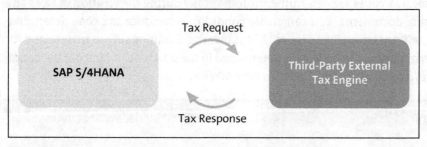

Figure 12.25 External Tax Engine Integration with SAP S4HANA

Some third-party external tax providers hook their external tax engines into SAP S/4HANA user exits. They also add configuration settings such as pricing procedures, tax calculation schemas, condition types, and tax codes in SAP S/4HANA for seamless integration. The connection between SAP S/4HANA and the external tax engine is established using a proxy. Request and response payloads use XML messages, and these messages go through the proxy.

Refer to Section 12.2.3 for the configuration of external tax engine integration with SAP S/4HANA.

The following are the key benefits of integrating an external tax engine with SAP S/4HANA:

- An external tax engine provides automated and real-time indirect tax determination capabilities for sales and purchasing processes in SAP S/4HANA in most countries across the globe.
- Complex indirect tax calculation rules are maintained by the third-party tax provider in the external tax engine based on regulatory requirements.
- External tax engines provide tools and capabilities to maintain business-specific custom rules for tax determination for both sales and purchasing processes.
- These engines offer increased indirect tax calculation control and accuracy.
- They also offer reduced maintenance costs for maintaining ever-changing regulatory and tax rules from tax authorities across multiple countries.
- Companies benefit from enhanced tax reporting capabilities and efficient monthly tax filings and reporting.
- These engines facilitate compliance and reduce the risk of underpayments and overpayments due to inaccurate tax calculations.
- External tax engines provide a search utility to check historical payloads for analysis.

12.2.2   External Tax Calculation for Purchase Orders and Supplier Invoice Documents

External tax calculation for purchase orders and accounts payable invoices happens smoothly in real time. Figure 12.26 shows the procure-to-pay process and the data flow between the external tax engine and SAP S/4HANA.

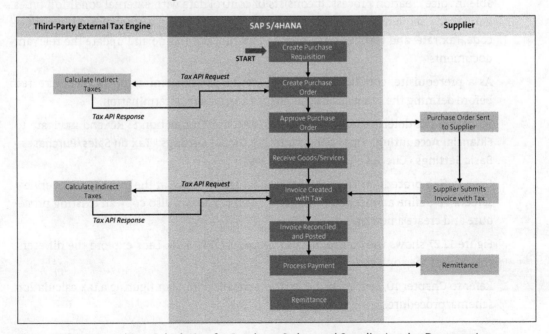

Figure 12.26 External Tax Calculation for Purchase Orders and Supplier Invoice Documents

In the procure-to-pay process, tax calculation happens at two points: at the time of creation of the purchase order, and during incoming invoice processing. A real-time tax calculation request will be communicated by sending an XML message (preferred method) to the external tax engine. The XML message is mapped to the corresponding target fields in the external tax engine to calculate taxes from the set of tax rules in the external tax engine. Once the taxes are determined, a tax code, tax rate, and tax amount in the transactional currency along with the transactional data (purchase order/supplier invoice data) will be sent as a tax response using an XML message to SAP S/4HANA. The incoming XML message is mapped to the corresponding target fields in SAP S/4HANA, and the transaction document is updated with the tax details.

12.2.3   Configuration of External Tax Engine Integration

External tax determination in SAP S/4HANA leverages the condition technique, partly to establish the integration with the external tax engine and collect the taxes from it. An external tax determination procedure is set up to collect taxes for purchasing documents, accounts payable invoice documents, and accounts receivable invoice documents comprising condition types, access sequences, and condition records. Let's dive deep into these settings.

Define Tax Determination Procedures

An external tax determination procedure ensures accurate tax determination during the purchase order outline agreement, accounts payable invoice, and accounts receivable invoice creation process. It consists of control data with external condition types defined in a sequential order that allows the system to collect tax details such as the tax code, tax rate, and tax amount from the external tax engine and update the relevant documents.

As a prerequisite, condition types for external tax determination must be created before defining the calculation schema for external tax determination.

To define tax determination procedures, execute Transaction SPRO and navigate to **Financial Accounting • Financial Accounting Global Settings • Tax on Sales/Purchases • Basic Settings • Check Calculation Procedure.**

Click **Define procedures** from the **Activity** screen to arrive at the procedures overview screen. To define a procedure, click **New Entries.** You can also copy an existing procedure and create a new one.

Figure 12.27 shows the control data of procedure **TAXGBX.** Let's explore the different columns of the control data.

Refer to Chapter 10, Section 10.2.4 for more details about configuring a tax calculation schema/procedure.

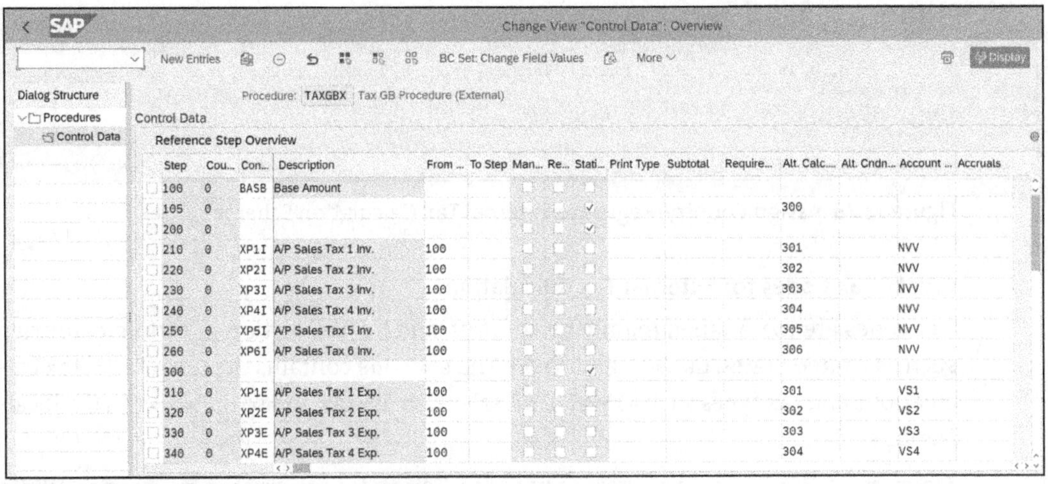

Step	Cou...	Con...	Description	From ...	To Step	Man...	Re...	Stati...	Print Type	Subtotal	Require...	Alt. Calc....	Alt. Cndn...	Account ...	Accruals
100	0	BASB	Base Amount												
105	0							✓				300			
200	0							✓							
210	0	XP1I	A/P Sales Tax 1 Inv.	100								301		NVV	
220	0	XP2I	A/P Sales Tax 2 Inv.	100								302		NVV	
230	0	XP3I	A/P Sales Tax 3 Inv.	100								303		NVV	
240	0	XP4I	A/P Sales Tax 4 Inv.	100								304		NVV	
250	0	XP5I	A/P Sales Tax 5 Inv.	100								305		NVV	
260	0	XP6I	A/P Sales Tax 6 Inv.	100								306		NVV	
300	0							✓							
310	0	XP1E	A/P Sales Tax 1 Exp.	100								301		VS1	
320	0	XP2E	A/P Sales Tax 2 Exp.	100								302		VS2	
330	0	XP3E	A/P Sales Tax 3 Exp.	100								303		VS3	
340	0	XP4E	A/P Sales Tax 4 Exp.	100								304		VS4	

Figure 12.27 External Tax Calculation Schema (Illustration Only)

The condition types to collect the taxes from the external tax engine are added to the purchase order pricing procedure to copy the tax code, tax rate, tax amount, jurisdiction code (United States only), and so on over to the purchase order document. This is a mandatory step, and the purchase order pricing procedure must be configured to add the external tax calculation condition types.

> **Note**
>
> Condition types and calculation schema settings are configured based on the instructions from the third-party external tax provider. They will publish a configuration guide to set up condition types, pricing procedures for purchasing documents, and the tax calculation schema (tax procedure). Figure 12.27 shows an illustration of the calculation schema for an external tax calculation.

Assign Country/Region to Calculation Procedure

An external tax determination/calculation procedure must be assigned to each country. The system determines the external tax procedure automatically during the creation of a purchase order and other documents based on the country of the company code in which the transactional document is being created.

To assign an external tax determination procedure to a country/region, execute Transaction SPRO and navigate to **Financial Accounting • Financial Accounting Global Settings • Tax on Sales/Purchases • Basic Settings • Assign Country/Region to Calculation Procedure**.

Figure 12.28 shows procedure **TAXGBX** assigned to country **GB**. You can assign the tax determination procedure directly to the country/region in this configuration setting.

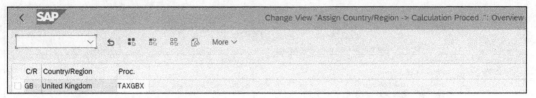

Figure 12.28 Assign Country/Region to External Tax Calculation Schema

Create Tax Codes for External Tax Calculation

Tax codes are maintained mainly by country, and by jurisdiction in certain countries such as United States, Canada, and so on. The tax code contains the tax rate for tax calculations and tax types such as VAT, GST, sales tax, use tax and so on based on the localizations.

Transaction FTXP is used to create tax codes. Tax codes are created by country. While creating a tax code, specify the country in the initial screen. The system automatically derives the calculation schema (tax procedure) assigned to that country/region. Tax codes are the main drivers for connecting SAP S/4HANA with external tax engines, and these are also called external tax codes. These tax codes are used to collect the taxes returned from an external tax engine in SAP S/4HANA. In general, the following tax codes are created in SAP S/4HANA for sales and purchasing processes:

- I1

  Input tax code to represent overtax or accurate tax with no tax accruals for purchasing process

- U1

  Input tax code to represent undercharged tax with tax accruals for purchasing process

- O1

  Output tax code for sales process

Figure 12.29 shows input tax code **I1** created for external tax calculation for country **GB** and for external tax procedure (calculation schema) **TAXGBX**. Here, tax type **V** indicates the input tax code. Input tax codes are used in the procure-to-pay process.

As another example, Figure 12.30 shows output tax code **O1** created for external tax calculation for country **GB** and for external tax procedure (calculation schema) **TAXGBX**. Here, tax type **A** indicates the output tax code. Output tax codes are used in the sales process.

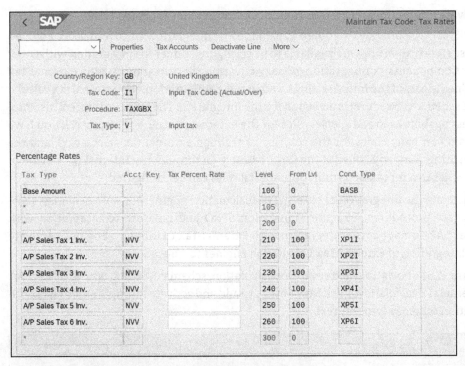

Figure 12.29 Input Tax Code for External Tax Calculation

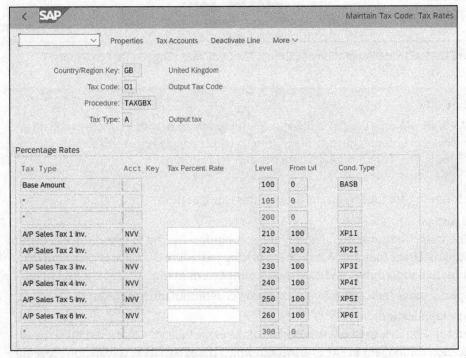

Figure 12.30 Output Tax Code for External Tax Calculation

Activate Integration of External Tax Calculation Engines for Procedures

SAP S/4HANA provides standard capabilities to integrate with external tax engines. These standard settings are available to be configured under the **Integration with Other SAP Components** configuration node. Activating the integration with external tax engines is part of the initial settings, and in this configuration, the external tax calculation schema (procedure) is activated for the integration. The standard system also provides capabilities to add special rules for the external tax call. For instance, if you have certain company codes for the country go through external tax calculation, and the remaining company codes of the same country go through the internal tax engine of SAP S/4HANA for tax determination, special rules can be configured.

To activate the integration of external tax calculation engines for tax calculation schemas (tax procedures), execute Transaction SPRO and navigate to **Integration with Other SAP Components • Integration with External Tax Calculation Engines • Activate the Integration of External Tax Calculation Engines for Procedures**.

Figure 12.31 shows that external tax calculation schema **TAXGBX** was activated for external tax calculation. Click **New Entries** to add and activate multiple external tax calculation schemas (procedures).

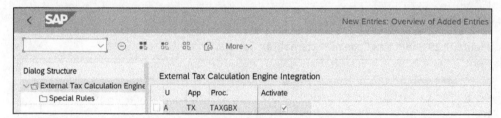

Figure 12.31 Activate Integration of External Tax Calculation Engines for Procedures

> **Note**
>
> SAP S/4HANA provides flexibility and control to integrate with multiple external tax engines. For instance, you can integrate and activate a separate external engine for the United States, and a different external engine for the UK.

The following are the important fields of this configuration:

- **U (usage)**
  The usage indicator denotes the usage condition for the external tax calculation schema. Usage indicator **A** indicates a pricing procedure, **B** indicates an output determination procedure, **C** indicates an account determination procedure, and so on.
  Specify usage indicator **A** for an external tax determination procedure.

- **App (application)**
  An application is used as a subdivision of the usage indicator, and it indicates the application component in SAP S/4HANA in which the external tax calculation procedure is being used. Specify **TX** (taxes) as the application for external tax determination.

- **Proc. (procedure)**
  Specify the external tax calculation schema that was configured to collect the taxes from an external tax engine in this field.

Figure 12.32 shows the special rules added to the activated external tax calculation schema, **TAXGBX**. To add the special rules, select the external tax calculation schema and click **Special Rules** in the **Dialog Structure**. In this setting, you can add rules to activate certain company codes for external tax calculation for the country that is enabled, and the remaining company codes for internal tax calculation for the same country. This setting is only required if you want to segregate company codes within the same country for internal tax calculation and external tax calculation.

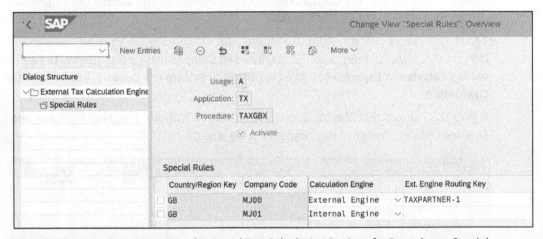

Figure 12.32 Activate Integration of External Tax Calculation Engines for Procedures: Special Rules

The following are the important fields of this configuration:

- **Country/Region Key**
  Specify the country or region keys for which the external tax calculation schema has been enabled and business requirements to segregate company codes from that country for internal and external tax calculation. Refer to our earlier discussion of assigning a country/region to the calculation procedure for more details.

- **Company Code**
  Enter a company code that belongs to the same country in which the external tax calculation schema has been enabled to add special rules.

- **Calculation Engine**
  Select an external or internal engine for the combination of country/region key and company code for external or internal tax calculation.

- **Ext. Engine Routing Key**
  A routing key is a technical data element added to the tax request payload that is transmitted to an external tax engine. It is used to determine which external tax engine is

activated for the combination of country/region key and company code. This is only required for external tax engines; it is not required for internal tax engines.

Assign a Classification System to a Country/Region for Tax Calculation

The product taxonomy is classified based on the products and services for e-commerce. Products and services are classified in a hierarchical manner and can have up to five levels. In SAP S/4HANA, an external material group can be assigned in the **Basic data 1** view of the material master data. The external material group uses a code from a classification system. The United Nations Standard Products and Services Code (UNSPSC) is a widely used classification system for products and services. The classification system ID (e.g., the UNSPSC number) will be sent to an external tax engine via the tax request payload for tax calculations.

To assign a classification system to a country for tax calculation, execute Transaction SPRO and navigate to **Integration with Other SAP Components • Integration with External Tax Calculation Engines • Assign a Classification System to a Country/Region for Tax Calculation**.

Figure 12.33 shows that classification system **United Nations Standard Products and Services Code** was assigned to countries **US**, **GB**, and **CA**.

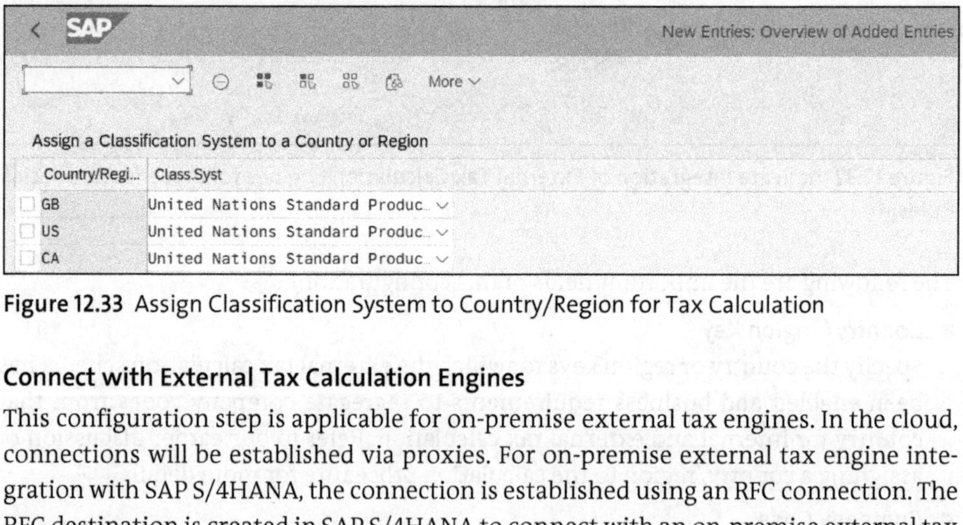

Figure 12.33 Assign Classification System to Country/Region for Tax Calculation

Connect with External Tax Calculation Engines

This configuration step is applicable for on-premise external tax engines. In the cloud, connections will be established via proxies. For on-premise external tax engine integration with SAP S/4HANA, the connection is established using an RFC connection. The RFC destination is created in SAP S/4HANA to connect with an on-premise external tax engine using Transaction SM59.

To connect with on-premise external tax calculation engines, execute Transaction SPRO and navigate to **Integration with Other SAP Components • Integration with External Tax Calculation Engines • Connect with External Tax Calculation Engines**.

To enable the communication between an external tax engine (on-premise) and SAP S/4HANA Cloud, click **New Entries** and specify a standard scenario ID defined for external tax engine connections and the RFC destination. Figure 12.34 shows that an RFC destination is assigned to the scenario ID for the on-premise external tax engine.

This configuration is required to integrate SAP S/4HANA Cloud with on-premise external tax engines. The communication (tax request and tax results) between SAP S/4HANA Cloud and the on-premise external tax system happens via standard scenario ID **SAP_COM_0249**. The standard scenario ID enables the communication with the on-premise system using RFCs.

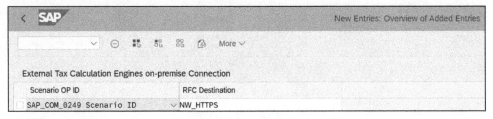

Figure 12.34 Connect with External Tax Calculation Engines

12.3   Summary

In this chapter, we introduced the integration of SAP S/4HANA with external business partners using electronic data interchange, and SAP S/4HANA integration with external tax engines for tax calculation. In Section 12.1, we explained the EDI standards, the EDI integration for materials management functions, and the configuration of EDI interfaces. In Section 12.2, we explained the external tax engine integration for tax calculation for purchasing and invoicing processes of materials management and the setup of external tax engine integrations with SAP S/4HANA.

The Author

Murthy KP is a senior manager at Deloitte and an accomplished supply chain leader, with extensive experience delivering complex global supply chain transformation projects through SAP solutions. He has successfully led multiple digital transformation projects, driving efficiency and innovation. Specializing in SAP S/4HANA materials management, he possesses extensive cross-functional expertise. He brings incredible value with his unique blend of SAP functional and technical knowledge. With a proven track record of leveraging SAP S/4HANA and SAP Ariba capabilities, he empowers organizations to unlock efficiencies, align technology with business strategy, and achieve sustainable growth.

Index

Interested in reading more?

Please visit our website for all new book
and e-book releases from SAP PRESS.

www.sap-press.com